PHYSICAL DATA FOR THE ELEMENTS (Concluded)

Element	Symbol	Atomic number	Atomic weight (amu)	Melting point (°C)	Density of solid, 20°C (gm/cm³)	Crystal structure, 20°C
Lanthanum	La	57	138.91	918	6.17	Hex.
Lead	Pb	82	207.2	327.502	11.34	FCC
Lithium	Li	3	6.941	180.6	0.533	BCC
Magnesium	Mg	12	24.31	650	1.74	HCP
Manganese	Mn	25	54.94	1246	7.47	Cubic
Mercury	Hg	80	200.59	−38.836	—	
Molybdenum	Mo	42	95.94	26.23	10.22	BCC
Neon	Ne	10	20.18	−248.587	—	
Nickel	Ni	28	58.71	1455	8.91	FCC
Niobium	Nb	41	92.91	2469	8.58	BCC
Nitrogen	N	7	14.01	−210.0042	—	—
Oxygen	O	8	16	−218.789	—	—
Phosphorus	P	15	30.97	44.14	1.82	Ortho.
Platinum	Pt	78	195.09	1769	21.44	FCC
Potassium	K	19	39.1	63.71	0.862	BCC
Silicon	Si	14	28.09	1414	2.33	Dia. cub.
Silver	Ag	47	107.87	961.93	10.5	FCC
Sodium	Na	11	22.99	97.8	0.966	BCC
Sulfur	S	16	32.06	115.22	2.09	Ortho.
Tin	Sn	50	118.69	231.9681	7.29	BCT
Titanium	Ti	22	47.9	1670	4.51	HCP
Tungsten	W	74	183.85	3422	19.25	BCC
Uranium	U	92	238.03	1135	19.05	Ortho.
Xenon	Xe	54	131.3	−111.7582	—	—
Zinc	Zn	30	65.38	419.58	7.13	HCP

THE SCIENCE AND DESIGN OF ENGINEERING MATERIALS

SECOND EDITION

James P. Schaffer
Lafayette College

Ashok Saxena
Georgia Institute of Technology

Stephen D. Antolovich
Washington State University

Thomas H. Sanders, Jr.
Georgia Institute of Technology

Steven B. Warner
University of Massachusetts, Dartmouth

The McGraw-Hill Companies, Inc.

Taipei New York San Francisco Washington, D.C. Auckland Bangkok Bogota Caracas
Hamburg Hong Kong Jakarta Lisbon London Madrid Manila Mexico Milan
Montreal New Delhi Paris San Juan Sao Paulo Seoul Singapore Sydney Tokyo
Toronto

WCB/McGraw-Hill

A Division of The **McGraw-Hill** *Companies*

THE SCIENCE AND DESIGN OF ENGINEERING MATERIALS
International Editions 1999

2 3 4 5 6 7 8 9 0   P H W   9

ISBN   0-07-116762-5

When ordering this title, use ISBN   0-07-116762-5

THE AUTHORS

James P. Schaffer

James P. Schaffer is an associate professor of Chemical Engineering at Lafayette College in Easton, Pennsylvania. After receiving his B.S. in mechanical engineering (1981) and his M.S. (1982) and Ph.D. (1985) in materials science and engineering from Duke University, he taught at the Georgia Institute of Technology for five years before moving to Lafayette in 1990. He has taught an introductory materials engineering course to more than 1200 undergraduate students using the integrated approach taken in this text.

Dr. Schaffer's field of research is the characterization of atomic scale defects in materials using positron annihilation spectroscopy along with associated techniques. Professor Schaffer holds two patents and has published more than 30 papers. He has received a number of teaching awards including the Ralph R. Teetor Educational Award (SAE, 1989), Jones Lecture Award (Lafayette College, 1994), Distinguished Teaching Award (Middle Atlantic Section of ASEE, 1996), Superior Teaching Award (Lafayette Student Government, 1996), Marquis Distinguished Teaching Award (Lafayette College, 1996), and the George Westinghouse Award (ASEE, 1998). He is a member of ASEE, ASM International, TMS, Tau Beta Pi, and Sigma Xi.

Ashok Saxena

Ashok Saxena is currently professor and chair of the School of Materials Science and Engineering at the Georgia Institute of Technology. Professor Saxena received his M.S. and Ph.D. degrees from the University of Cincinnati in materials science and metallurgical engineering in 1972 and 1974, respectively. After eleven years in industrial research laboratories, he joined Georgia Tech in 1985 as a professor of materials engineering. He assumed the chairmanship of the school in 1993. From 1991 to 1994, he also served as the director of the Campus-Wide Composites Education and Research Center.

Dr. Saxena's primary research area is mechanical behavior of materials, in which he has published over 125 scientific papers and has edited several books. His research in the area of creep and creep-fatigue crack growth has won international acclaim; he was awarded the 1992 George Irwin Medal for it by ASTM. He is also the recipient of the 1994 ASTM Award of Merit. Professor Saxena is an ASTM Fellow, a Fellow of ASM International, and a member of ASEE, TMS, Sigma Xi, and Alpha Sigma Mu.

Stephen D. Antolovich

Stephen D. Antolovich is currently a professor of Mechanical and Materials Engineering at Washington State University, where he also serves as director of the School of Mechanical and Materials Engineering. He received his B.S. and M.S. in metallurgical engineering from the University of Wisconsin in 1962 and 1963, respectively, and a Ph.D. in materials science from the University of California–Berkeley in 1966. He joined the Georgia Institute of Technology in 1983, where he served as professor of materials engineering, director of the Mechanical Properties Research Laboratory (MPRL), and director of the School of Materials Science and Engineering.

In 1988 Dr. Antolovich was presented with the Reaumur Medal from the French Metallurgical Society. In 1989 he was named Professeur Invite by CNAM University in Paris. In 1990 he was presented with the Nadai Award by the ASME. Dr. Antolovich regularly makes presentations to learned societies in the United States, Europe, Canada, and Korea and has carried out funded research/consultation for numerous government agencies. Dr. Antolovich has published over 100 archival articles in leading technical journals. His major research interests are in the areas of deformation, fatigue, and fracture, especially at high temperatures. He is a member of ASME, ASTM, and AIME, and a Fellow Member of ASM International.

Thomas H. Sanders, Jr.

Thomas H. Sanders, Jr., is currently Regents' Professor in the School of Materials Science and Engineering at the Georgia Institute of Technology. Professor Sanders received his B.S. and M.S. in ceramic engineering from Georgia Tech in 1966 and 1969, respectively. In 1974 he completed his research for his Ph.D in metallurgical engineering at Georgia Tech and joined the Physical Metallurgy Division of Alcoa Technical Center, Alcoa Center, Pennsylvania. While at Alcoa Center his major research efforts were directed toward developing and implementing processing microstructure–properties relationships for high-strength aluminum alloys used in aerospace applications. He was on the faculty in Materials Science and Engineering at Purdue University from 1980 to 1986 and joined the faculty at Georgia Tech in 1987. He was awarded the W. Roane Beard Outstanding Teacher Award for 1994.

Dr. Sanders's primary research area is physical metallurgy of materials with primary emphasis on aluminum alloys. He has published approximately 100 scientific papers and has edited several books. He was awarded a Fulbright grant in 1992 to conduct research at Centre National de la Recherche Scientifique (ONERA), Châtillon, France. Professor Sanders is a member of TMS and a Fellow of ASM.

Steven B. Warner

Steven B. Warner is Professor and Chairperson of the Textile Sciences Department, University of Massachusetts, Dartmouth. Dr. Warner earned his combined S.B. and S.M. degrees in metallurgy and ceramics in 1973 from the Massachusetts Institute of Technology. In 1976 he was awarded an Sc.D. from the Department of Materials Science and Engineering at MIT. He was a research scientist from 1976–1982 at Celanese Research Co. and from 1982–1988 at Kimberly-Clark Corp. In 1987 he joined Georgia Institute of Technology as Adjunct Professor in Chemical Engineering; in 1988 he became Associate Professor in Materials Engineering; and from 1990–1994 he was a faculty member in Textile and Fiber Engineering.

Dr. Warner's research interests are the structure-property relationships of materials, especially polymers. He has published more than 30 scientific papers, holds six U.S. patents, and is the author of *Fiber Science*. In addition he has been a technical expert in a number of patent cases.

FOREWARD

If one's technical library were to contain only a single book on materials, this is the book to have. The authors have succeeded in covering the field of materials science and engineering in even its broadest aspects. They have captured both the science of the discipline as well as the engineering and design of materials. All classes of materials are treated; metals, semiconductors, ceramics, and polymers, as well as composites made of combinations of these. As urged in the National Research Council's recent study of materials science and engineering, processing and synthesis also are included, as are the subjects of machinability and joining. (No material, however outstanding its properties, is likely to be very useful if it can't be produced, shaped, or attached to other components.)

The breadth of *The Science and Design of Engineering Materials*, which reflects the varied fields of expertise of the authors, makes it an ideal text for a survey course for students from all fields of engineering. Because of the depth as well as the breadth with which the topics are treated, the text also is an excellent choice for introductory courses for materials science and engineering majors. Graduates of these introductory and survey classes will value *The Science and Design of Engineering Materials* as a resource book for years to come. The clear explanations and frequent examples allow the practicing engineer, on his or her own, to become acquainted with the materials field or update his/her knowledge of it. Care and skill have been exercised in the choice of illustrations. Numerous drawings and graphs augment explanations in the text, and clearly reproduced micrographs provide real-life examples of the phenomena being described. The examples and questions are especially noteworthy. While a portion of the questions are of the "one right answer" kind, and are intended to reinforce and clarify the material in the text, others are of the open-ended, design type that require creative thought and more closely resemble real-life situations. They can form the bases for useful and provocative class discussions.

This new edition of *The Science and Design of Engineering Materials* is a valuable addition to the materials literature. It will contribute to the materials education of engineers and scientists for years to come.

Julia Weertman
Walter P. Murphy Professor of Materials Science and Engineering
Northwestern University

A society's ability to develop and use materials is a measure of both its technical sophistication and its technological future. This book is devoted to helping all engineers better understand and use materials to ensure the future of technology.

THE INTENDED MARKET

The book is intended for undergraduate students from all engineering disciplines. It assumes a minimal background in calculus, chemistry, and physics at the first-year college level. The text has been used successfully in a variety of situations including:

- A traditional 40- to 42-lecture single-semester/quarter course
- A yearlong course sequence
- A foundation course for materials engineering majors
- A service course with students from multiple engineering disciplines
- A service course targeted at a specific audience (for mechanical or electrical engineers only)
- A section composed of only first- and second-year students
- As a refresher course for materials engineering graduate students with a B.S. degree in another engineering discipline.

Though only some of the chapters might be used in a single-semester/quarter course, experience suggests that students benefit from reading the entire text. The authors have intentionally made no effort to mark optional sections or chapters, since topic selection is a function of many factors, including instructor preferences, the background and needs of the students, and the course sequence at a specific institution.

THE AUTHOR TEAM

The field of materials engineering is so vast that no single individual can master it all. Therefore, a team was assembled with expertise in ceramics, composites, metals, polymers, and semiconductors. The author team has the collective expertise to explain clearly all the important aspects of the field in a single coherent package. The authors teach or have taught in chemical, materials, mechanical, and textile engineering departments. We teach at small colleges, where the engineering program is within a liberal arts setting, as well as major technological universities. Just as a composite combines the best features of its constituent materials, this book combines the varied strengths of its authors.

THE INTEGRATED APPROACH

The book is organized into four parts. Part I, Fundamentals, focuses on the structure of engineering materials. Important topics include atomic bonding, thermodynamics and kinetics, crystalline and amorphous structures, defects in crystals, and strength of crystals. The concepts developed in these six chapters provide the foundation for the remainder of the course. In Part II, Microstructural Development, the important processing variables of temperature, composition, and time are introduced, along with methods for controlling the structure of a material on the microscopic level. Part III focuses on the engineering properties of the various classes of materials. It builds upon the understanding of structure developed in Part I and the methods used to control structure set forth in Part II. It is in the properties section of the text that our approach, termed the integrated approach, differs from that of most of the competing texts.

Traditionally, all the macroscopic properties of one type of material (usually metals) are discussed before moving on to describe the properties of a second class of materials. The process is then repeated for ceramics, polymers, composites, and semiconductors. This traditional progression offers several advantages, including the ability to stress the unique strengths and weaknesses of each material class.

As authors, we believe most engineers will be searching for a material that can fulfill a specific list of properties as well as economic, processing, and environmental requirements and will want to consider all classes of materials. That is, most engineers are more likely to "think" in terms of a property class rather than a material class. Thus, we describe the mechanical properties of all classes of materials, then the electrical properties of all classes of materials, and so on. We call this the *integrated approach* because it stresses fundamental concepts applicable to all materials first, and then points out the unique characteristics of each material class. During the development of the book the authors found that there were times when "forcing integration" would have degraded the quality of the presentation. Therefore, there are sections of the text where the integrated approach is temporarily suspended to improve clarity and emphasize the unique characteristics of specific materials.

The fourth and final part of the book deals with processing methods and with the overall materials design and selection process. These two chapters tie together all the topics introduced in the first three parts of the book. The goal is for the student to understand the methods used to select the appropriate material and processing methods required to satisfy a strict set of design specifications.

EMPHASIS ON DESIGN AND APPLICATIONS

Students are better able to understand the theoretical aspects of materials science and engineering when they are continually reinforced with applications and examples from their personal experiences. Thus, we have made a substantial effort to include both familiar and technologically important applications of every concept introduced in the text. In many cases we begin a discussion of a topic by describing a familiar situation and asking why certain results occur. This approach motivates the students to learn the details of the quantitative models so that they can solve problems, or understand phenomena, in which they have a personal interest.

The authors believe that most engineering problems have multiple correct solutions and must include environmental, ethical, and economic considerations. Therefore, our homework problems include both numerical problems with a single correct answer and design problems with multiple valid solution techniques and "correct" answers. The sample exercises within the text are divided into two classes. The **Examples** are straightforward applications of concepts and equations in the text and generally have a single correct numerical solution. In contrast, **Design Examples** are open-ended and often involve selecting a material for a specific application.

We have used a **Case Study** involving the design of a camcorder as a continuous thread throughout the manuscript. Each of the four parts of the text—Fundamentals, Microstructural Development, Properties, and Design—begins with the identification of several materials issues associated with the camcorder that can only be understood using concepts developed in that portion of the text. This technique allows students to get a view of the forest before they begin to focus on individual trees. The ongoing case also permits us to form bridges between the important aspects of the course within a context that is familiar to most students.

The authors' belief in the importance of materials design and selection is underscored by the inclusion of an entire chapter on this subject at the end of the book. We recommend strongly that the instructor have the students read this chapter even if the schedule does not permit its inclusion in lecture. We find that it "closes the loop" for many of our students by helping them to understand the relationships among the many and varied topics introduced in the text. The design chapter contains 10 case studies and addresses issues such as life-cycle cost analysis, material and process selection, nuclear waste disposal, inspection criteria, failure analysis, and risk assessment and product liability.

CHANGES TO THE SECOND EDITION

Five new features have been added to the second edition of the text:

1. Each chapter begins with a motivational insert called Materials in Action. This feature is designed to introduce the reader to the important ideas in the chapter through an interesting real-world situation. Examples include a description of how adding 0.4 weight percent carbon to iron increases the strength of the material by two orders of magnitude, a discussion of why directionally solidified nickel-based turbine blades are worth their weight in gold in some aerospace applications, and an illustration of the false economy of using less expensive machining operations if they have a negative influence on fatigue crack initiation. This new feature extends our emphasis on design and applications, which was one of the most popular attractions of the first edition.

2. We have developed a new Materials in Focus CD-ROM to enhance the textbook presentation. The CD-ROM contains a phase diagram tool and over 30 animations designed to help the reader gain an understanding of some of the visual concepts in the book. Examples include "three-dimensional" views of unit cells and polymer molecules, the movement of dislocations through crystals, changes in the population of electron energy levels in semiconductors with temperature, illustrations of polarization mechanisms, and examples of processing operations. In addition, the CD-ROM contains all of the photomicrographs in the text, and a series of interactive example problems. For example, in the portions of Chapter 7 on phase diagrams students can select a state point on a phase diagram and have the software help them determine the phases present, the compositions of the phases, and their relative amounts. Every illustration on the CD-ROM is directly linked to an illustration, concept, or problem in the text. In fact, every location in the text that has a link to a CD-ROM animation or example is clearly indicated by the presence of a "CD-ROM" icon in the margin of the text.

3. Over 225 new homework problems have been added throughout the text. The majority of the chapters contain several design problems (i.e., problems with multiple correct solutions). These homework problems are marked with a "Design Problem" icon in the margin of the text.

4. We have added an eight-page full color insert near the center of the book. This feature allows us to illustrate several important applications of materials science and engineering that simply are not easily described with either words or two-color illustrations.

5. The entire book has been redesigned for enhanced readability. In particular, the use of the icons illustrated below permits the reader to quickly identify several important features of the second edition:

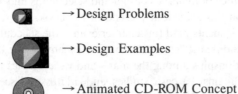

→ Design Problems

→ Design Examples

→ Animated CD-ROM Concept

We have made a determined effort to improve the quality of the photomicrographs and to eliminate errors that were present in the first edition. We would like to express our sincere thanks to those of you who spotted problems and pointed them out to us. The book is better for your efforts, and if you have additional suggestions for how to improve the text we would be happy to hear them.

6. A Web site for the book can be found at http://www.mhhe.com. It contains information about the book and its supplements, Web links, and teaching resources.

ACKNOWLEDGMENTS

This book has undergone extensive revision under the direction of a distinguished panel of colleagues who have served as reviewers. The book has been greatly improved by this process and we owe each reviewer a sincere debt of gratitude. The reviewers for the first edition were:

John R. Ambrose,	*University of Florida*
Robert Baron,	*Temple University*
Ronald R. Bierderman,	*Worcester Polytechnic Institute*
Samuel A. Bradford,	*University of Alberta*
George L. Cahen, Jr.,	*University of Virginia*
Stephen J. Clarson,	*University of Cincinnati*
Diana Farkas,	*Virginia Polytechnic Institute*
David R. Gaskell,	*Purdue University*
A. Jeffrey Giacomin,	*Texas A&M University*
Charles M. Gilmore,	*The George Washington University*
David S. Grummon,	*Michigan State University*
Ian W. Hall,	*University of Delaware*
Craig S. Hartley,	*University of Alabama at Birmingham*

Phillip L. Jones,	*Duke University*
Dae Kim,	*The Ohio State University*
David B. Knorr,	*Rensselaer Polytechnic Institute*
D. Bruce Masson,	*Washington State University*
John C. Matthews,	*Kansas State University*
Masahiro Meshii,	*Northwestern University*
Robert W. Messler, Jr.,	*Rensselaer Polytechnic Institute*
Derek O. Northwood,	*University of Windsor*
Mark R. Plichta,	*Michigan Technological University*
Richard L. Porter,	*North Carolina State University*
John E. Ritter,	*University of Massachusetts*
David A. Thomas,	*Lehigh University*
Peter A. Thrower,	*Pennsylvania State University*
Jack L. Tomlinson,	*California State Polytechnic University*
Alan Wolfenden,	*Texas A&M University*
Ernest G. Wolff,	*Oregon State University*

The reviewers for the second edition are:

Bezad Bavarian,	*California State University–Northridge*
David Cahill,	*University of Illionois*
Stephen Krause,	*Arizona State University*
Hillary Lackritz,	*Purdue University*
Thomas J. Mackin,	*University of Illinois–Urbana*
Arumugam Manthiram,	*The University of Texas at Austin*
Walter W. Milligan,	*Michigan Technological University*
Monte J. Pool,	*University of Cincinnati*
Suzanne Rohde,	*University of Nebraska–Lincoln*
Jay Samuel,	*University of Wisconsin–Madison*
Shome N. Sinha,	*University of Illinois–Chicago*

The authors would also like to thank the members of the editorial team: Tom Casson, publisher; Scott Isenberg; Kelley Butcher, developmental editor; and Gladys True, project manager. We would also like to thank James Mohler of the Department of Technical Graphics, Purdue University, the developer of the Materials in Focus CD-ROM.

SUPPLEMENTS

We have devoted considerable effort to the preparation of a high-quality solutions manual. Our approach is to employ a common solution technique for every homework problem. The procedure includes the following steps:

1. **Find:**   (What are you looking for?)
2. **Given:**   (What information is supplied in the problem statement?)
3. **Data:**   (What additional information is available, from tables, figures, or equations in the text, and is required to solve this problem?)

4. Assumptions: (What are the limits on this analysis?)
5. Sketch: (What geometrical information is required?)
6. Solution: (A detailed step-by-step procedure.)
7. Comments: (How can this solution be applied to other similar situations and what alternative solution techniques might be appropriate?)

The solutions manual is available to adopters of the text. Also, the authors have gained considerable experience using the "integrated" approach in the classroom and are available to discuss implementation strategies with interested colleagues at other institutions.

James P. Schaffer   **Thomas H. Sanders, Jr.**
Ashok Saxena   **Steven B. Warner**
Stephen D. Antolovich

BRIEF CONTENTS

CONTENTS

THE SCIENCE AND DESIGN OF ENGINEERING MATERIALS

SECOND EDITION

MATERIALS SCIENCE AND ENGINEERING

MATERIALS IN ACTION   Building Blocks of Technology

Materials are at the core of all technological advances. Mastering the development, synthesis, and processing of materials opens opportunities that were scarcely dreamed of a few short decades ago. The truth of this statement is evident when one considers the spectacular progress that has been made in such diverse fields as energy, telecommunications, multimedia, computers, construction, and transportation. Travel by jet aircraft would be impossible without the materials that were developed specifically for the jet engine, and there would be no computers as we know them without solid-state microelectronic circuits. Indeed, it has been stated that the transistor has had the most far-reaching impact of *any* scientific or technological discovery to date. The centrality of materials to advanced technical societies was recognized in a recent report to the U.S. Congress authored by some of the most distinguished educators and scientists in the country. In that report it was stated that

advanced materials and advanced processing of materials are critical to the nation's quality of life, security, and economic strength. Advanced materials are the building blocks of advanced technologies. Everything Americans use is composed of materials, from semiconductor chips to flexible concrete skyscrapers, from plastic bags to a ballerina's artificial hip, or the composite structures on spacecraft. The impact of materials extends beyond products, in that tens of millions of manufacturing jobs depend on the availability of high-quality specialized materials.

In that same report it was further stated that

advanced materials are the building blocks of technology. When processed in particular ways, they enable the technological advances that constitute progress. Advanced materials and processing methods have become essential to the enhancement of [the] quality of life, security, industrial productivity and economic growth. They are the tools for addressing urgent problems, such as pollution, declining natural resources and escalating costs.

The ability to develop and use materials is fundamental to the advancement of any society. In this text we will explore how that is done by engineers to improve the well-being of mankind.

Source: Reprinted from *Materials Science and Engineering for the 1990s: Maintaining Competitiveness in the Age of Materials,* National Research Council, Washington, D.C. (National Academy Press, 1989).

1.1 INTRODUCTION

Our purpose in this book is to examine the way in which materials impact society and to show how they are produced, processed, and used in all branches of engineering to advance the well-being of society. In doing this, we will emphasize the relationship between the structure of a material and its underlying properties, and we will develop general principles applicable to all materials. Our goal in following this approach is to enable students to develop a fundamental understanding of material behavior that will help prepare them for a rapidly changing, and sometimes bewildering, environment. Since engineering is essentially an applied activity, practical examples that build on and amplify the fundamentals will also be emphasized for all topics and materials that are considered. The final chapter presents case studies in which the principles and practical information developed in the preceding chapters are integrated into the solution of real, materials-based engineering problems.

In the remainder of this chapter, we will review the fundamental relationship between a society's economic well-being and its ability to understand and convert materials into usable forms. We will introduce the importance of the relationships between structure, properties, and processing for all classes of (solid) materials in all branches of engineering. The chapter concludes with examples of some of the exciting opportunities and challenges that lie ahead in the areas of mechanical, aerospace, electrical, and chemical engineering.

1.2 THE ROLE OF MATERIALS IN TECHNOLOGICALLY ADVANCED SOCIETIES

Throughout history, most major breakthroughs in technology have been associated with the development of new materials and processes. For example, consider the materials-processing innovations that led to the development of the Damascus sword. Two methods were used to fabricate such swords. In one process, alternating layers of soft iron and steel (in this case Fe with about 0.6% C) were hammered together at high temperatures to produce a blade that had an edge of hard steel to retain a sharp cutting surface and a body of iron that provided resistance to fracture. In Japan, similar results were obtained by hammering steel into a thin sheet and then folding it back upon itself many times. A finished Japanese sword is shown in Figure 1.2–1; the variations in structure are quite

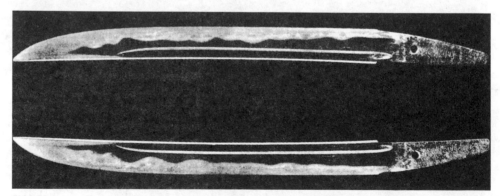

FIGURE 1.2–1 Photographs of the front and back sides of a Japanese sword forged by Hiromitsu in the mid-16th century. The smoothly waving outline was produced by polishing and the contrast enhanced by lighting. The structure of the hard and soft areas (mottled regions) can be seen along the edge from the tip to the midpoint. *(Source: Cyril Stanley Smith,* A Search for Structure: Selected Essays on Science, Art, and History, *MIT Press, Cambridge, MA, copyright 1992.)*

clear. The result of either processing method was a novel layered metal structure. Weapons produced from metals with this new structure gave their possessors a great advantage in battle. Similarly fabricated weapons in the Middle East provided one basis for the spread of the Syrian empire.

This example illustrates one of the key principles of materials science and engineering—the intimate link between structure, properties, and processing. The structure of the metal resulting from innovative processing methods provided new combinations of properties that offered significant advantage to those who developed the technology. Thus, these swords represent one of the first engineered materials.

More recently, development of processes to obtain precise compositional and structural control has made miniaturized transistor technology possible. The result has been an electronics revolution that produced products such as computers, cellular phones, and compact disk players that continue to affect all aspects of modern life.

Another area where materials provide the springboard for advance is the aerospace industry. Light, strong alloys of aluminum and titanium have fostered the development of more efficient airframes, while the discovery and improvement of nickel-base alloys spurred development of powerful, efficient jet engines to propel these planes. Further improvements are being made as composites and ceramics are substituted for conventional materials.

The role of materials in the exploration of space is of central importance. One prominent example lies with the U.S. space shuttle. During reentry, extremely high temperatures develop as a result of friction between the earth's atmosphere and the shuttle. These temperatures, which can exceed 1600°C, would melt any metal currently used in airframes. Ceramic tiles, which have the ability to withstand extremely high temperatures and have excellent insulating properties, provide a method for protecting the aluminum frame of the spacecraft.

The approximate temperature distribution developed on the surface of the space shuttle during reentry is shown in Figure 1.2–2. Those regions in which the temperature ranges

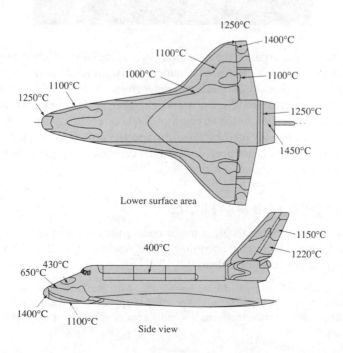

Lower surface area

Side view

FIGURE 1.2–2

Surface temperatures of U.S. space shuttle during reentry into the earth's atmosphere. *(Source: G. Lewis,* Selection of Engineering Materials, *Prentice Hall, Inc., Englewood Cliffs, NJ, 1990.)*

between 400 and 1260°C have been protected with about 30,000 silica tiles. The tiles are coated with a layer of black borosilicate glass to both insulate the surface and radiate thermal energy from the shuttle. In those regions that may reach 1600°C, coated reinforced carbon/carbon composites (materials composed of carbon fibers surrounded by a carbon matrix) are used. Without such materials, it is doubtful that a reusable space vehicle would be possible. This is an example of the way our highest aspirations are realized through our practical ability to develop and work with advanced materials.

Another example of materials providing the vehicle to technological breakthrough occurs in telecommunications. Information that was once carried electrically through copper wires is now being carried optically, through high-quality transparent SiO_2 fibers as shown in Figure 1.2–3. The optical properties of the fibers are deliberately and precisely varied across the fiber diameter to provide for maximum efficiency. Using this technology has increased the speed and volume of information that can be carried by orders of magnitude over what is possible using copper cable. Moreover, the reliability of the transmitted information has been vastly improved. In addition to these benefits, the negative effects of copper mining on the environment have been reduced, since the materials and processes used to produce glass fibers have more benign environmental effects.

FIGURE 1.2–3

Optical fiber preform used to manufacture lightguides. The rings represent areas having different indices of refraction. When the preform is drawn, the final fiber diameter is about 125×10^{-6} m. *(Source: Permission of AT&T Archives.)*

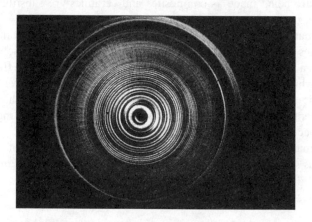

The centrality of materials to the economic well-being of the United States has been pointed out in the National Research Council study entitled *Materials Science and Engineering for the 1990s—Maintaining Competitiveness in the Age of Materials.* This document states that "materials science and engineering is *crucial* to the success of industries that are important to the strength of the U.S. economy and U.S. defense." A similar position has been adopted by Japan, where the ability to develop, process, and fabricate advanced materials has been declared the cornerstone of the nation's strategy to maintain a leading technological position.

1.3 THE ENGINEERING PROFESSION AND MATERIALS

In one way or another, materials are a major concern in all branches of engineering. In fact, the definition of engineering according to the Accreditation Board for Engineering and Technology makes this point clearly:

Engineering is the profession in which a knowledge of the mathematical and natural sciences gained by study, experience, and practice is applied with judgment to develop ways to utilize, economically, the materials and forces of nature for the benefit of mankind.

If this definition is accepted, we can see that engineering is a profoundly human activity that touches upon the life of all members of society. We can also see that an engineer is not only an applied scientist but much more. The engineer must have a good business sense, including an understanding of economics.

Important differences exist between the functions and approaches of engineers and scientists. Engineering is essentially an integrating activity, while science is a reductionist activity. The engineer often employs an intuitive, global (and, frequently, empirical) approach as opposed to that of the scientist, who breaks a problem down into its most basic elements to elucidate fundamental principles. In other words, an engineer is frequently required to solve problems by synthesizing knowledge from various disciplines and to produce items without a complete fundamental understanding of what he or she is dealing with. In such cases an engineer must define the operating conditions and develop a test program, based on his or her intuition, that will allow the project to move ahead in a safe, orderly, and economical manner.

In carrying out a job, the engineer will be faced with an almost infinite number of materials from which to choose. In some cases the materials will be put into service with little or no modification required, while in other cases additional processing will be necessary to obtain the desired properties. In choosing the best material for the job, the best approach is to determine the properties that are required and to then see what material will meet those properties at the lowest cost.

It is important to have a clear understanding of what is meant by the word *cost*. It does not simply refer to the initial cost of an item. Something may have a high initial cost, yet over the lifetime of the part, the total cost, taking all factors into account, may be low. An approach that considers the lifetime of the component or assembly is commonly referred to as life-cycle cost analysis. Factors such as reliability, replacement cost, the cost of downtime, the cost of environmental cleanup or disposal, and many others must all be considered. Materials play a key role in the life-cycle cost of a part. For example, consider tennis rackets or skis fabricated from composites, or macroscopic mixtures, of carbon fibers embedded in an epoxy matrix. While the initial cost of these items is relatively high, they are very durable and over their (significantly longer) lifetime are much less expensive than the metal or wood items they replaced.

It is also important for the engineer to realize that choice of materials cannot be made on the basis of a single property. For example, if an electrical engineer is designing a component in which the ability to conduct electricity is the principal property, he or she must remember that the material must be capable of being economically fabricated into the required form, be able to resist breaking, and have long-term stability so that the properties will not change significantly with time. Thus, in the majority of cases, choice of a material involves a complex set of trade-offs (including economic factors), and there is seldom one single solution that is "right" for the given application. Alternatively stated, there are often multiple "correct" solutions to a materials-selection problem; engineers must investigate several alternate solutions before making a final selection.

In addition, as we have seen in the case of the space shuttle, the materials selected must function together as a system. While each material is selected for specific properties to fulfill a given need, the materials must also be capable of operating together without degrading the properties of one another.

1.4 MAJOR CLASSES OF MATERIALS

The major classes of engineering materials are considered to be: (1) metals, (2) ceramics, (3) polymers, (4) composites, and (5) semiconductors. Metals with which you are probably familiar include iron, copper, aluminum, silver, and gold; common ceramics include sand, bricks and mortar, (window) glass, and graphite; examples of familiar polymers are cellulose, nylon, polyethylene, Teflon, Kevlar, and polystyrene; we have already discussed mixtures of materials known as composites such as carbon/carbon composites used in tiles on the space shuttle and carbon fibers in an epoxy matrix used in tennis rackets and skis; and the simplest semiconductors are silicon and germanium. By understanding the similarities and differences among these classes of materials, you will be in a position to make intelligent materials choices that can meet the challenges of modern technology.

Why are materials arranged in the groups listed above? Many materials have similar atomic structures or useful engineering properties or both that make it convenient to classify them into these five broad groups. It should be recognized that these classifications are somewhat arbitrary and may change with new discoveries and advances in technology. Composites, also sometimes called "engineered materials," provide an excellent example of a new classification. These materials are made by combining other (often conventional) materials, using advanced technology, to obtain properties that could not be obtained from the existing classes of materials.

In our discussion in this chapter and throughout the book we will emphasize that the properties of a material are related to its structure. We will deal with structure at many size scales ranging from the atomic scale ($\sim 0.1 \times 10^{-9}$ m or 0.1 nm) through the microscopic scale ($\sim 50 \times 10^{-6}$ m or 50 μm), and up to the macroscopic scale ($\sim 10^{-2}$ m or 1 cm). In the next chapter we will see that the material structure on each of these size scales can be used to understand and explain certain materials properties.

While the properties of a material are related to its structure, it is important to understand that the way in which a material is processed affects the structure and hence the properties. As an example of this important concept, consider the dramatic effect that thermal processing can have on the properties of steel. If slowly cooled from a high temperature, steel will be relatively soft and have low strength. If the same steel is quenched (i.e., rapidly cooled) from the same high temperature, it will be extremely hard and brittle. Finally, if it is quenched and then reheated to some intermediate temperature, it will have an excellent combination of strength and toughness. While we will study this example in depth later in the text, the major point to be made here is that each of the three thermal processes has produced a different structure in the same material, which in turn gives rise to different properties.

Each of the five classes of materials, together with some elementary structure-property relationships, is discussed briefly in the following sections.

1.4.1 Metals

Metals form solids in which the atoms are located in regularly defined, repeating positions throughout the structure. These regular repeating structures, known as crystals and discussed in detail in Chapter 3, give rise to specific properties. Metals are excellent conductors of electricity, are relatively strong, are dense, can be deformed into complex shapes, and are resistant to breaking in a brittle manner when subjected to high-impact forces. This set of mechanical and physical properties makes metals one of the most important classes of materials for both electrical and structural applications. Extensive (and in some cases exclusive) use of metals occurs in automobiles, airplanes, buildings, bridges, machine tools, ships, and many other applications where a combination of high strength and resistance to brittle fracture is required. In fact, it is largely the excellent combination of strength and toughness (i.e., resistance to fracture) that makes metals so attractive as structural materials.

The basic understanding of metals and their properties is advanced, and they are considered to be mature materials with relatively little potential for major breakthroughs. However, significant improvements have been and continue to be made as a result of advances in processing. Two examples are:

▮ Higher operating temperatures in jet engines have been attained through the use of turbine blades that are produced by controlled solidification processes. The blades are made of alloys (atomic-scale mixtures of atoms) of nickel or other metals and are in wide commercial use. Improvements will continue as processes are refined through use of advanced sensors and real-time computer control.

▮ Frequently parts are fabricated from metal powders by compacting them into a desired shape at high temperature and pressure in a process known as powder metallurgy (PM). An important reason for using PM processing is reduced fabrication costs. While some improvement in properties can be obtained through PM, a major benefit is the reduced variation in properties, which will allow the operating loads to be safely increased. Reduced production costs through PM will continue to impact the aerospace and automotive fields.

1.4.2 Ceramics

Ceramics are generally composed of both metallic and nonmetallic atomic species. Many (but not all) ceramics are crystalline, and frequently the nonmetal is oxygen, as in Al_2O_3, MgO, and CaO, all of which are typical ceramics. One significant difference between ceramics and metals is that in ceramics, bonding is ionic and/or covalent. As a result there are no "free" electrons in ceramics. They are generally poor conductors of electricity, but are frequently used as insulators in electrical applications. One familiar example is spark plugs, in which a ceramic insulator separates the metal components.

Ionic and covalent bonds are extremely strong. As a result, ceramic materials are intrinsically stronger than metals. However, because of their more complex structure, the ions or atoms cannot easily be displaced as a result of applied forces. Rather than bend to accommodate such forces, ceramics tend to fracture in a brittle manner. This brittleness generally limits their use as structural materials, although recent improvements have been made by incorporating ceramic fibers into a ceramic matrix and other innovative techniques. Ceramics' rigid bond structure confers other advantages, including high temperature stability, resistance to chemical attack, and resistance to absorption of foreign substances. They are thus ideal in high-temperature applications such as the space shuttle, as containers for reactive chemicals, and as bowls and plates for foods where surface contamination is undesirable.

Some ceramics are not crystalline. The most common example is window glass, which is composed primarily of SiO_2 with the addition of various metal oxides. Optical properties are of major importance in glass and may be controlled through composition and processing. In addition the thermal and mechanical properties of glass can also be controlled. Safety glass is simply glass that has been subjected to a thermal cycle that leaves the surface in a state of compression and thereby resistant to cracking. In fact, glass treated in this way is even difficult to crack when struck with a hammer!

Some current and potential applications for ceramic materials with a large economic impact are listed below:

▮ In the automotive industry the thermal and strength properties of ceramics make them very attractive for engine components. For example, there are over 60,000 autos in Japan with ceramic turbochargers, which increase the efficiency of the automobile. The materials in this application are Si_3N_4 or SiC processed to have some ability to resist brittle fracture.

■ Ceramics based on compounds such as $YBa_2Cu_3O_7$ and $Ba_2Sr_2CaCu_2O_x$ have increased critical superconducting temperatures to > 95 K. This means that superconducting films may be used as liners in microwave devices and as wires for all kinds of applications. Improving the current-carrying capacity and connection technology are essential for widespread application of these materials.

■ Next-generation computers will have ceramic electro-optic components that will give increased speed and efficiency.

1.4.3 Polymers

Polymers consist of long-chain molecules with repeating groups that are largely covalently bonded. Common elements within the chain backbone include C, O, N, and Si. An example of a common polymer with a simple structure, polyethylene, is shown in Figure 1.4–1. The bonds within the backbone are all covalent, so the molecular chains are extremely strong. Chains are usually bonded to each other, however, by means of comparatively weak secondary bonds. This means that it is generally easy for the chains to slide by one another when forces are applied and the strength is thus relatively low. In addition, many polymers tend to soften at moderate temperatures, so they are not generally useful for high-temperature applications.

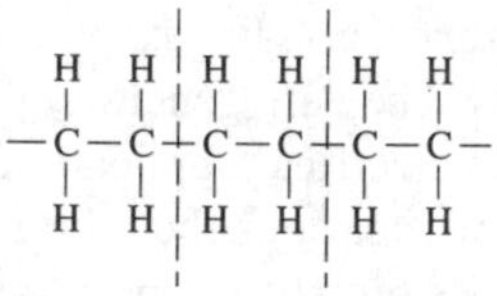

FIGURE 1.4–1   Schematic of the structure of polyethylene. The mer or basic repeating unit in the polymer is the $-C_2H_4-$ group.

Polymers, however, have properties that make them attractive in many applications. Since they contain common elements and are relatively easy to synthesize, or exist in nature, they can be inexpensive. They have a low density (in part because of the light elements from which they are constituted) and are easily formed into complex shapes. They have thus replaced metals for molded parts in automobiles and aircraft applications, especially where the load-bearing requirements are modest. Because of these properties, as well as their chemical inertness, they are used as beverage containers and as piping in plumbing applications.

Like metals and ceramics, their properties can be modified by compositional changes and by processing. For example, substitution of a benzene ring for one in four hydrogen atoms converts polyethylene, shown in Figure 1.4–1, to polystyrene, Figure 1.4–2. Polyethylene is pliable and is used for applications such as "squeeze bottles." In polystyrene, the comparatively large benzene side group restricts the motion of the long-chain molecules and makes the structure more rigid. If the benzene group in polystyrene is replaced with a Cl atom (intermediate in size between H and the benzene ring), polyvinylchloride is produced. The Cl atom will restrict the chain mobility more than an H atom but less than a benzene ring. A leathery material is produced with somewhat intermediate properties between polyethylene and polystyrene. These three polymers illustrate the fundamental principle, applicable to all materials, of the relationship between material structure and properties.

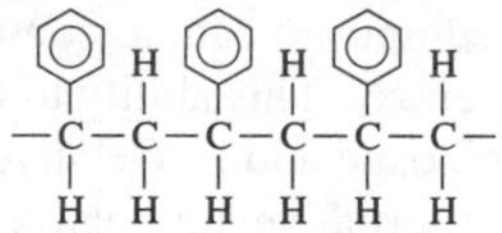

FIGURE 1.4–2   Schematic of the structure of polystyrene. This polymer has the same basic structure as the polyethylene shown in Figure 1.4–1 except that a benzene ring (C_6H_5) has been substituted for one of the four H atoms. As a result of the larger side group, which hinders the sliding motion of adjacent polymer chains, polystyrene is stiffer than polyethylene.

Some current and potential applications for polymers include the following:

■ The development of biodegradable polymers offers the potential for minimizing the negative impact on our environment that results from the tremendous amount of waste our society generates.

■ Advances in liquid-crystal polymer technology may permit development of lightweight structural materials.

■ Electrically conducting polymers may be able to replace traditional metal wires in weight-critical applications such as electrical cables in aerospace vehicles.

1.4.4 Composites

Composites are structures in which two (or more) materials are combined to produce a new material whose properties would not be attainable by conventional means. Examples include plywood, concrete, and steel-belted tires. The most prevalent applications for fiber-reinforced composites are as structural materials where rigidity, strength, and low density are important. Many tennis rackets, racing bicycles, and skis are now fabricated from a carbon fiber–epoxy composite that is strong, light, and only moderately expensive. In this composite, carbon fibers are embedded in a matrix of epoxy, as shown in Figure 1.4–3. The carbon fibers are strong and rigid but have limited ductility. Because of their brittleness, it would not be practical to construct a tennis racket or ski from carbon alone. The epoxy, which in itself is not very strong, plays two important roles. It acts as a medium to transfer load to the fibers, and the fiber-matrix interface deflects and stops small cracks, thus making the composite better able to resist cracks than either of its constituent components.

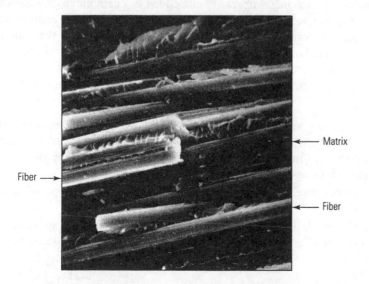

FIGURE 1.4–3
A cross-sectional view of a carbon-epoxy composite showing the strong and stiff graphite fibers embedded in the tough epoxy matrix. *(Source: Bhagwan D. Agarwal and Lawrence J. Broutman,* Analysis and Performance of Fiber Composites, *2nd ed., copyright © 1990 by John Wiley & Sons, New York. Reprinted by permission of John Wiley & Sons, Inc.)*

The strength and rigidity of a composite can be controlled by varying the amount of carbon fiber incorporated into the epoxy. This ability to tailor properties, combined with the inherent low density of the composite and its (relative) ease of fabrication, makes this material an extremely attractive alternative for many applications. In addition to the sporting goods described above, similar composites are used in aerospace applications such as fan blades in jet engines (where the operating temperatures are low) and for control surfaces in airframes. The use of composites in the F-18 fighter aircraft is shown in Figure 1.4–4.

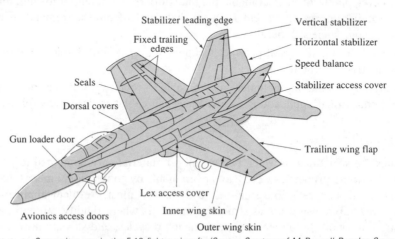

FIGURE 1.4–4 Composites use in the F-18 fighter aircraft. *(Source: Courtesy of McDonnell Douglas Corporation.)*

Composites can also be fabricated by incorporating strong ceramic fibers in a metal matrix to produce a strong, rigid material. An example is SiC fibers embedded in an aluminum matrix. Such a composite, known as a metal matrix composite, finds application as an airframe material for components in which moderate loads are encountered, such as in the skin of the fuselage.

Composites in which metal fibers are embedded in a ceramic matrix (ceramic-matrix composites) are produced in an attempt to take advantage of the strength of the ceramic while obtaining an increase in the toughness from the metal fibers that can deform and deflect cracks. When a crack is deflected, more load is required to make it continue to propagate, and the material is effectively tougher.

Some exciting new developments and possibilities for composites include the following:

▮ There is great potential to reduce the weight and increase the payload of airplanes. Initial uses are for lightly loaded parts such as vertical stabilizers and control surfaces made from carbon fiber–epoxy, but metal-matrix composites will play an increasingly important role.

▮ High-temperature ceramic-matrix composites will increase operating temperatures of engines.

▮ A significant challenge in increasing the use of composites is to learn to design with materials having totally different modes of failure than do conventional materials.

1.4.5 Semiconductors

The major semiconducting materials are the covalently bonded elements silicon and germanium as well as a series of covalently bonded compounds including GaAs, CdTe, and InP, among others. In some ways semiconductors are a subclass of ceramics, since their bonding characteristics and mechanical properties are similar to those previously described for ceramics. The commercial importance of semiconductors, however, warrants their consideration separately. For these materials to exhibit the level of reproducibility of properties required by the microelectronics industry, semiconductors must be processed in ways that permit precise control of composition and structure. In fact, the processing techniques for semiconductors are among the most highly developed of those used for any materials class. For example, impurity levels are routinely controlled in the parts-per-billion range (i.e., a few impurity atoms for every billion host atoms).

The previous discussion on composites focused on materials used for structural applications. It should be understood that microelectronic devices are essentially composites in which a host of radically different property requirements means that different classes of materials (metallic conductors, active semiconducting elements, and ceramic insulators) must be used in close proximity. One of the major challenges in the area of microelectronics lies in miniaturization and fabrication of these devices. The extremely fine scale of present-day microelectronic devices is shown in Figure 1.4–5. Here it is clear that many of the components of this composite structure are of submicron size!

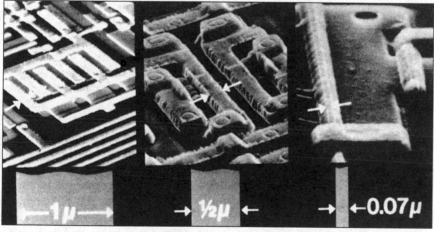

FIGURE 1.4–5 Microelectronic circuits. Note the very small size of some of the features on these devices. *(Source: Reprinted with permission from* Materials Science and Engineering for the 1990s: Maintaining Competitiveness in the Age of Materials. *Copyright 1989 by the National Academy of Sciences. Courtesy of the National Academy Press, Washington, DC.)*

Some present and future applications for semiconductors and microelectronic devices are listed below:

▪ The dominant mode of information transfer is changing from electrical to optical signals. While the technology for optical communication has already been developed, the materials and devices for optical computing arc still in the research stage. It is believed, however, that the developing technology will result in much faster and therefore more powerful computational devices.

▌ The size scale of microelectronic devices continues to decrease. While a typical "chip" in a 486-computer contains about 1 million devices, it is anticipated that by the year 2000, chips will contain on the order of 100 million devices. This will result in smaller, faster, more powerful electronic devices of all kinds.

▌ Micromachining is a relatively new technology in which mechanical components, such as miniature motors, are incorporated directly into the silicon chip. In this way the electrical and mechanical components are intimately linked in a manner that leads to decreased size as well as increased reliability and device performance. An example of this technology is the device used to trigger air bags in many automobiles. The mechanical component (an accelerometer) recognizes the rapid deceleration and initiates an electrical signal that results in the deployment of the air bag.

1.5 MATERIALS PROPERTIES AND MATERIALS ENGINEERING

Virtually all engineers are concerned with the selection of materials as a part of their job assignment. The materials used are selected on the basis of properties that are particularly important for the intended application. Thus, mechanical, aerospace, and civil engineers are often concerned with the mechanical properties of a material, chemical engineers with corrosion properties, and electrical engineers with electrical and magnetic behavior. Materials engineers frequently function as part of an interdisciplinary design team or serve as consultants to other engineers in the selection of materials. They are also often involved with the development of new materials. In the following sections, examples of some of the engineering properties that will be studied in more depth in subsequent chapters are introduced in the context of engineering applications.

Mechanical Properties

Many engineers must design structures that will be subjected to mechanical loads. For example, in the design and fabrication of bridges, automobiles, airplanes, and pressure vessels, the forces that are encountered must not cause the parts to collapse as a result of overloading, and impact loads must not lead to catastrophic failure. It is interesting to note that ~5% of the GNP in the United States and other industrialized societies is lost each year because of fracture! This is well over $150 billion. Furthermore, the materials that are selected must be able to resist the corrosive effects of the environment in which they are applied. (A similar amount of money to that lost from catastrophic failure is also lost from corrosion.) One of the most basic parts of the mechanical design process involves choosing a material that has sufficient strength, stiffness, toughness, and ductility for the intended structural application.

Electrical Properties

Perhaps the most basic electrical property of a material is its conductivity. The conductivity is essentially a normalized measure of the amount of charge that will flow per unit of time in response to an applied electrical field. Electrical conductivities are given in Table 1.5–1 for some common metals, ceramics, polymers, and semiconductors. Note the enormous range of materials and electrical properties that are available to an engineer.

The conductivity of a material can be changed significantly by the addition of impurities. For metals, impurities decrease conductivity, since the impurity atom interferes with the motion of the "free" electrons. Thus, when metals are used to conduct electricity, they are usually used in as pure a form as possible to reduce the resistance. The situation is quite different for semiconductors such as silicon. The addition of even a small amount of phosphorus can increase the conductivity by many orders of magnitude. As shown in Table 1.5–1, the addition of just two phosphorus atoms in 1 million silicon atoms can cause the conductivity to increase by a factor of approximately 5 million! The controlled addition of small amounts of impurities to elements such as Si and Ge is the basis for producing modern semiconductors. These semiconductors are used to fabricate electrical devices such as transistors that are revolutionizing the electronics and telecommunications industries.

Conductivities of some common materials at room temperature.

Material	Conductivity $[(\Omega\text{-m})^{-1}]$
Metals	
Cu	6.0×10^7
Ag	6.8×10^7
Al	3.8×10^7
Ceramics	
Al_2O_3	10^{-12}–10^{-10}
Porcelain	10^{-12}–10^{-10}
Polymers	
Polyethylene	10^{-17}–10^{-13}
Polystyrene	$< 10^{-14}$
Polyacetylene doped with AsF_5	10^5
Semiconductors	
Si (pure)	4×10^{-4}
Si (2×10^{-14} at.% P)	2240
Ge (pure)	2.2

Source: W. D. Callister, *Materials Science and Engineering: An Introduction*, 2nd ed., Copyright © 1991 John Wiley & Sons, New York. Reprinted by permission of John Wiley & Sons, Inc.

The electrical properties of polymers can be largely affected by impurities, either added impurities, naturally occurring impurities, or impurities on the surface. For example, the application of an antistatic coating or the addition of ions into the bulk of a polymer can change the conductivity of textile fibers by five orders of magnitude. Uncoated fibers can lead to serious static discharge problems, like the failure of a parachute to open.

Effects of the Environment

The environment in which materials are used is a factor that must always be kept in mind, since it can have a pronounced effect on a material's properties and the way they can change. In this context, environment refers to factors such as temperature, load, or contact with aqueous media.

Increasing temperature usually decreases the strength of most engineering materials. While there are some important exceptions to this general rule, it is largely true for the vast majority of engineering materials. Increased temperature also usually has the effect of speeding up surface reactions with materials, many of which degrade properties. An example of oxide formation and oxide penetration into a bulk material is shown in Figure 1.5–1.

FIGURE 1.5–1

Oxidation in a Ni-base alloy. Note that the oxide has formed on the surface and penetrated into the underlying metal. The oxide is brittle and degrades the mechanical properties of the alloy. *(Source: Claude Bathias and J.-P. Baïlon, eds.,* La Fatigue des Matériaux, © *Les Presses de L'Université de Montréal, and Éditions Maloine, Paris. Used with permission.)*

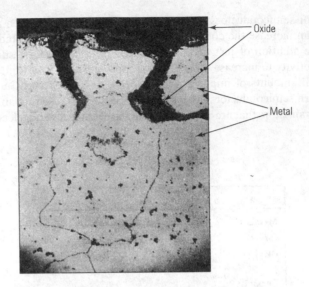

Oxide

Metal

Corrosion is a very complex phenomenon that manifests itself in a number of ways. In one mechanism, the material is attacked by particular ionic species in the medium. A well-known example of this occurred during the British rule of India. Cartridges whose shells were made of brass were stored in damp areas and failed when they were put into use. This usually occurred during the monsoon season and came to be known as "season cracking." The moisture in the air and minute amounts of gunpowder combined to form ammonium hydroxide (NH_4OH). The NH_4^+ ion attacked the brass, rendering the shells unusable.

In Chapter 15 we will learn that there are many other ways in which materials interact with their environment. Some, like the corrosion mechanism described above and the degradation of polymers by ultraviolet light, have a negative impact on material properties. Others, like the controlled oxidation of some ceramics, can significantly improve the properties of materials. Engineers must be concerned not only with the influence of the environment on their materials but also with the influence of their materials on the environment. Issues such as pollution and recycling will be recurring themes throughout this text.

1.6 THE INTEGRATED APPROACH TO MATERIALS ENGINEERING

To help you understand the structure, property, and processing relationships in materials, we have organized this textbook into four parts. Part I, Fundamentals, focuses on the structure of engineering materials. Important topics include atomic bonding, thermodynamics and kinetics, crystal structures, defects in crystals, strength of crystals, and noncrystalline structures. The concepts developed in these five chapters provide the foundation for the remainder of the text.

In Part II, Microstructural Development, we introduce the important processing variables of temperature, composition, and time. These two critical chapters develop the methods for controlling the structure of a material on the microscopic level. The concepts of phase diagrams and transformation kinetics are the central themes in this part of the text.

The third part of the book focuses on the engineering properties of the various classes of materials. It builds upon the understanding of structure developed in Part I and the

TABLE 1.6–1   The matrix of combinations of material classes and property classes discussed in most introductory materials engineering courses.

Property class	Materials class				
	Metals	Ceramics	Polymers	Semiconductors	Composites
Mechanical	XXXX	XXXX	XXXX	XXXX	XXXX
Electrical	XXXX	XXXX	XXXX	XXXX	XXXX
Dielectric and optical	XXXX	XXXX	XXXX	XXXX	XXXX
Magnetic	XXXX	XXXX	XXXX	XXXX	XXXX
Thermal	XXXX	XXXX	XXXX	XXXX	XXXX
Environmental interaction	XXXX	XXXX	XXXX	XXXX	XXXX

methods used to control structure set forth in Part II. It is in the properties section of the text that the integrated approach to materials engineering becomes most apparent.

Table 1.6–1 lists the matrix of topics included in Chapters 9 through 15. The five classes of materials are listed across the top of the grid and six classes of properties (mechanical, electrical, dielectric and optical, magnetic, thermal, and environmental interactions) are listed in the left-hand column. While the majority of the entries in this matrix of topics are covered in most introductory materials engineering courses, the order in which the topics are covered varies considerably. Traditionally, the topics are introduced by column. That is, all of the properties of one type of material (usually metals) are discussed before moving on to describe the properties of a second class of materials. The process is repeated class by class until all the topics have been introduced. This progression offers several advantages, including the ability to stress the unique strengths and weaknesses of each material class. It is also convenient for those students planning to specialize in a single class of materials. Ceramists, for example, may well benefit from a topic sequence that isolates the properties of this interesting class of materials.

We believe, however, that most engineers will be searching for a material that can fulfill a specific list of properties, as well as economic, processing, and environmental requirements, and will want to consider all classes of materials. That is, we feel that engineers are more likely to think in terms of rows of the matrix rather than columns. For this reason we have elected to package the material in Part III by property class rather than by material class. We will first describe the mechanical properties of all classes of materials, then the electrical properties of all classes of materials, and so on through the grid.[1] We call this an integrated approach to materials engineering because it attempts to teach fundamental concepts that are applicable to all materials first, and then point out the unique characteristics of each material class.

The fourth and final part of the book deals with processing methods for the various classes of materials and with the overall materials design and selection process. These two chapters attempt to tie together all of the diverse topics introduced in the first three parts of the book. The goal is for the reader to understand the methods used to select the appropriate material and processing methods required to satisfy a strict set of design specifications.

---

[1] The exception to this procedure is that composites are treated in isolation primarily because of their highly anisotropic nature (i.e., the properties of most composites containing fibers depend on the direction in which they are measured). This characteristic adds some mathematical complexity to the problem, which we have elected to treat in a separate chapter.

1.7 ENGINEERING PROFESSIONALISM AND ETHICS

Society has placed engineers in a position of trust, since products manufactured under an engineer's supervision have the potential to do great harm if they are not manufactured and used properly. For this reason, an engineer is expected to adhere to high ethical standards. Various branches of engineering have developed codes of ethics to address fundamental issues. The code of ethics of the Institute of Electrical and Electronics Engineers (IEEE) is reprinted here.

IEEE Code of Ethics

Preamble

Engineers, scientists and technologies affect the quality of life for all people in our complex technological society. In the pursuit of their profession, therefore, it is vital that IEEE members conduct their work in an ethical manner so that they merit the confidences of colleagues, employers, clients and the public. This IEEE Code of Ethics represents such a standard of professional conduct for IEEE members in the discharge of their responsibilities to employers, to clients, to the community and to their colleagues in this Institute and other professional societies.

Article I

Members shall maintain high standards of diligence, creativity and productivity, and shall:

1. Accept responsibility for their actions;
2. Be honest and realistic in stating claims or estimates from available data;
3. Undertake technological tasks and accept responsibility only if qualified by training or experience, or after full disclosure to their employers or clients of pertinent qualifications;
4. Maintain their professional skills at the level of the state of the art, and recognize the importance of current events in their work;
5. Advance the integrity and prestige of the profession by practicing in a dignified manner and for adequate compensation.

Article II

Members shall, in their work:

1. Treat fairly all colleagues and co-workers, regardless of race, religion, sex, age or national origin;
2. Report, publish and disseminate freely, information to others, subject to legal and proprietary restraints;
3. Encourage colleagues and co-workers to act in accord with this Code and support them when they do so;
4. Seek, accept and offer honest criticism of work, and properly credit the contributions of others;
5. Support and participate in the activities of their professional societies;
6. Assist colleagues and co-workers in their professional development.

Article III

Members shall, in their relations with employers and clients:

1. Act as faithful agents or trustees for their employers or clients in professional and business matters, provided such actions conform with other parts of this Code;
2. Keep information on the business affairs or technical processes of an employer or client in confidence while employed, and later, until such information is properly released, provided such actions conform with other parts of this Code;

3. Inform their employers, clients, professional societies or public agencies or private agencies of which they are members or to which they may make presentations, of any circumstance that could lead to a conflict of interest;

4. Neither give nor accept, directly or indirectly, any gift, payment or service of more than nominal value to or from those having business relationships with their employers or clients;

5. Assist and advise their employers or clients in anticipating the possible consequences, direct and indirect, immediate or remote, of the projects, work or plans of which they have knowledge.

Article IV
Members shall, in fulfilling their responsibilities to the community:

1. Protect the safety, health and welfare of the public and speak out against abuses in those areas affecting the public interest;

2. Contribute professional advice, as appropriate, to civic, charitable or other nonprofit organizations;

3. Seek to extend public knowledge and appreciation of the profession and its achievements.

For the most part, engineers have practiced their profession in adherence to these ethical canons and enjoy high respect in society.

SUMMARY

Materials may be classified into broad groupings depending on their fundamental structure. The major classes of materials are metals, ceramics, polymers, composites, and semiconductors. There is an intimate link between the chemical and physical structure of materials and their mechanical, electrical, thermal, optical, and magnetic properties. Engineers who understand the link between material structure, properties, and processing can select appropriate materials for use in any application.

It is no overstatement to say that a society's ability to develop and use materials is a measure of both its technical sophistication and its technological future. This book is devoted to helping scientists and engineers to better understand and use materials. To achieve this goal, the approach will be to develop broad, unifying principles applicable to the major classes of materials.

PART I

FUNDAMENTALS

n Part I we will be concerned with how atoms bond together in an infinite array to form engineering materials. These materials possess specific properties that are a direct result of their atomic scale structures. We will learn that in some solids, called crystalline materials, the atomic arrangement is in the form of a periodic array of atoms. In other solids, called noncrystalline or amorphous materials, the atoms are arranged in a near random fashion. Since crystals contain large numbers of atoms—one gram contains on the order of 10^{23} atoms—defects in the atomic arrangement are inevitable. These defects have a significant impact on the properties of the crystal.

It is important for engineers to understand the atomic scale structures of crystalline and noncrystalline materials if they are to use materials effectively in the design and construction of commercial products. As an example, consider the case of a camcorder. Camcorders must be light and compact yet sturdy, have automatic controls to compensate for light intensity and distance, and have high-quality optics and sound-recording capability, a pleasing appearance, and a competitive price. Engineers must satisfy these requirements by using a variety of materials with specific properties to achieve an optimum design.

Let us examine the major components of a camcorder. The lenses must transmit light without optical aberrations to achieve a high-quality picture. Lenses are made from glasses, which owe their transparency to their noncrystalline structure. The hardness of glass, which makes it suitable for precision grinding to achieve the proper lens shape, is a consequence of its three-dimensional bond network, as discussed in Chapters 2 and 6.

The camcorder housing parts must be light and easily formed into intricate shapes. These components are made from polymers composed of molecular chains of light elements that are loosely packed together resulting in a low-density material. Since the molecular chains are attached to each other by weak secondary bonds, which disintegrate easily upon heating, these materials have low softening points, which permit them to be shaped inexpensively. The structure of polymers will be discussed in Chapters 2 and 6.

A camcorder also contains a large number of electronic components, including: (1) electrically conducting wires made from

metals such as copper or gold, (2) precision resistors fabricated from carbon or nickel-chromium alloys, (3) semiconductors such as silicon doped with phosphorus or boron, and (4) ceramic or polymer insulators that isolate electrical signals between active elements. The tremendous variation in the electrical properties of these and other materials is a result of their atomic and crystal structures, as described in Chapters 2 and 3.

The quality of the connections between the electronic devices has a significant influence on the reliability of the circuits and therefore the performance of the camcorder. This quality can be assured only if sufficient diffusion (atomic scale mixing) occurs between the materials being bonded together. As discussed in Chapter 4, the diffusion required to produce a strong bond between materials would not be practical without the presence of point defects in crystals. Camcorders also contain small, intricately machined metal parts such as gear and sprocket assemblies that could not have been produced economically without the presence of line defects known as dislocations. As described in Chapter 5, dislocations make metals easy to form and machine into complicated shapes.

In sum, a camcorder, or any other complex commercial product, is composed of many components, each requiring a different combination of properties such as strength, formability, and response to electric fields. The job of the materials engineer, or any other engineer, is to select (or develop) the right material for each component.

ATOMIC SCALE STRUCTURES

MATERIALS IN ACTION   Bonding

Forces are exerted between atoms and groups of atoms that are in near proximity. In many cases these forces are attractive in nature and the atoms group together through the formation of bonds. These bonds have different characteristics depending on the atoms or groups of atoms in question. The bond character determines the physical, mechanical, and chemical properties including the state of aggregation (in other words, gas, liquid, or solid) as well as the structure (for example, crystalline or amorphous) under a given set of conditions. The fact that the "valence" electrons in a piece of metal are not localized on individual atoms means that they are free to conduct electricity under an applied electric field. This accounts for the excellent electrical conductivity of metals. On the other hand, most polymers (typically consisting of carbon, hydrogen, and oxygen in various combinations) are rather poor conductors of electricity. We shall see that in the bonds that constitute polymers, the electrons are localized and not free to move. These materials thus are excellent insulators as a result of their bond character.

An example of the way in which the bond character determines the structure and properties is found with carbon. It turns out that the bonds of carbon may have a different character, depending on the conditions under which the bonds were formed. In the case of graphite, the primary bonds exist within a plane; the carbon atoms lying within a single plane are strongly bonded in a two-dimensional network while the forces holding adjacent planes together are rather weak. A structure is thus formed in which adjacent planes can easily slide over one another and the material in this form exhibits good lubricating properties. On the other hand, if high temperature and pressure are exerted upon this solid, the character of the bonds changes to a three dimensional structure. The bonds of the carbon atoms take on a shape akin to toy jacks. These "jacks" bond together in a three-dimensional network to form diamond, the hardest material known to man. It is possible to produce diamonds commercially and such diamonds are used in drill bits for drilling through very hard minerals as well as for other kinds of cutting tools.

In this chapter we shall study the basic principles of bonding and examine some of the relationships between the underlying bond character and material properties.

2.1 INTRODUCTION

Have you wondered why metals are good electrical conductors while ceramics and plastics are generally good insulators? Why does rubber stretch while metals are relatively rigid in response to an applied force? Why is diamond able to cut through any other material? Why are steels able to resist considerable impact loading (rapid hammerlike blows) while ceramics shatter under relatively small impact loads? Why are oxide and polymer glasses transparent while other materials are opaque? Why do some materials expand more than others when heated?

We often take for granted the intrinsic properties of engineering materials. If we are to alter the structure of materials to enhance their properties, however, it becomes necessary to develop a thorough understanding of why materials behave the way they do.

Although the properties of materials depend on all levels of structure, many properties are determined by the **atomic scale structure** alone. By atomic scale structure we mean: (1) the types of atoms present, (2) the types of bonding between the atoms, and (3) the way the atoms are packed together.

In this chapter we will briefly review the structure of the atom and some basic principles of thermodynamics and kinetics. We will then describe the characteristics of the various types of bonds between atoms and the way in which small groups of atoms are spatially arranged. We will also explore the relationships between atomic scale structure and engineering properties and attempt to answer the questions we have posed above.

2.2 ATOMIC STRUCTURE

All matter is composed of atoms. The atoms, in turn, consist of electrons, protons, and neutrons. The properties of an atom are determined by many factors including: (1) the atomic number Z that corresponds to the number of electrons or protons in a neutral atom, (2) the mass of the atom, (3) the spatial distribution of the electrons in orbits around the nucleus, (4) the energy of the electrons in the atom, and (5) the ease of adding or removing one or more electrons from the atom to create a charged ion. The latter three factors, involving electrons, can be influenced by external conditions such as mechanical forces, electromagnetic fields, and temperature. Since we are interested in the influence of exactly these variables on the properties of solids, the bulk of our discussion will focus on the characteristics of electrons in atoms. An understanding of electron behavior, however, requires some understanding of quantum mechanics. Quantum mechanics theory, or QMT, is a mathematical framework developed by physicists in the early part of the 20th century to describe the interaction of electrons, protons, and neutrons in atoms and molecules.

One of the key features of QMT is the recognition that an electron exhibits both particle and wave characteristics. The properties of an electron in an atom can be best modeled by treating the electron as an energy wave. The equation describing the electron wave motion was developed by Erwin Schrödinger in 1925 and is known as the Schrödinger equation. Both this equation and its solution are beyond the level of this text. However, we will state without derivation some results of Schrödinger's equation, and of QMT, as they relate to the atomic structure of materials.

Only certain types of electron wave motion can satisfy the constraints of Schrödinger's equation. The valid solutions to this equation can be numbered for identification purposes. That is, each solution is identified by a set of three integer values (n, l, and m) known as

quantum numbers.[1] While the wave equation is extremely successful in explaining a great many aspects of the properties of electrons, the physicist Wolfgang Pauli found that a complete description of an electron required the specification of an additional value associated with electron "spin." This fourth quantum number, m_s, can only take on values of $\pm\frac{1}{2}$. Another outcome of Pauli's work, known as the **Pauli exclusion principle,** states that no two interacting electrons may have the same four values for their quantum numbers. Thus, each electron in an atom has a unique set of four quantum numbers that completely describe its characteristics.

One of the main reasons for introducing quantum numbers in this text is that they can be used to characterize the energy levels and arrangement of the electrons in an atom. The energy of an electron is primarily a function of its n and l quantum numbers (with m having a weaker influence). As shown in Figure 2.2–1, in the planetary or Bohr model of an atom the electrons are arranged in subshells in which all electrons have the same n and l values and, therefore, approximately the same energy.

Bohr model for a Zn atom

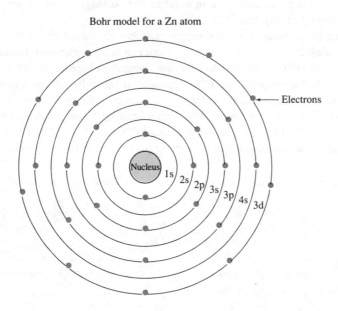

FIGURE 2.2–1 A schematic illustration of the Bohr model of a Zn atom showing the arrangement of electrons in the subshells.

The electron subshells are identified using an alphanumeric code in which the number represents the value of n and the letter gives the value of l using the following convention: s for $l = 0$, p for $l = 1$, d for $l = 2$, and f for $l = 3$. The Pauli exclusion principle can be used to show that the maximum number of electrons permitted in any subshell is

---

[1] For a potential energy function with spherical symmetry, as in the case of the hydrogen atom, in which the negatively charged electron is electrostatically attracted to the positively charged nucleus, the wave equation is solved more easily in spherical (r, θ, ϕ) coordinates. The principal quantum number n is associated with the boundary conditions on r; the angular momentum quantum number l is derived from the boundary conditions on θ; and the magnetic quantum number m is associated with the ϕ coordinate. More detailed quantum calculations show that while n can assume any integer value greater than or equal to 1, the values of the quantum numbers l and m are restricted as follows: l can only have integer values from 0 to $n - 1$, and m can only assume the integer values from $-l$ to $+l$.

determined by the value of the quantum number l and is given by $2(2l + 1)$. Thus, the maximum numbers of electrons in an s, p, d, and f subshell are respectively 2, 6, 10, and 14.

The **electron configuration** represents the distribution of electrons within the permissible energy levels. In the **ground state,** an atom's electrons occupy the lowest-energy subshells consistent with the Pauli exclusion principle. The subshells can be arranged in order of increasing energy as follows:

$$1s, 2s, 2p, 3s, 3p, 4s, 3d, 4p, 5s, 4d, 5p, 6s, 4f, 5d, 6p, 7s, 5f, 6d \ldots$$

In this notation, the number of electrons in each subshell is indicated using an integer superscript on the corresponding letter. For example, a half-filled subshell with quantum numbers $n = 3$ and $l = 2$ would be designated as $3d^5$.

How can we use this notation to describe the ground-state electron configuration for an oxygen atom that contains eight electrons? In the ground state the subshells will "fill" in the order 1s, 2s, 2p . . . and the maximum number of electrons in s and p subshells will be two and six, respectively. Thus, the ground-state electron configuration for oxygen is $1s^2 2s^2 2p^4$, indicating two electrons in each of the (filled) 1s and 2s subshells and four electrons in the (partially filled) 2p subshell.

In addition to the quantization of energy, another key result of the wave model is that the exact position of an electron within an atom can never be known. Instead, probability density functions (PDFs) are used to describe the spatial location of electrons. As shown in Figure 2.2–2, the shape of the PDF depends on the value of the quantum number l. Note that not all the distribution functions are radially symmetric. The consequence of a nonsymmetric PDF is that definite bond angles can be found in structures such as diamond, organic molecules, and polymeric chains. We will see that these specific bond angles influence the macroscopic engineering properties of the corresponding materials.

FIGURE 2.2–2

A highly schematic illustration of the probability density functions for electrons in certain subshells of an atom. Note that the s subshells are radially symmetric while the p subshells (and all other subshells) are highly directional.

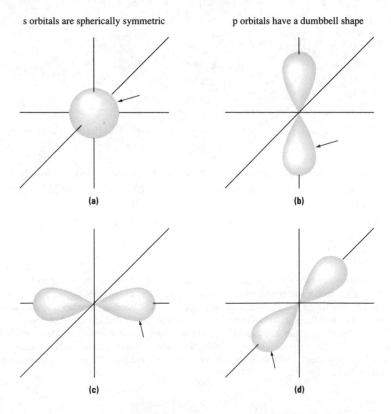

s orbitals are spherically symmetric

p orbitals have a dumbbell shape

(a)

(b)

(c)

(d)

PERIODIC TABLE

GROUP $\longrightarrow$

I A	II A	III A	IV A	V A	VI A	VII A	$\leftarrow$ VIII A $\rightarrow$			I B	II B	III B	IV B	V B	VI B	VII B	VIII B
1 **H** 1.0079												Metals | Nonmetals					2 **He** 4.0026
3 **Li** 6.941	4 **Be** 9.012											5 **B** 10.811	6 **C** 12.011	7 **N** 14.007	8 **O** 15.999	9 **F** 18.998	10 **Ne** 20.180
11 **Na** 22.99	12 **Mg** 24.305			*d* Transition Elements								13 **Al** 26.982	14 **Si** 28.086	15 **P** 30.974	16 **S** 32.066	17 **Cl** 35.453	18 **Ar** 39.948
19 **K** 39.098	20 **Ca** 40.078	21 **Sc** 44.956	22 **Ti** 47.88	23 **V** 50.942	24 **Cr** 51.996	25 **Mn** 54.938	26 **Fe** 55.847	27 **Co** 58.933	28 **Ni** 58.69	29 **Cu** 63.546	30 **Zn** 65.39	31 **Ga** 69.723	32 **Ge** 72.610	33 **As** 74.921	34 **Se** 78.960	35 **Br** 79.904	36 **Kr** 83.80
37 **Rb** 85.468	38 **Sr** 87.620	39 **Y** 88.906	40 **Zr** 91.224	41 **Nb** 92.906	42 **Mo** 95.940	43 **Tc** (97.907)	44 **Ru** 101.07	45 **Rh** 102.906	46 **Pd** 106.42	47 **Ag** 107.87	48 **Cd** 112.41	49 **In** 114.82	50 **Sn** 118.71	51 **Sb** 121.75	52 **Te** 127.60	53 **I** 126.90	54 **Xe** 131.29
55 **Cs** 132.91	56 **Ba** 137.33	57 **La*** 138.91	72 **Hf** 178.49	73 **Ta** 180.95	74 **W** 183.85	75 **Re** 186.21	76 **Os** 190.20	77 **Ir** 192.22	78 **Pt** 195.08	79 **Au** 196.97	80 **Hg** 200.59	81 **Tl** 204.38	82 **Pb** 207.20	83 **Bi** 208.98	84 **Po** (208.99)	85 **At** (209.99)	86 **Rn** (222.02)
87 **Fr** (223.02)	88 **Ra** (226.03)	89 **Ac*** (227.03)	104 **Unq** (261.11)	105 **Unp** (262.11)	106 **Unh** (262.12)												

Metals \| Nonmetals

Gas — | 34
Se
78.96 | — Atomic number / — Atomic mass (g mol⁻¹)
Liquid —

f Transition Elements

*Lanthanides (Rare Earths)

58 **Ce** 140.12	59 **Pr** 140.91	60 **Nd** 144.24	61 **Pm** (144.91)	62 **Sm** 150.36	63 **Eu** 151.97	64 **Gd** 157.25	65 **Tb** 158.93	66 **Dy** 162.50	67 **Ho** 164.94	68 **Er** 167.26	69 **Tm** 168.93	70 **Yb** 173.04	71 **Lu** 174.97

**Actinides

90 **Th** 232.04	91 **Pa** (231.04)	92 **U** (238.05)	93 **Np** (237.05)	94 **Pu** (244.06)	95 **Am** (243.06)	96 **Cm** (247.07)	97 **Bk** (247.07)	98 **Cf** (242.06)	99 **Es** (252.08)	100 **Fm** (257.10)	101 **Md** (258.10)	102 **No** (259.10)	103 **Lr** (260.11)

FIGURE 2.2–3   The periodic table of the elements.

Another less abstract consequence of QMT is a rational explanation of the periodic table of the elements, which was originally developed on the basis of experimental observations (see Figure 2.2–3). The elements were placed in order of increasing atomic number and arranged in a series of vertical columns, or groups, so that all the elements in a group display similar chemical properties. An explanation for the regularity of atomic properties within a group is obtained from the electron configurations of the elements. Elements within a group have the same number of electrons in their outer, or **valence,** shells that participate most strongly in chemical reactions.

....

EXAMPLE 2.2–1

Determine the electron configurations for a silicon atom ($Z = 14$) and a germanium atom ($Z = 32$). Explain why these two elements display similar characteristics.

Solution

Since the maximum number of electrons in a subshell is given by the equation $2(2l + 1)$, the corresponding numbers for an s shell ($l = 0$), a p shell ($l = 1$), and a d shell ($l = 2$) are respectively 2, 6, and 10. Combining this result with the observation that the order of the subshells is given by 1s, 2s, 2p, 3s, 3p, 4s, 3d, 4p allows one to determine that the electron configurations for silicon and germanium are $[1s^2 2s^2 2p^6]3s^2 3p^2$ and $[1s^2 2s^2 2p^6 3s^2 3p^6 3d^{10}]4s^2 4p^2$. Both elements have a valence electron structure of the form $xs^2 xp^2$ where x is 3 for Si and 4 for Ge. Since the valence electron distributions are similar for these elements, we should expect them to exhibit similar properties.

....

2.3 THERMODYNAMICS AND KINETICS

Thermodynamics is the study of the relationships between the thermal properties of matter and the external system variables such as pressure, temperature, and composition. Thermodynamic considerations are fundamental in determining whether chemical and physical reactions can occur. **Kinetics,** on the other hand, determines how rapidly reactions can proceed. The reactions and their rates determine the structure of the resulting material, which, in turn, determines the properties. Therefore, a working knowledge of thermodynamics and kinetics is necessary for understanding material behavior.

Basic thermodynamic principles show that a reaction occurs only if it results in a reduction in the total energy of the system. In the absence of energy reduction the reaction does not occur; however, the converse is not true. That is, even if a reaction is thermodynamically possible, it may not actually occur. Thus, thermodynamics establishes the necessary condition but not the sufficient condition for a reaction to occur.

Many chemical processes involve the breaking of atomic bonds and the subsequent formation of new bonds in materials. These changes in bond structure can alter the properties of the material. Since changes in bond structure involve energy changes, structural alterations are controlled by thermodynamic considerations. Examples of common bond-alteration processes include the extraction of metals such as aluminum, iron, and titanium from their ores, the melting of a solid to form a liquid, chemical attack of materials in aggressive environments, the formation of a solid from a collection of isolated atoms, the degradation of polymers in ultraviolet light, and the failure of materials resulting from the application of excessive force that breaks atomic bonds.

It is important to note that changes in the system variables can result in a change in the thermodynamically favored state. For example, consider the change in bond structure that occurs upon melting or freezing of a pure material. Figure 2.3–1 illustrates the relative energies of the two thermodynamic states, liquid and solid, as a function of temperature. At temperatures greater than the melting temperature, the liquid represents a lower energy state than the solid. It is the thermodynamically favored state. As the temperature is decreased below the melting temperature, the solid becomes the favored state and there is a thermodynamic driving force for solidification, which includes bond rearrangement, of the material. We will see that many of the important reactions in materials engineering can be understood by examining the influence of changes in system variables on the relative energies of different atomic scale structures.

Thermodynamics provides information about which processes may occur; however, it gives no information about the rate of the process. The science that deals with reaction

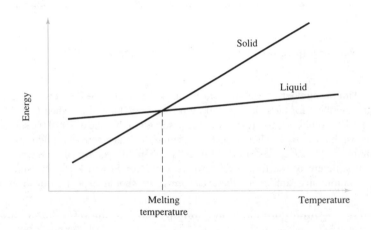

rates is kinetics. In many cases, the kinetic factors are more important than the thermo-dynamic factors. For example, ordinary window glass is not a thermodynamically stable structure. An investigation of the kinetics of the problem, however, shows that at *room temperature* the rate at which window glass moves toward the lowest-energy, or equi-librium, state is such that glass can exist in its nonequilibrium form for thousands of years. This example illustrates the important concept that a reaction or structural change will occur only if both the thermodynamics and kinetics are favorable.

The phrase *room temperature* was highlighted in the previous paragraph to emphasize the importance of temperature in the kinetics of a typical chemical process. In the vast majority of cases the rate of a chemical reaction increases exponentially with an increase in temperature. That is,

$$\text{Reaction rate} = C \exp\left(-\frac{Q}{RT}\right) \tag{2.3-1}$$

where C is a constant, R is the gas constant, T is absolute temperature, and Q is the activation energy for the process. Any reaction that obeys Equation 2.3–1 is termed an **Arrhenius process,** or a thermally activated process. We will see that Arrhenius processes are very common in the study of materials and that Equation 2.3–1 is used to model a wide variety of chemical and physical processes.

You will find that there are a great many equations introduced in this textbook. One key to success in any quantitative field is to gain an understanding of the terms in each equation. Several features of the Arrhenius equation are important. First, note that R is a universal constant, that is, it always has the same value. Second, T represents the absolute temperature of the system, usually reported in kelvins. Third, the preexponential term C and the activation energy Q are constants that are characteristic of any individual reaction, but their values change as the reaction changes (i.e., they are not universal constants). Fourth, the activation energy can be thought of as the amount of energy that must be supplied to the system in order for the reaction to proceed. For example, a block resting on the top of a staircase can lower its (potential) energy by moving to the bottom of the stairs. In order for this "reaction" to occur, however, the block must be given a push. The amount of energy required for this push can be envisioned as the activation energy for the process. The concept of activation energy is illustrated schematically in Figure 2.3–2. The units of Q are typically kilojoules per mole (kJ/mol). Fifth, from a

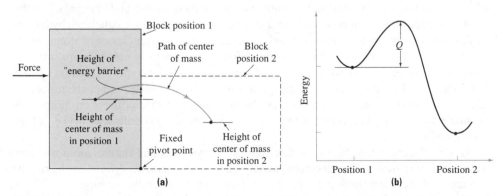

(a)   (b)

FIGURE 2.3–2 The concept of activation energy. **(a)** When a block is "pushed over," the height of its center of gravity initially increases and then decreases to a final value less than its initial value. The difference between the maximum and initial heights represents the "energy barrier" to this process. **(b)** The more general concept of an energy barrier of height Q is illustrated schematically. In moving from position 1 to position 2, the overall energy of the system decreases, but energy in the amount Q must be "supplied" in order for the transition to occur.

dimensional point of view the reaction rate and the preexponential constant C must have the same units (since the ratio inside the exponential term must be dimensionless). Finally, the units for Q and R must be consistent.

....

EXAMPLE 2.3–1

Typical activation energies for the reactions described in this text range from 30 to 300 kJ/mol. Using the value of 30 kJ/mol, calculate the change in the reaction rate when the temperature increases from 25°C to 100°C. Repeat this calculation using an activation energy of 300 kJ/mol.

Solution

Since the problem deals with reaction rates, we must use the Arrhenius equation, 2.3–1. If $k(T)$ represents the reaction rate at temperature T, then the ratio of the reaction rates at any two temperatures can be found from the following equation:

$$\frac{k(T_1)}{k(T_2)} = \frac{C \exp(-Q/RT_1)}{C \exp(-Q/RT_2)}$$

$$= \exp\left\{-\left(\frac{Q}{R}\right)\left[\left(\frac{1}{T_1}\right) - \left(\frac{1}{T_2}\right)\right]\right\}$$

For $Q = 30$ kJ/mol we find:

$$\frac{k(100°C)}{k(25°C)} = \exp\left\{-\left[\frac{30,000 \text{ J/mol}}{8.314 \text{ J/(mol-K)}}\right]\left(\frac{1}{373 \text{ K}} - \frac{1}{298 \text{ K}}\right)\right\}$$

$$= \exp(+2.435) \approx 11.4$$

Repeating the calculation for $Q = 300$ kJ/mol gives:

$$\frac{k(100°C)}{k(25°C)} = \exp(+24.35) \approx 3.76 \times 10^{10}$$

This calculation demonstrates the extreme sensitivity of the reaction rate to changes in either temperature or activation energy.

....

2.4 PRIMARY BONDS

Having reviewed the structure of the atom and the thermodynamic and kinetic factors that control reactions, we can proceed with a discussion of atomic scale structures by investigating the types of bonds that form between atoms. The two major classes of atomic bonds are primary and secondary bonds. Primary bonds are generally one or more orders of magnitude stronger than secondary bonds. The three major types of primary bonds are ionic, covalent, and metallic bonds. All primary bonds involve either the transfer of electrons from one atom to another or the sharing of electrons between atoms. The formation of primary bonds allows atoms to achieve a noble gas configuration (i.e., a filled valence shell).

One of the important factors in determining the type of bond that an atom will form is its electronegativity. The **electronegativity** (EN) of an element, defined as the relative tendency of that element to gain, or attract, an electron, generally increases as you read the periodic table from left to right (see Appendix A). Elements with comparatively high values of EN are said to be electronegative while those with comparatively low EN values are termed electropositive. Another important quantity in determining the type of bond formed in a compound involving two types of atoms is the electronegativity difference, ΔEN, between the atoms.

2.4.1 Ionic Bonding

The most common type of bond in a compound containing both electropositive and electronegative elements is the **ionic bond.** This bond involves **electron transfer** from the electropositive atom to the electronegative atom. A high ΔEN between the atoms favors the formation of ionic bonds. Some of the more common ionic compounds involve the electropositive Group I elements (Na and Li) combining with electronegative elements from either Group VII (Cl or F) or Group VI (O) to form alkali halides (LiF, NaCl) and oxides (Li_2O and Na_2O).

As shown in Figure 2.4–1, an ionic bond is formed in NaCl when the single electron from the nearly empty valence band of the Na atom ($1s^22s^22p^63s^1$) is transferred to the nearly filled valence band of the Cl atom ($1s^22s^22p^63s^23p^5$). The result is the formation of a Na cation, Na^+, and a Cl anion, Cl^-, each with a completely filled and therefore stable valence electron shell. Once charge transfer has occurred, a force of attraction develops between the ions. The magnitude of the force is a function of both the valence of the ions and their separation:

$$F_a(x) = \frac{|Z_1 Z_2| q^2}{4\pi\,\varepsilon_0 x^2} \qquad\qquad (2.4\text{–}1)$$

where Z_i is the valence of the ion, q is the charge of an electron, ε_0 is the permittivity of vacuum ($= 8.85 \times 10^{-12}$ C^2/N-m^2), and x is the separation distance between the ions. A force of the form given in Equation 2.4–1 is known as a **coulombic force.**

The coulombic force draws the ions together until their filled electron shells begin to overlap. Until the point of overlap, the electrons associated with the cation are independent of those of the anion. When the electron shells of the ions begin to impinge upon one another, their electrons begin to interact and they can no longer be considered to be independent. The Pauli exclusion principle requires that some of the interacting electrons be promoted to higher energy levels so that no two electrons will have the same four quantum numbers. Since this process requires the energy of the system to increase, a repulsive force develops in order to minimize the overlap of the electron shells of adjacent

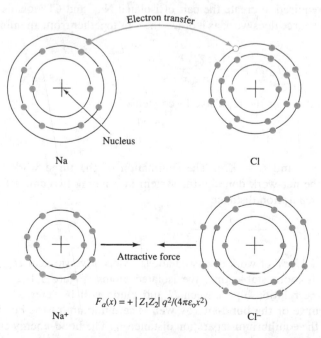

Electron transfer

Nucleus

Na   Cl

$$F_a(x) = +\,|Z_1 Z_2|\, q^2/(4\pi\varepsilon_0 x^2)$$

Attractive force

Na^+ Cl^-

FIGURE 2.4–1

An example of an ionic bond showing electron transfer from Na to Cl to form the Na^+ cation and Cl^- anion pair.

ions. An alternate explanation for the development of a repulsive force is that the ions occupy a finite amount of space (the hard sphere model) and resist being forced to occupy a smaller volume.

The magnitude of the repulsive force increases rapidly as the ions are forced closer and closer together. Expressing this in one of many possible formalisms:

$$F_r(x) = -\frac{K}{x^m} \tag{2.4-2}$$

where K and m are constants and $m > 2$ (a common experimental value for m is 12). Comparison of Equations 2.4–1 and 2.4–2 shows that the repulsive force is dominant at small values of x and the attractive force dominates at larger separation distances.

The equilibrium separation distance x_0 can be found by setting the sum of the forces equal to zero. That is, equilibrium occurs at the value of x for which

$$F_a(x) + F_r(x) = 0 = \frac{|Z_1 Z_2| q^2}{4\pi \varepsilon_0 x^2} - \frac{K}{x^m} \tag{2.4-3}$$

Note that Equation 2.4–3 is valid only in the directions in which ions are in contact with one another. Figure 2.4–2a shows the relationship between the competing forces and x_0. This figure, and others like it, will be referred to as the **bond-force curve.** Since x_0 represents the average center-to-center distance between atoms, it is also known as the **bond length.**

Consider the energy changes that occur during the formation of an ionic compound. The three important factors are the energy, or work, necessary to create the ions from neutral atoms, the work done by the attractive force (work done by the system) in drawing the ions together from an infinite separation distance, and the work done against the repulsive force (work done on the system) in bringing the ions together.

The energy required to remove an electron from an isolated neutral atom is referred to as its **ionization potential.** The ionization potential of $Na \rightarrow Na^+$ is 5.14 eV. The energy released when an isolated neutral electronegative atom gains an electron is termed its **electron affinity.** The electron affinity of $Cl \rightarrow Cl^-$ is 4.02 eV. Therefore, the net energy (work) required to create the pair of isolated Na^+ and Cl^- ions is 1.12 eV.

The attractive force does work as it draws the ions together from an infinite separation:

$$U_a(x) = \int_x^\infty F_a dx = -\frac{|Z_1 Z_2| q^2}{4\pi \varepsilon_0 x} \tag{2.4-4}$$

A similar integration for the repulsive force yields

$$U_r(x) = \int_x^\infty F_r dx = \frac{C}{x^n} \tag{2.4-5}$$

where $n = m - 1$ and $C = K/n$. The summation of the three work terms gives an expression for the net work done by the system in bringing two atoms from an infinite separation to a separation distance x:

$$U(x) = U_i - \frac{|Z_1 Z_2| q^2}{4\pi \varepsilon_0 x} + \frac{C}{x^n} \tag{2.4-6}$$

where U_i is the 1.12 eV of work discussed above. A negative value of $U(x)$ indicates that the compound is more stable than the isolated atoms. Figure 2.4–2b is a schematic illustration of the relationship between U and x and will be referred to as either the **bond-energy curve** or the **bond-energy well.** The minimum of the bond-energy curve corresponds to the equilibrium separation distance x_0. The bond-energy curve contains a

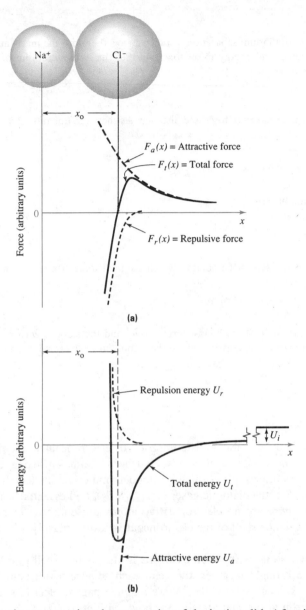

(a)

(b)

FIGURE 2.4–2
(a) The bond-force curve
showing the location of
the equilibrium separation
distance x_0. Note the approximately linear slope of
the total force curve in the
vicinity of x_0. **(b)** The
bond-energy curve for the
ionic compound NaCl
showing the location of
the equilibrium separation
distance x_0.

wealth of information concerning the properties of the ionic solid. After investigating the other types of atomic bonds, we will return to the bond-energy curve to extract this information.

··

EXAMPLE 2.4–1

Describe the electron transfer process that occurs in the formation of the ionic compound Li_2O.

Solution

Appendix B shows that the electron configuration for Li is $1s^2 2s^1$ and for O it is $1s^2 2s^2 2p^4$. Thus, Li has one valence electron and is electropositive. Oxygen has six valence electrons and is electronegative. Li can obtain a filled valence shell by transferring its lone valence electron to the electronegative O atom. This results in the formation of a Li^+ ion and an O^- ion. The O^- ion, however, still does not contain a filled valence shell. If a second Li atom transfers its lone valence electron to the O^- ion, the result is a stable group of ions composed of two Li^+ and one O^{2-} (i.e., Li_2O).

··

EXAMPLE 2.4–2

Use the concept of equilibrium separation distance x_0 together with the mathematical relationship among force, distance, and energy to demonstrate that the minimum point on the bond-energy curve corresponds to x_0.

Solution

Energy, or work, is the product of force and distance. As shown in Equation 2.4–4, the total energy is

$$U(x) = \int_x^\infty F \, dx$$

Alternatively we might write

$$F = \frac{dU}{dx}$$

Since equilibrium is defined as the point at which the total force is zero, we may write

$$F = \frac{dU}{dx} = 0 \text{ at equilibrium}$$

dU/dx represents the slope of the bond-energy curve, and the slope is zero at the bottom of the bond-energy curve. Hence, x_0 corresponds to the minimum of the bond-energy curve.

··

2.4.2 Covalent Bonding

Covalent bonds form in compounds composed of electronegative elements, especially those with four or more valence electrons. Hydrogen also has a strong tendency to form covalent bonds. Examples of covalently bonded materials include the elements in Group IV (C, Ge, Si), the diatomic gases (H_2, Cl_2, F_2), and essentially all carbon-based molecules. Since there are no electropositive atoms present, the "extra" electrons required to fill the valence shell of the electronegative atoms must be obtained by *sharing* electrons.

Consider the Cl atom, with seven electrons in its valence shell ($1s^2 2s^2 2p^6 3s^2 3p^5$). If two Cl atoms are brought together, they can each acquire a full complement of eight electrons in their valence shell if they share a single pair of electrons. Each atom contributes one electron to the shared pair. As shown in Figure 2.4–3, the shared electrons are spatially localized in the region between the Cl atoms. Although a complete description of the force holding the Cl_2 molecule together requires a quantum mechanics explanation, it can be envisioned as a coulombic attraction between the positively charged nuclei and the negatively charged shared electron pair localized between the two Cl atoms.

As with the ionic bond, a repulsive force develops when the **core electrons,** the electrons in the filled shells "below" the valence shell, begin to overlap. Although the form of the equation for the attractive force is slightly different from that given in Equation 2.4–1, x_0 can be found using a similar equation:

$$F_a + F_r = 0 = \frac{A'}{x^p} - \frac{B'}{x^q} \tag{2.4–7}$$

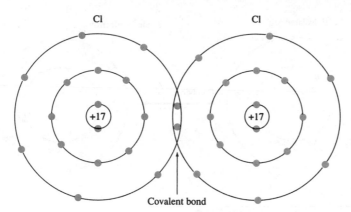

FIGURE 2.4–3   A schematic illustration of a covalent bond formed between two Cl atoms resulting in the formation of a Cl_2 molecule.

where A' and B' are positive constants and $p < q$. The bond-force and bond-energy curves for covalently bonded materials have the same general shape as those for ionic materials.

EXAMPLE 2.4–3

Describe the valence electron distribution in the covalently bonded F_2 molecule.

Solution

The electron configuration for F is $1s^2 2s^2 2p^5$. Thus, each F atom requires one additional electron to achieve a filled valence shell. Two F atoms each contribute a single electron to a shared electron pair, forming a single covalent bond. An illustration of the electron distribution for the F_2 molecule would look like that shown for Cl_2 in Figure 2.4–3 except that the covalent bond would involve a pair of electrons from the $n = 2$ shell rather than from the $n = 3$ shell.

2.4.3 Metallic Bonding

Solids composed primarily of electropositive elements containing three or fewer valence electrons are generally held together by **metallic bonds.** As mentioned above, the electropositive elements can obtain a stable electron configuration by "giving up" their valence electrons. Since there are no electronegative atoms present to receive the "extra" electrons, they are instead donated to the structure in general. That is, they are *shared by all of the atoms* in the compound.

Figure 2.4–4 is a schematic representation of this bond in which the valence electrons are delocalized from the nucleus and core electrons. The valence electrons form a "cloud" or "sea" of electrons that surrounds the ion cores. The force holding the metal together is the attraction between the positively charged ion cores and the negatively charged electron cloud. The electrons are shared, but they are not spatially localized.

As with the previous bond types, the Pauli exclusion principle results in a repulsive force that becomes significant when the filled electron shells associated with the ion cores begin to overlap. The sum of the attractive and repulsive forces leads to both bond-force and bond-energy curves similar in shape to those described for ionic and covalent bonds.

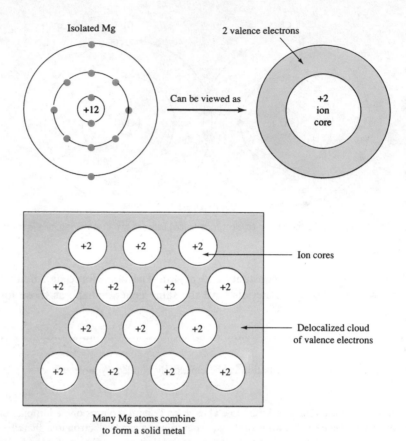

FIGURE 2.4–4 A schematic representation of the metallic bond in solid Mg. The valence electrons form an electron cloud that surrounds the ion cores.

EXAMPLE 2.4–4

Predict the most likely type of primary bond in each of these materials: (*a*) Cu, (*b*) KCl, (*c*) Si, and (*d*) CdTe.

Solution

To predict primary bond type, one must know the electronegativity values and number of valence electrons for each of the elements involved. From Appendix B, EN(Cu) = 1.90, EN(K) = 0.82, EN(Cl) = 3.16, EN(Si) = 1.90, EN(Cd) = 1.69, and EN(Te) = 2.1. Cu and K have one valence electron, Cd has two, Si has four, Te has six, and Cl has seven valence electrons.

    a. Since Cu bonds with itself, ΔEN = 0 and the bond is not ionic. Cu has an EN value that is neither highly electronegative nor highly electropositive, so we must consider the number of valence electrons. Since Cu has fewer than three valence electrons, it will have metallic bonds.

    b. For KCl, ΔEN = 3.16 − 0.82 = 2.34. Using the table in Appendix A, this corresponds to a bond that is ~75% ionic. Therefore, we predict the bonding in KCl can be considered chiefly ionic.

    c. Si (with ΔEN = 0) is not ionic. Like Cu, EN(Si) is neither highly electropositive nor highly electronegative. Since Si has four valence electrons, it is likely to display covalent bonds.

d. For CdTe, $\Delta EN = 2.1 - 1.9 = 0.2$. Using the table in Appendix A, this corresponds to a bond that is ~1% ionic. Therefore, the bonding in CdTe is either metallic or covalent. Since the average number of valence electrons in CdTe is $(2 + 6)/2 = 4$, we predict the bonding in CdTe is likely to be covalent.

2.4.4 Influence of Bond Type on Engineering Properties

At this point we can make a few preliminary observations concerning some of the mechanical and electrical properties of solids as a function of bond type.

Consider the difference in the response of a metal and an ionic solid (ceramics and oxide glasses are examples of solids with considerable ionic character) when each material is struck with a blow from a hammer. Atoms in the metal can slip and slide past one another without regard to electrical-charge constraints in response to the applied force and thus absorb the impact without breaking. This phenomenon is called **ductile** behavior of metals. On the other hand, as illustrated in Figure 2.4–5a, in an ionic solid, each ion is surrounded by oppositely charged ions. Thus, ionic slip may lead to like charges moving into adjacent positions, causing coulombic repulsion (see Figure 2.4–5b). This makes slipping much more difficult to achieve, and the material responds by breaking. This is one of the reasons why ceramics and oxide glasses fracture easily. Such behavior is known as **brittle** behavior.

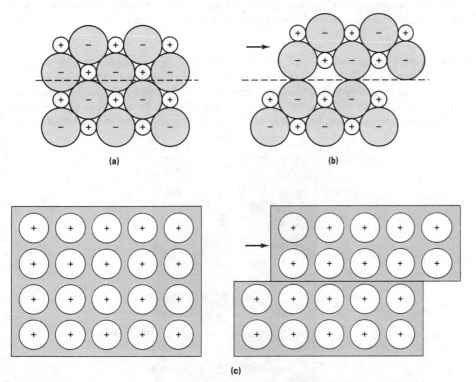

(a)   (b)

(c)

FIGURE 2.4–5   A comparison of the difference in the atomic scale response of a metal and an ionic solid to a hammer blow. **(a)** In an ionic solid before the hammer blow each ion is surrounded by oppositely charged ions. **(b)** When the ions attempt to slip past one another in response to the applied force, strong repulsive forces develop and lead to cracking. **(c)** In contrast, in a metal the electron cloud shields the positively charged atomic cores from each other so that the repulsive forces do not develop.

Next we consider the response of a solid to an applied electric field. Electrical conduction involves the directed motion of charges in response to an applied field. The electrical conductivity of materials depends on three factors: (1) the type of charge carrier in the materials (electrons or ions), (2) the spatial density of charge carriers, and (3) the charge carrier mobility, or ease with which the charge carriers can move through the atomic scale structure of the material. The charge carriers in metals are the electrons in the cloud that surrounds the atomic cores. In part because of their small size, the loosely bound or free electrons can move relatively unimpeded through the metal, resulting in a high charge-carrier mobility. The combination of high mobility and high concentration of charge carriers leads to high electrical conductivities for most metals.

In contrast, each ion in an ionic solid has a filled valence shell. Hence, none of the electrons can be easily removed from its host ion. Instead, charge motion and, hence, electrical conduction often require movement of entire ions. Since such motion is comparatively difficult and slow, and the density of mobile ions is considerably less than the density of mobile electrons in metals, ionic solids are generally characterized as electrical insulators rather than electrical conductors.

DESIGN EXAMPLE 2.4-5

Electric utility companies must transport electricity from the power generation plant to consumers. As shown in Figure 2.4–6, one of the common transmission methods is to use above-ground wires suspended between structural support towers. The towers and transmission wire are often fabricated from metals, but the "spacers" between the transmission lines and the towers are usually made from ionic solids. Explain the choice of materials for these three applications.

Solution

To minimize the power loss associated with the resistance of the transmission lines, it is necessary to select a material with a high electrical conductivity. Thus, metals rather than ionic compounds are used for this application. The structural support material is selected for its mechanical rather than its electrical properties. It would be unwise to fabricate these towers from a brittle material, since the structure may experience high loads from many sources including gusts of wind and

FIGURE 2.4–6   A photograph of power transmission lines showing the metal support structure, the ceramic insulators, and the wires carrying the electric current.   *(Source: Courtesy of Commonwealth Edison.)*

possible collisions with motor vehicles. In addition, because of their high ductility, metals are more easily formed into large and complex shapes. Finally, we must consider the purpose of the "spacers." If the metal transmission lines are in close proximity to the metal support structure, the possibility of a short circuit that will electrify the entire structure exists. Therefore, the wires must be insulated from the support structure by using a high–electrical-resistance ionic solid as the "spacer."

2.5 THE BOND-ENERGY CURVE

The bond-energy curve was introduced in Section 2.4.1. Several important macroscopic material properties can be obtained directly from this curve. Specifically, one can estimate the bond energy, the average bond length, the elastic modulus, and the coefficient of thermal expansion. Each of these characteristics of the curve will be discussed in some detail in this section.

Recall that the function U can be interpreted as the amount of work done by the system when two ions or atoms are brought together from infinite separation or, equivalently, the amount of energy that must be supplied to the system in order to completely separate the atoms from one another. Therefore, the magnitude of $U(x)$ at x_0, or the depth of the energy well, is a measure of the inherent strength of the bond, or the **bond energy.** In addition, we have already pointed out that the equilibrium separation distance x_0 corresponds to the bond length.

The bond-energy curve represents the variation of energy with position when the only forces acting on the atoms or ions are the atomic forces of attraction and repulsion. If other external forces or sources of energy act on the system (for example, the application of external loads, an electromagnetic field, or temperature changes), quantities such as bond length and effective bond energy may be altered. In fact, one of the key issues in materials engineering is to be able to understand and model the interaction between external variables and atomic scale structures.

Consider the response of a system of atoms to an external load. Figure 2.4–2a showed that the bond-force curve is approximately linear near the equilibrium position. The slope of the curve, $(\partial F/\partial x)$, at x_0 is a measure of the force required to displace atoms from their equilibrium positions. Mathematically, near x_0 the displacement (Δx) is proportional to the force:

$$F \propto \Delta x \qquad \text{or} \qquad \frac{F}{\Delta x} = aE \qquad\qquad (2.5\text{–}1)$$

where a is a geometric factor and E is a material property known as **Young's modulus.** Young's modulus is a measure of the resistance of the material to relative atomic separation (stiffness). The steeper the slope of the force curve at the equilibrium separation distance (i.e., the higher the value of E), the greater the force required to move the atoms from their equilibrium positions. Thus, materials with large values of E are stiff, since they are better able to resist changes in length under applied loads.

EXAMPLE 2.5–1

The bond-force curves for two materials are shown in Figure 2.5–1. If identically shaped beams were fabricated from each material, which would show a greater deflection under equal applied loads?

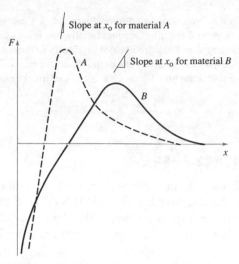

FIGURE 2.5–1   A comparison of the bond-force curves for two hypothetical materials, A and B.

Solution

As shown in Equation 2.5–1, Young's modulus is related to the slope of the bond-force curve at x_0. Figure 2.5–1 shows that the bond-force curve for material A is steeper than that for material B (at the point for which $F = 0$). Therefore, material A has a higher Young's modulus. Since Young's modulus is a measure of the inherent stiffness of the material, we should expect the beam fabricated from material A to be stiffer and therefore deflect less than an identical beam fabricated from material B.

As discussed previously the bond-force curve is the derivative of the bond-energy curve:

$$\frac{\partial U}{\partial x} = F \quad \text{and} \quad \frac{\partial^2 U}{\partial x^2} = \frac{\partial F}{\partial x} \tag{2.5–2}$$

Combining Equations 2.5–1 and 2.5–2 yields

$$\frac{\partial^2 U}{\partial x^2} = aE \tag{2.5–3}$$

Thus, the curvature of the bond-energy curve at x_0 is also proportional to Young's modulus. Note that the smaller the radius of curvature, the higher the stiffness. A physical interpretation of this relationship is that the steeper the sides of the energy well, the greater the amount of energy required to displace the atoms away from their equilibrium positions.

 Another macroscopic property with a strong connection to the bond-energy curve is the **coefficient of thermal expansion,** α_{th}. As temperature increases, the atoms gain energy and are able to "move up" the sides of the energy well. Figure 2.5–2a shows that at a temperature corresponding to the energy E^* there are two values of x with the same value of E^*. If the atom is assumed to vibrate between these two positions, then the mid-point represents an average separation distance. By drawing a series of constant-energy

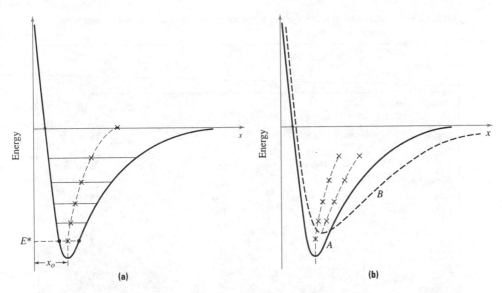

FIGURE 2.5–2 Relationship between the bond-energy curve and the coefficient of thermal expansion. **(a)** The dashed line passes through the midpoints of the constant-energy line segments and indicates that the average separation distance increases as energy (temperature) increases. **(b)** Comparison of the bond-energy curves for two materials: material A has a deeper and more symmetric bond-energy well so that the locus of midpoints on the constant-temperature line segments changes less rapidly for material A. This means that $\alpha_{th}(A) < \alpha_{th}(B)$.

line segments and connecting their midpoints, we see that the average atomic-separation distance increases as the temperature increases. For many materials this relationship can be represented over a limited temperature range by the linear equation

$$\frac{x_e - x_0}{x_0} = \alpha_{th}(T - T_0) \tag{2.5–4}$$

where x_e is the equilibrium spacing at temperature T, x_0 is the equilibrium spacing at the reference temperature T_0, and α_{th} is the coefficient of linear expansion.

Figure 2.5–2b shows bond-energy curves for two materials with approximately the same equilibrium separation distance. The curve for material A is nearly symmetric in the vicinity of x_0 while that for material B is highly asymmetric. The midpoints of the constant-temperature line segments for each curve show that the bond length changes more rapidly with temperature for material B than A. Since α_{th} is a measure of the normalized change in dimensions with temperature, the magnitude of α_{th} increases as the bond-energy curve becomes more asymmetric. We can combine this observation with another, that deeper energy wells tend to be more symmetric, and predict that materials with high bond energies—those with deep and symmetric wells—should have low α_{th} values. This prediction is supported by the data in Table 2.5–1. The elements with higher melting temperatures have stronger primary bonds and lower coefficients of thermal expansion.

It is important to recognize that the relationships between the bond-energy curve and macroscopic properties developed in this section show general trends. They are extremely helpful in understanding and predicting relative differences in properties between different materials. However, the constants in the corresponding equations are not known with

TABLE 2.5–1   Latent heat of fusion, melting temperatures, and coefficients of thermal expansion for some metallic elements.

Material	Latent heat of fusion (J/g)*	Melting temperature (K)	Coefficient of thermal expansion ($\times 10^{-6}$ °C)*
Row III metals			
Na	113	371	70
Mg	368	922	25
Al	397	933	25
Si†	1800	1685	3
Row IV metals			
K	63	336	83
Se	67	494	37
Zn	113	693	35
Cu	205	1358	17
Mn	268	1517	22
Fe	272	1809	12
Co	276	1768	12
Ni	297	1726	13
Cr	331	2130	6
V	410	2175	8
Ti	418	1943	9

*Adapted from the *CRC Handbook of Tables for Applied Engineering Science,* copyright CRC Press, Boca Raton, FL, 1979.
† Although silicon is not usually considered a metal, it is included here for comparison.

sufficient accuracy to facilitate calculation of the absolute values of bond length, bond energy, modulus of elasticity, and coefficient of thermal expansion. The values of these properties for engineering materials are usually directly measured in the laboratory.

DESIGN EXAMPLE 2.5–2

The bond-energy curves for two engineering materials are shown in Figure 2.5–3. Your task is to select the better material for use in each application described below.

FIGURE 2.5–3
A comparison of the bond-energy curves for two hypothetical materials, A and B.

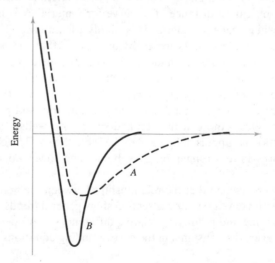

a. A beam that shows little deflection under moderate loads

b. A crucible to be used at a high service temperature

c. A device designed to sense changes in temperature by changing its dimensions

Solution

a. The application requires high stiffness. A material with a high value of Young's modulus is needed. The bond-energy curve for material B exhibits a smaller radius of curvature than does A. Hence, it has the higher modulus and should be selected for this application.

b. The component must operate at high temperatures. Since bond strength is related to the depth of the bond-energy well, we again select material B.

c. This application requires a material with a high value of α_{th}. Since α_{th} increases as the asymmetry of the bond-energy curve increases, we select the material with the more asymmetric bond-energy well. The better choice for this application is material A.

2.6 ATOMIC PACKING AND COORDINATION NUMBERS

As demonstrated above, several important macroscopic properties of materials can be estimated on the basis of only a knowledge of the atoms present, the type of bonding between atoms, and the shape of the bond-energy curve. Other properties, such as density, largely depend on the arrangement of the atoms in the solid. In this section we discuss the factors that influence the three-dimensional packing of atoms.

The arrangement of atoms within a solid can be principally characterized by the number of nearest neighbors, or **coordination number** (CN), of each atom in the structure. In turn, the coordination number is influenced primarily by the type of bonding present and by the relative sizes of the atoms or ions.

Consider the ionic compound CsCl. As discussed in Section 2.4.1, each ion can be approximated as a hard sphere. Because of the nature of the coulombic attractive force, we expect each Cs^+ cation to be surrounded by as many negatively charged Cl^- anions as is geometrically possible. The total energy for the system is minimized when the number of oppositely charged nearest neighbors for each ion is maximized. The geometric factor responsible for determining the CN of ions is the ratio of the radii of the ions.

Using the procedure outlined in Section 2.4.1, the equilibrium separation distance for CsCl, $x_0(CsCl)$, can be found from the bond-energy curve. It is then possible to write

$$r(Cl^-) + r(Cs^+) = x_0(CsCl) \tag{2.6-1}$$

where the terms on the left are the radii of the ions. Similar equations may be written for a variety of compounds, and the resulting set of simultaneous equations may be solved for the respective radii. Using this procedure, a self-consistent set of ionic radii can be calculated. The result justifies the use of a rigid sphere model for describing ionic solids. The atomic and ionic radii of the elements are given in Appendix C.

Let r represent the radius of the smaller ion, usually the cation, and let R represent the radius of the larger ion, usually the anion. The relationship between the ratio of the radii and the resulting CN can be determined using the following constraints: (1) cations "touch" anions, (2) the number of anions surrounding a given cation will be as high as geometrically possible, and (3) the ions cannot overlap.

Consider the geometry when a small cation is bonded to a much larger anion. It will always be possible to place two anions in contact with the cation, but as shown in Figure 2.6–1, it may not be possible to place a third anion in contact with the cation

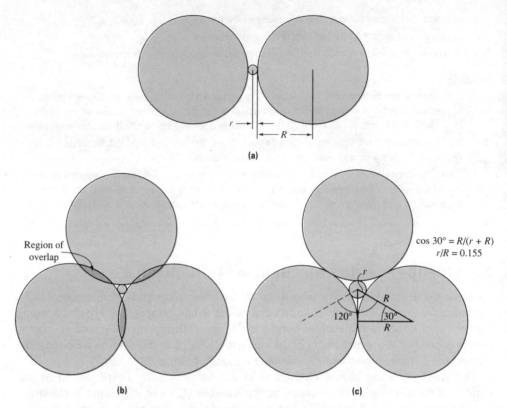

FIGURE 2.6–1 Geometry for CN = 2: **(a)** two anions "fit" but **(b)** a third would require overlap and **(c)** the minimum r/R value for a CN = 3 is shown to be 0.155.

without forcing the anions to overlap. The repulsive force between the ions prevents this overlap. Since it is impossible for the smaller ion to have CN = 3, this ratio of r/R results in $CN_{cation} = 2$.

If the radius of the cation is gradually increased while holding the radius of the anion constant, a value of r/R is eventually reached for which CN = 3 is possible (see Figure 2.6–1c). As r/R continues to increase, CNs of 4, 6, 8, and eventually 12 become possible. The critical radius ratio for each coordination number is given in Table 2.6–1. Note that these are minimum values for each CN. Although a given CN is geometrically possible for any r/R ratio greater than the value given in Table 2.6–1, it will be energetically favorable only until the minimum value of the next highest CN is reached. Thus, the maximum value of r/R for CN = 8 is the minimum r/R value for CN = 12.

Examination of Appendix C shows that the radius of an anion is generally larger than the radius of the corresponding neutral atom, and the radius of a cation is usually smaller than that of the neutral atom. This behavior can be readily understood on the basis of electron-electron and electron-proton interactions. In cations, the magnitude of the electronic repulsions decreases with the loss of electrons, and the positive charge in the nucleus is able to attract the remaining electrons more closely. The result is $r^+ < r^0$. Using the inverse argument, it can be shown that r^- will be greater than r^0.

Since r/R must always be less than or equal to 1, and since for most ionically bonded compounds $r(cation) < r(anion)$, the appropriate radius ratio is usually $r(cation)/R(anion)$. If, however, $r(anion) < r(cation)$, then the r/R ratio should be used to

TABLE 2.6–1   The critical (r/R) ratio for each coordination number. (Note that the drawings are not to scale.)

Coordination number	Critical (r/R) value	(r/R) Stability range	Geometry
2	0	$0 < r/R < 0.155$	Always possible
3	0.155	$0.155 \leq r/R < 0.225$	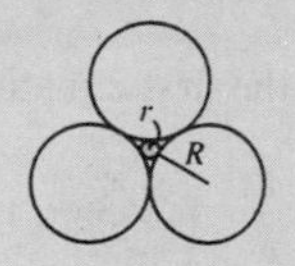
4	0.225	$0.225 \leq r/R < 0.414$	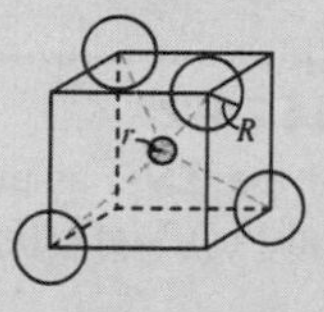
6	0.414	$0.414 \leq r/R < 0.732$	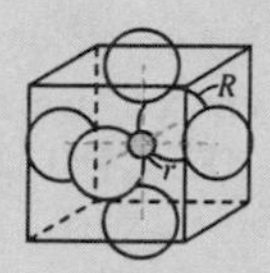
8	0.732	$0.732 \leq r/R < 1$	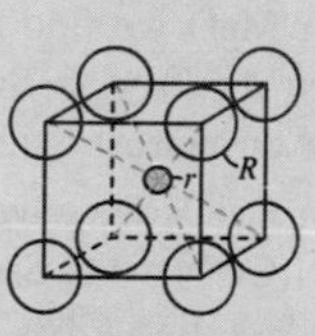
12	1	$r/R = 1$	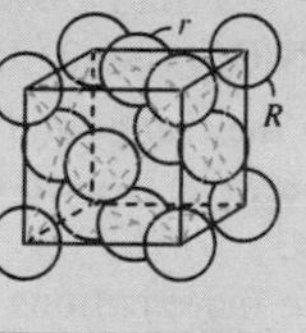

estimate the CN of the anion. Once the CN of the smaller ion is known, the CN of the larger ion can be determined based on the cation : anion ratio, or the stoichiometry of the compound.

....

EXAMPLE 2.6–1

Table 2.6–1 gives the ionic radius ratio range for CN = 6 as $0.414 \leq (r/R) < 0.732$. Derive these limiting values by investigating the critical geometry for CNs of 6 and 8.

Solution

The geometry for the critical (minimum) r/R ratio for CN = 6 is shown in Table 2.6–1. If a represents the length of the edge of the cube, then when all of the ions are just touching each other

$$r + R = \frac{a}{2} \quad \text{and} \quad R + R = \frac{a}{\sqrt{2}}$$

Dividing the first equation by the second equation yields

$$\frac{r + R}{2R} = \frac{1}{\sqrt{2}}$$

Solving for the desired quantity yields

$$\frac{r}{R} = \sqrt{2} - 1 = 0.414$$

Since the maximum value for CN $= 6$ corresponds to the minimum value for CN $= 8$, we must repeat the procedure for the critical geometry for CN $= 8$. In this case, we find

$$r + R = \frac{a\sqrt{3}}{2} \quad \text{and} \quad R + R = a$$

Dividing the first equation by the second and solving for r/R yields

$$\frac{r}{R} = \sqrt{3} - 1 = 0.732$$

EXAMPLE 2.6–2

Calculate the CNs, assuming ionic bonding, for each of the elements in each of the following compounds: (a) MgO, (b) Cr_2O_3, (c) K_2O.

Solution

From Appendix C we find that the relevant ionic radii are $r(Mg^{2+}) = 0.078$ nm, $r(Cr^{3+}) = 0.064$ nm, $r(K^+) = 0.133$ nm, and $r(O^{2-}) = 0.132$ nm. The stable ranges of r/R for each CN are given in Table 2.6–1.

a. For MgO we find $r(Mg^{2+})/R(O^{2-}) = 0.078/0.132 = 0.59$, which corresponds to $CN(Mg^{2+}) = 6$. Since the anion : cation ratio is $1 : 1$, each anion will have the same number of nearest neighbors as each cation so that $CN(O^{2-}) = 6$.

b. For Cr_2O_3 we have $r(Cr^{3+})/R(O^{2-}) = 0.064/0.132 = 0.485$, which corresponds to $CN(Cr^{3+}) = 6$. Since there are more anions than cations in the structure, $CN(O^{2-})$ must be less than $CN(Cr^{3+})$. We use the anion : cation ratio of $3 : 2$ to find that $CN(O^{2-}) = (2/3)CN(Cr^{3+}) = 4$.

c. In the compound K_2O the radius of the anion is less than the radius of the cation. Therefore, r/R is used to predict the CN of the anion as follows: $r(O^{2-})/R(K^+) = 0.132/0.133 = 0.992$, which corresponds to $CN(O^{2-}) = 8$. Since there are more cations than anions in the structure, $CN(K^+)$ must be less than $CN(O^{2-})$. We use the anion : cation ratio of $1 : 2$ to find that $CN(K^+) = (1/2)CN(O^{2-}) = 4$.

In contrast to ionic materials, for which CN is determined by geometry, the number of nearest neighbors in a covalently bonded material is determined by the number of electrons in the valence shell of each atom. The governing equation for the electronegative elements in Groups IV through VII is

$$N_B = (8 - N_v) \tag{2.6–2}$$

where N_B is the number of covalent bonds formed and N_v is the number of valence electrons in the neutral atom. In most covalent solids CN is equal to N_B. The exceptions to this rule are the covalent compounds containing double or triple bonds. It should also be recognized that the monovalent electronegative element hydrogen often forms a single covalent bond.

Let us consider as examples several compounds involving H ($1s^1$) and C ($1s^2 2s^2 2p^2$). Since an H atom requires only one additional electron to achieve a full valence shell, each H atom bonds to one additional H atom to form diatomic hydrogen. Since both atoms in the H_2 structure have filled valence shells, this diatomic molecule is stable and will not form any additional primary bonds.

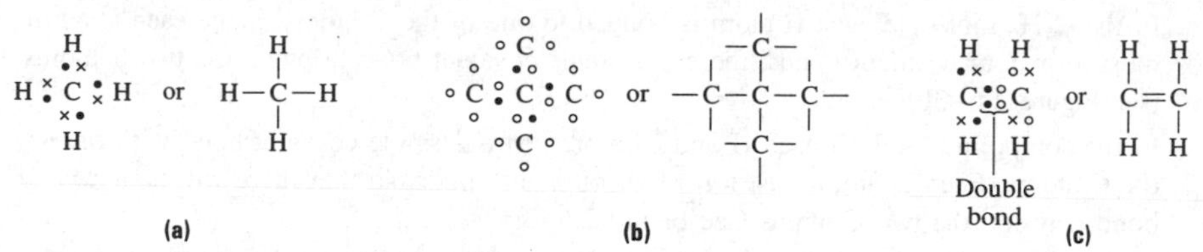

FIGURE 2.6–2 Schematic illustrations of the covalent bond structure in a series of compounds: **(a)** CH_4, **(b)** pure carbon in the diamond structure, and **(c)** C_2H_4. The x's represent the electrons from the H atoms and the •'s and ○'s represent those from the C atoms. Note the double bond in the compound C_2H_4.

The bond structure for carbon is more complex. Since C has only four electrons in its outer shell, it must share four pairs of electrons in order to achieve a full complement of eight valence electrons. In the compound CH_4, the C atom shares one electron with each H atom (see Figure 2.6–2a). The resulting methane molecule is stable. Next, consider the bonding arrangement for pure carbon. Each C atom may form a single covalent bond with four other C atoms, giving rise to the highly stable three-dimensional diamond structure shown in Figure 2.6–2b.

What is the bond structure in the molecule C_2H_4? Each H can satisfy its bonding requirements by forming a single covalent bond with one of the C atoms. In turn, each C atom will be covalently bonded to two H atoms. Each C atom, however, must still form two additional covalent bonds in order to achieve a filled valence shell. This can be accomplished if the C atoms share two pairs of electrons. The C_2H_4 molecule is shown schematically in Figure 2.6–2c. This molecule is fundamentally different from either H_2 or CH_4 in that it contains a double bond. C_2H_4 is an *unsaturated* hydrocarbon. We will see later that the double bond allows many C_2H_4 molecules to react to form the polymer polyethylene.

EXAMPLE 2.6–3

Use either of the shorthand electron notations shown in Figure 2.6–2 to depict the covalent bonding arrangement in each of the following materials: (a) H_2O, (b) C_2H_6, (c) C_2H_3Cl, and (d) Si.

Solution

The number of covalent bonds formed is directly related to the number of valence electrons (N_v) in an atom (Equation 2.6–2). From Appendix A, $N_v(H) = 1$, $N_v(C) = 4$, $N_v(O) = 6$, $N_v(Cl) = 7$, and $N_v(Si) = 4$.

a. In the H_2O molecule the O atom will form one covalent bond with each of the H atoms. This arrangement of electrons, shown in Figure 2.6–3a, allows all three atoms to obtain filled valence shells.

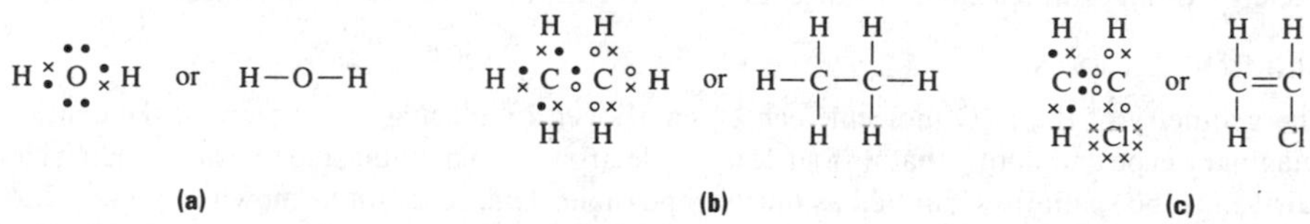

FIGURE 2.6–3 Schematic illustrations of the covalent bond structure in a series of compounds: **(a)** H_2O, **(b)** C_2H_6, and **(c)** C_2H_3Cl. The x's represent the electrons from the H and Cl atoms and the •'s and ○'s represent those from the C and O atoms. Note the double bond in the compound C_2H_3Cl.

b. In the C_2H_6 molecule each H atom is bonded to one of the C atoms. Since each C atom must form four covalent bonds, there is a single covalent bond bridging the two C atoms (see Figure 2.6–3b).

c. In the compound C_2H_3Cl each H and Cl atom forms a single covalent bond with one of the C atoms. Each C atom must form four covalent bonds, so that there will be a *double* bond between the two C atoms (see Figure 2.6–3c).

d. In silicon, each atom must be bonded to four other Si atoms, and the resulting structure is similar to the diamond structure described previously (see Figure 2.6–2b).

Covalent bonds are directional and are characterized by specific **bond angles.** The bond angles can be determined by the geometry of the structure or vice versa. Shared electrons, or bond pairs, and lone electron pairs constitute mutually repulsive negative-charge centers that tend to separate as much as possible. As shown in Figure 2.6–4a, the bond angle in a tetrahedral structure such as diamond is 109.5°, which places nearest-neighbor C atoms (and their associated shared electron pairs) as far apart as possible in space while satisfying the valency requirements. In contrast, when carbon is bonded to only three other atoms (one of which involves a double bond), the resulting structure is planar with a bond angle of about 120°, as shown in Figure 2.6–4b. The existence of specific bond angles in covalent molecules is important in understanding the properties of polymers.

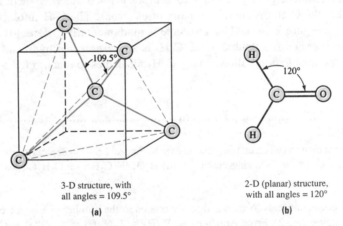

3-D structure, with
all angles = 109.5°

(a)

2-D (planar) structure,
with all angles = 120°

(b)

FIGURE 2.6–4 A schematic illustration of covalent bond angles in two compounds: **(a)** the bond angle in a tetrahedral structure such as diamond is 109.5°; **(b)** when the C is bonded to only three other atoms (one of which involves a double bond), the resulting structure is planar with a bond angle of ~120°.

EXAMPLE 2.6–4

Sketch the three-dimensional arrangement of covalent bonds in the H_2O molecule.

Solution

The geometry of the H_2O molecule can be envisioned by placing the O atom at the center of an imaginary cube and noting that its four pairs of electrons, two bonding and two nonbonding electron pairs, must be spatially separated as much as possible. This separation, shown in Figure 2.6–5, is obtained by placing the electron pairs along directions pointing to an alternating set of four corners of the imaginary cube. The H atoms are positioned at two of the cube corners associated with the

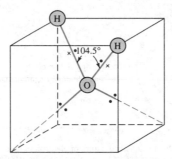

FIGURE 2.6–5 A schematic illustration of covalent bond angles in water. Note that the bond angle is 104.5°, which is slightly less than the tetrahedral angle of 109.5°.

bonding electron pairs. The structure of H_2O deviates slightly from this model, since nonbonding electron pairs repel each other slightly more than bonding electron pairs. The result is that the H—O—H bond angle is 104.5°—slightly less than the predicted 109.5°.

The shared electrons in a metallic bond are delocalized. Thus, the CN of an atom in a metallic solid is determined primarily by geometrical considerations. Indeed, many pure metals (e.g., Al, Cu, and Ni), for which $r/R = 1$, have structures with a CN of 12; however, several common pure metals such as Fe, Cr, and W have CNs of only 8, even in their purest forms.

Coordination numbers are useful because they describe the **short-range order,** defined as the number and type of nearest neighbors, associated with a particular solid structure. All solids exhibit short-range order. As we expand the consideration to include second- and higher-order neighbors, we find that there are two distinct types of solids. Those that exhibit both short-range order (SRO) and **long-range order** are called **crystalline materials** while those with SRO only are termed **amorphous,** or **noncrystalline, materials.**

2.7 SECONDARY BONDS

Secondary bonds are fundamentally different from primary bonds in that they involve neither electron transfer nor electron sharing. Instead, attractive forces are produced when the center of positive charge is different from the location of the center of negative charge. The resulting electric dipole can be either temporary, induced, or permanent and can occur in atoms or molecules. As shown in Figure 2.7–1 for Ar, a **temporary dipole** is formed when the electrons, which are constantly in motion, are momentarily arranged so as to produce an asymmetric charge distribution. The temporary dipole can then induce another dipole in an adjacent Ar atom. The two dipoles then experience a coulombic force of attraction. This type of bonding is responsible for the condensation of noble gases at low temperatures and is known as **van der Waals** (or van der Waals–London) bonding. Van der Waals bonds can also occur between symmetric molecules such as CH_4 and CCl_4. The strength of the van der Waals bond generally increases as the number of atoms in the compound increases. Hence, large molecules can have a large net attractive force. This phenomenon explains why the melting temperatures of the hydrocarbons with chemical formulas C_nH_{2n+2} increase as n increases.

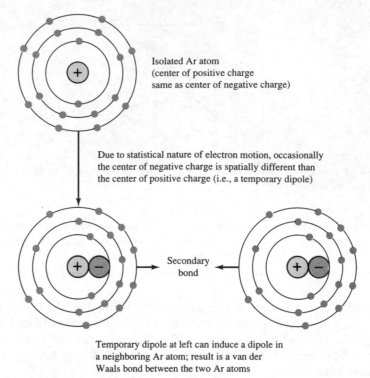

FIGURE 2.7–1   Formation of a temporary dipole in an Ar atom can induce a dipole in an adjacent Ar atom. This type of secondary bond is known as a van der Waals bond.

Figure 2.7–2 shows the charge distribution in H_2O, H_2S, and NH_3. These molecules are **permanent dipoles,** because their center of positive charge (indicated by the symbol δ^+) is always different from their center of negative charge (δ^-). Permanent dipole bonds are generally stronger than van der Waals bonds. One especially important type of permanent dipole bond is the **hydrogen bond,** which occurs whenever a hydrogen atom can be shared between two strongly electronegative atoms such as N, O, F, or Cl. The hydrogen bond is the strongest type of secondary bond, but it is still significantly weaker than a primary bond. Hydrogen bonds hold the wood fibers in a sheet of paper together.

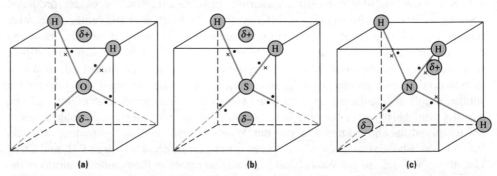

FIGURE 2.7–2   A schematic illustration of three permanent dipole molecules: **(a)** H_2O, **(b)** H_2S, and **(c)** NH_3. The x's represent the valence electrons from the H atoms and the •'s represent those from either O, S, or N. The δ^+ and δ^- symbols represent the spatial centers of positive and negative charge for the molecule. Note that nonbonding electron pairs are local regions of negative charge and the isolated nucleus of an H atom is a local region of positive charge.

EXAMPLE 2.7–1

Use the concept of secondary bond strength to predict which member of each pair of materials below has a higher melting temperature.

 a. C_2H_4 or $C_2H_2F_2$
 b. H_2O or H_2S
 c. Propane (C_3H_8) or dodecane ($C_{12}H_{26}$)

Solution

The two key factors are the type of secondary bond and the size of the molecules involved.

 a. C_2H_4 is a symmetric molecule, so that it is not a permanent dipole. Instead C_2H_4 molecules are held together by only weak van der Waals bonds. In contrast, the replacement of two of the H atoms with highly electronegative F atoms makes $C_2H_2F_2$ a permanent dipole. Thus, we expect $C_2H_2F_2$ to display the higher melting temperature. Indeed, the melting temperature of $C_2H_2F_2$ is $-84°C$ while that of C_2H_4 is $-169°C$.

 b. H_2O and H_2S are both permanent dipoles; however, O is more electronegative than S, [EN(O) = 3.44, EN(S) = 2.58]. Consequently, water forms stronger hydrogen bonds. The melting point of H_2O is $0°C$, while that for H_2S is $-85.5°C$.

 c. C_3H_8 and $C_{12}H_{26}$ are organic molecules composed of carbon and hydrogen. As such, the intermolecular attraction is only weak van der Waals forces. Since $C_{12}H_{26}$ is a larger molecule, it can form a larger number of dipoles. Thus, the melting temperature of $-9.6°C$ for $C_{12}H_{26}$ is higher than the $-189.7°C$ melting temperature of C_3H_8.

Secondary bonds control properties such as melting point and elastic modulus in solids where primary bonds do not form a three-dimensional (3-D) network. The nondirectional nature of both ionic and metallic bonds, coupled with the high coordination numbers for these solids (CN $\geq$ 4), usually results in the formation of a 3-D primary bond network. Hence, there are no regions in which primary bonds are absent, and therefore the properties of metallic and ionic solids are dominated by primary bonds.

In contrast, the directional nature of covalent bonds, coupled with the typically lower CNs for these structures (CN $\leq$ 4), can result in groups of atoms (molecules) that have fulfilled their primary bond requirements without the establishment of a 3-D primary bond network. In these materials, secondary bonds play a dominant role in determining the engineering properties. It is important to note that the strongest and stiffest directions in covalent solids are ones in which a continuous primary bond structure exists. In these directions covalent solids are every bit as strong and stiff as metallic and ionic solids. It is only in the directions in which primary bonds are absent that the properties are controlled by secondary bonds. The directional dependence of properties in covalent materials is utilized effectively in producing fibers that are very strong along their length direction (i.e., covalent bond direction) from polymeric materials. The high strength and stiffness along the fiber axis are obtained, however, at the expense of properties in the perpendicular directions.

2.8 MIXED BONDING

As mentioned previously, the type of primary bond formed in a compound depends on the electronegativities of the atoms involved. In compounds involving more than one element, ionic bonds are favored when the difference in electronegativities, ΔEN, is large, and covalent bonds are favored when ΔEN is small. The transition from pure ionic to pure covalent bonding is gradual, and many compounds display a bond with mixed ionic/

covalent characteristics. Calculations by Pauling show that a ΔEN of about 1.7 results in a bond that is ~50% ionic and 50% covalent. Appendix A includes a chart that relates the magnitude of ΔEN to the percent ionic characteristic of the bond. In fact, one of the reasons for the failure of the r/R ratio in predicting the correct CN in some "ionic" compounds is the mixed ionic/covalent characteristics of the bonds in these materials.

Many ceramics contain primary bonds with mixed ionic/covalent characteristics. Consider the bonding in the compound SiO_2. Each Si atom (EN = 1.9) shares one electron pair with each of its nearest-neighbor O atoms (EN = 3.44). The greater electronegativity of the O atoms, however, results in a partial transfer of the shared electrons from the Si atoms to the O atoms. The shared electrons imply covalent bonding while the electron transfer gives the bond some ionic characteristics. The ΔEN of 1.54 suggests that the Si—O bond is ~45% ionic and ~55% covalent.

Another class of materials that may exhibit an intermediate bond type is the metals. The bonds in the Group IA metals are predominately metallic, but as the valence of the elements increases (i.e., moving from left to right in the periodic table), the bonds begin to take on some covalent characteristics. The increase in covalent characteristic of the bonds in metals is part of the reason the density of metals generally decreases as one moves to the right in the periodic table in a given row.

When some metals are mixed together, they form stoichiometric compounds with characteristic metal atom ratios. Examples include AlLi, Ni_3Al, Al_3V, AlSb, CuZn, Ti_3Al, and Mg_2Si. These compounds are referred to as **intermetallics.** Most of these compounds exhibit either mixed metallic/covalent or mixed metallic/ionic bond characteristics, depending on the ΔEN values of the elements involved. Although brittle, these materials often have good high-temperature resistance and a high strength-to-weight ratio. As a result, Ni_3Al and Ti_3Al are finding usage in the aerospace industry.

In previous examples the term mixed bonding referred to the intermediate characteristics of an individual bond. Another type of "mixed" bonding occurs in materials having both primary and secondary bonds. As mentioned in the previous section, secondary bonds influence the macroscopic material properties only when there is an absence of primary bonds in certain regions of a material. Many of the polymers described in the next section display properties dominated by secondary bonds.

2.9 THE STRUCTURE OF POLYMER MOLECULES

An important class of materials consists of large molecules called macromolecules. A subset of macromolecules are those that are formed by linking together a large number of identical units, or **mers.** These molecules, called **polymers,** for "many mers," offer a variety of useful properties. Although all polymers contain covalent bonds within the molecules, they may have either primary or secondary bonds bridging the macromolecules. We have elected to introduce the concept of a polymer in the atomic scale structure chapter since the distinction between the two major subclasses of polymers is based on the type of bond between adjacent polymer chains.

The structurally simplest polymers are synthetic, or man-made. As an example, consider the linear polymer polyethylene, or PE, for which the monomer is the C_2H_4 molecule. As shown in Figure 2.9–1, the PE polymer chain is formed by opening the double bond between the C atoms in an individual monomer and linking a series of monomers together to form the linear macromolecule. An examination of bond energies associated with single and double covalent bonds between carbon atoms shows that the breaking of one double bond and the formation of one single bond (per monomer) result in a decrease in the free

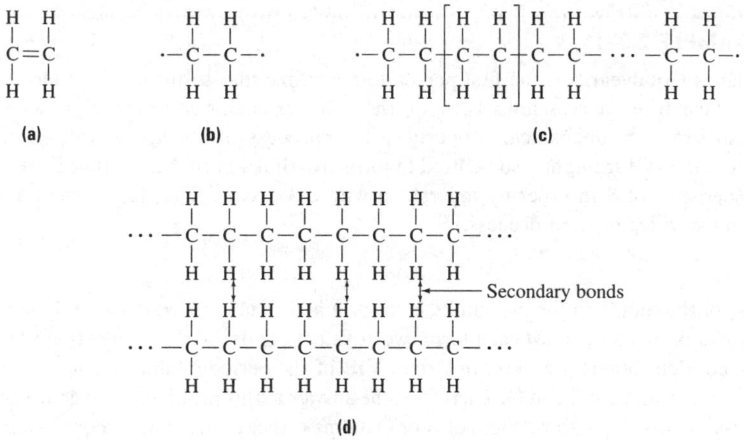

FIGURE 2.9-1 The structure of polyethylene, PE: **(a)** the basic building block for PE is the C_2H_4 monomer; **(b)** the double bond in the monomer is "opened" so that **(c)** many monomers can be linked together to form the PE polymer chain; **(d)** since the polymer chains are saturated, the only type of bond that can form between PE chains is the secondary bonds.

energy of the system. Thus, the formation of a PE polymer chain from a collection of identical monomers is a thermodynamically favored reaction. Note that in contrast to the monomer, the PE polymer chain is saturated, so there are no additional sites for primary bond formation. Thus, the only mechanism that remains for bond formation between PE chains is secondary bond formation. Linear polymers that form melts upon heating, such as PE, are called **thermoplastic polymers.**

The structure of rubber is fundamentally different from that of the thermoplastic polymers. Careful examination of the generic hydrocarbon rubber structure in Figure 2.9–2a shows that the polymer chains contain an unsaturated double bond. The existence of this double bond within the macromolecule permits the formation of additional primary bonds between chains (Figure 2.9–2b). The primary bonds between rubber chains formed by the opening of the unsaturated double bonds are known as **crosslinks.** When the crosslink density is low, only a small fraction of the double bonds have been opened, and the individual polymer chains retain their identity. There are only a "few" primary bonds between chains. As the crosslink density increases, the individual chains lose their identity and the structure begins to resemble a three-dimensional network of primary bonds. This 3-D primary bond structure is characteristic of many polymers that do not form a melt, or **thermoset polymers.**

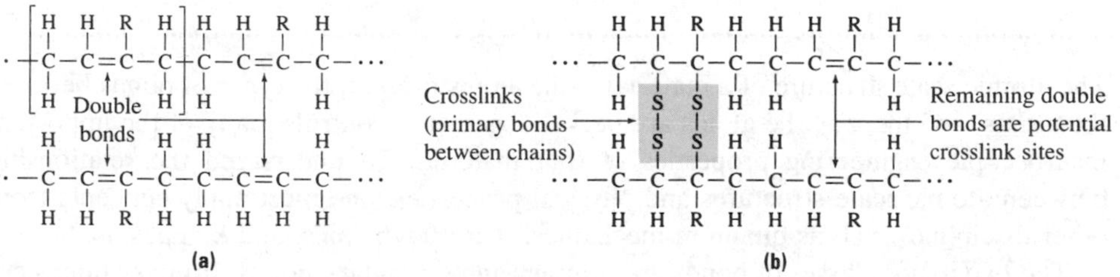

FIGURE 2.9-2 The structure of crosslinked rubber. The existence of double bonds along the length of the polymer chains shown in part **(a)** permits the formation of crosslinks between chains, as shown in part **(b)**. Note that in this case the crosslinks are composed of short chains of sulfur atoms.

DESIGN EXAMPLE 2.9–1

In 1844 Charles Goodyear was the first person to recognize that sulfur can be added to rubber to promote the formation of crosslinks between the polymer chains in order to produce a polymer (rubber) with superior mechanical properties. This process is known as vulcanization. What element other than sulfur might also be used to form crosslinks in rubber? (Hint: First consider the bond characteristics of S in order to understand why Goodyear found this element to be the key ingredient in the vulcanization process.)

Solution

The purpose of the atoms in the crosslinks is to form a "bridge" between two polymer chains. As such, the crosslinking atoms must each form two covalent bonds. The elements that are most likely to form two covalent bonds are those in Group VIB of the periodic table. The two most common elements in this column are S and O. Therefore, the answer to this problem is oxygen. The crosslinks in rubbers usually involve a short chain of S or O atoms rather than a single S or O atom. The level of crosslinking in an automobile tire is not great, and there are large numbers of unsaturated double bonds remaining. Therefore, when atmospheric sulfur or oxygen attacks the tire in use, the sidewalls become brittle and crack.

As mentioned above, the key difference between thermoplastic (TP) and thermoset (TS) structures is that TSs generally have a 3-D primary bond network while TPs have weak secondary bonds between the covalently bonded polymer chains. This difference has a profound influence on the properties of polymers.

The change in structure of a TS polymer as a result of an increase in temperature can be compared to that of a raw egg during cooking. In either case, the raw material—monomer, or uncooked egg—can be formed easily into almost any desired shape by "pouring" the raw material into a suitably shaped mold. The initial application of heat changes the structure by supplying the energy necessary to form a three-dimensional bond network. Upon cooling from the "fabrication" temperature, the material retains the shape of the mold. It is not possible, however, to repeat this process. Upon reheating, both the egg and the TS polymer will degrade—char or burn—rather than revert to their original fluidlike structure. Thus, TS polymers can be formed initially into almost any desired shape, but they cannot be re-formed at a later time.

In contrast, the behavior of a typical TP polymer is much closer to that of candle wax. Because of the comparative ease with which the secondary bonds between polymer chains can be temporarily suspended with heat, TPs can be reheated and re-formed over and over again. As a result, TPs are generally much easier to recycle than their TS counterparts.

SUMMARY

The atomic scale structure of a material—the atoms present, the types of bonds between the atoms, and the way the atoms are packed together—controls many of the important macroscopic engineering properties of that material. To understand the relationship between atomic scale structures and physical properties, one must apply concepts from other disciplines such as quantum mechanics, thermodynamics, and kinetics.

The two major classes of bonds are primary and secondary bonds. Primary bonds are one to two orders of magnitude stronger than secondary bonds. The three types of primary bonds are ionic, covalent, and metallic bonds.

The formation of an ionic bond involves electron transfer from the electropositive element to the electronegative element. This charge transfer results in a coulombic force of attraction between the ions. Ionic solids are comparatively brittle and are poor conductors of electricity.

Covalent bonds are favored when all the atoms are electronegative. Each atom obtains a filled valence shell by sharing electrons with its nearest-neighbor atoms. These shared electrons are spatially localized, and therefore, covalent bonds have specific bond angles associated with them. Examples of covalently bonded solids include Si, Ge, diamond, most polymers, and most organic molecules.

Metallic bonds are favored when all the atoms are electropositive and when the atoms have three or fewer valence electrons. These elements also share electrons, but the shared electrons are delocalized and form an electron cloud around the metal ion cores. The delocalized valence electrons lead to high electrical conductivities and ductile behavior for most metals.

Regardless of the type of primary bond formed between two atoms or ions, there is a competition between the atomic forces of attraction and repulsion. The separation distance for which the total force between the atoms or ions is zero is defined to be the equilibrium separation distance, or bond length.

The bond-energy curve, which is the integral of the bond-force curve, may be used to explain several important macroscopic properties of a solid, including the bond length, the modulus (or stiffness) of the material, the bond strength, and the linear coefficient of thermal expansion.

The coordination number (CN) of an atom represents its number of nearest neighbors. In covalently bonded materials CN is determined by the valence of the atoms involved. The CN in ionic and metallic solids is determined from simple geometrical considerations (i.e., the radius ratio).

Secondary bonds occur as a result of the formation of an electric dipole within certain atoms or molecules. The dipole can be temporary, induced, or permanent.

Some solids have primary bonds with mixed properties. For example, many ceramics exhibit mixed ionic and covalent characteristics. Other materials display properties typical of a mixture of primary and secondary bonding.

Thermoplastic polymers have secondary bonds between chains, and thermosetting polymers have a three-dimensional covalent bond network. As a result, thermoplastic polymers can be repeatedly reheated and reshaped, while thermosetting polymers retain their original shape and degrade rather than melt when they are reheated.

KEY TERMS

amorphous material	coordination number	ground state	polymers
Arrhenius equation	core electrons	hydrogen bond	quantum number
atomic scale structure	coulombic force	ionic bond	short-range order
bond angles	covalent bonds	ionization potential	temporary dipole
bond energy	crosslinks	intermetallics	thermodynamics
bond-energy curve (bond-energy well)	crystalline materials	kinetics	thermoplastic polymer
	ductile	long-range order	thermoset polymer
bond-force curve	electron affinity	metallic bonds	valence electrons
bond length	electron configuration	monomers	van der Waals bond
brittle	electron transfer	Pauli exclusion principle	Young's modulus
coefficient of thermal expansion	electronegativity	permanent dipole	

HOMEWORK PROBLEMS

SECTION 2.2
Atomic Structure

1. How many valence electrons do elements in Group IIIB have? How many valence electrons do elements in Group VB have?

2. In Example Problem 2.2–1 we showed why Si and Ge are similar. What element is next in this series? What is its electron configuration?

3. Would you expect Ca and Zn to exhibit similar properties? Why or why not?

4. What would be some of the consequences if electron energies were *not* quantized?

5. How many electrons does Cu have? Protons? Neutrons?

6. Write the electronic structure of C. How can C form four equal bonds? Covalent bonds are regions of high electron density and, hence, repel one another. Anticipate the bonding geometry (bond angle) of the four bonds in covalently bonded carbon.

SECTION 2.3
Thermodynamics and
Kinetics

7. The oxides of most metals are energetically "downhill" from the energy of the pure metal. Metals that oxidize generally gain weight in the early stages of oxidation. Do you want this behavior for "the gold standard," which is a metal stored for long periods of time and used to calibrate dollars?

8. Is it possible to have pure liquid water at $-1°C$ (30°F)?

9. The flow rate of molasses can be described as an Arrhenius process with an activation energy of about 50 kJ/mol. How much does the flow rate change when the temperature changes from 10°C to 25°C?

10. In order to form a polymer, many identical small molecules are linked together in a chemical reaction. The polymerization reaction is exothermic, or heat producing, and it can be described as an Arrhenius process. The activation energy is typically on the order of 80 kJ/mol. How much does the reaction rate change if the temperature increases 10°C?

11. Why are top-of-the-line electronic cable ends made using gold (contacts) rather than steel, aluminum, or copper?

12. It is energetically an uphill battle to reduce Al_2O_3 to 2Al and $(1.5)O_2$. Since the thermodynamics favor the oxide rather than the separate elements, how can a process exist to reduce bauxite (the oxide of aluminum) to pure aluminum?

SECTION 2.4
Primary Bonds

13. For each of the primary bond types answer the following questions:
    a. Does this type of primary bond usually involve electronegative atoms, electropositive atoms, or both types of atoms?
    b. Are the bonding electrons shared or transferred?
    c. If the bonding electrons are shared, are they spatially localized or delocalized?

14. Determine the most likely type of primary bond in each of the following materials: O_2, NaF, InP, Ge, Mg, CaF_2, SiC, $(CH_2)_n$, MgO, CaO.

15. Show how the four unpaired electrons in C form covalent bonds, say, with four H atoms. That is, show electron sharing and application of the "filled valence shell rule."

16. How does the coulombic force change with distance between charges? What other well-known force shows this behavior? At what charge separation is dF (coulombic)$/dx$ a maximum? A minimum?

17. Define what is meant by the terms *ionization potential* and *electron affinity*.

18. Why do covalent bonds form only between electronegative elements?

19. If O forms a covalent bond with itself and Si forms a covalent bond with itself, then is it reasonable to suggest that Si—O is a covalent bond?

20. Is it possible that a ceramic or ionic solid might be a reasonable conductor of electricity? What needs to occur?

21. Explain the physical significance of the requirement that exponents in Equation 2.4–7 obey the inequality $p < q$.

22. The bond-energy curve can be used to gain information about three of the four following materials properties. Which type of information cannot be determined from the bond-energy curve?
    a. Bond energy
    b. Equilibrium separation distance
    c. Primary bond type
    d. Vaporization temperature

SECTION 2.5
The Bond-Energy Curve

23. Of the two materials shown in Figure 2.5–1, which do you anticipate will have a higher melting temperature?

24. Given that the coefficient of expansion for Al is $25 \times 10^{-6}°C^{-1}$ and for SiC is $4.3 \times 10^{-6}°C^{-1}$, predict which material has the higher melting temperature.

25. Explain why it is difficult to find a material with both a high stiffness and a high coefficient of thermal expansion.

26. Many, many oxide ceramics or ionic solids have moduli of elasticity of $\sim6.9 \times 10^4$ MPa, independent of composition. Why?

27. The modulus of elasticity of window or oxide glass is 6.9×10^4 MPa and that of a "plastic" glass is 6.9×10^3 MPa. How much more will the plastic glass deflect than the oxide glass under a given load?

28. Sketch a bond-energy and bond-force curve for a material with a negative thermal expansion coefficient.

29. The center-to-center separation of two atoms in pure Ti is 2.94 Å. The thermal expansion coefficient of Ti metal is $9 \times 10^{-6}°C^{-1}$. What percent increase in atomic separation is realized in going from 25°C to 625°C? Repeat the problem for copper, using data from Table 2.5–1 and Appendix C.

30. How might a scientist or an engineer measure Young's modulus in the laboratory? How about the thermal expansion coefficient?

31. Consider the ionic compound CaF_2. Given that $r(Ca) = 0.197$ nm, $r(Ca^{2+}) = 0.106$ nm, $r(F) = 0.06$ nm, and $r(F^-) = 0.133$ nm, estimate the coordination numbers for each of the elements in this compound.

SECTION 2.6
Atomic Packing and
Coordination Numbers

32. Determine the coordination number of the cation for NiO, ZnS, and CsI.

33. Draw the structure of CH_4, $-\overset{\overset{\displaystyle O}{\|}}{C}-$, and $-\overset{\overset{\displaystyle H}{|}}{N}-$ along the lines of Figure 2.6–5.

34. Discuss the reasons why most covalent solids are low-density materials, ionic solids are intermediate density, and metals are high density.

35. Under what conditions would you use the (r/R) concept to predict the coordination numbers in a covalent solid?

36. What coordination number would you predict for germanium, which is a Group IV covalently bonded material?

37. Calculate the minimum and maximum r/R ratios for CN = 4.

38. A newly discovered crystalline compound has the chemical formula A_2B. You are given the following information: $r(A) = 0.12$ nm, $r(A^-) = 0.13$ nm, $r(B) = 0.15$ nm, and $r(B^+) = 0.14$ nm.
    a. Explain why $r(A^-) > r(A)$ but $r(B^+) < r(B)$.
    b. Assuming purely ionic bonding, calculate the coordination number (CN) of both the cation and the anion.

c. If you were told that the observed CN for the anion was 6, what would you conclude about the nature of the bond?

39. Two elements are combined and the coordination number of each type of atom is 12. What is the relative size of each atom type?

40. How many bonds form with C? How many with Si? Is it reasonable to expect that Si can be substituted for C in most organic structures, so that silicon chemistry has at least some similarities to carbon chemistry?

41. Show the correct geometrical structure of C_2H_6.

42. Why are there no solids with coordination number 5, 7, or 9?

43. Where would you put four tacks in a cubic room if you wanted each of them to be as far from the others as possible?

44. Discuss the value of the bond angle (H—N—H) in ammonia, NH_3.

45. If you add a proton, H^+, to ammonia, you get NH_4^+. If this material is introduced to Cl^-, what is the nature and strength of the bond formed? (This sort of bond occurs in a type of material called an ionomer.)

SECTION 2.7
Secondary Bonds

46. Predict which member of each pair of materials below has a higher melting temperature.
    a. C_2F_4 or $C_2H_2F_2$
    b. CH_4 or NH_3 (predict the higher vaporization temperature)
    c. $C_{14}H_{30}$ or C_5H_{12}

47. Secondary bonds are present and extremely important in polymer behavior. Natural polymers are either animal- or vegetable-based. Animal fibers contain —NH bonds. Vegetable fibers contain —OH bonds. Do you expect natural polymers to be moisture sensitive, i.e., capable of bonding with H_2O?

48. What binds one molecule to another in a molten mass of polymer?

49. Which chemical group is capable of larger secondary bonding forces, one that contains $C \equiv N$ groups or one that contains CH_3 groups?

50. Secondary bonding involves electronic attraction of dipoles, or partial charges. Primary bonding also involves electronic interactions, but of whole charges. Estimate the relative magnitude of secondary bonding forces and compare them with primary bond strengths. In so doing you have estimated how much stronger a polymer can be when all the covalent bonds are aligned in one direction, such as in some fibers, and how weak they are normal to that direction.

SECTION 2.8
Mixed Bonding

51. Give three examples of materials that exhibit mixed bonding at the primary bond level and three examples of materials that exhibit both primary and secondary bonding.

52. Suppose a material is partly ionic and partly covalently bonded. If an external environment can shift the balance of charge sharing and charge transfer, how will this affect electrical conductivity? If the material were partly covalent and partly metallic, how would your answer change? Of what use would such a material be?

53. Some covalently bonded materials can be doped with electron acceptors, allowing charge to jump from the covalent bond to the acceptor and back, perhaps into another covalent bond. How will the presence of dopant affect electrical conductivity.

54. Ni_3Al is an intermetallic that retains good mechanical properties at high temperature. This characteristic is typically ascribed to ionic solids. Is this material at least partially ionic?

SECTION 2.9
The Structure of Polymer Molecules

55. Human hair is not a thermoplastic material but rather a thermoset material. Much as in the case of vulcanized rubber, the crosslinks are sulfur bridges. Crosslinks give hair molecules a memory of their most stable position. Curly hair, for example, is always

curly. What do you think is necessary to occur in order to give hair a new "perm"(anent) memory? Why does a perm "smell"?

56. Is silk a thermoplastic or a thermoset polymer? Wool? Hair? Cotton? All natural polymers? What is the underlying reason? Recall secondary bonding.

57. As mentioned in the text, breaking of a double bond to produce a single bond can create a polymer and lots of heat. It can be a thermodynamically favorable process. What happens when the process proceeds quickly on a huge batch of material?

58. Suppose you are going to install some plastic piping in your home to transport hot water from the water heater to the shower. Would you be better off using polyvinylchloride (based on the mer $C_2H_2Cl_2$) pipe or polyethylene (based on the mer C_2H_4) pipe for this project? Why?

59. Are all plastic (polymer) components recyclable? Why or why not?

60. How would you expect the stiffness of rubbery material to change as the amount of sulfur (or oxygen) in the material increases? Why?

C H A P T E R 3

CRYSTAL STRUCTURES

MATERIALS IN ACTION Crystal Structure

Many materials crystallize in a regular array with basic building blocks being repeated at regular intervals. Such materials are said to be crystalline. Their actual structure is determined by the bond character and the requirement that the energy of the system be minimized. Crystals may be formed in metals, ceramics, and even polymers. The properties of materials depend on their bond character of which the crystal structure is an important manifestation. Because of the high symmetry of metallic crystals they possess good ductility and are used in structural applications where large-scale deformation is required. Also because they tend to have less directionality in their bonds than do other materials, they are excellent electrical conductors. In ceramics with more than one atomic species being present, the bonding is usually a mix of covalent and ionic. Thus, size and bond directionality become important considerations and the crystals are less symmetrical than in metals. These materials are thus harder, but less ductile. They find application where wear resistance and ultra high strength are important. Because of their localized and directional bonds, their electrical conductivity is usually not as good as that of metals. In crystalline polymers, the crystal structure is usually determined by geometry and secondary bond forces (van der Waals, dipoles). Unoriented polymers are weaker than either metals or ceramics. Pound for pound, oriented polymers can be every bit as strong, or stronger, than metals or ceramics.

In many ways crystal structure is a major determinant of the properties of a material. In this chapter we shall study crystal structures and those factors that determine crystal structure and to a degree, the physical, electrical, and mechanical properties of materials.

3.1 INTRODUCTION

In the previous chapter we saw that macroscopic properties of materials are influenced strongly by atomic scale structure. In previous discussions we focused our attention on an individual atom or ion and looked first at its spatial relationship to a single neighboring atom, specifically, at its equilibrium separation distance; and then at its spatial relationship to a small group of atoms known as its nearest neighbors, as manifested in its coordination number and bond angles. As mentioned in Section 2.6, the local arrangement of nearest-neighbor atoms about a central atom is known as short-range order (SRO).

In this chapter we expand our view of materials to incorporate larger numbers of atoms. Materials that exhibit order over distances much greater than the bond length are said to have long-range order (LRO). In fact, materials can be classified on the basis of the extent of LRO they exhibit: *amorphous* solids show SRO in three dimensions, but no LRO; *crystalline* solids exhibit both SRO and LRO in three dimensions. As a general rule, most metals are crystalline, while ceramics and polymers may be either crystalline, amorphous, or a combination of the two.

Since an understanding of SRO and LRO is central to this chapter, it is appropriate to pause and present a few examples. In the case of a noble gas, under most conditions the interaction between atoms is minimal. Consequently, there is no significant positional relationship between one gas atom and another. The material shows neither SRO nor LRO. This is not true in most condensed phases. In a liquid, for example, nearest neighbors are positioned at well-defined distances and SRO is established. The order within most liquids, however, does not persist beyond nearest-neighbor distances. The same type of order, SRO but no LRO, can occur in solids. As you might expect, the structure of amorphous solids, or glasses, is similar to that of a "frozen" liquid. A detailed discussion of amorphous solids is the subject of Chapter 6.

Many solids are crystalline. The establishment of LRO requires that atoms be arranged on a three-dimensional array that repeats in space. The 3-D framework is known as a **crystal lattice.** The details of the lattice pattern strongly influence the macroscopic properties of crystalline engineering materials.

The organization of this chapter is as follows. First we describe the concept of a crystal lattice and then present a few simple crystal structures. We will see that even for simple structures, a language is needed to describe specific points, directions, and planes in crystals. After introducing the appropriate nomenclature, we describe several methods to quantify various characteristics of crystal lattices. We then turn our attention to a description of more complex crystal structures, those structures typically associated with ionic, covalent, or molecular crystals. Next, we describe the fundamental relationship between crystal structure and macroscopic properties, a relationship that is emphasized throughout the textbook. Finally, we conclude the chapter with an introduction to X-ray diffraction, the technique most commonly used to characterize crystal structure.

3.2 BRAVAIS LATTICES AND UNIT CELLS

A **lattice** can be defined as an indefinitely extended arrangement of points each of which is surrounded by an identical grouping of neighboring points. Before proceeding to a discussion of three-dimensional (3-D) crystal lattices, we will introduce some of the important characteristics of a lattice using a 2-D analogy.

Wallpaper is a common example of a 2-D lattice. The smallest region that completely describes the pattern is known as the **unit cell.** Once the unit cell is established, the entire

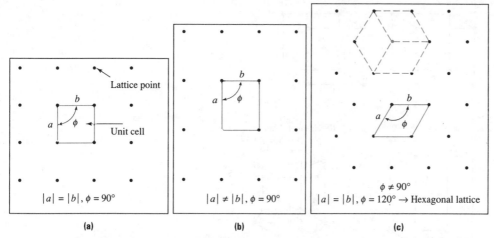

FIGURE 3.2–1   The possible unit cells of wallpaper, or any other 2-D pattern, include: **(a)** a square, **(b)** a rectangle, or **(c)** a parallelogram. The restrictions on the lattice parameters, *a* and *b*, and the angle between the edges of the unit cell, *φ*, are listed for each type of cell.

extended pattern can be generated by translating the unit cell in 2-D. As shown in Figure 3.2–1, the only permissible shapes for 2-D unit cells are a square, a rectangle, or a parallelogram. The reason for this is that the unit cell must have a shape that permits it to be arranged in a way that completely fills space. The vertices of the unit cell are known as **lattice points.** The lengths of the unit-cell edges are known as the **lattice parameters.** The angles and lengths within the repeat unit determine the class to which the lattice cell belongs.

Note that all three patterns in Figure 3.2–2 have exactly the same rectangular lattice, yet the patterns are distinguishable. Thus, the specification of a lattice alone is not sufficient to uniquely define a pattern. In addition, one must describe exactly what is located at each lattice point. The name given to the "group of things" located on a lattice point is the **basis.** For wallpaper the basis is one or more illustrations, while for a 3-D crystal the basis is one or more atoms. This idea can be formalized by the relationship

$$\text{Lattice} + \text{Basis} = \text{Crystal structure} \qquad (3.2\text{–}1)$$

Having introduced the important properties of a lattice in 2-D, we can focus on our main objective—the description of 3-D crystal lattices.

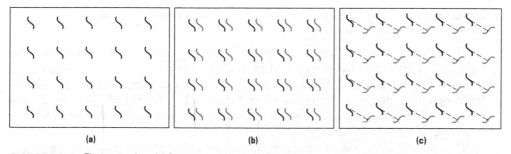

FIGURE 3.2–2   Three examples of 2-D patterns all created using the same rectangular lattice but each having a different basis: **(a)** the basis is a single character; **(b)** the basis contains a repeated character, and **(c)** the basis contains two characters with different orientations.

I. Cubic lattices $a = b = c$; $\alpha = \beta = \gamma = 90°$

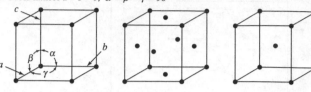

II. Tetragonal lattices $a = b \neq c$; $\alpha = \beta = \gamma = 90°$

III. Hexagonal lattices $a = b \neq c$; $\alpha = \beta = 90°$; $\gamma = 120°$

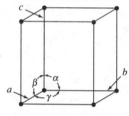 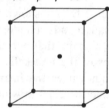 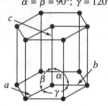

IV. Orthorhombic lattices $a \neq b \neq c$; $\alpha = \beta = \gamma = 90°$

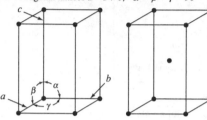

V. Rhombohedral lattices $a = b = c$; $\alpha = \beta = \gamma \neq 90°$

VI. Monoclinic lattices $a \neq b \neq c$; $\alpha = \gamma = 90° \neq \beta$

VII. Triclinic lattices $a \neq b \neq c$; $\alpha \neq \beta \neq \gamma$

 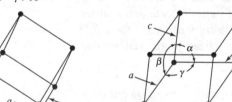  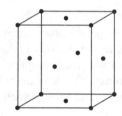 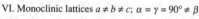 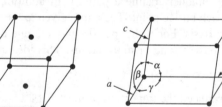

FIGURE 3.2–3   The 14 Bravais lattices grouped into the 7 lattice types. The restrictions on the lattice parameters *a, b,* and *c* and the angles between the edges of the unit cell α, β, and γ are listed for each unit cell.

There are 14 valid 3-D lattices, on which the basis—atoms or groups of atoms—can be placed. They are called Bravais lattices and are shown in Figure 3.2–3. Each of the lattice points is equivalent; that is, the lattice points are indistinguishable. The equivalence of lattice points is demonstrated in Figure 3.2–4 for the body-centered cubic crystal, in which the axis system is redrawn so that the "new" origin corresponds to the center of the "original" cube.

The unit cell is the smallest volume that shows all the characteristics of the system. Each of the cells repeats indefinitely in all directions, to the physical limits of the crystal. The properties of a unit cell are the same as those of the crystal. Hence, a unit cell is a convenient representative structure that can be used to calculate theoretical properties of a crystal, such as density.

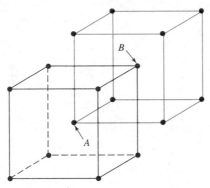

FIGURE 3.2–4 The equivalence of the lattice points is demonstrated for the body-centered cubic crystal by redrawing the axis system so that the "new" origin corresponds to the center of the "original" cube. Note that atom A is at the center of the black cube and the front left bottom corner of the colored cube. Atom B is at the back right top corner of the black cube and the center of the colored cube.

As in the previous 2-D example, 3-D unit cells are described in terms of the cell parameters—the lengths of the cell edges and the angles between axes. Consider, for example, the cell with the highest symmetry, the cubic cell. The axes in the cubic system are orthogonal (all angles 90°) and the lengths of the sides of the cube are equal. Hence, a cubic crystal is completely characterized by a single lattice parameter a_0. The lattice parameter a_0 is not equivalent to the equilibrium separation distance x_0. The former quantity represents the length of a cubic unit-cell edge and the latter represents the distance between the centers of adjacent nearest-neighbor atoms. At the other extreme, description of a triclinic crystal requires the specification of three lengths (a, b, and c) and three angles (α, β, and γ).

We will see in the next section that many metals have cubic structures and some have hexagonal structures. Materials with ionic bonds typically have larger, more complex crystal structures than metals. Since there is more than one type of atom present in ionic solids, the complexity of the basis increases. Polymers also have complex bases and crystal structures with large unit cells.

3.3 CRYSTALS WITH ONE ATOM PER LATTICE SITE AND HEXAGONAL CRYSTALS

In this section we examine and develop important concepts regarding unit cells for metals. The simplest unit cell, called the **simple cubic (SC)** structure, has atoms located in each of the cell corners. We will not dwell on the SC structure here, since no important metals have this structure. Instead, we will investigate the more common metal structures.

3.3.1 Body-Centered Cubic Crystals

The simplest cells are those with cubic symmetry and one atom per lattice position (i.e., the basis is a single atom). Consider the structure of tungsten, shown in Figure 3.3–1. An atom lies at each corner of the cube and one in the center. This is the **body-centered cubic (BCC)** structure. Each corner atom touches the central atom, but the corner atoms do not touch each other. Other metals with the BCC structure at room temperature include chromium, iron, molybdenum, and vanadium.

Three important characteristics of a cubic unit cell are the length of its lattice parameter a_0, the number of atoms in the unit cell, and the coordination number of each atom. Although values of a_0 for many materials are available in the technical literature, it is important to recognize that estimates for a_0 can be obtained from a knowledge of the

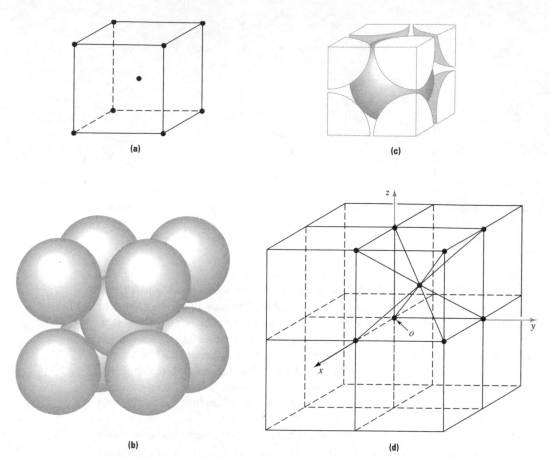

FIGURE 3.3–1   The structure of tungsten, a body-centered cubic metal, is illustrated in three different ways: **(a)** the point model shows only the locations of the atom centers, **(b)** the full solid sphere model shows all five atoms associated with the unit cell, and **(c)** the partial solid sphere model shows just the fractions of each atom contained within the unit cell. **(d)** An illustration of eight adjacent unit cells showing that some of the atoms in the BCC unit cell are shared among several adjacent unit cells.

radius of the atoms involved, r, and the geometry of the unit cell. To obtain the relationship between the lattice parameter and the atomic radius (the a_0-r relationship), find a direction in which atoms are touching, and equate the expression for atom center-to-center distance in terms of a_0 to the equivalent distance in terms of r.

Figure 3.3–2a shows that for a cube with edge length a_0, the length of any face diagonal is $a_0\sqrt{2}$ and the length of any body diagonal is $a_0\sqrt{3}$. Therefore, the distance between adjacent atoms in the BCC structure is $a_0\sqrt{3}/2$ (see Figure 3.3–2b). The repeat distance in terms of r is $2r$, so that the a_0-r relationship is

$$\frac{a_0\sqrt{3}}{2} = 2r \quad \text{or} \quad a_0(\text{BCC}) = \frac{4r}{\sqrt{3}} \tag{3.3–1}$$

Next, we determine the number of atoms per cell. Examination of Figure 3.3–1d shows that nine atoms are associated with each cell but some atoms are shared among several cells. Notice that each corner atom is shared by eight cells. Therefore, only 1/8 of any corner atom is associated with each cell. In contrast, a center atom is totally contained within its cell. Thus, the total number of atoms per unit cell is two $[(8 \times 1/8) + (1 \times 1)]$. Examination of Figure 3.3–1 shows that each atom in the BCC structure has eight nearest neighbors [CN(BCC) = 8].

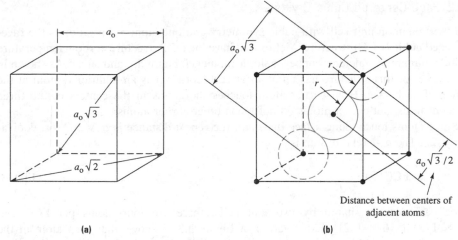

(a)   (b)

FIGURE 3.3–2   **(a)** For a cube with edge length a_0, the length of any face diagonal is $a_0\sqrt{2}$ and the length of any body diagonal is $a_0\sqrt{3}$. **(b)** The distance between adjacent atoms in the BCC structure may be expressed as either $a_0\sqrt{3}/2$ or, equivalently, $2r$.

EXAMPLE 3.3–1

Using data on atomic weight and atomic radius (in Appendices B and C), calculate the density of BCC Fe.

Solution

Density is defined to be mass divided by volume. Since the unit cell completely describes the crystal structure, we can calculate the density of iron based on the mass and volume of its unit cell. The mass of the unit cell, M_{uc}, can be determined by noting:

$$M_{uc} = \frac{\text{Number of atoms}}{\text{Unit cell}} \times \frac{\text{Mass}}{\text{Atom}}$$

The volume of the unit cell, V_{uc}, is just a_0^3. For iron,

$$M_{uc} = \left(\frac{2 \text{ atoms}}{\text{BCC unit cell}}\right)\left[(55.85 \text{ g/mol Fe})\left(\frac{1 \text{ mol Fe}}{6.023 \times 10^{23} \text{ atoms Fe}}\right)\right]$$

$$= 1.85 \times 10^{-22} \text{ g/(unit cell)}$$

and

$$V_{uc} = a_0^3 = \left(\frac{4r}{\sqrt{3}}\right)^3 = \left[\frac{4 \times (1.24 \times 10^{-8} \text{ cm})}{\sqrt{3}}\right]^3$$

$$= 2.35 \times 10^{-23} \text{ cm}^3/(\text{unit cell})$$

Therefore, the density of iron is calculated as

$$\rho = \frac{1.85 \times 10^{-22} \text{ g/(unit cell)}}{2.35 \times 10^{-23} \text{ cm}^3/(\text{unit cell})}$$

$$= 7.87 \text{ g/cm}^3$$

The measured density of Fe is also 7.87 g/cm^3. In general, however, we should not expect perfect agreement between the density estimate obtained using this method and the measured density of a solid, since our model assumes that the crystal is "perfect" while real crystals contain defects. These defects will be discussed in the next two chapters.

3.3.2 Face-Centered Cubic Crystals

Another common unit cell with cubic symmetry and one atom per position is the **face-centered cubic (FCC)** structure. Metals that have the FCC structure at room temperature include aluminum, calcium, copper, gold, lead, nickel, platinum, and silver. As shown in Figure 3.3–3, this structure has an atom at each corner plus an additional atom at the center of each face. Each corner atom touches the atoms in the centers of the three adjacent faces, but corner atoms do not touch other corner atoms.

Since atoms touch along a face diagonal, the repeat distance is $a_0\sqrt{2}/2$, or, equivalently, $2r$. The a_0-r relationship is:

$$a_0(\text{FCC}) = \frac{4r}{\sqrt{2}} \tag{3.3–2}$$

Since each face is shared by two unit cells, there are four atoms per FCC cell $[(8 \times 1/8) + (6 \times 1/2)]$. Examination of Figure 3.3–3 shows that each atom in the FCC structure has 12 nearest neighbors, or CN(FCC) = 12. Consider, for example, the

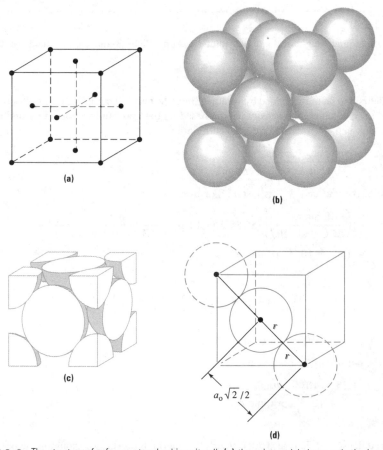

FIGURE 3.3–3   The structure of a face-centered cubic unit cell: **(a)** the point model shows only the locations of the atom centers, **(b)** the full solid sphere model shows all 14 atoms associated with this unit cell, and **(c)** the partial solid sphere model shows just the fractions of each atom contained within this unit cell. **(d)** The distance between adjacent atom centers in FCC can be expressed as either $a_0\sqrt{2}/2$ or, equivalently, $2r$.

atom in the center of the top surface of the cube. Its nearest neighbors are the four atoms in the centers of the vertical faces of the unit cell, the four corner atoms on the top surface of the cell, and the four atoms on the vertical faces of the unit cell that is positioned just above this unit cell.

EXAMPLE 3.3–2

Calculate the density of aluminum. Then discuss why the aerospace industry prefers aluminum-based alloys to iron-based alloys and why iron-based alloys are preferred over aluminum alloys in structural members of bridges and buildings.

Solution

Using the procedure in Example 3.3–1, we find

$$\text{Density} = \frac{M_{uc}}{V_{uc}}$$

with

$$M_{uc} = \left(\frac{\text{Number of atoms}}{\text{Unit cell}}\right) \times \left(\frac{\text{Mass}}{\text{Atom}}\right)$$

and

$$V_{uc} = a_0^3$$

For FCC aluminum,

$$M_{uc} = \left(\frac{4 \text{ atoms}}{\text{FCC unit cell}}\right)\left[(26.98 \text{ g/mol Al})\left(\frac{1 \text{ mol Al}}{6.023 \times 10^{23} \text{ atoms Al}}\right)\right]$$

$$= 1.79 \times 10^{-22} \text{ g/(unit cell)}$$

and

$$V_{uc} = a_0^3 = \left(\frac{4r}{\sqrt{2}}\right)^3 = \left[\frac{4 \times (1.43 \times 10^{-8} \text{ cm})}{\sqrt{2}}\right]^3$$

$$= 6.62 \times 10^{-23} \text{ cm}^3/(\text{unit cell})$$

Therefore, the density of aluminum is calculated as

$$\rho = \frac{1.79 \times 10^{-22} \text{ g/(unit cell)}}{6.62 \times 10^{-23} \text{ cm}^3/(\text{unit cell})}$$

$$= 2.70 \text{ g/cm}^3$$

The measured density of aluminum is also 2.70 g/cm^3. A comparison of the density of aluminum with that of iron ($\rho = 7.87$ g/cm^3) partially explains why aluminum alloys are preferred to iron alloys in the aerospace and other industries where minimizing the weight of structures is a critical design parameter. For the same cross-sectional area, steels (iron-based alloys) can sustain higher load levels and can also provide more stiffness against bending compared with aluminum alloys. Thus, if weight is not a consideration but volume of material is important, steel members have a higher load-bearing capacity than aluminum alloys. Other advantages of steel include cost and its weldability compared with aluminum.

3.3.3 Hexagonal Close-Packed Structures

The structure of a hexagonal system is most easily visualized by considering three unit cells arranged to form one larger cell, as shown in Figure 3.3–4a. The larger cell is not a unit cell, since the structure can be completely characterized by an even smaller volume

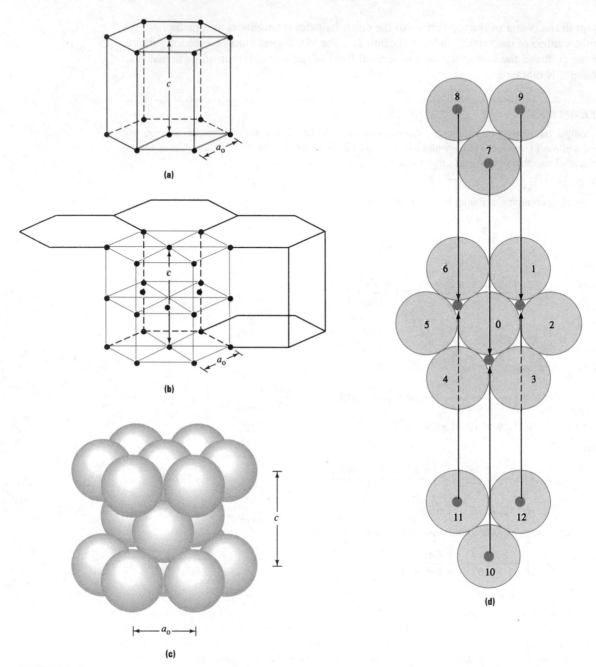

FIGURE 3.3–4   The structure of simple hexagonal (SH) and hexagonal close-packed (HCP) unit cells: **(a)** a point model of an SH unit cell showing the geometric relationship between the smaller primitive unit cell and the more convenient "large" unit cell; **(b)** a point model of the "large" unit cell for the HCP structure showing several neighboring unit cells; **(c)** the full solid sphere model for the HCP structure; and **(d)** an illustration of the 12 nearest neighbors for an atom in an HCP unit cell.

of material. Nevertheless, it is often more convenient to visualize and solve problems using this "big" cell. The upper and lower surfaces of the large cell are hexagons, and the six side faces are rectangles. In the simple hexagonal (SH) structure, atoms are positioned at each corner and in the center of the hexagonal faces. Although the SH structure is not a common crystal structure, one of its variations, the **hexagonal close-packed (HCP)** structure, is characteristic of many metals, including cadmium, cobalt, magnesium, titanium, yttrium, and zinc at room temperature. The HCP crystal structure is shown in Figure 3.3–4b. Note that there are six atoms at the corners of the top and bottom planes, each shared by six unit cells; one atom in the center of the upper and lower basal planes, each shared by two cells; and three atoms in the midplane. Thus, the total number of atoms in the large HCP cell is six $[(12 \times 1/6) + (2 \times 1/2) + (3 \times 1)]$. Since each large cell consists of three unit cells, each unit cell contains two atoms.

As shown in Figure 3.3–4c, each of the six corner atoms in the top and bottom hexagonal planes, known as the **basal planes,** touches the central atom. If a_0 is defined to be the length of the unit-cell edge, then the a_0-r relationship is simply $a_0 = 2r$. A complete description of the dimensions of the cell, however, requires an expression for its height, or the perpendicular distance between the basal planes. In the ideal HCP unit cell, the height, c, is related to a_0, and hence r, through the expression

$$c = \left(\frac{4}{\sqrt{6}}\right)a_0 = 1.633a_0 = 3.266r \qquad (3.3-3)$$

This relationship assumes that the atoms are perfect rigid spheres. Since this assumption is not always satisfied, many real HCP metals display a c/a_0 ratio significantly different from 1.633 (see Table 3.3–1). The volume of a "large" HCP cell is

$$V_{uc}(\text{large HCP}) = \left(\frac{\sqrt{3}}{2}\right)a_0^2 c \qquad (3.3-4)$$

The number of nearest neighbors in the HCP system is 12, or CN(HCP) = 12. This can be seen by considering the central atom in the lower basal plane. As shown in Figure 3.3–4a, this atom has six nearest neighbors in its own plane and three nearest neighbors in the parallel planes above and below.

TABLE 3.3–1   The c/a ratios for selected HCP metals at room temperature.

Metal	c/a ratio
Cd	1.886
Zn	1.856
"Ideal" HCP	1.633
Mg	1.624
Co	1.621
Zr	1.593
Ti	1.587
Be	1.568

3.4 MILLER INDICES

In this section we introduce the notation known as **Miller indices,** which is the most common convention used to describe specific points, directions, and planes in the crystal-lattice systems. Before proceeding with the details of the convention, however, we show the need for indexing using an example. Consider the geometry shown in Figure 3.4–1. In the next chapter we will need to know the angle between the direction from left to right along the bottom back edge of the cube and the direction from the bottom back left corner through the center of the cube. The Miller index notation not only simplifies the

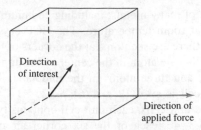

FIGURE 3.4–1 Examples of two directions, in a cubic unit cell. The direction of the applied force can be described as a projection from the bottom back left corner of the cube through the bottom back right cube corner. The cumbersome nature of this description provides some of the motivation for the development of a more concise nomenclature for naming points, directions, and planes in crystals.

description of directions, but also permits simple vector operations like the dot and cross products.

3.4.1 Coordinates of Points

The first step in the description of crystal structures is to select a coordinate system. We have elected, as is customary, to use a right-hand Cartesian coordinate system throughout the text. The next step is to orient the coordinate system in the unit cell. As shown in Figure 3.4–2 for a cubic unit cell, the most common orientation is to align the three coordinate axes with the edges of the unit cell, with the origin at a corner of the cell. It is important to note, however, that the choice of an origin is arbitrary, and the selection of an origin is a matter of convenience for each problem under consideration.

Having defined a coordinate system, points within the lattice are written in the form h, k, l, where the three indices correspond to fractions of the lattice parameters a, b, and c. Recall that the lattice parameters a, b, and c correspond to the length of the unit-cell edge in the x, y, and z directions. Hence, with reference to Figure 3.4–2, the three corners along the axes, marked A, B, and C, are 1, 0, 0; 0, 1, 0; and 0, 0, 1. Across the body diagonal at point D, the position is 1, 1, 1. Across the face diagonal the coordinates at E, F, and G are 1, 1, 0; 0, 1, 1; and 1, 0, 1.

FIGURE 3.4–2

Miller indices—h, k, l—for naming points in a crystal lattice. The origin has been arbitrarily selected as the bottom left back corner of the unit cell.

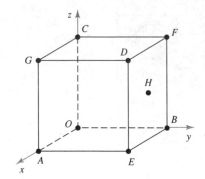

Position	Coordinate
O	0, 0, 0 (Origin)
A	1, 0, 0
B	0, 1, 0
C	0, 0, 1
D	1, 1, 1
E	1, 1, 0
F	0, 1, 1
G	1, 0, 1
H	1/2, 1, 1/2

EXAMPLE 3.4–1

Sketch a cubic unit cell and answer these questions:

   a. What are the coordinates of the points located at the centers of the six faces?

   b. What are the coordinates of the point at the center of the cube?

   c. Locate the point 1/4, 3/4, 1/4.

Solution

A cubic unit cell is sketched in Figure 3.4–3. The first step is to select an origin and orient the coordinate axes. We have elected to use the bottom back left corner of the cube as our origin.

   a. As shown in the figure, the six face centers have coordinates 0, 1/2, 1/2; 1/2, 0, 1/2; 1/2, 1/2, 0; 1, 1/2, 1/2; 1/2, 1, 1/2; and 1/2, 1/2, 1.

   b. The cube center has coordinates 1/2, 1/2, 1/2.

   c. The point 1/4, 3/4, 1/4 can be located by starting at the origin and moving out a distance of 1/4 of a lattice parameter in the *x* direction, then 3/4 in the *y* direction, and finally 1/4 in the *z* direction. This procedure is shown in Figure 3.4–3b.

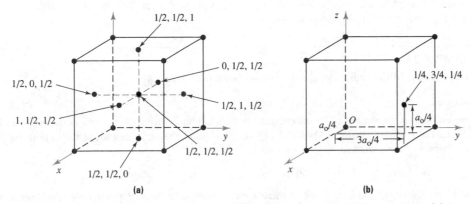

FIGURE 3.4–3   A cubic unit cell with the origin, point *O*, located at the bottom back left corner showing **(a)** the coordinates of the six face centers and the center of the cube, and **(b)** the location of the point 1/4, 3/4, 1/4, which is found by starting at the origin and moving a distance $a_0/4$ in the *x* direction, then $3a_0/4$ in the *y* direction, and finally $a_0/4$ in the *z* direction.

3.4.2 Indices of Directions

Miller indices for directions are obtained using the following procedure:

   1. Determine the coordinates of two points that lie in the direction of interest— h_1, k_1, l_1 and h_2, k_2, l_2. The calculation is simplified if the second point corresponds with the origin of the coordinate system.

   2. Subtract the coordinates of the second point from those of the first point: $h' = h_1 - h_2$; $k' = k_1 - k_2$; and $l' = l_1 - l_2$.

   3. Clear fractions from the differences—h', k', and l'—to give indices in lowest integer values, h, k, and l.

   4. Write the indices in square brackets without commas: $[h\ k\ l]$.

   5. Negative integer values are indicated by placing a bar over the integer. For example, if $h < 0$, we write $[\bar{h}\ k\ l]$.

Several examples are shown in Figure 3.4–4. Again we have used a cubic unit cell with the origin at a cube corner, although the technique is not restricted to this simple geometry. The nearest cube corners are in the directions [1 0 0], [0 1 0], and [0 0 1] as well as [$\bar{1}$ 0 0], [0 $\bar{1}$ 0], and [0 0 $\bar{1}$]. The body diagonal is [1 1 1]. The face diagonals are [0 1 1], [1 0 1], and [1 1 0].

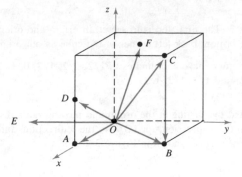

Direction	Indices $[h\,k\,l]$
$\overrightarrow{OA}$	$[1\,0\,0]$
$\overrightarrow{OB}$	$[1\,1\,0]$
$\overrightarrow{OC}$	$[1\,1\,1]$
$\overrightarrow{OD}$	$[2\,0\,1]$
$\overrightarrow{OE}$	$[0\,\bar{1}\,0]$
$\overrightarrow{OF}$	$[1\,1\,2]$
$\overrightarrow{CB}$	$[0\,0\,\bar{1}]$

FIGURE 3.4–4   Miller indices—$[h\,k\,l]$—for naming directions in a crystal lattice. The origin has been arbitrarily selected as the bottom left back corner of the unit cell.

EXAMPLE 3.4–2

Determine the Miller indices of the directions shown in Figure 3.4–1.

Solution

The problem can be simplified by selecting the bottom left back corner of the cube as the origin and defining the x, y, and z axes as shown in Figure 3.4–4. Since the coordinates of the bottom right back corner of the cube are 0, 1, 0, we find:

$$h_1, k_1, l_1 = 0, 1, 0 \qquad \text{and} \qquad h_2, k_2, l_2 = 0, 0, 0 \text{ (the origin)}$$

and $h' = 0$, $k' = 1$, and $l' = 0$. Since there are no fractions to clear, the indices for the direction along the bottom back edge are $[0\,1\,0]$. The second direction can be named in a similar way with reference to the point at the center of the cube (i.e., $h_1, k_1, l_1 = 1/2, 1/2, 1/2$). In this case we find $h' = k' = l' = 1/2$. Is the direction then $[1/2\ 1/2\ 1/2]$? No, this notation is invalid, since we have agreed that the indices for directions will always take on integer values. We must clear the fractions in $[1/2\ 1/2\ 1/2]$ by multiplying each term by 2 in order to obtain the correct notation, $[1\ 1\ 1]$.

If the properties of a crystal are measured in two different directions and found to be identical, then those directions are termed equivalent. For example, the properties of a cubic crystal measured along $[1\ 0\ 0]$ are the same as those along $[0\,\bar{1}\,0]$ or $[0\ 0\ 1]$. Similarly, all the face diagonals are equivalent, and all the body diagonals are equivalent. We refer to these **families of directions** using angle brackets: $\langle h\,k\,l \rangle$. Thus, the edges of a cube comprise the family of directions $\langle 1\ 0\ 0 \rangle$, the face diagonals comprise $\langle 1\ 1\ 0 \rangle$, and the body diagonals comprise $\langle 1\ 1\ 1 \rangle$. Note that for cubic unit cells the individual members of a family can be generated by taking all of the permutations of the symbols h, k, and l, using both positive and negative integers.

EXAMPLE 3.4–3

List the individual members of the family of directions $\langle 1\ 1\ 0 \rangle$ for a cubic unit cell.

Solution

We must find all of the permutations, both positive and negative, of the values 1, 1, and 0: $[1\ 1\ 0]$, $[1\ 0\ 1]$, $[0\ 1\ 1]$, $[\bar{1}\ 1\ 0]$, $[\bar{1}\ 0\ 1]$, $[0\ \bar{1}\ 1]$, $[1\ \bar{1}\ 0]$, $[1\ 0\ \bar{1}]$, $[0\ 1\ \bar{1}]$, $[\bar{1}\ \bar{1}\ 0]$, $[\bar{1}\ 0\ \bar{1}]$, and $[0\ \bar{1}\ \bar{1}]$. It may be useful for you to sketch these 12 directions and convince yourself that they are in fact equivalent.

It is frequently necessary to determine the angle between directions. After you have drawn the directions of interest in a unit cell, it is sometimes possible to determine the angle by inspection. Alternatively, in cubic crystals only, the angle between directions can be determined by taking the vector dot product. If

$$\mathbf{A} = u\mathbf{i} + v\mathbf{j} + w\mathbf{k} \qquad \text{and} \qquad \mathbf{B} = u'\mathbf{i} + v'\mathbf{j} + w'\mathbf{k}$$

then

$$\mathbf{A} \cdot \mathbf{B} = |\mathbf{A}||\mathbf{B}| \cos \theta \tag{3.4-1}$$

where θ is the angle between the two vectors. Solving for θ yields:

$$\theta = \cos^{-1}\left[\frac{uu' + vv' + ww'}{(u^2 + v^2 + w^2)^{1/2}(u'^2 + v'^2 + w'^2)^{1/2}}\right] \tag{3.4-2}$$

Note that because of step 3 in the algorithm used to define directions, in which fractions were cleared to result in indices with integer values, directions do not contain information about length. Consequently, directions are not true vectors. Fortunately, it is still possible to perform vector algebra operations on crystallographic directions, as long as only directional information is utilized. Information related to magnitude has no physical meaning for crystallographic directions.

....

EXAMPLE 3.4–4

Determine the angle between the directions $[0\ 1\ \bar{1}]$ and $[0\ 0\ \bar{1}]$.

Solution

By inspection of the sketch shown in Figure 3.4–5 we can see that the angle between the cube edge and the face diagonal is 45°. The solution can also be obtained using Equation 3.4–2:

$$\theta = \cos^{-1}\left[\frac{0 \times 0 + 1 \times 0 + (-1) \times (-1)}{\sqrt{(0 + 1 + 1)}\ \sqrt{(0 + 0 + 1)}}\right]$$

$$= \cos^{-1}\left(\frac{1}{\sqrt{2}}\right) = 45°$$

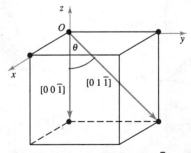

FIGURE 3.4–5 A sketch of a cubic unit cell showing the directions $[0\ 1\ \bar{1}]$ and $[0\ 0\ \bar{1}]$ and the angle θ between them. The origin was selected as the top left back corner of the cube so that both directions could be drawn within the boundaries of the unit cell.

....

3.4.3 Indices of Planes

Miller indices for planes are obtained using the following procedure:

1. Identify the coordinate intercepts of the plane, that is, the coordinates at which the plane intersects the x, y, and z axes. If the plane is parallel to one of the axes, the intercept is taken as infinity (∞). If the plane passes through the origin, consider an equivalent plane in an adjacent unit cell or change the location of the origin used to name the plane.
2. Take the reciprocal of the intercepts.
3. Clear fractions, but do not reduce to lowest integers.
4. Cite planes in parentheses—(h k l)—again placing bars over negative indices.

Several examples of planes are shown in Figure 3.4–6. The (1 0 0), (0 1 0), and (0 0 1) planes—the cube faces—are mutually orthogonal. Similarly, (1 1 0) and ($\bar{1}$ 1 0)—the planes that connect opposite edges through face diagonals—are orthogonal. **Families of planes** are expressed in braces: $\{h\ k\ l\}$. All planes in a family are equivalent in that they contain exactly the same arrangement of atoms. In cubic systems the members of a family of planes can be listed by taking all possible permutations of the indices. For example, the members of $\{1\ 0\ 0\}$ are (1 0 0), (0 1 0), and (0 0 1) and their negatives ($\bar{1}$ 0 0), (0 $\bar{1}$ 0), and (0 0 $\bar{1}$).

Working with the indices for directions and planes will familiarize you with several important features and relationships:

1. Planes and their negatives are equivalent. The negatives of directions are not equivalent but rather point in opposite directions.
2. Planes are not necessarily equivalent to their multiples. Directions are invariant to a multiplier.
3. In cubic crystals, a plane and a direction with the same indices are orthogonal.

FIGURE 3.4–6
Miller indices—(h k l)—for naming planes in a crystal lattice. The origin has been arbitrarily selected as the bottom left back corner of the upper unit cell in part (a) and the bottom left back corner of the unit cell in part (b).

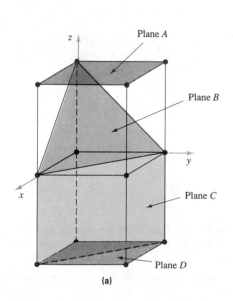

(a)

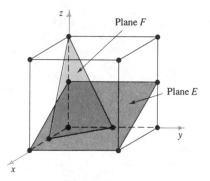

Plane	Intercepts	Indices
A	∞, ∞, 1	(0 0 1)
B	1, 1, 1	(1 1 1)
C	1, 1, ∞	(1 1 0)
D	∞, ∞, −1	(0 0 $\bar{1}$)
E	1, ∞, 1/2	(1 0 2)
F	1/2, 1/2, 1	(2 2 1)

(b)

EXAMPLE 3.4–5

Determine the indices of the planes labeled A and B in Figure 3.4–7a. Then sketch plane (1 $\bar{3}$ 0).

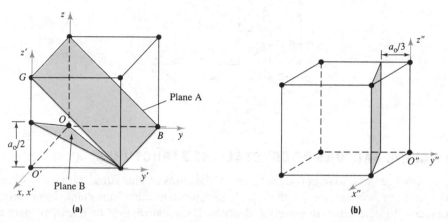

FIGURE 3.4–7   The cubic unit cells, coordinate systems, and planes referred to in Example 3.4–5.

Solution

Plane A can be named by choosing the bottom back left corner of the cube as the origin. Plane A is parallel to the x axis (intercept $= \infty$) and intersects both the y and z axes at 1. Taking reciprocals of the intercepts, plane A is the (0 1 1) plane with respect to origin O. Since plane B passes through O, it cannot be named using the same coordinate axes. What point should we use for the new origin? Although there is no unique answer to this question, a good choice is the front left bottom corner. The intercepts for this origin (O' in the figure) are $x' = -1$, $y' = 1$, and $z' = 1/2$. Taking reciprocals of the intercepts, plane B is the ($\bar{1}$ 1 2) plane with respect to O'. To sketch the plane (1 $\bar{3}$ 0), we note that the reciprocals of the indices are the intercepts: $x = 1$, $y = -1/3$, and $z = \infty$. The infinite intercept in the z direction means that the plane is parallel to the z axis. Since the x intercept is positive and the y intercept is negative, a good choice for the origin is the bottom back right corner (O'' in the figure). After plotting the x and y intercepts, draw lines through each intercept parallel to the z axis. The two lines define the plane (1 $\bar{3}$ 0).

3.4.4   Indices in the Hexagonal System

The notation used to describe points, directions, and planes in hexagonal lattices is similar to that used in cubic systems. As shown in Figure 3.4–8, there are four crystallographic axes in the hexagonal solid, which are most often referenced with respect to an origin located in the center of the basal plane. The three a axes are contained within the basal plane and the c axis is perpendicular to the basal plane. Since the hexagonal lattice has four coordinate axes rather than the three axes characteristic of the cubic systems, the Miller index convention for hexagonal systems is a bit more complex. Since we will be able to describe most of the important characteristics of HCP crystals without the aid of the numerical naming convention, we have elected to simplify our presentation by omitting the development of such a convention. Instead, we will simply refer to the important directions and planes as the a and c directions and the basal planes.

FIGURE 3.4–8
The important directions and
planes in hexagonal unit cells.

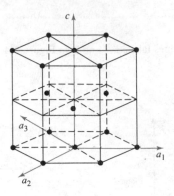

3.5 DENSITIES AND PACKING FACTORS OF CRYSTALLINE STRUCTURES

The previous sections have introduced some of the terms and notations of crystallography. In the next few sections we use these tools to determine some important characteristics of crystals. In particular, this section deals with the calculation of linear, planar, and volumetric densities.

3.5.1 Linear Density

The **linear density** ρ_L is the number of equivalent lattice points per unit length along a direction. Thus, ρ_L is defined as:

$$\rho_L = \frac{\text{Number of atoms centered along direction within one unit cell}}{\text{Length of the line contained within one unit cell}} \qquad (3.5\text{–}1)$$

As an example, consider the $[1\ 1\ 0]$ direction in an FCC crystal, as shown in Figure 3.5–1a. Atoms lie at the endpoints and the center of the face diagonal. Thus, there are three atoms associated with this direction. The two corner atoms, however, are shared with the continuation of the face diagonal in neighboring unit cells. Since some of the atoms are shared, we use weighting factors to determine the actual number of atoms along a face diagonal within a single unit cell. Since this is a linear calculation, the appropriate weighting factors can be envisioned using a linear sketch. Using Figure 3.5–1b, we see that we are interested in the fraction of the atomic diameter that lies within a single unit cell. In the case of ρ_L for $[1\ 1\ 0]$ in FCC, therefore, the number of atoms is 2 [i.e., $(2 \times 1/2) + (1 \times 1)$], and the length of the line is $4r$. Thus, using Equation 3.5–1, ρ_L for $[1\ 1\ 0]$ in FCC is $1/(2r)$.

The $\langle 1\ 1\ 0 \rangle$ family of directions has special significance in the FCC structure, since these are the directions in which atoms are in direct contact. As such, $\langle 1\ 1\ 0 \rangle$ directions have the highest ρ_L of any directions in the FCC system. In any crystal system, the directions with the highest ρ_L are termed the **close-packed directions.**

Note that the weighting factor for a corner atom or any other shared atom depends on the dimension of the calculation. In this linear calculation the corner atom contributed a factor of $1/2$ (of a diameter). In contrast, its weighting factor was $1/8$ (of a sphere) when we were counting the number of atoms in an FCC unit cell in Section 3.3.2 for a 3-D calculation.

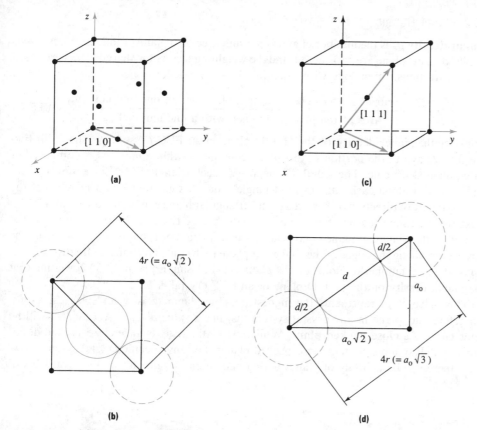

FIGURE 3.5–1   (a) The portion of the [1 1 0] direction contained within a single FCC unit cell is sketched in 3-D; (b) the same direction sketched in 1-D [shown within the (0 0 $\bar{1}$) plane]; (c) the portion of the [1 1 1] direction contained within a single BCC unit cell is sketched in 3-D; (d) the same direction sketched in 1-D [shown within the (1 $\bar{1}$ 0) plane].

EXAMPLE 3.5–1

Calculate the linear density along [1 1 1] in a BCC material. Repeat the calculation for the [1 1 0] direction in BCC.

Solution

The appropriate sketches are shown in Figure 3.5–1c and d. Using Equation 3.5–1, we find:

$$\rho_L = \frac{(2 \text{ corner atoms} \times 1/2) + (1 \text{ body-centered atom} \times 1)}{4r}$$

Hence, the linear density along [1 1 1] in a BCC material is also $1/(2r)$. Since this is the direction in which atoms are in direct contact, the $\langle 1\ 1\ 1 \rangle$ directions are the close-packed directions in the BCC structure. For the [1 1 0] direction:

$$\rho_L = \frac{2 \text{ corner atoms} \times 1/2}{a_0\sqrt{2}} = \frac{\sqrt{3}}{4r\sqrt{2}}$$

where we have made use of the a_0-r relationship for BCC given in Equation 3.3–1. Thus, in BCC the linear density in the close-packed directions is about 63% higher than in the $\langle 1\ 1\ 0 \rangle$ directions.

3.5.2 Planar Density

Planar density ρ_P is the number of atoms per unit area on a plane of interest. Only atoms centered on the plane are considered, and the weighting factors for shared atoms are based on area fractions. In analogy with Equation 3.5–1, we define ρ_P as:

$$\rho_P = \frac{\text{Number of atoms centered on a plane within one unit cell}}{\text{Area of the plane contained within one unit cell}} \qquad (3.5\text{–}2)$$

Consider the planer density of the (1 1 1) plane in an FCC crystal. As shown in Figure 3.5–2a and b, the portion of this plane contained within a unit cell is composed of an equilateral triangle. The length of the side of each triangle is $4r$, since atoms are in direct contact along each edge of the triangle (note that each edge is a member of the $\langle 1\ 1\ 0 \rangle$ family of directions). The area of this triangular plane can be calculated as $4r^2\sqrt{3}$. Next we determine the number of atoms in the plane. Each of the three atoms at the corners of the triangular plane contributes an area fraction of $1/6$ (i.e., $60°/360°$), and the three atoms along the edges of the triangular plane each contribute an area fraction of $1/2$. Thus, the total number of atoms on the plane is two. Using Equation 3.5–2, we find that the planar density on the (1 1 1) plane of an FCC crystal is $1/(2\sqrt{3}r^2)$.

This value of ρ_P represents the highest possible planar density for spherical atoms. Therefore, any plane in any crystal system that has a value of $\rho_P = 1/(2\sqrt{3}r^2)$ will be referred to as a **close-packed plane.** While all crystal systems have close-packed directions, not all systems contain close-packed planes. For any specific crystal system, however, there will be a family of planes with a maximum ρ_P value for that system. We will

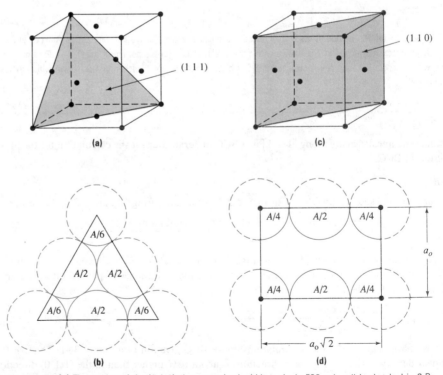

(a)

(b)

(c)

(d)

FIGURE 3.5–2   **(a)** The portion of the (1 1 1) plane contained within a single FCC unit cell is sketched in 3-D; **(b)** the same plane sketched in 2-D; **(c)** the portion of the (1 1 0) plane contained within a single FCC unit cell is sketched in 3-D; **(d)** the same plane sketched in 2-D.

TABLE 3.5-1   The close-packed directions and highest-density planes in the BCC, FCC, and HCP crystal structures.

Crystal structure	Close-packed directions	Highest-density planes	Are the highest-density planes close-packed?
BCC	$\langle 1\ 1\ 1 \rangle$	$\{1\ 1\ 0\}$	No
FCC	$\langle 1\ 1\ 0 \rangle$	$\{1\ 1\ 1\}$	Yes
HCP	a	Basal	Yes

refer to these planes as the **highest-density planes.** By our definitions all close-packed planes are highest-density planes, but not all highest-density planes are close-packed planes. Table 3.5–1 lists the close-packed directions and the highest-density planes for the BCC, FCC, and HCP crystal structures.

EXAMPLE 3.5-2

Determine the planar density of (1 1 0) in an FCC crystal.

Solution

The appropriate sketches are shown in Figure 3.5–2c and d. Using Equation 3.5–2, we find:

$$\rho_P = \frac{(4\text{ corners} \times 1/4) + (2\text{ face centers} \times 1/2)}{a_0 \times a_0\sqrt{2}} = \frac{2}{a_0^2\sqrt{2}}$$

Using $a_0(\text{FCC}) = 4r/\sqrt{2}$, as given in Equation 3.3–2, we find:

$$\rho_P = \left(\frac{2}{\sqrt{2}}\right)\left(\frac{\sqrt{2}}{4r}\right)^2 = \frac{1}{4\sqrt{2}r^2}$$

The planar density of (1 1 0) in FCC is $1/(4\sqrt{2}r^2)$. It is interesting to compare this value with the corresponding ρ_P value for the (1 1 1) plane in FCC $[= 1/(2\sqrt{3}r^2)]$. ρ_P on the (1 1 0) planes is only about 61% of that on the close-packed planes.

EXAMPLE 3.5-3

The highest-density planes in the BCC structure are the (1 1 0) planes. Determine the planar density of (1 1 0) in a BCC crystal, then compare this value with the planar density of the close-packed planes in the FCC structure.

Solution

Using the same procedure as in the previous example, we find:

$$\rho_P = \frac{(4\text{ corners} \times 1/4) + (1\text{ body center} \times 1)}{a_0 \times a_0\sqrt{2}} = \frac{2}{a_0^2\sqrt{2}}$$

Using $a_0(\text{BCC}) = 4r/\sqrt{3}$, as given in Equation 3.3–1, we find:

$$\rho_P = \left(\frac{2}{\sqrt{2}}\right)\left(\frac{\sqrt{3}}{4r}\right)^2 = \frac{3}{8\sqrt{2}r^2}$$

Thus, ρ_P of (1 1 0) in BCC is $3/(8\sqrt{2}r^2)$. Comparing this with ρ_P of (1 1 1) in FCC $[= 1/(2\sqrt{3}r^2)]$, we find that the value for the highest-density planes in BCC is about 92% of that for the close-packed planes in FCC. We will learn later that this result explains in part the difference in the mechanical properties of BCC and FCC metals.

3.5.3 Volumetric Density

Volumetric density (ρ_V) is the number of atoms per unit volume. The weighting factors for shared atoms are based on volume fractions.

Consider an FCC crystal. As shown previously, there are four atoms per unit cell. The cell volume is $V = a_0^3 = (4r/\sqrt{2})^3 = 16\sqrt{2}r^3$. Thus, ρ_V for FCC is $4/(16\sqrt{2}r^3) = 1/(4\sqrt{2}r^3)$. This is the highest volumetric density possible for spherical atoms. Crystal structures that display this value of ρ_V will be referred to as **close-packed structures.** As the name suggests, HCP crystals also have $\rho_V = 1/(4\sqrt{2}r^3)$.

....

EXAMPLE 3.5-4

Verify that $\rho_V = 1/(4\sqrt{2}r^3)$ for HCP.

Solution

The volume of the "big" HCP unit cell is equal to the area of the basal plane multiplied by the height of the cell. As shown in Figure 3.3–4b, the basal plane is composed of six equilateral triangles. The side of each triangle is $2r$, since atoms are centered at each apex and touch at the midpoint of each side. The area of each triangle can be calculated as:

$$\text{Area} = \left(\frac{1}{2}\right)b \times h = \left(\frac{1}{2}\right)(2r)(r\sqrt{3}) = r^2\sqrt{3}$$

The area of the basal plane is then $6r^2\sqrt{3}$. Combining this result with the $c:r$ relationship, $c = (8/\sqrt{6}r)$, yields:

$$V(\text{HCP}) = (6\sqrt{3}r^2)\left(\frac{8}{\sqrt{6}}\right)r = 24\sqrt{2}r^3$$

The number of atoms in the "big" HCP unit cell has been shown previously to be six. Therefore,

$$\rho_V(\text{HCP}) = \frac{6}{24\sqrt{2}r^3} = \frac{1}{4\sqrt{2}r^3}$$

....

3.5.4 Atomic Packing Factors and Coordination Numbers

The ratio of the volume occupied by the atoms to the total available volume is defined to be the **atomic packing factor (APF)** for the crystal structure. Thus, APF is calculated using any of the following equivalent expressions:

$$\text{APF} = \frac{\text{Volume of atoms in the unit cell}}{\text{Volume of the unit cell}} \tag{3.5-3a}$$

or

$$\text{APF} = \frac{(\text{Number of atoms in cell}) \times (\text{Volume of an atom})}{\text{Volume of the unit cell}} \tag{3.5-3b}$$

or

$$\text{APF} = \rho_V\left[\left(\frac{4}{3}\right)\pi r^3\right] \tag{3.5-3c}$$

For the crystal structures discussed previously: APF(SC) = 0.52, APF(BCC) = 0.68, and APF(FCC) = APF(HCP) = 0.74. It should not be surprising to find that the highest APFs occur for the two close-packed structures.

Why is it that most metals have either the FCC or the HCP structure whereas no important metals have the SC structure? It has been shown previously that it is energetically favorable for atoms to arrange themselves in a way that leads to the tightest packing and highest coordination numbers possible within the constraints of atom size (and valence in the case of covalent bonds). Combining the above information about APFs with a listing of the CNs for the various metal structures [CN(SC) = 6, CN(BCC) = 8, and CN(FCC) = CN(HCP) = 12] provides the answer to this question. In most cases one of the two close-packed structures represents the energetically favored arrangement of atoms. The SC structure is the least favorable atomic arrangement, and BCC is intermediate. In the next section we will investigate the differences between FCC and HCP structures.

EXAMPLE 3.5–5

Use a calculation to verify that the APF for the BCC structure is 0.68.

Solution

As discussed in Section 3.3.1, there are two atoms per cell in the BCC structure, and the $a_0 : r$ relationship is $a_0(\text{BCC}) = 4r/\sqrt{3}$. Substituting these values into the definition of APF given in Equation 3.5–3b yields:

$$\text{APF(BCC)} = \frac{2[(4/3)\pi r^3]}{(4r/\sqrt{3})^3} = \frac{\pi\sqrt{3}}{8} = 0.68$$

3.5.5 Close-Packed Structures

Although the FCC and HCP structures are similar, they have important fundamental differences. Both structures are characterized by an APF of 74% and a CN of 12. Both have sets of planes with the highest possible planar density, the close-packed planes; and both have directions with the highest possible linear density, the close-packed directions. The difference between the structures is in the arrangement of their close-packed planes.

Stacking sequence can be visualized by looking along the c direction in the HCP structure and along $[1\,\bar{1}\,1]$ in the FCC structure, as shown in Figure 3.5–3a and b. Consider the arrangement of atoms on one of the close-packed planes. Each atom is surrounded by six nearest neighbors in that plane. When a second layer of close-packed atoms is placed on top of the first layer, the atoms in the second layer do not lie directly on top of the atoms in the first layer. Rather, they lie in wells directly above the centers of the "holes" in the first layer (see Figure 3.5–3c).

There are two viable options for the placement of the third layer. One is to position each atom in the third layer directly above an atom in the first layer. This arrangement is shown in Figure 3.5–4b and corresponds to the HCP structure. The other option is to place the atoms in the third layer above the "holes" in the first layer that were not covered by atoms in the second layer. This arrangement, shown in Figure 3.5–4a, corresponds to the FCC structure. Thus, the distinction between HCP and FCC is in the placement of the third layer of close-packed atoms.

If we refer to any close-packed layer with its atoms in the positions associated with the first layer as an "A" layer, those layers with atoms positioned as in the second layer as "B" layers, and those with atoms positioned above the holes in both the A and B layers as "C" layers, then the stacking sequence in HCP is ABABAB . . . and the sequence in FCC is ABCABC . . . (see Figure 3.5–4).

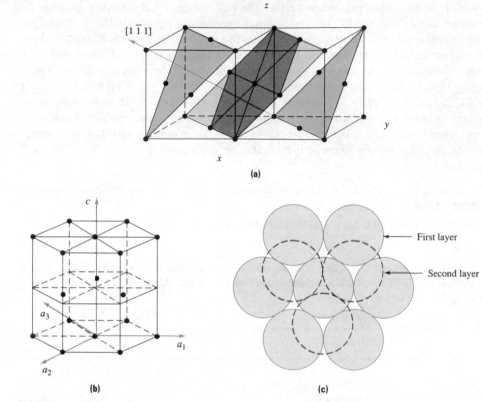

FIGURE 3.5–3 The stacking sequence of close-packed planes as viewed along **(a)** the $[1\,\bar{1}\,1]$ direction in an FCC unit cell and **(b)** the c direction in an HCP unit cell. **(c)** The atoms in the second layer lie in wells directly above the centers of the "holes" in the first layer.

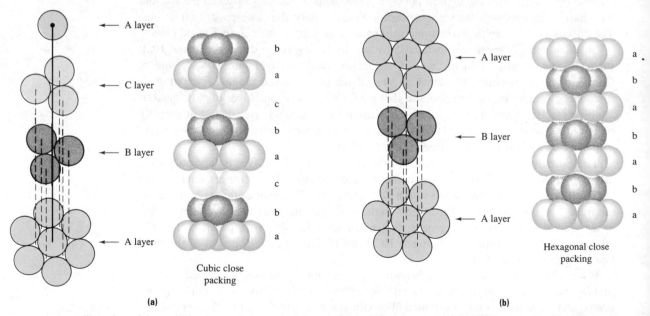

Illustrations of the stacking sequence of close-packed planes in **(a)** the FCC structure and **(b)** the HCP structure. Note that the stacking sequence is ABCABC . . . in FCC and ABABAB . . . in HCP.

3.6 INTERSTITIAL POSITIONS AND SIZES

The 3-D lattices that we have described are not completely filled with atoms (APF < 1.0 for all crystals). There are spaces around all and between some of the atoms. In this section we will investigate the size and position of the largest holes, or **interstices,** in the FCC, BCC, and HCP systems. We will then discuss crystals that contain more than one type of atom. As a rule, small atoms are positioned in the larger interstices. In ionic structures, positioning of the smaller ions depends on the stoichiometry of the compound, the charge on the ions, and the radius ratio.

3.6.1 Interstices in the FCC Structure

The largest hole in an FCC structure is at the center of the unit cell, as shown in Figure 3.6–1a. It is at this position that the largest sphere can be fitted into the cubic close-packed structure. The CN of an atom placed in this position is 6, since the atoms in the center of each face are equidistant. The polyhedron that connects the equidistant atoms can be used to describe the geometry of the interstitial site. In this case it has eight sides. Hence, this type of interstice is called an **octahedral site.** Equivalent sites are located at the center of each edge. These edge sites are shared among four unit cells. Hence there are four [i.e., $(12 \times 1/4) + (1 \times 1)$] octahedral sites per FCC unit cell.

The size of the octahedral holes is defined as the radius of the largest sphere that can be placed within it. It can be calculated from the geometry of the structure. As shown in Figure 3.6–1, the distance from the center of the top face to the center of the bottom face is equal to the lattice parameter a_0 for the unit cell. If κ is the radius of the hole, then $a_0 = 2r + 2\kappa$. Using the a_0-r relationship in FCC $[a_0(\text{FCC}) = 4r/\sqrt{2}]$ and solving for the radius ratio yields $\kappa/r = 0.414$. Thus, an atom roughly 40% of the size of the host atoms can "fit" into an octahedral interstitial position in the FCC structure.

The FCC structure also contains **tetrahedral sites,** as shown in Figure 3.6–1b. These sites are bounded by four atoms and lie completely within the cell in the $l/4$, $m/4$, $n/4$ positions, where l, m, and n are 1 or 3. Each cell contains eight of these $1/4$, $1/4$, $1/4$–type tetrahedral sites. Using arguments similar to those employed above, it can be shown that the κ/r ratio for tetrahedral sites is 0.225. This means that atoms up to ~20% of the size of the host atoms can "fit" in the tetrahedral interstitial positions in FCC structures. Note the similarities between the κ/r ratios for sizing interstitial holes in the FCC structure and the critical r/R ratios for determining CNs in ionic compounds (see Table 2.6–1). This should not be surprising, since the relevant geometries are identical. In summary, while there are twice as many tetrahedral sites as there are octahedral sites, each tetrahedral site is only about half the diameter of an octahedral site in the FCC structure.

....

EXAMPLE 3.6–1

Determine whether it is possible for a hydrogen atom to fit into either a tetrahedral or an octahedral site in FCC aluminum.

Solution

Using Appendix C, we find that $r(\text{H}) = 0.046$ nm and $r(\text{Al}) = 0.143$ nm. The important radius ratio is $r(\text{H})/r(\text{Al}) = 0.046/0.143 = 0.32$. Comparing this value with the critical κ/r ratios in the text (i.e., $\kappa/r \leq 0.414$ for octahedral sites and $\kappa/r \leq 0.225$ for tetrahedral sites) shows that H can fit only into the octahedral interstitial positions in Al.

....

FIGURE 3.6–1

The locations of the interstitial sites in the common
crystal structures: **(a)** octahedral sites in FCC, **(b)** tetrahedral sites in FCC,
(c) octahedral sites in BCC,
(d) tetrahedral sites in BCC,
(e) octahedral sites in HCP,
and **(f)** tetrahedral sites in
HCP.

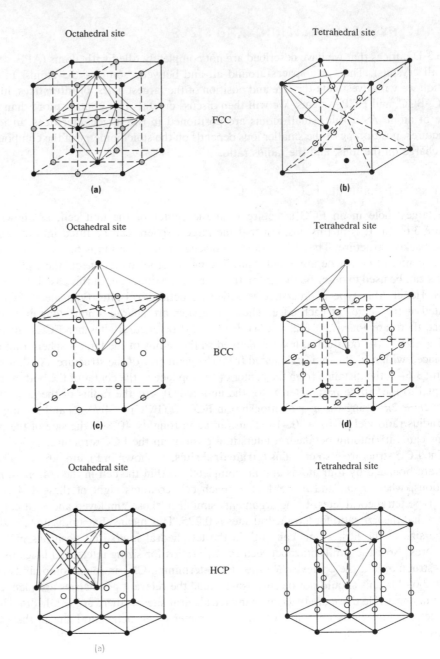

Octahedral site

Tetrahedral site

FCC

(a)

(b)

Octahedral site

Tetrahedral site

BCC

(c)

(d)

Octahedral site

Tetrahedral site

HCP

(a)

3.6.2 Interstices in the BCC Structure

Like the FCC structure, the BCC structure also contains both octahedral and tetrahedral
sites. As shown in Figure 3.6–1c, the octahedral sites are located in the center of each face
and the center of each edge, giving a total of six sites per unit cell. The diameter of the
octahedral site cannot be determined by examination of the face diagonal. The BCC
structure is not a close-packed structure, and the atoms that surround the interstitial site
are not all equidistant neighbors. When the largest possible atom occupies the octahedral
position, the atoms touch only along $\langle 1\ 0\ 0 \rangle$ as measured from one central atom to

another. Hence, $a_0 = 2r + 2\kappa$. Using the a_0-r relationship in BCC $[a_0(\text{BCC}) = 4r/\sqrt{3}]$ and solving for the radius ratio yields $\kappa/r = 0.155$.

The tetrahedral sites in BCC structures are located in the $1/4, 1/2, 0$–type positions, which are on the $\{1\ 0\ 0\}$ faces, as shown in Figure 3.6–1d. There are 24 such sites, each shared with another cell, for a total of 12 tetrahedral sites per unit cell. The diameter of the largest atom that just fits into the tetrahedral site can be calculated by considering atomic packing along $\langle 2\ 1\ 0 \rangle$ and is found to be $\kappa/r = 0.291$. Note that the tetrahedral sites in the BCC structure are more numerous and larger than the octahedral sites.

....

EXAMPLE 3.6–2

Determine whether it is possible for a carbon atom to dissolve in BCC iron.

Solution

Using Appendix C, we find: $r(\text{C}) = 0.077$ nm and $r(\text{Fe}) = 0.124$ nm. Hence, $r(\text{C})/r(\text{Fe}) = 0.077/0.124 = 0.62$. Comparing this with the critical κ/r ratios given in the previous section ($\kappa/r \leq 0.155$ for octahedral sites and $\kappa/r \leq 0.291$ for tetrahedral sites), we find that C does not fit easily into either type of interstitial position in BCC Fe. We will learn later that this is one of the key reasons for the high strength of steel alloys, which are composed of a mixture of small amounts of C in Fe. In addition, this "misfit" limits the solubility of C in the Fe crystal structure.

....

3.6.3 Interstices in the HCP Structure

As anticipated, the HCP structure also contains both octahedral and tetrahedral interstices. The positions of the interstices are shown in Figure 3.6–1e and f. There are 6 octahedral sites per "big" cell or 2 sites per unit cell and 12 tetrahedral sites per big cell or 4 per unit cell. Since both FCC and HCP are close-packed crystal structures, the relative sizes of the interstitial sites are the same in these two types of crystals. Table 3.6–1 summarizes the number and size of the tetrahedral and octahedral interstitial sites in the BCC, FCC, and HCP structures.

TABLE 3.6–1   The size and number of tetrahedral and octahedral interstitial sites in the BCC, FCC, and HCP crystal structures. The sizes of the interstitial sites are given in terms of the radius ratio (κ/r) where κ is the radius of the largest atom that can "fit" into the interstitial position and r is the radius of the host atoms. The number of interstitial sites is given in terms of both the number of sites per cell and, in parentheses, the number of sites per host atom.

Crystal structure	Size of tetrahedral sites	Size of octahedral sites	Number of tetrahedral sites per unit cell (per host atom)	Number of octahedral sites per unit cell (per host atom)
BCC	$\kappa/r = 0.291$	$\kappa/r = 0.155$	12 (6)	6 (3)
FCC	$\kappa/r = 0.225$	$\kappa/r = 0.414$	8 (2)	4 (1)
HCP	$\kappa/r = 0.225$	$\kappa/r = 0.414$	12 (2)	6 (1)

3.7 CRYSTALS WITH MULTIPLE ATOMS PER LATTICE SITE

In this section we expand our list of crystal structures to include systems with a basis of two or more atoms, that is, with multiple atoms per lattice point. In some cases the atoms are the same; in other cases the atoms are different. Examples of a 2-D lattice with a basis containing multiple characters can be found in the prints by the artist M. C. Escher shown in Figure 3.7–1.

3.7.1 Crystals with Two Atoms per Lattice Site

The Cesium Chloride Structure

Cesium chloride is an ionic solid. Its crystal structure, shown in Figure 3.7–2a, is composed of a simple cubic lattice with two ions, one of each type, per lattice position. One of the ions is centered on each lattice point, and the other is positioned at a distance of $a_0\sqrt{3}/2$ in the $[1\ 1\ \bar{1}]$ direction with respect to each lattice position. This structure is not a BCC structure, because there are two different atoms present. In this case the center position is not equivalent to a corner position. As shown in Figure 3.7–2b, the CsCl structure can be envisioned as a pair of interwoven SC lattices. This model clearly demonstrates the symmetry of the CsCl structure. Either atom can reside in cube corners with the other atom at the cube center.

The coordination number for each ion in the CsCl structure is 8, and there is one ion of each type per unit cell. The lattice constant can be readily determined by noting that ions touch along the body diagonal. Thus,

$$\frac{a_0\sqrt{3}}{2} = r + R \quad \text{or} \quad a_0(\text{CsCl}) = \frac{2(r + R)}{\sqrt{3}} \tag{3.7–1}$$

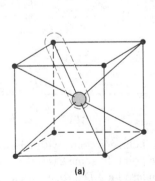

(a)

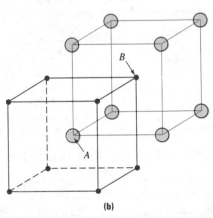

(b)

FIGURE 3.7–2   The CsCl crystal structure: **(a)** A simple cubic lattice with two different atoms per lattice point; **(b)** Alternatively, the structure can be viewed as a pair of interwoven simple cubic lattices.

Although there are other ionic solids with the CsCl structure, such as CsBr and CsI, the structure also occurs in other types of materials. For example, when equal numbers of copper and zinc atoms are mixed, they can crystallize under certain conditions into the CsCl structure.

....

EXAMPLE 3.7–1

Calculate the theoretical density of CsCl.

Solution

From Appendix B we find that the atomic weights of Cs and Cl are 132.9 g/mol and 35.45 g/mol. From Appendix C, $r(Cs^+) = 0.167$ nm and $R(Cl^-) = 0.181$ nm. Using the same procedure as in Example 3.3–1, we find:

$$\text{Density} = \frac{M_{uc}}{V_{uc}}$$

with

$$M_{uc} = \left(\frac{\text{Number of ions}}{\text{Unit cell}}\right) \times \left(\frac{\text{Mass}}{\text{Ion}}\right)$$

For CsCl,

$$M_{uc} = \left(\frac{1\ Cs^+}{\text{Cell}}\right)\left(\frac{\text{At. wt. Cs}}{\text{Avogadro's number}}\right) + \left(\frac{1\ Cl^-}{\text{Cell}}\right)\left(\frac{\text{At. wt. Cl}}{\text{Avogadro's number}}\right)$$

$$= (132.9\ \text{g/mol Cs})\left(\frac{1\ \text{mol}}{6.023 \times 10^{23}\ \text{atoms}}\right) + (35.45\ \text{g/mol Cl})\left(\frac{1\ \text{mol}}{6.023 \times 10^{23}\ \text{atoms}}\right)$$

$$= 2.80 \times 10^{-22}\ \text{g/(unit cell)}$$

Using Equation 3.7–1:

$$V_{uc} = a_0^3 = \left\{\frac{2[r(Cs^+) + R(Cl^-)]}{\sqrt{3}}\right\}^3 = \left[\frac{2(1.67 + 1.81) \times 10^{-8}\ \text{cm}}{\sqrt{3}}\right]^3$$

$$= 6.49 \times 10^{-23}\ \text{cm}^3/\text{(unit cell)}$$

Therefore,

$$\rho(CsCl) = \frac{2.80 \times 10^{-22}\ \text{g/cell}}{6.49 \times 10^{-23}\ \text{cm}^3/\text{cell}} = 4.31\ \text{g/cm}^3$$

The experimental density of CsCl is 3.99 g/cm³.

....

The Sodium Chloride Structure

As shown in Figure 3.7–3a, the NaCl structure is composed of an FCC arrangement of anions complemented with cations located in all of the octahedral positions. A more precise definition of the NaCl structure is an FCC lattice with a basis of two different atoms—one ion type is located on the FCC positions and the other ion type is positioned at a distance of $a_0/2$ in the [1 0 0] direction with respect to each lattice position. Figure 3.7–3b shows that the NaCl structure can be envisioned as a pair of interwoven FCC lattices.

There are four ions of each type in the NaCl structure, and the a_0-r relationship is $a_0(NaCl) = 2(r + R)$. Ions touch along the cube edge. Other compounds with this

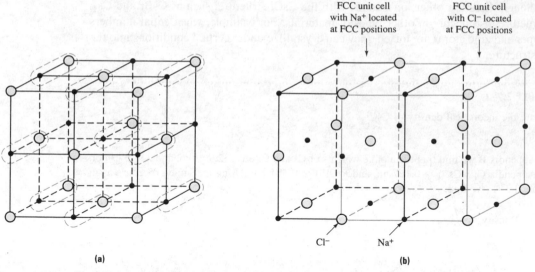

FCC unit cell
with Na+ located
at FCC positions

FCC unit cell
with Cl− located
at FCC positions

Cl−   Na+

(a)   **(b)**

FIGURE 3.7–3   The NaCl crystal structure: **(a)** An FCC lattice with two different atoms per lattice point; **(b)** Alternatively, the structure can be viewed as a pair of interwoven FCC lattices.

structure are all ceramics and include a number of metal oxides (MgO, CaO, SrO, FeO, BaO, MnO, NiO) and alkali halides (KCl).

The Diamond Cubic Structure

One of the crystalline forms of pure carbon is diamond. In the diamond cubic crystal structure, shown in Figure 3.7–4a, the atoms are arranged on an FCC lattice with additional atoms in half of the tetrahedral or 1/4, 1/4, 1/4–type positions. Alternatively, this structure can be described as having two atoms per site — one centered on each FCC position and the other located at a distance of $a_0\sqrt{3}/4$ in the $[1\ 1\ \bar{1}]$ direction with respect to each lattice position. Note that there are eight atoms per unit cell.

Why would carbon atoms assume this structure rather than any of those previously mentioned? The answer is that it must represent a lower-energy arrangement than any of the other structures. But why is this so? As discussed in Chapter 2, the covalent bonding in diamond requires that CN = 4 and that the C-C-C bond angle be 109.5°. While none

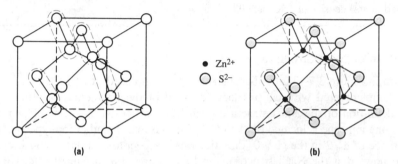

● Zn²⁺
○ S²⁻

(a)   **(b)**

FIGURE 3.7–4   **(a)** The diamond cubic crystal structure is composed of an FCC lattice with two atoms per lattice point. One atom from each pair is centered on each lattice point, and the second atom is positioned at $(a_0\sqrt{3}/4)$ $[1\ 1\ \bar{1}]$. **(b)** The zinc blende crystal structure is similar to the diamond cubic structure, except that the basis is composed of two different atoms.

of the previous structures satisfy these requirements, a examination of Figure 3.7–4a shows that the diamond cubic crystal structure fulfills the bonding criteria. The a_0-r relationship is a_0 (diamond cubic) $= 8r/\sqrt{3}$. Atoms touch along 1/4 of the body diagonal. Carbon is not the only material to have this structure. Silicon and germanium, both covalent semiconductors, crystallize in the diamond cubic structure.

The Zinc Blende Structure

As shown in Figure 3.7–4b, the zinc blende structure is similar to the diamond cubic structure. Atoms are located in the same positions, but here the two atoms per lattice site are different. The structure is named after an ionic compound ZnS, although a number of covalently bonded semiconductors such as GaAs and CdTe have this structure.

Why are only half of the tetrahedral sites filled? Why not fill all eight of the sites? The answers to these questions are related to the stoichiometry of the compound. The chemical formula implies that there is an equal number of each atom type in the compound. Since there are four FCC sites per cell and eight tetrahedral sites per cell, the one-to-one atom ratio demands that half the tetrahedral sites be empty. The zinc blende structure has a coordination number of 4, there are four atoms of each type per cell, and the a_0-r relationship is a_0(zinc blende) $= 4(r + R)/\sqrt{3}$. Atoms are in contact along 1/4 of a body diagonal.

We have examined three crystal structures that an ionic material with equal numbers of anions and metal cations, that is, compounds with MX stoichiometry, might assume. How does the material "choose" its crystal structure? The key concepts are the r/R ratio and stoichiometry. For example, consider MgO. It might crystallize into the NaCl structure, the CsCl structure, or the zinc blende structure. The ratio $r(Mg^{2+})/R(O^{2-})$ is 0.59. Using the stability criteria summarized in Table 2.6–1, one can determine that the most stable coordination number is 6. Consequently, MgO forms crystals of the NaCl type.

....

EXAMPLE 3.7–2

Predict the most likely crystal for (*a*) CsI and (*b*) GaAs.

Solution

We must first predict the CNs for each compound and then use these values to predict the crystal structure, noting that CN(zinc blende) $= 4$, CN(NaCl) $= 6$, and CN(CsCl) $= 8$. We begin by determining the primary bond type in each compound. Using electronegativity values in Appendix B, CsI is ionic and GaAs is covalent.

   a. Since CsI is ionic, the ratio $r(Cs^+)/R(I^-) = (0.167 \text{ nm})/(0.220 \text{ nm}) \approx 0.76$, with Table 2.6–1, shows that $CN(Cs^+) = CN(I^-) = 8$. Thus, we predict CsI has the CsCl structure.

   b. Since GaAs is covalent, the CNs are determined by the $8 - N_{ve}$ rule. In this compound, and many similar compounds, it is the average number of valence electrons that determines the structure. Ga has three valence electrons and As has five valence electrons, so the average number of valence electrons is four. GaAs has the zinc blende crystal structure.

For these two compounds the observed crystal structures are in agreement with predictions.[1]

....

---

[1] The simplified approach presented here, however, does not always yield the correct result. Recall that our model assumes the atoms to be rigid spheres. This assumption is not always valid. The interested reader should compare the predicted and observed crystal structures for the compound CaO.

3.7.2 Crystals with Three Atoms per Lattice Site

Compounds with a two-to-one ratio of atoms, such as CaF_2, SiO_2, and Li_2O, generally have crystal structures with a basis of three atoms. We will see in the following discussion that the complexity of the crystal structure tends to increase with the complexity of the basis.

The Fluorite Structure

Calcium fluoride (CaF_2) and a number of other materials with the MX_2 formula crystallize into a structure in which the M ions are located in the FCC positions and the X ions fill all the tetrahedral sites. The structure is shown in Figure 3.7–5. The stoichiometry dictates that there be double the number of X atoms as M atoms per unit cell. The M ions have CN(M) = 8 and the X ions have CN(X) = 4. Other compounds with this structure include UO_2, ThO_2, and ZrO_2.

Several compounds with the formula M_2X, including Li_2O, Na_2O, and K_2O, crystallize in the antifluorite structure. This structure is simply the inverse of the fluorite structure with the X ions at the FCC positions and the M ions filling all of the tetrahedral positions.

FIGURE 3.7–5

The fluorite crystal structure. The structure is composed of an FCC lattice with three atoms per lattice point. The M ions are located at the FCC positions and the X ions occupy all of the tetrahedral interstitial positions.

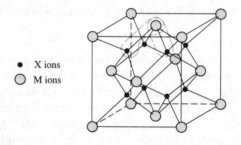

● X ions
○ M ions

The Crystobalite Structure

While SiO_2 (silica) has three atoms per lattice site, it is much easier to visualize the structure of crystobalite, an important crystalline form of silica, in a different fashion. The short-range order of any compound of Si and O requires that each Si atom form single covalent bonds with its four nearest neighbor O atoms. In turn, each O atom is covalently bonded to two Si atoms. Hence, as shown in Figure 3.7–6a, the basic building block for all Si-O compounds is the negatively charged $(SiO_4)^{4-}$ tetrahedron. The crystobalite crystal structure, shown in Figure 3.7–6b, can be envisioned as the diamond cubic structure with an $(SiO_4)^{4-}$ tetrahedron positioned on each lattice site. Since each O atom

FIGURE 3.7–6

The crystal structure of crystobalite is an FCC lattice with six atoms per lattice site. It can be envisioned as being composed of an $(SiO_4)^{4-}$ tetrahedron of atoms centered on each of the diamond cubic positions.

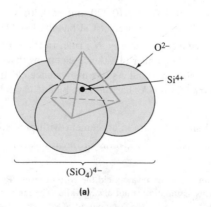

O^{2-}

Si^{4+}

$(SiO_4)^{4-}$

(a)

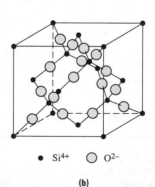

● Si^{4+} ○ O^{2-}

(b)

is shared by two tetrahedra, each tetrahedron contains one Si and two O atoms, or one SiO_2 group. Thus, crystobalite has an FCC lattice with six atoms, or two tetrahedra, per lattice site.

3.7.3 Other Crystal Structures

The structures presented to this point are some of the simplest crystal structures found. They serve to illustrate the essential salient concepts associated with crystallography, yet they represent the structures of a number of important materials. You will note that most metals are either cubic or hexagonal. Most nonmetals are neither cubic nor hexagonal. In the following discussion we will add to our list of important crystal structures. The basic concepts remain unchanged with these structures; however, the structures are more complex.

The Perovskite Structure

Calcium titanate ($CaTiO_3$) crystallizes with the perovskite structure, as shown in Figure 3.7-7a. The Ca^{2+} ions are located at the cube corners, the O^{2-} ions are at the center of each face, and a Ti^{4+} ion resides at the center of the cube. Thus, there are a total of five ions in the unit cell. As mentioned previously, the choice of an origin for the coordinate axis is arbitrary. It is equally valid to draw the perovskite unit cell with a Ti^{4+} ion located at the origin. (See Figure 3.7-7b.) Note that when the origin coincides with one of the Ti^{4+} ions, the O^{2-} ions are at the edge centers rather than the face centers.

Similar materials, such as $BaTiO_3$, have similar structures. In the case of barium titanate, however, the structure is simple tetragonal rather than cubic. What this means is that the lengths of the edges of the unit cell are not equal. As shown in Figure 3.7-8, in the case of $BaTiO_3$ the difference is small: $a = b = 0.398$ nm, $c = 0.403$ nm. Careful inspection of this figure shows that the central Ti^{4+} ion does not lie in the same plane as the four oxygen atoms in the side faces of the tetragonal unit cell. This charge displacement gives $BaTiO_3$ some important electrical properties. The shift in the relative positions of the central Ti cation and the surrounding O anions results in the formation of a local electric dipole. The strength of the dipole, which is related to the magnitude of the atomic displacement, can be altered by either an applied force or an electric field. The result, which will be discussed in more detail in Chapter 11, is that a barium titanate crystal can be used as a transducer, to convert electrical voltages into mechanical energy and vice versa. This leads to applications in telephone receivers and phonograph cartridges.

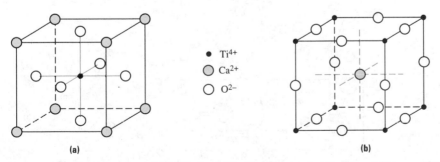

FIGURE 3.7-7   The perovskite unit cell for $CaTiO_3$ drawn (a) with the origin coinciding with a Ca^{2+} ion, and (b) with the origin coinciding with a Ti^{4+} ion.

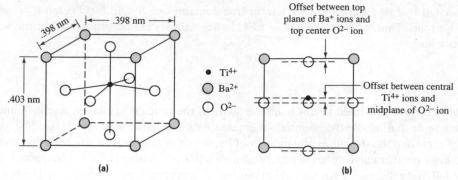

FIGURE 3.7–8   The tetragonal unit cell of $BaTiO_3$ shown **(a)** in 3-D, and **(b)** in 2-D.

The Structure of Methane

Methane (CH_4) is a gas at room temperature. Hence, it is not a material that is often used in solid form. The chief purpose of including it here is to show that the structures of low–molecular-weight organic crystals are similar in all respects to those of inorganic crystals. Below $-183°C$, methane crystallizes into an FCC structure. On each lattice site are placed five atoms. Similar to the case of crystobalite, it is perhaps easiest to envision the structure as having a carbon atom tetrahedrally coordinated by hydrogen atoms at each lattice site.

The Structure of Polyethylene

The unit cells of macromolecules (polymers) are more complex than those of metals, ionic compounds, or small-molecule organic crystals. In general, polymers crystallize only partially or not at all, and it is the linear ones that may be semicrystalline. The discussion below focuses on only those parts of the polymer that have been laid down to form crystals. We will revisit the subject of polymer crystallization in Chapter 6.

FIGURE 3.7–9

Crystalline polyethylene is composed of polyethylene chains arranged in an orthorhombic unit cell. *(Adapted from C. W. Bunn, Chemical Crystallography, 1945, by permission of Oxford University Press.)*

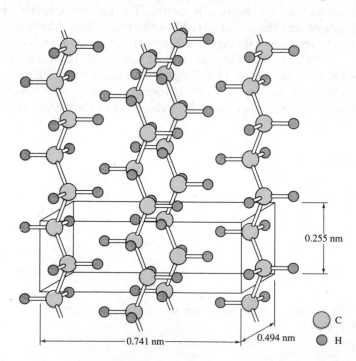

As mentioned in the previous chapter, polyethylene is a chain of CH_2 groups linked together with covalent bonds. Because of its simple symmetrical structure, polyethylene, in the scheme of polymers, crystallizes readily. In crystals, segments of the polymer chains line up parallel with one another in order to maximize intermolecular interactions, that is, secondary bond formation. The structure of a polyethylene crystal is shown in Figure 3.7–9. The unit cell is orthorhombic, meaning that no two edges have the same length but all angles are 90°. It is customary to use the c direction as the chain direction. Unit-cell dimensions for polymers are typically larger than those for metals; however, complex ionic crystals also have large cell parameters.

3.8 LIQUID CRYSTALS

As discussed in the introduction to this chapter, the order in most liquids is short range only. The term **liquid crystal** is given to fluids that show some degree of long-range order. The extent of the long-range order in these materials is intermediate between that of a crystalline solid and a liquid with SRO only. The order in one type of liquid crystals consists essentially of an association of molecules, giving regions of near parallel alignment of molecules as shown in Figure 3.8–1. In another type of liquid crystal the molecules twist cooperatively. This latter class of liquid crystals has found household applications as liquid crystal displays (i.e., the illuminated numbers in digital clocks and other electronic devices).

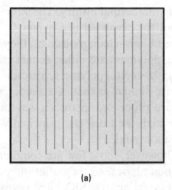
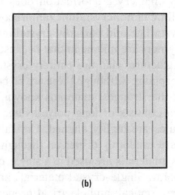

(a)   (b)

FIGURE 3.8–1   Schematic illustrations of the structure of some liquid crystals: **(a)** chain ends are unaligned; **(b)** chain ends are aligned.

3.9 SINGLE CRYSTALS AND POLYCRYSTALLINE MATERIALS

We have been concerned exclusively with the order of crystalline materials in this chapter. Indeed, many materials are crystalline but not quite as perfect as you may have been led to believe thus far. Virtually all metals form crystalline structures under normal conditions of cooling from the molten liquid; however, an entire part cast from the melt is rarely a single crystal. Rather, the casting is made up of a number of crystals with identical structures but different orientations. Most metals form **polycrystalline** structures, as shown in Figure 3.9–1. The grains are small crystals that are typically on the order of 0.5–50 μm across, but they may be up to a centimeter in diameter. The grain boundaries are internal surfaces of finite thickness where crystals of different orientations meet. It is possible that you have seen grains and grain boundaries on old brass doorknobs that have been polished and etched with perspiration through years of use.

FIGURE 3.9–1

A schematic illustration of a polycrystalline sample. The polycrystal is composed of many grains separated by thin regions of disorder known as grain boundaries. Note that the unit-cell alignment within grain A (shown in the high-magnification insert) is different from that in grain B.

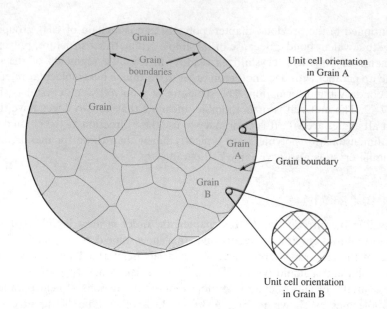

Many ceramic materials are also in the form of polycrystalline solids. With some inorganic solids, such as silica, it can be relatively easy to cool the material sufficiently quickly that crystal formation does not occur. Hence, these materials may be either crystalline or noncrystalline (amorphous), depending on thermal history. The structure of noncrystalline and partially crystalline materials will be discussed in Chapter 6.

Polymers are unique in that because of the nature of long-chain molecules, they rarely form structures that are entirely crystalline. Hence, polymers are either semicrystalline or amorphous. Although there are no commercial single-crystal polymers, Spectra® fiber, one of the strongest materials known, has a structure similar to that of a single crystal. Spectra consists of long polyethylene chains that are processed in such a way that the molecules are highly aligned. Crystallinity is very high, and defects, principally chain ends, are randomly dispersed through the continuous crystal.

Few materials are used in a single-crystal form; however, those few are commercially significant. Single-crystal materials have no grain boundaries, so they offer unique mechanical, optical, and electrical properties. Single-crystal quartz (SiO_2) and perovskites are used as transducers in a variety of applications, such as in high quality receivers and pickups (phonograph cartridges). Single-crystal germanium and silicon are used extensively in the microelectronics industry. Single-crystal nickel alloys are used in turbine blades in high-performance jet aircraft. Sapphire (Al_2O_3) and diamond (C) single crystals are precious stones.

3.10 ALLOTROPY AND POLYMORPHISM

Many materials can exhibit crystal structures that change from one unit cell to another at specific temperatures. Elements that exhibit this behavior are said to be allotropic while compounds that exhibit this behavior are **polymorphic.** Iron, for example, has a BCC structure at room temperature. As the material is heated, the structure changes to FCC at 912°C, and then reverts to a different BCC structure at 1394°C. These solid-phase transitions are reversible with temperature. Many ceramic materials, such as silica (SiO_2), alumina (Al_2O_3), and titania (TiO_2), undergo several changes in crystal structure with temperature. Property changes accompany structural changes. For example, volume, and

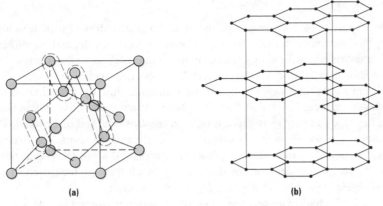

FIGURE 3.10–1   A comparison of the polymorphs of carbon: (a) diamond and (b) graphite.

therefore density, may increase or decrease. Many brittle materials cannot withstand the internal forces that develop as a result of these volume changes. These materials fail at the transformation temperature. An example is zirconia, ZrO_2. Tetragonal zirconia undergoes a polymorphic transition (upon cooling) at ~1000°C to form monoclinic zirconia. The accompanying volume change can crack the material. Similarly, ice on ponds cracks largely because of phase changes that occur with temperature.

Another allotropic material is carbon. As shown in Figure 3.10–1, it can exist in crystalline form as either diamond or graphite. The diamond structure has a 3-D tetrahedral network of covalent bonds with CN = 4, as expected for a Group IV element. The exceptionally high strength of the 3-D covalent network gives diamond the highest melting temperature of any of the elements. In contrast, the thermodynamically favored form of C at room temperature is graphite, which has a hexagonal two-dimensional layered structure in which each C atom has only three nearest neighbors in the plane of the layer. The bonds within each layer are covalent bonds, but the interlayer forces are comparatively weak secondary bonds. The relative ease with which these secondary bonds are broken gives graphite its excellent lubricating properties.

....

EXAMPLE 3.10–1

Calculate the percent volume change that accompanies the polymorphic transformation of ZrO_2 as it is cooled through 1000°C. The unit cell for the low-temperature phase is monoclinic with $a = 0.5156$ nm, $b = 0.5191$ nm, $c = 0.5304$ nm, and $\beta = 98.9°$. The unit cell for the high-temperature polymorph is tetragonal with $a = 0.5094$ nm and $c = 0.5177$ nm.

Solution

The calculation can be performed on a per–unit cell basis and is equivalent to finding the volume of each of the two unit cells. According to Figure 3.2–3, the volume of a tetragonal unit cell is simply $a^2 c$ while that for a monoclinic unit cell is $ac(b \sin \beta)$. The latter expression represents the area of the base of the unit cell multiplied by the projected length of the b axis in the perpendicular direction. Using these two expressions and the data given in the problem:

$$V(\text{tet.}) = (0.5094 \text{ nm})^2 (0.5177 \text{ nm}) = 0.134 \text{ nm}^3$$

$$V(\text{mono.}) = (0.5156 \text{ nm})(0.5304 \text{ nm})(0.5191 \text{ nm})(\sin 98.9°)$$

$$= 0.140 \text{ nm}^3$$

Therefore, the change in volume upon cooling through 1000°C is an expansion of approximately 4%. The experimentally observed values for this expansion are in the range 3–5%.

....

3.11 ANISOTROPY

When the properties of a material are independent of the direction in which they are measured, the material is termed **isotropic.** When properties depend on direction, the material is **anisotropic.** Anisotropy can occur at the microscopic or the macroscopic level.

We have shown in Section 2.5 that materials' properties, such as modulus of elasticity and coefficient of thermal expansion, can be estimated directly from the bond-energy curve. It should be recognized, however, that these estimates are valid only in the specific directions for which the curve is derived (i.e., in the close-packed directions). The separation distance between atoms in any direction other than the close-packed directions will be greater than x_0. Therefore, the bond-energy curves, and all of the properties derived from these curves, will depend on the direction in which they are measured. It is for this reason that single crystals exhibit some degree of anisotropy.

An example of a highly anisotropic crystalline material is graphite. Above 300°C its coefficient of thermal expansion along the c axis is more than 25 times greater than that in any direction parallel to the basal planes. This anisotropy is related to the relative strength of the primary bonds within planes and the secondary bonds between planes. Similarly, polymers in which the molecules are aligned are highly anisotropic.

While single crystals are anisotropic, most polycrystals are nearly isotropic on the macroscopic scale. The reason is that the properties of the polycrystal represent a statistical average of those of its randomly oriented constituent grains.

Anisotropy can also occur on a larger structural scale. Common examples include wood, steel-reinforced concrete, carbon fiber–reinforced epoxy, and oriented polymers. This anisotropy is related to the structure of these materials at the microscopic or even macroscopic levels (e.g., reinforced concrete) and will be discussed in later chapters.

3.12 X-RAY DIFFRACTION

You may have wondered how crystal structures are determined. The chief tool for investigation of structure on the scale of atomic dimensions is X-ray diffraction (XRD). Certain physical laws show that the wavelength of the electromagnetic radiation must be of the order of the size of the feature to be investigated. Hence, in order to investigate atomic dimensions in the range of 0.5–50 Å, electromagnetic radiation with an energy characteristic of X rays is required. X rays were discovered in the 19th century, and von Laue and Bragg developed the theory of XRD in the early 20th century. It is a relatively simple technique, yet it can provide a wealth of accurate information on interplanar spacings and atomic positions.

Let us review the wave interaction mechanism known as interference illustrated in Figure 3.12–1. It is based on the observation that amplitudes of interfering waves add to yield the amplitude of a composite wave. When waves of equal amplitude are in phase, their maxima and minima are aligned, and the amplitude of the composite wave is twice that of its components (Figure 3.12–1b). This is known as constructive interference. If, however, as shown in Figure 3.12–1c, one of the two waves is shifted by a distance of half a wavelength, the amplitude of the composite wave is equal to zero at all points. This phenomenon is known as destructive interference and is the operative principle of antireflective coatings (used, for example, on a camera lens).

The inteference phenomenon is central to understanding the interaction of X rays with crystals. Consider a beam of parallel X rays impinging upon the surface of a crystal. To the X-ray beam, the crystal lattice appears to be a collection of atoms located in planes,

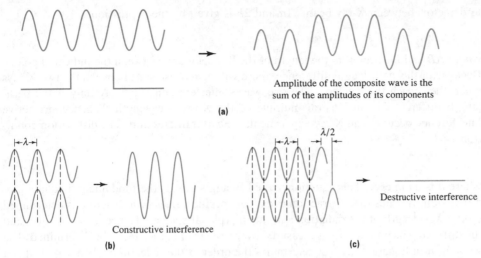

FIGURE 3.12–1 The phenomenon of wave interference: **(a)** the general case, **(b)** constructive interference, and **(c)** destructive interference.

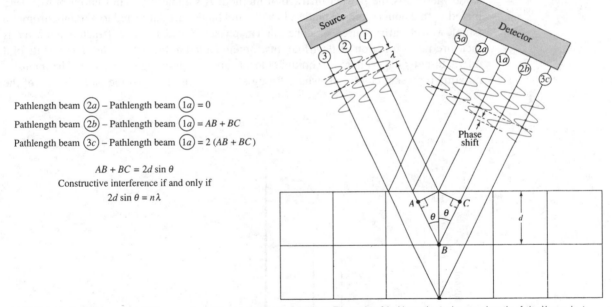

Pathlength beam $\widehat{2a}$ – Pathlength beam $\widehat{1a}$ = 0

Pathlength beam $\widehat{2b}$ – Pathlength beam $\widehat{1a}$ = AB + BC

Pathlength beam $\widehat{3c}$ – Pathlength beam $\widehat{1a}$ = 2 (AB + BC)

$$AB + BC = 2d \sin \theta$$
Constructive interference if and only if
$$2d \sin \theta = n\lambda$$

FIGURE 3.12–2 The geometry of X rays reflected from parallel atomic planes. The angle of incidence is θ, the wavelength of the X rays is λ, and the spacing between parallel planes of atoms is d. The important point is that there is a path-length difference between X rays reflected from parallel planes.

as shown in Figure 3.12–2. The wavelength of the radiation is λ, the angle of incidence is θ, and the perpendicular distance between parallel planes, the interplanar spacing, is d. Consider two parallel waves of X rays. The first wave is reflected from the top plane of atoms, and the second is reflected from the next lower parallel plane of atoms. These are waves 1a and 2b in Figure 3.12–2. The difference in the distance traveled, from source to detector, between X-ray beams $1a$ and $2b$ is given by the expression:

$$AB + BC = 2d \sin \theta \qquad (3.12\text{--}1)$$

where AB and BC represent the length of the line segments between the indicated points. Because of this path-length difference, there will be a phase shift between the two X rays when they reach the detector. The two waves interfere constructively only if the path-length difference is an integral multiple of the X-ray wavelength. When constructive interference occurs in an X-ray experiment, we call it **diffraction.** The diffraction condition is:

$$2d \sin \theta = n\lambda \qquad (3.12\text{--}2)$$

where n is an integer. This equation, called **Bragg's law,** is the fundamental equation of diffraction. It states that diffraction occurs at specific values of θ, where θ is determined by the wavelength of the radiation and the interplanar spacing. Hence, if λ is known and the diffracted beam intensity is measured over a range of θ, then d can be determined. The value of n in Equation 3.12–2 determines the order of the reflection; when $n = 1$, it is a first-order reflection; $n = 2$ is a second-order reflection; and $n = i$ is the ith order reflection. Using XRD, lattice parameters can be determined with high accuracy—in angstroms to four or five digits to the right of the decimal place!

There are literally dozens of techniques used in XRD. Perhaps the most easily understood method is the powder diffraction method. A sample is ground into a powder and mixed with a noncrystalline binder. Powder and binder are subjected to a monochromatic (single-wavelength) source of coherent (in-phase) X radiation. Diffracted intensity is measured as a function of θ with an instrument called a diffractometer. The results of a diffractometer experiment on sodium chloride are shown in Figure 3.12–3. The intensity maxima represent conditions where Bragg's law is satisfied for some plane in one of the

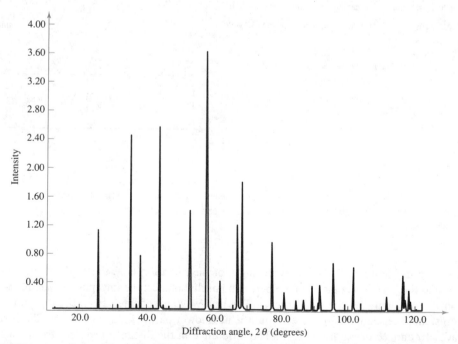

FIGURE 3.12–3 A plot of X-ray diffraction data, in the form of beam intensity versus incidence angle θ, for polycrystalline NaCl. *(Source: Courtesy of P. Desai.)*

many ground crystal fragments. There are many families of planes in a crystal—{1 0 0}, {1 1 0}, {1 1 1}, {2 0 0}, {2 1 0}, {2 1 1}, {2 2 0}, {2 2 1}, {2 2 2}, and so on. Each peak must be identified or indexed according to the plane and d spacing that produce the diffraction. Once all the d spacings are determined, the crystal structure can be assigned.

The d spacing, or interplanar spacing of crystals, can be determined geometrically for a known crystal. For example, the interplanar spacing for (1 1 1) in an FCC crystal can be determined by inspection of Figure 3.5–3a. The normal to the plane intersects three parallel members of the {1 1 1} family of planes per unit cell. Hence, the interplanar spacing is:

$$d(1\ 1\ 1) = a_0 \frac{\sqrt{3}}{3} \tag{3.12–3}$$

where in this notation $a_0 = d(1\ 0\ 0)$. In cubic systems, the geometrical relations can be expressed relatively simply:

$$d(h\ k\ l) = \frac{a_0}{(h^2 + k^2 + l^2)^{0.5}} \tag{3.12–4}$$

where $d(h\ k\ l)$ is the d spacing for plane $(h\ k\ l)$. More complex equations have been developed for other crystal systems.

....

EXAMPLE 3.12–1

a. Calculate the interplanar spacing of the {1 1 1} family of planes in an FCC material with atomic radius r.

b. Use the result of part a to verify that the c/a ratio in the HCP structure is equal to 1.633.

Solution

a. Using Equation 3.12–4 for a cubic crystal,

$$d(1\ 1\ 1) = \frac{a_0}{(1^2 + 1^2 + 1^2)^{0.5}} = \frac{a_0}{\sqrt{3}}$$

In the FCC structure, the a_0-r relationship is $a_0 = 4r/\sqrt{2}$. Combining these two equations yields:

$$d(1\ 1\ 1) = \frac{4r}{\sqrt{6}}$$

b. The quantity $d(1\ 1\ 1)$ represents the distance between two close-packed planes. In the HCP structure the height of the unit cell is equal to twice the distance between close-packed planes of atoms. That is,

$$c = 2d(1\ 1\ 1) = 2\left(\frac{4r}{\sqrt{6}}\right) = \frac{8r}{\sqrt{6}}$$

In the HCP structure, the a_0-r relationship is simply $a_0 = 2r$. Therefore, we find:

$$c = \frac{8r}{\sqrt{6}} = \frac{4a_0}{\sqrt{6}} \Rightarrow \frac{c}{a_0} = \frac{4}{\sqrt{6}} = 1.633$$

....

Crystallographers have had a great deal of experience indexing diffraction patterns. One way to expedite the process is through the use of "missing" diffraction peaks. Consider the {0 0 1} diffraction from a simple cubic structure as shown in Figure 3.12–4a. An intensity

FIGURE 3.12–4

The geometry of diffraction for several families of crystal planes: **(a)** the {0 0 1} family in simple cubic; **(b)** the {0 0 1} family in BCC, neglecting the contributions from the parallel {0 0 2} family; and **(c)** the complete {0 0 2} family in BCC.

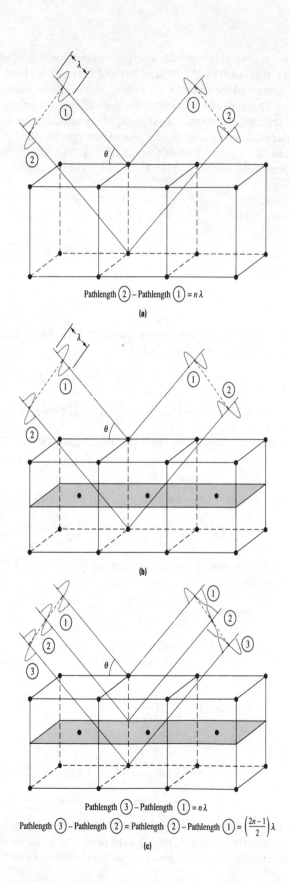

Pathlength ② – Pathlength ① $= n\lambda$

(a)

(b)

Pathlength ③ – Pathlength ① $= n\lambda$

Pathlength ③ – Pathlength ② = Pathlength ② – Pathlength ① $= \left(\dfrac{2n-1}{2}\right)\lambda$

(c)

peak will occur at an angle in accordance with Bragg's law. Now consider {0 0 1} diffraction from a BCC material. Assume conditions are appropriate so that the X rays interfere constructively, as shown in Figure 3.12–4b. The atoms in the center of this and other cells create a plane of equal density to those of the {0 0 1}. These are the {0 0 2} planes. These planes fall midway between successive {0 0 1} planes. At the angle appropriate for constructive interference of {0 0 1}, the {0 0 2} produce destructive interference (see Figure 3.12–4c). Hence, no intensity maximum is observed for {1 0 0} in BCC structures.

The concept of "missing" diffraction peaks from certain crystal structures can be extended to develop some general rules for diffraction. Two of the important rules are:

1. The sum $(h + k + l)$ must be even for diffraction to occur from $(h\ k\ l)$ in BCC structures, and

2. h, k, and l must be all even or all odd for diffraction to occur from $(h\ k\ l)$ in FCC structures.

X-ray diffraction is an extremely powerful tool for investigating the structure of crystalline materials. We have presented only the fundamental relationships between X rays and crystal structure. A number of complex X-ray techniques have been developed, but all the techniques rely on the basic concepts presented above.

EXAMPLE 3.12–2

Calculate the spacing of planes responsible for an intensity peak at 32.2°, assuming first-order diffraction: The wavelength of the radiation is 1.54 Å.

Solution

This is a Bragg's law problem, so the relevant equation is 3.12–2:

$$2d \sin \theta = n\lambda$$

Solving for d:

$$d = \frac{n\lambda}{2 \sin \theta}$$

Inserting the values $n = 1$, $\lambda = 1.54$ Å, and $\theta = 32.2°$ yields:

$$d = \frac{1.54\ \text{Å}}{2 \sin 32.3°} = 1.445\ \text{Å}$$

SUMMARY

Most solids exhibit short- and long-range order and are called crystalline solids. The atoms that make up crystals are arranged on a crystallographic lattice. There are 14 Bravais lattices, and all crystals belong to one and only one lattice type. A material that changes lattice type with temperature is called polymorphic. Some materials change crystal structure many times prior to melting. Unit cells are arranged in a periodic fashion and repeat indefinitely in three demensions. Miller indices are a shorthand notation used to refer to directions and planes in a unit cell.

The simplest crystals are those containing one element, or pure metals. Many metal crystals are based on a cubic unit cell: FCC or BCC. However, some are based on hexagonal or tetragonal unit cells. The repeat distance, or lattice parameter, of most metals is small—on the order of a few angstroms.

Metals are close-packed if the atoms occupy the maximum fraction of space possible in the unit cell. The atomic packing factor of a close-packed crystal is 74% and the coordination number is 12. Both FCC and HCP structures are as dense as possible—with close-packed planes and close-packed directions. One difference between HCP and FCC is the stacking sequence of the close-packed planes.

The crystallography of two-element metals or ceramics is more complex. Typically one component is positioned in SC, BCC or FCC positions, and the other component resides in the octahedral or tetrahedral holes in the unit cell. The lattice parameter of most two-component solids, such as an MX ceramic, is typically of the order of 5 to 10 Å.

Polymers crystallize only partially at best. Polymer crystals are characterized by large unit cells, perhaps as large as 20 Å in one direction. There are parts of four or more mers and, hence, many atoms in a unit cell. Polymeric crystals are typically of low symmetry.

While most materials are polycrystalline, some materials are specially grown in the form of single crystals. Since they lack periodic interruptions in structure—grain boundaries—single crystals have unique and useful properties.

X-ray diffraction is a powerful technique for investigating the atomic structure of materials. Bragg's law, $n\lambda = 2d \sin \theta$, is the fundamental equation that governs the interaction of X rays with crystallographic planes. Using X-ray diffraction, we can determine the crystal structure of a material and the precise positions of the atoms in the unit cell.

KEY TERMS

anisotropic

atomic packing factor (APF)

basal plane

basis

body-centered cubic (BCC)

Bragg's law

close-packed directions

close-packed planes

close-packed structures

crystal lattice

diffraction

face-centered cubic (FCC)

families of directions

families of planes

hexagonal close-packed (HCP)

highest-density plane

interstices

isotropic

lattice

lattice parameter

lattice point

linear density (ρ_L)

liquid crystal

Miller indices

octahedral site

planar density (ρ_P)

polycrystalline

polymorphic

simple cubic (SC)

tetrahedral site

unit cell

volumetric density (ρ_V)

HOMEWORK PROBLEMS

SECTION 3.1
Introduction

1. Discuss the effect of temperature on short- and long-range order. Include the effect of melting.
2. Go find a printed textile fabric or wallpaper. Sketch the pattern. Does it have LRO? Does a mural have LRO?
3. How does short-range order change when a crystal melts?

SECTION 3.2
Bravais Lattices and Unit Cells

4. Sketch tetragonal and orthorhombic cells.
5. Based on a sample of printed textile fabric, show the unit cell, the lattice points, and the basis, and measure the lattice parameters.

6. Draw a base-centered cubic structure. How is this crystal system included in Figure 3.2–3?

7. Redraw the FCC structure with one of the atoms in the center of a face at the origin.

8. Do the corner atoms in the BCC touch one another? Do those of the FCC structure?

9. Optical microscopy is commonly used by mineralogists to identify composition. The macroscopic form of a crystal often, but not always, reflects the symmetry of the unit cell. Quartz crystals are pointed hexagons. Which cell might characterize this crystalline form of SiO_2?

10. Calculate the length of a face diagonal in Cr, a BCC metal.

11. Calculate the density of FCC copper using data based only on atom measurements. The measured value is 8.96 g/cm^3.

12. Calculate the density of magnesium, an HCP metal, from atomic data, and compare the result with the measured value, 1.74 g/cm^3. Account for differences.

13. Show why the c/a ratio in HCP materials should be close to 1.63. Why does c/a in some metals differ from 1.63?

14. Gadolinium changes from HCP to BCC upon heating at 1260°C. The lattice parameters for the HCP are $a = 0.36745$ nm and $c = 1.18525$ nm. The lattice parameter of the BCC structure is 0.4060 nm. Calculate the volume change associated with the change in crystal structure.

15. List the five metals with the lowest densities. These materials are prime candidates for high ratios of strength or stiffness to weight (density). Note which of these metals are ones that you are familiar with.

16. If atoms are located at all the 1, 0, 0– and 1/2, 1/2, 0–type positions in a cubic system, what is the Bravais lattice?

17. Identify the plane in an FCC metal formed by the intercepts 1, −1, −2. What is the normal to this plane?

18. Calculate the angle between [1 0 0] and [1 1 1] in Al.

19. Construct a coordinate system at the center of a cubic unit cell with the axes parallel to the ⟨1 0 0⟩ directions. Determine the tetrahedral angle, the angle between directions from the origin to two ends of any face diagonal.

20. What is the angle between [1 1 0] and [1 1̄ 1] in a cubic crystal? Between (1 1 0) and (1 1 1)? Between (1 1 0) and [1 1 1]?

21. Give the indices of the points, directions, and planes in the cubic cells shown in Figure HP3.1.

SECTION 3.3
Crystals with One Atom per Lattice Site and Hexagonal Crystals

SECTION 3.4
Miller Indices

FIGURE HP3.1

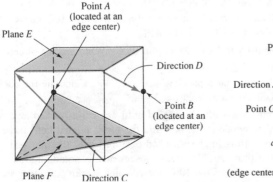

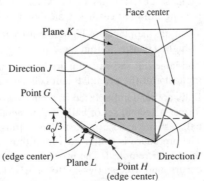

22. Give the indices of the points, directions, and planes in the cubic cells shown in Figure HP3.2.

FIGURE HP3.2

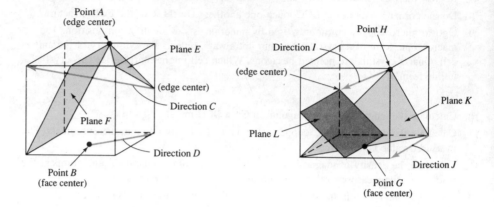

23. Identity the plane formed by the intercepts $1/2$, -1, 2 in a cubic cell. What is the normal to this plane?

24. Sketch all the members of the {1 0 0} family of planes in a tetragonal unit cell. Now sketch all the members of the {0 0 1} family in a similar tetragonal unit cell.

SECTION 3.5
Densities and Packing
Factors of Crystalline
Structures

25. Calculate the planar density of atoms on the (1 0 0), (1 1 0), and (1 1 1) planes in an FCC unit cell.

26. Calculate the planar density of (1 0 0) and (1 1 0) in a BCC structure.

27. Sketch the atoms in the closest-packed direction on the closest-packed plane in Cr.

28. What is the packing factor of the unit cell shown in Figure 3.7–1?

29. Calculate the separation of close-packed planes in FCC and HCP structures.

30. Compare the linear density of [1 0 0] and [1 1 1] in FCC crystals.

31. Fiber-reinforced composites are made from fibers and resin. Determine the maximum volume fraction of monosize-diameter fibers that can be put into a composite by calculating the packing factor of close-packed rods.

32. What is the "coordination number" of the rods described in the previous problem?

33. Why do some metals crystallize into BCC structures when FCC is more dense and has a higher CN?

34. Suppose you discover a new material that is characterized by close-packed planes stacked in the sequence ABAC, repeating indefinitely. Is this discovery significant? Can you calculate the atomic packing factor of the new material?

35. Suppose a crystal of sodium, which has a BCC structure, has 1 in every 100 atoms missing from lattice sites. Calculate the effect on the APF.

36. What are the planes of highest density in FCC, HCP, and BCC? What are the directions of highest density within these planes?

37. The large cell shown in Figure 3.3–4b has a packing factor of 74%. Calculate the packing factor of the primitive unit cell shown in Figure 3.3–4a.

38. Balls are randomly placed in a bin. The balls are jiggled until one close-packed layer forms. A second set of balls is then added to the bin, and the bin is jiggled until these balls assume a close-packed plane sitting in close-packed positions over the first set. The procedure is continued with a third set. The stacking sequence is noted and the third layer removed. The procedure is repeated 100 times. Which stacking sequence occurs more frequently—ABC or ABA? Why?

39. Fill the octahedral hole in an FCC structure with the largest sphere that will fit. Do the same for a BCC structure. How many neighbors does the sphere touch in each case?

40. If all the octahedral and tetrahedral sites in the FCC structure were filled with atoms, what would the APF be?

41. Why is the octahedral site in the BCC structure smaller than the tetrahedral site?

42. Compare the number and size of tetrahedral sites in the BCC structure with those of the octahedral sites in the FCC structure.

43. An impurity with a radius of 0.07 nm is introduced into crystalline Ca. How many impurities can be put into the structure? Give your answer in terms of a multiple of the number of calcium atoms..Repeat for an impurity with a radius of 0.04 nm.

44. Carbon is added to iron to form steel. The BCC structure of iron, called ferrite, is stable to 912°C, whereupon the structure becomes FCC, called austenite. Which form of iron do you anticipate is capable of dissolving more carbon? Explain your answer.

SECTION 3.6
Interstitial Positions and
Sizes

45. Calculate the lattice parameter of silicon.

46. Figure 3.7–4a shows the diamond cubic structure, which is characteristic of a number of covalent crystals. Calculate the theoretical density of diamond and silicon.

47. Calculate the density of the $(SiO_4)^{4-}$ tetrahedron, the basic building block of silicates.

48. Calculate the lattice parameter of MgO. One ion is cubic close-packed and the octahedral sites are filled by the other ion. What structure is this commonly known as?

49. Redraw Figure 3.7–4b with Zn^{2+} at the origin.

50. Sketch (1 1 0) and (1 1 1) in MgO, which has the NaCl structure. Indicate the locations of the tetrahedral and octahedral sites.

51. Predict the structure of NiO.

52. Predict the structure of SiC.

53. Predict the structure of CaO.

54. Calculate the density of CsBr.

55. Show why materials such as ThO_2, TeO_2, and UO_2 crystallize into the fluorite structure.

56. Calculate the density of UO_2.

57. Calculate the density of CaF_2. Which direction is most dense? What plane is most dense?

58. Calculate the O—Si—O bond angle in crystobalite.

59. What information do you need to facilitate calculation of the density of SiO_2 in the form of crystobalite?

60. Calculate the density of polyethylene, realizing from Figure 3.7–9 that there are two mers (each—CH_2—CH_2—) per unit cell. Explain why most commercial samples of polyethylene have a lower density, about 0.97 g/cm³.

61. What is the coordination number of each ion in the perovskite structure?

SECTION 3.7
Crystals with Multiple
Atoms per Lattice Site

62. Speculate what occurs when *liquid crystals* are cooled below the temperature at which they solidify. Will they form crystals?

63. Sketch the structure of a liquid crystal used in a numeric display in a digital clock.

64. Referring to Figure 3.8–1, do you anticipate that polymer liquid crystals are of the type shown in part a or part b of the figure? Recall that in a polymer sample there are many chains and no two are identical in length.

SECTION 3.8
Liquid Crystals

65. Metals are always opaque because the electrons absorb incident (visible) radiation. Polymers and ceramics may be transparent. In general, to be transparent a ceramic or

SECTION 3.9
Single Crystals and
Polycrystalline Materials

polymer must be either single-crystal or completely noncrystalline (amorphous), or the crystals must be very small. Give several examples of transparent materials and note whether they are noncrystalline or single crystals.

66. How might grain boundaries affect density?

67. Kevlar® is a polymer fiber that is formed from a liquid crystalline phase. Its morphology is quite similar to that of Spectra® polyethylene. Pound for pound, both fibers are considerably stronger than steel. Sketch the molecular arrangement in a Kevlar fiber.

68. Two samples of unoriented polyethylene are subjected to X-ray diffraction. Diffraction from (2 0 0) in sample A is 1/3 as much as that from sample B, the standard, which is 80% crystalline. What is the crystallinity of sample A? How do the densities of the samples compare?

SECTION 3.10
Allotrophy and
Polymorphism

69. In Chapter 2 you learned a little about thermodynamics.
    a. Use your knowledge of this subject to explain the nature of the driving force for reversible changes in crystal structures.
    b. Crystal structure changes usually are brought about by changes in temperature. Suggest other driving forces for crystal structure changes.

70. Water shows anomalous behavior by expanding when it changes phase from liquid to crystal (ice). Do an experiment using your freezer to estimate the volume change of H_2O upon freezing. Give several consequences of the volume change that occurs when water is frozen (hint: "on the rocks," ice skating or string cutting through ice, rock splitting).

71. Discuss the changes that occur in bonding when graphite transforms to diamond.

72. Boron nitride has comparable crystal structures to those of carbon. One form of BN has excellent lubricating properties. Sketch the other crystal form of BN.

73. Stress can induce a crystal structure change in polymers. Human hair is a good example. At a few percent strain, human hair changes unit cell from a short one (α) to a long one (β) along the fiber axis. How might this transformation affect the amount of stretch human hair can endure before failing?

SECTION 3.11
Anisotropy

74. Why is plywood stronger than ordinary wood?

75. The electrical conductivity is measured along each of the principal axes of a large single crystal of silicon. The values are identical within experimental error. Can we safely conclude that the material is isotropic?

76. Single-crystal alumina, which has a hexagonal unit cell, has a thermal expansion coefficient of $8.3 \times 10^{-6}/°C$ normal to the c axis and $9.0 \times 10^{-6}/°C$ along the c axis. Estimate or bracket the expansion coefficient of polycrystalline alumina?

77. Polystyrene is an amorphous polymeric material, a glass. Glasses have a structure similar to that of a frozen liquid. Is polystyrene isotropic?

78. Polycrystalline crystobalite has a thermal expansion coefficient of about $11 \times 10^{-6}/°C$, whereas that of amorphous or glassy silica, called fused silica, is about $0.5 \times 10^{-6}/°C$. Did you anticipate the values would be more similar? Use bond-energy concepts to rationalize the difference.

SECTION 3.12
X-ray Diffraction

79. A cubic crystal shows a diffraction maximum from copper radiation, $\lambda = 1.54$ Å at $\theta = 33°$, corresponding to diffraction from (1 3 0). Calculate the lattice parameter.

80. What is the angle at which diffraction occurs in Au from (1 0 1) using incident radiation with a wavelength of 1.54 Å?

81. At what angle does diffraction occur in Ag from (1 1 1) using incident radiation with a wavelength of 1.54 Å?

82. What plane is responsible for diffraction in NaCl at $2\theta = 32.04°$ using incident radiation with a wavelength of 1.54 Å? What about $2\theta = 15.86°$?

83. Calculate the d spacing for planes that show diffraction at $\theta = 42.0°$, when subjected to radiation of wavelength 1.54 Å. At what angle will the diffraction occur if the wavelength is 3.29 Å?

84. Referring to Figure 3.7–9, calculate the angle at which diffraction will occur from (0 0 1) when using incident radiation with a wavelength of 1.54 Å. Note that the c-axis is perpendicular to (0 0 1).

85. A cubic metal shows diffraction from (1 1 1) at $\theta = 22.62°$. May the metal be Al, Cr, or Cu? The wavelength is 1.54 Å.

86. In polymers, structure on the scale of 100–200 Å is critical to the mechanical properties of the material. How might X-ray diffraction be used to probe this scale?

87. Explain how X-ray diffraction might be used to identify unknown samples.

POINT DEFECTS AND DIFFUSION

MATERIALS IN ACTION Point Defects and Impurities

We have discussed atomic bonding and the solid-state structures that are closely related to the nature of the bonds. In this chapter we will see that it is actually impossible for a crystal to be "perfect" in all respects. In fact, based on fundamental laws of physics all crystalline materials must have a small fraction of atoms or ions that are missing from their normal lattice sites. We will see later that these missing atoms or ions can have a profound effect upon the mechanical, chemical, and electronic properties of materials. In addition to the defects that must be present, there are defects that while not fundamentally required to be present, nonetheless have an overwhelming probability of being present. Such "defects" may be impurities which may or may not be intentionally present. These defects have an influence on the properties of materials way out of proportion to their concentration. Perhaps the most obvious example of the effect of impurities is found in steel, a mixture of iron and a very small concentration of carbon. The carbon atoms, which typically might constitute only 0.4 weight percent of steel (i.e., approximately 1 carbon atom for every 50 Fe atoms) can have the effect of increasing the strength of iron from about 15 MPa to more than 1,500 MPa! The superb structural properties of steel, which derive from the impurity carbon atoms, are the reason for world-wide use of steel that exceeds hundreds of billions of dollars.

Another example of the effect of impurities on the properties of a material is seen in semiconductors, which are the basis of the electronics "revolution." A good example of this occurs with the addition of P to Si. We have already seen in the previous chapter that Si is a covalently bonded crystal. As such there are no free electrons to carry current and Si is essentially an insulator. However, when a minute amount of P is added to Si the situation changes dramatically. The P, which has a valence of 5, substitutes for Si atoms in the crystalline lattice. However, Si has a valence of 4. Thus there is one extra unbonded electron for every P atom that is added. This extra "free" electron then provides for electrical conduction. When Si is "doped" in this way, an n-type semiconductor is produced. When Si is doped with B (valence 3) there is a "missing" electron or a "hole." This hole, which is positively charged, can migrate from atom to atom thereby providing p-type conduction. When an n-type semiconductor is sandwiched between two p-type semiconductors a p-n-p transistor is produced. The production of such electronic devices accounts for billions of dollars in economic activity world-wide.

In this chapter, we will gain an appreciation for various types of defects and impurities and will be able to relate the defect state to properties of the material.

4.1 INTRODUCTION

In the previous chapter we investigated the structure of crystalline materials on a unit-cell level. We must recognize, however, that engineering materials are composed of very large numbers of atoms in small volumes. For example, in BCC Fe, with $a_0 = 0.287$ nm and two atoms per unit cell, there are approximately 8.5×10^{22} atoms per cm³! Because of the large number of atoms involved, we should expect a high probability for some "mistakes" to occur in the atomic arrangement. In crystalline solids these mistakes are called *defects,* and they may occur in several different forms, including single atoms (point defects), rows of atoms (linear defects), planes of atoms (planar defects), and small three-dimensional clusters of atoms (volume defects). We will see that it is *impossible* to avoid defects in materials.

It is observed that to obtain 100% purity of any material is extremely difficult. Of course, in many cases it is *desirable* to add additional atom species to an otherwise "pure" material. For purposes of discussion in this chapter, we consider the existence of additional atom species incorporated within the major atomic species to be a defect, whether or not such species are intentionally added. We call this defect an impurity. We show that it is impossible to obtain a material, in engineering quantities, in the pure form.

For all materials, it is necessary to consider the existence of defects and impurities, collectively known as imperfections, as they affect material properties. Imperfections at the atomic level significantly affect properties. Imperfections exist at the macroscopic level as well. Macroscopic defects, with dimensions at least six orders of magnitude larger than the atomic size, can be seen by the naked eye or with a low-power microscope. For example, macroscopic cracks can arise from improper processing or may be formed during service. In contrast to atomic scale defects, macroscopic imperfections always result in a degradation of properties.

The purpose of this chapter is to examine point defects in crystals at the atomic scale and to discuss their influence on diffusion in solids. Other types of defects and their engineering significance are discussed in the next chapter.

4.2 POINT DEFECTS

In this section we examine the types of point defects that can occur in pure crystalline materials. For purposes of this discussion, a **point defect** may be defined as an imperfection that involves a few atoms at most. Point defects occur in all classes of materials, and unifying principles govern their occurrence and behavior. The materials that will be considered may be ionic (e.g., NaCl or KCl), covalent (e.g., Si or Ge), mixed ionic/covalent (e.g., Al_2O_3 or SiO_2), metallic (e.g., Al or Fe), or molecular (e.g., polyethylene or nylon). Point defects have profound effects on the mechanical and physical properties of most engineering materials.

4.2.1 Vacancies and Interstitials in Crystals

A two-dimensional lattice is shown in Figure 4.2–1. Here we see that at certain positions, there are missing atoms (a normally occupied position is vacant) while in other places atoms are in "wrong" positions (atoms are located in normally unoccupied positions). The former are called **vacancies** and the latter are termed **interstitials.** Both are examples of point defects.

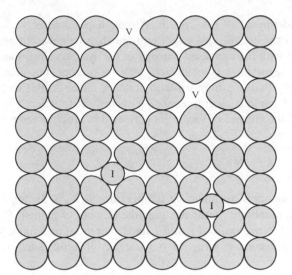

FIGURE 4.2–1
Vacancies and interstitials
in a crystalline material.

Examination of Figure 4.2–1 shows that both types of defects alter the equilibrium spacing between atoms in their immediate vicinity. As discussed in Chapter 2, moving atoms away from their equilibrium positions requires an increase in the energy of the system. The increase in energy that arises as a result of the change in bond length is known as strain energy. Strain energy may be viewed as analogous to the energy increase that occurs when a spring (in this case the atomic bond) is either stretched or compressed. In addition, either type of point defect may alter the coordination number of the surrounding atoms, which also results in an energy increase. Thus, we can conclude that the presence of vacancies or interstitials increases the internal energy of the crystal.

One question that arises is, If the internal energy of the system increases, why should such defects form in the first place? To answer this question we must look a bit more closely at the concept of energy minimization introduced in Chapter 2. There are several different types of energy, including potential energy, kinetic energy, and work (e.g., force multiplied by distance). Other forms of energy with which you may be less familiar include, for example, the rest-mass energy of a particle ($E = mc^2$) and Gibbs free energy. It is the Gibbs free energy, G, that is important for our discussion of defects in crystals. The key point is that if a group of atoms can be arranged in two different ways, then the arrangement with the lower G value will be the thermodynamically favored state.

The next step is to understand the factors that control the magnitude of G. The free energy of a crystal is a balance between enthalpy (a measure of the internal energy of the crystal) and entropy (a measure of the randomness of the crystal). G increases with either an *increase* in internal energy or a *decrease* in entropy.

As mentioned above, point defects increase the internal energy of the crystal through an increase in strain energy. If this were their only contribution to G, then the defects would be energetically unstable. However, they also have an effect on the entropy of the crystal.

The entropy of a crystal increases whenever the number of distinct possible arrangements of atoms increases. Relating this idea to the presence of point defects, we see that if the system contained no imperfections, there would be only one possible atomic configuration—each lattice point would be occupied by an identical atom. In contrast, by changing the positions of the point defects, it is possible to form many distinct atomic arrangements of equivalent energy. Thus, the presence of point defects greatly increases the entropy (randomness) of the system. Although point defects usually increase the internal energy of the crystal, this effect is more than balanced by the decrease in free energy associated with the increased entropy of the system. Thus, thermodynamic arguments not only suggest that point defects may be present, but actually demand their presence and imply that it is impossible to create a stable crystal without point defects.

The concept of minimizing free energy with respect to defect concentration can be used to show that the number of vacancies is related to absolute temperature through the following equation:

$$N_v = N_T \exp\left(-\frac{Q_{fv}}{RT}\right) \tag{4.2-1}$$

where N_v is the number of vacancies at temperature T, N_T is the total number of lattice sites (equal to the number of vacancies plus the number of atoms), Q_{fv} is the activation energy for vacancy formation, and R and T have their usual meanings. We can also define the vacancy concentration in terms of the fraction of lattice sites that are vacant, C_v:

$$C_v = \frac{N_v}{N_T} = \exp\left(-\frac{Q_{fv}}{RT}\right) \tag{4.2-2}$$

where C_v is a dimensionless quantity. A similar equation can be obtained for the interstitial concentration by substituting the appropriate value for interstitial formation energy, Q_{fi}.

···

EXAMPLE 4.2-1

Calculate the concentration of vacancies in Cu at room temperature (298 K) and just below the melting point (1356 K), given that the activation energy is approximately 83,600 J/mol and the gas constant R is 8.31 J/mol-K.

Solution

The vacancy concentration at each temperature can be determined using Equation 4.2-2 with the values for Q_{fv} and R as given above. At 1356 K this leads to:

$$C_v = \exp\left(-\frac{Q_{fv}}{RT}\right) = \exp\left(-\frac{83,600}{8.31 \times 1356}\right)$$

$$= \exp(-7.419) = 6 \times 10^{-4}$$

Similarly, at 298 K the vacancy concentration is:

$$C_v = \exp(-33.759) = 2.2 \times 10^{-15}$$

Note the dramatic increase in the vacancy concentration with increasing temperature and also that, even at the highest temperature, only 6 out of 10,000 lattice sites are vacant. This latter number may be converted to the spacing between vacancies, with the result that vacancies are approximately eight atom spacings apart from one another, a short enough distance for vacancies to affect the properties of the material.

···

While the arguments presented above are valid for both metallic and covalent solids, the situation in crystals involving macromolecules (polymers) is more complex. A vacancy in a polymer crystal may be envisioned as the space between the end of one polymer chain and the beginning of the next chain. There are many types of defects in polymer crystals, but they are usually of little consequence in the properties of engineering polymers, which contain vast amounts of noncrystalline material.

In summary, we have identified two possible types of point defects in crystals: vacancies and interstitials. It has been demonstrated that not only are such defects possible but their existence is thermodynamically necessary. Point defects play a major role in determining several properties of crystals, and their effects will be discussed throughout this and subsequent chapters.

4.2.2 Vacancies and Interstitials in Ionic Crystals

Just as vacancies and interstitials occur in nonionic crystals, similar point defects are also found in ionic solids. However, an extremely important difference between ionic and nonionic crystals arises with respect to the generation of vacancies. In ionic crystals, isolated single vacancies do not occur, since the removal of a single ion would result in an electrical charge imbalance in the crystal. Instead, vacancies can occur only in small groups with the cation : anion vacancy ratio such that the crystal as a whole remains electrically neutral. These electrically neutral cation-anion vacancy clusters are called **Schottky defects.** Examples of Schottky defects in NaCl and $MgCl_2$ are shown in Figure 4.2–2. In NaCl, in which there are equal numbers of cations and anions, electrical neutrality requires equal numbers of cation and anion vacancies, and the Schottky defect is composed of a single anion/cation vacancy pair. In contrast, in $MgCl_2$, for every Mg^{2+} vacancy there are two Cl^- vacancies to maintain electroneutrality.

The Schottky defect is not the only point-defect cluster that can maintain charge neutrality. Another possibility involves the formation of a vacancy/interstitial pair. These small point-defect clusters are known as **Frenkel defects** and are shown schematically in Figure 4.2–3 for the compound AgCl. Generally speaking, Frenkel defects involving cation interstitials are far more common than those involving anion interstitials, since most interstices are too small to accommodate the larger ions, usually anions.

The relative concentrations of the various point defects are determined by the energies of defect formation and the requirement of electroneutrality. Using thermodynamic

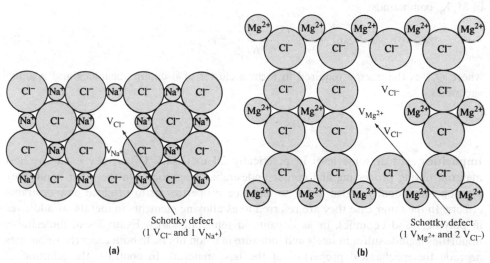

FIGURE 4.2–2   Schottky defects in **(a)** NaCl and **(b)** $MgCl_2$. The vacancies are created so that overall charge neutrality is maintained. The figure shows the vacancies associated (close together); however, the cation and anion vacancies need not be located spatially close to one another.

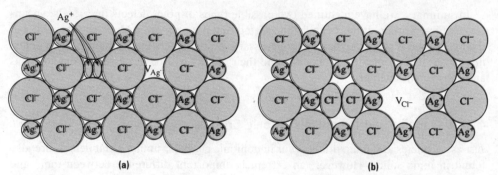

FIGURE 4.2–3 Frenkel defects in AgCl. The defect consists of vacancy/interstitial pairs. Frenkel defects involving cations **(a)** are more common than those involving anions **(b)**, since cations are usually smaller.

reasoning similar to that used to develop the expression for the vacancy concentration in nonionic solids, the concentrations of vacancies and interstitials in ionic solids can be estimated.

For Frenkel defects:

$$C_v = \frac{N_v}{N_T} = \exp\left(-\frac{Q_{fvi}}{2RT}\right) = \frac{N_i}{N_T} = C_i \tag{4.2–3}$$

where C_v and C_i are respectively the concentrations of vacancies and interstitials and Q_{fvi} is the energy required to form a vacancy and an interstitial. The factor of 2 in the denominator of the exponential occurs because two defects are created in the process.

For Schottky defects in an MX compound:

$$C_{v,\text{cat}} = \frac{N_{v,\text{cat}}}{N_T} = \exp\left(-\frac{Q_{fvp}}{2RT}\right) = \frac{N_{v,\text{an}}}{N_T} = C_{v,\text{an}} \tag{4.2–4a}$$

where $C_{v,\text{cat}}$ and $C_{v,\text{an}}$ are respectively the concentrations of cation and anion vacancies and Q_{fvp} is the energy required to form a cation/anion vacancy pair. For Schottky defects in M_nX_p compounds:

$$pC_{v,\text{cat}} = (np) \exp\left[-\frac{Q_{fvc}}{(n^2 + p^2)RT}\right] = nC_{\text{an}} \tag{4.2–4b}$$

where Q_{fvc} is the energy required to form a cluster of n cation vacancies and p anion vacancies.

4.3 IMPURITIES

Impurities exist at some level in practically all materials. Often they enter the host material during processing and confer undesirable properties. On the other hand, they may be intentionally added either to enhance properties or to produce new desirable effects. In the latter case they are referred to as alloying elements in metals, as additives in polymers and ceramics, or as dopants in semiconductors. Examples of undesirable impurities include sulfur in steels and moisture in nylon fibers. In both cases the impurities degrade the mechanical properties of the host material. In contrast, the addition of phosphorus to silicon to confer desirable electrical (semiconducting) characteristics, carbon to iron to increase strength, methyl acrylate to polyacrylonitrile to give a dyeable polymer, and metallic ions to window glass to produce desirable colors are examples of beneficial impurity additions. In this section we describe the different types of impurities, demonstrate the basis for their existence, and discuss their practical effects.

4.3.1 Impurities in Crystals

Just as vacancies and interstitials in pure materials lower the free energy, in many cases impurities dissolved in a material also lower the total free energy. While the internal energy is generally increased as a result of adding impurity atoms, the entropy is also increased, and at any temperature there will be a certain number of impurities for which the free energy is a minimum. The equilibrium concentration of impurity atoms at any temperature is determined by a trade-off between increases in the entropy and increases in the internal energy.

Atoms of the primary atomic species are called **solvent atoms,** while the impurities are usually referred to as **solute atoms.** In crystalline solids, impurity atoms can be either present in the spaces between the solvent atoms or substituted for the solvent atoms. When the impurities lie in the spaces between the solvent atoms, they are called interstitials, and the mixture formed by the two atomic species is called an **interstitial solid solution.** When impurities substitute for the solvent atoms, they are called **substitutional** atoms, and the mixture of the two species is called a **substitutional solid solution.** Examples of interstitial and substitutional solid solutions are shown in Figures 4.3–1 and 4.3–2, respectively.

There are certain requirements for the formation of either interstitial or substitutional solid solutions. For example, it is possible for atoms of the host species to exist in interstitial positions as a result of high-energy impacts, such as those received during irradiation. However, this is an unusual case: interstitial solid solutions normally form only when the interstitial atoms are significantly smaller than the solvent atoms and are

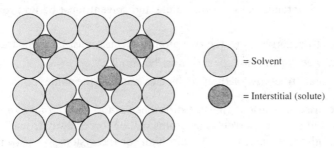

FIGURE 4.3–1   Interstitial solid solution. The solute atom is positioned in the void spaces between solvent atoms, causing strain in the lattice.

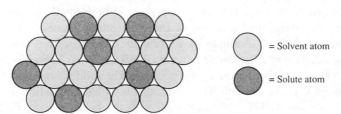

FIGURE 4.3–2   Substitutional solid solution. The solute and solvent atoms must have sizes (radii) within 15% and must have similar bond characteristics.

comparable in size to the interstitial sites they occupy.[1] While these are generalizations, it is not possible to formulate precise rules as to when an interstitial solid solution will be formed. Indeed, for the case of C in Fe, the radius of a C atom is significantly larger than the interstitial position in which it is located (see Example 3.6–2). One reason for this seeming discrepancy lies in the fact that atoms are usually considered to be perfect spheres. While often a useful approximation, the assumption of spherical atoms is an oversimplification for C, which results in the apparent contradiction. Detailed consideration of the shape of atoms and the way in which they can fit into the interstitial positions is a topic of current research.

The example of an interstitial solid solution of C in BCC iron is particularly relevant, since it represents the technologically important metal alloy known as steel. The equilibrium concentration of carbon in iron can be computed using Equation 4.2–2 with an appropriate value for Q, and is found to be 0.022% at 727°C and much less at room temperature. As we shall see later in this section, carbon atoms in solution increase the strength of steel significantly. Other examples of interstitial atoms in Fe include H, O, N, and S (note that these are all small atoms). These impurities may have different effects. For example, H in steel sometimes results from certain electroplating operations and can result in a degradation of the mechanical properties of the steel (to be discussed in Chapter 15). Nitrogen, on the other hand, can combine with Al (present in substitutional form) to form particles of AlN, which improve the steel's ability to be formed into items such as fenders for cars and refrigerator doors. The improved malleability is achieved by the effect of AlN on preventing grain growth during hot-working operations, as discussed in Chapter 5.

The requirements for the formation of substitutional solid solutions are clearly defined and can be expressed as the well-known **Hume-Rothery rules,** named after the person who proposed them. In order to form a substitutional solid solution over a wide range of compositions, the following conditions must be met:

1. The size difference between the solute and solvent must be no greater than ~15%.
2. The electronegativities of the two atomic species must be comparable.
3. The valences of the two species must be similar.
4. The crystal structures of the two species must be the same.

The first rule arises because if the size difference is too great, a large amount of strain energy will result (i.e., the surrounding atoms will be forced away from their equilibrium positions) and the free energy will be too high for the system to exist in the form of a substitutional solid solution. Note that the extent of solid solubility increases as the size difference decreases. The second and third rules ensure that the solute and solvent will have similar bond characteristics. The fourth rule is applicable only if solid solutions are to be formed for *all* proportions of the two atomic species. It can be ignored for dilute solid solutions (i.e., those in which the solute is present in only small amounts).

The previously mentioned example of the addition of P to Si to produce desirable electrical properties is an example of a commercially important substitutional solid solution. The addition of appropriate amounts of Ni and Cr to Fe results in an important example of a metallic substitutional solid solution. The solid solution formed in this way stabilizes the FCC phase at room temperature. It is called austenitic stainless steel, and a typical composition is 18% Cr and 8% Ni with the balance being Fe (so-called 304

---

[1] Recall that we calculated the size of the interstitial holes in the BCC, FCC, and HCP crystal structures in Section 3.6. The results of those calculations are summarized in Table 3.6–1.

stainless steel). The Ni and Cr additions result in a steel with the ability to survive in aggressive environments that would rapidly degrade the properties and appearance of steel containing just C and Fe (plain carbon steel). For example, austenitic stainless steels can be used at moderate temperatures without oxidizing, whereas plain carbon steels oxidize rapidly. Other applications of stainless steels are found in the chemical industry, in biomedical implants, and in cutlery, where corrosion resistance is important.

EXAMPLE 4.3–1

Copper and nickel form a solid solution in all proportions. Predict this result using the Hume-Rothery rules.

Solution

The atomic radii of nickel and copper are 0.128 and 0.125 nm (Appendix C). The difference in their radii is 2.4%, which is significantly less than the 15% allowed by the Hume-Rothery rules. The electronegativities of copper and nickel are 1.90 and 1.91 (Appendix B), respectively, which are also very similar. The most common valences are $+1$ or $+2$ for copper and $+2$ for nickel (Appendix C). Finally, both metals have the FCC crystal structure. Thus, the Cu-Ni system obeys all four Hume-Rothery rules and is expected to show solid solution behavior over the entire compositional range.

Impurities also occur in polymeric crystals. Sometimes impurities are incorporated directly into the polymer chain. Copolymers are examples of this type of impurity, as shown in Figure 4.3–3. The incorporation of a small amount of a second mer into the polymer chain does not necessarily imply inclusion of the second mer into the crystalline regions of the polymer. If the characteristics (size, shape, and bond structure) of the second mer are similar to those of the bulk polymer, then it may be incorporated directly into the crystalline regions of the polymer, as shown schematically in Figure 4.3–4a. If this is not possible, either the polymer will not crystallize or the crystalline regions will form in such a way as to exclude most, if not all, of the second mer. In the latter case, the impurity will reside in the surrounding noncrystalline regions (see Figure 4.3–4b).

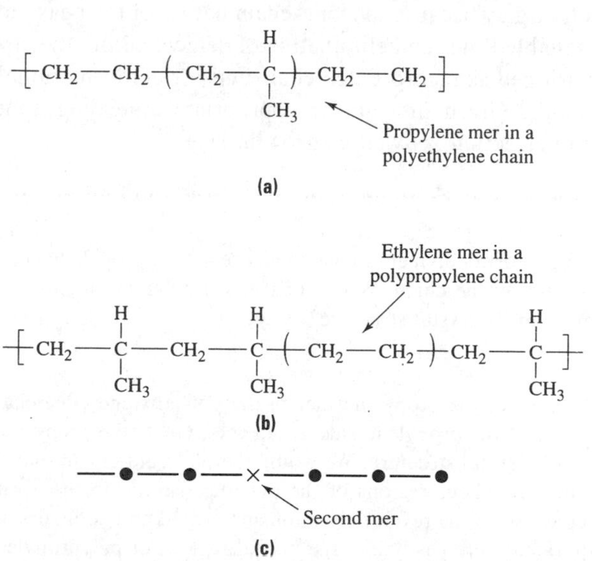

FIGURE 4.3–3 Examples of a second mer in a polymer chain: **(a)** a propylene mer in a polyethylene chain, and **(b)** an ethylene mer in a polypropylene chain. **(c)** A schematic of a second mer is provided.

FIGURE 4.3–4

Polymer crystals containing a second type of mer (denoted by the symbol x). In **(a)** the second type of mers are included in the crystal while in **(b)** they are largely excluded.

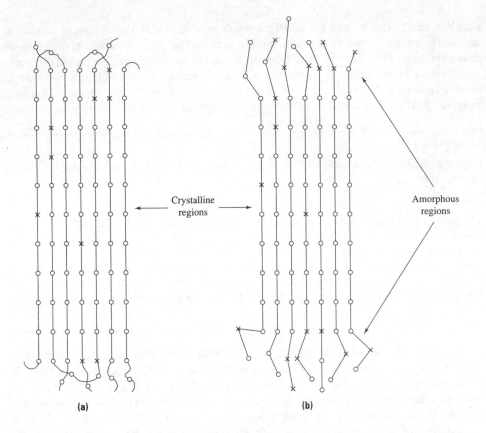

(a)   (b)

Crystalline regions

Amorphous regions

The partially crystalline structure of polymers is described in more detail in Chapter 6. The specifics of incorporation of an impurity into a growing polymer crystal are similar in principle to those of other growing crystals; i.e., principles of size difference and bond compatibility are equally applicable.

Even "pure" polymer crystals formed under ordinary processing conditions are generally quite imperfect, largely because the long-chain nature of the beast precludes reaching equilibrium in reasonable times and elimination of defects within the crystal. Most of the defects may be envisioned as incorporated chain ends or conformational defects, such as chain folds in crystals. Given time under appropriate annealing conditions, polymer crystals improve in perfection, as well as grow in size.

....

EXAMPLE 4.3–2

Consider the two examples of copolymers shown in Figure 4.3–3, propylene in a polyethylene chain and ethylene in a polypropylene chain. Which of the two polymers is more likely to be able to include the comonomer in its crystal structure? Why?

Solution

First we consider the case of the propylene mer in the polyethylene (PE) chain. Because of the comparatively large size of the propylene side group (i.e., CH_3), the propylene mer is not easily incorporated into the PE crystal structure. We would therefore anticipate that the CH_3 side group would be found in the amorphous regions of the polymer (similar to the arrangement shown in Figure 4.3–4b). In contrast, in the reverse situation an isolated and comparatively small ethylene mer could be incorporated more easily into the crystal regions of polypropylene.

....

4.3.2 Impurities in Ionic Crystals

Impurities occur in ionic solids just as they do in other materials. Anions exist primarily in the substitutional form—they are generally too large to "fit" in the interstices. Cations, on the other hand, can be present at both the substitutional and interstitial positions. As was the case for interstitials and vacancies in ionic crystals, electroneutrality must be maintained within the crystal. This requirement results in some interesting effects. For example, suppose that divalent metallic Mg^{2+} ions are found as impurities in NaCl. The Mg^{2+} ions will substitute for the Na^+ ions, resulting in one "extra" positive charge. How does the crystal compensate for this additional positive charge? Since no additional anions are available, the most frequent result is the effective removal of a single positive charge through the formation of a Na^+ vacancy (i.e., one V_{Na} is formed for each Mg^{2+} substitutional impurity ion). Thus, in addition to the vacancies that would normally be present (called thermal vacancies), there are extra vacancies related to the impurity content. This effect is illustrated in Figure 4.3–5. Similar reasoning would apply to impurity cations having different charges or to anions.

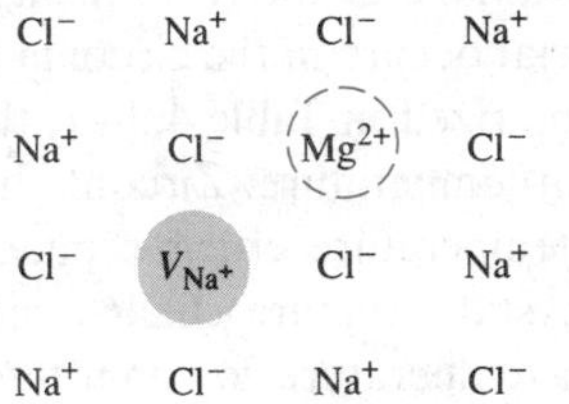

FIGURE 4.3–5   Substitution of Mg for Na in NaCl. The divalent Mg requires that a Na vacancy (V_{Na^+}) be created in order to maintain electroneutrality.

EXAMPLE 4.3–3

Consider a sample of MgO containing 0.2 weight percent Li_2O as an impurity. Compute the additional vacancy concentration that arises because of the presence of the impurity.

Solution

As a basis of calculation we assume 100 grams of material. We first determine the number of moles of Li^+, Mg^{2+}, and O^{2-} ions present. The number of moles of Li_2O is calculated as:

$$\text{Moles } Li_2O = \frac{\text{Grams } Li_2O}{2(\text{Mol. wt. Li}) + (\text{Mol. wt. O})}$$

$$= \frac{0.2 \text{ g}}{(2 \times 6.941 \text{ g/mol}) + (16 \text{ g/mol})} = 6.7 \times 10^{-3} \text{ moles}$$

Therefore, there are $2(6.7 \times 10^{-3}) = 1.34 \times 10^{-2}$ moles of Li^+ and 6.7×10^{-3} moles of O^{2-} in the sample. Using a similar calculation, the number of moles of MgO is:

$$\text{Moles MgO} = \frac{99.8 \text{ g}}{(24.31 + 16) \text{ g/mol}} = 2.48 \text{ moles}$$

Since MgO is a 1 : 1 compound, there are 2.48 moles of Mg^{2+} ions and 2.48 moles of O^{2-} ions. The total number of each type of ion is:

$$N_{Li} = 1.34 \times 10^{-2} \text{ moles}$$

$$N_{Mg} = 2.48 \text{ moles}$$

$$N_O = 6.7 \times 10^{-3} + 2.48 = 2.4867 \text{ moles}$$

The total number of moles of ions in the system is:

$$N_T = N_{Mg} + N_{Li} + N_O = 1.34 \times 10^{-2} + 2.48 + 2.4867 = 4.9801 \text{ moles}$$

Each substitution of a Li^+ for Mg^{2+} results in the loss of one positive charge. If no interstitials are created, this loss of positive charge must be balanced by the creation of anion vacancies. Charge neutrality requires one oxygen vacancy created for every two Li^+ impurity ions. Therefore, the number of oxygen vacancies created is:

$$N_{v,O} = \frac{N_{Li}}{2} = \frac{1.34 \times 10^{-2}}{2} = 0.67 \times 10^{-2} \text{ moles}$$

The concentration of oxygen vacancies is:

$$\frac{N_{v,O}}{N_T} = \frac{0.67 \times 10^{-2} \text{ moles}}{4.9801 \text{ moles}} = 1.35 \times 10^{-3}$$

This concentration of vacancies resulting from the addition of even a modest amount of impurity is far greater than the thermal vacancy concentration.

....

An industrially important example of the relationship between impurities, vacancies, and structure in an ionic material occurs in the zirconia (ZrO_2) system. As mentioned in the previous chapter and summarized in Table 4.3–1, this material may exist in several crystalline forms depending on temperature. Zirconia has excellent thermal properties, which make it an ideal high-temperature crucible material. Unfortunately, the volume change associated with the crystal structure change can cause cracking when the part is cooled from the fabrication temperature to room temperature. The cracking can be avoided by suppressing the tetragonal-to-monoclinic transition to below room temperature (i.e., if the tetragonal phase is stabilized down to room temperature). This has been achieved by adding materials such as CaO, MgO, and Y_2O_3. The resulting product, called stabilized zirconia, is widely used as a crucible material.

TABLE 4.3–1   Structure of ZrO_2 as a function of temperature.

Temperature range	Structure
Above 2680°C	Liquid
2370–2680°C	Cubic solid
1170–2370°C	Tetragonal solid
Below 1170°C	Monoclinic solid

4.4 SOLID-STATE DIFFUSION

Thus far we have implied that the point defects described in the previous section are located at fixed positions within the solid. These defects, however, can and do move around within the solid. One mechanism by which atoms or molecules move is known as diffusion.

If a drop of ink is placed in a glass of still water, the ink moves (diffuses) throughout the water. Similarly, if a layer of carbon atoms is placed on the surface of a hot iron plate, the carbon migrates (diffuses) into the plate. These are examples of diffusion, the first in a liquid and the second in a solid. Diffusion can also occur in a gas, but the effects of diffusion are usually masked by free convection of air. Thus, if a bottle of perfume is opened in a room, the fragrance can be detected everywhere in the room after only a short time. Certainly diffusion has taken place, but to a much greater degree the perfume molecules have been transported by convection.

Diffusion is a process of mass transport that involves the movement of one atomic species into another. In its simplest form, it occurs by random atomic jumps from one position to another and takes place in the gaseous, liquid, and solid states for all classes of materials. Since diffusion involves atomic jumps, it becomes more rapid at higher temperatures because of the higher thermal energy of the atoms (the rate of diffusion can be modeled using the Arrhenius equation). The diffusion rate also depends on the openness of the structure, being generally more rapid for less densely packed structures.[2] Consequently diffusion is most rapid in the gaseous state and least rapid in the solid state. Even though it is least rapid in the solid state, solid-state diffusion has extremely important practical consequences for engineering materials. In addition to its importance from a scientific point of view, many industrial processes having significant economic consequences involve diffusion. In this regard it should be noted that diffusion can occasionally lead to a degradation in the engineering properties and thus also has important economic implications. For example, diffusion of oxygen into jet engine components can cause severe embrittlement and failure.

In the following sections we will review some industrially important examples of diffusion, develop a fundamental physical description of the diffusion process, examine the relationship between the diffusion process and point defects, and finally apply our knowledge of diffusion to a quantitative study of important engineering applications.

4.4.1 Practical Examples of Diffusion

The controlled diffusion of P, B, or other dopants (beneficial impurities) into Si wafers is of fundamental importance to the microelectronics industry. The controlled diffusion of oxygen and carbon dioxide through a silicon-based rubber membrane in a heart-lung machine permits surgeons to operate on the human heart. The controlled diffusion of oxygen through a ZrO_2 membrane is the basis of the sensing devices used to monitor and control air/fuel mixtures in internal-combustion engines. These are but a few of the many commercial applications of diffusion.

Another important industrial process involving diffusion is the carburization of steels. Carburization increases the carbon content in the near-surface region. The extra carbon combines with the iron to form strong iron carbide particles having excellent wear resistance. The carburized surface is ideal for applications such as gears and crankshafts in automotive engines, where there is considerable friction and where the surface must be wear-resistant. One method by which steels are carburized is to heat them to high temperatures in a CO/CO_2 atmosphere in which the ratio of these gases is adjusted to maintain a desired surface concentration of C. Under these conditions, carbon atoms are deposited on the surface, and with time they move into the body of the part being carburized. By allowing sufficient time, a carbon-rich layer near the surface can be produced. Intuitively, we expect the thickness of this layer to depend on factors such as time, temperature, and the carbon concentration at the surface. Figure 4.4–1a shows an example of a carburized surface.

As a final example of diffusion, consider the coating of turbine blades used in jet engines. Turbine blades operate at temperatures as high as 1100°C in an environment containing highly corrosive combustion products. They are fabricated from Ni-base alloys, which, although strong and relatively resistant to oxygen attack, would nonetheless be completely oxidized over time if not protected. One technique that is widely used for

---

[2] This is an oversimplification. Diffusion within Si is generally slow even though the basic Si lattice is rather open. This is because of covalent nature of the Si bond and the effects that it has on the low number of vacancies and on migration of other species.

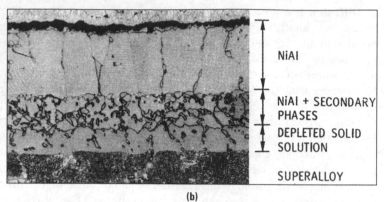

(a)   (b)

FIGURE 4.4–1 Example of surface layers formed by solid-state diffusion: **(a)** A carburized surface layer on a steel part. The carbon has been diffused into the surface region from the left-hand side and has transformed into iron carbide particles. **(b)** A coating on a typical Ni-base superalloy. The coating is formed by interdiffusion of Al, Ni, and other elements in the superalloy. The phases formed as a result of the coating are indicated on the micrograph. The coating is added to protect the base metal from oxidation. Note the dark oxide phase at the top of the coating. *(Source:* **(a)** *Metals Handbook Desk Edition, 1985, ASM International, Materials Park, OH.* **(b)** The Superalloys, *edited by C. T. Sims and W. C. Hagel, © 1972 by John Wiley & Sons. Reprinted by permission of John Wiley & Sons, Inc.)*

coating turbine blades is to deposit Al on the surface of the superalloy via dissociation of AlF_3 gas. The Al and Ni in the superalloy interdiffuse and react to form NiAl (plus other secondary phases). This diffusion coating is highly resistant to oxidation and protects the base alloy. Figure 4.4–1b shows a section of a typical coated turbine blade.

There are also examples of diffusion processes that lead to a degradation of engineering properties. For example, consider the microelectronic circuit, composed of a silicon wafer containing diffused impurities, described above. The performance of such circuits depends critically on the spatial distribution of the impurity atoms in the diffusion zone. While the circuit may originally have the ideal impurity distribution, it is possible for this distribution to change from the ideal during use of the device through a combination of driving forces such as temperature and electric fields. For example, the heating that occurs as a result of current flow within these devices can result in additional unwanted diffusion of the impurity atoms and attendant loss of function.

4.4.2 A Physical Description of Diffusion (Fick's First Law)

The diffusion process can be understood in terms of a simple physical description. Consider two adjacent atomic planes of a solid solution of A and B atoms, such as shown in Figure 4.4–2. Diffusion can be modeled as the jumping of atoms from one plane to another. Our job is to calculate the net number of A atoms moving from plane 1 to plane 2 per unit area and unit time. This quantity is called the diffusion flux with units of atoms/(cm²-s). Let the concentration of A atoms on planes 1 and 2 be denoted by C_1 and C_2 atoms/cm³, and let the spacing between the adjacent planes be Δx.

What factors might affect the rate of atomic motion (i.e., the diffusion rate)? Some of the important variables are:

1. The concentration difference, $C_1 - C_2$, between the two planes. As the concentration difference increases, we expect the diffusion flux to increase.
2. The jump distance, Δx (this variable is a function of the crystal structure). As the jump distance decreases, we expect the diffusion flux to increase.

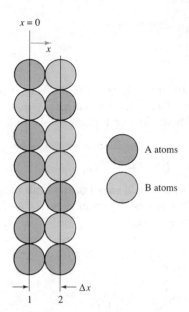

$x = 0$

1   2 $\leftarrow \Delta x$

A atoms

B atoms

FIGURE 4.4–2

Two adjacent atomic planes with different concentrations of A and B atoms.

3. The frequency at which atoms attempt to jump from one plane to another (this variable is an exponential function of temperature). As the jump frequency increases, we expect the diffusion flux to increase.

Combining these observations yields the following mathematical model for the net diffusion flux J of A atoms in the $+x$ direction (at constant temperature):

$$J = D\left(\frac{C_1 - C_2}{\Delta x}\right) \tag{4.4–1a}$$

where D is the **diffusion coefficient** (with units of cm²/s). As discussed in more detail below, the diffusion coefficient contains the information about the temperature dependence of the jump frequency. The term in the parentheses is just the negative of the concentration gradient, dC/dx, so that the equation can be rewritten in the form:

$$J = -D\left(\frac{dC}{dx}\right) \tag{4.4–1b}$$

Equation 4.4–1 is called **Fick's first law** of diffusion. Its use is restricted to problems for which the concentration gradient does not change with time. The minus sign in Fick's first law indicates that mass flows "down" the concentration gradient (i.e., from regions of high concentration to regions of low concentration). It is completely analogous to heat transfer through a barrier of thickness x in which the hot surface is at T_h and the cold surface is at T_c. In this case, the net heat flux per unit area per unit time (watts/m²) through the plate is proportional to the temperature gradient. In mathematical form this is:

$$q = \frac{k(T_h - T_c)}{x} = -k\left(\frac{dT}{dx}\right) \tag{4.4–1c}$$

Equation 4.4–1c is known as Fourier's law.

The diffusion coefficient D contains the temperature dependence of the jump frequency as well as the information about interplanar distances, which depend on crystal structure. It has the mathematical form

$$D = D_0 \exp\left(-\frac{Q}{RT}\right) \tag{4.4-2}$$

where D_0 is a constant with units cm²/s, Q is the activation energy in J/mol for the diffusion process, and R and T have their usual meanings.

Taking the logarithm of both sides of Equation 4.4–2 yields:

$$\ln D = \ln D_0 + \left(-\frac{Q}{R}\right)\left(\frac{1}{T}\right) \tag{4.4-3}$$

This equation predicts that $\ln D$ should vary linearly with reciprocal temperature, $1/T$. A schematic plot of $\ln D$ versus $1/T$ is shown in Figure 4.4–3. The slope of the curve is $(-Q/R)$, and the y intercept is $\ln D_0$.

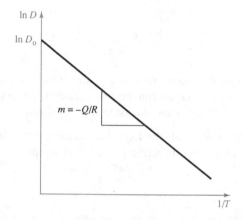

FIGURE 4.4–3   Variation of diffusion coefficient with temperature. The slope is proportional to the activation energy.

We have seen that diffusion is a process based on random atomic jumps. Thus, in a homogeneous solid solution there will be no net flux of atoms in any direction, since the concentration of atoms is the same everywhere. When there is a concentration gradient, however, the flux of atoms will be *down the concentration gradient*. Recall the examples used to open this discussion—ink in water and carbon in iron. In each case the diffusing species moves down the concentration gradient from regions of high concentration to regions of low concentration.

··

EXAMPLE 4.4–1

Assume that a thin plate of BCC Fe is heated to 1000 K. One side of the plate is in contact with a CO/CO₂ gas mixture that maintains the carbon concentration at the surface at 0.2 weight percent (wt. %) C. The other side is in contact with an oxidizing atmosphere that maintains the surface concentration at 0 wt. % C. Compute the number of carbon atoms transported to the back surface per second through an area of 1 cm². The plate is 0.1 cm thick. The density of BCC Fe is approximately 7.9 g/cm³, and the diffusion coefficient is 8.7×10^{-7} cm²/s at 1000 K.

Solution

Since the concentration gradient is constant, we can use Fick's first law (Equation 4.4–1) to determine the carbon flux. To do this we first calculate the concentration gradient in terms of (carbon atoms/cm^3)/cm. The concentration of carbon atoms at either surface is calculated as follows:

$$C = \left(\frac{\text{wt. \% C} \times \text{density of BCC Fe}}{\text{Mol. wt. C}} \right) \times (\text{Avogadro's Number})$$

$$C_1 = \left[\frac{(0.002)(7.9 \text{ g/cm}^3)}{12.01 \text{ g/mol}} \right] (6.02 \times 10^{23} \text{ atoms/mol})$$

$$= 7.92 \times 10^{20} \text{ atoms/cm}^3$$

$$C_2 = 0$$

The concentration gradient is then:

$$\frac{dC}{dx} \approx \frac{C_2 - C_1}{\text{Thickness of sample}}$$

$$= -\frac{7.92 \times 10^{20} \text{ (atoms/cm}^3)}{(0.1 \text{ cm})} = -7.92 \times 10^{21} \text{ atoms/cm}^4$$

The number of atoms per second per cm^2 transported through the plate is obtained by multiplying the concentration gradient by the diffusion coefficient and inserting a minus sign:

$$J = -D\left(\frac{dC}{dx} \right) = (8.7 \times 10^{-7} \text{ cm}^2/\text{s})(7.92 \times 10^{21} \text{ atoms/cm}^4)$$

$$= 6.9 \times 10^{15} \text{ atoms/(cm}^2\text{-s)}$$

....

EXAMPLE 4.4–2

For low concentrations of Zn in Cu, the diffusion coefficient of Zn has been measured to be 3.67×10^{-11} cm^2/s at 1000 K and 8.32×10^{-18} cm^2/s at 600 K. Determine the activation energy for this process and then determine the value of the diffusion coefficient at 450 K.

Solution

This problem can be solved using Equation 4.4–2. Using the given data, we can write the two equations

$$D(T_1) = 3.67 \times 10^{-11} \text{ cm}^2/\text{s} = D_0 \exp\left(-\frac{Q}{RT_1} \right)$$

and

$$D(T_2) = 8.23 \times 10^{-18} \text{ cm}^2/\text{s} = D_0 \exp\left(-\frac{Q}{RT_2} \right)$$

There are, of course, many techniques for finding the two unknowns D_0 and Q. We will demonstrate one general method. Dividing the expression evaluated at T_1 by that at T_2 and taking the natural logarithm of both sides of the resulting equation:

$$\ln\left[\frac{D(T_1)}{D(T_2)} \right] = \left(-\frac{Q}{R} \right)\left[\left(\frac{1}{T_1} \right) - \left(\frac{1}{T_2} \right) \right]$$

Solving for Q yields:

$$Q = -R\left\{ \ln\left[\frac{D(T_1)}{D(T_2)} \right] \right\}\left(\frac{1}{1/T_1 - 1/T_2} \right)$$

Substituting the data given in the problem statement yields:

$$Q = -(8.314)\left[\ln\left(\frac{3.67 \times 10^{-11}}{8.32 \times 10^{-18}}\right)\right] \times \left(\frac{1}{1/1000 - 1/600}\right)$$

$$= 190,800 \text{ J/mol}$$

Next, the equation for D at 1000 K can be solved for D_0 to yield:

$$D_0 = D \exp\left(+\frac{Q}{RT}\right) = 3.67 \times 10^{-11} \exp\left[\frac{190,800}{(8.314)(1000)}\right]$$

$$= 0.34 \text{ cm}^2/\text{s}$$

The diffusion coefficient at 450 K can be determined as:

$$D = D_0 \exp\left(-\frac{Q}{RT}\right) = (0.34 \text{ cm}^2/\text{s}) \exp\left[-\frac{190,800}{(8.314)(450)}\right]$$

$$D(450 \text{ K}) = 2.41 \times 10^{-23} \text{ cm}^2/\text{s}$$

....

EXAMPLE 4.4–3

A thin plastic membrane is used to separate hydrogen from a gas stream. At steady state the concentration of hydrogen on one side of the membrane is 0.025 mol/m^3, on the other side it is 0.0025 mol/m^3, and the membrane is 100 micrometers (μm) thick. Given that the flux of hydrogen through the membrane is 2.25×10^{-6} mol/(m^2-s), calculate the diffusion coefficient for hydrogen.

Solution

Since this is a steady-state membrane problem, it can be solved using Fick's first law:

$$J = -D\left(\frac{dC}{dx}\right)$$

Solving for the D and substituting the appropriate values from the problem statement yields:

$$D = -\frac{J}{(dC/dx)} \approx -\frac{J}{(C_2 - C_1)/\Delta x}$$

$$= -\frac{2.25 \times 10^{-6} \text{ mol/(m}^2\text{-s)}}{(0.0025 - 0.025 \text{ mol/m}^3)/(100 \times 10^{-6} \text{ m})}$$

$$D = 1 \times 10^{-8} \text{ m}^2/\text{s} = 1 \times 10^{-4} \text{ cm}^2/\text{s}$$

....

4.4.3 Mechanisms of Diffusion in Covalent and Metallic Crystals

So far our discussion has referred to the motion of atoms and ions without being very specific as to the detailed mechanism by which they can move in a material. The mechanism of diffusion determines the energy barrier that must be overcome (i.e., the activation energy Q) for the process to occur. Since energy is supplied thermally, the higher the temperature, the greater the probability that large numbers of atoms will have sufficient energy to overcome the energy barrier and the more rapid will be the diffusion process. The lattice geometry also affects the diffusion coefficient through the preexponential constant D_0 in Equation 4.4–2. Thus the mechanism and crystal structure exert strong influences on the rate at which diffusive processes occur.

We previously discussed carburizing of steels. In this process the diffusing species is carbon. It is known that carbon forms an interstitial solid solution with Fe, as mentioned

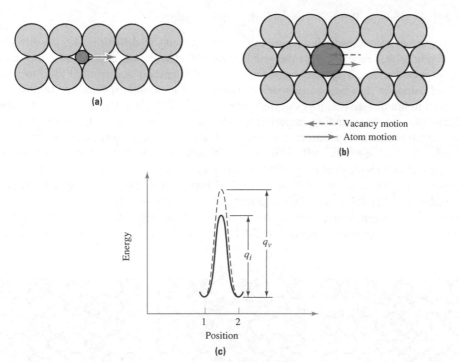

─ ─ ─ ─► Vacancy motion
────────► Atom motion

(b)

FIGURE 4.4–4 An illustration of the two main diffusion mechanisms in solids: **(a)** interstitial diffusion mechanism, and **(b)** the vacancy exchange mechanism of diffusion (note that the direction of atom motion is opposite to that of vacancy motion). **(c)** Schematic illustration of the activation energies associated with each type of diffusion.

in the section on impurities. Thus, carbon migrates by jumping from one interstitial position to another. As shown schematically in Figure 4.4–4a, the amount of energy required to "squeeze" the interstitial C atom between two Fe atoms so that it can move into an adjacent interstitial position corresponds to the activation energy q_i per atom (or Q_i per mole) for the interstitial diffusion process.

Diffusion in substitutional solid solutions (such as the Al and Cr diffusing into a Ni-base alloy mentioned previously) is also important, and this process is characteristically different from that of interstitial diffusion. In this case, an atom can move only if there is a vacant lattice site adjacent to it. As shown in Figure 4.4–4b, the diffusion mechanism involves an exchange between a vacancy and a diffusing atom. The probability of a jump in any given direction depends not only on the geometry of the crystal structure, but also on the probability that a vacancy lies adjacent to the atom being considered in the direction being considered. The probability of having a vacancy adjacent to an atom is proportional to the fraction of vacant lattice sites (i.e., C_v in Equation 4.2–2). Thus, the atomic mechanism involved in diffusion in substitutional solids is different from the case of C in Fe, in which interstitial sites are always adjacent to the diffusing atom.[3]

In general, the activation energies for vacancy-assisted diffusion Q_v are higher than those for interstitial diffusion Q_i. The reason is that the former mechanism requires energy to both form a vacancy and move an atom into the vacancy, while in the latter case energy is needed only to move the interstitial atom into the (always available) interstitial site.

---

[3] Strictly speaking, this is not precisely true for all compositions. For example, if the concentration of C atoms is high, then there is a significant probability that some C atoms will be adjacent to each other, thus occupying sites that would otherwise be available for interstitial diffusion.

4.4.4 Diffusion for Different Levels of Concentration

The movement of atoms within a pure material is termed **self-diffusion,** and the diffusion coefficient in this case is called the self-diffusion coefficient or, equivalently, the tracer diffusion coefficient. The latter term arises from the method that is often used to study self-diffusion. In this method radioactive isotopes of an atomic species are sandwiched between nonradioactive isotopes of the same species. With time, the radioactive isotopes diffuse into the surrounding material, as shown in Figure 4.4–5. By measuring the concentration profiles at different times, the tracer diffusion coefficient can be determined. Knowledge of the self-diffusion coefficients is important in the study of the mechanical and rheological properties of materials at high temperatures. This topic will be discussed in more detail in Chapter 9. Self-diffusion coefficients for selected materials are listed in Table 4.4–1 and shown graphically in Figure 4.4–6.

In contrast, when diffusion occurs by the movement of solute atoms in a dilute solution, the process is referred to as **impurity diffusion,** and the diffusion coefficient D is called

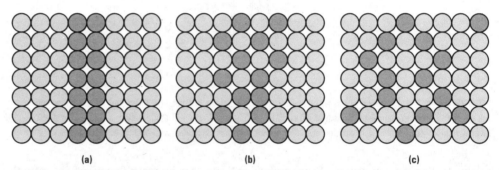

(a)   **(b)**   **(c)**

FIGURE 4.4–5   Diffusion of radioactive isotopes in a matrix of the same atomic species. Increasing time is indicated in **(a)**, **(b)**, and **(c)**. With time the radioactive species diffuse into the surrounding material. By measuring the concentration profiles at different times, the tracer diffusion constant can be determined.

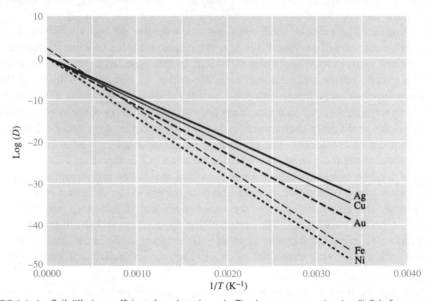

FIGURE 4.4–6   Self-diffusion coefficients for selected metals. The data were extrapolated to "infinite" temperature to demonstrate D_0. Data obtained from several sources.

TABLE 4.4–1 Diffusion coefficients for selected systems.

Material	D_0 (m²/s)	Q (kJ/mol)
Self-diffusion coefficients		
Ni	1.30×10^{-4}	279
Cu	2.00×10^{-5}	197
Ag	4.00×10^{-5}	184
Au	9.10×10^{-6}	174
Fe	1.18×10^{-2}	281
Si	1.80×10^{-1}	460
PE in melt	1.20×10^{-11}	28
Impurity diffusion coefficients		
Cu in Ag	1.20×10^{-4}	193
Cu in Al	1.50×10^{-5}	126
Zn in Ag	5.40×10^{-5}	174
Zn in Cu	3.40×10^{-5}	191
Ni in Cu	2.70×10^{-4}	236
Ni in Fe	7.70×10^{-5}	280
C in BCC Fe	2.00×10^{-6}	84
C in FCC Fe	2.00×10^{-5}	142
N in Fe	3.00×10^{-7}	76
Al in Al_2O_3	2.8×10^{-3}	477
O in Al_2O_3	1.9×10^{-1}	636
Mg in MgO	2.49×10^{-5}	330
O in MgO	4.3×10^{-9}	344
Ni in MgO	1.8×10^{-9}	202
O in SiO_2	2.7×10^{-8}	111
CO_2 in polyester (PET)	6.0×10^{-5}	51
CO_2 in PE	2.0×10^{-4}	38
CO_2 in PVC	4.2×10^{-2}	64
O_2 in PET	5.2×10^{-5}	47
O_2 in PE	6.2×10^{-4}	41
O_2 in PVC	4.1×10^{-3}	54

the impurity diffusion coefficient. Values of the impurity diffusion coefficient for selected systems are also given in Table 4.4–1.

When diffusion occurs in solid solutions having a significant amount of solute, the diffusion coefficient that is used in calculations involving both species is called the chemical diffusion coefficient. Extensive measurements of concentration profiles and thermodynamic quantities are necessary to determine the chemical diffusion coefficient. However, for dilute solid solutions, the chemical diffusion coefficient reduces to the impurity diffusion coefficient for the diffusing species.

EXAMPLE 4.4–4

Explain each of these observations:

a. The activation energy for the diffusion of H in FCC iron is less than that for self-diffusion in FCC iron.

b. The activation energy for the diffusion of H in BCC iron is less than that for the diffusion of H in FCC iron.

Solution

The key factors to consider are the diffusion mechanism (interstitial or vacancy exchange) and the crystal structure of the solvent.

    a. From Appendix C, we see that $r(H) \ll r(Fe)$, so we can correctly predict that H diffuses through FCC Fe via the interstitial mechanism. The self-diffusion of Fe involves the vacancy exchange mechanism. Since the activation energy for vacancy exchange is the sum of two terms—the energy required to form a vacancy and the energy required to promote atom/vacancy exchange—while the activation energy for interstitial diffusion is just the energy necessary to move an atom into a neighboring interstitial site, we expect Q_v to be greater than Q_i if all other factors are roughly equal.

    b. H diffuses through both FCC and BCC Fe via the interstitial mechanism. Why is the activation energy lower in the BCC structure? If we interpret the activation energy as the energy necessary to squeeze an interstitial atoms between solvent atoms as it moves to an adjacent interstitial position, and if we recall that the BCC structure is more open than the FCC structure (i.e., has a lower APF), we can predict that diffusion in less densely packed structures is generally easier than it is in close-packed structures when all other factors are roughly equal.

4.4.5 Mechanisms of Diffusion in Ionic Crystals

The diffusion mechanisms discussed for nonionic crystals are applicable to diffusion in ionic crystals with some modifications. One important difference is that at least two charged species (ions and charged vacancies) must be involved in diffusion in ionic solids in order to maintain electroneutrality. Another factor is that, as we noted for stabilized zirconia, the vacancy concentration may dramatically increase by the addition of impurities. The extent to which the additional vacancies influence diffusion depends on the degree to which they must remain closely associated with the impurity ions. If the vacancies are free to move away from their substitutional impurity atoms, then these vacancies can significantly increase the diffusion rates in the ionic crystal. Diffusion coefficients for selected ionic systems, in the absence of significant impurity concentrations, are listed in Table 4.4–1.

    In Chapter 2, we mentioned that ionic solids tend to be less efficient conductors of electricity than metals. One reason given was the limited mobility of the ions transporting the electric charge. Clearly this is a diffusion-related issue, and we should expect a correlation between the electrical conductivity of an ionic solid and the diffusion coefficients of the ions. In fact, much of the diffusion data for ionic crystals has been obtained from electrical conductivity measurements. Since the diffusion coefficients are a function of temperature and impurity concentration, we should not be surprised to find that the conductivity of ionic solids is also a strong function of these same variables. We will return to this issue when we focus our attention on the electrical properties of materials in Chapter 10.

EXAMPLE 4.4–5

When a sample of Si is exposed to air, a layer of SiO_2 grows on its surface. The exceptional properties of this oxide layer are one of the primary reasons for the success of Si in the semiconductor industry. The oxide growth mechanism is diffusion-controlled. Given that the growth rate of the oxide doubles as the temperature is raised from 1000°C to 1090°C, determine whether the oxide growth is controlled by the diffusion of O or the diffusion of Si through the SiO_2 layer. Assume that the oxide growth rate is an Arrhenius process.

Solution

We can use the data given in the problem to calculate an apparent activation energy for the oxide growth process and then compare this activation energy with the values listed in Table 4.4–1. If we let $r(T)$ represent the oxide growth rate at temperature T, we can obtain the expression:

$$\frac{r(1090°C)}{r(1000°C)} = 2 = \exp\left[-\frac{Q/R}{(1/1363 \text{ K}) - (1/1273 \text{ K})}\right]$$

Solving for Q and substituting $R = 8.314$ J/mol-K yields:

$$Q = (8.314 \text{ J/mol-K})(\ln 2)[(1/1273 \text{ K}) - (1/1363 \text{ K})]^{-1}$$

$$\approx 111{,}100 \text{ J/mol}$$

From Table 4.4–1 we find that the activation energy for the diffusion of O in SiO_2 is 112 kJ/mol. No corresponding value for the diffusion of Si through SiO_2 is given in the table. Nonetheless, the striking similarity of the Q values for the oxide growth rate and the diffusion of O in SiO_2 suggests that the growth of the SiO_2 layer is likely controlled by the diffusion of O through the oxide.

4.4.6 Mechanisms of Diffusion in Polymers

Diffusion in polymers can generally be described by Fick's first law. The mechanisms of diffusion, however, are significantly different in polymers from those in other materials. Since the individual atoms in polymer chains are not free to move independently, the concept of self-diffusion in a polymer must correspond to the motion of an entire thermoplastic polymer chain. Such motion is extremely difficult, if not impossible, in the crystalline regions of a polymer. Therefore, we will restrict our discussion to polymer chain motion in amorphous regions and in polymer melts. A simplified model for the process is to view the motion of the polymer chain as being constrained within a "bent pipe" (see Figure 4.4–7). The pipe is spatially constrained by physical entanglements with neighboring polymer molecules. Since the motion of the polymer through the pipe resembles the motion of a snake, the corresponding mathematical model is known as reptation theory. Rather than present the details of the theory, we will simply state without proof that the diffusion coefficient obeys the form given in Equation 4.4–2 with the preexponential constant D_0 varying inversely with the square of the length of the polymer chain.

Whereas impurity diffusion in metal, ionic, and covalent crystals takes place by the motion of individual atoms or ions through the lattice, whole molecules may diffuse

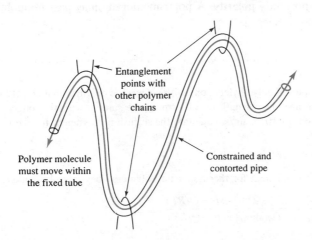

Polymer molecule
must move within
the fixed tube

Entanglement
points with
other polymer
chains

Constrained and
contorted pipe

FIGURE 4.4–7

Entanglements of large polymer chains limit self-diffusion in polymers. The molecule is constrained to move within the "pipe" that is illustrated schematically.

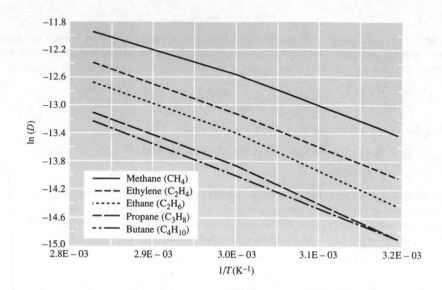

through the relatively open structure of the amorphous polymer regions. Thus it is possible for water molecules, for example, to be absorbed into a polymer by a diffusive process. The mechanism is similar to that for interstitial diffusion in the nonionic solids described above in that energy is required to move the diffusing species from one open-volume region within the polymer to another similar open-volume region. Thus, impurity diffusion coefficients for polymers also display an exponential variation of D with T as given in Equation 4.4–2. Like the motion of entire thermoplastic polymer chains, impurity diffusion is much faster in amorphous regions than in crystalline regions.

Because of their unique properties, polymers have been employed as very effective separating agents (filters). If two organic gaseous species need to be separated, and if the two species have significantly different sizes, then it is usually possible to find a polymer for which diffusion of the smaller species through a membrane is rapid and diffusion of the larger species does not occur at all. In this way virtually complete separation can be obtained. The effect of the size of the impurity molecule is illustrated in Figure 4.4–8, which shows the diffusivity of several hydrocarbons through natural rubber. The diffusion coefficients decrease as the size of the molecule increases (at constant T).

Polymer membranes separate not only on the basis of size, but also on the basis of other things, most importantly polarity. A polar membrane may pass nonpolar species but be a barrier to polar species.

EXAMPLE 4.4–6

Consider two samples of polyethylene, one with an average polymer chain length of 0.2 μm and the other with an average chain length of 3.0 μm. If the sample with the longer chains has a self-diffusion coefficient of 10^{-18} m²/s, calculate the diffusion coefficient for the other sample at the same temperature.

Solution

The self-diffusion coefficient for thermoplastic polymers obeys the Arrhenius equation, so that:

$$\frac{D(\text{short})}{D(\text{long})} = \frac{[D_0(\text{short})] \exp(-Q/RT)}{[D_0(\text{long})] \exp(-Q/RT)}$$

Assuming that the activation energy for self-diffusion is not a function of chain length, this ratio reduces to:

$$\frac{D(\text{short})}{D(\text{long})} = \frac{D_0(\text{short})}{D_0(\text{long})}$$

As mentioned in the text, D_0 varies inversely with the square of the chain length. Therefore,

$$\frac{D_0(\text{short})}{D_0(\text{long})} = \left(\frac{3.0\ \mu m}{0.2\ \mu m}\right)^2 = 225$$

Combining these results gives:

$$D(\text{short}) = 225 \times D(\text{long})$$
$$= 225 \times (10^{-18}\ m^2/s) = 2.25 \times 10^{-16}\ m^2/s$$

Thus far we have considered the fundamentals of diffusion. In particular, the effects of point defects on atom movements in solids have been studied, and Fick's first law has been developed for the rate of mass transport in a constant composition gradient. In the following sections some important engineering examples of diffusion are described in quantitative detail.

4.4.7 Fick's Second Law

Recall that Fick's first law assumes that the concentration gradient is independent of time. During many commercially important diffusion processes, however, the concentration of the diffusing species within the host solid is changing with time at any given location. Thus, we need to extend our mathematical model to account for the changes in concentration with time (i.e., we must include the term $\partial C/\partial t$).

The rate at which the concentration of an atomic species varies with time and position can be developed by use of Fick's first law. Consider the geometry shown in Figure 4.4–9. How does the number of "A" atoms within a thin slice of material of thickness dx and area dA change with time? The rate of change of the number of atoms in the slice is equal to the rate that atoms are entering the slice minus the rate that atoms are leaving the slice. Since concentration is just number divided by volume, and flux is rate divided by area,

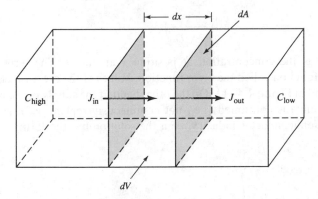

FIGURE 4.4–9   Diffusion through a solid bar. The derivation of Fick's second law involves the relationship between the time rate of change in the concentration of the diffusing species in the volume dV and the fluxes into and out of this differential volume.

we can write the equation[4]:

$$\left(\frac{\partial C}{\partial t}\right) dV = (J_{in} - J_{out})\, dA \tag{4.4-4}$$

Recognizing that $dV = dA \cdot dx$ and rearranging yields:

$$\frac{\partial C}{\partial t} = \frac{J_{in} - J_{out}}{\partial x} = -\frac{\partial J}{\partial x} \tag{4.4-5}$$

Substituting the expression for flux given in Fick's first law, $(J = -D(dC/dx))$, yields:

$$\frac{\partial C}{\partial t} = D\left(\frac{\partial^2 C}{\partial x^2}\right) \tag{4.4-6a}$$

where we have made the reasonable assumption that the diffusion coefficient is independent of position, which is equivalent to saying that the diffusion coefficient does not depend on composition. Equation 4.4–6a is known as **Fick's second law** of diffusion. This equation is quite general for diffusion taking place in one direction and can be generalized for diffusion in three dimensions.

Just as there is a heat transfer analogy for Fick's first law of diffusion, there is an exact heat transfer analogy for Fick's second law of diffusion. The change in temperature for non-steady-state conditions is governed by the equation:

$$\frac{\partial T}{\partial t} = D_{th}\left(\frac{\partial^2 T}{\partial x^2}\right) \tag{4.4-6b}$$

where D_{th} is the thermal diffusivity. This equation is known as the heat equation.

Solving Fick's second law yields solutions in which the concentration is a function of position and time, that is to say, $C = C(x, t)$. The form of the solution to Equation 4.4–6a depends on the geometry of the problem (more precisely, on the initial and boundary conditions for the differential equation). For diffusion into a thick plate (see Figure 4.4–10), the boundary conditions are that the concentration at the surface, C_s, and the initial bulk concentration, C_0, are constants. A constant concentration of C at the surface can be maintained through a balance between the rate at which C is deposited at the surface and the rate at which C diffuses into the base metal. Under the condition of a continuously replenished source, solution of Fick's second law yields:

$$\frac{C(x, t) - C_0}{C_s - C_0} = 1 - \text{erf}\left(\frac{x}{2\sqrt{Dt}}\right) \tag{4.4-7a}$$

where $C(x, t)$ is the concentration at position x at time t. The term erf refers to a mathematical function called the error function. Values of the error function are given in Table 4.4–2 and in Figure 4.4–11. On the other hand, if the supply of the diffusing species (β grams per cm^2) is consumed (i.e., not continuously replenished), as is the case for doping in the semiconductor industry, then the solution has the form:

$$C(x, t) = \left(\frac{\beta}{2\sqrt{\pi Dt}}\right) \exp\left(-\frac{x^2}{4Dt}\right) \tag{4.4-7b}$$

---

[4] Please note that in this discussion partial derivatives must be used since the concentration is a function of the variables, x and t.

From these examples we see that the concentration depends on both location and time, and that the exact form depends on the initial conditions and geometry.

As mentioned earlier in this chapter, carburizing of steel is frequently done to produce surfaces that are wear-resistant. This process is used extensively and is perhaps one of the most important examples of diffusion. The goal of gas carburizing is to produce a concentration profile of C that will yield the desired surface properties. Modeling this process requires the use of Fick's second law, as illustrated in the following example.

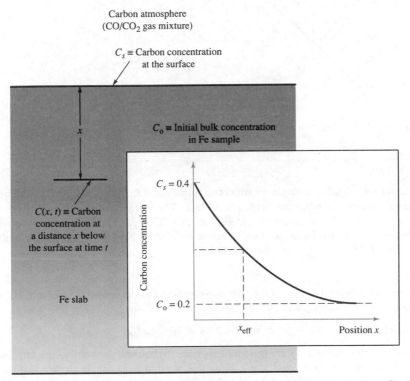

FIGURE 4.4–10   Diffusion of C into a large (i.e., semi-infinite) slab of Fe. Schematic concentration profile of C as a function of position at a particular instant in the carburizing cycle is also indicated.

TABLE 4.4–2   Selected values of the error function.

z	erf(z)	z	erf(z)
0	0	0.55	0.563
0.05	0.056	0.60	0.604
0.10	0.113	0.65	0.642
0.15	0.168	0.70	0.678
0.20	0.223	0.75	0.711
0.25	0.276	0.80	0.742
0.30	0.329	0.85	0.771
0.35	0.379	0.90	0.797
0.40	0.428	0.95	0.821
0.45	0.476	1.00	0.843
0.50	0.521	1.05	0.862

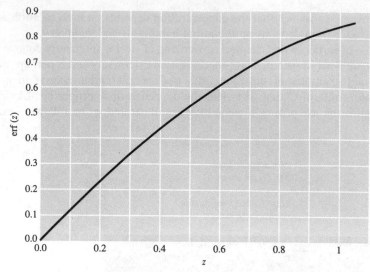

FIGURE 4.4–11   Graph of the error function.

EXAMPLE 4.4–7

Determine the time it takes to obtain a carbon concentration of 0.24% at a depth 0.01 cm beneath the surface of an iron bar when carburizing at 1273 K. The initial concentration of carbon in the iron bar is 0.20% and the surface concentration in a CO/CO$_2$ gas environment is maintained at 0.40%. In this example, the Fe has the FCC structure and from Table 4.4–1 the diffusion coefficient is $D = (2 \times 10^{-5} \text{ m}^2/\text{s})\{\exp[-(142,000 \text{ J/mol})/RT]\}$.

Solution

The first step is to obtain the value of the diffusion coefficient at the temperature of interest. Direct substitution into the given equation yields:

$$D(1273 \text{ K}) = 2.98 \times 10^{-11} \text{ m}^2/\text{s} = 2.98 \times 10^{-7} \text{ cm}^2/\text{s}$$

Equation 4.4–7a may be used to solve this problem, since the boundary conditions are satisfied. The left side of Equation 4.4–7a is equal to 0.2. From this, we find using simple algebra that the value of the error function is 0.8. Using Table 4.4–2 (or Figure 4.4–11), we find that when the error function is 0.8, then the argument is approximately 0.90, that is,

$$\frac{x}{2\sqrt{Dt}} = 0.9$$

In the above equation all terms are known except for t, the time. Solving for time gives:

$$t = \frac{\{x/[2(0.9)]\}^2}{D} = \frac{(0.01/1.8)^2}{2.98 \times 10^{-7}}$$

$$= 104 \text{ s} = 1.73 \text{ min}$$

Thus it will take just 1.73 minutes for the carbon concentration to reach 0.24% at a depth of 0.01 cm beneath the surface when carburizing at 1273 K.

EXAMPLE 4.4–8

The strength of many oxide glasses can be improved tremendously by exchanging larger ions for smaller ions in the surface of the glass, a process called ion stuffing. For example, immersion of a lithia-alumina-silicate glass body in molten potassium nitrate (KNO$_3$) at about 500°C for four hours allows the potassium ions to replace the smaller lithium ions, putting the surface in compression. Composition profiles show that the concentration of potassium at a depth of 20 μm increases from 0 to about 10 weight percent. During immersion in the salt the surface concentration of potassium is about 16 weight percent. Estimate the effective diffusion coefficient.

Solution

This is a Fick's second law problem with solution

$$\frac{C(x, t) - C_0}{C_s - C_0} = 1 - \text{erf}\left(\frac{x}{2\sqrt{Dt}}\right)$$

The initial concentration of potassium in the glass is 0 weight percent (wt. %) so that $C_0 = 0$. The surface concentration is maintained at $C_s = 16$ wt. %. After four hours ($t = 14,400$ s) the concentration at a depth of 20 μm ($x = 20 \times 10^{-6}$ m) is 10 wt. %, so that $C(x, t) = 10$ wt. %. Rearranging the above equation to solve for the erf yields:

$$\text{erf}\left(\frac{x}{2\sqrt{Dt}}\right) = 1 - \frac{C(x, t) - C_0}{C_s - C_0}$$

$$= 1 - \frac{10 - 0}{16 - 0} = 0.625$$

Linear interpolation between the values for $z = 0.60$ and $z = 0.65$ in Table 4.4–2 shows that for $\text{erf}(z) = 0.625$, $z = 0.6277$. Thus,

$$\frac{x}{2\sqrt{Dt}} = 0.6277$$

Solving this equation for D yields:

$$D = \frac{[(x/2)/0.6277]^2}{t}$$

$$= \frac{\{[(20 \times 10^{-6} \text{ m})/2]/0.6277\}^2}{14,400 \text{ s}}$$

$$= 1.76 \times 10^{-14} \text{ m}^2/\text{s} = 1.76 \times 10^{-10} \text{ cm}^2/\text{s}$$

In the preceding examples we have carried out rather precise calculations. However, engineers frequently find it necessary to make good estimates using less complicated approaches. One of the most useful approximations when dealing with diffusion involves the concept of an effective penetration distance. Let us define the **effective penetration distance** as the point at which the concentration of the diffusing species has a value equal to the average of the initial concentration and the surface concentration:

$$C(x_{\text{eff}}, t) = \frac{C_0 + C_s}{2} \tag{4.4–8}$$

Substituting this value into Equation 4.4–7 yields the result:

$$0.5 = \text{erf}\left(\frac{x_{\text{eff}}}{2\sqrt{Dt}}\right) \tag{4.4–9}$$

From Figure 4.4–11 we find that $\text{erf}(0.5) \approx 0.5$. Therefore, we have

$$x_{\text{eff}} \approx \sqrt{Dt} \tag{4.4–10}$$

The utility of this expression is that for a given effective penetration distance, it provides a simple method for determining diffusion times and temperatures (temperatures are determined through the diffusion coefficient). Equation 4.4–10 can be generalized to most diffusion problems with the introduction of a geometry-dependent constant:

$$x_{\text{eff}} = \gamma \sqrt{Dt} \tag{4.4–11}$$

The constant γ is 1 for a plate geometry and 2 for cylinders.

EXAMPLE 4.4–9

Suppose that an effective diffusion distance of 0.05 cm is required for carburizing a steel whose initial bulk concentration is 0.04% carbon. Suppose also that because of economic and time constraints, the time in the furnace cannot exceed one hour. Estimate the temperature at which the process must be carried out, assuming that the carburizing atmosphere is such that a surface concentration of 0.4% is maintained. Further assume that the parts being carburized are cylinders 5 cm in diameter.

Solution

We may use Equation 4.4–11 with a γ value of 2 (since this is the value corresponding to a cylindrical geometry) and $x_{\text{eff}} = 0.05$ cm. Solving the equation

$$x_{\text{eff}} = \gamma \sqrt{Dt}$$

for the diffusion coefficient yields:

$$
\begin{aligned}
D &= \left(\frac{x_{\text{eff}}}{\gamma}\right)^2 \left(\frac{1}{t}\right) \\
&= \left(\frac{0.05 \text{ cm}}{2}\right)^2 \left[\frac{1}{(1 \text{ h})(3600 \text{ s}/1 \text{ h})}\right] \\
&= 1.736 \times 10^{-7} \text{ cm}^2/\text{s} = 1.736 \times 10^{-11} \text{ m}^2/\text{s}
\end{aligned}
$$

From Table 4.4–1,

$$D = (2 \times 10^{-5} \text{ cm}^2/\text{s}) \exp\left(-\frac{142{,}000 \text{ J/mol}}{RT}\right)$$

Solving this expression for T and substituting the appropriate value for D yields:

$$
\begin{aligned}
T &= -\frac{142{,}000 \text{ J/mol}}{8.314 \text{ J/mol-K}} \bigg/ \ln\left(\frac{1.736 \times 10^{-11} \text{ m}^2/\text{s}}{2 \times 10^{-5} \text{ m}^2/\text{s}}\right) \\
&= 1224 \text{ K} = 951°\text{C}
\end{aligned}
$$

It is worth noting that diffusion problems can be solved by analogy to heat transfer. Solutions to the heat equation (Equation 4.4–6b) can be and are applied to diffusion situations with appropriate modifications. Thus whole classes of solutions and techniques are readily available for solving diffusion problems.

SUMMARY

In this chapter we have seen that point defects occur in all crystalline engineering materials. Although there are general rules that guide the formation of defects in all crystals, it is important to remember the additional requirement of electroneutrality in ionic crystals.

Point defects such as vacancies and interstitials are thermodynamically stable and usually have a significant influence on the mechanical, chemical, and electrical properties.

Their equilibrium concentrations can be determined using an analysis of the competing effects of changes in internal energy and configurational entropy with variations in the point-defect concentration.

We have seen that vacancies play an important role in diffusion of substitutional solid solutions, since atom movement occurs by atoms and vacancies exchanging positions. By considering random atomic jumps, it was demonstrated that diffusion will occur down a concentration gradient. Under steady-state conditions, the net flux of atoms is given by Fick's first law. In non-steady-state conditions, diffusion is governed by Fick's second law. The diffusion coefficient, knowledge of which is essential for solving problems involving mass transport, can be expressed in terms of a structure-related constant and an activation energy. Since diffusion is thermally activated, the diffusion coefficient increases exponentially with temperature. Diffusion in polymer melts is somewhat different. Polymer molecules diffuse or move most easily along their contour length. Movement in other directions is strongly hindered by a multiplicity of entanglements.

Impurities exist in all solids. Their locations within the crystal depend on several factors, including the size, crystal structure, and bond characteristics of both the solute and solvent atoms. The impurities can occupy interstitial positions if they are much smaller than the solvent atoms, or they can substitute for solvent atoms if they satisfy the Hume-Rothery rules. Impurities can interact with other defects so as to profoundly influence the engineering properties. Thermodynamic arguments show that it is impossible to obtain engineering quantities of perfectly pure materials.

The fundamental point of this chapter is that defects and impurities are always present in engineering materials and the properties of such materials can be understood largely in terms of the defect state. Through intelligent use of defects and impurities, materials can be engineered to achieve combinations of desirable properties.

KEY TERMS

diffusion	Fick's second law	interstitial	solute
diffusion coefficient	Frenkel defect	interstitial solid solution	solvent
effective penetration distance	Hume-Rothery rules	point defect	substitutional solid solution
	impurities	Schottky defect	vacancy
Fick's first law	impurity diffusion	self-diffusion	

HOMEWORK PROBLEMS

1. Provide an example of a material property that is dramatically affected by the presence of a small concentration of defects.

   SECTION 4.2
   Point Defects

2. In a certain crystalline material, the vacancy concentration at 35°C is twice that at 25°C. At what temperature would the vacancy concentration be one-half that at 25°C? [Note: An alternative statement of the problem is, Given $C_v(35°C) = 2C_v(25°C)$, find T such that $C_v(T) = 0.5C_v(25°C)$.]

3. In a certain crystalline material the vacancy concentration at 25°C is one-fourth that at 80°C. At what temperature would the vacancy concentration be 3 times that at 80°C?

4. When Ge is the solvent and N is the solute, an interstitial solid solution results. When Ge and Si are mixed in any proportion, a substitutional solid solution results. Using only the information given in this problem statement, determine what type of solid solution will result when N is the solvent and Si is the solute. Explain your answer.

   SECTION 4.3
   Impurities

5. Al-Zn alloys are used as corrosion-resistant coatings on steel sheets. In the solid state, Al and Zn form a metallic solid solution. Are the two elements completely soluble in this solid solution?

6. Using the Hume-Rothery rules for the formation of substitutional solid solutions, indicate whether the following systems would be expected to exhibit extensive solid solubility:

   a. Al in Ni   b. Ti in Ni   c. Zn in Fe
   d. Si in Al   e. Li in Al   f. Cu in Au
   g. Mn in Fe   h. Cr in Fe   i. Ni in Fe

7. Suppose one attempts to add a small amount of Ni to Cu in an effort to create a solid solution.

   a. Which element is the solvent and which is the solute?
   b. Do you think the resulting solid solution (S.S.) will be a substitutional S.S. or an interstitial S.S. Why?
   c. Do you think Ni and Cu will be completely soluble in each other? Why or why not?

8. Cu can be used to increase the strength of Al. How does Cu dissolve in Al?

9. What type of solid solution is likely to be formed when C is added to Fe?

10. In the text, we stated that the equilibrium concentration of C in Fe is 0.022% at 727°C. Calculate Q_{fv}.

11. Use a sketch to show both a Schottky defect and a Frenkel defect in the MgF_2 structure.

12. Explain why the following statement is incorrect: In ionic solids, the number of cation vacancies is equal to the number of anion vacancies.

13. What type of defects and how many of them are likely to be created when 2 moles of NiO are added to 98 moles of SiO_2? Assume the concentration of interstitials in the system is low enough to be neglected.

14. What type of defects and how many of them are likely to be created when 1 mole of MgO is added to 99 moles of Al_2O_3? Assume the concentration of interstitials in the system is low enough to be neglected.

15. An ionic solid contains cation vacancies, anion vacancies, and cation interstitials. The energy of formation of the cation vacancies is 20 kJ/mol. For anion vacancies it is 40 kJ/mol, and for the cation interstitials it is 30 kJ/mol. Compute the relative concentrations of all of the defect species. Assume that the cations and anions are monovalent.

16. Consider the ionic compound UO_2.

   a. Describe what a Schottky defect would look like in this compound.
   b. Would you expect to find more cation or anion Frenkel defects in this compound? Why?

17. Consider the ionic compound Li_2O.

   a. Describe what a Schottky defect would look like in this compound.
   b. Describe what a Frenkel defect would look like in this compound.

18. Consider the ionic compound Na_2O [with $r(Na^+) = 0.098$ nm and $r(O^{2-}) = 0.132$ nm]. Suppose a small sample of this material contains one Schottky defect and one Frenkel defect. Determine the most likely number of each of the following types of point defects:

   a. Interstitial Na^+ ions
   b. Interstitial O^{2-} ions
   c. Vacant Na^+ sites
   d. Vacant O^{2-} sites

19. Consider the possibility of solid solutions with Au acting as the solvent.

   a. Which element (N, Ag, or Cs) is most likely to form an interstitial solid solution with Au?
   b. Which element (N, Ag, or Cs) is most likely to form a substitutional solid solution with Au?

20. Under what condition can Fick's first law be used to solve diffusion problems?

SECTION 4.4
Solid-State Diffusion

21. Suppose that 1 weight percent of B is added to Fe.
    a. Would the B be present as an interstitial or a substitutional impurity?
    b. Compute the fraction of sites (either interstitial or substitutional) occupied by the B atoms.
    c. If the Fe containing the B were to be gas-carburized, would the process be faster or slower than for Fe that has no B? Explain.

22. Which type of diffusion do you think will be easier (have a lower activation energy)?
    a. C in HCP Ti
    b. N in BCC Ti
    c. Ti in BCC Ti
    Explain your choice.

23. At one instant in time there is 0.19 atomic % Cu at the surface of Al and 0.18 atomic % Cu at a depth of 1.2 mm below the surface. The diffusion coefficient of Cu in Al is 4×10^{-14} m^2/s at the temperature of interest. The lattice parameter of FCC Al is 4.049 Å. What is the flux of Cu atoms from the surface to the interior?

24. SiO_2 can be either a glass or a crystalline solid at room temperature. Is diffusion faster in vitreous (glassy) silica or crystalline silica? Why?

25. Most textile fibers are semicrystalline. That is, they contain both crystalline and non-crystalline regions. Do you expect dyes to penetrate faster in the crystalline or in the noncrystalline regions? (In fact, analysis of dyed fibers shows the dye has penetrated only one type of region.)

26. By what factor does the diffusion coefficient of Al in Al_2O_3 change when the temperature is increased from 1800 to 2000°C?

27. Consider the diffusion of C into Fe. At approximately what temperature would a specimen of Fe have to be carburized for 2 hours to produce the same diffusion result as at 900°C for 15 hours?

28. Helium gas is stored at 20°C under a pressure of 5×10^5 Pa in a glass cylinder 50 cm in diameter and 10 cm long. The glass walls are 3 mm thick. What is the initial rate of mass loss through the cylinder walls? (The diffusion coefficient of He through oxide glass at 20°C is about 4×10^{-4} m^2/s and the concentration of He on the inner surface is about 2.2×10^{-3} kmol/m^3.)

29. When food is wrapped in polyethylene, a nonpolar polymer, and placed in the freezer, the amount of ice within the package increases with time in the freezer. This observation is not true for Saran Wrap, which is made using a polymer that contains highly polar atoms. Explain the difference in behavior.

30. A Ti rod (HCP) is to be placed in a furnace in order to increase its carbon content. If the initial carbon content of the rod is 0.2 weight % and the carbon content in the furnace is the equivalent of 1.0 weight %, find the temperature required to yield a carbon content of 0.5 weight % at a depth of 0.4 mm below the surface of the rod in 48 hours.

31. Ge is covalently bonded while Cu possesses a metallic bond. Self-diffusion in each takes place by a vacancy interchange mechanism. For comparable temperatures, which would have the larger diffusion coefficient? Why?

32. Refer to Figure 4.4–5. Describe how the entropy and energy change in going from part a to b to c.

33. A candidate material for a turbine blade application oxidizes by diffusion of metal atoms through the oxide to the metal surface, where metal and oxygen react to form the oxide. After 10 hours at 550°C, an oxide layer 8 μm thick has formed. What will the thickness be after 100 hours?

34. A polyester (PET) carpet fiber 50 μm in diameter is immersed into a boiling dye bath at 100°C containing water, dye, and dye carrier (a fluid that is soluble in water). The diffusion coefficients of water, dye carrier, and dye are 1.0×10^{-12}, 1.0×10^{-13}, and 1.0×10^{-14} m^2/s, respectively.

a. Estimate the times required for the water, dye, and carrier to penetrate to the center of the fiber.

b. How would your answer change if the fiber diameter were doubled?

c. If the thermal diffusivity of PET is 8×10^{-8} watt/(m-K), how long will it take for the heat to penetrate to the center of a 50-μm-diameter fiber?

35. A 1/4-inch-thick plate of steel is case-carburized to 1/16 of an inch in 4 hours. How long will it take to make 1/4-inch-diameter steel chains of the same composition case-carburized to 1/16 of an inch under the same carburization conditions?

36. When cotton or wool fiber is moved into an environment that contains more or less humidity than where it has been, the moisture content of the fiber changes. It increases if the relative humidity is higher and decreases if the relative humidity is lower. A reasonable value for the diffusion coefficient of moisture through cotton or wool is 10^{-11} m²/s. How long will it take single cotton and wool fibers to come to equilibrium, given that the diameter of a wool fiber is about 25 μm and cotton has a ribbon shape with the small dimension about 2 μm? Repeat the calculation for a tightly packed cubic bale of fiber that is 1 m on a side.

37. Carbonated soft drinks come in oxide glass, plastic, and metal containers. It has been noted that soft drinks stored in PET bottles lose their fizz more rapidly than those stored in oxide glass or metal containers. Comment on the reasons for the high rate of fizz reduction, given that the diffusion coefficient of CO_2 through PET is very low, about 7×10^{-14} m²/s, and the concentration of CO_2 on the inner surface is low. (The diffusion of CO_2 through oxide glass is not less than 10^{-14} m²/s.)

38. Compare the diffusion coefficient of methane in rubber at 293 K with the diffusion of Cu in Ag at the same temperature. Is this result expected? To what do you attribute the large difference in diffusion rates?

39. Make a schematic plot of the diffusion rate in natural rubber as a function of temperature, from −50 to 100°C. Note that rubber freezes to a glass at about −5°C.

40. Polytetrafluoroethylene has been suggested as a membrane to separate water vapor from benzene. Will this work? If there are any problems with this approach, make a suggestion for an improved membrane material.

41. As mentioned in the text, one of the practical applications of diffusion is case hardening of steel. If you wish to increase the amount of carbon at a given depth below the surface, should you:

a. Increase or decrease the temperature (assume all other variables constant)?

b. Increase or decrease the carbon content in the furnace (assume all other variables constant)?

c. Increase or decrease the time the steel part remains in the carburization furnace (assume all other variables constant)?

42. Use the concept of an effective diffusion distance to estimate the following quantities:

a. How long does it take for a Cu atom to diffuse a distance of 1 nm through a Cu crystal lattice at room temperature? (Note that 1 nm is roughly the size of a single unit cell in Cu.)

b. How far can a Cu atom diffuse through a Cu crystal lattice in one hour at a temperature of 1000°C?

c. Do you think we need to be concerned about diffusion at room temperature in solids? What about diffusion at elevated temperatures?

43. In contrast to carburization, decarburization occurs when carbon diffuses from the interior of the steel to the surface and then enters the furnace atmosphere. How long will it take for a steel with an initial carbon content of 1.0% to obtain a carbon content of 0.832% at a distance of 0.2 cm below the surface if the furnace atmosphere is free of carbon and maintained at 1200°C?

44. Suppose it takes three hours for an Al atom to diffuse a distance x through an Al_2O_3 crystal at $T = 1000°C$.

    a. Estimate the time required for an Al atom to diffuse a distance of $9x$ through the same Al_2O_3 crystal.

    b. Calculate the temperature at which the diffusion rate of Al through Al_2O_3 will be 10 times that at 1000°C [i.e., find T such that $D(T) = 10D(1000°C)$].

45. Estimate the temperature required for a Zn atom to diffuse 50 μm into a Cu crystal. [Given: $D_0 = 34 \times 10^{-6}$ m^2/s, $Q = 191$ kJ/mol, and $R = 8.314$ J/(mol-K).]

46. A diffusion process is known to occur in 3 hours at 600°C.

    a. If the activation energy for this process is 80 kJ/mol, at what temperature must the diffusion process occur if it is to be completed in 1.5 hours (with the same result)?

    b. In an industrial setting, what additional information would you require in order to select the optimal temperature for this diffusion process?

C H A P T E R 5

LINEAR, PLANAR, AND VOLUME DEFECTS

MATERIALS IN ACTION   Line and Planar Defects

We have already seen that real crystals contain extremely small amounts of point defects. We have also seen that these defects have a very significant influence on the properties of crystalline materials. In addition to point defects, it was hypothesized in the 1930s and proven experimentally in the 1950s that crystalline materials also contain line defects. In a typical metallic material approximately only 5 atoms out of every 100 million lie along line defects. This extremely low concentration nonetheless has a major effect on the strength of such crystals. It has been experimentally shown that the existence of such defects can reduce the strength of metals by a factor of 10,000. While such a reduction in strength might at first seem to be very detrimental, it is these defects that are responsible for the good ductility of metals. Furthermore, by impeding the motion of these defects through changes in composition, heat treatment, and thermo-mechanical processing, the strength of metals and alloys can be increased by orders of magnitude.

Just as there are point and line defects, there are also planar defects. Most crystalline materials exist as aggregates of very small crystals that are "glued" together to form a solid. The regions that separate the small crystals have their atoms out of registry with the crystals on either side and hence have a higher energy than the surrounding regions. In a typical crystal whose volume is 1 cm^3, the internal area of such defects is 3000 cm^2 or 500 times more area than is on the surface of such a crystal. These internal boundaries strengthen metals, act as sites of corrosive attack, and are preferential regions for the formation of new phases.

5.1 INTRODUCTION

In the previous chapter we learned that point defects such as impurities, vacancies, and interstitials have a significant effect on engineering properties and diffusion in crystalline materials. Similarly, linear defects, planar defects, and volume defects also have a profound influence on mechanical and other properties of engineering materials. These defects and their characteristics are considered in this chapter.

Linear defects, known as dislocations, allow large amounts of plastic deformation in metals and make it possible for us to form them into complex shapes of engineering components. The existence of mobile dislocations can be used to explain the deviation between the theoretically predicted strength of ideal crystals and the strength actually observed. The concept of dislocations is used to explain the difference in the mechanical behavior of crystals resulting from differences in bond types and crystal structures. The chapter concludes with an introduction to various types of planar and volume defects, emphasizing their relationships to the physical properties of crystals.

The ideas discussed in this chapter apply equally to covalent, ionic, metallic, molecular, and "mixed" crystals. However, as briefly discussed in Chapter 2, engineering polymers always contain significant amounts of amorphous regions, or noncrystalline material. Since properties of these materials are largely determined by the highly imperfect crystalline regions acting in conjunction with the noncrystalline material, the existence of these sorts of defects within the crystalline regions of a polymer is often of limited engineering consequence. For this reason, most of the examples in this chapter come from the metal and ceramic (ionic and covalent) families of materials.

5.2 LINEAR DEFECTS, SLIP, AND PLASTIC DEFORMATION

An important category of defects is the line defect, defined as a one-dimensional (linear) region in the lattice characterized by local faults in the atomic arrangement. One important consequence of the inevitable presence of such defects, usually called **dislocations,** is that they enable atoms to slip and slide past one another under applied forces that are much lower than would otherwise be predicted. In contrast to the format used for point defects—describe the defect, then describe an application (diffusion)—in this section we will invert the procedure and begin by discussing the application (plastic deformation) before describing the structure of a dislocation. To accomplish this task we must first introduce some terminology.

5.2.1 The Shear Strength of Deformable Single Crystals

The permanent displacement of atoms within a crystal resulting from an applied load is known as **plastic deformation.** Plastic deformation occurs as a result of the application of a force of sufficient magnitude to displace (move) atoms from one equilibrium position to another. Thus, one can speak of a critical force per bond necessary to cause plastic deformation. In engineering, however, it is more convenient to discuss force per unit area, or **stress,** rather than force per bond.

While a force can be described by giving its magnitude and direction, the description of a stress requires four quantities: the magnitude and direction of the force, the area of the plane over which the force acts, and the direction of the plane normal. The two basic types of stresses are illustrated in Figure 5.2–1. **Normal stresses** result from a force in the direction parallel to the plane normal, N, while **shear stresses** result from a force in

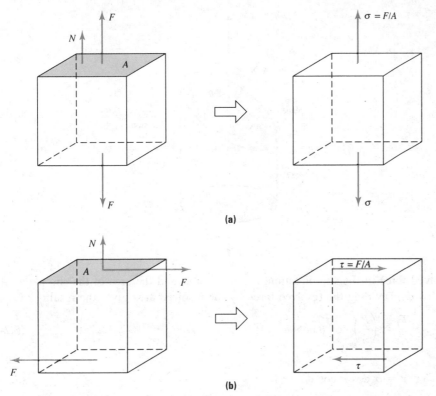

FIGURE 5.2–1 Illustration of **(a)** a normal stress, given the symbol σ and defined as F/A when F is parallel to the plane normal N; and **(b)** a shear stress, given the symbol τ and defined as F/A when F is perpendicular to N.

the direction perpendicular to N. We will use the symbols σ and τ to represent normal and shear stresses, respectively.

If single crystals are subjected to external forces and the plastic deformation of the crystals is carefully monitored, several observations emerge: (1) plastic deformation occurs as a result of shear stresses; (2) plastic deformation is anisotropic (i.e., not equal in all directions)—it occurs on high-density planes and in close-packed directions; and (3) for each crystalline material there is a critical value of shear stress on the favored plane in the favored direction at which plastic deformation will begin. We will investigate each of these observations in the following discussion.

The details of plastic deformation depend strongly on crystal type through the influence of the highest-density planes and close-packed directions. Before investigating this dependence on crystal structure, however, it is useful to discuss plastic deformation in more general terms. We therefore introduce the following definitions. The plane on which deformation occurs is the **slip plane,** and the direction within the slip plane along which deformation occurs is the **slip direction.** Combinations of slip planes and slip directions are called **slip systems.** Using these terms, we can simplify the discussion while retaining the ability to include the influence of crystal type at the appropriate time.

Let us return to the first experimental observation: plastic deformation occurs because of shear stresses. Does this mean that a normal stress applied on a single crystal does not produce any slip? The answer is no. For example, consider the geometry shown in Figure 5.2–2. In this figure, θ is the angle between the applied force F and the slip direction, and ϕ is the angle between the force and the normal to the slip plane. The

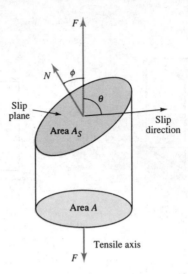

resolved force in the slip direction, F_s, is $F \cos \theta$, and the area of the slip plane, A_s, is $A/(\cos \phi)$. Dividing the resolved force by the resolved area gives the result:

$$\frac{F_s}{A_s} = \left(\frac{F}{A}\right) \cos \theta \cos \phi \tag{5.2–1a}$$

or

$$\tau = \sigma \cos \theta \cos \phi \tag{5.2–1b}$$

where τ is the shear stress in the slip direction on the slip plane and σ is the applied normal stress. Thus, a normal stress applied to a single crystal causes a shear stress on the slip plane along the slip direction. This component of stress, called the resolved shear stress, can cause slip if it is large enough.

The value of the shear stress at which plastic deformation occurs is known as the **critical resolved shear stress**, τ_{CR}, which is a constant for a given slip system in a given crystal. Solving Equation 5.2–1b for σ and inserting the value of τ_{CR} yields an expression for the magnitude of the critical normal stress, σ_c, necessary to cause plastic deformation:

$$\sigma_c = \frac{\tau_{CR}}{\cos \theta \cos \phi} \tag{5.2–2}$$

Equation 5.2–2 is known as **Schmid's law.** Note that the stress necessary to initiate slip, σ_c, is a function of the crystal orientation with respect to the direction of the applied force on the crystal axis. Its magnitude is not a material constant.

....

EXAMPLE 5.2–1

An FCC crystal yields under a normal stress of 2 MPa applied in the $[\bar{1}\ 2\ 3]$ direction. The slip plane is $(1\ 1\ 1)$ and the slip direction is $[\bar{1}\ 0\ 1]$. Determine τ_{CR} for this crystal.

Solution

To solve this problem we must find both $\cos \theta$ and $\cos \phi$ in Equation 5.2–2. This can be done using the vector dot product as described in Section 3.4–2:

$$\cos \phi = \frac{[\bar{1}\ 2\ 3] \cdot [1\ 1\ 1]}{|[\bar{1}\ 2\ 3]||[1\ 1\ 1]|}$$

$$= \frac{-1 + 2 + 3}{\sqrt{14}\sqrt{3}} = 0.617$$

$$\cos \theta = \frac{[\bar{1}\ 2\ 3] \cdot [\bar{1}\ 0\ 1]}{|[\bar{1}\ 2\ 3]||[\bar{1}\ 0\ 1]|}$$

$$= \frac{1 + 0 + 3}{\sqrt{14}\sqrt{2}} \doteq 0.756$$

Solving Equation 5.2–2 for τ_{CR} and substituting the data given in the problem statement yields:

$$\tau_{CR} = (2\ \text{MPa}) \times (0.617) \times (0.756) = 0.933\ \text{MPa}$$

The original theoretical model used to predict the value of τ_{CR} assumed that one atomic plane slides over another as a unit, as shown in Figure 5.2–3. Such a process would require sufficient force to simultaneously break all of the atomic bonds across the slip plane. This model leads to predicted values of τ_{CR} on the order of $E/10$, where E is the elastic modulus introduced in Chapter 2. In metals, the experimentally determined values of τ_{CR} are in many cases orders of magnitude lower than the theoretically calculated strengths of defect-free crystals. Typical values of the critical resolved shear stress and corresponding theoretical shear strengths for a number of metals are given in Table 5.2–1.

Since the calculations based on the assumption of a defect-free crystal failed to yield results consistent with experiments, new models specifically involving crystalline defects were investigated. To explain the much lower strengths of crystalline materials, a model in which slip can occur without simultaneously breaking all of the atomic bonds across a slip plane was considered.

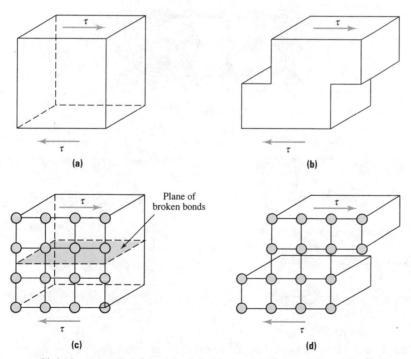

FIGURE 5.2–3 Model for computation of the theoretical critical resolved shear strength. Under the application of a shear stress τ, the top plane of atoms is assumed to slide over the bottom plane as a unit to produce permanent offset. Parts **(a)** and **(b)** represent a macroscopic view, while parts **(c)** and **(d)** represent an atomic scale view, of the process. As discussed in the text, this model requires that an entire plane of bonds be broken simultaneously.

TABLE 5.2–1 Experimental and theoretical critical resolved. shear stresses for metal crystals.

Metal	Critical resolved shear stress (MPa)	
	Experimental	Theoretical ($E/10$)
Cu	0.49	10.3×10^3
Ag	0.37	9.2×10^3
Al	0.78	7.0×10^3
Fe	27.44	21.0×10^3
Ti	13.72	11.0×10^3

5.2.2 Slip in Crystalline Materials and Edge Dislocations

In 1934 Sir Geoffrey Taylor, a British physicist and meteorologist, hypothesized that crystals could contain defects arising from the insertion of part of an atomic plane, as shown in Figure 5.2–4a. He called such a defect an **edge dislocation.** For simplicity, a simple cubic crystal has been used to illustrate the concept; however, the same principles apply to all crystal structures. Note that the crystal shown in Figure 5.2–4a is free of

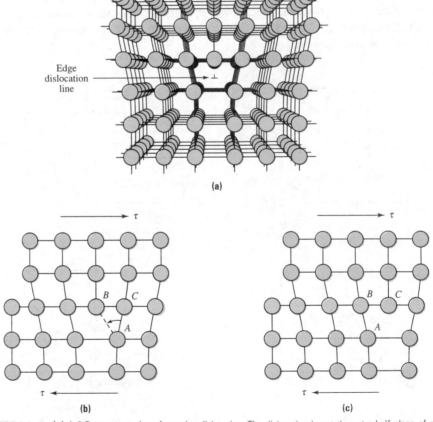

FIGURE 5.2–4 **(a)** A 3-D representation of an edge dislocation. The dislocation is not the extra half plane of atoms that has been inserted but rather the *line* that runs along the bottom of the extra half plane. Parts **(b)** and **(c)** illustrate the motion of an edge dislocation in response to the application of a shear stress τ. The details of this motion are described in the text.

defects in all regions except along the termination of the partial plane, where there is a row of atoms with less than the ideal number of nearest neighbors. As an idealization, the row of atoms along which atomic packing is imperfect defines a linear defect.

Although the idea of the edge of an extra half plane of atoms as a model for an edge dislocation is a useful concept, it is incorrect to think of the insertion of a half plane of atoms as a mechanism of dislocation formation. Instead, dislocations are introduced into a crystal in several ways including: (1) "accidents" in the growth process during solidification of the crystals, (2) internal stresses associated with other defects in the crystal, and (3) interactions between existing dislocations that occur during plastic deformation.

Next we consider the mechanisms of dislocation motion. If a shear stress τ is applied to the crystal as shown in Figure 5.2–4b, there is a driving force for breaking the bonds between the rows of atoms marked A and C and the formation of bonds between the atoms in rows A and B (see Figure 5.2–4c). Before the application of the shear stress, atoms in rows A and C have the correct coordination number while atoms in row B are unsatisfied (and therefore correspond to the dislocation line). After application of the stress, atoms in rows A and B have the correct CN but atoms in row C are unsatisfied. In effect, the dislocation has moved to the right by one atomic spacing.

The dislocation motion described in Figure 5.2–4b and c is known as **dislocation glide.** The process of breaking and reestablishing one row of atomic bonds may continue until the dislocation passes completely out of the crystal. As shown in Figure 5.2–5, when the dislocation leaves the crystal, the top half of the crystal is permanently offset by one atomic unit relative to the bottom half.

Since the permanent deformation (i.e., irreversible motion of atoms from one equilibrium position to another) via dislocation glide was produced by breaking only one row of atomic bonds at any one time, the corresponding theoretical τ_{CR} should be much lower than for the previous model, where all of the bonds were broken simultaneously. In fact, detailed calculations have shown that theoretical predictions of τ_{CR} for the dislocation glide model are in good agreement with the experimental values. In addition, this model is consistent with the two other experimental observations (i.e., the sensitivity to shear stress, and the crystallographic nature of slip).

Examination of Figure 5.2–5 suggests that two important geometric quantities are associated with a dislocation: the crystallographic direction in which it lies and the corresponding displacement vector (i.e., the magnitude and direction of the atomic displacement that results from the motion of the dislocation). Each of these characteristic can be described quantitatively.

A measure of the atomic displacement associated with the motion of a dislocation (see Figure 5.2–5c) can be obtained by making a circuit around the dislocation through

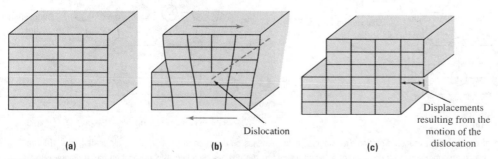

Displacements
resulting from the
motion of the
dislocation

Dislocation

(a)   **(b)**   **(c)**

FIGURE 5.2–5   Passage of a dislocation through a crystal: **(a)** the dislocation is just about to enter at the left, **(b)** the dislocation has proceeded halfway across the crystal, and **(c)** the dislocation has exited on the right-hand side and the top half has shifted (been displaced) by one atomic unit.

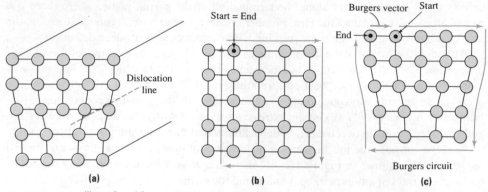

FIGURE 5.2-6   Illustration of Burgers circuit and Burgers vector for an edge dislocation. **(a)** A 3-D view of an edge dislocation. **(b)** A Burgers circuit closes upon itself when it surrounds a dislocation-free region of a crystal. **(c)** When the Burgers circuit surrounds a dislocation, the start and stop points are not coincident and the vector pointing from the stop point to the start point is defined to be the Burgers vector for the dislocation.

defect-free material. As shown in Figure 5.2–6, the circuit is made in a clockwise sense around the dislocation. It has an equal number of atomic steps on parallel sides such that the start and end of the circuit would be coincident if it did not surround a dislocation (see Figure 5.2–6b). When a similar circuit is made around a region of material that contains a dislocation, however, the circuit does not close upon itself (see Figure 5.2–6c). A vector is defined by joining the endpoint to the starting point. Such a circuit is called a **Burgers circuit,** and the resulting vector is called the **Burgers vector, b.**

Examination of Figure 5.2–7a reveals a problem associated with our definition of a "clockwise" Burgers circuit. One gets different Burgers vectors for the same dislocation depending on the direction in which the dislocation is viewed. This problem is solved by defining the **unit tangent vector t** that is locally tangent to the direction of a dislocation at the point of interest (see Figure 5.2–7b). The Burgers circuit is then taken in a clockwise direction while looking in the direction of the unit tangent vector. While the initial choice of the positive direction for the unit tangent vector is arbitrary, the key is that the chosen

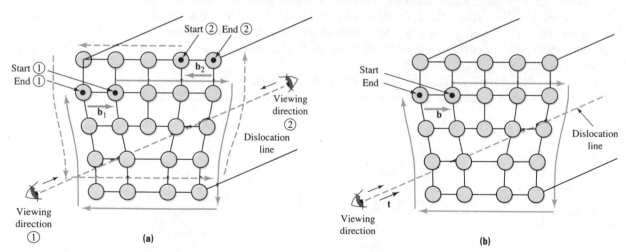

FIGURE 5.2-7   The concept of a positive dislocation direction. **(a)** Burgers vector depends on direction of viewing along dislocation line. **(b)** The choice of a positive direction and a corresponding unit tangent vector **t**. When viewed in the positive direction, this dislocation has a unique Burgers vector.

direction must be used consistently throughout all related calculations associated with any single dislocation.

Note that for an edge dislocation the Burgers vector is perpendicular to the dislocation line, as represented by the unit tangent vector (i.e., $\mathbf{b} \perp \mathbf{t}$). The slip plane for the edge dislocation contains both $\mathbf{b}$ and $\mathbf{t}$. In cubic crystals, the Miller indices of the slip plane can be obtained by the cross product of $\mathbf{b}$ and $\mathbf{t}$ (i.e., $\mathbf{b} \times \mathbf{t}$). When determining the critical resolved shear stress of a single crystal, the Burgers vector (because it specifies the slip direction) and the normal to the active slip plane must be used to determine the angular quantities in Equation 5.2–2, as illustrated in the following example.

EXAMPLE 5.2–2

Consider a dislocation in an FCC crystal with the following characteristics: $\tau_{CR} = 0.5$ MPa, $\mathbf{t} = (1/\sqrt{6})[\bar{1}\ \bar{1}\ 2]$, and $\mathbf{b} = (a_0/2)[\bar{1}\ 1\ 0]$.

a. Determine the slip plane for this dislocation.

b. Calculate the magnitude of the applied normal stress in the [0 1 0] direction necessary to cause the motion of this dislocation.

c. Repeat part b for a normal stress applied in the [0 0 1] direction.

Solution

a. The slip plane is determined by taking the cross product of $\mathbf{b}$ and $\mathbf{t}$. The resultant vector, $\mathbf{n}$, is normal to the plane defined by $\mathbf{b}$ and $\mathbf{t}$ (i.e., the slip plane). Thus:

$$\mathbf{n} = \mathbf{b} \times \mathbf{t} = \left(\frac{a_0}{2}\right)\left(\frac{1}{\sqrt{6}}\right)\begin{vmatrix} \mathbf{i} & \mathbf{j} & \mathbf{k} \\ -1 & 1 & 0 \\ -1 & -1 & 2 \end{vmatrix} = \left(\frac{a_0}{\sqrt{6}}\right)[\mathbf{i} + \mathbf{j} + \mathbf{k}]$$

where $\mathbf{i}$, $\mathbf{j}$, and $\mathbf{k}$, are unit vectors in the x, y, and z directions. Using Miller indices, $\mathbf{n} = (a_0/\sqrt{6})[1\ 1\ 1]$. Clearing the constant factor (recall that the magnitude of a crystallographic direction has no physical significance) and noting that in cubics a plane has the same indices as its normal, we see that the slip plane is (1 1 1).

b. Equation 5.2–2 can be used to calculate the critical normal stress required to cause plastic deformation (dislocation motion):

$$\sigma_c = \frac{\tau_{CR}}{\cos\theta\cos\phi}$$

In this case θ is the angle between the applied force direction [0 1 0] and slip direction [$\bar{1}$ 1 0]. Using the dot product, we find:

$$\cos\theta = \frac{[0 \times (-1)] + (1 \times 1) + (0 \times 0)}{(1)(\sqrt{2})} = \frac{1}{\sqrt{2}}$$

Similarly, ϕ is the angle between the normals to the two planes involved (i.e., [1 1 1] and [0 1 0]), so that:

$$\cos\phi = \frac{(1 \times 0) + (1 \times 1) + (1 \times 0)}{(\sqrt{3})(1)} = \frac{1}{\sqrt{3}}$$

Substituting these values into the expression for σ_c yields:

$$\sigma_c = \frac{0.5 \text{ MPa}}{(1/\sqrt{2})(1/\sqrt{3})} = 1.22 \text{ MPa}$$

c. This time θ is the angle between $[0\ 0\ 1]$ and $[\bar{1}\ 1\ 0]$ and ϕ is the angle between $[1\ 1\ 1]$ and $[0\ 0\ 1]$, so that:

$$\cos\theta = \frac{[0\times(-1) + (0\times1) + (1\times0)]}{(1)(\sqrt{2})} = 0$$

and

$$\cos\phi = \frac{(1\times0) + (1\times1) + (1\times0)}{(\sqrt{3})(1)} = \frac{1}{\sqrt{3}}$$

Substituting these values into the expression for σ_c yields:

$$\sigma_c = \frac{0.5\ \text{MPa}}{(0)(1/\sqrt{3})} = \infty$$

How do we interpret this result? Since $\cos\theta = 0$, $\theta = 90°$. This means that the applied stress direction $[0\ 0\ 1]$ is perpendicular to the slip direction $[\bar{1}\ 1\ 0]$ so that the applied force has no shear component lying in the slip direction. Therefore, stress applied along $[0\ 0\ 1]$ cannot cause dislocation motion in the $[\bar{1}\ 1\ 0](1\ 1\ 1)$ slip system.

....

While edge dislocation theory can be used to explain many of the important features of plastic deformation in crystals, it is necessary to invoke the presence of other types of dislocations to explain other aspects of deformation. The generalized dislocation theory can explain essentially every known feature of plastic deformation in crystals.

5.2.3 Other Types of Dislocations

Figure 5.2–8 shows an edge dislocation along with several other types of dislocations, including a screw dislocation, a mixed dislocation, and a dislocation loop. In the case of the **screw dislocation** (Figure 5.2–8b) the Burgers vector is parallel to the dislocation line (**b** ∥ **t**). A screw dislocation can also be envisioned as forming the axis of a helical ramp that runs through the crystal.[1] Recall that for any dislocation the plane that contains both **b** and **t** is a potential slip plane. Since **b** is parallel to **t** for a screw dislocation, any plane that contains the line defined by **b** (or, equivalently, **t**) is a potential slip plane. Therefore, the screw dislocation can glide (move) from one valid slip plane onto another valid slip plane. This is in contrast to the edge dislocation, which has a unique slip plane and which can glide on only that plane. For this reason, screw dislocations are generally more mobile than edge dislocations.

Figure 5.2–8c shows a curved dislocation. At position A the dislocation has an edge character (**b** ⊥ **t**), and at position B it has a screw character (**b** ∥ **t**). At intermediate points between A and B, the dislocation has a mixed character (i.e., neither edge nor screw). It is therefore known as a mixed dislocation. Note that Burgers circuits at points A and B each yield the same Burgers vector. In fact, the Burgers vector is an invariant for a dislocation, meaning that the Burgers vector is constant for any given dislocation.

Figure 5.2–8d shows a dislocation loop within a crystal. The loop has the same Burgers vector at all positions, but its character (i.e., edge, screw, or mixed) continuously changes. It is also possible to form dislocation loops that have the same orientation at all points.

---

[1] A reasonable mental image for a screw dislocation is the type of helical ramp that is often found in parking garages. Movement (displacement) along the ramp (dislocation) allows a car to move from one parking level (plane of atoms) to the next level.

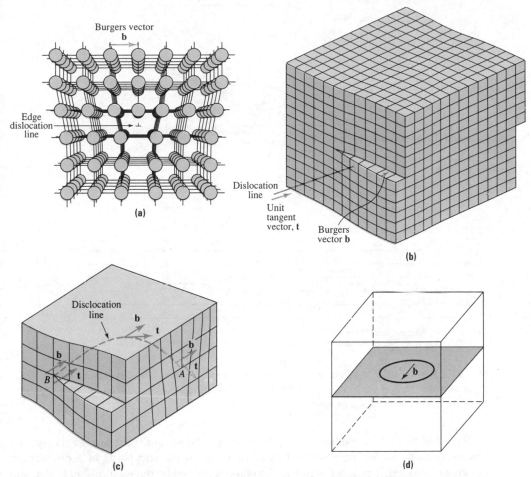

FIGURE 5.2–8 Illustrations of four types of dislocations: **(a)** an edge dislocation, **(b)** a screw dislocation, **(c)** a mixed dislocation, and **(d)** a dislocation loop. *(Source:* **(b)** *William D. Callister, Jr.,* Materials Science and Engineering, *2nd ed., Copyright © 1991 by John Wiley & Sons. Used with permission of John Wiley & Sons, Inc.* **(c)** *James F. Shackelford,* Introduction to Materials Science for Engineering, *3rd ed. Copyright © Macmillan Publishing Company, Inc. Used with permission of Macmillan College Publishing Company.)*

In the example shown in Figure 5.2–9, a portion of a (1 1 1) plane is removed from an FCC crystal. This might happen, for example, by the formation of a disk-shaped cluster of vacancies. Here the Burgers vector is perpendicular to the plane of the loop so that the dislocation has pure edge characteristics over its entire length.

The preceding examples illustrate several key features of dislocations:

1. The character of a dislocation is defined by the relationship between its Burgers vector and unit tangent vector. For edges $\mathbf{b} \perp \mathbf{t}$, for screws $\mathbf{b} \parallel \mathbf{t}$, and for mixed dislocations, $\mathbf{b}$ and $\mathbf{t}$ may form angle between 0° and 90°.

2. The plane on which a dislocation may slip contains both $\mathbf{b}$ and $\mathbf{t}$. The dislocation can glide (move) on this plane provided it is an admissible slip plane for the crystal type (i.e., a plane of high atomic density).

3. The Burgers vector is invariant. Thus, while the character of a dislocation may change from position to position, the Burgers vector is always the same.

4. A dislocation cannot end in the middle of a defect-free region of a crystal. It can end at the surface of the crystal, on itself, or on another dislocation.

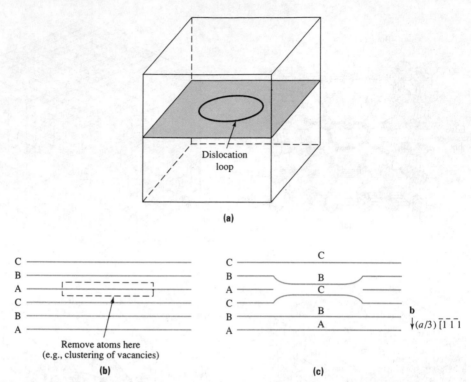

FIGURE 5.2–9 Formation of a dislocation loop by removal of a circular portion of a plane of atoms: **(a)** a 3-D view of the dislocation loop, **(b)** a 2-D view showing the section of the plane to be removed, and **(c)** the atomic displacement that occurs when the atoms are removed.

We need to further explain the significance of the second sentence in the second observation above. We have stated two restrictions on the slip plane of a dislocation: (1) from a geometric point of view the slip plane is defined by the relationship $\mathbf{b} \times \mathbf{t}$; and (2) from a crystallographic point of view the slip plane must be the highest–atomic density plane. A dislocation is free to move by the glide mechanism (i.e., the mechanism shown in Figure 5.2–4) only if its geometric slip plane coincides with an allowable crystallographic slip plane. The valid slip planes and directions for the three common metal crystal structures are discussed in the next subsection.

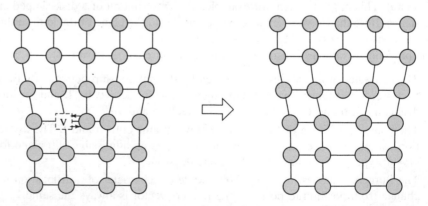

FIGURE 5.2–10 Climb of edge dislocation. The dislocation moves up one atomic spacing when it absorbs a vacancy.

Even if a dislocation cannot glide, it is possible for it to move by a different mechanism. This mechanism involves the diffusion of atoms, vacancies, or both toward the dislocation (see Figure 5.2–10). When a dislocation moves in this way, the process is called **dislocation climb.** When the atoms along the dislocation line exchange places with a nearby linear array of vacancies, the dislocation "moves up" one atomic plane. This motion occurs in a direction perpendicular to the plane defined by $\mathbf{b} \times \mathbf{t}$ and is therefore fundamentally different from dislocation glide. We will see in the chapter on mechanical behavior that dislocation climb is important at high temperatures, where the vacancy concentration is high and linear arrangements of vacancies are more likely.

5.2.4 Slip Planes and Slip Directions in Metal Crystals

We begin this section by attempting to explain why slip occurs in close-packed directions on highest-density planes. One interpretation of the critical resolved shear stress is that it represents the inherent lattice resistance, or inherent lattice friction, associated with slip on parallel crystal planes. We might, therefore, expect that slip will occur on the planes that display the least amount of lattice friction (i.e., the ones with the lowest values of τ_{CR}). Intuitively, the planes that offer the least resistance must be those that are smoothest on the atomic scale. The smoothest planes are those with the highest density of atoms.

The most favorable direction for slip may be inferred from energetic and crystallographic considerations. Figure 5.2–11 shows that a dislocation causes a local distortion of the lattice. This distortion, or change in the equilibrium separation distance between nearest-neighbor atoms, causes an increase in the energy of the system proportional to the volume of the distorted region. As an initial estimate, we can assume that the distorted volume has the shape of a cylinder with its axis coincident with the dislocation line (see Figure 5.2–11). This simplified model can be used to obtain two important results. First, the energy of the crystal increases as the total length of dislocations within the crystal increases. Second, the energy of a particular dislocation, on a per–unit length basis, is proportional to the cross-sectional area of the distorted cylinder of material. If we assume that the cylinder radius is proportional to the length of the Burgers vector, we can predict that the energy of the dislocation, E_{disl}, should scale with the square of the length of the Burgers vector. Mathematically,

$$E_{disl} \alpha |\mathbf{b}|^2 \tag{5.2–3}$$

where $|\mathbf{b}|$ is the magnitude or length of the Burgers vector.

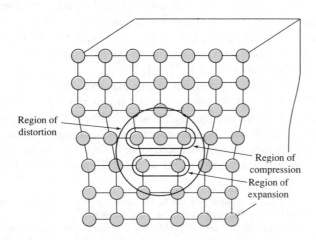

Region of distortion

Region of compression

Region of expansion

FIGURE 5.2–11

An illustration of the distorted region of a crystal lattice in the vicinity of an edge dislocation.

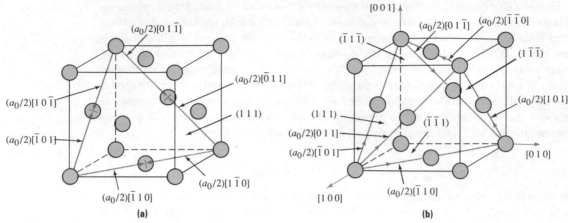

FIGURE 5.2–12   Burgers vectors and slip systems in the FCC structure: **(a)** an FCC unit cell showing the location of the (1 1 1) slip plane and the six valid Burgers vectors in the (1 1 1) plane. Note that the Burgers vectors occur in pairs, such that only three of the Burgers vectors are independent. **(b)** The tetrahedron formed by four members of the {1 1 1} family of planes in the FCC structure, including the three independent Burgers vectors in each plane.

Examination of Equation 5.2–3 shows that the energy associated with a dislocation is a minimum for the shortest Burgers vector. However, there is an additional requirement that the Burgers vector for unit dislocations (the most common dislocations) must join crystallographically equivalent positions in the lattice. That is, the motion of a dislocation must transport atoms from one equilibrium position to another. Thus, the most energetically favorable, and therefore the most common, Burgers vectors will be the shortest vectors that connect equivalent lattice positions. The result is that in simple crystal structures, the Burgers vectors are in the close-packed directions.

Consider the FCC crystal structure. Since the {1 1 1} planes are the close-packed planes, they represent the slip planes in FCC. The shortest vector joining equivalent lattice positions occurs in the close-packed directions and in FCC crystals corresponds to one-half of any face diagonal. The notation for such a vector is $(a_0/2)\langle 1\ 1\ 0\rangle$, where the term in the pointed brackets describes the family of directions and the term in parentheses gives the length of the vector.[2] Thus, the slip directions in FCC materials are members of the $\langle 1\ 1\ 0\rangle$ family of directions. Figure 5.2–12b shows that the three independent Burgers vectors contained within the (1 1 1) plane are $(a_0/2)[\bar{1}\ 0\ 1]$, $(a_0/2)[1\ \bar{1}\ 0]$, and $(a_0/2)[0\ 1\ \bar{1}]$. The combination of a slip direction and a slip plane constitutes a slip system. We can see from Figure 5.2–12b that each {1 1 1} plane contains three nonparallel $\langle 1\ 1\ 0\rangle$ directions, so there are three slip systems on each {1 1 1} plane. Since there are four nonparallel {1 1 1} planes, there are 12 slip systems in the FCC structure. Note that the slip planes intersect with one another in the FCC structure. The 12 slip systems in FCC are collectively represented as {1 1 1}$\langle 1\ 1\ 0\rangle$.

In the BCC system the situation in more complex, since there is no close-packed plane. However, the planes of highest atomic density are of the type {1 1 0}, and they are frequently observed to be the slip planes. These planes contain close-packed directions in the $\langle 1\ 1\ 1\rangle$ family, which are the slip directions. The Burgers vector is of the form $(a_0/2)\langle 1\ 1\ 1\rangle$, and the slip system is denoted by {1 1 0}$\langle 1\ 1\ 1\rangle$. In addition to the {1 1 0} planes, slip is also observed on the {1 1 2} and {1 2 3} planes[3] in BCC systems. Both of

---

[2] For example, $(a_0)\langle 1\ 1\ 0\rangle$ corresponds to a full-face diagonal, $(a_0/2)\langle 1\ 0\ 0\rangle$ represents one-half of a cube edge, and $(a_0/4)\langle 1\ 1\ 1\rangle$ describes one-quarter of a body diagonal.

[3] The planar densities on the {1 1 2} and {1 2 3} planes are only slightly lower than that on the {1 1 0} planes.

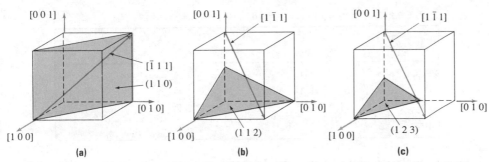

FIGURE 5.2–13 Slip planes and directions in the BCC structure: **(a)** a member of the {1 1 0}⟨1 1 1⟩ system, **(b)** a member of the {1 1 2}⟨1 1 1⟩ system, and **(c)** a member of the {1 2 3}⟨1 1 1⟩ system.

these families contain the ⟨1 1 1⟩ directions, so the additional slip systems are represented as {1 1 2}⟨1 1 1⟩ and {1 2 3}⟨1 1 1⟩. The slip planes and directions in the BCC system are shown in Figure 5.2–13. When all possible slip planes and directions are considered, there are 48 slip systems in the BCC structure. Comparing the FCC and BCC structures, we find that in both cases the slip directions are the close-packed directions, and the slip planes intersect with one another. In contrast to FCC, however, the slip planes in BCC are not close-packed.

The situation is yet more complex for the HCP structure. As discussed in Section 3.5.5, the HCP and FCC structures both contain close-packed planes and close-packed directions. The difference between the structures is in the stacking sequence of the close-packed planes. In the HCP structure the stacking sequence is ABABAB. . . . This stacking sequence results in only one set of parallel close-packed planes (i.e., the basal planes). This leads to the important result that in the ideal HCP structure the close-packed slip planes do not intersect.

With only one set of slip planes containing three slip directions, there are a total of only three slip systems in the ideal HCP structure (see Figure 5.2–14). Slip on the basal planes in the *a* directions is referred to as **basal slip.** However, basal slip cannot operate at all times. For example, if the basal plane is either parallel or perpendicular to the loading direction, the resolved shear stress is zero (recall the result of Example 5.2–2). In this case dislocations will not move on the basal plane, and other slip systems become important. The other systems that operate in HCP crystals depend on the shape of the atoms, which is measured by the axial ratio, c/a_0. For all slip systems other than the basal system, the slip planes intersect, but they are not close-packed. The slip planes and directions for the three metal crystal structures are summarized in Table 5.2–2.

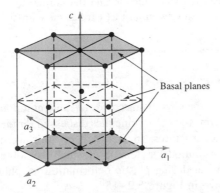

FIGURE 5.2–14 The slip planes and directions in the HCP structure are the basal planes and the *a* directions.

TABLE 5.2–2   Primary slip systems in the BCC, FCC, and HCP systems.

Crystal structure	Slip planes	Slip directions	Number of slip systems
FCC	{1 1 1}	⟨1 1 0⟩	12
BCC	{1 1 0}	⟨1 1 1⟩	12
	{2 1 1}	⟨1 1 1⟩	12
	{3 2 1}	⟨1 1 1⟩	24
HCP	Basal	a	3

In general, the ductility of a pure metal depends on the number of slip systems, whether they intersect, and the planar density of the slip planes. Since the FCC structure has 12 independent and intersecting slip systems, all of which involve close-packed planes, materials with this structure generally exhibit ductile behavior.

The BCC materials also exhibit good ductility at higher termperatures but in certain cases may exhibit poor low-temperature ductility because of lack of close-packed planes. This is especially true for BCC Fe below room temperature.

Hexagonal metals are much more complicated. For example, if a Zn single crystal is oriented for basal slip, it will show extensive plasticity, since the close-packed planes are active. However, in the ploycrystalline form, not all of the grains will be suitably aligned. The result is that polycrystalline HCP materials may crack rather than deform plastically when loaded. This is why Zn has poor ductility compared with FCC metals such as Al, Cu, and Ag.

....

EXAMPLE 5.2–3

Which of the following are valid slip systems in an FCC metal crystal?

  a. $(a_0/2)[1\ 1\ 1](1\ 0\ \bar{1})$
  b. $(a_0)[1\ \bar{1}\ 0](1\ 1\ 1)$
  c. $(a_0/2)[1\ 0\ 1](1\ 1\ 1)$
  d. $(a_0/2)[1\ 0\ 1](\bar{1}\ 1\ 1)$

Solution

In FCC, the valid slip systems are of the form $(a_0/2)\langle 1\ 1\ 0\rangle\{1\ 1\ 1\}$. System a is invalid, since the plane and direction do not belong to the appropriate families. System b is invalid because the magnitude of the vector is incorrect. $(a_0)[1\ \bar{1}\ 0]$ corresponds to an entire face diagonal rather than half of a face diagonal, which is the shortest distance between equivalent positions. System c is of the correct form, but it too is invalid, since the slip plane must contain the slip direction and the $[1\ 0\ 1]$ direction is not contained within the $(1\ 1\ 1)$ plane. We know that $[1\ 0\ 1]$ is not in $(1\ 1\ 1)$, since the dot product between the slip direction and the normal to the slip plane is not zero. System d does represent a valid slip system in FCC. It has the correct form, and the $[1\ 0\ 1]$ lies within $(\bar{1}\ 1\ 1)$.

....

5.2.5 Dislocations in Ionic, Covalent, and Polymer Crystals

As mentioned in Chapter 3, the crystal structures of ionic compounds are more complex than those of metals; however, the same principles discussed for metals apply equally well to dislocations in complex structures. In fact, LiF crystals were used by Gilman and Johnson in a classic study as a model system upon which they obtained an understanding of various atomic phenomena during plastic deformation. A photomicrograph showing dislocations in KCl is shown in Figure 5.2–15.

FIGURE 5.2–15   Dislocations in KCl. The KCl crystal is transparent and the dislocations (white lines) have been "decorated" by impurities to make them visible. The dislocations form a network.   *(Source: Reprinted from Amelinckx, Acta Metallurgica 6 (1958), p. 34, with kind permission from Elsevier Science Ltd., The Boulevard, Langford Lane, Kiblington OX5 IGB, UK.)*

As with point defects, the main factor that complicates the concept of a dislocation in ionic solids is the need to maintain local charge neutrality. Charge neutrality manifests itself in several ways, including: (1) the shortest vectors between crystallographically equivalent positions do not point in the directions in which ions are in contact; (2) in ionic solids the slip planes (i.e., highest-density planes) are not close-packed planes; and (3) the number of viable slip systems is limited because of the potential problem associated with placing ions of like charge next to each other (see Section 2.4.4).

Consider the unit cell for NiO shown in Figure 5.2–16. This ionic compound has the NaCl structure composed of an FCC lattice with two ions per lattice position (see Section 3.7–1). Three vectors are shown in Figure 5.2–16. Which one represents a valid Burgers vector in the NiO unit cell? We might be tempted to conclude that V_1 is the best choice, since it represents the shortest distance between ions. This is incorrect, however, since a Burgers vector must connect equivalent lattice points, and vector V_1 connects a Ni^{2+} site with an O^{2-} site. In contrast, vectors V_2 and V_3 both represent valid Burgers vectors. Since the length of vector V_3 is $a_0\sqrt{2}/2$ while the length of vector V_2 is a_0, V_3 represents the most common Burgers vector for this crystal structure. Note that this

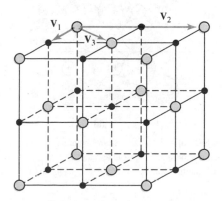

FIGURE 5.2–16

The NiO structure showing three possible choices for the Burgers vector. V_3 is the preferred vector, since it is the shortest vector (i.e., lowest-energy Burgers vector) that connects crystallographically equivalent positions.

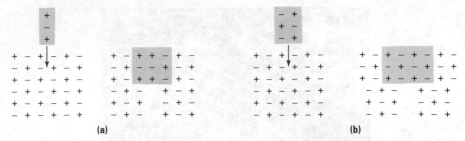

FIGURE 5.2–17 Burgers vectors in ionic crystals. **(a)** An edge dislocation is created by inserting an extra half plane. However, this process brings like charges into direct contact, which results in high energy. **(b)** An edge dislocation is created by inserting two extra half planes, and like charges are not in contact.

Burgers vector is of the form $(a_0/2)\langle 1\ 1\ 0 \rangle$, which is the Burgers vector for an FCC lattice. Also note that the $\langle 1\ 1\ 0 \rangle$ directions are not the directions in which ions are in direct contact (ions touch along $\langle 1\ 0\ 0 \rangle$ in the NaCl structure).

The necessity for longer Burgers vectors in ionic crystals can be illustrated using the extra half plane of atoms model for an edge dislocation. As shown in Figure 5.2–17, the insertion of a single plane of ions results in the placement of like charges adjacent to one another. If two planes of ions are inserted, the alternating cation-anion arrangement is preserved.

Just as the slip directions are not truly close-packed directions, the slip planes in ionic solids are not close-packed planes. Which are the highest-density planes in the NiO structure? Consider the $(1\ 0\ 0)$ planes shown in Figure 5.2–18. If we count the total number of ions of either type, we find that the $\{1\ 0\ 0\}$ family has the highest density. If, however, the crystal is plastically deformed on $(0\ 0\ 1)$, as shown in Figure 5.2–18b, slip results in the placement of atoms of like charge next to one another. The result is a brittle failure of the crystal. Examination of Figure 5.2–18c suggests another possible slip system. The Burgers vector shown in this figure is $(a_0/2)[0\ 1\ 1]$, and if the slip occurs by moving the parallel $(0\ 1\ 1)$ planes, the mutual repulsion of like ions never occurs. The most common slip system in NiO is in fact $\{1\ 1\ 0\}\langle 1\ 1\ 0 \rangle$. For NiO, and for ionic crystals in general, neither the slip directions nor the slip planes are close-packed. As a result, the resistance to dislocation motion is higher in ionic crystals than in metal crystals. In addition, the possibility of like ions coming in contact further restricts the number of slip systems in ionic crystals. All these factors contribute to a decrease in the mobility of dislocations in ionic crystals, leading to a tendency for brittle fracture.

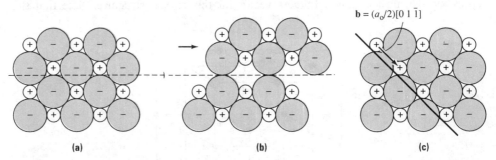

$$\mathbf{b} = (a_0/2)[0\ 1\ \bar{1}]$$

(a)   **(b)**   **(c)**

FIGURE 5.2–18 Deformation in NiO. **(a)** The crystal is shown before any dislocation has passed over the $(0\ 0\ 1)$ plane. **(b)** A dislocation with Burgers vector $(a_0/2)[0\ 1\ 0]$ is assumed to have passed over the plane, resulting in like charges adjacent to one another. **(c)** Slip has occurred in the $(0\ 1\ 1)$ plane, and the preferred charge arrangement is maintained.

EXAMPLE 5.2–4

Determine the direction and length of the lowest-energy (i.e., shortest) Burgers vector in CsCl.

Solution

As discussed in Section 3.7.1, the unit cell for CsCl has one type of ion at the simple cubic positions and the other ion at the $1/2, 1/2, 1/2$ position. The unit cell for this structure, together with two candidate Burgers vectors, is shown in Figure 5.2–19. Although vector $\mathbf{V}_1$ is the shorter of the two vectors, it is not a valid Burgers vector because it does not connect equivalent lattice positions. Instead, the shortest valid Burgers vector is $\mathbf{V}_2$ (i.e., a vector of the form $a_0\langle 1\ 0\ 0\rangle$).

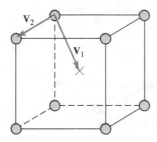

FIGURE 5.2–19

The CsCl structure with two candidate Burgers vectors. $\mathbf{V}_2$ is the preferred Burgers vector, since it connects equivalent atomic positions.

Dislocations also occur in covalent crystals. An example of dislocations in Si is shown in Figure 5.2–20. While covalent crystals do not suffer from the same charge-neutrality constraints as ionic crystals, the slip systems in covalent compounds have some of the same characteristics as those in ionic solids. The low coordination numbers in covalent crystals lead to comparatively low atomic packing factors and correspondingly low values for linear and planar densities. As with ionic crystals, this results in comparatively long Burgers vectors, and high values of inherent lattice resistance to dislocation motion. In addition, high-energy covalent bonds must be broken when a dislocation moves, and this contributes to the high strength. The most commonly observed slip system in covalent crystals such as Si and Ge is $\{1\ 1\ 1\}\langle 1\ 1\ 0\rangle$. The significance of dislocations in Si and Ge crystals lies in their influence on the electrical and growth characteristics of the crystal

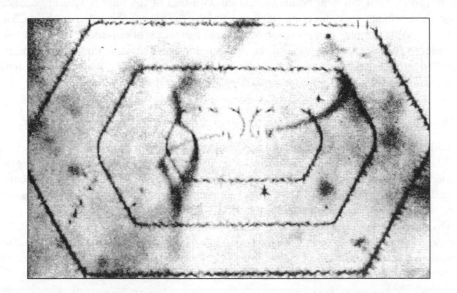

FIGURE 5.2–20

Dislocations (dark lines) in Si follow well-defined crystallographic directions in order to minimize the energy of the system. *(Source: Courtesy of Dunod Editeur from Jacques Friedel, Dislocations.)*

rather than in their role in plastic deformation. We will return to our discussion of dislocations in covalent crystals in Chapter 10, on the electrical properties of materials.

....

EXAMPLE 5.2–5

Figure 5.2–21 shows two candidate Burgers vectors for the diamond cubic structure of Si. Explain why slip occurs along vector $V_1 = (a_0/2)[1\ 1\ 0]$ in this structure even though V_1 is longer than the vector $V_2 = (a_0/4)[1\ 1\ 1]$.

FIGURE 5.2–21

Potential Burgers vectors in diamond cubic Si. The actual Burgers vector is $(a_0/2)[1\ 1\ 0]$.

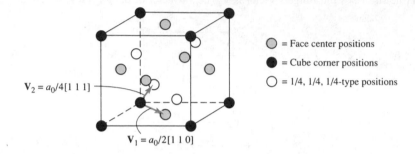

$V_2 = a_0/4[1\ 1\ 1]$

◯ = Face center positions
● = Cube corner positions
◯ = 1/4, 1/4, 1/4-type positions

$V_1 = a_0/2[1\ 1\ 0]$

Solution

Vector V_2 is not a valid Burgers vector. When a dislocation passes through a crystal, each atom must be transported from one equilibrium lattice position to another equivalent position. Although vector V_2 does move some of the atoms in the structure to another equivalent position, it does not do so for every atom in the structure. For example, if the atom originally located at the position 1/4, 1/4, 1/4 were translated by the vector V_3, it would end up at the position 1/2, 1/2, 1/2, which is not a normally occupied lattice position in the diamond cubic structure. Vector V_1 is the shortest lattice vector that moves all of the atoms in the unit cell onto equivalent lattice positions.

....

Dislocations also occur in the crystalline regions of polymers. The geometry is more complex, and the Burgers vectors are considerably larger than in other crystal structures because of the size of the polymer unit cells. As a result of the higher relative strength of the intramolecular covalent bonds compared with that of the intermolecular secondary bonds, the principal slip directions tend to be along the axis of the polymer chains. Since dislocations do not play a dominant role in the deformation of polymers, they will not be considered further. A description of the deformation mechanisms in polymers requires a knowledge of structure of the amorphous and partially crystalline regions of a polymer. Therefore, analysis of polymer deformation is postponed until the next chapter.

5.2.6 Other Effects of Dislocations on Properties

In addition to the crucial role played by dislocations in deformation behavior, they also significantly influence other properties. For example, the dilated (expanded) regions around an edge dislocation provide an easy path for diffusion. Diffusion along dislocations is called pipe diffusion, and it becomes particularly important at low temperatures because of its low activation energy.

Dislocations also influence the electrical, optical, and magnetic properties of materials. For example, we will show in the chapter on electrical properties that there is a need to minimize the concentration of dislocations in crystals engineered for optimum electrical properties.

Dislocations play an important role in the processing of engineering materials. For example, during the growth of crystals from the vapor phase, the presence of a screw dislocation at the surface can significantly increase the crystal growth rate.

5.3 PLANAR DEFECTS

We have now considered point and line defects and their effect on engineering properties of materials. In this section we will discuss the structure of two-dimensional or planar defects. The relationship between these defects and macroscopic properties is developed in Section 5.5 and in several later chapters.

5.3.1 Free Surfaces in Crystals

All materials share one common type of defect—their free or external surfaces. Why are free surfaces considered to be defects? As shown in Figure 5.3–1, surface atoms have fewer nearest neighbors, and therefore higher energy, than atoms with the optimum CN situated inside the crystal. The extra energy associated with the free surface is called **surface tension** and is given in units of energy/area (or equivalently, force/length). Free surfaces are often sites for chemical reactions, since the extra energy is available to drive the reactions. For example, in a very humid room, water droplets form on mirrors, door handles, and plumbing fixtures. Why does this happen? In part, the adsorption of the water molecules onto the surface decreases the energy of the system by better satisfying the bonding requirements of the surface atoms.

Reconsider the hypothetical crystal surface shown in Figure 5.3–1. Suppose additional atoms attempt to attach themselves to this surface. Where are the most likely locations for

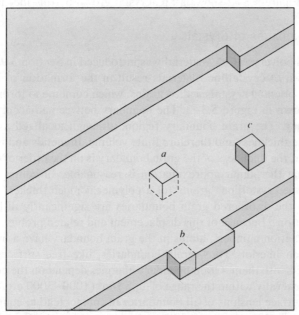

FIGURE 5.3–1 Schematic illustration of atoms (assumed to be cubes) adsorbed at a free surface. The lowest-energy location for adsorption is at *a*, since the atom is attached with a reduction in the surface area and, hence, a reduction in the surface energy. At position *b* the surface energy remains unchanged, while at *c* the energy increases, since the surface area increases. The site at *b* is a "repeatable step," since the process may be repeated continually with no change in energy.

FIGURE 5.3–2   Crystal growth from the vapor phase around a screw dislocation in SiC. Note the spiral growth step, which provides favorable positions for the adsorption of atoms. *(Source: Courtesy of Dunod Editeur from Jacques Friedel,* Dislocations.*)*

attachment? The most energetically favorable positions are those that reduce the number of unsatisfied bonds. For example, the incorporation of an atom into a "hole" in the surface results in a reduction of the number of local unsatisfied bonds from five to one (location *a* in the figure). If, instead of filling a hole, the attachment takes place at a ledge (location *b* in the figure), the number of unsatisfied bonds and the surface energy remain unchanged. Finally, consider the adsorption of an atom onto a flat region of the surface (point *c* in the figure). This process increases the number of unsatisfied bonds and is energetically unfavorable.

In some cases the ledge is formed from the intersection of a screw dislocation with the crystal surface. Figure 5.3–2 shows a SiC crystal grown by this mechanism.

5.3.2 Grain Boundaries of Crystals

The concept of a polycrystalline material was introduced in Section 3.9. Most techniques for the production of crystalline materials result in the formation of large numbers of small, randomly oriented crystals, called grains, which combine to form a polycrystalline aggregate, as shown in Figure 5.3–3. The boundary between adjacent crystals is called a **grain boundary.** The grain boundary region, although idealized as a planar defect, actually has finite thickness and therefore finite volume. In metals and ionic and covalent crystals, however, the thickness of the grain boundary is on the order of only a few atomic diameters, so that the planar approximation is reasonable. In contrast, the disordered region between the crystalline "grains" of a polymer is much thicker.

The atoms in the disordered grain boundaries are significantly displaced from their equilibrium positions. Because of this displacement and related problems associated with improper coordination numbers, atoms in the grain boundary have a higher energy than those in the grain interior. Thus, grain boundaries, like free surfaces, have a surface tension associated with them. Grain boundary energies depend on the composition of the crystal but are generally within the range of 1–3 J/m^2 (1000–3000 erg/cm^2). Under most conditions, the surface tensions of all boundaries in contact tend to equilibrate, and three grains will come together to form ~120° angles (refer to Figure 5.3–3).

As a result of their high energy, grain boundaries are regions in which chemical reactions and changes in structure may take place. For these and other reasons, grain boundaries play an important role in determining the mechanical, electrical, optical, and magnetic properties of materials.

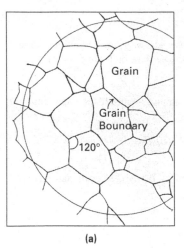

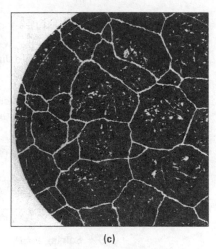

(a)   (b)   (c)

FIGURE 5.3–3 Grain structures in metals: **(a)** a schematic representation, **(b)** the grain structure in brass, and **(c)** the grain structure in austenitic steel. In **(b)**, the planar regions inside the grains are twins, which are discussed in Section 5.3.5.   *(Source:* ASTM Annual Book of Standards *(1990),* Section 3, "Metals Test Methods and Analytical Procedures," E 112, pp. 283–84, Figures 1, 3, 4. Copyright ASTM. Reprinted with permission.)

5.3.3 Grain Size Measurement

Since the size of the individual grains has a significant effect on many engineering properties, it is important to have a consistent method for determining grain size. The most widely accepted measure of the grain size is the American Society for Testing and Materials (ASTM) grain size number N, which is defined through the equation:

$$n = 2^{N-1} \tag{5.3–1}$$

where n is the number of grains per in^2 at a magnification of $100\times$ (i.e., the number of grains per square inch measured on a photograph taken at $100\times$ magnification). The grain size N is essentially an index of the fineness of the structure; as N increases the structure becomes increasingly fine. Since N is defined as an exponent, the grain size does not vary linearly with N. Typically, a fine-grained material will have an ASTM grain size of 8 to 10, while coarse-grained materials have grain sizes on the order of 2 or 3. Table 5.3–1 gives the average grain diameter for each of the common ASTM grain sizes.

TABLE 5.3–1 Average grain diameter for each of the common ASTM grain sizes.

ASTM grain size	Grain diameter (μm)
0	359
1	254
2	180
3	127
4	90
5	64
6	45
7	32
8	22.4
9	15.9
10	11.2
11	7.94
12	5.61
13	3.97
14	2.81

EXAMPLE 5.3–1

Estimate the average grain size, in microns, of a material whose ASTM grain size numbers are 2 and 8 for two different conditions. (Compare your results with the data in Table 5.3–1.)

Solution

For ease of calculation we will assume that the grains have a square shape. If the edge length at $100\times$ is D_{100}, then the number of grains per in² at $100\times$ is:

$$n = \frac{1}{(D_{100})^2}$$

This may be equated to the number of grains/in² as determined by the ASTM grain size number:

$$\frac{1}{(D_{100})^2} = n = 2^{N-1}$$

Solving for D_{100} yields:

$$D_{100} = 2^{(1-N)/2}$$

By definition, the grain size at $1\times$ is $1/100$ of D_{100}. Thus,

$$D = \frac{2^{(1-N)/2}}{100} \text{ in}$$

Unit conversion to microns gives:

$$D = \frac{2^{(1-N)/2}}{100} \text{ in} \times (2.54 \text{ cm/in}) \times (10^4 \ \mu\text{m/cm})$$

$$D = (2^{(1-N)/2})(254) \ \mu\text{m}$$

For ASTM #2 and ASTM #8 we obtain the following results:

$$D_{\text{ASTM}\#2} = 180 \ \mu\text{m} \quad \text{and} \quad D_{\text{ASTM}\#8} = 22.5 \ \mu\text{m}$$

5.3.4 Grain Boundary Diffusion

In Chapter 4 we considered diffusion as a process that occurs in the bulk of a crystalline material either by an interstitial or a vacancy exchange mechanism. Earlier in this chapter, we discussed diffusion along a dislocation line called pipe diffusion. Diffusion can also be aided by the presence of grain boundaries, which have a structure that is more open (less densely packed) than the defect-free regions of the crystal. The lower packing density in the vicinity of a grain boundary results in additional room for atomic motion and a correspondingly lower activation energy for diffusion.

When extended defects (i.e., linear or planar defects) are present, the overall diffusion rate is the sum of the weighted contribution from diffusion through the bulk and diffusion through defects. The relative magnitude of the two contributions is determined by three factors: (1) the diffusion coefficients for bulk and defect diffusion, (2) the fraction of material associated with the extended defects, and (3) the temperature.

The activation energy for diffusion along grain boundaries has been found empirically to be roughly half of that in the bulk. Thus, at constant temperature the rate of diffusion along grain boundaries is always higher than that in the bulk. The total mass transferred by each diffusion mechanism is, however, determined not only by the rate of diffusion, but also by the volume fraction of material affected by each process. At most temperatures of interest the sheer volume of bulk material is the dominant factor, and the majority of mass transport occurs in the bulk. As the temperature decreases, however, the relative contribution from the planar defects increases rapidly. This idea is illustrated mathematically in the following example problem.

....

EXAMPLE 5.3–2

The activation energy for bulk diffusion of C in BCC Fe is 84,000 J/mol. If the activation energy for grain boundary diffusion is one-half that of bulk diffusion and if both types of diffusion have the same D_0 value, compute the ratio of the diffusion coefficients first at 1000 K and then at 300 K.

Solution

Expressing the diffusion coefficient in the form given by Equation 4.4–2, it is possible to write the ratio of the diffusion coefficients as:

$$\frac{D_{gb}}{D_{bulk}} = \frac{D_0 \exp(-Q_{gb}/RT)}{D_0 \exp(-Q_{bulk}/RT)}$$

Since D_0 is assumed to be constant and $Q_{gb} = (1/2)Q_{bulk}$, we find:

$$\frac{D_{gb}}{D_{bulk}} = \exp\left(+\frac{Q_{bulk}}{2RT}\right)$$

$$\text{At 1000 K: } \frac{D_{gb}}{D_{bulk}} = \exp\left[\frac{84,000}{(2)(8.31)(1000)}\right] = 1.57 \times 10^2$$

$$\text{At 300 K: } \frac{D_{gb}}{D_{bulk}} = \exp\left[\frac{84,000}{(2)(8.31)(300)}\right] = 2.07 \times 10^7$$

The relative importance of grain boundary diffusion increases by many orders of magnitude as the temperature is lowered.

....

5.3.5 Other Planar Defects

While free surfaces and grain boundaries are perhaps the most obvious types of planar defects, other 2-D defects also exist. In fact, we have already seen another example in a different context. If we refocus our attention on Figure 5.2–9, which illustrates the formation of a dislocation loop via the collapse of a disk-shaped cluster of vacancies, we see that a 2-D fault in the stacking sequence of close-packed planes has been introduced. This is an example of the class of planar defects known as **stacking faults.** In addition to the collapse of a vacancy cluster, stacking faults may be introduced either by insertion of a cluster of interstitials or as a result of complex dislocation interactions. Stacking faults also occur during the growth of a crystalline material. Depending on the crystalline structure, the stacking fault energy can range from 0.01 to 0.3 J/m² (10–300 erg/cm²) and can be varied through alloy additions. For example, the stacking fault energy of pure Cu is about 0.1 J/m² (100 erg/cm²), but when Cu is alloyed with Al, the fault energy may be as low as 0.01 J/m² (10 erg/cm²). Since the density of stacking faults in a metal is inversely proportional to the energy of the fault, the addition of Al to Cu increases the stacking fault density in the metal.

Note that the amount of extra energy associated with the stacking fault is less than that for a grain boundary. Why is this so? In a stacking fault each atom maintains the appropriate number of nearest neighbors. The packing "errors" associated with the fault are related to second- and higher-order nearest neighbors. As such, the perturbation in the crystal structure is less severe than that in a grain boundary, and the corresponding defect energy is lower.

EXAMPLE 5.3–3

The stacking fault shown in Figure 5.2–9 results in a local region of HCP characteristics within an FCC structure. Describe a mechanism for the formation of a local region of FCC characteristics within an HCP crystal.

Solution

The stacking sequence of close-packed planes in FCC is ABCABCABC. Therefore, the removal of a section of the first C plane via the vacancy mechanism described in the text results in a local packing sequence of **AB_AB**CABC. Note that the first four planes appear to have the characteristic ABABAB sequence of the HCP structure. The question is, How can one alter the ABABAB sequence to obtain a local FCC character? The answer involves the formation of a stacking fault by the insertion of a disk-shaped cluster of interstitials. If the interstitials reside between the first B and the second A layers, the local stacking sequence becomes **ABCAB**AB, and the first five layers exhibit the characteristic FCC sequence.

FIGURE 5.3–4

Illustration of tilt and twist boundaries in crystals: **(a)** a low-angle tilt boundary made up of a vertical pile-up of edge dislocations of like sign, and **(b)** a 3-D representation of a low-angle twist boundary.

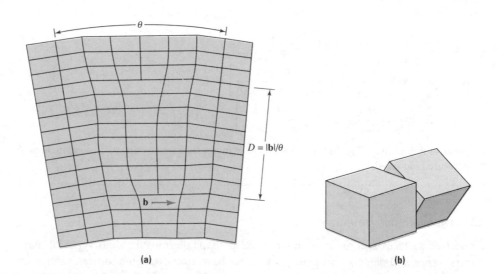

(a)   (b)

Another type of two-dimensional defect is a low-angle **tilt boundary.** As shown in Figure 5.3–4a, this defect can be defined as an ordered arrangement of edge dislocations. The name tilt boundary arises because the edge dislocations act as a wedge that tilts the two halves of the grain about an axis that lies in the plane of the boundary.[4] As shown in Figure 5.3–4b, a similar type of planar defect, known as a low-angle **twist boundary,** results from an ordered arrangement of screw dislocations. If the density of dislocations making up either type of boundary is increased so that the misorientation angle θ is greater than about 20°, the structure loses its geometric interpretation as an arrangement of dislocations and instead takes on the characteristics of a generalized grain boundary discussed previously.

Like all crystalline defects, low-angle boundaries possess extra energy, since they represent local regions of lattice distortion. The magnitude of the boundary energy increases as the misorientation angle increases (for $\theta < 20°$). This can be interpreted in either of two ways: (1) as the angle increases, so too does the dislocation density; or (2) as

_____

[4] A low-angle tilt boundary can glide as a unit on the slip plane, and indeed it was the reversible motion of such boundaries that provided the first experimental verification of the existence of dislocations.

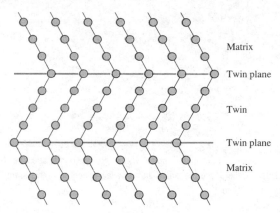

FIGURE 5.3–5   Schematic illustration of a twinned region in a crystal. The atom positions in the twin are mirror images of those outside the twin. The basic crystal structure is the same inside and outside the twin.

the angle increases, the misorientation between the two crystals becomes more and more severe, and therefore the structure deviates further from the lowest-energy atomic arrangement.

The final type of planar defect we will consider is the twin boundary or twin plane illustrated in Figure 5.3–5. A **twin** is a crystalline region in which there is a mirror image of the crystal structure across a boundary. Twins can be formed either during plastic deformation (deformation twins) or during cooling from high temperatures (annealing twins) by somewhat complex dislocation mechanisms. The amount of extra energy associated with a twin boundary is comparable to that of a stacking fault. Note that twins are visible in Figure 5.3–3b as a planar regions within the grains.

5.4 VOLUME DEFECTS

As the name implies, the term *volume defect* is reserved for three-dimensional regions in which the long-range order characteristics of a crystalline material have been lost. Volume defects range in size and complexity from small clusters of point defects to extensive amorphous regions. Like all crystalline defects, volume defects possess additional energy and are capable of influencing the engineering properties of materials.

Just as a cluster of linear dislocations resulted in the formation of a planar low-angle boundary, so too can a cluster of point defects result in the formation of a volume defect. Common examples include **voids,** which are 3-D clusters of vacancies, and **precipitates,** which are 3-D clusters of interstitial impurities, substitutional impurities, or both.[5]

Voids, also known as pores, are open volume regions in a material. They can be formed by the agglomeration of diffusing vacancies, from the bombardment of a crystal by high-energy radiation, or as a result of many processing operations that produce gas bubbles. Their size can range from a few nanometers to centimeters or larger. Pores have a significant detrimental effect on the mechanical properties of materials.

The size of precipitates can also range from atomic to macroscopic dimensions. Often, the precipitate takes on its own crystal structure that is different from that of the host lattice. In this case it is perhaps more appropriate to consider the two-dimensional

---

[5] Precipitates are often intentionally incorporated into a crystal to improve the properties of the material. Nonetheless, we have elected to include precipitates in our list of defects, since they represent a deviation from the "defect-free" regions of the host phase microstructure.

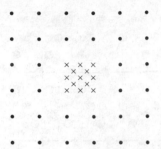

FIGURE 5.4–1 A 2-D illustration of a precipitate formed within the host crystal structure. The atoms within the precipitate are indicated by the symbol x and the matrix atoms by the symbol ●. In this example, the crystal structures of the matrix and precipitate are different.

interface between the matrix and the precipitate as the "defect" in the crystal structure. A schematic illustration of a precipitate with a different crystal structure from that of the matrix is shown in Figure 5.4–1.

The ultimate extension of the concept of a 3-D crystalline defect is the formation of a comparatively large volume of material in which all traces of LRO are removed, that is, an amorphous region within a polycrystalline aggregate. Materials composed of either predominantly amorphous structures of mixtures of amorphous crystalline regions are discussed in the following chapter.

EXAMPLE 5.4–1

Classify each of the following defects as point, line, planar, or volume defects:

    a. A screw dislocation

    b. A low-angle twist boundary composed of screw dislocations

    c. A vacancy

    d. A spherical cluster of approximately 50 vacancies

    e. A local region of atoms with a BCC structure within an FCC lattice

    f. The boundary between the BCC and FCC regions in part *e*

Solution

    a. Any dislocation is an example of a linear or 1-D defect.

    b. An ordered arrangement of linear defects results in the formation of a 2-D defect. Therefore, a low-angle twist boundary is a planar defect.

    c. A vacancy is a point defect.

    d. A spherical cluster of vacancies, however, begins to take on some 3-D qualities and becomes a volume defect.

    e. A local region of BCC structure within an FCC crystal represents a precipitate, which may be treated as a volume defect.

    f. The boundary between a precipitate and the matrix is a 2-D surface, so it is a planar defect.

5.5 STRENGTHENING MECHANISMS IN METALS

Previously we learned that plastic deformation in crystals occurs by the movement of dislocations. The relative ease of dislocation motion in metals accounts for their far lower strength than would otherwise be predicted. This simple statement has tremendous engineering consequences. Several strengthening mechanisms for metals will be introduced in this section.[6] The basic concept, however, is simple—to strengthen a metal one must either eliminate the dislocations (which is possible only in small volumes of metal such as thin whiskers) or impede their motion.

Consider the design of a car axle. Since the axle will fail to serve its intended purpose if it bends or plastically deforms under an applied load, the metal used for the axle should have a high resistance to plastic deformation. If the presence of mobile dislocations decreases the strength of a metal, then perhaps we can use the inverse argument and alter the structure of a metal so that dislocation motion, and hence plastic deformation, is more difficult. In fact, many of the commercially important metal-strengthening operations are based on this fundamental idea.

The discussion in this section is limited to metals because dislocation motion in ionic and covalent crystals is already difficult, which accounts for their brittle characteristics. Therefore, strengthening of these materials is of no practical significance. In fact, the challenge in these materials is quite the opposite: to create additional valid slip systems to get more ductility in these systems. This is a topic of research outside the scope of this introductory text.

5.5.1 Alloying for Strength

The calculation of shear stress required for dislocation glide assumes that the dislocation moves through a defect-free region of a crystal. When foreign atoms are added to a crystal, however, the defect-free assumption is invalid, and the stress required to cause dislocation motion, known as the **flow stress,** may be greater than anticipated.

How does the presence of impurity atoms, either interstitial or substitutional, impede the motion of dislocations? As mentioned in Section 4.3.1, the incorporation of interstitial atoms into a crystal results in a slight shifting of the neighboring solvent atoms away from their equilibrium positions and a corresponding increase in the strain energy of the crystal. The region of material over which the impurity atom exerts its influence is known as its strain (or stress) field. Similarly, there is some local atomic distortion in the vicinity of a dislocation. This distortion results in a dislocation having excess energy in proportion to the square of the length of its Burgers vector and a corresponding stress field. Depending on the character of the impurity atom, it can lower its energy by being located in either the expanded or the compressed region of the dislocation. If the dislocation is moved away from the impurity by a shear stress, the system energy increases. The increased energy must be supplied by increasing the stress beyond what it would be for equivalent motion with no impurity atom present. Thus, the strength is effectively increased.

One type of solute strengthening, known as solute drag, can be used to explain why C is such a potent strengthening element in Fe. Examination of Figure 4.3−1 shows that the lattice in the vicinity of an oversized interstitial atom is locally compressed. In contrast, Figure 5.2−11 shows that the region just below the extra half plane of atoms associated with an edge dislocation is locally expanded. If an oversized C interstitial atom diffuses to the expanded region of an edge dislocation, it can result in a neutralization of both of their lattice distortions and a reduction in the defect energy of the crystal. After this cooperative defect structure is established, considerable work must be done to move the dislocation away from the C atom. A high stress is required to do this work, and the Fe is strengthened.

---

[6] These same concepts will be revisited in Chapter 9.

The magnitude of the solute-strengthening effect depends on the characteristics of the solute atom strain field. The most potent strengthening effect comes from interstitial elements occupying the tetrahedral positions in BCC metals. Since substitutional solute atoms are about the same size as solvent atoms, the corresponding lattice distortion is less extensive, and the strengthening effect may be less pronounced.

EXAMPLE 5.5–1

Which alloying element, N or Co, is more likely to have a pronounced effect on the strength of pure Fe?

Solution

The key is to determine the type of solid solution formed by each alloy in Fe. From Appendix C we find $r(Fe) = 0.124$ nm, $r(N) = 0.071$ nm, and $r(Co) = 0.125$ nm. This suggests that Co may form a substitutional solid solution (SSS) with Fe. Examination of Appendix B shows that Co and Fe have similar electronegativity values and valences. Since the first three Hume-Rothery rules are satisfied, Co forms a limited SSS with Fe. Since $r(N) \ll r(Fe)$, N may occupy interstitial positions within the BCC Fe unit cell. Table 3.6–1 shows that the size ratio for interstitial atoms in the tetrahedral BCC sites is $\kappa/r = 0.291$. The ratio of $r(N)$ to $r(Fe)$ is $0.071/0.124 = 0.57$. Thus, the N interstitials are significantly oversized and will produce a large stress field. We should therefore expect that N is a more potent strengthening element than Co in Fe.

5.5.2 Strain Hardening

Just as an increase in the concentration of interstitial or substitutional atoms increases the flow stress of a crystal, an increase in the concentration of dislocations also increases the strength of a crystal. The reason for this strengthening effect is very much the same — extra work must be done to drag the stress field of a dislocation through the combined stress fields of the other dislocations in the crystal.

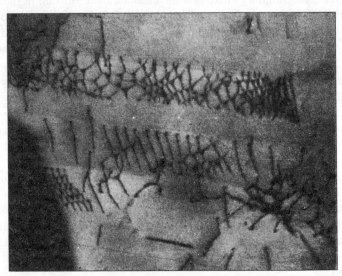

FIGURE 5.5–1   Dislocations in stainless steel deformed 10% in tension (40,000×).   *(Source: Hirsch, Partridge, and Segall, Philosophical Magazine 4 (1958), p. 721. Taylor and Francis, Inc. Washington, DC. Reproduced with permission. All rights reserved.)*

Once we recognize the direct relationship between dislocation density and the flow stress, and note that during plastic deformation the number of dislocations in a crystal increases dramatically, we can conclude that continued plastic deformation requires continually higher and higher stresses. The increase in the flow stress resulting from plastic deformation is usually referred to as **strain hardening.**

To quantify the relationship between flow stress and dislocation concentration, we must define the dislocation density ρ_{disl} using the equation:

$$\rho_{disl} = (\text{cm of dislocation})/(\text{cm}^3 \text{ of material}) \qquad (5.5\text{--}1)$$

Typically, the dislocation density of undeformed material is on the order of 10^8 cm^{-2} while the density of heavily deformed material may be as high as 10^{12} cm^{-2}. The distribution of dislocations in a stainless steel deformed 10% in tension is shown in Figure 5.5–1.

For many metals the relationship between the flow stress τ_{flow} and the dislocation density is of the form:

$$\tau_{flow} = \tau_0 + k\sqrt{\rho_{disl}} \qquad (5.5\text{--}2)$$

where τ_0 and k are constants for a given material.

....

EXAMPLE 5.5–2

The flow stress of a copper alloy increases from 2 MPa to 55 MPa when the dislocation density increases from 10^7 cm^{-2} to 10^{10} cm^{-2}. Calculate the flow stress for a similar heavily deformed copper alloy with a dislocation density of 10^{12} cm^{-2}.

Solution

The relationship between dislocation density and flow stress is given by Equation 5.5–2. Using the data in the problem statement, we can write the simultaneous equations

$$\tau_{flow} = 55 \text{ MPa} = \tau_0 + k\sqrt{10^{10} \text{ cm}^{-2}}$$

and

$$\tau_{flow} = 2 \text{ MPa} = \tau_0 + k\sqrt{10^7 \text{ cm}^{-2}}$$

Subtracting the second equation from the first yields:

$$53 \text{ MPa} = k(\sqrt{10^{10} \text{ cm}^{-2}} - \sqrt{10^7 \text{ cm}^{-2}})$$

Solving for k gives:

$$k = \frac{53 \text{ MPa}}{\sqrt{10^{10} \text{ cm}^{-2}} - \sqrt{10^7 \text{ cm}^{-2}}} = 5.47 \times 10^{-4} \text{ MPa-cm}$$

Substituting this k value into the equation for the flow stress of alloy with $\rho_{disl} = 10^7$ cm^{-2} and solving for τ_0 yields:

$$\tau_0 = (2 \text{ MPa}) - (5.47 \times 10^{-4} \text{ MPa-cm})(\sqrt{10^7 \text{ cm}^{-2}}) = 0.27 \text{ MPa}$$

Thus, the variation of flow stress with dislocation density for this copper alloy is:

$$\tau_{flow} = (0.27 \text{ MPa}) + (5.47 \times 10^{-4} \text{ MPa-cm})(\sqrt{\rho_{disl}})$$

For a sample with $\rho_{disl} = 10^{12}$ cm^{-2}, we find:

$$\tau_{flow} = (0.27 \text{ MPa}) + (5.47 \times 10^{-4} \text{ MPa-cm})(\sqrt{10^{12} \text{ cm}^{-2}})$$
$$= 547.3 \text{ MPa}$$

....

5.5.3 Grain Refinement

So far our discussion of plastic deformation has focused on the motion of dislocations in single crystals. We know, however, that many engineering materials are polycrystalline. Like point and line defects, the presence of grain boundaries impedes the motion of dislocations and, therefore, increases the stress necessary to promote dislocation motion (i.e., plastic deformation).

How can we quantify the influence of grain boundaries on dislocation mobility? In single crystals, the important quantity is the resolved shear stress necessary to promote dislocation motion. We discuss resolved shear stresses rather than normal stress in order to avoid the anisotropy issue and recognize that the shear stresses can be converted to normal stresses using the Schmid factor analysis in Section 5.2.1. Since polycrystals are isotropic, it is often more convenient to describe plastic deformation directly in terms of normal stresses. The magnitude of the normal stress required to promote plastic deformation in a polycrystal is known as the **yield stress,** σ_{ys}.

Since grain boundaries are effective obstacles to dislocation motion, and since small-grained materials will have a higher density of grain boundaries per unit volume, we find that the yield stress of a polycrystal generally increases with decreasing grain size. This relationship can be expressed analytically by the **Petch-Hall** equation:

$$\sigma_{ys} = \sigma_0 + \frac{k'}{\sqrt{d}} \qquad (5.5\text{--}3)$$

where σ_0 and k' are material constants and d is the average grain size, or grain diameter, of the polycrystal. Note the similarity to Equation 5.5–2. In both cases there is an intrinsic resistance to dislocation motion (τ_0 or σ_0) and a strengthening term resulting from the presence of the defects ($k\sqrt{\rho_{disl}}$ or $k'/\sqrt{d}$).

....

EXAMPLE 5.5–3

The strength of a low-carbon steel is 622 MPa for ASTM grain size #2 and 663 MPa for ASTM grain size #8. What will the strength be for ASTM grain size #10 ($d = 11.2$ μm)?

Solution

The average grain diameters for ASTM grain sizes #2 and #8 were found in Example 5.3–1 to be 180 μm, respectively. The relationship between grain size and strength is given by Equation 5.5–3. Using the data in the problem statement, we can write the simultaneous equations

$$\sigma_{ys}(\#8) = 663 \text{ MPa} = \sigma_0 + \frac{k'}{\sqrt{22.5 \text{ μm}}}$$

and

$$\sigma_{ys}(\#2) = 622 \text{ MPa} = \sigma_0 + \frac{k'}{\sqrt{180 \text{ μm}}}$$

Substracting the second equation from the first yields:

$$41 \text{ MPa} = k'\left(\frac{1}{\sqrt{22.5 \text{ μm}}} - \frac{1}{\sqrt{180 \text{ μm}}}\right)$$

Solving for k' gives:

$$k' = \frac{41 \text{ MPa}}{(1/\sqrt{22.5 \text{ μm}}) - (1/\sqrt{180 \text{ μm}})}$$
$$= 301 \text{ MPa-}\sqrt{\text{μm}}$$

Substituting this k' value into the equation for the yield strength of the ASTM #2 steel and solving for σ_0 yields:

$$\sigma_0 = (663 \text{ MPa}) - \left(\frac{301 \text{ MPa-}\sqrt{\mu\text{m}}}{\sqrt{22.5 \ \mu\text{m}}}\right) = 599.5 \text{ MPa}$$

Thus, the Petch-Hall equation for this steel is:

$$\sigma_{ys} = (599.5 \text{ MPa}) + \frac{301 \text{ MPa-}\sqrt{\mu\text{m}}}{\sqrt{d \ \mu\text{m}}}$$

For an ASTM #10 sample with $d = 11.2 \ \mu$m, we find:

$$\sigma_{ys} = (599.5 \text{ MPa}) + \frac{301 \text{ MPa-}\sqrt{\mu\text{m}}}{\sqrt{11.2 \ \mu\text{m}}}$$

$$= 689.4 \text{ MPa}$$

5.5.4 Precipitation Hardening

Another way to strengthen metals is by the inclusion of precipitates in the crystal. Dislocation motion is impeded by the lattice distortion in the vicinity of a precipitate, so that the flow stress is higher than when no precipitates are present. This mechanism of hardening, called **precipitation hardening,** is widely used in commercial aluminum alloys used in aircraft structures, nickel alloys used in jet engines, steels, and many other alloy systems. An example of dislocation motion being impeded by precipitates is shown in Figure 5.5–2.

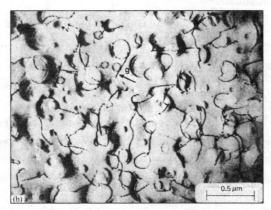

FIGURE 5.5–2   A photomicrograph of dislocations being impeded by precipitates in a Ni-base alloy. In this example the dislocations can bypass the particles by looping around them.   *(Source: B. A. Lerch, N. Jayaraman, and S. D. Antolovich, Materials Science and Engineering 66, Elsevier Science Publishing Company, 1984, pp. 151–66.)*

Note the similarity between the precipitation hardening mechanism and all the previously mentioned strengthening techniques. In all cases the insertion of an obstacle that impedes the motion of dislocations results in an increase in the strength of the crystal.

SUMMARY

Linear, surface, and volume defects occur in all crystalline materials and are directly involved in determining important engineering properties. Linear defects in crystals are called dislocations. Although not thermodynamically stable, dislocations occur in all crystalline materials as a result of growth "accidents" and internal stresses in the crystals. The character of a dislocation is defined by the relationship between its Burgers vector **b** and the unit tangent vector **t**. Two distinct orientations are edge (**b** ⊥ **t**) and screw (**b**‖**t**). While a dislocation may change its character, its Burgers vector is invariant. The open volume associated with an edge dislocation permits diffusion to occur more rapidly in these regions. Dislocations allow plastic deformation to occur at stresses that are much lower than stresses that would be required if dislocations were not present. This means that crystalline materials (and in particular, metals) may be strengthened by impeding the motion of dislocations. Strengthening can be achieved through: (1) addition of impurities that bind themselves to dislocations, (2) an increase in the dislocation concentration within the crystal, (3) an increase in the number of internal surfaces such as grain boundaries, and (4) the insertion in the lattice of particles (precipitates) that have a different crystal structure from that of the host material.

Planar defects occur as free surfaces, grain boundaries, stacking faults, low-angle boundaries (tilt and twist), and twin planes. In all cases, these boundaries possess excess energy, since the atoms are either severely out of register with the lattice (grain boundaries) or have second nearest-neighbor violations (twin boundaries and stacking faults). Because the boundary regions are characterized by higher energy, they represent preferential areas for chemical reactions. They also represent barriers to dislocation motion and, as such, strengthen the material. Grain boundaries have the most potent strengthening effect. The relationship between grain size and yield strength is known as the Petch-Hall equation.

An understanding of the cause and effect of linear, planar, and volume defects present in materials gives scientists and engineers the ability to manipulate the defects to optimize properties for each application.

KEY TERMS

basal slip	edge dislocation	screw dislocation	surface tension
Burgers circuit	flow stress	shear stress	tilt boundary
Burgers vector b	grain boundary	slip direction	twin
critical resolved shear stress (τ_{CR})	normal stress	slip plane	twist boundary
	Petch-Hall equation	slip system	unit tangent vector t
dislocation	plastic deformation	stacking fault	void
dislocation climb	precipitation hardening	strain hardening	yield stress
dislocation glide	Schmid's law	stress	

HOMEWORK PROBLEMS

SECTION 5.2:
Linear Defects, Slip, and Plastic Deformation

1. A fiber with a diameter of 25 μm is subjected to an elongation load of 25 g along the fiber axis. What is the stress on the fiber? Is the applied stress a shear stress or a tensile tress?

2. Consider the FCC metal Cu, which has a lattice parameter a of 0.362 nm.
    a. Calculate the length of the lowest-energy Burgers vector in this material.
    b. Express this length in terms of the radius of a Cu atom.
    c. On which family of planes will slip occur in this material? Why?

3. The lowest-energy Burgers vector in FCC Ag has length 0.288 nm. Find the length of the units cell edge in Ag.

4. Use sketches to predict the most common Burgers vector in the FCC and NaCl crystal structures.

5. A molybdenum (Mo) crystal has a Burgers vector of length 0.272 nm. If the lattice parameter a in Mo is 0.314 nm, determine whether Mo has the BCC or the FCC crystal structure.

6. Section 5.2.5 discusses slip in ionic crystals. If you force an ionic solid to shear and it breaks, will the fracture surface be atomically smooth or rough? Why?

7. The lowest-energy Burgers vector in BCC Cr has length 0.25 nm. Find the length of the unit cell edge in BCC Cr.

8. A normal stress of 123 MPa is applied to BCC Fe in the [1 1 0] direction. What is the resolved shear stress in the [1 0 1] direction on the (0 1 0) plane?

9. Calculate the magnitude of the applied stress in the [1 2 3] direction necessary to promote slip in the [1 1 1] direction on the (1 $\bar{1}$ 0) plane in a BCC crystal with $\tau_{CR} = 5.5$ MPa.

10. Why do Burgers vectors usually lie in closest-packed directions?

11. Why are the closest-packed planes usually the slip planes?

12. Consider the (1 $\bar{1}$ 1) plane of an FCC crystal and a dislocation whose Burgers vector is parallel to [$\bar{1}$ $\bar{1}$ 0]. The dislocation itself is parallel to the intersection of (1 $\bar{1}$ 1) and (1 1 1).
    a. Give the Burgers vector of the dislocation.
    b. Indicate the character of the dislocation.

13. Show that the dislocation reaction given below is both vectorially proper and energetically favorable for a BCC metal:

$$(a/2)[1\ 1\ 1] + (a/2)[1\ \bar{1}\ \bar{1}] \rightarrow a[1\ 0\ 0]$$

14. A dislocation in an FCC solid has a Burgers vector of $(a/2)[1\ 0\ 1]$. The dislocation is parallel to [$\bar{1}$ 1 0]. Determine the character of this dislocation and its slip plane. Will this dislocation be able to move by glide?

15. Why are FCC metals more ductile than either HCP or BCC metals?

16. Consider the simple cubic crystal structure. In which family of directions will the lowest-energy Burgers vectors lie?

17. MgO has the same crystal structure as NaCl and has a density of 3.65 g/cm^3. Use these data to calculate the length of a Burgers vector in this material. The atomic weight of oxygen is 16.00 and of magnesium is 24.32.

18. Suppose a metal single crystal is loaded (stressed) in the [1 0 1] direction.
    a. If the critical resolved shear stress for this material is 0.34 MPa, what magnitude of applied stress is necessary for dislocations to begin to move in the (1 1 1)[$\bar{1}$ 1 0] slip system?
    b. If the slip system listed in part a is valid, does this metal have the FCC, HCP, or BCC crystal structure?

19. Suppose you have an FCC metal single crystal which is known to have a critical resolved shear stress of 55.2 MPa.
    a. Find the largest normal stress that could be applied to a bar of this material in the [1 1 2] direction before dislocations begin to move in the [1 0 $\bar{1}$] direction on the (1 1 1) plane.
    b. How does your answer change if it is a BCC crystal and the slip system is [1 1 1](1 0 $\bar{1}$) (with the same τ_{CR})?

20. An Al crystal slips on the (1 1 1) plane in the [1 $\bar{1}$ 0] direction when 3.5 MPa is applied in the (1 $\bar{1}$ 1) direction. Compute the critical resolved shear stress.

21. Discuss the similarities and differences between edge and screw dislocations.

22. The critical resolved shear stress of a BCC metal is 7 MPa. A single crystal of this alloy is stressed along the [0 0 1] direction.
    a. What slip systems will be activated?
    b. What normal stress will cause plastic deformation?

23. *a.* Calculate the critical resolved shear stress in crystal if a stress of 170 MPa in the $[1\ 0\ 0]$ direction is required to move a dislocation in the $[1\ 1\ \bar{1}]$ direction on the $(1\ 0\ 1)$ plane.

   b. Is this a BCC or an FCC crystal? Why?

24. Calculate the magnitude of the applied stress in the $[1\ 2\ 3]$ direction necessary to promote slip in the $[1\ 1\ 1]$ direction on the $(1\ \bar{1}\ 0)$ plane in a BCC crystal with $\tau_{CR} = 5.5$ MPa. [It may or may not help you to know that a certain FCC single crystal with $\tau_{CR} = 0.55$ MPa slips in the $[1\ \bar{1}\ 0]$ direction on the $(1\ 1\ 1)$ plane under an applied stress of 9.63×10^{-2} MPa applied in the $[1\ 2\ 3]$ direction.]

25. Consider the crystal structure for $CaTiO_3$ (the perovskite structure) in which Ca^{2+} ions are located at the simple cubic positions, O^{2-} ions are located at the six face centers, and the Ti^{4+} ion is at the 1/2, 1/2, 1/2 position. Determine the direction and length of the lowest-energy Burgers vector for this compound.

26. *a.* Calculate the critical resolved shear stress in a crystal if a stress of 1.7 MPa in the $[1\ 0\ 0]$ direction is required to move a dislocation in the $[1\ 0\ 1]$ direction on the $(1\ 1\ 1)$ plane.

   b. Is this a BCC or an FCC crystal? Why?

   c. This problem is slightly flawed. Can you identify the reason for this statement?

27. Sketch a two-dimensional arrangement of atoms that contains a dislocation. Show two Burgers circuits, one that contains the dislocation and one that encloses a defect-free region of the crystal.

SECTION 5.3:
Planar Defects

28. Surface tension is energy divided by area. Show that this is equivalent to force divided by length.

29. Two processes are used to produce polycrystalline Al_2O_3. In the first process the grain size is 10 μm and in the second case it is 40 μm.

   a. Compute the ASTM grain size number for each of these structures.

   b. Compute the total grain boundary area for each structure.

30. The strength of a low-carbon steel is 200 MPa at an ASTM grain size of 4. It is 300 MPa at ASTM grain size #6. What will the strength be at ASTM grain size #9?

31. A BCC metal has a lattice parameter of 0.25 nm. A tilt boundary is formed with an angular difference of 2.5°. Compute the dislocation density in the tilt boundary.

32. Why do water droplets suspended in the air or in oil tend to be spherical?

33. The notation along the side of Figure 5.3–4a states that $D = |\mathbf{b}|/\theta$. Prove this equality.

SECTION 5.4:
Volume Defects

 34. Suppose you have a sample of a material. You suspect there are numerous large voids in the material, randomly spaced. How can you test your hypothesis? What if there were precipitates rather than voids?

 35. You suspect a material contains both crystalline and noncrystalline regions. How might you test your hypothesis?

SECTION 5.5:
Strengthening
Mechanisms in Crystals

36. Give five examples of material applications that require little or no plastic deformation during use, i.e., applications in which it is imperative that the material behave in a purely elastic manner.

37. How do changes in the dislocation density affect the strength and ductility of metals?

38. How would an increase in the point defect concentration of a metal affect its ductility (i.e., its ability to deform plastically)? Why?

39. Does the strength of a metal increase or decrease as its grain size decreases? Explain your answer.

40. List three ways in which a metal can be strengthened by hindering the motion of disloca-tions. Discuss these strengthening mechanisms from a physical point of view and indicate whether or not the indicated mechanism would be strongly temperature-dependent.

41. The dislocation density of a specific low-carbon steel is $10^{12}/cm^2$. One carbon atom per atomic site along a dislocation is required to "lock" a dislocation. What is the minimum carbon concentration that will cause the alloy to impede the motion of all dislocations?

42. Your friend shows you a bar of pure metal, say, tin. It has been well annealed so that the initial defect density is low. He bends the bar with his bare hands, hands it to you, and asks you to straighten it out. You know that you are stronger than your friend, but you cannot bend the bar. Why?

C H A P T E R   6

NONCRYSTALLINE AND SEMICRYSTALLINE MATERIALS

MATERIALS IN ACTION Polymers

Polymers are a class of organically based materials that consist of long chain molecules, usually with a carbon "backbone" and with other elements or structural units attached as side groups. The structure of these molecules can be manipulated to produce a wide range of desired properties. In general, the most desirable properties of polymers are their light weight, relatively high strength, good corrosion resistance, and electrical insulating properties. It has been estimated that there are over 60,000 commercially available polymers with over 30 million tons of plastics being manufactured each year. The economic impact of plastics is huge. Polymers are relatively easy to form into useful shapes by molding and extrusion processes, making their production economical compared to that of most other materials.

Polymers are not new. In the 16th century the Mayas had discovered how to use rubber trees to create polymers from which toys such as rubber balls were made. In the 19th century, one of the major breakthroughs in polymers was made when Goodyear discovered that rubber could be hardened by adding sulfur to natural rubber and reacting it at about 135° C. This process is called vulcanization. Vulcanized rubber is used in all tires manufactured worldwide.

One of the most widely used polymers is polyvinylchloride (PVC). It is used in plumbing applications for drain pipes and as such has virtually eliminated copper for such applications. In fact the tonnage of PVC piping worldwide has exceeded that of copper for several decades. PVC is also used as the coating of choice for electrical wires because of its excellent insulating properties and because it does not support combustion.

Finally, polymers are the matrix material of choice for many composite applications such as sporting goods and automotive applications. One interesting automotive application is found in one-piece driveshafts for light trucks. In this application, a proprietary vinyl resin is mixed with glass and graphite fiber and formed into the shape of a driveshaft. The final unit eliminates the need for center bearings, reduces vibration and noise, is corrosion resistant and is 60% lighter than its two-piece steel predecessor, thereby improving the fuel economy of the part. The potential for such units in GM vehicles alone is over 500,000.

6.1 INTRODUCTION

The structure of crystalline materials was presented in Chapter 3. Although many materials are crystalline or partially crystalline, some solids do not have their atoms arranged on a lattice that repeats periodically in space. Such solids are **noncrystalline,** or **amorphous.** Examples of noncrystalline materials include most thermoset plastics, transparent polymers, rubbers, and oxide and metallic glasses. Our emphasis in this chapter will be on noncrystalline solids; however, discussions of liquids are needed because most amorphous solids are formed by cooling from the melt and the structure of a glass is essentially that of a frozen liquid. In fact, theoretically any material will form an amorphous solid if the cooling rate from the melt is sufficiently rapid to suppress crystal formation.

Noncrystalline solids are unique in that they have no grain boundaries. Although in some cases grain boundaries improve the properties of a solid, in many cases they are detrimental. In Chapters 9–15, we will learn that grain boundaries can limit electrical, magnetic, and mechanical properties, and since they are high-energy surfaces, they can serve as initiation sites for corrosion and other forms of environmental attack. Thus, amorphous metals are used as transformer cores because of their low power loss in response to alternating magnetic fields, and metallic glass razor blades corrode much more slowly than crystalline ones. Grain boundaries also scatter light, making polycrystalline ceramics and semicrystalline polymers opaque (nontransparent). Amorphous polymers and oxide glasses can indeed often be readily identified by their transparency to visible light. Both oxide and polymer glasses are used in fiber-optic devices.

6.2 THE GLASS TRANSITION TEMPERATURE

Most noncrystalline solids are either rubbers or glasses. A modern working definition of **glass** is a material that lacks long-range order and is below the temperature at which atomic or molecular rearrangements can occur on a time scale similar to that of the experiment. In contrast, **rubber** is an amorphous solid for which molecular rearrangements can occur on the time scale of the experiment. Hence, glasses and rubbers are structurally similar, the chief difference between the two being the ability to rearrange molecularly. For example, compare the room-temperature responses of a piece of window glass and a rubber band to a blow from a hammer. The glass will break (i.e., it is brittle), while the rubber band will stretch to absorb the energy. When a rubber band is immersed in liquid nitrogen, however, it becomes a brittle glass. For any amorphous solid, the critical temperature that separates glassy behavior from rubbery behavior, on the time scale of the experiment, is known as the **glass transition temperature.**

The glass transition temperature is a characteristic of all noncrystalline materials regardless of whether they are organic polymers, metals, or inorganic oxide glasses. It is most clearly demonstrated by considering the density or volume changes associated with heating or cooling a material. Consider the processes that may occur during cooling of a material from the melt, as shown in Figure 6.2–1a. In this figure, the specific volume, which is the inverse of the mass density (i.e., volume per unit mass), is plotted as a function of temperature. As the temperature is lowered, the specific volume of the liquid decreases. The slope of the line, normalized to the sample volume V, is called the **volumetric** (or **bulk**) **thermal expansion coefficient** α_v:

$$\alpha_v = \left(\frac{1}{V}\right)\left(\frac{dV}{dT}\right) \tag{6.2–1}$$

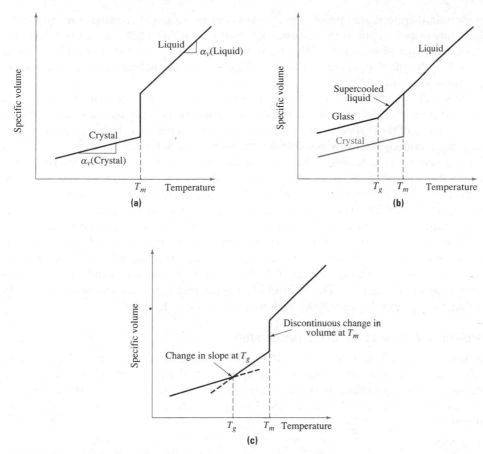

FIGURE 6.2–1 Specific volume as a function of temperature for a series of materials. **(a)** The liquid-to-crystalline solid transformation. A discontinuous change in volume occurs at the melting temperature T_m. **(b)** The liquid-to-glass transformation (the liquid-to-crystal curve is shown for reference). The temperature range in which the slope of the liquid-glass curve changes is the glass transition temperature T_g. **(c)** Specific volume versus temperature for a semicrystalline material. The discontinuous change in volume occurs at T_m, and a change in slope occurs at T_g.

This expression is analogous to the definition of the linear thermal expansion coefficient, α_{th}, given in Equation 2.5–4. It can be shown that $\alpha_v \approx 3\alpha_{th}$.

When the cooling rate is low, a sample may crystallize at a fixed temperature, called the melting point T_m. A sudden reduction in volume occurs at this temperature as a result of the change in atomic packing from that of the liquid to that of the crystalline solid. Below T_m, the specific volume once again decreases approximately linearly with temperature, reflecting the thermal expansion coefficient of the solid, which is roughly 1/3 that of the liquid for many materials.

All materials tend to crystallize when cooled below T_m because the energy of a collection of atoms packed in a crystalline array is lower than that of any other arrangement of the atoms. The crystal represents the lowest-energy state and therefore the most stable state of the material. Crystal formation occurs over a period of time because the establishment of long-range order requires atomic rearrangement by diffusion. In most materials, therefore, it is possible to avoid crystal formation by cooling at a sufficiently high rate so as to suppress the diffusion necessary to establish LRO in the crystal. In this case, as shown in Figure 6.2–1b, the volume of the collection of atoms continues to decrease with the slope characteristic of the liquid below the melting temperature, forming a **supercooled liquid.**

Since for most materials, $\alpha_v^{liq} > \alpha_v^{sol}$, if the supercooled liquid continued to change its volume at the rate characteristic of the liquid, it would eventually have a specific volume less than that of the corresponding crystal at the same temperature. In most circumstances it is not thermodynamically possible for the more loosely packed supercooled liquid to have a specific volume lower than that of the crystal, the slope of the curve for the supercooled liquid must eventually decrease to at least that of the crystal. As shown in Figure 6.2–1b, the temperature at which this slope change occurs is the glass transition temperature T_g. Since several physical property changes are associated with T_g, this phenomenon has been thoroughly characterized and studied in amorphous materials.

Many polymers are semicrystalline, containing both crystalline and amorphous regions. What would a sketch of specific volume versus temperature look like for a semicrystalline solid? As shown in Figure 6.2–1c, a semicrystalline solid exhibits two defining temperatures, T_m and T_g. The crystalline regions undergo a discontinuous change in volume at T_m, and the amorphous regions cause a slope change at T_g.

Molecular Motion at the Glass Temperature

Below T_g, large-scale molecular motion is not possible, because the material is frozen. Above T_g, molecular motion on the scale of the repeat unit occurs. By repeat unit, we mean a mer in a polymeric macromolecule—or, in the case of SiO_2, the repeat unit can be envisioned as a $(SiO_4)^{4-}$ tetrahedron. Examples of these two types of repeat units are illustrated in Figure 6.2–2.

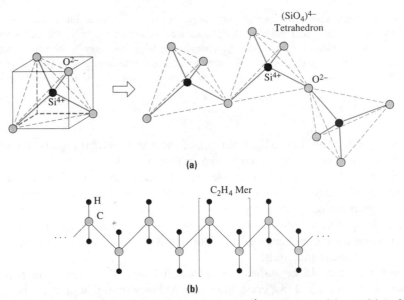

FIGURE 6.2–2 Repeat units in two different structures: **(a)** the $(SiO_4)^{4-}$ tetrahedron in SiO_2, and **(b)** the C_2H_4 mer in polyethylene. In this figure we show the polymer molecule in a simplified linear (stretched-out) orientation. In fact, as discussed later in the chapter, polymer molecules rarely look like this. When drawn in the linear fashion, however, the backbone carbon atoms are in the plane of the paper but the side groups (H, etc.) stick out of, or go into, the plane of the paper.

Consider the case of the polymer polystyrene. Below T_g one can envision the structure of this polymer as a 3-D tangled array of rigid molecules much like a plate of frozen cooked spaghetti. Above T_g, however, it is better to think of the polymer as a collection of entangled snakes with each snake (chain) able to wiggle all parts of its body. In the case of the oxide glasses, above T_g the individual $(SiO_4)^{4-}$ tetrahedra are able to "wiggle," while below T_g they are almost entirely rigid.

The enabling of molecular motion accounts for the softening of the material in the glass transition temperature range. Amorphous materials at temperatures below T_g are called glasses and exhibit brittle behavior. Between the glassy and liquid states, amorphous materials are essentially supercooled liquids. In the specific case of polymers, the supercooled liquid state is called rubber. Supercooled liquids, including polymers and oxides (but not metals), exhibit viscous behavior. This type of deformation is defined and discussed in the next section. In addition, the elastic modulus of amorphous solids decreases by several orders of magnitude as the temperature increases above T_g.

Our definition of T_g included the phrase *on the time scale of the experiment*. What influence does time have on the mechanical behavior of amorphous solids? The key, once again, is to recognize that molecular motion requires time. If a load is applied slowly to a sample, at a temperature near T_g, there may be sufficient time for molecular motion.[1] In contrast, a rapid loading rate may not allow sufficient time for significant motion regardless of the temperature. Therefore, since the loading rate influences the time available for molecular motion, it also influences the effective glass transition temperature of the solid. During a high–loading-rate experiment (i.e., when the time for molecular motion is very short), the effective T_g is higher than it would be in a low–loading-rate (slow) experiment.

Those of you familiar with the children's toy Silly Putty may have experienced the influence of loading rate on T_g. The T_g for this material is near room temperature. If the material is pulled rapidly it snaps in two, but if it is loaded more slowly it can be pulled down to a fine diameter. In this example the high loading rate causes the material to behave in a glass-like fashion at room temperature. In contrast, the low loading rate causes the material to behave in a fluid-like fashion at room temperature, so the material does not exhibit brittle behavior.

···

EXAMPLE 6.2–1

Do you think the glass transition temperature for ordinary window glass is above or below room temperature? What about the glass transition temperature for the polymer in a rubber band?

Solution

Our everyday experiences with window glass tell us that it is a brittle solid at all common temperatures. Since glasses exhibit brittle behavior when the temperature is below T_g, we can conclude that the glass transition temperature for window glass must be well above room temperature. In contrast, a rubber band is not brittle at room temperature. Therefore, we can conclude that the T_g for this polymer is below room temperature.

···

---

[1] Recall that the effective diffusion distance is given by the expression $x_{eff} = \sqrt{Dt}$. This implies that both time and temperature (through its influence on D) control the extent of diffusion. In much the same way, time (disguised as loading rate) and temperature determine whether an amorphous material will exhibit brittle behavior.

6.3 VISCOUS DEFORMATION

Above the glass transition temperature and below the thermodynamic melting temperature the structure of noncrystalline materials is that of a supercooled liquid. Consequently, several properties of these materials resemble those of liquids. These properties are distinct from the elastic properties of crystalline solids. For example, the response of amorphous material above T_g to an applied force is time-dependent, while the corresponding response below T_g is independent of time.

Recall that the concept of shear stress was introduced in Section 5.2.1. Consider a rectangular block of material with a top surface area A and height dx, subject to a shearing force F, as shown in Figure 6.3–1a. When the block is a solid, the shearing force is directly proportional to the deformation dy (i.e., $F \propto dy$). To make this relationship independent of sample dimensions, the force F is divided by the area A, and the displacement dy is divided by the height dx, yielding the relationship:

$$\left(\frac{F}{A}\right) \propto \frac{dy}{dx} \tag{6.3–1a}$$

Recognizing that F/A is the definition of shear stress τ, and defining the quantity dy/dx as the shear strain γ, leads to a more useful equivalent form of the equation:

$$\tau \propto \gamma \tag{6.3–1b}$$

This proportionality can be converted to an equality by inserting a constant known as the shear modulus G to give the equation:

$$\tau = G\gamma \tag{6.3–2}$$

When, as shown in Figure 6.3–1b, the solid is replaced with a fluid, the shearing force no longer produces a characteristic displacement (strain). Instead, the displacement continues with time, and it is the rate of displacement $d(dy/dt)$ that is proportional to the shearing force (i.e., $F \propto d(dy/dt)$). Normalizing the force by the area and the displacement rate by the height yields the expression:

$$\tau \propto \frac{d(dy/dt)}{dx} \tag{6.3–3a}$$

Since dx is a constant, we can rearrange the equation to yield:

$$\tau \propto \frac{d(dy/dx)}{dt} \tag{6.3–3b}$$

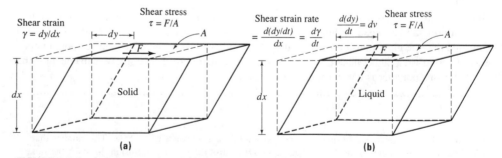

FIGURE 6.3–1   A comparison of the response of solids and liquids to the application of a shear stress: **(a)** for a solid, a shear stress τ results in a constant shear strain γ; and **(b)** for a liquid, a shear stress τ results in a constant shear strain rate $d\gamma/dt$.

Recognizing that dy/dx is the shear strain γ and converting the proportionality into an equality yields:

$$\tau = \eta\frac{d\gamma}{dt} = \eta\dot{\gamma} \qquad (6.3\text{--}4)$$

where the constant η is known as the viscosity. The viscosity is a measure of the work done on a material undergoing shear deformation. Equation 6.3–4 is known as Newton's law of viscosity.

Viscosity is an important property, and the study of flow is called rheology. This and other similar relationships are used extensively during the forming of glass and polymers into complex shapes by processes such as injection molding, fiber forming, or glass blowing (to be discussed in Chapter 16). The units of viscosity are poise, P $[= g/(cm\text{-}s)]$. Water at room temperature and liquid metals have a viscosity of about 0.01 P or 1 centipoise (cP). Molasses has a viscosity at room temperature of about 50 P, caramel about 10^5 P, and window glass about 10^{25} P. Sometimes it is easier to understand viscosity in terms of fluidity. Fluidity, ϕ, is simply the reciprocal of viscosity, $1/\eta$. Thus, as a material becomes more fluid, its viscosity decreases.

Temperature Dependence of Viscosity

Viscous flow is a thermally activated process. Thus, the temperature dependence of fluidity, like the diffusion coefficient discussed in Chapter 4, is governed by the Arrhenius equation:

$$\phi = \frac{1}{\eta} = \phi_0\exp\left(-\frac{Q}{RT}\right) \qquad (6.3\text{--}5a)$$

where ϕ_0 is a material constant and Q is the activation energy for viscous deformation. The relationship can also be written:

$$\eta = \eta_0\exp\left(+\frac{Q}{RT}\right) \qquad (6.3\text{--}5b)$$

Either equation fits the data reasonably well at temperatures above T_g. As an example, the data in Figure 6.3–2 show the effect of temperature on the viscosity of the most widely used glassy material, ordinary window or container glass (i.e., $Na_2O\text{-}CaO\text{-}SiO_2$ or soda-lime-silicate glass).

....

EXAMPLE 6.3–1

Calculate the viscosity of molasses at 100°C assuming an activation energy of 30 kJ/mol.

Solution

This problem can be solved using Equation 6.3–5b, $\eta = \eta_0\exp(+Q/RT)$, and by noting that in the text it was stated that $\eta(25°C) = 50$ P. Setting up a ratio of the viscosities at two temperatures gives:

$$\frac{\eta(T_2)}{\eta(T_1)} = \frac{\eta_0\exp(+Q/RT_2)}{\eta_0\exp(+Q/RT_1)}$$

$$= \exp\left[\frac{Q}{R}\left(\frac{1}{T_2} - \frac{1}{T_1}\right)\right]$$

Solving the equation for $\eta(T_2) = \eta(100°C)$ and substituting the appropriate values gives:

$$\eta(100°C) = (50\text{ P})\exp\left[\left(\frac{30{,}000\text{ J/mol}}{8.314\text{ J/mol-K}}\right)\left(\frac{1}{373\text{ K}} - \frac{1}{298\text{ K}}\right)\right] = 4.41\text{ P}$$

....

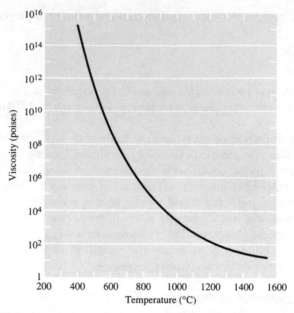

FIGURE 6.3–2   The effect of temperature on the viscosity of a soda-lime-silicate glass.   *(Source: W. D. Kingery et al.,* Introduction to Ceramics, *2nd ed. Copyright © 1976 by John Wiley & Sons. Reprinted by permission of John Wiley & Sons, Inc.)*

In this section we have introduced the concept of viscous deformation in amorphous materials at temperatures above T_g. Semicrystalline polymers contain both crystalline and amorphous regions. In these materials we must be concerned with the elastic nature of crystal deformation and the viscous nature of the rubbery material (noncrystalline polymer above T_g). Thus, these polymers show viscoelastic behavior, which is a combination of time-independent and time-dependent behavior. In Chapter 9, we will learn that crystalline solids can also exhibit time-dependent deformation or viscoelastic behavior when they are stressed for long times at high temperatures.

6.4 STRUCTURE AND PROPERTIES OF AMORPHOUS AND SEMICRYSTALLINE POLYMERS

In the previous section we discussed the common features of noncrystalline materials as a whole without attempting to distinguish between polymers, glasses, and rubbers. We now focus our attention on the atomic-level structure and characteristics of the various types of amorphous solids. This section describes the molecular structure of amorphous polymers, focusing on those that do not crystallize rapidly, the completely amorphous polymers.

6.4.1 Polymer Classification

Although the structure of a polymer is perhaps best illustrated with a drawing, it is often necessary to describe a polymer's structure in a less cumbersome manner. For example, the chemical structure of polyethylene (illustrated in Figure 6.2–2b) can be described using the notation:

$$\ldots CH_2 - CH_2 - CH_2 - CH_2 - CH_2 - CH_2 - CH_2 - CH_2 \ldots$$

which can be abbreviated as:

$$-[-CH_2 - CH_2 -]_n -$$

The n represents the number of repeat units (or monomers) in the polymer and is known as the **degree of polymerization.** Typically, n is very large, ~50,000 to 500,000 for polyethylene. If n is small, the polyethylene is a paraffin wax or oil.

This notation can be used to introduce several other commercially important synthetic polymers. For example, the vinyl polymers are based on the polyethylene structure with one of the H atoms in each polyethylene mer replaced with a different atom, or group of atoms. Thus, the generic formula for a vinyl polymer is:

$$-[-CH_2 - CH -]_n - \\ | \\ R$$

where R represents the side group. For example, when R is a CH_3 group, the result is the commercially important polymer polypropylene. Because of its low cost, excellent formability, and chemical resistance, polypropylene is used extensively in disposable diapers, for disposable clothing such as the gowns worn in operating rooms, and for laboratory equipment such as plastic beakers. Some of the other members of the vinyl family are listed in Table 6.4–1.

Another important family of polymers is the hydrocarbon rubbers, which share the generic formula:

$$-[-CH_2 - CH = CR - CH_2 -]_n -$$

where the $=$ symbol represents a covalent double bond in the C backbone. For example, when R is an H atom, the rubber is polybutadiene. Several other members of the rubber family are listed in Table 6.4–1.

The polymers mentioned above contain only carbon atoms in the backbone, or main chain. A polymer consisting of only C and H is called a polyolefin. There are, however, other polymers that contain different atoms in the chain backbone. One way to classify these polymers is by naming the characteristic group of atoms within the main chain. For example, as shown in Table 6.4–1, all polyesters contain the ester linkage,

$$-O - C - . \\ \| \\ O$$

One of the most important commercial polyesters is poly(ethylene terephthalate), or PET, which has the structure:

$$-[-O - C - C_6H_4 - C - O - CH_2 - CH_2]_n - \\ \| \qquad \| \\ O \qquad O$$

Mylar, Dacron, Fortrel, and Terylene are some of the common trade names for this polymer. Polyesters in general, and in particular PET, are used in large quantities as film and as fiber. Specific applications include the transparency material used for projectors, a reinforcement in rubber such as in tires and belts, a woven cloth for sails, and in denim and corduroy jeans. PET is the largest-volume fiber in the United States.

The polyamides, also known as nylons, all contain the amide linkage, $-\underset{\underset{H}{|}}{N}-\underset{\overset{||}{O}}{C}-$.

Specific examples are shown in Table 6.4–1. Note that the commercially important polymer nylon 6,6 is composed of six C atoms located on either side of the N in the characteristic amide linkage. Nylon is an important engineering material. One major application is in the area of carpet fiber. You are familiar with many other polyamides; silk, wool, and hair are natural fibers based on amino acids that are linked with the amide group. In fact, all protein fibers are polyamides.

TABLE 6.4–1   The structure of some of the more common engineering polymers.

Polymer	Structure	Applications
Vinyls and related polymers		
Polyethylene	$CH_2=CH_2 \rightarrow \{CH_2-CH_2\}$	Clear film, flexible bottles
Polyvinylchloride	$CH_2=\underset{Cl}{\overset{H}{C}} \rightarrow \{CH_2-\underset{Cl}{\overset{H}{C}}\}$	Floors, pipes, hoses
Polystyrene	$CH_2=\overset{H}{\underset{\bigcirc}{C}} \rightarrow \{CH_2-\overset{H}{\underset{\bigcirc}{C}}\}$	Containers (clear or foam), toys
Polypropylene	$CH_2=\underset{CH_3}{\overset{H}{C}} \rightarrow \{CH_2-\underset{CH_3}{\overset{H}{C}}\}$	Sheet, pipe, film, containers
Polyacrylonitrile	$CH_2=\underset{C\equiv N}{\overset{H}{C}} \rightarrow \{CH_2-\underset{C\equiv N}{\overset{H}{C}}\}$	Fibers—synthetic wool
Polytetrafluoroethylene (Teflon)	$CF_2=CF_2 \rightarrow \{CF_2-CF_2\}$	Nonstick coatings, gaskets, seals
Polymethylmethacrylate (Plexiglas)	$CH_2=\underset{\underset{O-CH_3}{\overset{|}{C=O}}}{\overset{CH_3}{C}} \rightarrow \{CH_2-\underset{\underset{O-CH_3}{\overset{|}{C=O}}}{\overset{CH_3}{C}}\}$	Lenses, transparent enclosures, windows
Rubbers		
Polybutadiene	$\underset{H}{\overset{H}{C}}=\underset{}{\overset{H}{C}}-\underset{}{\overset{H}{C}}=\underset{H}{\overset{H}{C}} \rightarrow \{\underset{H}{\overset{H}{C}}-\overset{H}{C}=\overset{H}{C}-\underset{H}{\overset{H}{C}}\}$	Tires and molded parts
Polyisoprene (natural rubber)	$\underset{H}{\overset{H}{C}}=\underset{}{\overset{H}{C}}-\underset{}{\overset{CH_3}{C}}=\underset{H}{\overset{H}{C}} \rightarrow \{\underset{H}{\overset{H}{C}}-\overset{H}{C}=\overset{CH_3}{C}-\underset{H}{\overset{H}{C}}\}$	Tires and gaskets
Polychloroprene	$\underset{H}{\overset{H}{C}}=\underset{}{\overset{H}{C}}-\underset{}{\overset{Cl}{C}}=\underset{H}{\overset{H}{C}} \rightarrow \{\underset{H}{\overset{H}{C}}-\overset{H}{C}=\overset{Cl}{C}-\underset{H}{\overset{H}{C}}\}$	Belts, bearings, and foams
Polydimethylsiloxane (silicone rubber)	$Cl-\underset{CH_3}{\overset{CH_3}{Si}}-Cl\ \ H-O-H \rightarrow \{\underset{CH_3}{\overset{CH_3}{Si}}-O\}$	Gaskets, insulation, and adhesives

TABLE 6.4–1   *(concluded)*

Polymer	Structure	Applications
Polyesters Polyethyleneterephthalate	H—O—C—C₆H₄—C—O—H + H—O—CH₂—CH₂—O—H → H₂O ⎡O—C—C₆H₄—C—O—CH₂—CH₂⎤	Films (magnetic tape), fibers, and clothing
(Thermoset variation)	H—O—C—C₆H₄—C—O—H + H—O—CH₂—CH—CH₂—O—H (with OH branch) → H₂O	Boat and auto body parts (fiberglass), helmets, and chairs
Polyamides Nylon 6,6	H—N—(CH₂)₆—N—H + H—O—C—(CH₂)₄—C—O—H → H₂O ⎡N—(CH₂)₆—N—C—(CH₂)₄—C—O⎤	Carpets, parachutes, rope, gears, insulation, and bearings
Kevlar, or poly p-phenyleneterephthalamide (PPTA)	⎡N—⟨benzene⟩—N—C—⟨benzene⟩—C⎤	Fibers, bulletproof vests
Other common polymers Polyacetal	C=O → ⎡C—O⎤	Gears and machine parts
Polycarbonate	H—O—⟨benzene⟩—C(CH₃)₂—⟨benzene⟩—O—H + Cl—C—Cl → HCl ⎡O—⟨benzene⟩—C(CH₃)₂—⟨benzene⟩—O—C⎤	Lenses, helmets, lamp casings, machine parts
Phenolformaldehyde	(phenol + CH₂=O → crosslinked network) → H₂O	Casings, (motor and telephone), electrical components, distributor caps
Polyurethane [R, R' → complex molecules]	H—O—R—O—H + C=N—R'—N=C → ⎡O—R—O—C—N—R'—N—C⎤	Foam, sheet and tubing, in-line skate wheels
Epoxy	H—O—R—O—H + H—C—C—CH₂—Cl → HCl ⎡O—R—O—C—C—CH₂⎤ (with H, OH)	Adhesives, used in composites

....

EXAMPLE 6.4–1

Determine whether each of the following polymers is a vinyl, a rubber, an ester, or a nylon:

a. $[-CH_2-CH=C(CH_3)-CH_2-]$

b. $[-CH_2-CHCl-]$

c.
$$[-N-C_6H_4-N-C-C_6H_4-C-]$$
$$\quad | \qquad\quad | \ \ \| \qquad\qquad\quad \|$$
$$\quad H \qquad\quad H \ O \qquad\qquad\quad O$$

d.
$$[-O-(CH_2)_4-O-C-C_6H_4-C-]$$
$$\qquad\qquad\qquad\qquad \| \qquad\qquad\ \|$$
$$\qquad\qquad\qquad\qquad O \qquad\qquad\ O$$

Solution

A comparison of the four polymers with the generic structures for each class of polymers shows that polymer a is a rubber; polymer b is a vinyl, known as polyvinylchloride, or PVC; polymer c is a polyamide, known as poly(p-phenyleneterephthalamide) or PPTA, which in fiber form is Kevlar®; and polymer d is a polyester called poly(butylene terephthalate) or PBT.

....

In Chapter 2 we introduced the idea of classifying polymers on the basis of the type of intermolecular bond present. Thermoplastic (TP) polymers have secondary bonds between chains and thermoset (TS) polymers have primary bonds between chains (in fact, the concept of an individual polymer chain loses its meaning in TS polymers).

Recall that the rubber family of hydrocarbon polymers, introduced in Chapter 2 and illustrated in Table 6.4–1, contains an unsaturated double bond in the backbone of the polymer chain. This double bond can be opened (usually in the presence of oxygen or sulfur) to form crosslinks between adjacent chains. As the crosslink density increases, the structure of the rubber gradually changes from TP to TS. The concept of a transition from TP to TS structures because of crosslink formation is not restricted to rubbers, although there must be crosslinking to prevent viscous flow and enable good elastic recovery for rubbers to be useful. Consider the structures of the two polyesters shown in Figure 6.4–1a and b. PET has a linear structure and is a TP polymer. In contrast, the other polyester contains unsaturated linear chains. When styrene, a vinyl monomer with R representing a benzene ring, is added to the latter polyester, crosslinks are formed (Figure 6.4–1c). As the crosslink density increases, the polyester is transformed from TP to TS. This material is the polyester resin used in many fiberglass composites.

Not all TSs, however, are formed from linear macromolecules containing unsaturated double bonds. For example, one of the oldest thermosetting polymers is phenolformaldehyde, or Bakelite®. As shown in Figure 6.4–2, the three-dimensional primary bond network in this TS polymer results from the existence of three equivalent bonding sites associated with each phenol molecule. Epoxies are also examples of TS polymers In fact, to be TS simply requires a minimal number of mers with tri- or tetrafunctionality.

Synthetic polymers were formulated and made by chemists after years of studying natural macromolecules such as deoxyribonucleic acid (DNA) and cellulose. Cellulose is the chief component in cotton, wood, and all vegetable fibers. The structure of this important natural polymer is shown in Figure 6.4–3. Also shown in this figure is part of the structure of lignin, a natural polymer found extensively in wood and in the cell walls of plants.

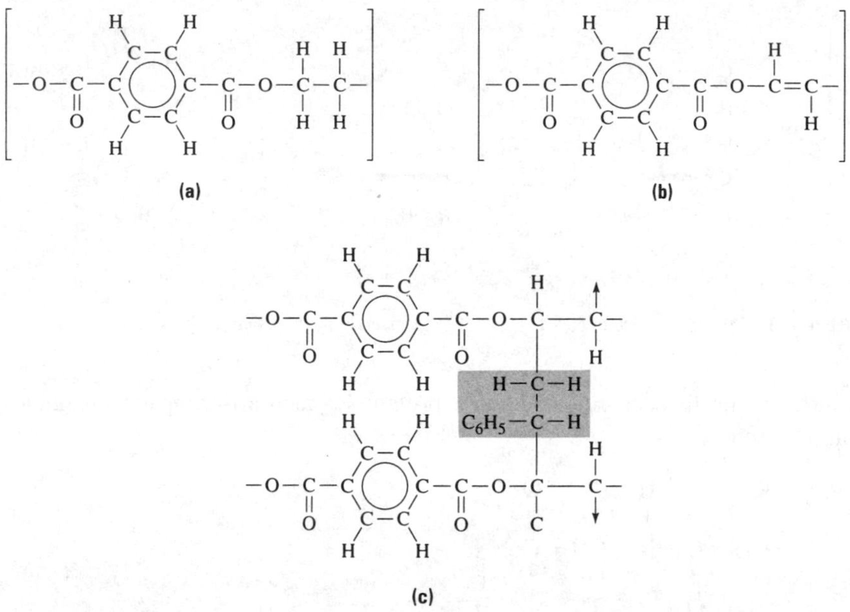

FIGURE 6.4–1 A comparison of two polyesters. **(a)** A schematic illustration of PET showing a saturated chain structure that results in the formation of a TP polymer. **(b)** The existence of the double bond in the backbone of this unsaturated polyester permits the formation of crosslinks. **(c)** In the unsaturated polyester the crosslinks formed by the polystyrene monomer $C_2H_3(C_6H_5)$ result in a 3-D primary bond network characteristic of a TS polymer.

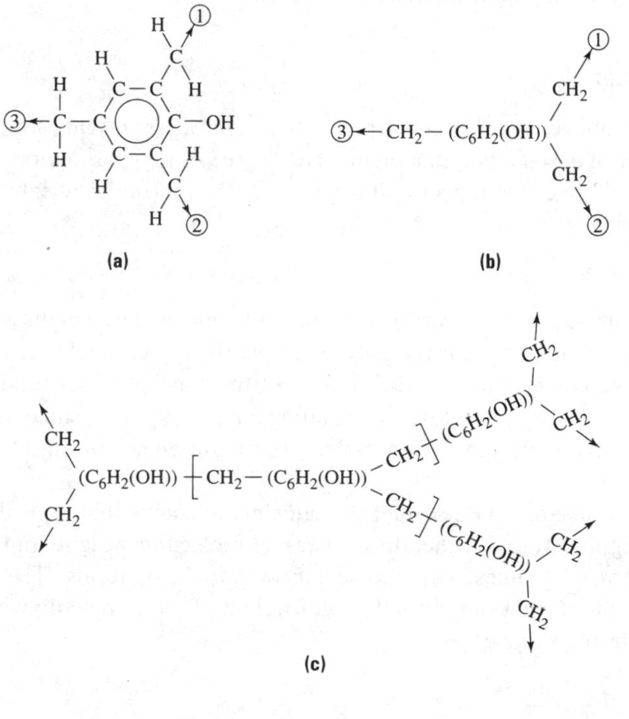

FIGURE 6.4–2

The structure of the TS polymer phenolformaldehyde (Bakelite®) showing: **(a)** the basic building block for the structure (the numbers 1–3 correspond to the three sites for primary bold formation with neighboring mers), **(b)** a simplified representation of the basic building block, and **(c)** a 2-D representation of the 3-D primary bond network in this TS polymer.

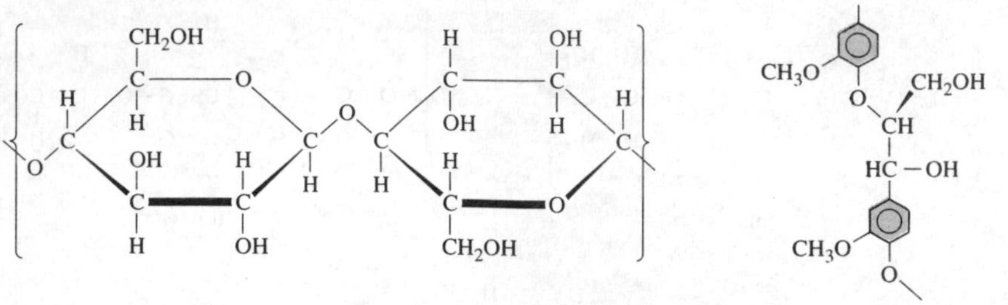

(a)   **(b)**

FIGURE 6.4–3   **(a)** The structure of cellulose, and **(b)** schematic of the structure of lignin.

Another naturally occurring polymer is protein. Proteins are complex polyamides, containing the unit:

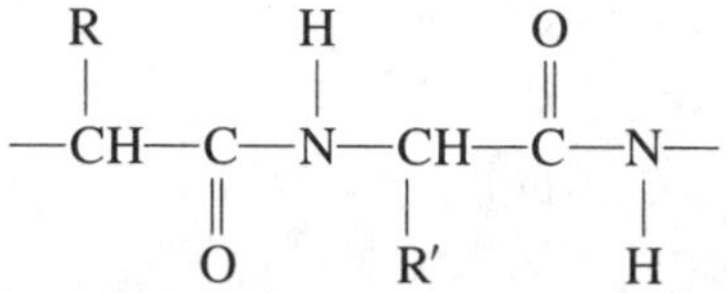

The difference among the various protein molecules, such as human muscle, silk, wool, and hair, is largely the result of differences in the amino acids that reside in the R and R′ locations.

At this point it is easy to be overwhelmed by the apparently endless number of available polymers and their chemistry. Although we have discussed only a few of the polymers that are currently available, those that we have mentioned on the preceding pages constitute more than 70% of the total market for polymers.

6.4.2 Molecular Weight

As mentioned above, the degree of polymerization, n, represents the number of mers linked together in a single polymer chain. The degree of polymerization multiplied by the molecular weight m of the repeat unit (mer) gives the molecular weight M of the macromolecular chain:

$$M = n \times m \tag{6.4–1}$$

For example, if n for PET were 100, then the molecular weight would be 100×192 g/mol $= 19{,}200$ g/mol for the polymer. This is a typical molecular weight for commercial PET. As you may have realized already, the concept of degree of polymerization is reserved for TP polymers. It loses its meaning for TS polymers, since they are in essence one single massive molecule with a molecular weight equal to that of the macroscopic polymer part.

An actual sample of a TP polymer contains many chains that typically have different molecular weights. We must, therefore, speak of molecular weights in terms of averages and be concerned at times with molecular weight distributions. The number-average molecular weight $\overline{M}_n$ of a sample is the total weight of the sample divided by the number of molecules in the sample:

$$\overline{M}_n = \frac{\Sigma_i(N_i M_i)}{\Sigma_i(N_i)} \tag{6.4–2}$$

where N_i is the number of molecules with molecular weight M_i. The weight-average molecular weight $\overline{M}_w$ of a sample is:

$$\overline{M}_w = \frac{\Sigma_i(w_i M_i)}{\Sigma_i(w_i)} \tag{6.4-3}$$

where $w_i = N_i M_i$.

The ratio of the two measures of average molecular weight is known as the polydispersity (PD) and is given by the expression:

$$PD = \frac{\overline{M}_w}{\overline{M}_n} \tag{6.4-4}$$

A polydispersity of 1.0 indicates that all the molecules have the same molecular weight. As polydispersity increases, the breadth of the molecular weight distribution increases.

Why do we need two measures of molecular weight? We will attempt to answer this question with an analogy. Suppose we chose at random five people each with some change. Consider the following two questions: (1) What is the average number of coins per person? and (2) What is the average amount of money per person? To answer the first question we simply count the total number of coins and divide by five. For this calculation a penny and a quarter make equal contributions. In contrast, to answer the second question we must distinguish between the values of the coins. A quarter will make a larger contribution than a penny. Relating this analogy to the two measures of molecular weight, we find that the number-average molecular weight treats all polymer chains equally, while the weight-average molecular weight permits the longer and heavier chains to make a proportionally larger contribution to the calculation. While $\overline{M}_n$ is perhaps conceptually less complex, the properties of polymers often better correlate with $\overline{M}_w$.

··

EXAMPLE 6.4–2

Consider a sample of polyvinylchloride (PVC) that is composed of only two types of chains. Ninety percent of the chains in this sample have a degree of polymerization (n) of 10,000, and 10% of the chains have $n = 100,000$. Calculate the polydispersity for this polymer sample.

Solution

First we must calculate the mass of a single PVC mer (m); then we calculate the molecular weight of each type of chain (M_i). Next we determine the two measures of average molecular weight ($\overline{M}_n$ and $\overline{M}_w$), and finally we can determine the polydispersity of the sample.

$$m(PVC) = (2 \times \text{at. wt. of C}) + (3 \times \text{at. wt. of H}) + (\text{at. wt. of Cl})$$

$$= 2(12) + 3(1) + 35.45 = 62.45 \text{ (g/mol)/mer}$$

$$M_1 = n_1 \times m(PVC) = 10,000 \text{ mers} \times 62.45 \text{ (g/mol)/mer}$$

$$= 624,500 \text{ g/mol}$$

$$M_2 = n_2 \times m(PVC) = 100,000 \text{ mers} \times 62.45 \text{ (g/mol)/mer}$$

$$= 6,245,000 \text{ g/mol}$$

Using Equation 6.4–2 we find:

$$\overline{M}_n = \frac{\Sigma_i(N_i M_i)}{\Sigma_i(N_i)}$$

$$= \frac{(0.9 \times 624,500) + (0.1 \times 6,245,000)}{0.9 + 0.1}$$

$$\overline{M}_n = 1,186,550 \text{ g/mol}$$

Using Equation 6.4–3 with the definition $w_i = N_i M_i$, we find:

$$\overline{M}_w = \frac{\Sigma_i(w_i M_i)}{\Sigma_i(w_i)} = \frac{\Sigma_i(N_i M_i^2)}{\Sigma_i(N_i M_i)}$$

$$= \frac{(0.9 \times 624{,}500^2) + (0.1 \times 6{,}245{,}000^2)}{(0.9 \times 624{,}500) + (0.1 \times 6{,}245{,}000)}$$

$$\overline{M}_w = 3{,}582{,}658 \text{ g/mol}$$

Finally, use Equation 6.4–4 to calculate the polydispersity:

$$PD = \frac{\overline{M}_w}{\overline{M}_n} = \frac{3{,}582{,}658}{1{,}186{,}550} \approx 3.02$$

Polydispersity of polymers typically ranges between 2 and 20.

6.4.3 Polymer Conformations and Configurations

We have seen that there are a great number of covalent bonds in polymeric macro-molecules. In Section 2.6 we found that covalent bonds are characterized by specific bond angles. The bond angles are only rarely 180°, so polymer chains are nonlinear. As shown in Figure 6.4–4a, a carbon atom in polyethylene (PE) forms four covalent bonds. These bonds are separated by equal angles in space, the so-called tetrahedral angle of 109.5°. The C—C bonds in PE are free to rotate about an axis that preserves the 109.5° angle (see Figure 6.4–4b). Although the chain can be stretched out so that the C—C bonds all lie in the same plane (the planar zigzag conformation, Figure 6.4–4c), it is more typical of a polymer chain to assume a contour similar to that of a piece of cooked spaghetti (Figure 6.4–4d). This is called the random coil conformation (to be described in Section 6.6.5), and the end-to-end separation can be calculated using random walk statistics. The conformation of a polymer can be changed by simple bond rotation. The random coil conformation is a relatively low energy state for a polymer chain. Polymer molecules in solution, melts, and noncrystalline solids seek to be random coils. Only the crystalline state has lower energy.

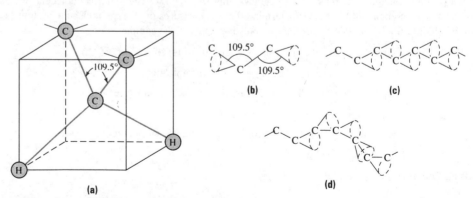

FIGURE 6.4–4   The 3-D structure (conformation) of a polyethylene chain. **(a)** The C—C—C bond angle in PE. **(b)** The bond angle does not define the location of the neighboring C atoms but only restricts their location to a specific cone of rotation. **(c)** If all of the C atoms in the chain backbone lie in the same plane, the planar zigzag conformation results. **(d)** The more common (lower-energy) conformation of PE is the random coil structure.

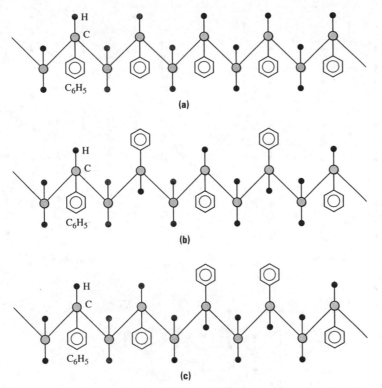

FIGURE 6.4–5 A planar illustration of the **(a)** isotactic, **(b)** syndiotactic, and **(c)** atactic configurations of polystyrene. Note: The C_6H_5 rings are actually much larger than indicated in these sketches.

Polymer configuration is a term used to describe the relationship of side groups to one another. The three most common configurations are atactic, isotactic, and syndiotactic. They are illustrated in Figure 6.4–5 for polystyrene. When the side groups are all on the same side of the molecule, the polymer is **isotactic;** if on alternating sides, it is **syndiotactic;** if in random positions, it is **atactic.** Tacticity is unchanged by simple bond rotations. Tacticity, also known as stereochemistry, is determined during polymer formation. The Nobel Prize discovery of catalysts that direct polymerization enabled the production of stereoregular polymers (i.e., isotactic or syndiotactic polymers that display a regular repeating structure).

EXAMPLE 6.4–3

Sketch the three different configurations of polyvinylchloride (use the planar zigzag conformation). Show several chains of each configuration lined up parallel to one another and describe the secondary bond interactions between molecules.

Solution

Figure 6.4–6 shows the isotactic, syndiotactic, and atactic configurations, all in the planar zigzag conformation, for polyvinylchloride. In all three cases the potential exists for relatively strong secondary bond formation between the electronegative Cl side groups and the electropositive H side groups.

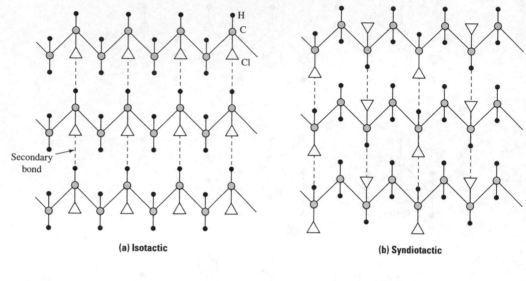

(a) Isotactic   **(b) Syndiotactic**

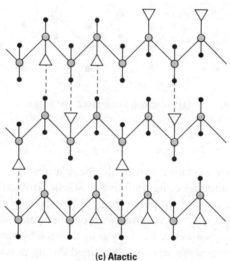

(c) Atactic

FIGURE 6.4–6 An illustration of the **(a)** isotactic, **(b)** syndiotactic, and **(c)** atactic configurations of polyvinylchloride. (See Example 6.4–3.)

6.4.4 Factors Determining Crystallinity of Polymers

Thermosetting polymers do not crystallize. The ability of a thermoplastic polymer to partially crystallize is determined by the ease with which the molecules can move and be efficiently packed together to create long-range order. Several factors influence the efficiency of packing polymer chains, including:

1. The size of side groups.
2. The extent of chain branching.
3. Tacticity.
4. The complexity of the repeat unit.
5. The degree of secondary bonding between parallel chain segments.

Each of these factors is discussed in more detail below.

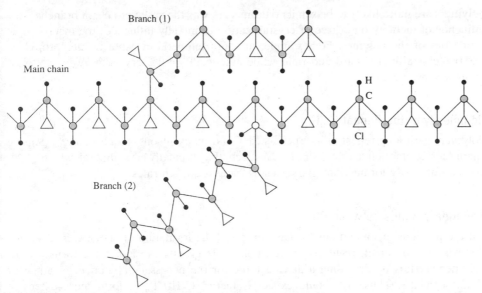

Branch (1)

Main chain

H
C
Cl

Branch (2)

FIGURE 6.4–7   A schematic illustration of chain branching in polyvinylchloride.

Size of Side Groups

Polymers with large, bulky side groups cannot be packed efficiently and, therefore, have difficulty forming crystals. Thus, one can predict that polypropylene, $[C_2H_3(CH_3)]_n$, is less likely to crystallize than polyethylene, $[C_2H_4]_n$, but more likely to crystallize than polystyrene, $[C_2H_3(C_6H_5)]_n$, since the CH_3 group is larger than a single H atom but smaller than a C_6H_5 group.

Chain Branching

As shown in Figure 6.4–7, a chain branch is a location on the main-chain backbone where a side group has been removed and replaced with another "branch" of backbone atoms. This may be a "defect" that occurred during polymerization or it may have been intentional. Just as it is far more difficult to efficiently stack branched tree limbs than it is to stack two-by-fours, it is far more difficult to form crystals with branched chains than with unbranched chains.

The degree of crystal formation in polyethylene $(C_2H_4)_n$ is strongly influenced by the density of chain branches. In turn, the density, strength, and modulus of PE all increase as the degree of crystallization increases. For example, a sample of PE with a moderate concentration of chain branches may be approximately 40–60% crystalline with a density of ~0.94 g/cm^3, a strength of 4–16 MPa, and a modulus of 100–260 MPa. In contrast, a sample of PE with a low concentration of chain branches may be > 95% crystalline with a density of ~1.0 g/cm^3, a strength of 20–40 MPa, and a modulus of 400–1200 MPa. The former type of PE with a lot of chain branches is commonly referred to as low-density PE, while the polymer with the low branch concentration is termed high-density PE. Because of its superior mechanical properties, high-density PE is used for structural parts, while the low-density material is preferred for applications such as food storage bags.

Tacticity

Atactic polymers with large side groups cannot be easily packed in such a way as to establish the long-range order required of a crystalline material. In contrast, it is far easier to establish LRO in isotactic and syndiotactic polymers, with the result that these polymers are more likely to be semicrystalline. As was the case with chain branches, the influence of tacticity on degree of crystallization ultimately influences the macroscopic properties of the polymer. For example, at room temperature isotactic polypropylene (~50% crystalline) is hard and rigid while atactic PP (~0% crystallinity) is a useless gummy substance.

Complexity of the Repeat Unit

Polymers with long repeat units, such as PET and many nylons, require more extensive chain segment motion to establish LRO. As such, they typically crystallize slowly, if at all, and can be easily formed into glasses using modest quench rates.

Secondary Bonds between Chains

All the previous factors on our list have hindered the formation of crystals. In contrast, the existence of small, regularly spaced polar side groups can aid in the formation of polymer crystals by providing a driving force for the necessary alignment of adjacent chain segments. Thus, we should expect isotactic $(C_2H_3Cl)_n$ to form more extensive crystalline regions than isotactic $(C_2H_3CH_3)_n$.

In sum, the factors that hinder crystal formation in polymers are an atactic configuration, bulky side groups, and chain branches. Polar side groups tend to aid in the formation of crystalline regions in polymers. Polymers with large repeat units are generally slow to crystallize. Polymers with simple symmetric structures are generally semicrystalline. Although the rules given in this section are incomplete, they lead to correct predictions in most common engineering situations. We pause to point out that the inability to form extensive crystalline regions is not necessarily a disadvantage; however, it means that the useful temperature range of the material may be determined by T_g rather than T_m. In fact, the reciprocal property, known as glass-forming ability, is often a highly desired material property.

DESIGN EXAMPLE 6.4–4

Predict which polymer in each pair listed is a better glass former:

    a. Isotactic $[C_2H_3(CH_3)]_n$ or syndiotactic $[C_2H_3F]_n$

    b. Atactic $[C_2H_3(CH_3)]_n$ or isotactic $[C_2H_3Cl]_n$

Solution

    a. Both configurations are symmetric and can form semicrystalline structures. The relatively bulky CH_3 group hinders crystal formation in $[C_2H_3(CH_3)]_n$. The secondary bonds associated with the polar F side group in $[C_2H_3F]_n$ favor crystal formation. Thus, we predict that isotactic $[C_2H_3(CH_3)]_n$ is a better glass former, and syndiotactic $[C_2H_3F]_n$ is more likely to form crystalline regions.

    b. Atactic $[C_2H_3(CH_3)]_n$ has two factors working against the establishment of extensive crystalline regions: the atactic configuration, and the bulky CH_3 side group. Isotactic $[C_2H_3Cl]_n$, however, has a regular structure, no bulky side groups, and relatively strong secondary bonds associated with its Cl atom. Therefore, we predict that atactic $[C_2H_3(CH_3)]_n$ is a better glass former and isotactic $[C_2H_3Cl]_n$ is more likely to form crystalline regions.

6.4.5 Semicrystalline Polymers

Even polymers for which factors are favorable for crystallization such as polyethylene are never fully crystalline. As such, they are referred to as semicrystalline, or partially crystalline, polymers. Because the macromolecules are highly entangled in the melt and diffusion rates are low, the chains do not have sufficient time to completely disentangle during solidification.

Recall that crystalline metals and ceramics are composed of grains of single crystals. Semicrystalline polymers are composed of **spherulites** (aggregates of crystalline and noncrystalline regions) as shown in Figure 6.4–8a. Note that the polymer chains are oriented normal to the radius of the spherulite and that the chains fold irregularly. The radial symmetry of spherulites gives them the characteristic "Maltese cross" appearance, shown in Figure 6.4–8b, when viewed under crossed polars.

It is often important to know the extent of crystallization in a semicrystalline polymer. This is most conveniently reported as the weight or volume percent of the polymer that is crystalline, and is known as the degree of crystallinity (χ). The amorphous regions in the semicrystalline polymer can be thought of as massive volume defects. In fact, one of the principal features distinguishing semicrystalline polymers from most other crystalline materials is the percentage of noncrystalline material. In semicrystalline polymers, typical values of χ range from 40 to 95%; in contrast, ceramic and metallic crystals are rarely less than 99% crystalline.

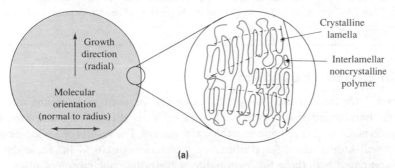

(a)

FIGURE 6.4–8

Spherulites in semicrystalline polymers: **(a)** a schematic of a growing spherulite, and **(b)** the characteristic Maltese cross pattern obtained by viewing magnified spherulites between crossed polars.

(b)

6.4.6 The Relationship between Structure and T_g

In section 6.4.4 we discussed on a qualitative level the relationship between glass-forming ability and molecular structure of polymers. Similar relationships can be developed for estimation of the glass transition temperature. These relationships are based on knowledge that the glass transition temperature indicates the onset of large-scale molecular motion. Since larger repeat units, bulkier side groups, chain branches, and polar side groups each tend to inhibit molecular motion, they also tend to increase the glass transition temperature.

It is interesting to note that for many polymers composed of symmetric monomers (e.g., C_2H_4, C_2F_4, and $C_2H_2Cl_2$) it is found that $T_g/T_m \approx 1/2$, while many polymers composed of unsymmetric monomers (e.g., the vinyls based on C_2H_3R) exhibit the relationship $T_g/T_m \approx 2/3$.

....

EXAMPLE 6.4–5

Explain on a structural basis the differences in the glass transition temperatures of polyethylene ($T_g = -140°C$), polyvinylchloride ($T_g = 50°C$), and polystyrene ($T_g = 100°C$).

Solution

The structures of the repeat units of these polymers are each shown in Table 6.4–1. The polymer backbone is identical in these three polymers. The bulkiness and polarity of the side group are the principal differences between the polymers. The C_6H_5 side group in PS is significantly larger than either the Cl in PVC or the H in PE, so PS has the highest T_g. The strong polar nature of Cl gives PVC a higher T_g than PE.

....

6.5 STRUCTURE AND PROPERTIES OF GLASSES

In Chapter 3 we learned that X-ray diffraction can be used to determine a number of important characteristics of crystalline materials. X-ray diffraction studies on noncrystalline materials can also provide valuable information. The results of a wide-angle X-ray scan on a well-known glass, fused silica, are shown in Figure 6.5–1a. These data can be usefully compared with those for crystobalite, a crystalline polymorph of silica, shown in Figure 6.5–1b. Sharp X-ray maxima are absent from the pattern obtained from the glass, but the position of the broad maximum is approximately the same as that of the first peak in the crystalline material. These results show that nearest-neighbor distances are essentially unchanged in the glass relative to the crystal. In other words, the short-range order (SRO) is the same in crystals and glasses. This result should not be surprising, since the SRO of a structure is uniquely determined by the types of atoms present and the types of bonds between those atoms.

The fundamental structural similarities and differences between glasses and crystals can be further illustrated with an example. The basic building block in silicate structures, amorphous and crystalline, is the $(SiO_4)^{4-}$ tetrahedron (see Figure 6.2–2). In the crystal the tetrahedra are arranged on a periodic lattice, as shown for crystobalite in Figure 3.7–6. In the glass the tetrahedra are joined at the corners in such a way that each oxygen is shared by two tetrahedra (thus satisfying the coordination number of 2 for each oxygen), but the resulting structure lacks the long-range three-dimensional order characteristic of the crystal.

The 3-D structure of silica glass is difficult to draw, so we will resort to a 2-D model. The 2-D model, known as the random network model, is shown in Figure 6.5–2a. The corresponding 2-D representation of a silica crystal is shown in Figure 6.5–2b. Although

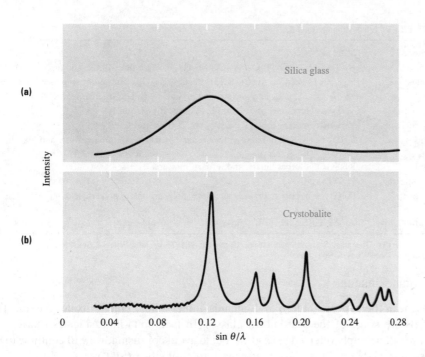

(a)

(b)

FIGURE 6.5–1

A comparison of the X-ray scans for **(a)** amorphous silica and **(b)** a crystalline polymorph of silica (crysto-balite).   *(Source: Adapted from B. E. Warren and J. Biscal, Journal of American Ceramic Society, 2149, 1938.)*

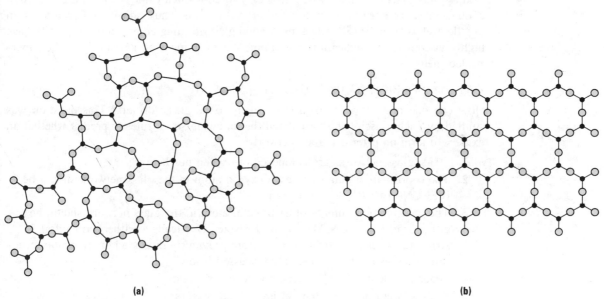

(a)   **(b)**

FIGURE 6.5–2   The 2-D representations of **(a)** silica glass and **(b)** a crystal of silica.

we have described the specific structure of only glassy silica, the structure of other noncrystalline materials is directly analogous. Several additional examples are provided below.

A wide variety of materials can form glasses, including materials from each of the three bonding classifications: metallic, ionic, and covalent. In fact virtually any material is capable of glass formation. The only requirement is that the material be cooled from the liquid rapidly enough that crystal structures are given insufficient time to develop. Table 6.5–1 lists a number of glass-forming compounds.

TABLE 6.5–1   Glass-forming systems.

Elements:	S, Se, P
Oxides:	B_2O_3, SiO_2, GeO_2, P_2O_5, As_2O_5, Sb_2O_3, In_2O_3, SnO_2, PbO_3, and SeO_2
Halides:	BeF_2, AlF_3, $ZnCl_2$, Ag(Cl, Br, I), Pb(Cl_2, Br_2, I_2), and multicomponent mixtures
Sulfides:	As_2S_3, Sb_2S_3, CS_2, and various compounds of B, Ga, In, Te, Ge, Sn, N, P, and Bi
Selenides:	Various compounds of Tl, Sn, Pb, As, Sb, Bi, Si, and P
Tellurides:	Various compounds of Tl, Sn, Pb, As, Sb, Bi, and Ge
Nitrides:	KNO_3-Ca(NO_3)$_2$ and many other binary mixtures containing alkali and alkaline earth nitrates
Sulfates:	$KHSO_4$ and other binary and ternary mixtures
Carbonates:	K_2CO_3-$MgCO_3$
Polymers:	Polystyrene, polymethylmethacrylate, polycarbonate, polyethylene terephthalate, and nylon
Metallic alloys:	Au_4Si, Pd_4Si, (Fe-Si-B) alloys

Source: Robert H. Doremus, *Glass Science,* 1st ed., Copyright © 1973 by John Wiley & Sons. Reprinted by permission of John Wiley & Sons, Inc.

6.5.1 Ionic Glasses

Entries near the top of Table 6.5–1 include ionic solids, chiefly oxide glasses. These materials are some of the best-known glasses. In fact, soda-lime-silicates (Na_2O-CaO-SiO_2) are often simply referred to as glass, but to avoid confusion, we will continue to refer to these materials as oxide glasses. It is easy to cool silica (SiO_2) to form a glass. High cooling rates are not necessary to avoid crystallization. Thus, silica is referred to as an excellent glass former. Like silica, most good glass-forming compositions typically have highly viscous melts, indicating that atomic rearrangement occurs slowly in the super-cooled melt.

Zachariasen's Rules for Oxide Glass Formation

How can one predict which oxides are likely to be glass formers? This question was studied by Zachariasen, who formulated the following set of rules to predict whether an oxide will form an extensive glass network:

1. Oxide glass networks are composed of oxygen polyhedra.
2. The coordination number of each oxygen atom in the glass network should be 2 [i.e., CN(O) = 2].
3. The coordination number of each metal atom in the glass network should be either 3 or 4 [i.e., CN(M) = 3 or 4]. Note: This leads to either tetrahedral structures such as the $(SiO_4)^{4-}$ structure previously discussed or triangular structures such as $(BO_3)^{3-}$, which are discussed below.
4. Oxide polyhedra share corners, not edges or faces.
5. Each polyhedron must share at least three corners.

Some of the common oxides that satisfy Zachariasen's rules are SiO_2, GeO_2, B_2O_3, P_2O_5, and As_2O_5. Since we have already illustrated the basic structure of glassy SiO_2, we will investigate the consequences of Zachariasen's rules for B_2O_3. The first step is to determine the shape of the basic building block (i.e., the shape of the oxide polyhedron). This is done by noting that CN(B) = 3 and CN(O) = 2. As shown in Figure 6.5–3a, the polyhedron that satisfies these dual requirements is a triangular structure with a boron atom at the center and an oxygen at each corner. The CNs of the oxygens are satisfied if each of the three corners is shared by two triangles. Note that this is consistent with rule 5. An examination of the isolated building block shows that it contains one B^{3+} ion and three O^{2-} ions, resulting in the formula $(BO_3)^{3-}$.

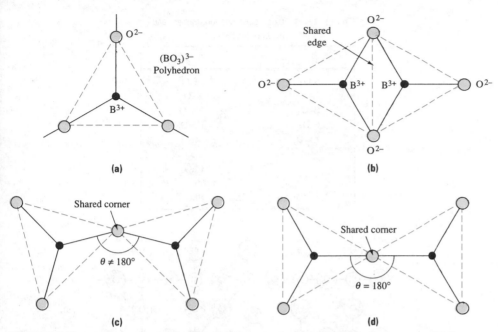

FIGURE 6.5–3   The structure of amorphous B_2O_3: **(a)** the basic building block, a triangular polyhedron $(BO_3)^{3-}$; **(b)** two polyhedra shown sharing an edge; **(c)** two polyhedra sharing a corner, with the B—O—B angle slightly different from 180°; and **(d)** two polyhedra sharing a corner, with the B—O—B angle equal to 180°.

Figure 6.5–3 also shows three possible ways of connecting two adjacent $(BO_3)^{3-}$ polyhedra together. A comparison of the arrangements in parts b and c of this figure can be used to interpret rule 4. Note that the B^{3+} ions are much closer together when the polyhedra share edges. Since the energy of the system is minimized when the separation distance between like charges is maximized, the arrangement with the shared corners is more stable than the one with shared edges (or shared faces). Thus, the arrangement in Figure 6.5–3b is not found in real oxide glass networks. Now consider the difference between the arrangements of the polyhedra shown in parts c and d of Figure 6.5–3. The principal difference is the value of the B—O—B angle. Sincc the B—O bond is partially covalent, a bond angle of 180° is predicted. A bond angle of exactly 180°, however, would lead to a regular and repeating atomic arrangement—that is, a crystal. In contrast, the arrangement shown in Figure 6.5–3c with an average bond angle of ~180° corresponds to the glass structure. The variation of this bond angle gives substantial width to the X-ray peak in glasses.

Because the oxides that satisfy Zachariasen's rules form extensive three-dimensional glass networks, they are referred to as network-forming oxides, or more simply as **network formers.** Other oxides, known as **network modifiers,** are incapable of forming an extended primary bond network. Table 6.5–2 lists the common network modifier oxides, including most of the Group IA and IIA oxides, as well as the five network formers.

The commercial significance of the network modifiers is in the changes they induce in the bond-structure of a glass when they are combined with one or more of the network-forming oxides. As shown in Figure 6.5–4, the network modifiers tend to break up the 3-D primary bond network and, therefore, decrease the primary bond density. As the primary bond density in the glass decreases, the glass transition temperature decreases. Note the

TABLE 6.5–2   Oxide glass network formers and network modifiers.

Network formers	Network modifiers
SiO_2	Li_2O
GeO_2	K_2O
B_2O_3	Na_2O
P_2O_5	Cs_2O
As_2O_5	MgO
	BaO
	CaO
	ZnO
	PbO

FIGURE 6.5–4

A highly schematic 2-D illustration of the way in which the network modifier Na_2O breaks up the 3-D primary bond network in silica glass.

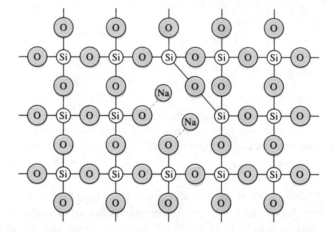

similarity between the glasses with and without network modifiers and the properties of polymers with and without crosslinks. In both cases, as the primary bond density decreases, the material loses its rigidity at a lower temperature.

An understanding of the role of network modifiers permits us to appreciate why glass containers, jars, and windows are made from soda-lime-silica glass while glass crucibles that must be able to retain their shape at high service temperatures are fabricated from vitreous silica (pure SiO_2). In the former application, the glass will not experience high temperatures during service. Thus, it need not have an excessively high T_g. This is important because the fabrication of a glass component must be performed at a temperature above T_g, where the lower viscosity makes shape changes possible. Since a large fraction of the cost of a glass object is associated with the cost of energy necessary to raise the temperature of the glass above T_g for the forming operations, a lower T_g glass composition offers significant economic advantages. In the case of the glass crucible material, the more demanding high-temperature service environment dictates the use of a glass composition free of network modifiers even though such a composition is more difficult and more expensive to process.

Pyrex, a glass composition containing the network formers SiO_2 and B_2O_3, and the network modifiers Na_2O and CaO, offers properties intermediate to those of soda-lime-silicate glass and vitreous silica. It has about three times better thermal shock resistance than the Na_2O-CaO-SiO_2 glass composition, but does not require the excessively high processing temperatures of vitreous silica. Pyrex is often the material of choice for laboratory glassware that must be able to survive changes in temperature ranging from that of ice water to that of a gas burner flame.

Other Ionic Glasses

Appearing next on the list of glass-forming systems in Table 6.5–1 are halide glasses, followed by chalcogenide glasses. The former compositions contain chlorine, fluorine, iodine, or the like. The compositions of the latter include materials containing sulfur, selenium, or tellurium. The halides and the chalcogenides, also falling in the class of ceramic materials, have useful optical, electrical, and magnetic properties. For example, arsenic sulfide glass (a chalcogenide), which has a high resistance to attack by moisture and is transparent to radiation in the infrared region, is used as a "window" material in the nose section of some radar-guided missiles.

6.5.2 Covalent Glasses

Further down the list in Table 6.5–1 are organic glasses, including the common polymer glasses. Like the oxide glasses, polymers form viscous melts and are good glass formers. In some cases the inherent molecular structure of the polymer prohibits crystallization. In other cases, it is necessary to rigidify the structure and reduce molecular motion to prevent devitrification or crystal formation, as is the case in oxide glasses. We have learned how to recognize and build polymer structures that are inherent glass formers. Table 6.5–3 compiles the melting and glass transition temperatures of several useful polymers. In general, the simple polymers are characterized by a glass transition temperature about 1/2 their absolute melting temperature, and the more complex polymers have a glass transition temperature of about 2/3 their absolute melting temperature.

Another class of covalent glasses is the amorphous semiconductors. The useful electrical properties of semiconductors will be discussed in Chapter 10. At this point we simply note that although the vast majority of semiconducting devices are fabricated from crystals, some applications favor the use of amorphous structures. As an example, in solar cell applications, amorphous silicon offers several advantages, including significant cost savings during production and the elimination of property-degrading grain boundaries.

TABLE 6.5–3   Observed melting and glass transition temperatures for selected polymers.

Polymer	T_m (°C)	T_g (°C)
High-density polyethylene	137	−120
Polyvinylchloride	—	87
Polypropylene	170	−16
Polystyrene	240	110
Polyacrylonitrile	320	107
Polytetrafluoroethylene	327	—
Polychlorotrifluoroethylene	220	—
Polymethylmethacrylate	—	100
Acetal	181	−85
Nylon 6,6	265	50
Cellulose acetate	230	—
Polycarbonate	230	145
Poly(ethylene terephthalate)	255	90
Silicone	—	−123
Polybutadiene	120	−90
Polychloroprene	80	−50
Polyisoprene	30	−73

6.5.3 Metallic Glasses

Appearing toward the bottom of Table 6.5–1 are metallic glasses. Metals are perhaps the most difficult materials to prevent from crystallizing because they have small building blocks (single atoms) and simple crystal structures, and the melts are quite fluid. The required cooling rates for forming metallic glasses are in excess of 10^8 K/s. Techniques used to form pure glassy metals include vapor deposition onto a cold substrate and melt quenching of thin ribbons.

Metallic glass-forming systems are multicomponent alloys with relatively low melting temperatures. Typical glass-forming metal alloys contain about 80 atomic percent (at. %) metal and about 20 at. % of a semimetal.[2] Glass formation from these sorts of compositions requires cooling rates on the order of about 10^5 K/s. Two examples of compositions that can be used to form metallic glasses are (80% Au–20% Si) and (78% Fe–9% Si–13% B). The presence of the semimetal reduces the melting temperature of the alloy below that of the pure metal, thereby reducing the temperature range over which the material needs to be rapidly cooled. The semimetal also increases the size of the unit cell, requiring atoms to diffuse longer distances and hence requiring longer times in order to crystallize. Finally, the semimetal may also raise the energy of the crystal by requiring the crystal to distort slightly to accommodate the different-sized substitutional alloying element into the crystal lattice. Each of these effects enhances formation of an amorphous metal.

Glassy metals are unstable to heat. They crystallize readily upon heating to a temperature where the atoms become mobile (i.e., in the range $T_g < T < T_m$). Thus, their use is limited to relatively low temperatures, on the order of a few hundred °C.

The chief experimental difficulty in obtaining a glassy metal is to remove the heat at such a high rate that crystal formation is avoided. Most pure metals such as aluminum or iron have not yet been made amorphous by quenching because heat cannot be removed quickly enough. Even with the best glass-forming metal alloy systems, only ribbons or other thin samples can be made using current technology.

Metallic glasses have interesting magnetic properties (to be discussed in Chapter 12) and are used in transformer cores, in heads for magnetic recording devices, and as security strips in library books. Metallic glasses also exhibit high strength, since they are free of dislocations (a crystalline concept) and are occasionally used as reinforcing fibers in pressure vessels and tires. Finally, metallic glasses are by definition free from grain boundaries. This gives them excellent corrosion resistance (to be discussed in Chapter 15) and has led to such applications as razor blades.

6.6 STRUCTURE AND PROPERTIES OF RUBBERS AND ELASTOMERS

Rubbers are noncrystalline materials for which the glass transition temperature is below the ambient temperature. Most glassy materials crystallize when they are heated above T_g, when molecular motion becomes possible. However, some polymers possess unique molecular architectures that preclude crystallization. In these materials, the polymer chains are highly irregular and asymmetric. They simply cannot pack together to form a crystal, or the energy required to do so is prohibitive.

---

[2] The term *semimetal* is used to describe an element with behavior intermediate between the metals and the nonmetals. One way of distinguishing the three classes of elements is based on their electronegativity (EN) values. Metals typically have EN < 1.8, nonmetals have EN > 2.2, and the semimetals have 1.8 < EN < 2.2. The most common semimetals are B, Si, Ge, As, Sb, and Te.

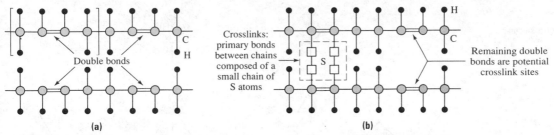

FIGURE 6.6−1   The structure of polybutadiene: **(a)** as a thermoplastic polymer, and **(b)** as a crosslinked rubber.

6.6.1 Thermoset Elastomers

When a thermoplastic polymer is heated above its glass transition temperature and placed under load or stress, the molecules flow or slip past one another. To make an elastomer, the molecules must be prevented from slipping past one another. This is generally achieved by lightly crosslinking the polymer. The crosslinks, which are covalent bonds between chains, provide a memory to the chains, so that molecular segments cannot slide past one another irrecoverably. The crosslinks allow the structure to return to its original coiled configuration after the stress is released.

An example of crosslinking is shown in Figure 6.6−1. Each mer in polybutadiene contains one unsaturated bond, a carbon-carbon double bond. After the linear polymer is formed, the double bonds can be reacted with sulfur, causing crosslinking between molecules. If only 1 in 1000 mers is reacted, the crosslinking is light, and a flexible rubber is formed—perhaps a rubber band. If the crosslinking is 1 in 10 mers, then the crosslinking is heavy, and a hard, brittle material is produced. An automobile tire is an example of a moderately crosslinked material.

....

EXAMPLE 6.6−1

How much weight can an initially 10-kg sample of polybutadiene rubber gain by complete reaction with oxygen? Assume that on average a single crosslink consists of a chain of two oxygen atoms.

Solution

Figure 6.6−1 shows that there is one double bond per monomer, so that the maximum number of crosslink sites is equal to the number of mers in the polymers. (Each double bond opens to form a pair of crosslinks, but each crosslink is shared between two mers so that the ratio of crosslinks to mers is 1:1.) The number of moles of mers in the sample, N, is found by dividing the mass of the sample by the molecular weight of a mer to yield:

$$N = \frac{10,000 \text{ g}}{4(12) + 6(1) \text{ g/mol}} = 185 \text{ moles}$$

Therefore, there are 185 moles of crosslink sites in this polymer. Since there are approximately two oxygen atoms per crosslink, we require $(2 \times 185) = 370$ moles of oxygen to completely crosslink the rubber. The total weight gain during the crosslinking process would be:

$$(370 \text{ mol O}) \times (16 \text{ g/mol}) = 5920 \text{ g} = 5.92 \text{ kg}$$

....

All rubbers are polymers. The most well-known rubbers are **elastomers,** materials that can be deformed several hundred percent and recover substantially completely. The most common elastomers are of the thermoset type described in the preceding paragraphs, polymers that are lightly crosslinked. These materials do not melt to a fluid state. A quick glance back at the bottom of Table 6.5-3 indicates that these rubbers have glass transition temperatures well below room temperature. An example of another class of crosslinked

FIGURE 6.6–2
The structure of a silicone
rubber.

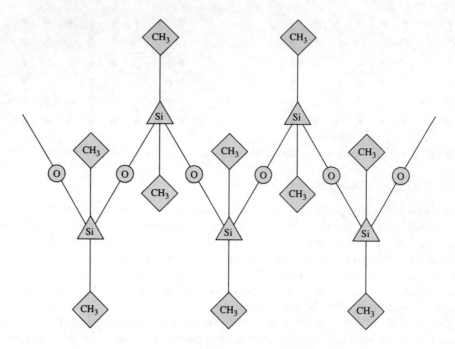

rubbers, based on silicon rather than carbon, is shown in Figure 6.6–2. Silicone rubber, polydimethylsiloxane, contains Si—O backbone bonds and organic (methyl, CH_3) side groups. You may be familiar with silicone rubbers in the form of caulking compounds used to waterproof windows, doors, and bathtubs around the home.

6.6.2 Thermoplastic Elastomers

The first thermoplastic elastomers were developed about 25 years ago. Unlike thermoset materials, thermoplastic polymers have the property that they can be repeatedly melted, shaped, and solidified. The molecules in thermoplastic polymers are not crosslinked. Instead, thermoplastic elastomers have relatively complex chemical structures. As shown in Figure 6.6–3a, each chain consists of blocks of mers, and hence, these materials are called

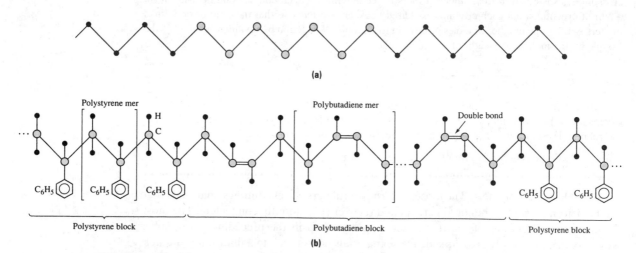

FIGURE 6.6–3　Copolymers and thermoplastic rubbers: **(a)** a schematic illustration of a block copolymer, and **(b)** the structure of the triblock copolymer thermoplastic rubber polystyrene-polybutadiene-polystyrene.

block copolymers. The solid polymer consists of regions where the segments of one type of mer cluster and regions where the segments of the other type of mer cluster. In this copolymer, one set of segments, called the hard segments, are rigid, either a crystal or a glass, at use temperature. The other segments, called the soft segments, are highly flexible, or fluid-like, at use temperature. The soft segments are elastomeric, and the hard segments prevent the long soft segments from flowing without bound under stress. Thus, the polymer consists of islands of hard segments in a sea of soft segments.

An example of a thermoplastic elastomer is the triblock copolymer of polystyrene-polybutadiene-polystyrene (S-B-S). The structure of the material is shown in Figure 6.6–3b. S-B-S has found commercial use because it can be repeatedly heated to a state where both the hard and soft segments are in the melt and processed using standard techniques for thermoplastic polymers. Applications for thermoplastic elastomers include gaskets, tires, adhesives, and wire insulation.

6.6.3 Crystallization in Rubbers

The molecules in a polymer melt, glass, or rubber can be accurately envisioned as entangled and coiled upon one another. Upon stretching, the molecules become aligned with the direction of the applied load. Such molecular alignment, illustrated in Figure 6.6–4, reduces the barrier to crystal formation. Recall from Chapter 3 that the molecules in polymer crystals consist of aligned sections of molecules. Recall from Chapter 4 that diffusion in polymers is slow, so movement of molecules into the conformation suitable for crystal formation is slow. Manually aligning the molecules with stress can increase the rate of crystal formation by up to 100 orders of magnitude.

Several years ago natural rubber was used to make rubber bands. The polymer crystallized upon elongation, producing heat. If the heat were removed, the polymer would stay elongated. Recovery from stretching would occur only when heat was provided to melt the crystals. The composition of a rubber band was changed to prevent the "lazy contraction." The effect of crystallization was put to use as an advantage in the mid 1980s. Using elastomeric random block copolymers containing a soft block of poly(tetramethylene oxide), [—CH_2—CH_2—CH_2—CH_2—O—], and a hard block of nylon, a polymer with soft segments that crystallize when the length is increased more than a factor of 4 was produced. In the stretched and crystallized state, the polymer could be easily applied to fabrics. The elasticity of the material is recovered by melting the crystals with a heat gun. This is one method by which "elastic" gathering has been incorporated into clothing and diapers.

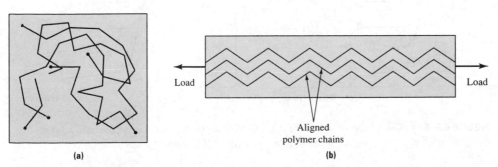

(a)   **(b)**

FIGURE 6.6–4   Changes in polymer structure as a result of an applied load: **(a)** unaligned chains in an unloaded polymer, and **(b)** chains aligned under an applied load.

6.6.4 Temperature Dependence of Elastic Modulus

Technically, rubbers need not be elastomers. Glassy polymers that are not crosslinked soften to a viscous state upon heating to temperatures above the glass transition temperature. If the polymer is lightly crosslinked, the polymer assumes a rubbery state upon heating. This effect is best illustrated by plotting the elastic modulus as a function of temperature, as shown in Figure 6.6–5a. Glassy polymers are characterized by a modulus of about 7000 MPa. Rubbers are characterized by a modulus of 0.7–70 MPa. In the absence of crosslinking, these materials have little use above T_g because of viscous flow. When the rubbery plateau occurs at room temperature and the molecules are tied down at points, the materials are elastomers. The temperature range over which the modulus drops from a value characteristic of a glass to one of a rubber is the glass transition temperature.

As shown in Figure 6.6–5b and c, the modulus of the plateau at temperatures greater than T_g increases with increasing crystallinity or crosslink density. At very high levels of crosslinking or crystallinity the material does not have a glass transition, and the elastic modulus remains high until degradation in the case of a crosslinked polymer, or melting

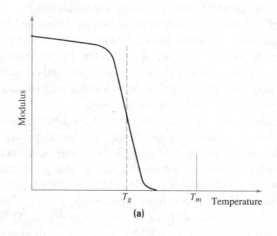

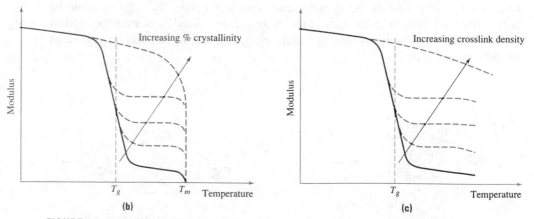

FIGURE 6.6–5   The relationship between modulus and temperature. **(a)** An amorphous uncrosslinked polymer. **(b)** The influence of increasing percent crystallinity. **(c)** The influence of increasing crosslink density.

in the case of a semicrystalline polymer. Many glassy thermoset polymers, such as epoxies and phenolics, do not soften prior to degradation.

Returning attention to Figure 6.6–5a, we note that as the temperature of an entangled amorphous polymer is increased to a value greater than that characteristic of the rubbery plateau, the modulus of the material falls to zero. This is the point where all secondary bonds between molecules are surpassed by the average thermal energy available to the atoms in the system. A force exerted on the material no longer produces a finite deformation or strain but rather produces a continuous deformation or strain rate. This is the region viscous flow.

6.6.5 Rubber Elasticity

An elastomer is capable of sustaining a tensile deformation perhaps 10 or more times its original length, from which it recovers almost completely. The mechanism of extension and recovery characteristic of most materials, as described in Chapter 2, is the stretching of atomic bonds. If the deformation of elastomers simply involved stretching atomic bonds, then the extent of deformation would be substantially less than that observed. The rubber deformation mechanism is not simple bond stretching and flexing; rather it involves the uncoiling of molecular chains. As shown in Figure 6.6–6a, the molecules in an unstretched rubber are in a random coil conformation. When the material is stretched, the chain segments uncoil and the end-to-end separation increases, as shown in Figure 6.6–6b. Upon removal of the stretching force, the molecules return to a random coil conformation.

To appreciate the difference between the elastic deformation that occurs by bond stretching (energy elasticity) and that which occurs by the uncoiling of polymer chains (entropy elasticity), it is useful to compare the magnitudes of the maximum elongation resulting from each mechanism. First consider the bond-stretching mechanism. The bond-force diagram introduced in Chapter 2 is recreated in Figure 6.6–7. The equilibrium separation distance x_0 is roughly 0.2 to 0.3 nm. Figure 6.6–7 shows that an increase in separation distance of magnitude δ would cause the atoms to become unbonded. The magnitude of δ is typically 0.03 to 0.05 nm. By dividing the change in separation distance

L = Distance between chain ends in coiled conformation

(a)

(b)

L_{ext}

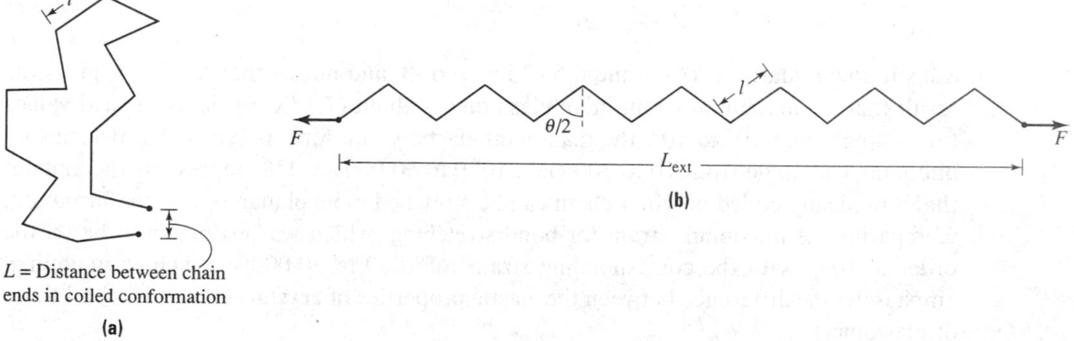

FIGURE 6.6–6   A comparison of (a) the random coil conformation of an unstretched rubber, and (b) the stretched, or planar zigzag, conformation in which the separation distance between the chain ends is maximum.

FIGURE 6.6–7

The bond-force diagram. The quantity x_0 is the equilibrium separation distance, and δ represents the displacement necessary to cause the two atoms to "fly apart."

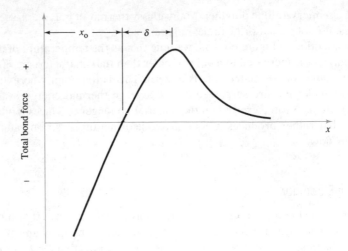

necessary to cause decohesion by the equilibrium separation distance, we find that the maximum elastic strain in a solid that deforms by bond stretching is on the order of 10%. In most materials, it is in fact much smaller.

In comparison, with reference to Figure 6.6–6b, the length of a fully extended hydrocarbon polymer chain, L_{ext}, is:

$$L_{ext} = ml \cos\left(\frac{\theta}{2}\right) \tag{6.6–1}$$

where m is the number of carbon-carbon bonds in the backbone of the polymer chain, l is the length of a carbon-carbon bond, and θ is 180° minus the C—C—C bond angle (109.5°). The end-to-end separation distance for the coiled polymer chain shown in Figure 6.6–6b, which we present here without derivation, is:

$$L \approx l\sqrt{m}\left(\frac{1 + \cos\theta}{1 - \cos\theta}\right) \tag{6.6–2}$$

As before, the maximum permissible elastic strain is found by dividing the change in separation distance by the original separation distance:

$$\text{Maximum strain} = \frac{L_{ext} - L}{L} = \left(\frac{L_{ext}}{L}\right) - 1 \tag{6.6–3}$$

Substituting Equations 6.6–1 and 6.6–2 into 6.6–3, and noting that $\sqrt{m} \gg 1$, gives the result that the maximum strain in an elastomer is about $(1.15\sqrt{m})$. Since typical values for m range from 10^2 to 10^5, the maximum elastic strain for a polymer that deforms by uncoiling can range from 10 to 300 (i.e., 1000 to 30,000%). This represents the amount that a randomly coiled polymer chain can be stretched to its planar zigzag conformation. Comparing the maximum strain for bond stretching, which was estimated to be on the order of 10%, with the corresponding strains of 1000 to 30,000% for chain uncoiling, emphasizes the difference between the elastic properties of crystals and glasses and those of elastomers.

The entropy associated with a polymer chain of given length is directly related to the number of ways the chain can be arranged while satisfying the required bond angles and maintaining a constant separation distance between chain ends. As shown in Figure 6.6–8a, only one conformation is possible for the planar zigzag arrangement. In

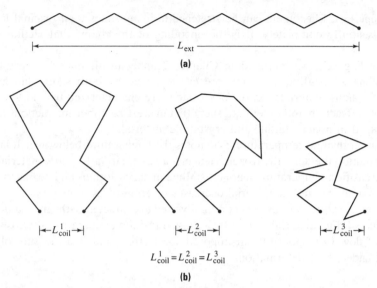

FIGURE 6.6–8   A comparison of the number of equivalent conformations for **(a)** the planar zigzag conformation and **(b)** a partially coiled chain.

contrast, as shown in Figure 6.6–8b, as the ends of the chain are brought closer together, the number of equivalent conformations increases. Thus, as a polymer chain is stretched, its number of equivalent conformations decreases and its entropy decreases. Since the lower-entropy conformations are less stable than those with higher entropy (see the discussion on free energy in Section 4.2.1), a stretched polymer chain is less stable than a coiled polymer chain. Outside the crystal (which can have a prohibitively high activation energy for formation) the most stable structure is the random coiled structure with an end-to-end separation given by Equation 6.6–2. Thus, a positive force (or a tensile stress) is required to extend the length of the sample, as is consistent with our everyday experiences. Removal of the force results in a tendency for the rubber to return to its coiled conformation.

Another consequence of the dominance of high conformational entropy in elastomers is found in their peculiar thermal behavior. While most crystalline materials expand when their temperature is raised, elastomers shrink when their temperature is raised because of their increased mobility and, hence, ability to coil.

SUMMARY

In this chapter we have developed a basic understanding of the structure and properties of amorphous, or noncrystalline, materials. Essentially any material—polymers, ceramics, metals—can be made into a glass by rapidly cooling the material from the melt in such a way that crystal formation is prevented. With some materials, crystal formation is very slow, and these materials can be formed into glasses easily. Examples include many polymers and silicate oxide glasses. Polymers that are good glass formers are those that have large unit cells, large monomers, or random copolymers. Atactic polymers with large, bulky side groups or extensive chain branching may not be able to form crystalline regions under any conditions. These materials not only form glasses, but can be heated to above the glass transition temperature without crystal formation. Only polymers can form rubbers, since the long-chain nature of the material gives this class of materials unique properties. Elastomers can be stretched to many times their original length and recover essentially completely. It is the uncoiling of the chains that facilitates rubber elasticity.

The bonding schemes outlined in Chapter 2 apply to all materials, crystalline or noncrystalline. Crystalline materials are characterized by both short- and long-range order. Amorphous materials, on the other hand, are characterized by short-range order only. X-ray diffraction and other data show that nearest-neighbor interactions are essentially identical in noncrystalline and crystalline materials.

The glass transition temperature T_g denotes the temperature below which large-scale molecular motion is essentially frozen. Hence, glasses are usually brittle materials. Above the glass transition temperature, a noncrystalline material softens and begins to flow. The modulus of a noncrystalline material decreases several orders of magnitude at the glass transition temperature. Viscosity is commonly used to characterize the mechanical behavior of noncrystalline materials above their glass transition temperature. Viscosity reflects the ease of flow in response to an imposed shear stress and is a measure of the heat dissipated under shear deformation.

KEY TERMS

amorphous	glass	nylon	vinyl
atactic	glass transition	polyamide	viscosity
block copolymer	temperature T_g	polyester	volumetric thermal
configuration	isotactic	rubber	expansion coefficient α_v
conformation	network formers	spherulites	
degree of polymerization	network modifiers	supercooled liquid	
elastomers	noncrystalline	syndiotactic	

HOMEWORK PROBLEMS

SECTION 6.1
Introduction

1. Compare the structure of an amorphous solid to that of the same material in the liquid form. Is the density the same? What are the units for reciprocal density?

2. The text states that if cooling is rapid enough, then glasses can be formed from any material. What process are you trying to "beat" by cooling rapidly?

3. How do you expect the specific heat of an amorphous material to change as it is heated through its glass transition? Specific heat is the amount of heat necessary to raise the temperature of a fixed mass of material 1°.

SECTION 6.2
The Glass Transition
Temperature

4. Consider the following thermal treatment of oxide glass. The molten glass is shaped into a slab about 3 mm thick; then both surfaces of the glass are rapidly cooled to room temperature. The center of the slab cools slowly to room temperature. How do the glass transition temperature and specific volume of the material near the surface of the slab compare with those of the material near the center of the slab? This cooling rate effect is used to strengthen the glass.

5. Estimate the volumetric thermal expansion coefficient of ordinary window glass, given that the melt has a linear thermal expansion coefficient of $(10 \times 10^{-6})°C^{-1}$.

6. Derive an approximate relationship between the linear and volumetric thermal expansion coefficients. (Hint: Consider a cube of material, side length s, that expands to $s + \delta s$ upon heating to a certain temperature. Calculate the expanded volume and discard small terms.)

7. Sketch the volume-temperature curve for polystyrene, realizing that it is glassy at room temperature and has a T_g of about 110°C.

8. Sketch the volume-temperature curve for pure Al through the melting temperature.

9. Is the glass transition temperature a well-defined temperature, or is it actually a range of temperatures? Why?

10. The glass transition temperature of a polymer is defined as the temperature at which mers become free to move upon heating or frozen upon cooling. It requires excess energy to rise above the glass transition temperature because the mers absorb energy at the glass transition temperature. It is possible that units smaller than mers are capable of energy absorption. Such transitions are capable of providing high impact resistance to polycarbonate and enable characterization of polymers. Will such transitions occur at temperatures below or above the glass transition temperature? Why?

11. Suppose you work for NASA and your job is to design a flexible, rubberlike material for use in outer space. What are the difficulties of this assignment? (Recall that the *Challenger* exploded in 1986 because of a faulty rubber seal.)

12. Provide examples of materials that behave like Silly Putty.

SECTION 6.3
Viscous Deformation

13. Motor oil manufacturers sometimes boast that their oil does not "thin" (decrease in viscosity) as the engine temperature is raised. Is this normal behavior, or do they really have a unique feature that they can effectively use in promotional literature?

14. An experiment is conducted using high- and low-viscosity oils. Two drops of oil are placed on a glass slide and the oil is covered with another slide. In this way the slides are separated by 0.1 mm of oil. The upper slide is pushed parallel to the lower slide at 0.05 cm/s. Is it more difficult to push the slide with the higher or lower viscosity oil? Why? Scotch tape consists of a thin layer of oil on a polymer film. The oil never dries. How does Scotch tape work?

15. Honey and water at room temperature are both forced through a capillary, which is a tube with a small hole. Five cubic centimeters of water flow through the capillary under the force of gravity in just two seconds. Pressure is applied to the honey to force it through the hole, also in two seconds. Immediately after the fluid is passed through the capillary, the temperature of the fluid is measured. What do you expect to find?

SECTION 6.4
Structure and Properties
of Amorphous and
Semicrystalline Polymers

16. Draw the atactic and isotactic configurations of polyvinylalcohol. One form is more likely to be semicrystalline, whereas the other is amorphous. Which is which, and why?

17. Sketch the possible configurations of polyvinylcyanide, otherwise known as polyacrylonitrile, $[—CH—CH_2—]$.
    $$\begin{array}{c} | \\ C \equiv N \end{array}$$

18. Which polymer is more likely to be crystalline: $[—CH_2—CF_2—]_n$ or $[—CH_2—CHF—]_n$? Explain your answer.

19. Given that PET melts at about 255°C, estimate its glass transition temperature, and compare the value to that given in Table 6.5–3.

20. Calculate the number of mers in a commercial sample of polypropylene with a molecular weight of 150,000 g/mol.

21. Calculate the molecular weight of a mer of cellulose. If the molecular weight of cotton were 9000 g/mol, how many mers would be joined?

22. Pentaerythritol is a tetrafunctional monomer that is occasionally combined in small quantities with the monomers used to synthesize PET. How will the addition of pentaerythritol affect crystallinity? As more and more is added, how does the glass transition temperature change?

23. Radiation, especially high-energy radiation, breaks the covalent bonds in polymers. How will high-energy radiation affect the crystallinity of a semicrystalline polymer?

24. When polypropylene is observed under crossed polars, the viewing screen is completely filled with Maltese cross patterns. Does this indicate that the polymer is completely crystalline? If so, why? If not, why not?

25. When a transparent PET film is stretched slowly at room temperature, it often remains transparent; however, when it is stretched at 130°C, it may become opaque. Why?

26. Cellulose contains a number of oxygen atoms. How many O atoms are in each mer, and how many OH groups are in each mer? Note that the OH groups provide cellulose and all vegetable fibers with an important characteristic—the ability to absorb water.

SECTION 6.5
Structure and Properties of Glasses

27. Why are CaO and Na_2O added to SiO_2 in most commercial oxide glasses, such as window and beverage glasses?

28. Use Zachariasen's rules to predict whether lead oxide is a good glass former.

29. Epoxies and polyesters are often used as the matrix material in fiber-reinforced composites. Both matrix materials are hard and brittle and heavily crosslinked. Are they crystalline or glassy? Why?

30. In the text we discussed the formation of glassy metals. When a glassy metal is heated above room temperature, it will crystallize at some temperature. Design a method or experiment that would allow you to detect, or measure, when crystallization occurs.

31. Do you expect to have a good glass former when you combine lead oxide, which has a high density and a high refractive index, with silicon dioxide, given that lead oxide is completely soluble in the melt of silicon dioxide? "Crystal" is the result, which is really a high-lead silica glass.

32. A silica glass containing soda is placed in a salt melt to effect exchange of ions. Potassium ions exchange with sodium ions. Will the glass transition temperature increase or decrease as a result of the exchange? What if lithium ions are exchanged for the sodium? Hint: check ion size.

33. Allied-Signal Corporation is the only domestic manufacturer of amorphous metals, Metglas. Its biggest-selling item is an iron-based film used in magnetic devices. The composition of the alloy is 78 wt. % Fe, 13 wt. % B, and 9 wt. % Si. Comment on the roles of boron and silicon in the alloy.

SECTION 6.6
Structure and Properties of Rubbers and Elastomers

34. Plot the effect of aligning molecules along the fiber axis on the modulus of a polymer fiber. What happens when the molecules are completely aligned?

35. What polymer would you use to make a flexible seal for a liquid nitrogen tank (liquid nitrogen boils at 77 K)?

36. Consider a polymer that consists of molecules that bond at 180°. Calculate the equilibrium end-to-end separation of such a macromolecule with a degree of polymerization of 100 and a mer length of 15 Å. Compare this value with the stretched-out length. This material forms a liquid crystal polymer.

37. A sandwich bag is made using polyethylene. Spectra 1000 fiber is also made using only polyethylene. The modulus of the bag is roughly 5 GPa, and that of Spectra is roughly 150 GPa. Account for the difference.

38. Examine the chemical structure of poly(dimethylsiloxane). What is unique about this polymer, and how does this uniqueness show in properties?

39. Calculate the equilibrium end-to-end separation of polystyrene with a degree of polymerization of 5000.

40. A glassy sample of unoriented atactic polystyrene has a molecular weight of 150,000 g/mol. What is the approximate separation of a molecule's chain ends?

41. Sketch the change in modulus with temperature of a polymer that is semicrystalline. The crystallinity is estimated to be about 50%, the glass transition temperature about 0°C, and the melting temperature 160°C. The polymer is isotactic polypropylene.

42. A given epoxy has a molecular weight between crosslinks of about 1000 g/mol. How will the mechanical properties of the epoxy change if the epoxy is reformulated so that the molecular weight between crosslinks (a) increases or (b) decreases?

43. A lightly crosslinked rubber band is stretched several hundred percent. If the restraining force were removed, then the rubber band would recover. The stretched rubber band is, however, placed in a freezer. In one case the freezer is below the glass transition temperature and in the other case it is not. In either case, the rubber band stays stretched when the restraining force is removed. What might be the reason for this behavior?

44. A garment containing synthetic fibers is washed and dried. During the process the garment shrinks. Use your knowledge of entropy to explain the shrinkage behavior of the fibers.

45. Why is the modulus of a rubber so much less than that of a ceramic or oxide glass?

46. Calculate the end-to-end separation of a 150,000 g/mol chain of polypropylene that is (a) coiled on itself or (b) stretching out as much as possible.

MICROSTRUCTURAL DEVELOPMENT

t is rare for a pure element to be used in an engineering application. Most materials of practical interest are a mixture or compound of two or more elements and often contain multiple solid phases. A solid phase is a chemically homogeneous portion of a material. Physical properties of engineering materials such as strength, hardness, toughness, and fatigue and corrosion resistance depend on both chemical composition and microstructure. Microstructural features can range in size from a few hundred angstroms ($\sim 10^{-8}$ m) to a few hundred micrometers ($\sim 10^{-4}$ m). In Chapters 7 and 8 we will discuss the concepts of phase equilibria and the kinetics of structural transformations such as those that occur during solidification. This knowledge provides us with a conceptual framework for predicting the phases present at a given temperature, their amounts, and their distribution in the microstructure.

Let us return to the example of camcorders to see how our knowledge of phase equilibria and kinetics of microstructural evolution can assist us in the design and manufacture of high-quality camcorders. The compactness and therefore weight of handheld camcorders depends on successfully making hundreds of electrical interconnections to efficiently use space on the circuit board. These connections must be reliable, which means that they must be able to withstand mechanical forces and conduct electrical signals without attenuation. An example of such interconnections is shown in the accompanying figure. Solders, which are low-melting-point metal alloys that contain primarily tin and lead, are used to make these connections.

Metal alloys satisfy the requirement for good electrical conductivity. The low melting point of solders helps make connections without exposing the electronic assembly to excessive heat and the risk of failure in the components. The Sn contained in solders easily reacts with metals such as Cu, Au, Ag, and Al used for making electrical lead wires. The reaction product is an intermetallic compound that participates in making the bonds on either side of the solder.

Sn and Pb melt at 232°C and 327°C, respectively. However, an alloy containing 61.9% Sn and 38.1% Pb melts at 183°C, which is lower than the melting point of either metal. Therefore, this composition is used as a common soldering alloy. In Chapter 7 we discuss melting and the microstructure produced during solidification. The microstructures of the interconnect joints can be

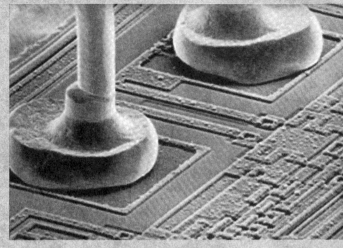

Magnified picture of an electronic package showing soldered interconnections. *(Dr. Jeremy Burgess/Science Photo Library)*

further optimized by making small compositional changes and by controlling cooling rates during solidification of the solder. The mechanical strength and corrosion resistance of the joint are influenced by its microstructure. These topics will be explored in Chapter 8.

Another fascinating application of phase diagrams in electronic component packaging such as in camcorders is the attachment of a silicon chip to a ceramic substrate. This bond is important in providing adequate protection to the chip from shock loading. The back of the chip is attached to the substrate with the help of a thin layer of gold. Au melts at a temperature above 1000°C, and Si at a temperature of approximately 2550°C, but a mixture of Au and 3% Si melts at an astonishingly low temperature of 363°C. Thus, a thin layer of gold is first applied to the surface of the cavity where the chip is to be attached on the substrate. The chip is then placed in the cavity and heated to approximately 400°C. Si diffuses into the gold at the interface and creates a melted region, which solidifies when cooled to create a strong bond between the chip and the substrate. This bond ensures that the integrated chip remains firmly in place when the camcorder is handled during manufacture, shipping, and use.

The discovery of solders and the Si-Au attachments would not have been possible without an understanding of concepts of phase equilibria. Similarly, improvements in the strength of the soldered joints would not be possible without understanding how microstructures evolve during solidification. Part II of the book deals with these topics.

PHASE EQUILIBRIA AND PHASE DIAGRAMS

MATERIALS IN ACTION   Phase Diagrams

The structure and state of aggregation in which a material exists may depend upon the temperature, pressure, and proximity of other materials. Examples of this abound in our common, everyday experience. For example, we all know that oil and water do not mix. We can add water to oil and shake vigorously in an attempt to mix them but the best we can do is to produce a dispersion of water in oil which soon separates out into the two components. In this case the oil doesn't affect the water and vice versa. Each exists independently of the other. However, the situation is entirely different for sugar and water. If we attempt to add a teaspoon of sugar to iced water (e.g., in iced tea) we find that only a small amount of the sugar dissolves in the iced tea and the rest can be seen at the bottom of the glass. However, if we attempt to add the same amount of sugar to hot tea the sugar goes into solution readily. In this case the solubility of the sugar depends on the temperature and both the water and the sugar are affected by the other. Another example of the relationship between temperature, pressure, and state of aggregation is seen in ice skating. Careful measurements have shown that ice is not really slippery and actually has a rather high coefficient of friction. Yet an ice skater is able to gracefully glide across the ice. How can this be? The answer lies in the pressure/temperature relationship between ice and water. At the normal temperatures of ice, only relatively modest pressure is required to convert the ice to water. More than enough pressure is applied through the blade of a skate to form a thin film of water which "lubricates" the ice/blade combination and allows the skater to glide. Two solids in close proximity can react to form a liquid. Solid water and solid road "salt" are not stable as solids in close proximity for temperatures above about −6°C. Thus when salt is spread on an icy road a liquid solution (which is the equilibrium state at this temperature for these materials) results and the ice is removed.

Another example of this phenomenon is found in the microelectronics industry where silicon chips are attached to ceramic substrates.

In all of these examples, a material or combination of mateirals exist in equilibrium for a given combination of temperature, pressure, and composition. In this chapter we will study the factors which govern the equilibrium state of materials (called phase equilibria) and develop "maps" which incorporate temperature, pressure, and composition (called phase diagrams) which are used to present such information.

7.1 INTRODUCTION

One of the recurring themes in materials engineering is that the properties of a material depend strongly on its underlying structure. As a macroscopic example, consider an electrical cable composed of an inner cylindrical core of one material surrounded by one or more concentric layers of different materials. The overall properties of the cable depend on several factors, including:

1. The number of concentric layers
2. The composition (and corresponding properties) of each layer
3. The thickness of each layer (i.e., the relative amount of each material)
4. The spatial arrangement of the layers (the properties of a cable with a metal core surrounded by plastic are different from those of one in which the materials exchange positions)

Similarly, in steel-reinforced concrete the properties of the mixture depend on the size, shape, number, and spatial distribution of the steel-reinforcing rods. Although both the cable and steel-reinforced concrete are examples of macroscopic mixtures of materials, the same dependence of properties on (micro) structure occurs in engineering alloys composed of mixtures of multiple types of materials on the microscopic scale.

In this chapter (and the next) we will learn how to predict the microstructure of a material as a function of its composition and thermal processing history. Armed with this knowledge, we can not only understand the structure-property relationship but, perhaps more important, develop methods to alter the microstructure of a material in order to achieve desirable combinations of properties. For example, we will learn how changes in composition and thermal processing can be used to vary the properties of Al alloys from an extremely soft and ductile material like aluminum foil to the high-strength alloys used as structural components in aircraft.

We begin by defining the term **phase** as a chemically and structurally homogeneous region of a material. Thus, a single-phase material is one that has the same composition and structure at every point. Many materials, however, are composed of two or more phases. For example, ice water is a mixture of a solid phase and a liquid phase. Most plain carbon steels are mixtures of one phase that is almost pure iron and a second phase that contains a significant amount of carbon. For polyphase microstructures, the overall properties of the material depend on: (1) the number of phases present, (2) their relative amounts, (3) the composition and structure of each phase, and (4) the size and spatial distribution of the phases.

The microstructure of a material can have great political and strategic significance. In World War II, German army food supplies were packaged in tin cans. During the Winter Campaign in Russia, German soldiers found that the tin cans had crumbled and their food was completely contaminated with small bits of metal! The reason for this can be found in the temperature dependence of the structure of tin. At moderate temperatures, tin has a body-centered tetragonal (BCT) structure. At low temperatures, however, it undergoes a phase transformation from BCT to a diamond cubic (DC) structure. The DC polymorph has a greater specific volume (27% larger) than the BCT polymorph, so upon cooling the polycrystalline tin cans experienced local expansion, causing internal stresses and failure.

The microstructure of a material depends upon its overall composition as well as variables such as temperature and pressure. As the composition becomes more complex, (e.g., involves more elements), the number of phases that may be present increases. Due to the large variety of commercial materials of complex composition, a systematic method is needed to characterize materials as a function of temperature and pressure. Phase diagrams provide a convenient method for achieving this goal. A **phase diagram** is a graphical representation of the phases present and the ranges in composition, temperature, and pressure over which the phases are stable.

All phase diagrams presented in this chapter are equilibrium phase diagrams. They pertain to the lowest-energy state of the system, in which the compositions, structures, and amounts of the phases do not vary with time at a given temperature and pressure. In general, the time required to reach equilibrium decreases as the temperature increases, since the rate of atomic rearrangement (diffusion) increases with temperature. In Chapter 8 we will deal with the kinetics of phase transformations (i.e., the influence of time on the rate of approach to equilibrium).

We have elected to restrict our discussion to relatively simple phase diagrams. We note, however, that the concepts developed using these simple systems can be applied directly to the more complex materials generally found in engineering products.

Before concluding our introduction to phase diagrams we must explain the conspicuous absence of polymer diagrams in this chapter. Phase diagrams are of limited use in polymer science for three reasons:

1. Equilibrium is very slowly achieved, so slow in fact that most melt-processed polymers never reach equilibrium. In general, only polymers in dilute solutions can be expected to achieve equilibrium.
2. Polymers consist of many molecules, each with a different molecular weight. (Hence, we use molecular weight averages.) Since each molecule behaves differently, modeling polymer systems is rather complex.
3. Experimental observations and recent calculations show that polymers are usually incompatible (i.e., they do not mix) unless they have very similar chemical structures.

Thus, at this time polymer phase diagrams have limited practical importance and will not be discussed further.

7.2 THE ONE-COMPONENT PHASE DIAGRAM

In the previous discussion we referred to the complexity of the composition of a material. This complexity is characterized by stating the number of components from which the material is composed. A **component** can be loosely defined as a chemically distinct and essentially indivisible substance. The most common components are elements (e.g., Fe, Si, or C) and stoichiometric compounds (e.g., NaCl, Li_2O, or Si_3N_4). Examples of single-component systems include any pure element or molecular compound such as H_2O. Examples of two-component systems include plain carbon steel (Fe + C) and seawater (NaCl + H_2O). Common window glass (SiO_2 + Na_2O + CaO) is an example of a three-component system.

Because of the importance of the terms *phase* and *component,* it is appropriate to pause and clarity their relationship using some examples. Pure FCC Al is a single-component (Al), single-phase (FCC) system. A mixture of pure ice and pure water is a single-component (H_2O) system composed of two phases. A mixture of BCC iron and FCC iron is also a single-component system composed of two phases. However, a solid solution of Cu and Ni is a two-component single-phase system. Similarly, a solid solution of NiO and MgO is also a two-component single-phase system.

Let us focus our attention on the least complex class of phase diagrams—those that involve a single component. In these systems the important thermodynamic variables are

FIGURE 7.2–1

Schematic of part of the $\mathcal{P}$-V-T surface for H_2O projected onto the $\mathcal{P}$-T plane.

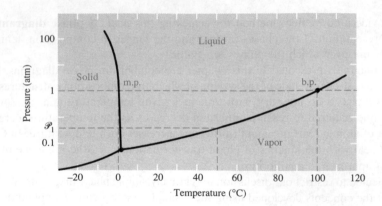

temperature T and pressure $\mathcal{P}$. The phase relationships may be represented on a pressure-temperature diagram. Figure 7.2–1 shows such a diagram, known as a one-component phase diagram, for the H_2O system. The phase diagram is composed of regions of pressure and temperature where only a single phase is stable. Note that if pressure and temperature are known, the equilibrium state of the system (i.e., the equilibrium phase) is established. On the other hand, if it is known that two phases are present at 50°C, the pressure is uniquely determined by the point $\mathcal{P}_1$ on the diagram. The solid line separating the liquid and vapor phase fields indicates pressure-temperature combinations for which these two phases coexist. Such a line is called a univariant line, since specification of one of the external variables and the phases present automatically fixes the other variable.

Phase equilibria can be described by a simple yet powerful rule called **Gibbs phase rule,** which expressed mathematically is:

$$F = C - P + 2 \tag{7.2–1}$$

This equation relates the number of **degrees of freedom,** F, at equilibrium to the number of components, C, in the system, the number of phases in equilibrium, P, and the two state variables temperature and pressure. The number of degrees of freedom available to the system at equilibrium is the number of variables (pressure, temperature, or composition) that can be independently adjusted without disturbing equilibrium. In many applications, pressure is constant at 1 atmosphere and Gibbs phase rule reduces to:

$$F = C - P + 1 \tag{7.2–2}$$

In the case of the H_2O system we may ask ourselves how much freedom there is to change variables ($\mathcal{P}$ and/or T) if two phases are present. Since we have one component and two phases. Equation 7.2–1 gives F = 1 − 2 + 2 = 1. This means that if *one* variable (T or $\mathcal{P}$) is changed there will still be two phases present. Once this variable is changed, however, the other is automatically fixed, since all of the "freedom" has been used up. For example, we may choose to change the temperature while still maintaining equilibrium between ice and water. If the temperature is increased, the pressure is fixed at the corresponding point on the solid-liquid phase boundary line. On the other hand, if we require three phases to be in equilibrium, then the phase rule tells us that $F = 0$. This means that there is no freedom in specifying variables for three-phase equilibrium in a single-component system. Three-phase equilibrium occurs at a fixed temperature and pressure. Equilibrium in which there is no freedom to change variables is called **invariant.** The invariant point is called a triple point, since three phases are in equilibrium.

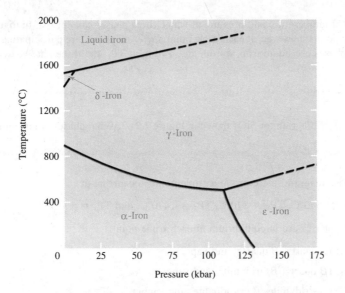

FIGURE 7.2–2

The equilibrium tempera-
ture-pressure diagram for
iron. (*Source: D. A. Porter
and K. E. Easterling*, Phase
Transformations in Metals
and Alloys, *Chapman &
Hall, 1981, p. 9. Reprinted
by permission of Chapman
& Hall.*)

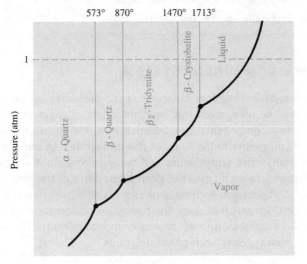

FIGURE 7.2–3

Pressure-temperature phase
diagram for SiO_2. (*Source:
Robert T. DeHoff*, Thermodynamics in Materials Science, *McGraw-Hill, 1973.
Reproduced with permission
of McGraw-Hill, Inc.*)

Many single-component systems have phase diagrams that are more complex than the pressure-temperature regime shown for water. Figures 7.2–2 and 7.2–3 show the temperature-pressure equilibrium diagrams for iron and silica (SiO_2), respectively. In these diagrams more than one triple point occurs. It is important to realize that even for single-component diagrams, the complexity depends on the component being considered and the range of temperature and pressure considered.

··

EXAMPLE 7.2–1

Consider the $\mathcal{P}/T$ diagram for iron shown in Figure 7.2–2. Two triple points are indicated on the diagram. Explain why the presence of two triple points does not violate the phase rule.

Solution

The phase rule (Equation 7.2–1) indicates that three-phase equilibrium in a one-component system occurs at one point (i.e., $F = C - P + 2 = 1 - 3 + 2 = 0$). That is, there is a unique combination of temperature and pressure at which three phases can coexist. In Figure 7.2–2 there are two triple points; however, at each triple point, *different* phases are participating in equilibrium. At TP_1, the phases are liquid iron, δ iron, and γ iron; at TP_2, the three phases are γ iron, α iron, and ε iron.

EXAMPLE 7.2–2

Consider the $\mathcal{P}/T$ diagram for SiO_2 shown in Figure 7.2–3. What phases are in equilibrium at each triple point?

Solution

From the phase diagram, there are four triple points, occurring at

$$TP_1 = 573°C, \ TP_2 = 870°C, \ TP_3 = 1470°C, \ and \ TP_4 = 1713°C$$

The following phases are in equilibrium at each triple point:

TP_1: (α quartz, β quartz, and vapor)

TP_2: (β quartz, β_2 tridymite, and vapor)

TP_3: (β_2 tridymite, β crystobalite, and vapor)

TP_4: (β crystobalite, liquid, and vapor)

7.3 PHASE EQUILIBRIA IN A TWO-COMPONENT SYSTEM

While the one-component systems discussed in the preceding section do have practical significance, it must be recognized that most materials of engineering interest are composed of at least two components. The addition of a composition variable introduces a modest degree of complexity to the study of phase diagrams. In most practical situations, the important variables are temperature and composition, since the pressure is usually fixed at 1 atmosphere. In such cases the appropriate form of the phase rule is that given in Equation 7.2–2. If pressure is eliminated as a variable, it is possible to construct two-dimensional phase diagrams that show the regions of composition and temperature over which any phase or combination of phases is in equilibrium. The following sections consider various prototypes of such phase diagrams.

7.3.1 Specification of Composition

Before beginning our discussion of phase diagrams, it is important to show the various ways to specify composition. It is often useful to specify compositions in terms of atomic percentages or atomic fractions. On the other hand, in many practical situations, weight percent or weight fraction of a given component is used to specify the composition. Therefore, it is sometimes necessary to convert back and forth between weight percent and atomic percent. If the composition is given in weight percent, conversion to atomic percent is done as follows:

$$at. \% \ A = \frac{(wt. \% \ A)/(at. \ wt. \ A)}{[(wt. \% \ A)/(at. \ wt. \ A)] + [(wt. \% \ B)/(at. \ wt. \ B)]} \times 100\%$$

$$(7.3–1)$$

where (at. % A) is the atomic percent of component A, (wt. % B) is the weight percent of component B, and (at. wt. A) is the atomic weight of component A. Conversely, if the composition is expressed in atomic percent, the corresponding weight percent is given by:

$$\text{wt. \% A} = \frac{(\text{at. \% A}) \times (\text{at. wt. A})}{[(\text{at. \% A}) \times (\text{at. wt. A})] + [(\text{at. \% B}) \times (\text{at. wt. B})]} \times 100\%$$

$$(7.3–2)$$

EXAMPLE 7.3–1

Calculate the atomic fraction of copper in aluminum for a two-component alloy containing 5 wt. % copper.

Solution

This is a two-component system composed of 5 wt. % Cu and 95 wt. % Al. Equation 7.3–1, along with the atomic weight data in Appendix B, can be used to solve for the at. % Cu as follows:

$$\text{at. \% Cu} = \frac{(\text{wt. \% Cu})/(\text{at. wt. Cu})}{(\text{wt. \% Cu})/(\text{at. wt. Cu}) + (\text{wt. \% Al})/(\text{at. wt. Al})} \times 100\%$$

$$= \frac{5/63.54}{(5/63.54) + (95/26.9)} \times 100 = 2.18 \text{ at. \% Cu}$$

The at. % Al can be calculated either by using the appropriate form of Equation 7.3–1 or, more simply, by subtracting the at. % Cu from 100%. That is:

$$\text{at. \% Al} = (100 \text{ at. \%}) - (2.18 \text{ at. \%}) = 97.82 \text{ at. \% Al}$$

In ceramic systems components are frequently compounds, and composition is usually expressed as mole fractions. One mole of Al_2O_3, for instance, contains 2 times Avogadro's number (6.023×10^{23}) of aluminum atoms and 3 times Avogadro's number of oxygen atoms.

If a material contains n_A moles of component A and n_B moles of component B, the mole fraction of component A, N_A, in the phase is:

$$N_A = \frac{n_A}{n_A + n_B}$$

$$(7.3–3)$$

When the components are elements, mole fractions and atomic fractions are equivalent. For a closed system—that is, when material cannot enter or leave the system—the sum of the mole fractions equals unity. Thus, for a two-component system:

$$N_A + N_B = 1$$

$$(7.3–4)$$

The concept of a closed system is important when dealing with phase equilibria. As temperature is changed, the elements can move back and forth between phases to achieve their equilibrium compositions, but there is a fixed amount of each component in the system.

Another important idea is the concept of solubility. We are all familiar with solubility in a system such as sugar and water. Increasing the temperature increases the amount of sugar that can be dissolved. When dissolved, the sugar and water form a two-component single-phase liquid. Decreasing the temperature decreases the solubility and, depending upon the initial concentration and temperature, may result in the formation of solid sugar crystals (rock candy). That is, lowering the temperature results in the formation of a two-phase mixture of liquid (containing some dissolved sugar) and a crystalline solid phase. Many metallic and ceramic systems behave in much the same way. That is, temperature and the similarity of the atomic structures of the components generally affect the solid solubility of one component in another.

7.3.2 The Isomorphous Diagram for Ideal Systems

The simplest two-component system is the **isomorphous system.** As the name suggests, complete solubility can occur over the entire composition range in both the liquid and solid states. Recall from Section 4.3.1 that complete solid solubility occurs only for well-defined conditions known as the Hume-Rothery rules. Figure 7.3–1a illustrates an ideal isomorphous binary (two-component) diagram for the system A-B. Pressure is assumed to be constant. The vertical axis is temperature and the horizontal axis represents

FIGURE 7.3–1

The idealized binary (A-B) isomorphous system: **(a)** the composition-temperature phase diagram with associated definitions, and **(b)** a similar diagram showing the liquidus and solidus temperatures for a specific alloy of composition X_1.

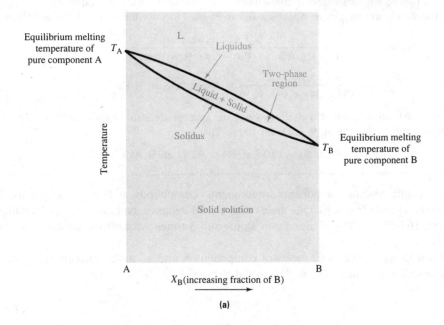

(a)

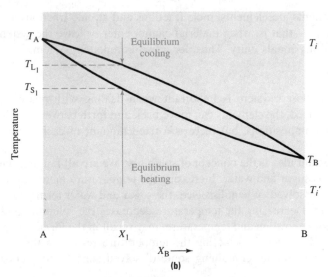

(b)

the composition of the alloy in weight, atomic, or mole fraction of component B. Therefore, the left edge of the phase diagram represents pure component A ($X_B = 0$) and the right edge corresponds to pure component B ($X_B = 1$). On these two edges, and only on these edges, the system is composed of a single component. For all other points on the phase diagram, however, two components (A and B) are present.

Although the variables are composition and temperature, in most engineering applications the composition of the system is fixed, and we are primarily concerned with the variation in the phases present, their composition, and their relative amounts as a function of temperature. We investigate the influence of temperature by drawing (either literally or figuratively) a vertical line on the phase diagram that intersects the composition axis at a value characteristic of the alloy. For example, if we investigate the properties of an alloy composed of 40% B and 60% A, we draw a vertical line that intersects the horizontal axis at $X_B = 0.4$.

Examination of Figure 7.3–1a shows that if pure A ($X_B = 0$) liquid is cooled, it solidifies at T_A, the melting point. Below this temperature pure A is 100% solid. If, however, a small amount of B is added to A ($0 < X_B \ll 1$), solidification begins at a somewhat lower temperature. In addition, solidification is not complete (i.e., the material is not 100% solid) until a significantly lower temperature is reached. For any mixture of A and B, solidification occurs over a temperature *range* rather than at a distinct temperature, as is the case of a pure component. As shown in the figure, the temperatures that mark the onset and completion of solidification are functions of the alloy composition. If all of the points marking the onset and completion of solidification are joined together to form two lines, a phase diagram is obtained that shows the equilibrium phase or phases present for any given combination of composition and temperature. The lines created in this way are called phase boundaries, and the areas enclosed by these lines are called phase fields. The phase boundary separating the single-phase liquid region from the two-phase (solid + liquid) region is called a **liquidus boundary.** The boundary separating the two-phase (S + L) region from the single-phase solid region is called a **solidus boundary.**

Figure 7.3–1b shows a binary isomorphous system with an alloy of composition X_1 indicated. The temperature coordinate of the point at which the vertical alloy composition line crosses the liquidus, T_{L_1}, is known as the liquidus temperature, and the corresponding value at the solidus boundary, T_{S_1}, is the solidus temperature. What happens if alloy 1 is slowly cooled from temperature T_i? Initially the alloy is 100% liquid (single phase). When the temperature reaches the liquidus temperature, a solid phase begins to appear. For temperatures between T_{L_1} and T_{S_1}, the liquid and solid phases coexist. Below the solidus temperature the alloy is 100% solid (single phase).

This example illustrates a very general principle. To determine the phases present for an alloy of composition X_0 at temperature T_0, locate the point on the phase diagram whose coordinates are X_0, T_0. Such a point is termed a **state point.** Once the state point is located, simply "read" the phases present directly from the phase diagram.

7.3.3 Phases in Equilibrium and the Lever Rule

Recall that the properties of a material depend not only on the phases present, but also on their compositions and relative amounts. How can one use a two-component phase diagram to determine this important information? The first step in determining phase compositions is to locate the corresponding state point. If the state point falls within a single-phase field, then the phase composition is identical to the alloy composition, and the material is 100% of that phase.

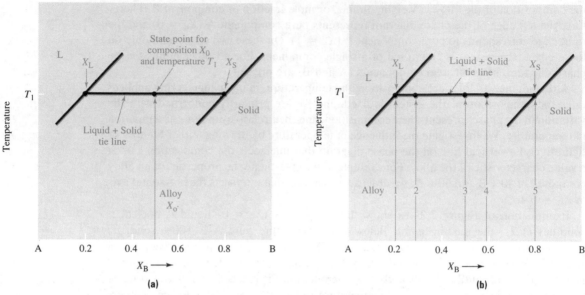

FIGURE 7.3–2 Graphical definitions of the tie line and the lever rule in a two-phase field: **(a)** the tie line through the state point defined by temperature T_1 and alloy composition X_0, and **(b)** the same tie line shared by all five alloy compositions at temperature T_1.

When the state point falls within a two-phase field, the phase rule tells us that there is only one degree of freedom. Thus, a change in temperature must be accompanied by a change in composition. In a two-phase field the phase compositions are found in the following manner:

1. Extend a line, known as a **tie line,** through the state point and parallel to the composition axis until it intersects a phase boundary in each direction.

2. The compositions of the two phases are given by the horizontal (composition) coordinates defined by the intersection of the tie line with the phase boundaries. These compositions can be read directly from the horizontal axis and will have units of weight, atomic, or mole fraction of component B.

This procedure is illustrated in Figure 7.3–2a, which shows an expanded section of a two-phase (liquid + solid) region. If the alloy composition is $X_0 = 0.5$ B and the temperature of interest is T_1, the state point lies within the liquid-plus-solid phase field, and a tie line must be constructed as shown. The compositions of the liquid and solid phases are given by X_L (≈ 0.2 B) and X_S (≈ 0.8 B), respectively. Note that the phase compositions are not the same as the alloy composition but that all compositions are given in units of fraction of component B.

Let us investigate how the phase compositions depend on the alloy composition while holding the temperature constant at T_1. As shown in Figure 7.3–2b, for any of the five alloys indicated, the tie line through T_1 is exactly the same. This means that as long as the composition of the alloy lies between 0.2 B and 0.8 B (i.e., the boundaries of the two-phase region at T_1), the compositions of the liquid and the solid in equilibrium at T_1 are fixed.[1]

---

[1] This observation is directly related to the phase rule, which states that for constant pressure, the number of degrees of freedom associated with a two-component two-phase system is $F = C - P + 1 = 2 - 2 + 1 = 1$. That is, as soon as the temperature is specified, the corresponding compositions in the two-phase field are uniquely determined.

What about the relative amounts of the two phases present at T_1 for the various alloy compositions shown in Figure 7.3–2b? For alloy 1, the state point is close to the liquid single-phase field, suggesting that this alloy is almost completely liquid at T_1. In contrast, for alloy 5 the state point is nearly in the solid single-phase field. Therefore, we can intuitively conclude that the fraction of solid present at T_1 progressively increases as the fraction of component B in the alloy increases. The relative amounts of the liquid and solid phases can be determined quantitatively using a mass balance as described below.

Returning to Figure 7.3–2a, for an alloy of overall composition X_0 at temperature T_1, component B is distributed in the liquid phase with composition X_L and in the solid phase with composition X_S. The mass of component B contained in any phase is equal to the total mass of that phase times the fraction of the phase that is composed of component B. For example, the mass of component B in the solid phase is given by $M_S X_S$ where MS is the mass of the solid phase. Since mass is conserved, the total mass of component B in the alloy must be equal to the sum of the mass of component B in the liquid and solid phases. Mathematically,

$$M_0 X_0 = M_S X_S + M_L X_L \tag{7.3–5a}$$

where M_L represents the mass of liquid and the other symbols have been defined previously. Dividing both sides of this equation by the total mass yields:

$$X_0 = \left(\frac{M_S}{M_0}\right) X_S + \left(\frac{M_L}{M_0}\right) X_L \tag{7.3–5b}$$

or

$$X_0 = f_S X_S + f_L X_L \tag{7.3–5c}$$

where $f_S \ (= M_S/M_0)$ and $f_L (= M_L/M_0)$ are the mass fractions of the solid and liquid, respectively. Noting that $f_S + f_L = 1$ and solving for f_L yields:

$$f_L = \frac{X_0 - X_S}{X_L - X_S} = \frac{X_S - X_0}{X_S - X_L} \tag{7.3–6}$$

Similarly, the fraction of the alloy that is solid is given by:

$$f_S = \frac{X_L - X_0}{X_L - X_S} = \frac{X_0 - X_L}{X_L - X_S} \tag{7.3–7}$$

Equations 7.3–6 and 7.3–7 are known as the lever rule. Equations of this type can be used to compute the relative amounts of phases in equilibrium independent of whether the phase mixtures are liquid-solid, solid-solid, or liquid-liquid. For each alloy in Figure 7.3–2b, the fraction of the two phases can be determined by a straightforward application of these equations:

Alloy 1

$$f_L = \frac{0.8 - 0.2}{0.8 - 0.2} = 1.0; \quad f_S = \frac{0.2 - 0.2}{0.8 - 0.2} = 0.0$$

Alloy 2

$$f_L = \frac{0.8 - 0.3}{0.8 - 0.2} = 0.833; \quad f_S = \frac{0.3 - 0.2}{0.8 - 0.2} = 0.167$$

Alloy 3

$$f_L = \frac{0.8 - 0.5}{0.8 - 0.2} = 0.5; \quad f_S = \frac{0.5 - 0.2}{0.8 - 0.2} = 0.5$$

FIGURE 7.3–3

A schematic illustration of the lever rule. The tie line represents a "lever" with its pivot point located at the alloy composition X_0, its left end fixed at "position" X_L, and its right end located at X_S with blocks of mass M_L and M_S resting on either end.

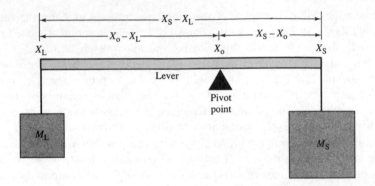

Alloy 4

$$f_L = \frac{0.8 - 0.6}{0.8 - 0.2} = 0.333; \quad f_S = \frac{0.6 - 0.2}{0.8 - 0.2} = 0.667$$

Alloy 5

$$f_L = \frac{0.8 - 0.8}{0.8 - 0.2} = 0.0; \quad f_S = \frac{0.8 - 0.2}{0.8 - 0.2} = 1.0$$

We conclude our discussion of the lever rule with a memory aid designed to help you use the lever rule correctly when determining mass fractions. As shown in Figure 7.3–3, one can envision the tie line as a lever with its pivot point located at the alloy composition X_0, its left end fixed at "position" X_L, and its right end located at X_S. We can also imagine two blocks of mass M_L and M_S resting on the left and right sides of the lever, respectively. In general, the lengths of the two lever arms will be unequal, and, therefore, the masses must also be unequal. For any such lever there is an inverse relationship between the mass and the lever arm length. That is, the heavier mass is on the side with the shorter lever. Symbolically, the relationship is given by:

$$\frac{M_{left}}{M_{total}} = \frac{\text{Right lever arm length}}{\text{Total lever length}}$$

Substituting the corresponding phase quantities yields:

$$\frac{M_L}{M_0} = f_L = \frac{X_S - X_0}{X_S - X_L}$$

which is, of course, the lever rule. The key is to remember that the mass fraction is proportional to the length of the *opposite* lever arm.

In summary, the lever rule gives us a method to determine the relative amount of each of the two phases present in a two-phase field. Thus, in concert with the tools provided in the previous few sections, we are now able to determine the phases present, the compositions of those phases, and their relative amounts for any state point on a binary phase diagram. This ability is tremendously powerful and forms a significant part of the foundation of materials science and engineering.

7.3.4 Solidification and Microstructure of Isomorphous Alloys

All alloys in an isomorphous system solidify under equilibrium conditions in a similar manner. The changes in microstructure as a function of temperature for an alloy of composition $X_0 = 0.6$ B are described in Figure 7.3–4a. Figure 7.3–4b is an expanded

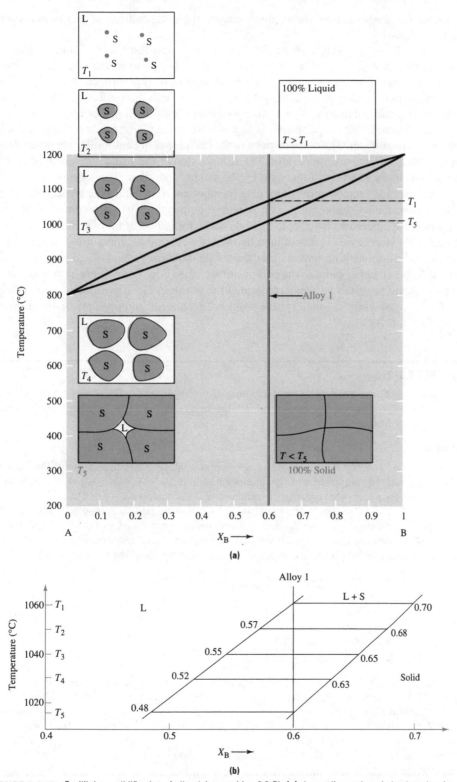

FIGURE 7.3–4 Equilibrium solidification of alloy 1 (composition 0.6 B): **(a)** the cooling path and sketches showing the development of the microstructure, and **(b)** an expanded section of part **(a)** showing the compositions of the liquidus and solidus boundaries in the range of 1010°C to 1060°C.

section of the diagram that shows more clearly the compositions of the liquidus and solidus boundaries.

Above T_1 the alloy is single-phase liquid, and the composition of the liquid is identical to that of the alloy (i.e., $X_L = X_0 = 0.6$ B). The alloy begins to freeze when the temperature of the molten alloy reaches T_1. The first solid crystals that form have a composition given by the intersection of the horizontal T_1 isotherm with the solidus boundary. When the temperature is changed to T_2, the composition of both the liquid phase and the solid phase must change. Once equilibrium is established, the temperature and composition in each phase are uniform. The compositions of the two phases in equilibrium are determined by the tie line at the new temperature. The composition of the liquid is determined by the liquidus boundary and that of the solid by the solidus boundary. The solid is enriched in the higher-melting temperature component (component B), whereas the liquid is enriched in the lower-melting temperature component (component A). As would be expected, dropping the temperature from T_1 to T_2 leads to the formation of more solid. The relative amount of the two phases is determined by the lever rule. As temperature is reduced, the changing solid composition moves along the solidus boundary, and at T_5, the composition of the solid is extremely close to the composition of the alloy. The composition of the very last bit of liquid is given by the intersection of the isotherm T_5 with the liquidus boundary. For $T < T_5$, the alloy is single-phase solid with uniform composition of 0.6 B (i.e., $X_S = X_0 = 0.6$ B).

....

EXAMPLE 7.3–2

Determine the compositions and mass fractions of the liquid and solid phases for alloy X_0 in Figure 7.3–4b at each of the five temperatures T_1 through T_5. Assume that the alloy is cooled under equilibrium conditions.

Solution

At temperature T_1 (1060°C) the alloy has just begun to solidify. The composition of the liquid is $X_{L_1} = 0.6$ B and the composition of the developing solid is $X_{S_1} = 0.7$ B. Since this state point is on the phase boundary, the system is essentially 100% liquid (i.e., $f_{L_1} \approx 1$ and $f_{S_1} \approx 0$). At temperature T_2 (1050°C) the state point is within the solid-plus-liquid two-phase field, and a tie line should be constructed. Using the tie line, the compositions are found to be $X_{L_2} = 0.57$ B and $X_{S_2} = 0.68$ B. The mass fractions are calculated using the lever rule. Since the liquid is on the "left side," the quantity in the numerator of the lever rule is the length of the right lever arm:

$$f_{L_2} = \frac{X_{S_2} - X_0}{X_{S_2} - X_{L_2}} = \frac{0.68 - 0.6}{0.68 - 0.57} = 0.73$$

Similarly,

$$f_{S_2} = \frac{X_0 - X_{L_2}}{X_{S_2} - X_{L_2}} = \frac{0.6 - 0.57}{0.68 - 0.57} = 0.27$$

Note that $f_{L_2} + f_{S_2} = 1.0$, which serves as a convenient verification of our calculations.

At T_3, $X_{L_3} = 0.55$ B, $X_{S_3} = 0.65$ B, $f_{L_3} = (0.65 - 0.6)/(0.65 - 0.55) = 0.5$, and $f_{S_3} = (0.6 - 0.55)/(0.65 - 0.55) = 0.5$.

At T_4, $X_{L_4} = 0.52$ B, $X_{S_4} = 0.63$ B, $f_{L_4} = (0.63 - 0.6)/(0.63 - 0.52) = 0.27$, and $f_{S_4} = (0.6 - 0.52)/(0.63 - 0.52) = 0.73$.

Finally, at T_5, $X_{L_5} = 0.48$ B, $X_{S_5} = 0.6$ B, $f_{L_5} \approx 0$, and $f_{S_5} \approx 1$.

....

7.3.5 Determination of Liquidus and Solidus Boundaries

When a phase change occurs in a material such as a metal, ceramic, or polymer, the energy content of the system changes. During solidification, the system generates heat. Consider the system shown in Figure 7.3–5. If pure component A is cooled from liquid to solid and temperature is continuously monitored as a function of time, a cooling curve similar to that shown in Figure 7.3–5b is obtained. The two breaks in the curve correspond to the beginning and end of solidification. The liberated heat (called the heat of transformation) balances the drop in temperature associated with cooling. Thus, solidification of a pure substance is indicated by a horizontal thermal arrest. This result is consistent with expectation, since freezing of a pure component takes place at a constant temperature.

For alloys (two-component mixtures), the shape of the cooling curve is different. Alloy 1 ($X_0 = 0.6$ B) has a freezing range. That is, there is a temperature interval between the liquidus and solidus. When the liquidus is reached, a small amount of solid is formed, and some heat is released. Consequently, the overall rate of cooling is reduced, and there is a break in the cooling curve, as shown in Figure 7.3–5c. It is important to recognize that since only a small amount of solid forms at this temperature, a much smaller amount of heat is liberated than was the case for the pure material. The process of heat liberation continues over the entire solidification range. Cooling curves aid in the determination of phase diagrams. The applications are generally restricted to liquid-to-solid (L → S) transformations rather than solid-to-solid (S → S) transformations, since energy changes associated with L → S transformations are larger than those of S → S transformations.

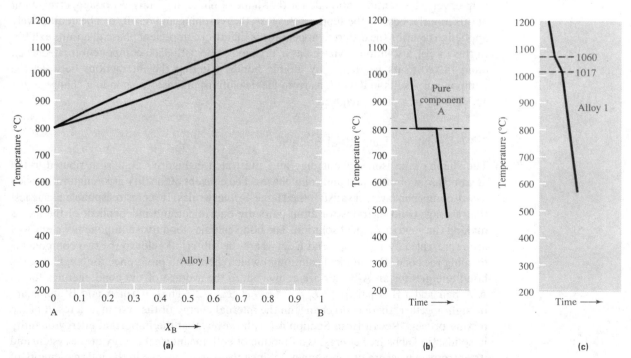

FIGURE 7.3–5 **(a)** A binary isomorphous phase diagram showing cooling curves for **(b)** pure component A, and **(c)** alloy 1.

7.3.6 Specific Isomorphous Systems

An isomorphous system is *only* possible for substitutional solid solutions. In the Cu-Ni system, Ni can substitute directly for the Cu atoms located on their FCC lattice positions and vice versa. The substitution occurs randomly on these lattice sites. This is because the Cu and Ni atoms are so similar that neither type of atom has any thermodynamic preference for either Cu or Ni nearest neighbors. Such a solid solution is said to be ideal. The Cu-Ni system has practical significance. These alloys have excellent corrosion resistance and are frequently used in applications such as water-cooled heat exchangers.

In the case of a ceramic system illustrated by the binary NiO-MgO system, there is again complete solubility in both liquid and solid states. Nickel oxide and magnesium oxide both have the NaCl crystal structure discussed in Section 3.7.1. Recall that the NaCl structure has an FCC basis with one cation and one anion per lattice point. In this context, the oxide structure may be visualized as two interpenetrating FCC sublattices—one of cations and the other of anions. In the case of the NiO-MgO solid solution, the substitution occurs on the cation sublattice.

Phase diagrams for four isomorphous systems are shown in Figure 7.3–6. Section 4.3.1 showed that the requirements for the formation of a substitutional solid solution are given by the Hume-Rothery rules. All the Hume-Rothery rules are satisfied for each of the component pairs illustrated in Figure 7.3–6. As might be expected, however, only a limited number of two-component systems strictly satisfy all four requirements set forth by Hume-Rothery and, therefore, exhibit ideal behavior. Deviations from ideal behavior are the norm, and most two-component phase diagrams do not correspond to the ideal isomorphous systems presented thus far.

An atomistic model, called the quasi-chemical theory of solutions, is often applied to help develop a scientific rationale for the shape of phase diagrams. Although a treatment of this model is beyond the scope of this text, the reasoning and results can be qualitatively applied to describe the different appearances that two-component phase diagrams exhibit. In this model, a solution is viewed as a large molecule with each component treated as an atom linked to its neighbors by atomic bonds. Altering the interactions between the components results in deviations from ideal solutions and a corresponding change in the appearance of a phase diagram.

7.3.7 Deviations from Ideal Behavior

The shape of the isomorphous diagrams illustrated in Figure 7.3–1 is attributed to the ideal behavior of the liquid and solid phases. The concept of ideality arises naturally when a two-component system satisfies the Hume-Rothery rules. It seems reasonable to suggest that average bond energies for atom pairs are equivalent and independent of the atoms making the bond in an ideal solution. The bond energies for a two-component system A-B are represented by E_{AA}, E_{BB}, and E_{AB}, where the subscripts refer to the two components forming the bond. For an ideal solution in which there is no preference for neighbors, the bond energies for an A-B pair are equivalent to the average of the bond energies for an A-A pair and a B-B pair [i.e., $E_{AB} = (E_{AA} + E_{BB})/2$]. Thus, when A and B atoms are brought together, there is no change in the internal energy of the system as a result of the mixing process. Recall from Section 4.2.1, however, that the important energy quantity in solids, the Gibbs free energy, is a function of both the internal energy of the system and its entropy (a measure of randomness). Since there is no change in the internal energy of the system during mixing, the Gibbs free energy of the system is minimized when the

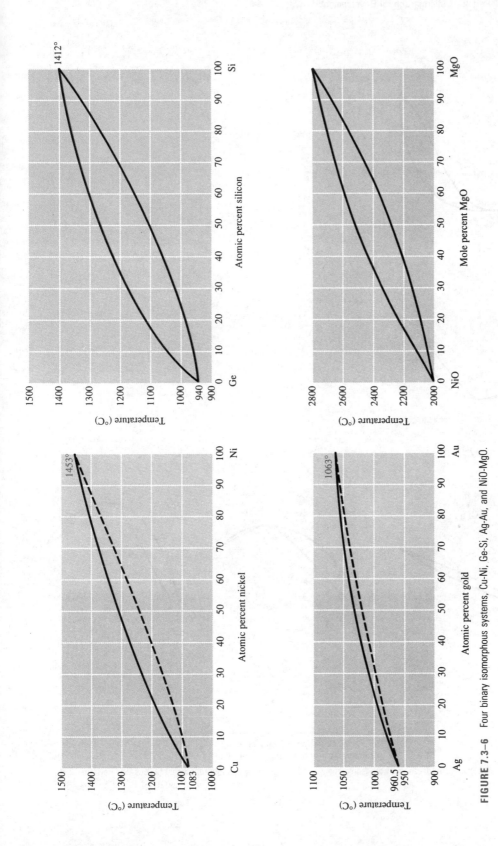

FIGURE 7.3−6 Four binary isomorphous systems, Cu-Ni, Ge-Si, Ag-Au, and NiO-MgO.

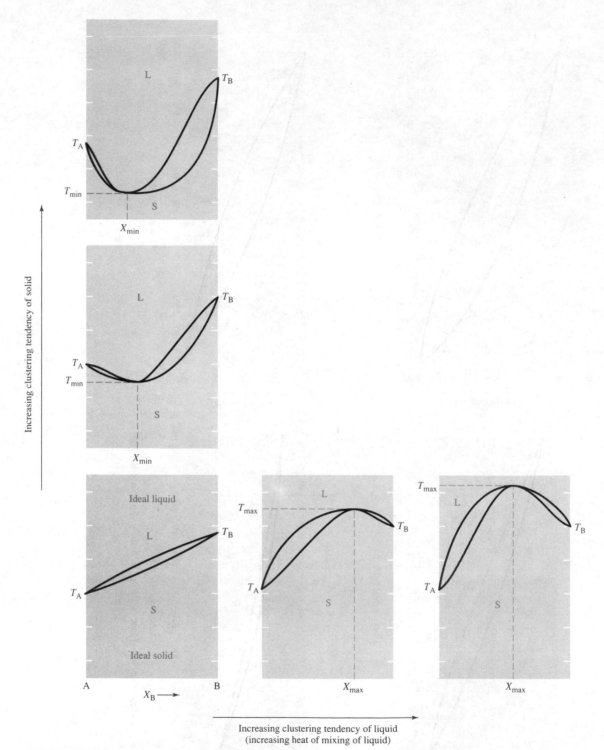

FIGURE 7.3–7 Progressive change in the form of the isomorphous phase diagram as the solid and liquid phases deviate from ideal behavior.

randomness of the system is maximized. Thus, in an ideal solution the lowest-energy configuration of the components is the one in which they are randomly distributed.

An important question to ask is, How does nonideality change the form or shape of a phase diagram? Figure 7.3–7 illustrates a sequence of phase diagrams that can occur when deviation from ideality increases in the liquid or solid phase. The causes of the deviations from ideal behavior are presented in the next section, but a few general remarks regarding the shapes of the phase diagrams in Figure 7.3–7 are appropriate. First, one type of deviation from ideality can lead to a melting point maximum, and the other to a melting point minimum. Second, increasing deviations from ideality increase the maximum or decrease the minimum. These melting point extremes are called **congruent melting** points, and at these points the compositions of the liquid and solid phases are the same. As illustrated by the phase diagrams in Figure 7.3–7, the liquidus and solidus boundaries touch. At the alloy composition corresponding to the congruent melting point, the alloy freezes at a single temperature rather than over a temperature range. Alloy compositions to the right or the left of the extreme composition, however, freeze over a range of temperatures.

....

EXAMPLE 7.3–3

Figure 7.3–8 shows a phase diagram with an alloy X_{max} melting congruently. Consider the two alloy compositions marked X_1 and X_2 in the figure. For each of these two alloys:

- *a.* Determine whether solidification occurs at a unique temperature or over a range of temperatures.
- *b.* Determine the composition of the first solid to form.
- *c.* Determine the composition of the last liquid to solidify.

Solution

- *a.* Since the vertical composition line for each of these two alloys crosses the liquid-plus-solid phase field, both alloys freeze over a temperature range. The boundaries of the freezing temperature range are functions of composition.
- *b.* The tie lines drawn at the liquidus temperatures for the two alloys give X_{S_1} and X_{S_2} for the compositions of the first solids to form for alloys X_1 and X_2, respectively.
- *c.* The tie lines drawn at the solidus temperature give the compositions for the last liquids to solidify, X_{L_1} and X_{L_2} for alloys X_1 and X_2, respectively.

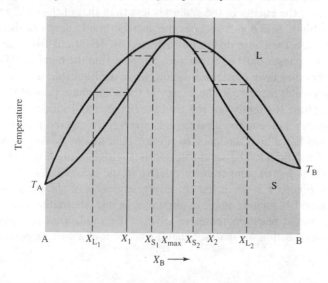

FIGURE 7.3–8

A binary phase diagram showing a congruently melting alloy of composition X_{max}. Two additional alloy compositions are shown.

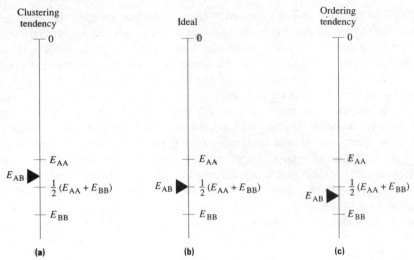

FIGURE 7.3–9   Illustration of the selective bond energies that lead to (a) clustering, (b) ideal, and (c) ordering tendencies in solutions.   (*Source: Robert T. DeHoff*, Thermodynamics in Materials Science, *McGraw-Hill, 1973. Reproduced with permission of McGraw-Hill.*)

The tendency to form congruent melting compositions can be understood in terms of the bond energies of the atoms that form the solution and the corresponding change in the free energy of the system. If E_{AB} is greater than the average of E_{AA} and E_{BB}, then the A component will "prefer" to be surrounded by other As and the B component will prefer to be surrounded by Bs. When As prefer As and Bs prefer Bs, a clustered solution is formed in which the number of A-B bonds is minimized. When mixing occurs, it increases the energy of the system, as shown in Figure 7.3–9a (i.e., there are more A-B bonds in the mixed solution than when the two components are isolated from one another, and the A-B bonds have a higher average energy).

On the other hand, if $E_{AB} < (E_{AA} + E_{BB})/2$, then the tendency is for the A component to prefer B nearest neighbors and vice versa. As shown in Figure 3.7–9c, the favored arrangement is an ordered solution in which the maximum number of A-B bonds is formed. In this case the energy of the system is less than that of the isolated components (i.e., the formation of A-B bonds reduces the average bond energy of the system).

To determine how the phase diagram is affected by deviations from ideality, the free energies of the two phases must be compared. Suppose that the bond energies in the solid phase are such that they favor cluster formation, while the liquid phase is ideal. In this case mixing causes a greater increase in the energy of the solid phase than of the liquid phase. Thus, the liquid phase is slightly favored, and the liquidus and solidus lines shift down to lower temperatures relative to the ideal case. Conversely, if the solid solution is ideal and the liquid solution has a high tendency to cluster, then mixing causes a greater increase in the energy of the liquid. The solid solution is slightly favored, and the liquidus and solidus lines shift to higher temperatures relative to the ideal case.

As shown in Figure 7.3–7, as the clustering tendency of the solid phase increases relative to the liquid, a melting point minimum appears. Alternatively, as the clustering tendency of the liquid phase increases relative to the solid, a melting point maximum appears. In the next section we shall see that the concept of a melting point minimum can be used to understand one of the most important and commercially significant types of phase diagrams, the eutectic system.

The quasi-chemical approach introduced in this section permits one to show that there is a thermodynamic basis for all features occurring in a phase diagram. We follow this approach as we introduce the different types of simple binary phase diagrams.

7.4 THE EUTECTIC PHASE DIAGRAM

As shown in Figure 7.4–1, as the clustering tendency of the solid increases, not only is there a depression in the liquidus and solidus temperatures, but a region of immiscibility (region of nonmixing) begins to develop at low temperatures. In effect, there are some combinations of temperature and composition for which the two components are no longer completely soluble in one another. For example, in Figure 7.4–1b, at T_1 a region

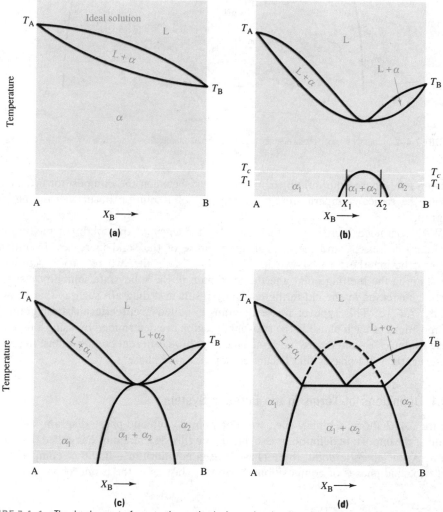

FIGURE 7.4–1   The development of a eutectic reaction by increasing the clustering tendency in the solid phase: **(a)** the diagram for an ideal system, and **(b–d)** increasing clustering tendency in the solid phase.

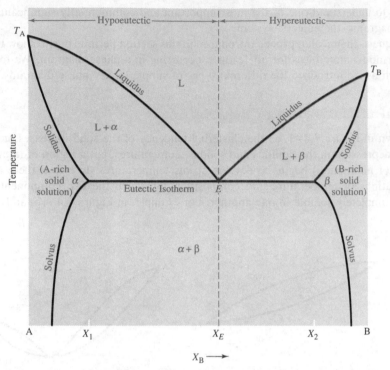

FIGURE 7.4–2   A binary eutectic phase diagram and the associated terms used to describe regions of a eutectic system.

of two–solid-phase equilibrium ($\alpha_1 + \alpha_2$) occurs between the compositions X_1 and X_2. Above the critical temperature T_c, a complete solid solution occurs across the phase diagram.[2]

With continued increases in the clustering tendency of the solid, the melting point minimum decreases, and the critical temperature of the solid increases. Eventually a diagram having the form shown in Figure 7.4–1c results, where curves defining the minimum in the melting point and the maximum in the solid-state solubility just touch. Further increases in the clustering tendency result in a diagram such as that shown in Figure 7.4–1d. This type of phase diagram is called a **eutectic** phase diagram. The compositions of each phase in the two-phase region are determined by using a tie line. In these examples α_1 and α_2 are the same phase but have different compositions. In general, however, the solid phases that occur are not the same phase.

7.4.1 Definitions of Terms in the Eutectic System

Figure 7.4–2 shows a binary (i.e., two-component) eutectic phase diagram, along with some of the important definitions associated with this system. Point E is called the eutectic point. At the eutectic point, three phases are in equilibrium—liquid of composition X_E and two solid phases of compositions X_1 and X_2. Just as in the isomorphous system, the

---

[2] One might reasonably ask why there is complete solid solubility at any temperature in a system that favors clustering. The answer is related to the temperature-dependent contribution of the entropy term to the free energy of the system. The complete explanation, however, is beyond the scope of this text.

boundary separating the liquid-phase field from the two-phase (S + L) field is called a liquidus boundary. In the eutectic system, however, there are two liquidus boundaries. One boundary is associated with the A-rich side of the diagram, and the other is associated with the B-rich side. The solidus boundaries separate the solid-phase regions from the two-phase (L + S) regions.

An additional type of boundary appears in the eutectic system that was not present in the ideal isomorphous system—a solvus boundary. The **solvus boundaries** separate single-phase solid regions from two-phase solid regions. There is a solvus boundary on each side of the diagram.

Figure 7.4–2 illustrates the convention used to name the solid solution regions on the phase diagram. These regions are usually labeled with Greek letters across the diagram from left to right, as α, β, γ, and so on. Note that for a binary phase diagram, any isotherm (other than one associated with an invariant reaction) that extends across the entire diagram will begin in a single-phase field (for pure component A) and then alternately pass through two-phase fields and single-phase fields until it terminates in a single-phase field for pure component B. For example, any isotherm below the eutectic temperature begins in the single-phase α field, then enters the two-phase $\alpha + \beta$ field, and ends in the single-phase β field. Similarly, an isotherm between the melting temperature of component B and the eutectic temperature begins in α, then enters $\alpha + L$, then L, then $L + \beta$, and ends in β.

The eutectic diagram is conveniently divided by designating compositions to the left of the eutectic point as **hypoeutectic** (meaning less of component B than the eutectic composition) and those to the right of the eutectic point as **hypereutectic.** Of course, this is arbitrary, since either the A component or the B component could be placed on the left side of the diagram. For purposes of discussion, however, the component on the left will always be referred to as component A.

7.4.2 Melting and Solidification of Eutectic Alloys

For an alloy of composition X_E at temperature T_1 in Figure 7.4–3, only one phase (liquid) is present, and the composition of the phase is the composition of the alloy, X_E. If the alloy is slowly cooled under equilibrium conditions to temperature T_E, three phases are then in equilibrium. The phases are liquid of composition X_E, and two solids, one of composition X_1 and the other of composition X_2. If the temperature is either lowered or raised by a differential amount ΔT, one or two of the phases will disappear. For example, increasing the temperature by an amount ΔT eliminates the two solid phases, leaving only the liquid of composition X_E behind. Alternatively, decreasing the temperature by an amount ΔT below T_E eliminates the liquid, leaving only two solid phases with compositions X_1' and X_2'.

Because the pressure is assumed to be fixed (usually at 1 atmosphere), the equilibrium between the liquid and the two solids at the eutectic composition is invariant (i.e., there are zero degrees of freedom associated with this state point). That means that the temperature as well as the compositions of the phases participating in equilibrium are fixed. Since the eutectic transformation takes place at a fixed temperature, we expect the cooling curve for an alloy of eutectic composition to exhibit an arrest on cooling associated with the formation of the two solids. The eutectic reaction can be represented symbolically as:

$$\text{Liquid} \rightleftarrows \text{Solid 1} + \text{Solid 2} \quad \text{or} \quad L \rightleftarrows \alpha + \beta \quad\quad (7.4\text{--}1)$$

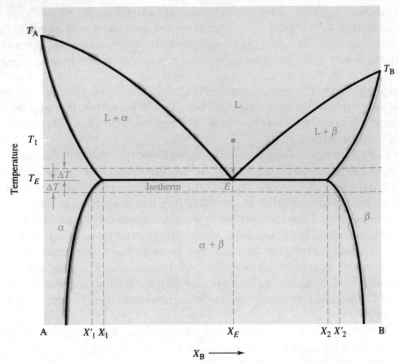

FIGURE 7.4–3   A binary eutectic equilibrium phase diagram showing the changes in composition of the phases present as the temperature is changed by an amount ΔT above and below the eutectic isotherm.

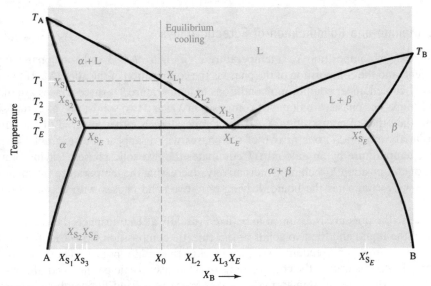

FIGURE 7.4–4   Equilibrium solidification of an off-eutectic alloy of composition X_0.

7.4.3 Solidification of Off-Eutectic Alloys

Suppose a hypoeutectic alloy X_0 is cooled under equilibrium conditions, as illustrated in Figure 7.4–4. At temperature T_1, the first solid (sometimes called the **primary,** or proeutectic, **phase**) appears. At T_1 two phases are present, solid of composition X_{S_1} and liquid of composition X_{L_1}. The fraction of each can be estimated using the lever rule:

$$f_{S_1} = \frac{X_{L_1} - X_0}{X_{L_1} - X_{S_1}} \quad \text{and} \quad f_{L_1} = \frac{X_0 - X_{S_1}}{X_{L_1} - X_{S_1}}$$

Similarly, at T_2,

$$f_{S_2} = \frac{X_{L_2} - X_0}{X_{L_2} - X_{S_2}} \quad \text{and} \quad f_{L_2} = \frac{X_0 - X_{S_2}}{X_{L_2} - X_{S_2}}$$

At a temperature just slightly above the eutectic temperature T_E, the composition of the solid is given as X_{SE}, the composition of the liquid is X_{LE} ($\approx X_E$), and the fractions of solid and liquid are:

$$f_{SE} = \frac{X_{LE} - X_0}{X_{LE} - X_{SE}} \quad \text{and} \quad f_{LE} = \frac{X_0 - X_{SE}}{X_{LE} - X_{SE}}$$

At the eutectic temperature, three phases are in equilibrium—the solid phases of composition X_{SE} and X'_{SE}, and the liquid phase of composition X_{LE}.

At temperatures below the eutectic isotherm, the alloy consists of primary α and a mixture of α and β that formed as a result of the eutectic transformation. Since the α and β in the eutectic form simultaneously, their morphology is usually different from that of the primary α (or β for a hypereutectic alloy). A morphologically distinct phase or mixture of phases is called a constituent. Thus, below the eutectic isotherm, the alloy consists of two constituents: primary α and eutectic $\alpha + \beta$.

Note that for any alloy with composition in the range $X_{SE} < X_0 < X'_{SE}$, at the eutectic temperature liquid of composition X_{LE} will transform to α and β of compositions X_{SE} and X'_{SE}, and that the relative amounts of α and β formed from this liquid will be constant. In fact, this is what makes the eutectic reaction invariant.

EXAMPLE 7.4–1

Figure 7.4–5 shows a hypothetical binary eutectic phase diagram on which we indicate an alloy of

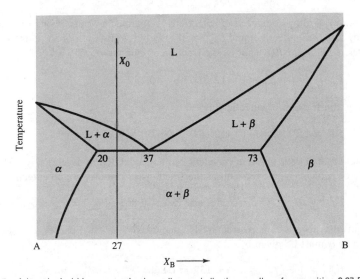

FIGURE 7.4–5   A hypothetical binary eutectic phase diagram indicating an alloy of composition 0.27 B.

composition 0.27 B. Calculate the following quantities:

a. The fraction of primary solid that forms under equilibrium cooling at the eutectic temperature.

b. The fraction of liquid with the eutectic composition that will transform to two solid phases below the eutectic isotherm.

c. The amount of α and β that will form from the liquid just below the eutectic isotherm.

d. The total amount of α phase in the alloy at a temperature just below the eutectic temperature.

Solution

a. The fraction of primary α, f_α^P, is determined by performing a lever rule calculation just above the eutectic temperature and using a composition corresponding to that of the alloy (i.e., $X_0 = 0.27$ B):

$$f_\alpha^P = \frac{X_L - X_0}{X_L - X_\alpha} = \frac{37 - 27}{37 - 20} = 0.588$$

b. Similarly, the fraction of liquid having the eutectic composition is:

$$f_L^{eut} = \frac{X_0 - X_\alpha}{X_L - X_\alpha} = \frac{27 - 20}{37 - 20} = 0.412$$

c. To determine the fraction of α that forms during solidification of the *eutectic liquid,* we must perform a lever rule calculation just below the eutectic temperature at a composition corresponding to that of the eutectic liquid. Therefore, the fraction of α in the eutectic constituent is:

$$f_\alpha = \frac{X_\beta - X_L^{eut}}{X_\beta - X_\alpha} = \frac{73 - 37}{73 - 20} = 0.679$$

Similarly, the fraction of β in the eutectic constituent is:

$$f_\beta = \frac{X_L^{eut} - X_\alpha}{X_\beta - X_\alpha} = \frac{37 - 20}{73 - 20} = 0.321$$

The fraction of alloy composed of eutectic α is obtained by multiplying the fraction of the alloy that was eutectic liquid ($f_L^{eut} = 0.412$) by the fraction of the eutectic liquid that becomes α ($f_\alpha = 0.679$). That is, the fraction of eutectic α, f_α^{eut}, is:

$$f_\alpha^{eut} = (f_L^{eut})(f_\alpha) = (0.412)(0.679) = 0.280$$

Similarly, the fraction of eutectic β is:

$$f_\beta^{eut} = (f_L^{eut})(f_\beta) = (0.412)(1.0 - 0.679) = 0.132$$

d. The total amount of α phase in the alloy can be calculated in several ways. The total fraction of α phase is just the sum of the fractions of primary α and eutectic α:

$$f_\alpha^{total} = f_\alpha^P + f_\alpha^{eut} = 0.588 + 0.280 = 0.868$$

Alternatively, since the microstructure is composed of just two phases, $\alpha + \beta$, the total fraction of α must be given by:

$$f_\alpha^{total} = 1 - f_\beta^{eut} = 1 - 0.132 = 0.868$$

Finally, the total amount of any phase at any temperature can be calculated directly by the lever rule evaluated at the corresponding state point. That is,

$$f_\alpha^{\text{total}} = \frac{X_\beta - X_0}{X_\beta - X_\alpha} = \frac{73 - 27}{73 - 20} = 0.868$$

An important commercial system that exhibits a eutectic is the Al-Si system shown in Figure 7.4–6. The bond characteristics of the components (Al and Si) are quite different, with Al being metallic and Si being covalent. Thus, there should be a tendency toward

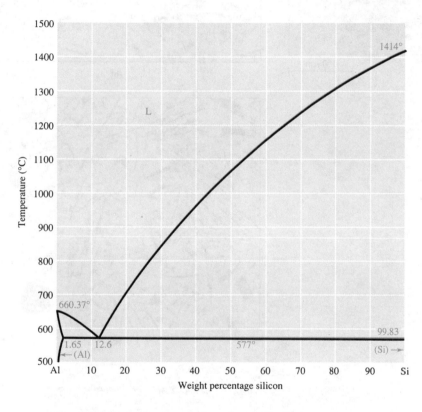

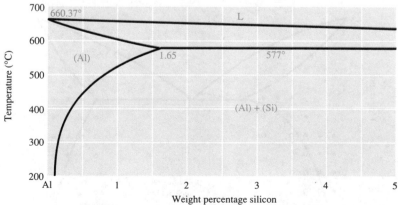

FIGURE 7.4–6

The Al-Si eutectic phase diagram. (*Source: F. Shunk, Constitution of Binary Alloys, McGraw-Hill. Reproduced with permission of McGraw-Hill.*)

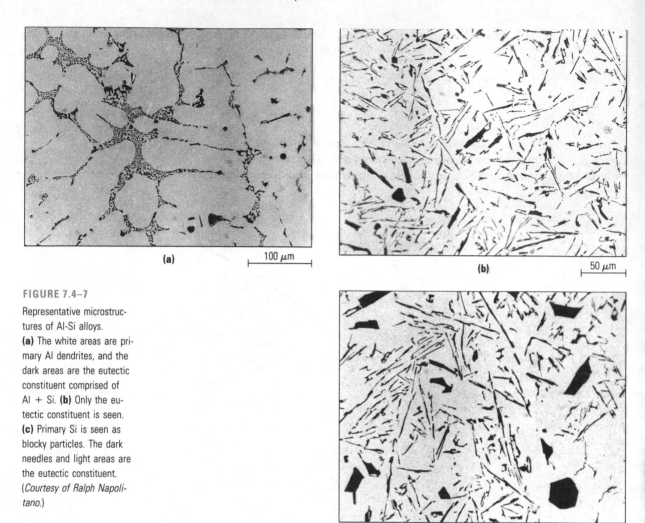

(a)

100 μm

(b)

50 μm

(c)

50 μm

FIGURE 7.4–7

Representative microstructures of Al-Si alloys.
(a) The white areas are primary Al dendrites, and the dark areas are the eutectic constituent comprised of Al + Si. **(b)** Only the eutectic constituent is seen. **(c)** Primary Si is seen as blocky particles. The dark needles and light areas are the eutectic constituent. (*Courtesy of Ralph Napolitano.*)

FIGURE 7.4–8

Optical metallography has been an effective technique to establish phase boundaries on a diagram. Filled circles represent regions containing only one phase, and two-phase regions are indicated by the black-and-white circles. The solvus boundary would thus represent the line separating microstructures containing one phase from those containing two phases.

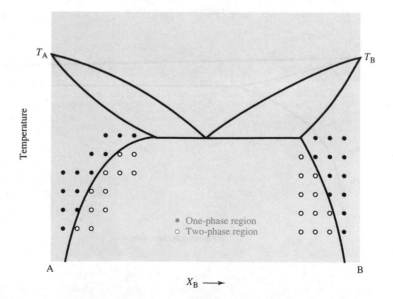

• One-phase region
○ Two-phase region

T_A

T_B

Temperature

A

B

$X_B \longrightarrow$

clustering in the solid state, and a eutectic is expected. Typical microstructures of a hypoeutectic alloy are shown in Figure 7.4–7. Figure 7.4–7a shows α-Al (white areas) with a mixture of α-Al and silicon between the large α grains. Figures 7.4–7b and c are higher magnifications showing the two-phase eutectic mixture of α-Al and silicon. Al-Si alloys of approximately the eutectic composition are widely used as piston materials in the automotive industry because of their excellent wear resistance (due to the hard Si particles), light weight (the eutectic is primarily Al), good strength, and relative thermal stability (there is relatively little solubility of one component in the other up to the eutectic temperature).

7.4.4 Methods Used to Determine a Phase Diagram

As shown in Figure 7.4–8, the solvus boundaries (i.e., the phase boundaries separating the single-phase field, α or β, from the two-phase $\alpha + \beta$ field) are nonvertical. Thus, if

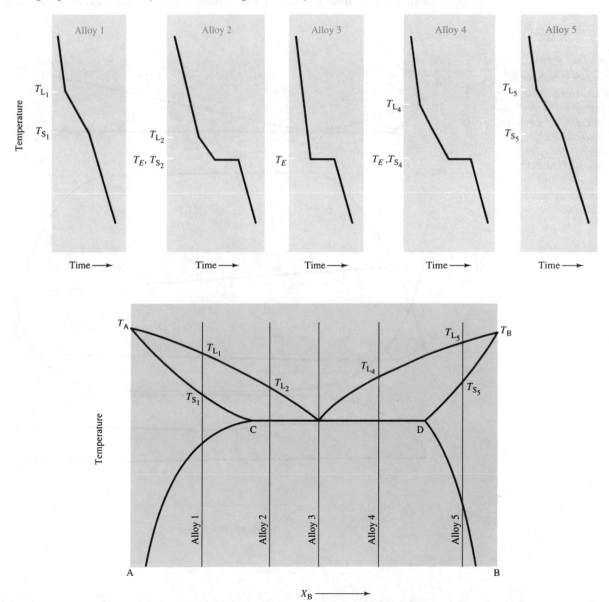

FIGURE 7.4–9   The use of cooling curves to establish the liquidus and solidus in a binary eutectic alloy cooled under equilibrium conditions.

an alloy with a composition that crosses one of the solvus boundaries is slowly cooled from the single-phase field (either α or β), there will be a specific temperature below which the alloy exhibits a two-phase microstructure. Since single-phase microstructures can be distinguished from two-phase microstructures using an optical microscope, the solvus boundary can be determined experimentally by slowly cooling a series of alloys and noting the temperature at which the two-phase microstructure appears. Figure 7.4–8 illustrates this procedure. On the left-hand side the filled circles represent a single-phase microstructure of α (A with specific amounts of component B homogeneously distributed throughout each grain of material). Similarly, on the right-hand side, the filled circles represent single-phase β (B with specific amounts of component A homogeneously distributed throughout each grain of material). In both cases, the open circles represent two-phase mixtures of α and β.

The upper part of the phase diagram can be determined by cooling curves such as those shown in Figure 7.4–9. Because alloys 1 and 5 solidify under equilibrium conditions and

FIGURE 7.4–10

Properties of annealed Ag-Cu alloys measured at room temperature. The annealing temperature was just below the eutectic isotherm. (*Source: F. Rhines,* Phase Diagrams in Metallurgy, *McGraw-Hill, 1957. Reproduced with permission of McGraw-Hill.*)

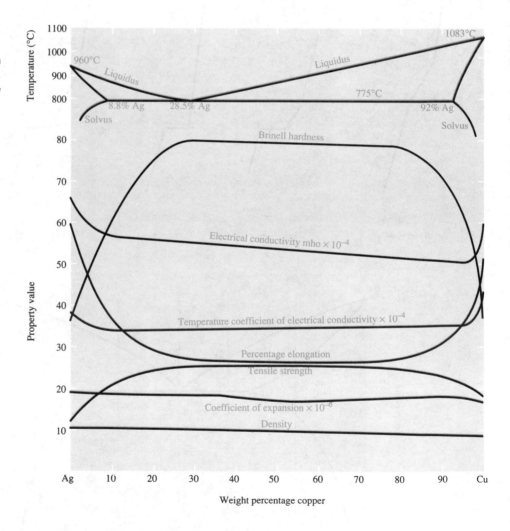

pass through a two-phase field during cooling, the cooling curves resemble those of an isomorphous system. Alloys between C and D solidify with a fraction of eutectic material that depends upon composition. Because a eutectic transformation is an invariant reaction, the liquid solidifying at the eutectic temperature does so at constant temperature, resulting in a horizontal arrest line. By comparing the cooling curves for alloys 2, 3, and 4, the relative amount of eutectic can be estimated.

Several other techniques can be used to locate phase boundaries. The key is to recognize that many material properties, including lattice parameter, density, electrical resistivity, and thermal expansion, are functions of the compositions and amounts of the phases present. Thus, experimentally measured changes in these properties can assist in establishing the location of phase boundaries.

Figure 7.4–10 shows the approximate variation in a number of properties for a series of Ag-Cu alloys homogenized just below the eutectic temperature (the phase diagram for this binary eutectic system is contained in the upper portion of this figure). Some, but not all, of the properties are sensitive to the composition of the phase present. Increasing the solute portion changes these properties continuously. However, once the solvus boundary is reached, the composition of the phases will remain constant, and only the relative amounts of the two phases change. Consequently, an abrupt change in properties is usually associated with the location of the solvus boundary. The principles described in this section are appropriate for the determination of any phase diagram.

7.4.5 Phase Diagrams Containing Two Eutectics

In a complex two-component system, it is possible to have numerous solid solution regions, as illustrated in Figure 7.4–11a. Again, the two components are A and B, but

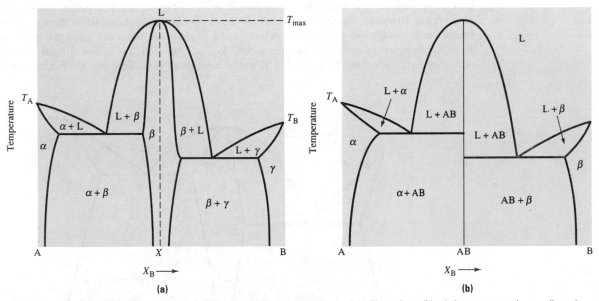

FIGURE 7.4–11   (a) A binary equilibrium phase diagram containing two eutectic reactions illustrating solid solution ranges, an intermediate phase β, and a congruent melting reaction. **(b)** When the solubility of the intermediate phase becomes limited, the line compound AB results.

there are two eutectic reactions and three solid solutions given by the Greek symbols α, β, and γ. The eutectic reactions can thus be represented symbolically as:

$$L \rightleftarrows \alpha + \beta \quad \text{and} \quad L \rightleftarrows \beta + \gamma \quad\quad\quad (7.4\text{--}2)$$

There is special significance to alloys of composition X in this figure. As liquid alloy X is cooled from temperature T_1 to T_2 through T_{max}, two phases are in equilibrium—liquid and solid. What is special about equilibrium at composition X is that either heating or cooling through T_{max} results in a phase change without a composition change. An alloy of composition X, therefore, is said to melt congruently. If an alloy on either side of alloy X is cooled through T_{max}, the composition of the solid that forms is not the composition of the liquid.

Figure 7.4–11b also shows a binary phase diagram with two eutectics, but in this case the single-phase field in the center of the diagram (i.e., $X_B = 0.5$) is a **line compound.** That is, this single phase displays essentially no solid solubility for either component A or B. Since this line compound contains 50 mol % A and 50 mol % B, its chemical formula is AB. Line compounds are generally labeled with their stoichiometric chemical formula rather than with a Greek letter.

....

EXAMPLE 7.4–2

Consider the phase diagram shown in Figure 7.4–12a. Label all of the phase fields and identify each of the invariant reactions.

Solution

The diagram is labeled by recalling that any isotherm (other than one associated with an invariant reaction) will begin in a single-phase field for pure component A, and then alternately pass through two-phase fields and single-phase fields until it terminates in a single-phase field for pure component B. The isotherm corresponding to T_1 in Figure 7.4–12b passes through five phase fields. Therefore, the first, third, and fifth fields contain single phases while the second and fourth fields contain two phases. By convention the single-phase fields are given the names α, β, and γ from left to right. The two-phase field located "between" the α and β fields contains $\alpha + \beta$ and the two-phase field between β and γ contains $\beta + \gamma$. Similarly, the isotherm at T_2

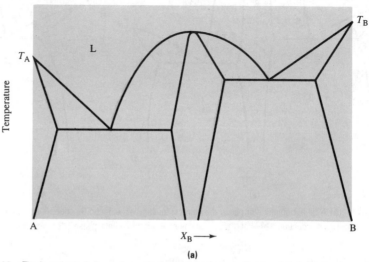

(a)

FIGURE 7.4–12   The hypothetical phase diagram referred to in Example Problem 7.4–2: **(a)** the phase diagram in the problem statement, and **(b)** the solution to the example problem.

FIGURE 7.4–12   Concluded

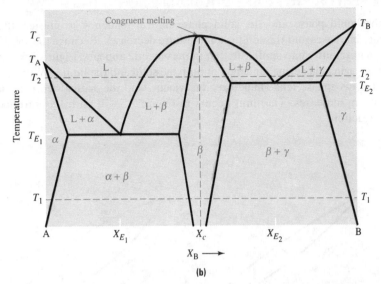

(b)

crosses nine phase fields—α, $\alpha + L$, L, $L + \beta$, β, $\beta + L$, L, $L + \gamma$, and γ. The three invariant reactions are:

1. Eutectic reaction at T_{E_1} and X_{E_1}: $L \leftrightarrows \alpha + \beta$
2. Eutectic reaction at T_{E_2} and X_{E_2}: $L \leftrightarrows \beta + \gamma$
3. Congruent melting at T_c and X_c: $L \leftrightarrows \beta$

When the α and β phase fields in a eutectic system are very narrow (i.e., the solubilities of A in B and B in A are negligible), the eutectic diagram may be simplified, as shown in Figure 7.4–13, which is similar to that of the Al-Si system discussed previously. This approximation is often done out of convenience, but it should be recognized that there must always be some solubility. This occurs because the free energy of a pure component can always be reduced by very small additions of a second component, which increases the randomness of the system (refer to the discussion in Section 4.3.1). Thus, regardless of how dissimilar the components are, there is always a small amount of mutual solubility.

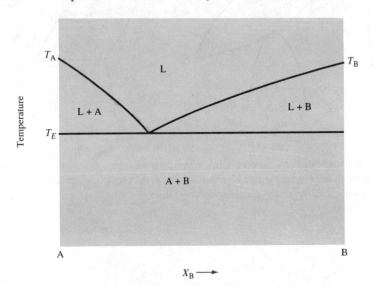

FIGURE 7.4–13

Binary eutectic system when the mutual solubilities of the two components in one another are extremely small.

7.5 THE PERITECTIC PHASE DIAGRAM

If both the liquid phase and the solid phase have a tendency to cluster, the liquidus temperature increases and the solidus temperature decreases. Because the solid is tending to cluster, a miscibility gap similar to that in the eutectic appears. Figure 7.5–1 shows the effect of increasing clustering tendency of both the solid and the liquid. With a progressive increase, the two-phase lens-shaped region widens, and the critical temperature for the miscibility gap increases. The limit occurs in Figure 7.5–1d with the introduction of a peritectic reaction.

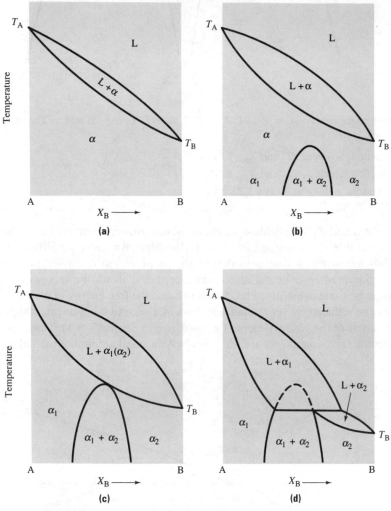

FIGURE 7.5–1 Development of a peritectic system by increasing the clustering tendencies of the solid and liquid phases. The clustering tendency for the solid is greater than that of the liquid. (*Source: Adapted from Albert Prince, Alloy Phase Equilibria, 1966. Permission granted from Elsevier Science.*)

The peritectic reaction and the associated definitions are shown in Figure 7.5–2a. The **peritectic reaction** can be written in symbolic form:

$$L + \alpha \leftrightarrows \beta \qquad\qquad (7.5\text{–}1)$$

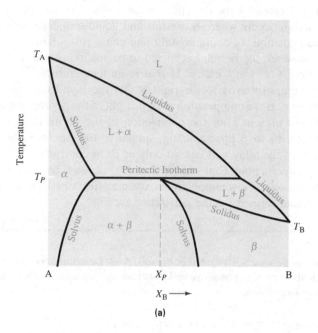

(a)

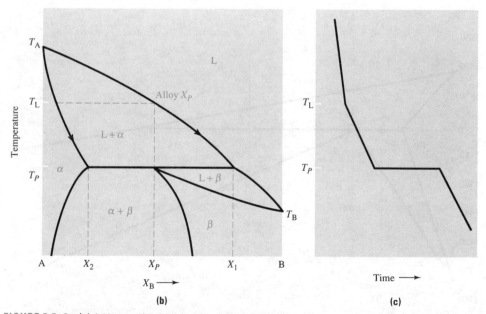

(b)   **(c)**

FIGURE 7.5–2   **(a)** A binary peritectic phase diagram and the associated terms used to describe regions of a peritectic system, **(b)** a simple peritectic system showing the equilibrium cooling of an alloy whose composition is the peritectic composition X_P, and **(c)** the corresponding cooling curve for alloy X_P.

As in the eutectic system, only alloys that cross the peritectic isotherm undergo the invariant reaction. Solidification of an alloy of the peritectic composition, X_P, proceeds as shown in Figure 7.5–2b. As the alloy is slowly cooled below the α liquidus line, primary crystals of α form. Their size and number increase as cooling continues. The compositions of the two equilibrium phases move along the liquidus and solidus boundaries, as indicated by the direction of the arrows on the figure. At the peritectic isotherm, a reaction occurs where α crystals and liquid transform into β crystals. Thus, the alloy of composition X_P contains only one phase just slightly below the peritectic isotherm, T_P. However, as the solid is cooled, α begins to precipitate from the β, so the material will consist of α in a matrix of β at room temperature.

If the alloy composition is to the right of X_1, the liquid transforms to β as in an isomorphous system. If the composition is to the right of the peritectic point but still below X_1, there is liquid present after the completion of the peritectic reaction. This liquid eventually transforms to β upon cooling, similar to the isomorphous system. Thus, at lower temperatures the alloy will consist only of β. In contrast, if the composition is to the left of the peritectic point (but greater than X_2), excess α remains after completion of the peritectic reaction. At low temperatures, the material consists of a mixture of primary α along with β that formed via the peritectic reaction.

EXAMPLE 7.5–1

Consider the peritectic system illustrated in Figure 7.5–3. Determine the composition and relative amounts of each phase present just above and just below the peritectic isotherm for each of the three alloy compositions indicated.

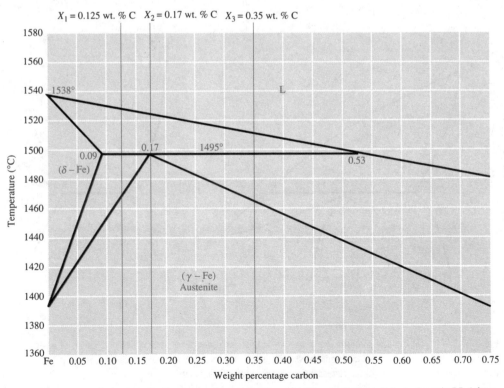

FIGURE 7.5–3 A simple peritectic diagram showing three specific alloy compositions. Refer to Example 7.5–1 for a discussion of this phase diagram.

Solution

First we determine the composition of each phase in each alloy. In all three cases, just above the peritectic isotherm the alloy consists of $L + \delta$. The phase compositions are given by the horizontal coordinate of the intersections of the tie line (through the state point) and the neighboring phase boundaries. Specifically, just above the peritectic isotherm all three alloys are composed of δ of composition $X_\delta = 0.09$ wt. % C and liquid of composition $X_L = 0.53$ wt. % C. Just below the peritectic, the situation is quite different. Alloy 1 consists of δ, of composition $X_{\delta_1} = 0.09$ wt. % C, and γ_1 of composition $X_{\gamma_1} = 0.17$ wt. % C. Alloy 2 exists as single-phase γ with composition $X_{\gamma_2} = 0.17$ wt. % C. Alloy 3 is a mixture of γ, with composition $X_{\gamma_3} = 0.17$ wt. % C, and liquid, with composition $X_{L_3} = 0.53$ wt. % C.

The relative amounts of each phase (by weight) are calculated using the lever rule, and the results of the calculations are summarized below.

	Above peritectic	Below peritectic
Alloy 1		
f_L	$\dfrac{0.125 - 0.09}{0.53 - 0.09} = 0.080$	0
f_δ	$\dfrac{0.53 - 0.125}{0.53 - 0.09} = 0.920$	$\dfrac{0.17 - 0.125}{0.17 - 0.09} = 0.563$
f_γ	0	$\dfrac{0.125 - 0.09}{0.17 - 0.09} = 0.437$
Alloy 2		
f_L	$\dfrac{0.17 - 0.09}{0.53 - 0.09} = 0.182$	0
f_δ	$\dfrac{0.53 - 0.17}{0.53 - 0.09} = 0.818$	0
f_γ	0	1.0 (single phase)
Alloy 3		
f_L	$\dfrac{0.35 - 0.09}{0.53 - 0.09} = 0.591$	$\dfrac{0.35 - 0.17}{0.53 - 0.17} = 0.500$
f_δ	$\dfrac{0.53 - 0.35}{0.53 - 0.09} = 0.409$	0
f_γ	0	$\dfrac{0.53 - 0.35}{0.53 - 0.17} = 0.500$

7.6 THE MONOTECTIC PHASE DIAGRAM

In some systems, the bond characteristics of the components are so different that a region of immiscibility develops in the liquid state. In essence, over some portion of the phase diagram two insoluble liquids form. A common example of this occurs with oil and water, which at ordinary temperatures and pressures are virtually insoluble in one another. Such behavior can also occur with metals and ceramics, one manifestation of which is associated with the **monotectic reaction.**

The monotectic reaction is characterized by the presence of a two-phase liquid field. Figure 7.6–1a shows a phase diagram that exhibits the monotectic reaction. As in the previous examples of phase equilibria, the description of a monotectic reaction can be written symbolically as:

$$\text{Liquid 1} \rightleftarrows \text{Liquid 2} + \alpha \qquad (7.6\text{–}1)$$

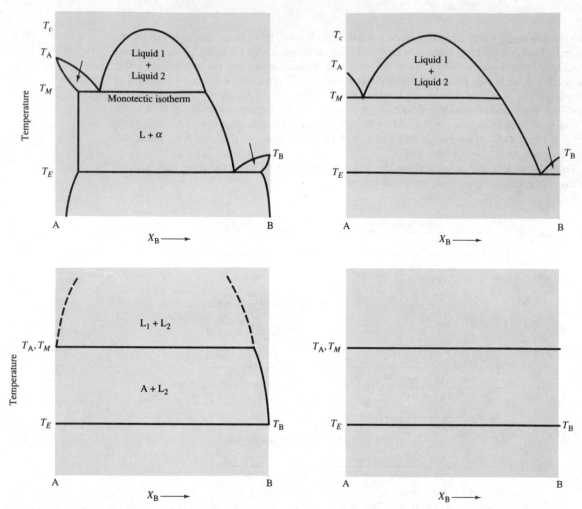

FIGURE 7.6–1 The limiting case of the monotectic reaction occurs when there is effectively no mutual solubility in either the liquid or solid phase.

Above the monotectic isotherm, two two-phase regions are present. One region, associated with component A, is the $(L_1 + \alpha)$ two-phase region. The other region, associated with the miscibility gap, is the $(L_1 + L_2)$ two-phase region. From the monotectic isotherm up to some critical temperature T_c, two liquid phases coexist. As the temperature increases to T_c, the compositions of the two phases in equilibrium change, moving closer together. At the critical temperature the two liquids have the same composition. Above T_c, and at all liquid compositions outside the miscibility gap, a single-phase liquid occurs

.   Evolution of the monotectic system to the limiting case of complete insolubility in the liquid and solid is illustrated in the sequence of diagrams in Figure 7.6–1. As the solubility in the liquid state decreases, the monotectic isotherm extends across the diagram. In the extreme, there is virtually no solubility in the liquid state. The resulting phase diagram is one in which the monotectic composition X_M approaches pure A and the monotectic temperature T_M approaches T_A. In addition, the eutectic composition X_E approaches pure B, and the eutectic isotherm T_E approaches T_B.

The W-Cu system is a commercially important example of this extreme behavior. Alloys of W-Cu are used in spot-welding operations. The tungsten is used because it imparts high-temperature strength to the welding electrode, and it is also refractory. The copper imparts its excellent electrical conductivity. The mutual insolubility in both the liquid and solid states for W-Cu suggests that alloys of this material could not be solidified conventionally for two reasons. First, the components cannot be mixed homogeneously in one another at reasonable temperatures, and second, there is a significant density differ-ence between tungsten and copper. The heavier liquid phase would simply sink to the bottom of the crucible. However, If finely divided copper and tungsten powders are mechanically mixed, "alloys" may be fabricated by hot-pressing and sintering these elemental powders together. These mechanical alloying processes are alternative mecha-nisms to alloy production by conventional solidification, which are addressed in Chap-ter 16, on processing.

Knowledge of phase equilibria can be invoked to suggest methods of alloy preparation. Such knowledge can also be used to determine when conventional alloy production of specific compositions through solidification is inappropriate and alternative approaches must be tried.

7.7 COMPLEX DIAGRAMS

In the diagrams presented so far, the complexity of the systems has been minimal. However, most diagrams of practical systems are more complex. Rather than having just one invariant reaction, or, as presented in Section 7.4.5, two of the same type of invariant reaction, many alloys contain combinations of invariant reactions. Figures 7.7−1 and 7.7−2 show complex diagrams containing two eutectics and a peritectic. The major difference between the two diagrams is that in the former all of the single-phase fields dis-play limited solubility while the latter diagram the single-phase fields are line compounds.

As shown in Figure 7.7−1b, such diagrams can be divided into three parts illustrating the various reactions participating in the equilibria. Breaking complex phase diagrams into parts simplifies their analysis. It should be recognized that, regardless of how compli-cated the phase diagram of a binary system may be, the principles in this chapter always

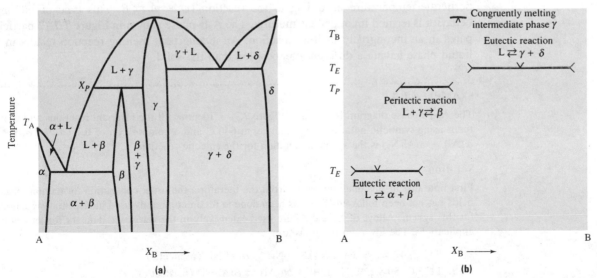

FIGURE 7.7−1 **(a)** Complex phase diagram containing a peritectic and two eutectic reactions, and **(b)** the invariant reactions in **(a)** emphasized along with their symbolic representations. When the β phase is heated to the peritectic temperature, an incongruent melting reaction occurs at T_P, the peritectic temperature.

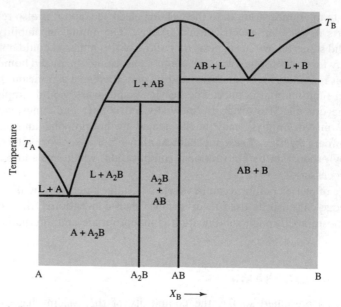

FIGURE 7.7-2   A diagram similar to Figure 7.7-1, but the β and γ are seen as line compounds. The α and δ phases are terminal solid solutions with essentially no solubility. Thus, they are simply labeled A and B, corresponding to the pure components.

apply. For example, when the point corresponding to the overall composition and temperature of interest lies in a two-phase field, the phase compositions are determined using a tie line, and their relative amounts are calculated using the lever rule. Three-phase equilibrium always occurs at a specific point for which the temperature and compositions of the phases participating in equilibrium are fixed.

Figure 7.7-1 shows an important distinction between the β and γ phases. The γ phase melts congruently according to the reaction $L \rightleftarrows \gamma$. In contrast, the β phase melts at the peritectic temperature according to the reaction $(L + \gamma) \rightleftarrows \beta$. This type of melting behavior is termed **incongruent melting.** The A_2B phase shown in Figure 7.7-2 participates in an incongruent melting reaction. An incongruent melting reaction results in a liquid phase having a different composition than the solid.

EXAMPLE 7.7-1

The Al-Ni phase diagram is shown in Figure 7.7-3. Identify all the invariant reactions and write them using symbolic notation. For example, at 640°C and a composition of 6.1 wt. % Ni, $L \rightleftarrows \alpha(\text{Al}) + \alpha(\text{Al-Ni})$ is the symbolic notation for the eutectic reaction.

Solution

First note that in this diagram, taken from the literature, the usual convention for naming phase fields has not been followed. This has been done to retain commonly used industrial designations for this system. Alloys of Ni and Al are used extensively in the aerospace industry for jet engine applications. The invariant reactions are:

1. 854°C, 42 wt. % Ni: $L + \beta(\text{Al-Ni}) \rightleftarrows \alpha(\text{Al-Ni})$ (peritectic)
2. 1133°C, 59 wt. % Ni: $L + \beta(\text{Ni-Al}) \rightleftarrows \beta(\text{Al-Ni})$ (peritectic)
3. 1638°C, 68.5 wt. % Ni: $L \rightleftarrows \beta(\text{Ni-Al})$ (congruent melting)
4. 1395°C, 85 wt. % Ni: $L + \beta(\text{Ni-Al}) \rightleftarrows \alpha(\text{Ni-Al})$ (peritectic)
5. 1385°C, 86.7 wt. % Ni: $L \rightleftarrows \alpha(\text{Ni-Al}) + \alpha(\text{Ni})$ (eutectic)

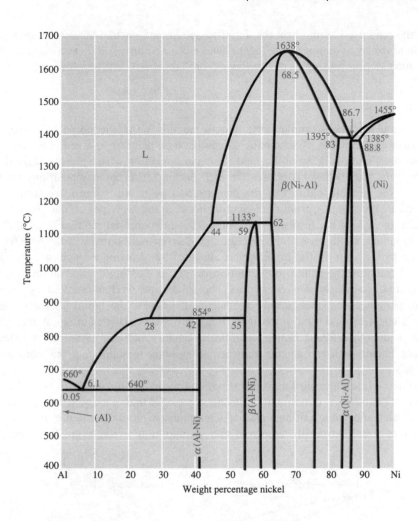

FIGURE 7.7–3

The Al-Ni binary phase diagram.

It is worth noting that α(Ni-Al) is often called γ' and is the basis of strengthening in Ni-base superalloys.

7.8 PHASE EQUILIBRIA INVOLVING SOLID-TO-SOLID REACTIONS

So far phase diagrams in which the invariant reactions involve a liquid phase have been emphasized:

$$L \leftrightarrows \alpha + \beta \text{ (eutectic)} \tag{7.8–1}$$

$$L + \beta \leftrightarrows \alpha \text{ (peritectic)} \tag{7.8–2}$$

and

$$L_1 \leftrightarrows L_2 + \alpha \text{ (monotectic)} \tag{7.8–3}$$

It is possible to replace the liquid phases in the above symbolic reactions with solid phases, creating a new set of invariant reactions:

$$\gamma \leftrightarrows \alpha + \beta \text{ (\textbf{eutectoid})} \tag{7.8–4}$$

$$\gamma + \beta \leftrightarrows \alpha \text{ (\textbf{peritectoid})} \tag{7.8–5}$$

and

$$\alpha_1 \leftrightarrows \alpha_2 + \beta \text{ (\textbf{monotectoid})} \tag{7.8–6}$$

All of the rules used to interpret the liquid-to-solid reactions in a binary diagram are applicable to the lower-temperature regime where these solid-to-solid reactions may occur. In the next section, examples of the three "-oid" reactions will be presented.

7.8.1 Eutectoid Systems

The eutectoid reaction is similar to the eutectic reaction in that one phase decomposes into two phases upon cooling. A schematic eutectoid phase diagram is given in Figure 7.8–1. It shows a region of solid solubility and a eutectoid reaction where the single solid phase γ transforms into two solid phases $(\alpha + \beta)$ having different compositions from the γ phase. The general features of the eutectic and eutectoid systems are similar; both reactions are invariant. A primary distinction between the two types of reactions concerns the scale, or size, of the resulting two-phase structures and the relative rates of the two transformations. In the case of the eutectic, the transformation involves the redistribution of solute by diffusion in the *liquid* phase ahead of the growing two-phase solid. In the eutectoid reaction, the solute redistribution occurs in the single *solid* phase directly in front of the two-phase $(\alpha + \beta)$ region. The solid state is characterized by a densely packed, periodically spaced atomic arrangement, whereas the liquid for the same substance is characterized by a more open, random network. Diffusion is enhanced in the open structure of the liquid. Hence, equilibrium can be achieved more rapidly in reactions involving liquids such as eutectic transformations than in solid-state reactions like the eutectoids.

Another distinction regarding transformations occurring in solids is the importance of prior grain boundaries. Grain boundaries are regions of high energy, which can promote the formation and growth of new phases. Consequently, in the formation of a two-phase structure of α and β from the eutectoid transformation of γ, the α and β phases tend to

FIGURE 7.8–1

A general eutectoid equilibrium phase diagram with associated definitions.

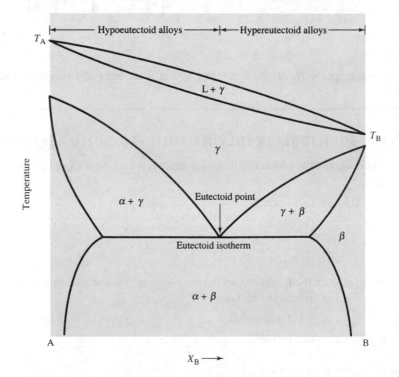

form preferentially at the original γ grain boundaries. These issues will be presented in more detail in Chapter 8.

Perhaps the most important commercial system that exhibits a eutectoid reaction is illustrated by the Fe-Fe$_3$C system. Figure 7.8–2 shows part of the Fe-Fe$_3$C system. Note that a eutectoid reaction is seen at 727°C, in addition to peritectic (1495°C) and eutectic (1148°C) reactions. Although strictly speaking the Fe-Fe$_3$C system is not an equilibrium system (carbon is more stable than Fe$_3$C), Fe$_3$C is quite stable, and for all practical purposes we may consider the diagram in Figure 7.8–2 to be an equilibrium diagram. When an alloy containing 0.77 wt. % carbon is cooled under conditions approaching equilibrium from the single-phase γ field called **austenite** (FCC Fe with dissolved C), the austenite decomposes into a two-phase mixture of α Fe and Fe$_3$C (**cementite**). The eutectoid constituent is called **pearlite;** the microstructure of pearlite is seen in Figure 7.8–3. This important system will be discussed more fully in Chapter 8. Most steels and cast irons for applications in automobiles, buildings, bridges, gears, landing gear, and so on are based on the diagram shown in Figure 7.8–2 or modifications thereof.

In addition to the eutectoid, there are also peritectoid and monotectoid reactions. Examples of these types of phase diagrams are illustrated in Figures 7.8–4 and 7.8–5.

The approach taken in this chapter—to introduce phase diagrams of two-component systems—was designed to emphasize the relative behavior of the solutions involved in forming the phase diagram. While the simple bond model used in this chapter is only an approximation, it helps to demonstrate that there are underlying thermodynamic factors that control the features of phase diagrams.

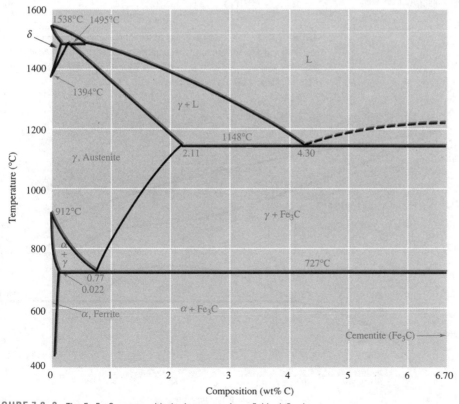

FIGURE 7.8–2 The Fe-Fe$_3$C system with the important phase fields defined.

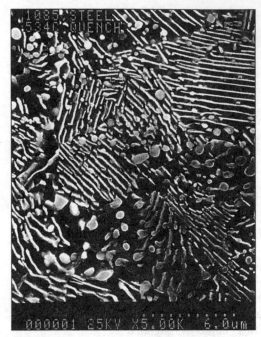

FIGURE 7.8–3 Structure of the eutectoid composition 0.77 wt. % C steel. The eutectoid constituent, called pearlite, has a lamellar (layered) morphology consisting of alternating plates of α Fe (light areas) and Fe_3C (dark areas).

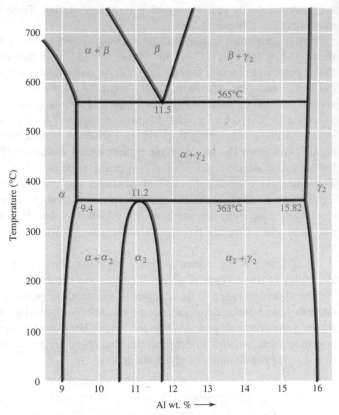

FIGURE 7.8–4 The copper-rich side of the Cu-Al system shows a peritectoid reaction in the approximate temperature range of 300–400°C. Symbolically the reaction is written $\alpha + \gamma_2 \rightleftarrows \alpha_2$. *(Source: Adapted from Albert Prince, Alloy Phase Equilibria, 1966. Permission granted from Elsevier Science.)*

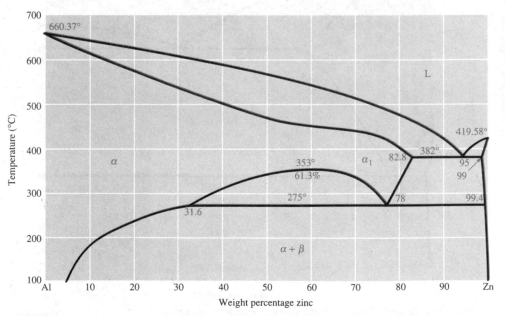

FIGURE 7.8–5 The Al-Zn systems contain a monotectoid reaction at 275°C. Symbolically the reaction is $\alpha_1 \rightleftarrows \alpha + \beta$.

7.9 PHASE EQUILIBRIA IN THREE-COMPONENT SYSTEMS

Over the years, the approaches taken to teach phase equilibria in metallurgical systems and ceramic systems have deviated significantly. Much of the work in metallic materials has centered on treating systems that exhibit a significant degree of solid-state solubility. For example, in the next chapter the principles of precipitation hardening will be presented. For a precipitation-hardening system, the objective is to increase the amount of solute present in the single phase and rapidly quench to a lower temperature, where the solubility is significantly reduced. By exercising controls on heat-treating temperatures and times, a fine dispersion of second-phase particles can be precipitated from the supersaturated solid solution. The fine distribution of phases increases strength while maintaining an acceptable level of ductility. Most commercial alloy systems contain numerous components, but the guiding principles of heat treating can be illustrated using a binary diagram. The emphasis in metallurgy, therefore, has been to develop the concepts of binary diagrams and generate understanding of more complex systems by analogies to conceptually simpler binary systems.

Ceramists, on the other hand, work with materials that tend to form line compounds with limited solubility of the various components in one another. Hence, by recognizing that fixing the composition of one of the phases in a two-phase field simplifies the interpretation of phase equilibria, ceramists have opted for more complex ternary (i.e., three-component) phase diagrams with reduced solubilities of solid phases.

Concepts of ternary equilibria, beyond introducing the basic principles of plotting compositions, introducing the lever rule in a two-phase region, and determining the fraction of phases in a three-phase region, are not appropriate areas for an introductory course in materials engineering. With the basic skills presented for binary diagrams, however, the remaining sections will provide the foundation that can be built upon in more advanced courses.

7.9.1 Plotting Compositions on a Ternary Diagram

The description of equilibrium in a two-component system at a fixed pressure is represented by a plane figure, since only two variables are required: temperature and composition. For a system at constant pressure in a ternary diagram, three composition variables (two of them independent) and a temperature variable must be represented. The description of phase equilibria relating composition and temperature to the phases present necessitates the use of a three-dimensional diagram (Figure 7.9.–1).

The 3-D representation contains all of the necessary information, but such a diagram is difficult to construct and interpret. However, a common way to represent a three-component system is by making use of isothermal sections, which are generated by making constant-temperature slices of the 3-D diagram shown in Figure 7.9–1. The liquidus and solidus surfaces can be projected onto a plane. The resulting temperature contours are then similar to a topographic map illustrating elevations using contours. Figure 7.9–2 shows a projection of the liquidus surface from Figure 7.9–1 onto the composition plane. Each temperature contour represents an isotherm. The spacing of the isotherms indicates the slope of the surface. The more closely spaced the isotherms are, the steeper the surface.

FIGURE 7.9–1
A space model showing a ternary system with a continuous series of liquid and solid solutions. The diagram is the ternary equivalent of an ideal system.

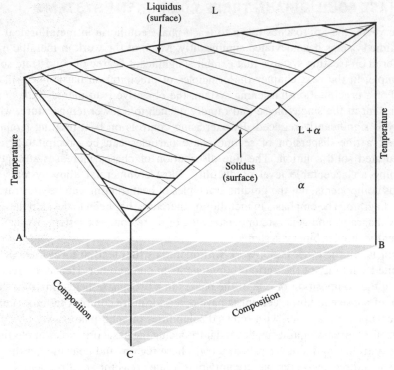

FIGURE 7.9–2
Projection of the liquidus surface of Figure 7.9–1 onto the composition plane.

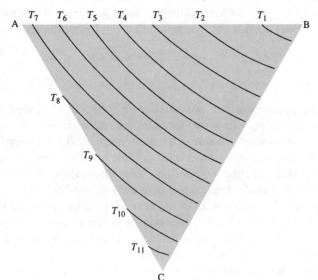

The compositions in Figures 7.9–1 and 7.9–2 have been plotted on an equilateral triangle. The percentage or fraction of each component can be determined by using the properties of an equilateral triangle and establishing proportionalities. In an equilateral triangle, the sum of the perpendiculars from any point in the triangle to the sides of the triangle is constant and equal to the height of the triangle.

To apply this property of an equilateral triangle to compositions in ternary alloys, consider Figure 7.9–3a. Each corner of the triangle is labeled with a specific (pure) component, A, B, or C. When a perpendicular runs from component B to side AC, the

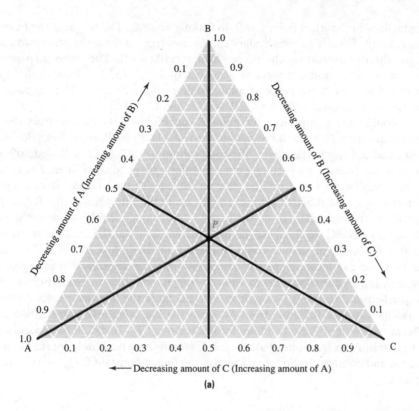

(a)

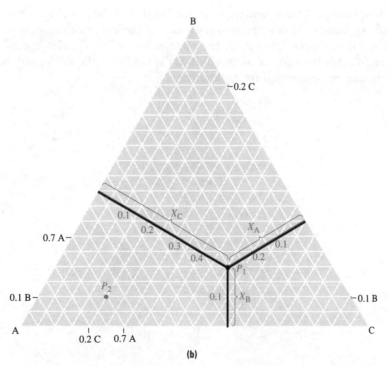

(b)

FIGURE 7.9–3   An equilateral triangle is used to plot compositions in a three-component system: **(a)** illustration of basic component fraction labeling, and **(b)** the location of an alloy composition composed of 30% A, 20% B, and 50% C.

horizontal lines are marked from 1 at B, to 0 along line AC. The height of the triangle is, therefore, unity. Similar perpendiculars can be constructed from the other two components and the distance along the perpendicular marked off. The three perpendiculars drawn from each component meet at distances of 1/3, 1/3, 1/3 from each edge. As required, the sum of these perpendicular distances must be unity. Hence, point P on Figure 7.9–3a is located at A = 1/3, B = 1/3, and C = 1/3.

To determine the composition of point P_1 on Figure 7.9–3b, construct the three perpendiculars from P_1 to the three sides of the triangle. Point P_1 is 0.3 from side BC, 0.2 from AC, and 0.5 from AB. Hence, the composition is 30% A, 20% B, and 50% C. Of course, it is not necessary to actually construct these perpendicular lines. The compositions can simply be read off from the scale on the sides of the triangle. To determine the fraction of C associated with P_1, for example, locate the line parallel to AB that passes through P_1. Then counting up from A (where $X_C = 0$) along the side AC, determine where this line intersects AC. In this case the intersection occurs at 0.5, and the fraction of C is thus 0.5. The same procedure could be carried out along line BC.

Similarly, we can locate a specific composition on the diagram. For example, locate the composition P_2 corresponding to 70% A, 10% B, and 20% C. Using Figure 7.9–3b, the perpendicular distance 0.7 away from side BC gives the line that marks the location of A, and the distance 0.1 away from AC gives the line where the composition B must lie. The intersection of these two lines locates the point P_2. It should be recognized that since the sum of the fractions of A, B, and C is unity, we need only locate the lines of constant A and constant B. For completeness, the line of constant C ($X_C = 0.2$) was also determined.

7.9.2 The Lever Rule in Ternary Systems

As in a two-phase field for a binary system, the fraction of the two phases in equilibrium in a ternary system can be determined using a form of the lever rule. Suppose that the two points α and β in Figure 7.9–4 mark the compositions of two phases in equilibrium at temperature T_1. For the alloy of composition X, the fraction of liquid, f_L, and the fraction of solid, f_S, can be determined by the lever principle.

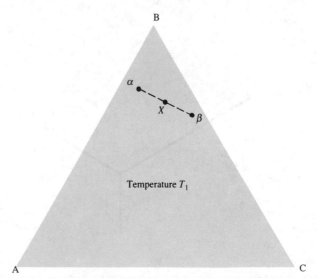

FIGURE 7.9–4   The use of the lever rule to determine the fraction of two phases α and β, for a ternary alloy of composition X.

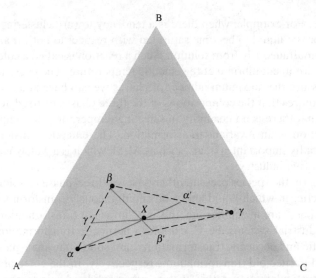

FIGURE 7.9–5 A diagram illustrating the application of a tie triangle to compute the fractions of three phases present at equilibrium.

When three phases are in equilibrium in a three-component system, they form a triangle. Figure 7.9–5 shows a schematic of three phases in equilibrium, α, β, and γ. The fraction of each component for an alloy of composition X can be determined by a "center-of-gravity" principle. As in the lever rule, the triangle may be considered as being supported at a fulcrum point at X and having masses at its corners proportional to the amounts of the phases present. At equilibrium, the position X is balanced when:

$$t_\alpha = \frac{\alpha' - X}{\alpha' - \alpha} \tag{7.9–1}$$

$$f_\beta = \frac{\beta' - X}{\beta' - \beta} \tag{7.9–2}$$

and

$$f_\gamma = \frac{\gamma' - X}{\gamma' - \gamma} \tag{7.9–3}$$

where the quantity $\alpha' - X$ represents the distance between the points α' and X, and the other differences are similarly defined.

SUMMARY

In this chapter we have demonstrated that equilibrium occurs at specific conditions of temperature, composition, and pressure. The phase rule provides a relationship between the number of phases, components, and other variables (such as pressure, temperature, and composition) that can be varied while maintaining specified equilibrium conditions.

The emphasis was placed on two-component systems at constant pressure, since they have extensive industrial significance. In the ideal case, when a second component is added freezing and melting take place over a range of temperatures. As conditions deviate from ideality (i.e., as the components exhibit stronger interactions), significant new features arise. For example, when there is a tendency toward clustering of like species, it is possible for the liquid to become saturated with respect to both α and β, so that they precipitate simultaneously from solution. Such a reaction is called a eutectic reaction, and three phases are in equilibrium at the eutectic temperature. The phase rule tells us that at constant pressure the maximum number of phases we can have at a given temperature is three, and requires that the compositions of the three phases be fixed at that temperature. This means that there is no flexibility in selecting temperature or composition and that a eutectic reaction is an invariant transformation. The eutectic transformation occurs in many industrially important systems, such as Al-Si, which is a widely used piston material in the automotive industry.

Depending on the species present, other types of three-phase reactions can occur, such as the peritectic, in which liquid and solid simultaneously transform to a different solid. Since three phases are in equilibrium, this is an invariant transformation. One of the most important industrial systems, Fe-C, exhibits a peritectic transformation.

The eutectic and peritectic transformations have solid-state analogues, which are called eutectoid and peritectoid transformations, respectively. In the eutectoid, one solid transforms at a particular temperature to two other solids. Again, since three phases are involved in equilibrium, the reaction is invariant. The Fe-C system also exhibits a eutectoid transformation that is of great industrial significance.

Regardless of their level of complexity, all phase diagrams share common features. All are maps that provide an indication of what is possible in terms of compositions and amounts of phases, equilibrium temperatures, and invariant transformations. They are particularly valuable, since they can be used to determine the phases present, the compositions of the phases, and the relative amounts of the phases. For two-component two-phase systems, compositions are given by the horizontal coordinates of the intersections of a tie line (through the state point) with the nearest phase boundaries on either side of the state point. Phase fractions are calculated using the lever rule, which states that the phase fraction is proportional to the length of the "opposite" lever arm.

The information about phase compositions and amounts is extremely important, since the properties of a polyphase material are linked to its microstructure. In addition, however, we must also understand the influence of time and thermal history on microstructure before we can begin to control the structure of a material to achieve desirable combinations of macroscopic properties. The next chapter investigates the mechanisms and kinetics of transformations in materials.

KEY TERMS

austenite	Gibbs phase rule	line compound	phase
cementite	hypereutectic (-oid) alloy	liquidus boundary	phase diagram
component	hypoeutectic (-oid) alloy	monotectic reaction	primary solid phase
congruent melting	incongruent melting	monotectoid reaction	solidus boundary
degrees of freedom	invariant reaction	pearlite	solvus boundary
eutectic reaction	isomorphous system	peritectic reaction	state point
eutectoid reaction	lever rule	peritectoid reaction	tie line

HOMEWORK PROBLEMS

Questions 1–8 pertain to the binary phase diagram and alloy X_0 in Figure HP7.1.

1. T_A and T_B represent the equilibrium melting temperatures of A and B, respectively. Label the regions of the phase diagram.

2. What is the name given to this type of equilibrium diagram?

3. If component A is FCC, what is the crystal structure of component B, and why?

4. Sketch equilibrium cooling curves for alloy X_0 and pure component B. Explain why they have different shapes.

5. What are the liquidus temperature and the solidus temperature of alloy X_0?

6. What is the composition of each phase in equilibrium at 1100°C for alloy X_0? Determine the fraction of liquid, f_L, and the fraction of solid, f_S, at 1100°C.

7. Sketch f_L and f_S for alloy X_0 as it is cooled under equilibrium temperatures from 1200°C to room temperature.

8. Using the phase diagram in Figure HP7.1, indicate how the compositions of the liquid and solid phases change during equilibrium cooling through the two-phase field.

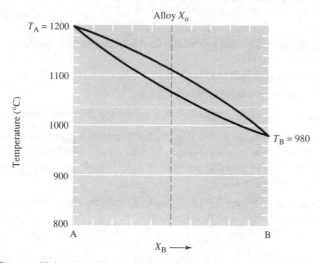

FIGURE HP7.1   Binary equilibrium phase diagram for the hypothetical two-component A-B system.

For a hypothetical two-component system A-B, consider the following problem.

9. From the following data construct a plausible equilibrium phase diagram. Component A melts at 800°C and B melts at 1000°C; A and B are completely soluble in one another at room temperature; and if solid α containing 0.3 B is heated under equilibrium conditions, the solid transforms to liquid having the same composition at 500°C.

Questions 10–14 pertain to the Cu-Ni phase diagram in Figure HP7.2.

10. For alloys containing 10, 22, 25, 27, and 40 wt. % Ni, determine the number of phases present and the composition of the phases in equilibrium at 1200°C.

11. Beginning with a statement of mass balance, derive the lever rule in a two-phase system.

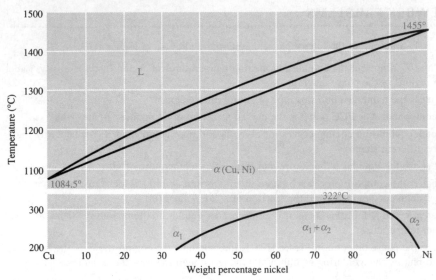

FIGURE HP7.2   The Cu-Ni phase diagram, an example of a binary isomorphous system.   *(Source: Donald T. Hawkins and Ralph Hultgren, "Constitution of Binary Alloys," Metals Handbook, Vol. 8, American Society for Metals, 1973, p. 24.)*

12. Discuss each of the factors that permit the Cu-Ni system to be isomorphous over the temperature range 350–1000°C.

13. What does the temperature 322°C represent?

14. Are α_1 and α_2 different crystal structures? Explain.

For a hypothetical two-component system A-B, consider the following problem.

15. Two alloys, alloy 1 containing 30% B and alloy 2 containing 50% B, when equilibrated at temperature T are in the same two-phase (L + S) region. The fraction of liquid in alloy 1 is 0.8 and the fraction of liquid in alloy 2 is 0.4. Calculate the equilibrium phase boundaries at temperature T.

Questions 16–23 pertain to Figure HP7.3. T_A and T_B represent the equilibrium melting temperature of pure components A and B, respectively.

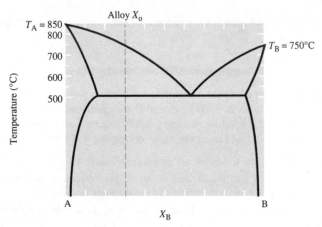

FIGURE HP7.3   Binary equilibrium phase diagram for a hypothetical two-component eutectic system.

16. Label all regions of the phase diagram and label the boundaries of monovariant equilibrium.

17. Sketch an equilibrium cooling curve from above the eutectic to room temperature for an alloy of eutectic composition.

18. Explain why the equilibrium cooling curves for alloys on either side of the eutectic composition will be different from the equilibrium cooling curve for a eutectic alloy.

19. What is the maximum solid solubility of B in A? Of A in B?

20. For an alloy of eutectic composition, determine the composition of the solid phases in equilibrium with the liquid.

21. Plot f_L, f_α, and f_β as a function of temperature for the equilibrium cooling of an alloy of eutectic composition.

22. For alloy X_0, calculate the fraction of α that forms as primary α and the fraction of α that forms by eutectic decomposition when the alloy is cooled from 850°C to room temperature, under equilibrium conditions.

23. What is the total fraction of α at room temperature for alloy X_0?

24. Consider a hypothetical phase diagram in which component A melts at 900°C, component B melts at 1000°C, and there is an invariant reaction at 600°C. The solubility of B in α is known to increase from almost nil at room temperature to a maximum of 10%. When an alloy containing 30% B is cooled under equilibrium conditions just above 600°C, a two-phase mixture is present, 50% α and 50% liquid. When the alloy is cooled just below 600°C, the alloy contains two solid phases, α and β. The fraction of α is 0.75. After cooling under equilibrium conditions to room temperature, the amount of α in the $\alpha + \beta$ mixture decreases to 68%. Sketch the equilibrium phase diagram.

25. The binary system A-B with $T_B > T_A$ is known to contain two invariant reactions of the type:

$$L \rightleftarrows \alpha + \beta \text{ at } T_1$$
$$L \rightleftarrows \beta + \gamma \text{ at } T_2$$

where $T_1 < T_A$ and $T_2 > T_A$. If $\beta(0.5 \text{ B})$ is a congruently melting phase at a temperature higher than T_B, sketch a possible phase diagram.

26. Shown in Figure HP7.4 are two binary phase diagrams of MgO with two other oxides, NiO and CaO. Based on ionic radii, which phase diagram is the MgO-NiO and which is the MgO-CaO? Label the regions on the diagram and identify the invariant reaction.

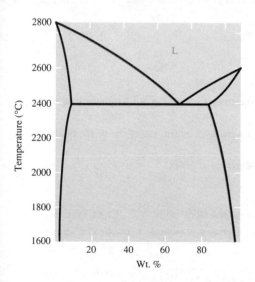

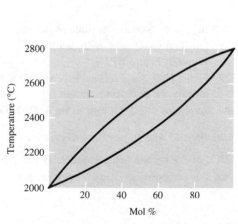

FIGURE HP7.4

Two binary phase diagrams with MgO. *(Source: Bergeron and Risbud, eds., An Introduction to Equilibria in Ceramics, 1984. Reprinted by permission of the American Ceramic Society.)*

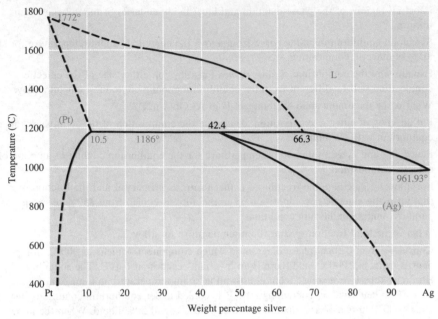

FIGURE HP7.5 The Ag-Pt phase diagram. *(Source: Donald T. Hawkins and Ralph Hultgren, "Constitution of Binary Alloys," Metals Handbook, Vol. 8, American Society for Metals, 1973.)*

Questions 27–30 pertain to Figure HP7.5.

27. What is the maximum solid solubility of Ag in Pt?

28. For an alloy of peritectic composition, what is the composition of the last liquid to solidify at 1186°C?

29. What is the range of alloy compositions that will peritectically transform during equilibrium cooling?

30. Plot the fraction of liquid, f_L, the fraction of α, f_α, and the fraction of β, f_β, as a function of temperature during equilibrium cooling from 1800 to 400°C for an alloy of peritectic composition.

31. Consider a hypothetical two-component system A-B with $T_A > T$ and two invariant reactions of the type:

$$L + \alpha \rightleftarrows \beta \text{ at } T_1$$
$$L \rightleftarrows \beta + \gamma \text{ at } T_2$$

If $T_1 > T_2$, sketch a possible phase diagram.

32. Label all phase fields and identify the invariant reactions in the Ag-Al phase diagram shown in Figure HP7.6.

33. Figure HP7.7 is a schematic of the V_2O_5-NiO phase diagram. Label the phase fields and identify the invariant reactions.

34. Apply the 1-2-1 . . . rule to the V_2O_5-NiO diagram at 600°C.

Questions 35 and 36 pertain to Figure HP7.8.

35. Label all phase fields and identify the invariant reactions in the V_2O_5-Cr_2O_3 system.

36. For compositions X and Y, plot the fraction of phases present as a function of temperature.

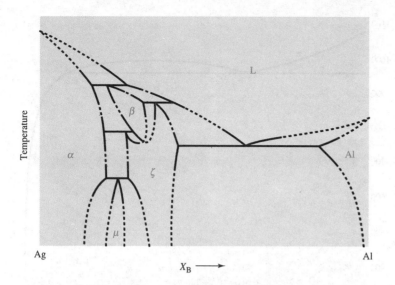

FIGURE HP7.6

A schematic of the Ag-Al phase diagram with the invariant reactions indicated by solid lines.   *(Source: Adapted from Albert Prince, Alloy Phase Equilibria, 1966. Permission granted from Elsevier Science.)*

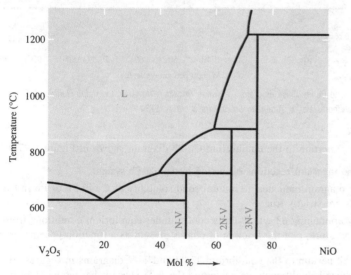

FIGURE HP7.7

A schematic of the V_2O_5-NiO phase diagram. *(Source: Bergeron and Risbud, eds., An Introduction to Equilibria in Ceramics, 1984. Reprinted by permission of the American Ceramic Society.)*

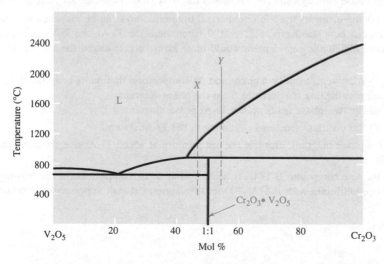

FIGURE HP7.8

A schematic of the V_2O_5-Cr_2O_3 phase diagram. *(Source: Bergeron and Risbud, eds., An Introduction to Equilibria in Ceramics, 1984. Reprinted by permission of the American Ceramic Society.)*

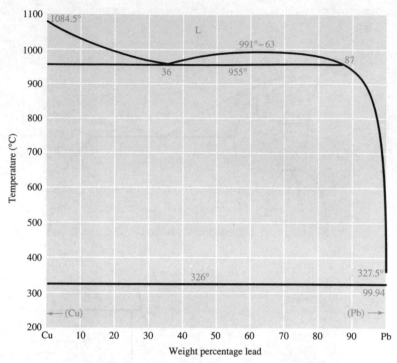

FIGURE HP7.9 The Cu-Pb phase diagram. *(Source: Donald T. Hawkins and Ralph Hultgren, "Constitution of Binary Alloys,"* Metals Handbook, *Vol. 8, American Society for Metals, 1973.)*

Questions 37–39 pertain to the equilibrium Cu-Pb diagram shown in Figure HP7.9.

37. Identify the invariant reactions occurring in the Cu-Pb system.

38. The phase diagram indicates no mutual solid solubility of Cu in Pb or Pb in Cu. Explain why this is not strictly true.

39. If an alloy containing 63 wt. % Pb is cooled under equilibrium conditions from 1100°C to room temperature, plot the fraction of phases present as a function of temperature.

Questions 40–52 pertain to the equilibrium Ti-Al and Ti-V diagrams in Figures HP7.10. Note that the α Ti phase is body-centered cubic and the β Ti phase is hexagonal close packed.

40. Most alloying elements used in commercial titanium alloys can be classified as either alpha stabilizers or beta stabilizers. Figure HP7.10 contains the Ti-Al and Ti-V equilibrium phase diagrams. Which alloying element would most likely be considered the alpha stabilizer? Explain.

41. In the Ti-Al system, identify a phase and its composition that melts congruently. Estimate the congruent melting temperature from the phase diagram.

42. Label all the two-phase fields on the Ti-Al phase diagram.

43. Label all the invariant reactions occurring in the Ti-Al system.

44. From the phase diagram, estimate the temperature at which Ti_3Al congruently transforms to α Ti.

45. When the line compound β $TiAl_3$ is heated, what is the composition of the first liquid that forms in equilibrium with β $TiAl_3$? Does this compound melt congruently? Explain.

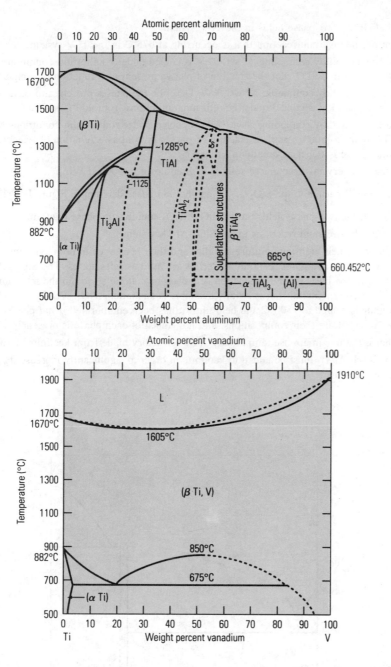

FIGURE HP7.10

The Ti-Al and Ti-V phase diagrams. *(Source: Hugh Baker, ed., ASM Handbook, Vol. 3, ASM International, Materials Park, OH, 1992.)*

46. The presence of Ti_3Al in Ti-Al alloys has a detrimental effect on the ductility. To control this problem, the amount of aluminium needs to be less than 6 wt. %. Consequently, all of the products that you are producing contain a maximum of 6 wt. % of aluminium. You have been told that additions of tin, zirconium, and oxygen (often present as an impurity) are all known to be alpha stabilizers. If there is a possibility that small additions of oxygen may enter the system as your alloy is melted, should you reduce the amount of aluminium in your alloy or not worry about it if ductility is an important property for your product? Explain your reasoning.

47. Label all the two-phase fields in the Ti-V system.
48. Identify the invariant reaction that is occurring at 675°C in the Ti-V system.
49. If a titanium alloy containing 52 wt. % V is cooled under equilibrium conditions from 900°C down to 500°C, plot the fraction of phases present as a function of temperature.
50. Consider an alloy containing 52 wt. % V. Describe the phases present and their compositions as the alloy is cooled under equilibrium conditions from 900°C to 500°C.
51. If a titanium alloy containing 5 wt. % V is cooled under equilibrium conditions from 900°C down to 500°C, plot the fraction of phases present as a function of temperature. Which phase is richer in vanadium, α or β?
52. What is the crystal structure of solid vanadium down to 500°C? Why?

Questions 53–57 pertain to the SiO_2-Al_2O_3 equilibrium phase diagram in Figure HP7.11.

53. Label all the two-phase fields on the SiO_2-Al_2O_3 equilibrium phase diagram.
54. Identify the invariant reactions in the SiO_2-Al_2O_3 system.
55. Because of their high-temperature stability and their resistance to hot environments that include liquids and gases, ceramic materials are used as refractories. The SiO_2-Al_2O_3 system forms the basis for the production of a class of materials used in the refractories industry. A starting material that is readily available is kaolinite, $Al_2O_3 \cdot 2SiO_2 \cdot 2H_2O$. If kaolinite is heated to 1500°C and held until it has been equilibrated, what phases will be present, what are their compositions, and how much of each phase is present?
56. What is the maximum use temperature for a refractory made from kaolinite? Explain.
57. How much Al_2O_3 must be added to kaolinite in order to significantly increase its maximum use temperature?

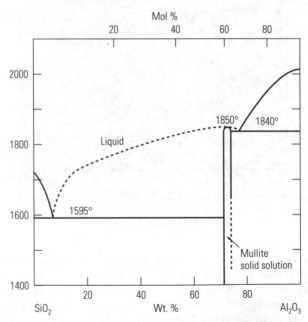

FIGURE HP7.11   The SiO_2-Al_2O_3 equilibrium phase diagram.   *(Source:* Phase Diagrams for Ceramists, *Vol. IV, compiled at the National Bureau of Standards, edited and published by The American Ceramic Society, Inc., 1981, Columbus, OH, 43214.)*

Problems 58–61 pertain to the Al-Ni equilibrium phase diagram of Figure 7.7–3 in the text.

58. As explained in Section 7.3.1, it is often convenient to express compositions in terms of atomic fractions rather than weight fractions. For the phases α(Ni-Al), β(Al-Ni), and α(Ni-Al), determine the compositions in atomic fractions and provide approximate chemical formulas for these phases. For α(Ni-Al) use the composition at the congruent melting temperature.

59. On the Ni-rich side of the Al-Ni diagram, liquidus boundaries of β(Al-Ni), α(Ni-Al), and solid solution Ni all appear to converge in a small temperature range, and thus this part of the diagram appears to be ambiguous. Assuming that the two invariant reactions are correct, provide an expanded view of the phase diagram in this vicinity. Illustrate how these two invariant reactions must be connected.

60. Label all two-phase fields on the Al-Ni diagram.

61. When an alloy containing 83 wt. % Ni is cooled under equilibrium conditions from 1395°C down to room temperature, show how the fraction of α(Ni-Al) changes as a function of temperature.

Problems 62–66 pertain to the Fe-Fe$_3$C diagram of Figure 7.8–2 in the text.

62. Three invariant reactions for alloys of Fe and carbon occur in the region covered by this diagram. Identify the three reactions.

63. What is the maximum solid solubility for carbon in the FCC phase?

64. For an alloy containing 0.77 carbon, what are the fractions of α Fe (ferrite) and Fe$_3$C (cementite) just below 727°C?

65. It is often useful to know the maximum amount of austenite that can transform to α Fe and Fe$_3$C as pearlite for a particular carbon content. Calculate the maximum amount of austenite that can transform as the eutectoid when an alloy containing 0.5 wt. % carbon is cooled from the single-phase austenite range to below 727°C.

66. Calculate the maximum amount of austenite that can transform as the eutectoid when an alloy containing 1.5 wt. % carbon is cooled from the single-phase austenite range to below 727°C.

CHAPTER 8

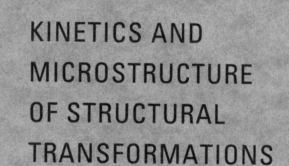

KINETICS AND
MICROSTRUCTURE
OF STRUCTURAL
TRANSFORMATIONS

MATERIALS IN ACTION   Phase Transformations

Perhaps the one phase transformation with which we are all familiar is the freezing of water, in which a liquid is transformed into a solid by extraction of heat. Other well-known examples of a liquid-to-solid phase transformation would include solidification of metals, which is frequently the first step in their production. In addition to the transformation of a liquid to a solid, many changes in internal structure occur entirely in the solid state as well. For example, when steel is heated to relatively high temperatures without melting, the carbon and iron atoms exist as a single phase with the iron atoms forming an FCC unit cell and the carbon atoms existing within the interstitial spaces in the lattice. When cooled to room temperatures, the structure completely changes to a mixture of iron carbide particles and BCC iron (with a small amount of carbon in the interstitial spaces). Many questions arise relative to phase transformations:

■ By what process do these transformations initiate?

■ What is the rate at which this process proceeds?

■ How do factors such as temperature, pressure, or the presence of impurities affect the transformation?

■ How stable is the resultant phase that forms?

■ How does the *mechanism* of transformation affect the microstructure and the properties?

These questions will be addressed in detail in this chapter. However, before proceeding let's consider a phase transformation of considerable economic importance in the jet engine industry and of great concern to all air travelers. This example relates to the production of turbine blades for jet engines. In the turbine section of an engine a fuel/air mixture is ignited at temperatures in excess of 1500 K! This causes rapid expansion of the gases, part of whose energy is converted by the turbine blades to rotation of the engine. The rotation of the compressor blades compresses the air in the front section of the engine and forces it to the back, where it is ignited and the process continues. The turbine blades operate under very exacting conditions: the temperatures are very high, the combustion environment is extremely corrosive, and the stresses are very high. Virtually the only class of materials that can withstand such conditions are called nickel-base superalloys, which in the simplest terms consist of a microstructure of an FCC matrix (essentially Ni plus other elements) and particles of $Ni_3(Al, Ti)$.

Like most metals, these Ni-base alloys were used in polycrystalline form. However, the grain boundaries are areas where excessive deformation and chemical attack (i.e., oxidation) occur and as such are preferential fracture sites. Fracture of turbine blades represents an unacceptable risk because of the damage that can be caused to the engine. Thus the operating conditions (e.g., stresses and temperatures) had to be held to relatively low levels when using polycrystalline turbine blades. It is especially desirable to eliminate grain boundaries that are normal to the applied stress, since these are the sites that will be preferentially attacked. The challenge in the jet engine industry was to remove these boundaries and thus allow for higher operating temperatures and stresses. This was successfully done by the directional extraction of heat. Thus, instead of allowing the blades to freely solidify with the formation of many randomly oriented grains, heat was *unidirectionally* extracted. Those crystals which formed and which were most favorably oriented for growth (as discussed in this chapter) dominated the solidification process, resulting in a structure of elongated grains with relatively few grain boundaries normal to the loading direction (which is parallel to the blade axis). This structure allowed for higher stresses and operating temperatures, and these improvements were clearly tied to an understanding of how to control the liquid/solid phase transformation of a multicomponent system. While this advance was very important, there were still grain boundaries, and these could be fracture sites.

The next step in the development of superior turbine blades was to eliminate all grain boundaries through the production of single-crystal blades. This was done by allowing only one grain to grow into the mold cavity. This step required great control of the temperature gradients, the solidification rates, and the composition of the alloy being used to produce this blade. It also required detailed knowledge of the crystal growth characteristics of the alloy. Most commercial engines produced contain single-crystal turbine blades, which have allowed significant increases in operating temperatures and stresses, with attendant economic benefits. While the total amount of materials used in jet engines worldwide is not large ($\sim$80,000 tons/yr), the value-added is extraordinary. For modern jet engines, the cost approximates its weight in silver, while for spacecraft, which share much of the engine technology, it approximates its weight in gold!

In this chapter we examine the way in which the structure of most materials (and hence their properties) is affected by the phase transformations that occur during production and processing.

8.1 INTRODUCTION

In Chapter 7 we presented the fundamental concepts of phase equilibria, the application of those concepts to the construction and interpretation of phase diagrams, and the application of phase diagrams to problems in materials engineering. An underlying assumption in Chapter 7 was that the system under investigation was in a state of thermodynamic equilibrium. However, materials are generally neither produced nor used in their equilibrium states. Oxide glasses, steels, and polymers are important commercial materials that represent examples of nonequilibrium systems. Composite materials, such as SiC fibers embedded in an Al matrix, also exist under nonequilibrium conditions. Their usefulness, to some degree, is dictated by the amount of deleterious interfacial phases that are present and the rate at which they develop.

Given sufficient time, all systems tend to their lowest free-energy state, as represented by the phase diagrams in the previous chapter. However, in practical engineering applications it is important to know the time required for a given transformation or reaction to occur. We have already seen an example of this in the sections on diffusion, where the time to produce a given diffusion layer was calculated. This was an example of a kinetic calculation. As applied to materials science, **kinetics** refers to the rate at which a process occurs. In any study of kinetics, the primary variables are time and temperature. If the time-temperature-microstructure relationships are known, the properties of a product can be tailored by controlling composition and heat treating cycles.

This chapter describes the kinetics of phase transformations. It begins with an intro-duction of scientific principles governing the kinetics of structural changes and then applies these principles to engineering systems. There were two criteria used to choose the specific systems described in this chapter: first, they have engineering significance, and second, they demonstrate and reinforce important scientific principles. Examples are provided from each class of materials. The emphasis, however, is on metals and alloys because heat treating is a key step in the production of an alloy microstructure, whereas heat treating per se has typically played a less important role in processing ceramics and polymers.

8.2 FUNDAMENTAL ASPECTS OF STRUCTURAL TRANSFORMATIONS

8.2.1 The Nature of a Phase Transformation

A phase transformation can range from simple to highly complex. The complexity de-pends upon the number and types of changes that occur during the reaction. For example, when water solidifies to form ice at 0°C, or when liquid copper solidifies at 1085°C to form solid copper, there is a change from the noncrystalline liquid state to the crystalline solid state. However, there is no change in composition. Since the liquid state adjusts to volume change occurring during solidification, strain within the original phase does not develop or affect the progress of the phase change in an open system.

Alternatively, as pointed out in the introduction to Chapter 7, when body-centered tetragonal tin transforms to the diamond cubic phase, there is a 27% increase in the specific volume. Unlike liquid-to-solid transformations, there is resistance to the phase change when a solid transforms to another solid having a different volume. Thus, transfor-mation stresses and their effects may be important.

Consider the aluminum-copper binary phase diagram shown in Figure 8.2−1. When an aluminum alloy containing 4.5% copper is rapidly cooled from 820 K (a single-phase

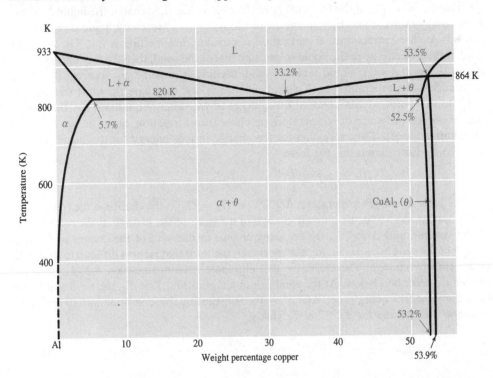

FIGURE 8.2−1

The aluminum-rich end of the Al-Cu binary phase dia-gram.

field) to 675 K (a two-phase field), the solid solution must transform to the two-phase mixture predicted by the phase diagram. Since the compositions of the resulting two phases differ from the composition of the higher-temperature single phase, the establishment of the equilibrium microstructure requires diffusion of both Al and Cu atoms through the FCC lattice. As discussed in Chapter 4, diffusion takes time. Other factors that complicate this phase transformation include the development of an interface (phase boundary) between the θ and α phases, and the volume changes associated with the transformation.

Most of the commercially important metal alloy systems that are heat-treated require numerous microstructural changes to take place before the material can be used in an engineering application. We will see in the next sections what effect changing the temperature has on various transformations. Our intuition tells us that increasing the temperature increases the rate of diffusion and speeds up atomic rearrangement. As demonstrated in the previous aluminum-copper phase transformation, however, other factors are important, and each of these factors has its own temperature dependence. Consequently, most microstructural changes have an optimum temperature at which the transformation proceeds at the fastest rate. An understanding of the process occurring during a transformation will assist us in choosing heat treatment conditions to give the best combination of properties in a reasonable time frame.

8.2.2 The Driving Force for a Phase Change

Before a qualitative description of a phase transformation can be presented, relevant questions must be addressed. For example, what is the reason for the phase change? Or, stated another way, what is the driving force causing the transformation? Also, is it possible to increase the driving force by changing temperature? The answers to these questions can be found by investigating the solidification of a pure substance.

Whether a phase change is possible depends solely on the change in free energy for the reaction. For example, if the free energy of the solid is less than that of the liquid at a particular temperature, then the free energy of the system is lowered by liquid transforming to solid. At that temperature solidification is possible, but melting is not. On the other hand, if the free energy of the liquid is less than that of the solid, then the free energy of the system is lowered when liquid replaces solid. Melting is possible but freezing is not. At the equilibrium temperature, both phases coexist because they have the same free energy.

Up to this point we have treated the concept of free energy qualitatively. In order to understand the influence of temperature on phase transformation, however, we require a quantitative model. The expression for the change in free energy for the liquid-to-solid (L $\rightarrow$ S) transformation has the form:

$$\Delta G^{L \rightarrow S} = \Delta H^{L \rightarrow S} - T\,\Delta S^{L \rightarrow S} \tag{8.2-1}$$

where T is the absolute temperature, $\Delta G^{L \rightarrow S}$ ($= G_S - G_L$) is the change in the free energy of the system, $\Delta S^{L \rightarrow S}$ is the entropy change (a measure of the change in the randomness of the system), and $\Delta H^{L \rightarrow S}$ is the enthalpy change (a measure of the change in internal energy of the system). For the present argument, the term that is most difficult to deal with is the entropy change. Therefore, we will eliminate it from Equation 8.2-1, using the procedure described below. At the equilibrium temperature, $T = T_E$, the solid and liquid have the same free energy (i.e., $\Delta G^{L \rightarrow S} = 0$). Substituting this condition into the previous equation and solving for $\Delta S^{L \rightarrow S}$ at T_E yields:

$$\Delta S^{L \rightarrow S} = \frac{\Delta H^{L \rightarrow S}}{T_E} \tag{8.2-2}$$

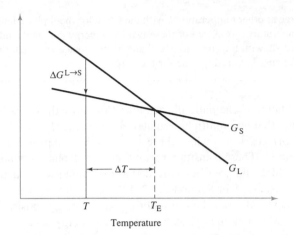

FIGURE 8.2–2

An approximation of the variation in free energy for the liquid and solid phases close to the equilibrium melting temperature.

Assuming the heat capacities of the two phases are about the same, this expression for the entropy change can be substituted into Equation 8.2–1 to yield an alternative expression for $\Delta G^{L \to S}$ (at any temperature of the form):

$$\Delta G^{L \to S} = \Delta H^{L \to S} - T\left(\frac{\Delta H^{L \to S}}{T_E}\right) \tag{8.2–3}$$

Rearranging Equation 8.2–3 and substituting $\Delta T = T_E - T$ yields:

$$\Delta G^{L \to S} = \left(\frac{\Delta H^{L \to S}}{T_E}\right) \Delta T \tag{8.2–4}$$

The next factors to consider is the sign of $\Delta H^{L \to S}$. When a phase transformation occurs, heat is either evolved (an exothermic reaction) or absorbed (an endothermic reaction). The solidification of a pure substance is an exothermic transformation, so that $\Delta H^{L \to S} = H^S - H^L < 0$. Since ΔT is positive for all temperatures below the equilibrium temperature, the fact that $\Delta H^{L \to S}$ is less than zero guarantees that $\Delta G^{L \to S}$ will also be less than zero for all temperatures below T_E. Thus, Equation 8.2–4 shows that the liquid-to-solid phase transformation is thermodynamically favored for all $T < T_E$. This equation also shows that the thermodynamic driving force for the phase transformation (i.e., the magnitude of ΔG) increases as ΔT increases. Thus, a *decrease in temperature results in an increase in the driving force for the solidification of a pure substance.*

Figure 8.2–2 illustrates the change in free energies with temperature for the solid and liquid phases and provides a graphical interpretation of the relationship between temperature and the driving force for the phase change. At the melting temperature, the free energies of liquid and solid phases are equal. The divergence of the two free-energy curves above and below the equilibrium freezing temperature represents the magnitude of the driving force for the specific reaction, $L \to S$ (freezing) or $S \to L$ (melting).

EXAMPLE 8.2–1

Consider the solid-state phase transformation γ-Fe $\to \alpha$-Fe that occurs at 912°C. Sketch the variation of Gibbs free energy over a narrow temperature range for the two phases.

Solution

At 912°C, both phases are in equilibrium, so that $G^\alpha(912°C) = G^\gamma(912°C)$. Since γ is the stable phase at higher temperatures, $G^\gamma(T) < G^\alpha(T)$ for $T > 912°C$. Similarly, for $T < 912°C$, the α phase is stable, so that $G^\alpha(T) < G^\gamma(T)$. The free-energy curves of the two phases must intersect at 912°C and diverge at other temperatures, with the curve for the α phase being below that for the γ phase at low temperatures. A sketch of these two free-energy curves would look much like that shown in Figure 8.2–2, with the curve for the liquid being replaced by that for the γ phase and the curve for the solid being replaced by one for the α phase.

....

Figure 8.2–3 shows a schematic illustration of the growth of a solid particle within a liquid matrix. Note that any multiphase material, including this solid-liquid system, must contain internal surfaces across which there are transitions in properties and structure between adjacent phases. These transition regions, known as phase boundaries, are typically only a few atoms thick and are often regarded as two-dimensional defects, much like the grain boundaries discussed in Section 5.3.2. Like grain boundaries, the atoms in phase boundaries are displaced from their equilibrium (low-energy) positions. As such, phase boundaries have additional energy, called interfacial energy, associated with them.

When a new phase forms from an existing phase, extra energy must be supplied to create the phase boundaries. Where does this energy come from? Recall from Figure 8.2–2 that as the degree of undercooling increases, there is more and more "extra" energy available to drive the phase transformation. Thus, by reducing the temperature— that is, increasing the undercooling—energy becomes available to create the phase boundaries. All other things being equal, the amount of undercooling necessary to create these interfaces is directly proportional to the interfacial energy of the boundaries. This concept will be particularly useful when we discuss solid-to-solid phase transformations later in the chapter.

Thus far, no mention has been made of the possible sites where a phase transformation might initiate. A new phase forms, or **nucleates,** at one of two types of locations. One possibility is nucleation at random locations in the parent phase. A second possibility is that the phase transformation nucleates at preferred sites. Preferred sites may be structural discontinuities in the parent phase (e.g., grain boundaries or dislocations) or foreign objects, such as the mold wall or an impurity particle trapped in the liquid. The distinction between these two types of nucleation events is important, and specific terms have been developed to identify the different mechanisms. When the new phase nucleates randomly in the parent phase, the process is called **homogeneous nucleation.** However, when the new phase nucleates preferentially at specific sites in the parent phase, the process is called **heterogeneous nucleation.**

8.2.3 Homogeneous Nucleation of a Phase

A simplified but helpful view of a liquid is to visualize the atoms comprising the liquid as a random distribution of hard spheres. The interatomic distances in the liquid are similar to those of the crystalline phase, but each atom on average has fewer nearest neighbors

FIGURE 8.2–3

Schematic showing **(a)** a region containing only liquid, and **(b)** a region containing liquid with a solid and a solid-liquid (S/L) interface.

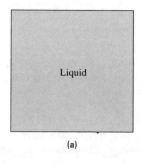

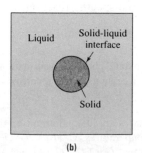

(a)

(b)

in the liquid than in the crystal. The structure is therefore more open and allows greater atomic mobility than does that of the solid state. Distributed throughout the liquid are small, closely packed atomic clusters having a packing arrangement similar to that of the solid. Because of the open structure, these clusters form and disperse quickly. The relationship between the size and the stability of the clusters depends on temperature.

For the purposes of conducting a simplified calculation relating size to stability, we make two assumptions. First, the energy at the interface between a cluster and the liquid can be regarded as isotropic, that is, independent of the specific crystal plane that forms the interface. Such a restriction implies that the transforming phase is spherical. Second, the interfacial energy per unit surface area is independent of the size of the solid phase. Although most transforming microstructures will not develop spherical products, adopting this simple geometry will lead to a model that can be applied, with minor modifications, to real systems.

Two specific components are associated with the free-energy change of the liquid-solid transformation. The first is the change in energy associated with the creation of the liquid-solid interface, and the second is the difference in bulk free energies of the liquid and solid phases. The total interfacial energy associated with a spherical particle is the product of the area of the interface ($4\pi r^2$) and the interfacial energy per unit are (γ_{SL}). Figure 8.2–4a shows the increase in the free energy of the system due to the presence of the interface as a function of the radius of the growing phase.

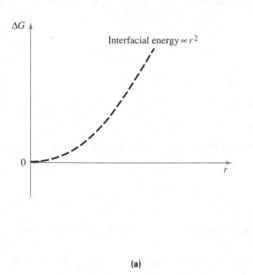

(a)

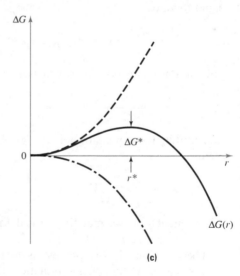

(c)

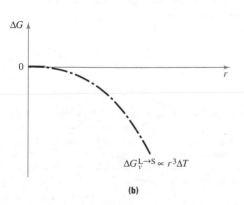

(b)

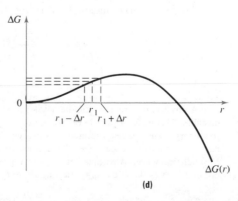

(d)

FIGURE 8.2–4

The dependence of the various energy terms associated with nucleation as a function of the radius of the growing phase: **(a)** the relationship between cluster radius and surface energy of growing spherical solid phase in a liquid, **(b)** the relationship between the cluster radius and $\Delta G_v^{L\rightarrow S}$, **(c)** the sum of the previous two curves, and **(d)** an annotated version of ΔG plotted versus r (see footnote 1 on page 294).

The second energy term is the bulk free energy. It is the difference in free energy of the solid and liquid phases multiplied by the volume of the particle. For $T < T_E$, the solid has a lower bulk free energy than the liquid and the L → S free-energy change is negative. As shown in Figure 8.2–4b, the product of the change in free energy per unit volume ΔG_v and the volume $(4/3)\pi r^3$ becomes more and more negative as the radius of the particle increases.

The total change in the free energy of the system is the sum of the interfacial energy term and the bulk energy term. The result of this summation is shown in Figure 8.2–4c. Note that the two energy terms act in opposite directions. Initially the r^2 area term dominates, and as the particle becomes larger, there is an increase in the net free energy. Once a **critical radius** r^* is reached, however, the r^3 volume term begins to dominate, and further increases in the particle size result in a reduction in the free energy of the system. Thus, r^* represents the size of a stable nucleus—smaller particles redissolve into the liquid phase while larger particles will continue to grow.[1]

The graphical method used thus far can be supported with an equivalent mathematical treatment. The change in free energy as a function of r is given by the expression:

$$\Delta G(r) = (4\pi r^2)\gamma_{SL} + \frac{4}{3}\pi r^3(\Delta G_v)$$

(8.2–5)

The value of r^* can be determined by taking the derivative of this equation with respect to r and setting the resultant expression for the slope of the energy curve in Figure 8.2–4c equal to zero:

$$\frac{d[\Delta G(r)]}{dr} = 0 = 8\pi r\gamma_{SL} + 4\pi r^2(\Delta G_v)$$

(8.2–6)

Solving this expression for r, which is r^*, gives:

$$r^* = \frac{-2\gamma_{SL}}{\Delta G_v}$$

(8.2–7)

Substituting the expression for ΔG_v given in Equation 8.2–4 into the previous equation yields:

$$r^* = \frac{(-2\gamma_{SL})T_E}{\Delta H_v \, \Delta T}$$

(8.2–8)

This equation shows that the critical radius decreases as the degree of undercooling increases, that is, $r^* \propto (1/\Delta T)$.

The final step in the process is to substitute the expression for r^* back into the expression for $\Delta G(r)$ given in Equation 8.2–5 to obtain the change in free energy necessary to form stable nuclei. Performing this substitution yields:

$$\Delta G^* = \frac{16\pi(\gamma_{SL})^3 T_E^2}{3(\Delta H_v)^2} \times \frac{1}{(\Delta T)^2}$$

(8.2–9a)

---

[1] A physical picture of r^* can be developed with the aid of Figure 8.2–4d. All particles with $r < r^*$ are unstable. That is, for a particle of size r_1 to grow an amount Δr, $r_1 \rightarrow (r_1 + \Delta r)$ represents an increase in free energy, which is not a spontaneous process. The spontaneous process would be the dissolution of a particle of radius r_1, since $r_1 \rightarrow (r_1 - \Delta r)$ lowers the free energy. All particles with $r < r^*$ are referred to as subcritical nuclei. A particle with $r = r^*$ represents a critical particle size, since either growth or dissolution decreases the free energy. Beyond r^*, only growth reduces the free energy.

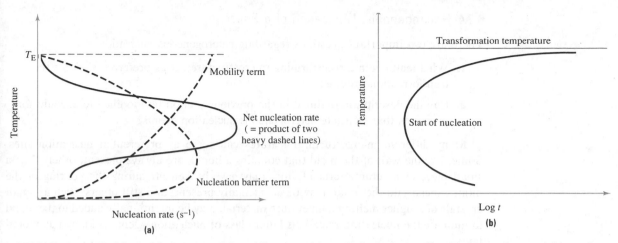

FIGURE 8.2–5 **(a)** The influence of temperature on the mobility term and the nucleation barrier term. The opposing processes result in a maximum in the nucleation rate at an intermediate temperature. **(b)** Since the time for nucleation is inversely related to the nucleation rate, the time curve exhibits a minimum at an intermediate temperature. Because of its shape, this curve is often referred to as a C curve.

or

$$\Delta G^* \propto \frac{1}{(\Delta T)^2} \qquad (8.2\text{–}9b)$$

It may be useful to offer a physical interpretation of the quantity ΔG^*. Returning to Figure 8.2–4c, we see that ΔG^* corresponds to the energy maximum in the plot of ΔG versus r. Since this is the amount of energy that must be supplied to the system in order to nucleate the new phase, ΔG^* can be thought of as the size of the energy "barrier" to the nucleation process. The crucial point is that *as the undercooling increases, the energy barrier to nucleation decreases.*

With larger undercoolings, both r^* and ΔG^* decrease, suggesting that simply lowering the temperature of the system allows nucleation to occur ever more readily. However, there are practical kinetic limits to this effect. For example, with decreased temperature there is a corresponding reduction in atomic mobility. The random fluctuation in the local arrangements of atoms is the process that provides the clusters. Since the formation of the clusters depends on atomic mobility, a reduction in the temperature reduces the rate of clustering. Thus, as shown in Figure 8.2–5a, the overall nucleation rate exhibits a maximum at an intermediate temperature. The maximum in the nucleation rate leads to a minimum in the time required to nucleate a phase, as shown in Figure 8.2–5b. Because of its shape, this curve is known as a **C curve.**

The picture of homogeneous nucleation that we just developed to describe solidification is relatively easy to present on paper, but it is extremely difficult to achieve in the laboratory. The primary reason is that the competitive process, heterogeneous nucleation, is more efficient and it can occur at temperatures very close to the equilibrium temperature. Thus, the extent of undercooling required for homogeneous nucleation is rarely reached in practice. As stated previously, heterogeneous nucleation occurs when a specific site in the system stimulates the transformation. In the next section we will present a qualitative discussion of heterogeneous nucleation so that we may compare the two processes.[2]

---

[2] When water boils in a pan on a stove, do the bubbles form within the liquid or on the walls of the pan? Why?

8.2.4 Heterogeneous Nucleation of a Phase

There are two important questions regarding heterogeneous nucleation:

1. What features in a transforming microstructure act as preferred sites for the transformation process?

2. Can the description outlined in the previous section be modified to account for processes that facilitate heterogeneous nucleation events?

Many different microstructural features can serve as preferential nucleation sites. Some, like the wall of the mold that contains a liquid, are always present; others, like a small piece of a ceramic crucible that may have broken off during the pouring of the molten metal into the mold, may be accidentally present; and still others, such as small crystals of a higher melting temperature material, may be intentionally added to the liquid to increase the nucleation rate. This latter class of nucleation agents is known as inoculants.

There are numerous practical examples of the use of inoculants to control phase transformations. One example is the use of dry ice crystals to seed clouds and promote the formation of raindrops. In industry, TiO_2 is often used to nucleate crystals in silicate glasses, resulting in the formation of crystalline solids called glass-ceramics. Nucleating agents are also used in polymer systems to facilitate crystallization and in cast iron to nucleate graphite nodules.

How can one predict whether a specific material will act as a heterogenous nucleation site for a specific solidification event? A major factor is a characteristic referred to as *wetting*. You are probably already familiar with the concept of wetting in the context of either car wax or wood-preserving treatments. The advertising compaigns for both types of products focus on the idea that when the underlying material is properly protected, water "beads up" on the surface. If the water spreads out as a thin film on the surface, then it is said to "wet" that surface. If, however, the water forms beads on the surface of the material, it does not wet that solid. In the present context, if a droplet of the molten metal does not wet a candidate nucleating agent, then that agent will not be an effective inoculant. In contrast, a material that is wet by the molten metal is likely to be an effective nucleating agent.

Consider the case of heterogeneous nucleation of a solid on the wall of a mold containing a liquid. Figure 8.2–6a is a schematic illustration of this geometry. The contact angle θ is a function of the energy of the liquid/mold interface γ_{LM}, the liquid/solid interface γ_{LS}, and the mold/solid interface γ_{MS}. These interfacial energies are equivalent to surface tensions acting along the direction of contact. The mathematical relationship that expresses a force balance in the horizontal direction is:

$$\gamma_{LM} = \gamma_{MS} + \gamma_{LS} \cos \theta \qquad (8.2\text{–}10a)$$

or

$$\cos \theta = \frac{\gamma_{LM} - \gamma_{MS}}{\gamma_{LS}} \qquad (8.2\text{–}10b)$$

For any combination of mold (or other solid nucleating agent), solid, and liquid, the wetting behavior will lie at or between the extremes exhibited in Figure 8.2–6b. When $\theta = 0°$, the nucleated solid completely wets the surface. When $\theta = 180°$, there is no wetting, the underlying surface has no effect on the formation of the solid cap, and the process is indistinguishable from homogeneous nucleation.

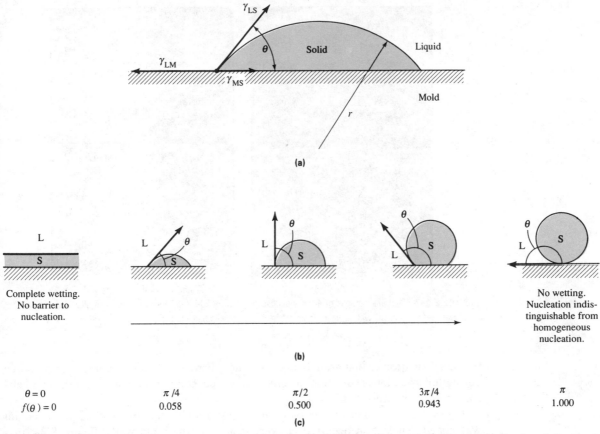

FIGURE 8.2–6 The wetting characteristics of a number of solid-liquid systems: **(a)** The angle θ, as defined in the illustration, is an indication of the wettability of the system; **(b)** as θ increase from $\theta = 0$ to $\theta = \pi$, the wettability decreases; **(c)** the function $f(\theta)$ varies from 0 to 1 as θ varies from 0 to π.

Next we turn our attention to extending the model for homogeneous nucleation to include the heterogeneous case. The homogeneous nucleation barrier ΔG^*_{hom} can be modified by taking into account the effectiveness of a particular feature for heterogeneously nucleating a new solid phase. The barrier to heterogeneous nucleation has the form:

$$\Delta G^*_{\text{het}} \propto (\Delta G^*_{\text{hom}}) f(\theta) \qquad (8.2\text{--}11)$$

where $f(\theta)$ is a function varying between 0 and 1. The value of $f(\theta)$ is related to the contact angle θ, as shown in Figure 8.2–6c.

Equation 8.2–11 shows that since $f(\theta)$ has a value between 0 and 1, the energy barrier for heterogeneous nucleation is *always* lower than that for homogeneous nucleation. Why is this so? As the solid grows, the solid/liquid interface area increases, the solid/mold area increases, and the liquid/mold area decreases. The energy released by the elimination of liquid/mold interface can provide energy to form the liquid/solid and solid/mold interface. Similarly, the energy released by the elimination of any defect acting as a nucleation site can be used to reduce the energy needed for nucleation in the absence of that defect. For example, in a solid-to-solid phase transformation, the nucleation and growth of a second-phase particle along a previously existing grain boundary is assisted by the energy released

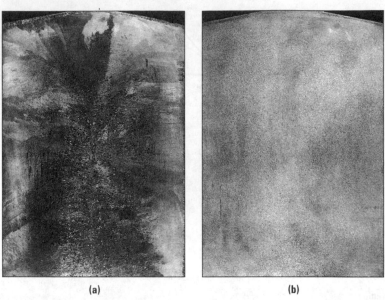

(a)   **(b)**

FIGURE 8.2–7 Photomicrographs of a transverse section from a 20-in × 54-in cast aluminum alloy: **(a)** with no ingot grain refiner, and **(b)** with an ingot grain refiner. Illustration is half the cross section.

by the elimination of that grain boundary. Since there is no corresponding energy source during homogeneous nucleation, heterogeneous nucleation is almost always the energetically favored process.

Figure 8.2–7a is a cross section of an as-cast aluminum alloy to which no inoculant was added. Nucleation occurred heterogeneously at the mold walls. Figure 8.2–7b is the same alloy cast under similar solidification conditions but containing an inoculant. A fine dispersion of TiB_2 particles distributed throughout the liquid was responsible for the uniform heterogeneous nucleation of the solid phase. The dramatic difference in the size and uniformity of the grain structure when a nucleating agent is present is evident.

In the preceding sections we presented the concepts of homogeneous and heterogeneous nucleation as applied to solidification. The same concepts can be applied directly to solid-state transformations that require solid-state diffusion as the mechanism for atomic rearrangement. One advantage of working with solid-state transformations is that diffusion rates are much slower than in liquids and it is often possible to rapidly quench below the transformation temperature and follow the changes in the microstructure as a function of time (at a constant temperature). The results of numerous experimental investigations of solid-state transformation involving structural changes, with or without associated compositional changes, have confirmed the types of behavior predicted in the previous sections. The dependence of the overall nucleation rate on temperature is a product of two factors—the energy barrier to nucleation, and the diffusion rate. The result is a maximum rate of nucleation and a corresponding minimum time for nucleation occurring at an intermediate temperature.

8.2.5 Matrix/Precipitate Interfaces

The nature of the solid/liquid interface and the contribution of the interfacial energy to the nucleation process were discussed previously. Interfacial energy considerations also play an important role in solid/solid transformations. However, some additional factors must be considered. In particular, we must present a description of the nature of interfaces that can develop when a second solid nucleates from a parent phase. There are three general types of interfaces between two solids: a coherent interface, a semicoherent interface, and an incoherent interface. The definitions of these interfaces are related to the continuity of atom planes across the interface.

A **coherent interface** arises when there is a one-to-one correspondence of atomic planes across the boundary separating two phases, as shown in Figure 8.2–8. Interfacial energy arises because unlike neighbors are present across the interface. This type of interface is most easily formed when the lattice parameters of the two phases are nearly identical (Figure 8.2–8a), but coherent interfaces can also form if the two phases have slightly different lattice parameters (Figure 8.2–8b). In the latter case, however, either one or both phases must distort in order to maintain continuity of the planes across the interface. The distortion gives rise to coherency strains. Fully coherent Al_3Li (δ') precipitates form in Al/Li alloys used as structural materials in the aerospace industry. Figure 8.2–8c is a high-resolution image of a fully coherent particle.

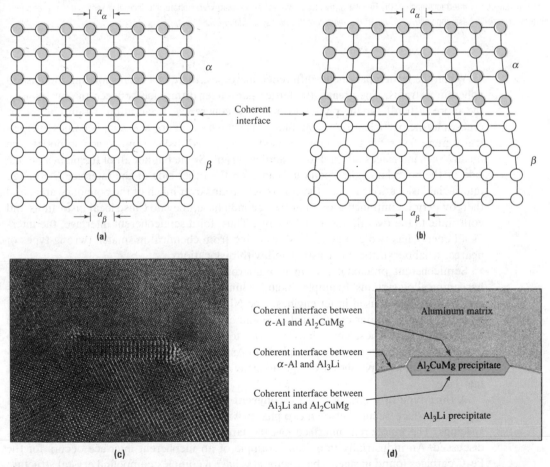

FIGURE 8.2–8   Schematic illustration of coherent interface between two phase α and β. **(a)** The crystal structures and lattice parameters are identical. **(b)** The crystal structures are identical but the lattice parameters are different. The difference in lattice parameters leads to coherency strains. **(c)** High-resolution image showing the coherent interface that exists between the aluminum matrix and two precipitates that form during artificial aging of an Al-Cu-Mg-Li alloy. The white dots are columns of atoms that form the phase. **(d)** A schematic of the photomicrograph in part c identifying the phases and the coherent interfaces.
(Source: Photograph provided by V. Radmilovic, National Center for Electron Microscopy, LBL, Berkeley, CA.)

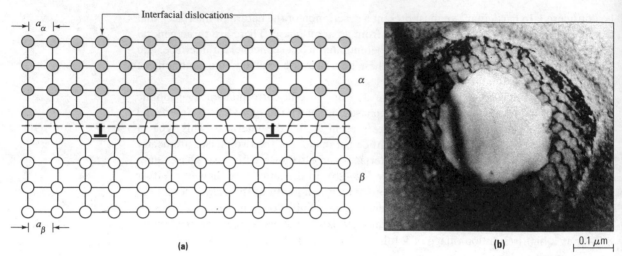

FIGURE 8.2-9 **(a)** Schematic of a semicoherent interface between two phases. The insertion of the periodically spaced dislocations eliminates coherency strains but increases the interfacial energy. **(b)** Regular array of dislocations at the matrix/precipitate interface of a semicoherent precipitate in Astroloy, a high-temperature Ni-base superalloy.   *(Source: Photomicrograph courtesy of Mario Luis Maciá).*

When the lattice parameter difference increases, there is a corresponding increase in coherency strains. The greater the lattice parameter difference between the two phases, the more difficult for the phases to maintain coherency. In addition, strain energy increases as the particle grows. Eventually, as shown in Figure 8.2–9, it becomes energetically more favorable to introduce periodically spaced dislocations to accommodate the misfit. Such interfaces are said to be **semicoherent.** Some but not all of the planes on one side of the boundary correspond to planes on the other side of the interface. Although interfacial dislocations reduce the coherency strains over much of the boundary area, and thereby permit continued growth of the second phase, they do cause local distortions and contribute to the overall interfacial energy. Thus, for a semicoherent interface, the interfacial energy has two components, one arising from chemical mismatch (wrong types of nearest neighbors), the other associated with dislocations.

Semicoherent precipitates form in numerous commerically important precipitation hardening alloy systems. Examples include aluminum alloys used for the skin of airplanes and Ni-base alloys used in jet engines. The Ni_3Al γ' precipitates that form in the FCC γ matrix have a similar crystal structure and a slightly different lattice parameter from the parent phase. These conditions give rise to a semicoherent interface. Figure 8.2–9b is an electron micrograph of a precipitate in Astroloy, a high-temperature Ni-base superalloy. The photograph shows a regular array of dislocations at the precipitate/matrix interface.

A third type of interface is the **incoherent interface,** shown in Figure 8.2–10. This type of interface occurs between two phases with different crystal structures and atomic spacings. The incoherent interface has the largest interfacial energy of the three types discussed. An industrially important example of an incoherent interface occurs for the Fe_3C carbides found in steel. The precipitates have a complex compound crystal structure and are embedded in a BCC matrix.

The three types of interfaces generally have interfacial energies on the order of 200 mJ/m² for coherent, 500 mJ/m² for semicoherent, and 1000 mJ/m² for incoherent

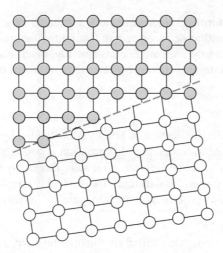

FIGURE 8.2–10   An incoherent interface occurs when the crystal structures and lattice parameters of two phases are different.   *(Source: D. A. Porter and K. E. Easterling,* Phase Transformations in Metals and Alloys, *Champman & Hall, 1981, p. 147. Reprinted by permission of Chapman & Hall.)*

interfaces. Equations 8.2–9 and 8.2–11, which describe the height of the homogeneous and heterogeneous nucleation barriers, show that ΔG^* is proportional to $\gamma_{\alpha\beta}^3$. Consequently, the barrier to the nucleation process is strongly dependent on the interfacial energy of the phase nucleating. Generally speaking, increasing $\gamma_{\alpha\beta}$ decreases the probability of homogeneous nucleation.

The relationships between interfacial energy and nucleation process are summarized in Table 8.2–1. The nature of the nucleation process dramatically influences the size and distribution of the precipitates. For homogeneous nucleation, sites are randomly distributed throughout the microstructure. Heterogeneous nucleation, on the other hand, occurs at specific locations, so the resulting distribution of nucleated phases is not random. Furthermore, diffusion rates are higher along grain boundaries and dislocations than in

TABLE 8.2.–1   Relationship among interfacial energy, the nature of the nucleation process, and the precipitate.

Type of interface	Interfacial energy	Nucleation process	Location of precipitates	Number of precipitates per unit volume
Coherent	Increasing	Homogeneous	Throughout the matrix	$\sim 10^{18}/cm^3$
Semicoherent		Heterogeneous	Mostly at dislocations	Orders of magnitude less than homogeneous nucleation
Noncoherent		Heterogeneous	At grain boundaries	Orders of magnitude less than homogeneous nucleation

the marix. Thus, once formed, heterogeneously nucleated phases tend to grow faster (higher diffusion rates) and to be more coarse (limited number of nucleation sites) than those that are homogeneously nucleated. Consequently, much can be inferred about the nucleation process by examining the precipitate size and distribution.

8.2.6 Growth of a Phase

Although uniformity in the distribution of phases is controlled by the nucleation process, the final structure and time to complete transformation is strongly affected by the growth rate of the nucleated phase. As in the nucleation step, there may also be competing processes that lead to a maximum growth rate at a particular temperature.

All phase changes that require atomic transport by diffusion share the following characteristics:

1. There is a time period, often called the incubation time, required to nucleate the phase; the incubation time is a function of undercooling. No measurable phase transformation occurs during the incubation time.

2. Once nucleated, the phase begins to grow and there is a rapid increase in the amount of new phase present.

3. Eventually, the growth rate of the new phase decreases because of either depletion of solute or physical impingement of the growing phase.

These characteristics can be described conveniently by a simple mathematical function plotted in Figure 8.2–11. The horizontal axis represents time and the vertical axis corresponds to the fraction of the parent phase that has been transformed. Notice that the range of the fraction transformed is from 0 to 1.0 on a linear scale. The behavior illustrated in Figure 8.2–11 is referred to as a sigmoidal relationship. Mathematically, this curve can be described using the function:

$$X = 1 - \exp[-(kt)^n] \qquad (8.2\text{–}12)$$

where X is the fraction of transformed material, k is similar to a rate constant in a chemical reaction and has the units of $(\text{time})^{-1}$, t is the time during which the material is transform-

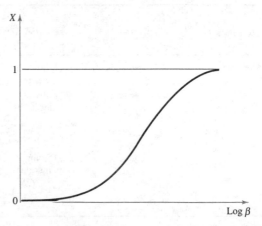

FIGURE 8.2–11   Sigmoidal behavior of the relationship $X = 1 - \exp(-\beta)$, where $\beta = (kt)^n$.

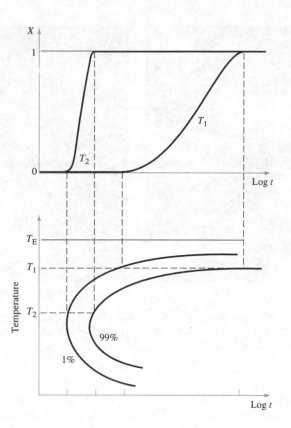

FIGURE 8.2–12

Construction of an isothermal transformation (IT) diagram.

ing, and n is a constant typically in the range of 0.5 to 5. Transformations in most materials and systems can be modeled using this equation.[3]

Note that X refers to the total amount transformed and thus includes nucleation effects as well as growth. In particular, the time at which $X = 0+$ (i.e., the instant at which a critical nucleus forms) is somewhat difficult to determine. While the concept of a critical nucleus is well defined, experimentally identifying a critical nucleus is much less straightforward. In subsequent chapters correlations between microstructure and properties will be developed. By following changes in a particular property, the kinetics of a phase transformation can be determined. For example, the electrical resistivity of an alloy may be sensitive to the amount of solute in solid solution. By quenching an alloy to a specific temperature and monitoring the isothermal change in electrical resistivity, we may be able to estimate the start of the reaction. The more sensitive the property being measured is to structure and composition, the easier it is to estimate the start of the nucleation process.

An isothermal transformation (IT) diagram can be constructed, as illustrated in Figure 8.2–12, by following the isothermal transformation kinetics at different temperatures. Since it is difficult to decide exactly when a reaction starts and is completed, the "start" and "finish" lines on an IT diagram are arbitrarily defined. In Figure 8.2–12 the start line

---

[3] This very powerful kinetic treatment was first presented by Johnson and Mehl in 1939 to describe the isothermal (constant-temperature) transformation kinetics of the decomposition of austenite (γ-Fe) into pearlite (α + Fe_3C). This model was later modified by Anderson and Mehl to describe the recrystallization of cold-worked aluminum. The equation was also first used by Avrami to describe polymer crystallization.

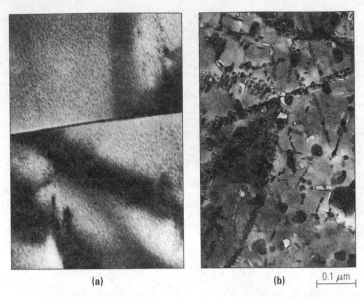

(a)   **(b)** $\overset{0.1\ \mu m}{\longmapsto}$

FIGURE 8.2–13 The distribution of precipitates in an aluminum alloy: **(a)** precipitates that homogeneously nucleated, and **(b)** those that heterogeneously nucleated. In part a, the precipitates are the fine, pepperlike features that are
scattered throughout.

was chosen to be when 1% of the parent phase transformed. The finish line was chosen
to be when 99% of the parent phase transformed. Intermediate amounts of transformed
product would lie between these lines, although not in direct proportion.

Figure 8.2–13 shows two electron micrographs of an aluminum alloy given different
heat treatments designed to produce different precipitate morphologies. These photomicrographs show the result of both the nucleation and growth processes. Figure 8.2–13a
shows precipitates along grain boundaries that are coarser and more widely spaced than
the finer precipitates away from the grain boundaries in the matrix. Homogeneous nucleation and growth were responsible for intergranular precipitates. The coarse distribution
along the grain boundaries, however, was due to heterogeneous precipitation at grain
boundaries followed by growth. An important difference between the two types of
precipitates is illustrated in this figure. Because heterogeneous nucleation occurs at a
higher temperature (due to the lower energy barrier) and diffusivity along grain
boundaries is higher than in the matrix, precipitates at the grain boundaries are always
coarser than those in the grain interiors. Figure 8.2–13b shows the same alloy heat-treated
to produce precipitates heterogeneously nucleated on dislocations within the grains.

This example illustrates two key points: (1) altering the heat treatment has a significant
effect on the size and spatial distribution of the precipitates through its influence on the
nucleation and growth rate, and (2) since macroscopic material properties depend
strongly on the size and spatial distribution of the phases, heat treatment is a powerful tool
for modifying the properties of engineering materials.

8.3 APPLICATIONS TO ENGINEERING MATERIALS

The basic principles regarding structural transformations introduced in the last few sections can be used to describe specific industrially important transformations. It is unreal-
istic to attempt to describe the vast number of materials and heat treatments currently used

by industry. By carefully choosing a limited number of examples where a wealth of data relating the effect of heat treatment on microstructure exists, however, we can develop an appreciation for and confidence in the application of the basic principles.

8.3.1 Phase Transformations in Steels

Steel is one of the most important structural materials in use today. It is also a widely studied alloy system because of the variety of microstructures that develop when Fe-C alloys are heat-treated. The Fe-C system displays most of the major types of phase transformations. Table 8.3–1 summarizes the terminology associated with different microstructures that commonly occur. Our examples will concentrate on a plain carbon steel, that is, steel in which carbon is the only alloying element.

Figure 8.3–1 is a section of the Fe-Fe$_3$C phase diagram in the vicinity of the eutectoid reaction. As described in the previous chapter, a eutectoid reaction is an invariant reaction in which a single-phase solid is replaced upon cooling by two different solid phases having

TABLE 8.3–1 Terminology for commonly occurring microstructures in iron-carbon alloys.

Microstructure name	Description
δ-ferrite	An interstitial solid solution of carbon in δ iron (BCC).
Austenite	An interstitial solid solution of carbon in γ iron (FCC).
α-ferrite	An interstitial solid solution of carbon in α iron (BCC).
Pearlite	Eutectoid of α-ferrite and cementite with a lamellar microstructure of alternate α-Fe and cementite plates.
Bainite	Eutectoid of α-ferrite and cementite. The α-ferrite either has a feathery appearance or occurs as plates. Carbide particles lie between the α-ferrite regions.
Spheroidite	Spherical particles of cementite in a matrix of α-ferrite.
Martensite	An interstitial solid solution of carbon in a body-centered tetragonal (BTC) Fe crystal structure.
Hypoeutectoid steels	Alloys with compositions to the left of the eutectoid reaction.
Hypereutectoid steels	Alloys with compositions to the right of the eutectoid reaction.
Proeutectoid ferrite	Ferrite that forms prior to the eutectoid ferrite.
Proeutectoid cementite	Cementite that forms prior to the eutectoid cementite.

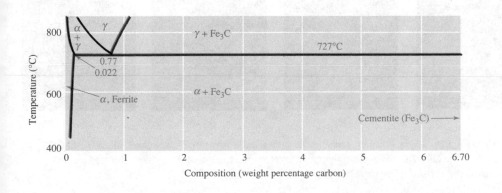

FIGURE 8.3–1

Section of the Fe-Fe$_3$C phase diagram showing the eutectoid region.

compositions different from the original phase. Written symbolically, the eutectoid reaction for plain carbon steel is γ-Fe $\rightarrow$ α-Fe + Fe$_3$C. Thus, upon cooling, the single-phase γ-Fe (austenite) decomposes into α-Fe (ferrite) and Fe$_3$C (cementite). The solubility of carbon in the FCC γ phase is greater than in the BCC α phase, but less than in the Fe$_3$C phase. The growth rate and the morphology of the α and Fe$_3$C phases are controlled by the diffusion of iron and carbon. When cooled to just below the eutectoid temperature, an alloy containing 0.77 wt. % C dissolved in the single-phase γ-Fe transforms into a two-phase mixture with a lamellar (layered) morphology of alternating α-Fe and Fe$_3$C plates. This two-phase lamellar material is called **pearlite.**

C curves for the beginning and end of pearlite formation are obtained by instantaneously quenching from the austenite phase field and isothermally holding at various temperatures below the eutectoid temperature to bring about changes in microstructure (austenite to pearlite) with time. Figure 8.3–2 shows the microstructure of pearlite formed at several hold temperatures, and Figure 8.3–3 is a diagram of the fraction of transformed pearlite as a function of time (note the sigmoidal shape).

The compositions of the α and Fe$_3$C phase in equilibrium are determined from the phase diagram by constructing a tie line in the two-phase field through the state point. The α-Fe is low in carbon and, for most calculations, can be considered pure iron, whereas cementite contains ~6.7 wt. % C. As shown in Figure 8.3–4, the redistribution of carbon required for this phase transformation can occur along the austenite-pearlite interface. During the transformation, the interface moves with constant velocity into the austenite, leaving behind parallel plates of ferrite and cementite. Since the formation of the pearlite structure depends upon the redistribution of carbon, the pearlite spacing (i.e., relative thickness of the α and Fe$_3$C layers) will depend on how far the carbon atoms can diffuse in a unit of time (i.e., on the diffusion rate). Thus, both growth rate and interlamellar spacing are strong functions of temperature. In general, the pearlite formed at higher temperatures (low nucleation rate and high growth rate) is coarser than pearlite formed at lower temperatures.

As the amount of undercooling is increased, an alternative morphology to pearlite becomes more energetically favorable. This two-phase mixture of α + Fe$_3$C is called **bainite** and occurs in a eutectoid steel transformed at temperatures less than 550°C. Like pearlite, bainite is the result of austenite decomposition. An exact mechanism describing all observations has not been identified, and to this day, the mechanisms proposed are still

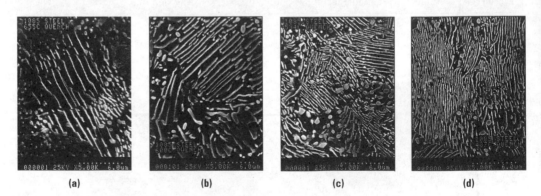

(a)   (b)   (c)   (d)

FIGURE 8.3–2 Microstructure of pearlite formed at different isothermal hold temperatures: **(a)** 655°C, **(b)** 600°C, **(c)** 534°C, and **(d)** 487°C. Notice that the morphologies of the two-phase structure are similar but their spacings decrease with decreasing isothermal hold temperature.

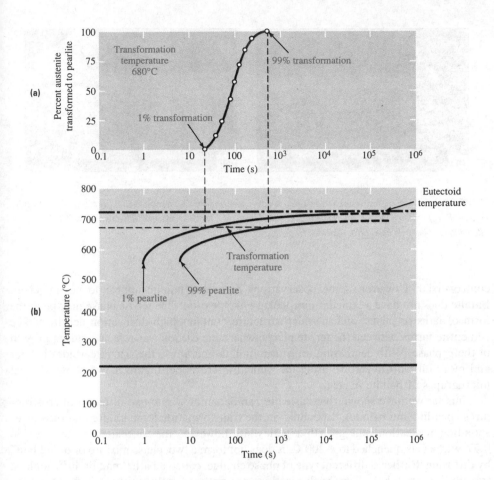

(a)

(b)

FIGURE 8.3–3

Fraction of transformed pearlite as a function of time at 680°C for a eutectoid alloy. *(Source:* Atlas of Isothermal Transformation Diagrams, *A.I.M.E. Reprinted by permission of the publisher.)*

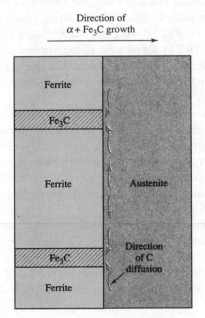

Direction of
α + Fe_3C growth

FIGURE 8.3–4

Schematic of the redistribution of carbon along the austenite-pearlite interface.

(a)   **(b)**

FIGURE 8.3–5   The microstructure of bainite in an alloy steel isothermally transformed at: **(a)** 495°C, and **(b)** 410°C. *(Source: R. F. Hehman, "Ferrous and Nonferrous Bainite Structures," Metals Handbook, 8th ed., pp. 194–96. ASM International, Materials Park, Ohio. Reprinted by permission of the publisher.)*

controversial. However, some observations regarding bainite appear to be universal. Bainite does not have a lamellar morphology like pearlite. The ferrite phase may be in the form of laths or plates, and the microstructure contains high dislocation densities. The cementite forms between the ferrite plates, where the carbon was rejected during growth of the α phase. With decreasing transformation temperature, the size of carbides is finer and more difficult to resolve using the optical microscope. Figure 8.3–5 shows optical micrographs of bainite in steel.

Thus far we have shown that austenite can decompose into two different microstructures (pearlite and bainite), depending on the transformation temperature and the corresponding nucleation and growth kinetics. If, however, the austenite of composition 0.77 wt. % C is quenched to ~200°C, it does not form a two-phase mixture of α and Fe_3C by diffusion. Rather, a different type of phase change, called an athermal, or diffusionless, transformation, takes place. Such reactions are not thermally activated and occur very rapidly and continuously during cooling. Thus, the amount of transformed phase depends *solely* on temperature, not time. In steels, the resulting phase is known as **martensite.**

The martensite transformation is of great importance, since the formation of the body-centered tetragonal (BCT) martensite phase formed by the rapid quenching of austenite is usually the first step in the heat treatment of steels. The heat treatment of steels is of significant commercial importance, and a large industrial sector is devoted to this activity.

Athermal transformations also occur in nonferrous systems such as Cu-Al and Au-Cd and in oxides such as the cubic-to-tetragonal $BaTiO_3$ transformation and the tetragonal-to-monoclinic transformation in ZrO_2. Ceramists and geologists have independently identified characteristics of the martensitic transformation in nonmetals and use the term **displacive** to describe transformations that a metallurgist would call martensitic.

On an isothermal transformation (IT) diagram, a martensitic phase change is drawn as a horizontal line. Figure 8.3–6 is an IT diagram for a eutectoid steel showing a typical C curve for a diffusional transformation at high temperatures and the athermal martensite reaction at low temperatures. The temperature at which the martensite reaction begins is labeled M_s, and the M_{50} and M_{90} lines refer to the temperatures at which the martensitic phase transformation is 50% and 90% completed. Figure 8.3–7 is a plot of M_s and M_f (the temperature at which the martensitic transformation is finished) as a function of carbon

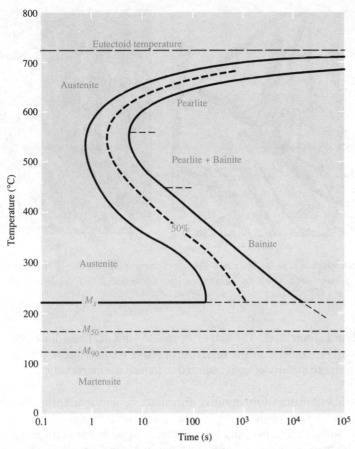

FIGURE 8.3-6 Schematic of an isothermal transformation (IT) diagram for a eutectoid steel.

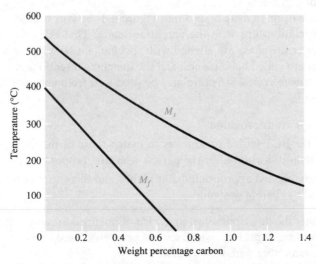

FIGURE 8.3-7 The effect of carbon on the M_s and M_f temperatures for a plain carbon steel.

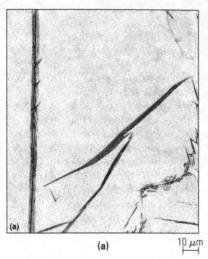
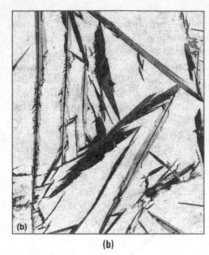

(a)   10 μm   (b)   (c)

FIGURE 8.3–8   Progress of athermal martensitic transformation in an Fe–1.8 wt. % C alloy after cooling to **(a)** 24°C, **(b)** −60°C, and **(c)** −100°C. *(Source: G. Kraus and A. R. Marder, The Morphology of Martensite in Iron Alloys. Metallurgical Transactions, Vol. 2, 1971, pp. 2343–57. Reprinted by permission of the publisher.)*

content for a plain carbon steel. The slopes of the M_s and M_f lines illustrates a strong dependence of carbon content on the martensitic phase transformation. As the carbon content increases, more undercooling is required to induce the martensitic phase transformation.

The martensitic phase transformation is diffusionless, and martensite has the same composition as the parent phase. Because the phase transformation is diffusionless, carbon atoms are not able to move out of their interstitial positions in the γ-Fe lattice as they do during the formation pearlite or bainite. Instead, the C atoms remain trapped in interstitial positions. These trapped carbon atoms, and their associated strain fields, are very effective obstacles to dislocation motion; therefore, martensite is the strongest of the common phases found in steels.

The microstructure of martensite is shown in Figure 8.3–8. This needlelike phase has specific orientation relationships with the parent austenite. That is, certain planes and directions in the martensite phase are aligned with specific planes in the austenite phase. The matching planes are called habit planes, and the matching directions are called habit directions. The martensite crystal structure may be produced from austenite by performing three operations:

1. An FCC to BCT transformation.
2. Distortion of the BCT lattice parameters to match those of the martensite. This is called the Bain distortion after the person who first proposed it.
3. Rotation to produce the appropriate habit planes and directions between the martensite and the parent austenite.

It should be noted that the description just given is not intended to represent the mechanism by which martensite forms. It is instead a set of geometric steps that can produce one lattice structure from another without diffusion.

The FCC/BCT transformation and Bain distortion are illustrated in Figure 8.3–9. This figure also shows the martensite crystal structure with lattice parameters a and c. Although the specific values of a and c depend on the carbon content of the alloy, volume

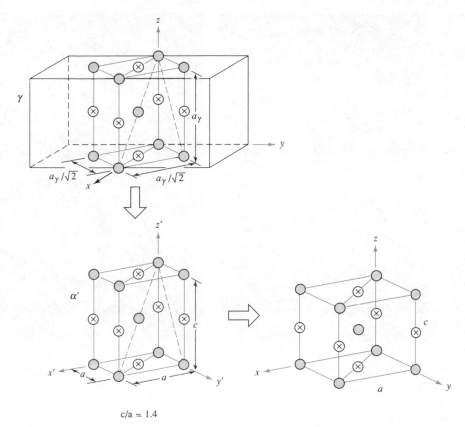

c/a ≃ 1.4

FIGURE 8.3–9   The relationship between the FCC γ phase and the BCT martensite phase. The possible sites for carbon atoms are located at the x's, and the iron atoms are located at the open circles. To obtain the α' phase from the γ unit cell, the c axis must contract about 20% and the a axis must expand about 12%.

always increases when austenite transforms to martensite. As a result of the volume change, significant internal stresses develop, and problems with cracking may occur if great care is not exercises during quenching.

Martensite is metastable with respect to the two-phase α-Fe/Fe$_3$C system. Thus, when a microstructure containing martensite is heated, Fe$_3$C begins to form. The excess free energy associated with the "trapped" carbon in the BCT structure is reduced as carbon diffuses out of the metastable BCT structure to the more stable carbide phase. As a result of the decrease in carbon content in the martensite, the c/a ratio approaches 1 and the BCT structure tends toward a BCC structure. The process by which carbide precipitates are formed in a martensitic matrix is called **tempering.** Because this transformation requires diffusion, the higher the temperature and the longer the time at temperature, the coarser the cementite particles will be. Because of the roughly spherical morphology of the carbide phase in tempered martensite, this microstructure is termed **spheroidite** (see Figure 8.3–10).

Spheroidized microstructures of carbides in ferrite can also be produced from pearlitic or bainitic microstructures. The slowest spheroidizing rates are associated with pearlite, especially from those microstructures with coarse spacings. Spheroidizing is more rapid if the carbides are initially in the form of fine carbides, which is the case for bainite. Spheroidizing a martensitic microstructure occurs even more rapidly than either bainitic or pearlitic microstructures.

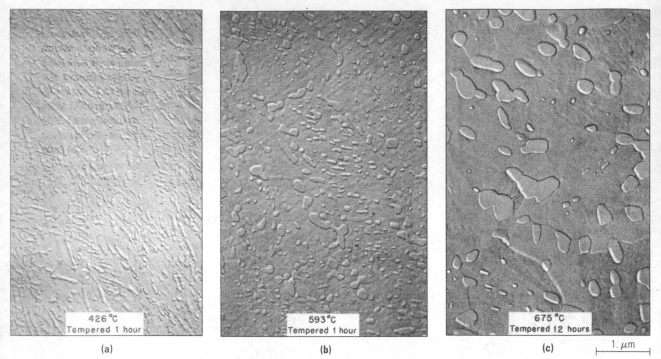

426°C Tempered 1 hour	593°C Tempered 1 hour	675°C Tempered 12 hours
(a)	(b)	(c)

1. μm

FIGURE 8.3–10 Microstructure of a tempered martensite (spheroidite) in a steel with 0.7 wt. % C.   *(Source:* Transactions of the Metallurgical Society *(212), 1958, a publication of the Minerals, Metals & Materials Society, Warrendale, PA. Reprinted by permission of the publisher.)*

Spheroidized microstructures are the most stable microstructures found in steels. Because the microstructure is composed of spherical carbides in a ferrite matrix, these microstructures are characterized by good ductility. This may be extremely important for microstructures such as low-carbon steels (less than 0.25 wt. % C) that are extensively cold-formed or for high-carbon steels (greater than 0.6 wt. % C) that are machined prior to heat treating to produce hard, wear-resistant surfaces.

Application of an IT diagram is illustrated in Figure 8.3–11. If an alloy of eutectoid composition is heated to a temperature above the eutectoid isotherm into the single-phase austenite region, instantaneously quenched to 600°C, and then held for 20 seconds (path 1), all the austenite will transform to pearlite. If, however, the alloy is instantaneously quenched to 650°C and held for 100 seconds (path 2), the resulting microstructure will still be pearlite, but the size scale of the pearlite will be larger (i.e., it will be composed of thicker layers of α and Fe_3C). This is because at the higher transformation temperature the nucleation rate is lower (fewer particles) and the diffusion rate is higher (bigger particles and larger spacings). In either case, once the diffusion-based $\gamma \rightarrow \alpha + Fe_3C$ transformation is complete, subsequent decreases in temperature have no effect on the microstructure.

Suppose that rather than holding at 600°C for 20 seconds, the hold time was reduced to 3 seconds and then the alloy was rapidly quenched to room temperature. Holding at 600°C for 3 seconds transforms about 50% of the austenite to pearlite (see Figure 8.3–11). A microstructure containing a mixture of ~50% untransformed austenite and ~50% transformed pearlite would result. Rapid quenching to room temperature would have no effect on the pearlite. However, there would be a significant effect on the untransformed austenite. By quenching to room temperature, the alloy passes through the M_s and M_f lines. Consequently, the austenite is transformed to martensite upon quenching to room

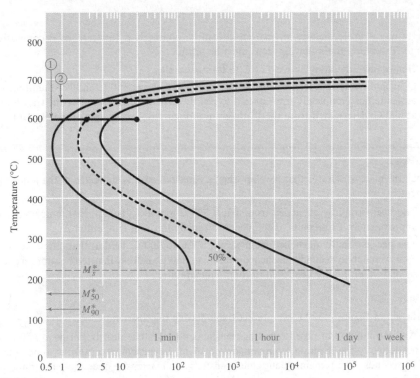

FIGURE 8.3–11 Examples of time-temperature paths on the isothermal transformation diagram of a eutectoid steel.

temperature. The final structure would thus consist of a mixture of 50% pearlite and 50% martensite.

EXAMPLE 8.3–1

When a eutectoid steel is heat-treated in the austenite phase field, quenched, and isothermally held at temperatures below 727°C, different morphologies of ferrite and cementite (the carbide phase) are possible. Describe these microstructures and how they are produced.

Solution

The microstructure that forms can be determined with the aid of the IT diagram shown in Figure 8.3–6.

a. Quenching from above 727°C and holding at temperatures below 727°C to about 550°C results in pearlite. Pearlite is a lamellar structure of ferrite and carbide. The lower the isothermal hold temperature, the finer the pearlite spacing.

b. Quenching from above 727°C and holding at temperatures below 550°C but above M_s results in a bainitic structure, which is fine carbides distributed in ferrite. Again, the lower the isothermal hold temperature, the finer the constituents of the microstructure.

c. Reheating either pearlitic, bainitic, or martensitic microstructures results in a distribution of spherical carbides in ferrite. The microstructures are termed spheroidite. The size of the carbides is a function of the starting microstructure and the time and temperature of the spheroidizing heat treatment.

EXAMPLE 8.3–2

Suppose you have been asked by your instructor to prepare a sample of pearlite for a demonstration in class. If the only available sample is a eutectoid steel with a bainitic structure, describe the procedure you would use to complete your assignment.

Solution

The only way pearlite, bainite, or martensite can be produced is through the decomposition of austenite. Therefore, the following steps must be taken:

    a. Homogenize the alloy above 727°C to form austenite.

    b. Quench directly to a temperature below 727°C and above 550°C and hold until all the austenite has completely transformed to pearlite. The time required for the transformation at 600°C, determined from the IT diagram in Figure 8.3–6, is approximately 10 s.

EXAMPLE 8.3–3

Describe the microstructures that develop when a eutectoid steel is subjected to the following heat treatments:

    a. The steel is rapidly quenched from above 727°C to 650°C, held at that temperature for approximately 12 seconds, then quenched rapidly to room temperature.

    b. The steel is rapidly quenched from above 727°C to 600°C, held at that temperature for approximately 3 seconds, quenched rapidly to 350°C, held at that temperature for 10^3 seconds, then finally quenched to room temperature.

Solution

With the aid of Figure 8.3–11, we predict the following microstructures:

    a. A 50-50 mixture of pearlite and martensite is present. The pearlite forms at 650°C, and according to the diagram, approximately 50% of the austenite transforms to pearlite after 12 seconds. The remaining austenite transforms to martensite once the temperature drops below M_{90}.

    b. A 50-50 mixture of pearlite and bainite is present. As in part *a* of this problem, 50% of the austenite transforms to pearlite at 600°C; the remaining austenite transforms to bainite at 350°C. The principle that applies here is that even though some austenite remains after the elevated temperature exposure, when the new lower temperature is reached, the remaining austenite transforms at that temperature as though it had never experienced time at 600°C. Consequently, the remaining austenite does not begin to transform to bainite until it is at 350°C for approximately 10 seconds, and the transformation is complete after crossing the bainite finish line.

Changing the composition of the alloy affects the appearance of the IT diagram. The diagram illustrated in Figure 8.3–11 is for a eutectoid composition steel. Alternatively, the IT diagram for a 0.5 wt. % C alloy, a hypoeutectoid composition, is shown in Figure 8.3–12. The primary difference between the IT diagrams for the eutectoid composition and the off-eutectoid composition of 0.5 wt. % C is the occurrence of the proeutectoid ferrite that forms as the alloy is cooled through the austenite/ferrite two-phase field. If an alloy is instantaneously quenched from the austenite single-phase field to 650°C and held at that temperature, the first phase formed is ferrite. Because it forms before the ferrite in pearlite, it is termed proeutectoid ferrite. The amount of proeutectoid ferrite that forms depends on the isothermal hold temperature. Once the pearlite start line is crossed, the remaining austenite begins to transform to pearlite. The proeutectoid and eutectoid phases are heterogeneously nucleated at the austenite grain boundaries. The maximum amount of proeutectoid phase that forms occurs when the alloy is slowly cooled through the two-phase field. This amount can be estimated using the lever rule.

Care must be exercised in using isothermal transformation curves to estimate microstructures that result from different thermal histories. Such diagrams were developed assuming that the material was held in a single-phase field, instantaneously quenched to

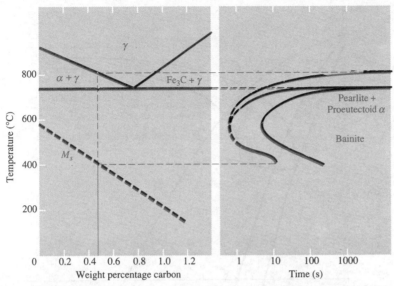

FIGURE 8.3–12   Relationship between the Fe-Fe$_3$C phase diagram and the IT diagram for a 0.5 wt. % carbon alloy.

a specific temperature, and held for various times to produce different amounts of the transformed product. They were not developed under conditions of continuous cooling and should not be rigorously used to estimate microstructures that result from such conditions. Thus, an IT diagram, while illustrating a number of important concepts, has limited application to a commercial operation in which a large, complex-shaped part would experience neither an instantaneous nor a uniform quench rate. A continuous-cooling transformation (CT) diagram, which is developed using procedures similar to those for an IT diagram, must be used when the quench rates are not instantaneous. Figure 8.3–13 shows a schematic of a CT diagram for a eutectoid steel. These diagrams are more versatile, since quench paths other than instantaneous quenches followed by isothermal holds can be analyzed.

Three quench paths are illustrated in Figure 8.3–13. Path 1 shows a slower quench rate than does path 2. For path 1, as the alloy cools along the segment O to a_1, the microstructure is 100% austenite. It is not until the quench path intersects the pearlite start line that pearlite begins to form. From point a_1 to point b_1 the fraction of transformed pearlite increases. Once the path crosses the pearlite finish line, all the austenite has decomposed into the two-phase pearlite microstructure, and further decreases in temperature have no effect.

Quench path 2, on the other hand, develops a different microstructure. Pearlite begins to form at point a_2, and continued cooling to point b_2 results in an increase in the pearlite fraction up to ~33%. When the microstructure passes point b_2 it contains a mixture of ~33% pearlite and ~67% austenite. The remaining austenite does not transform until the path crosses the M_s temperature at point c_2. Continued cooling through the martensite transformation range increases the amount of martensite. Thus, at room temperature, the microstructure for path 2 will contail ~33% pearlite and ~67% martensite. Quench path 3 represents the critical cooling rate necessary to obtain 100% martensite. All quench paths to the left of path 3 also result in 100% martensite.

To obtain a martensitic structure throughout the cross section of a part, the critical cooling rate must be exceeded in every section of the part. The surface, which is in contact with the quench medium, cools most quickly, and the rate of cooling decreases as the

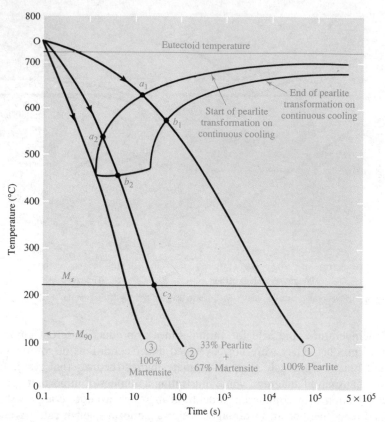

FIGURE 8.3–13   Schematic of a CT diagram for a eutectoid steel showing three quench paths.

depth below the surface increases. If the section is sufficiently thick, then there may be some point below the surface where the cooling rate is less than the critical cooling rate. From that point inward, the amount of martensite decreases. Steels with low critical cooling rates can form extensive amounts of martensite, while steels with high critical cooling rates are able to form only relatively thin shells of martensite on their surface. Since martensite is the strongest and hardest phase in steel, an alloy that can form extensive amounts of martensite is said to have high **hardenability.** Steels with higher critical cooling rates have lower hardenability and a thinner martensite shell. Because the composition of the alloy affects the rate at which the austenite decomposes, composition is the most important factor in controlling hardenability.

A rapid method of determining the hardenability of a steel is the **Jominy end-quench test,** illustrated in Figure 8.3–14. A standard-size specimen, which has been heated to the austenite single-phase field, is placed in a fixture that positions its lower end over a quenching-water pipe. The water is turned on, and the specimen cools rapidly from the bottom, with a cooling rate that decreases with distance from the quenched surface. The quench rate is fixed by the water temperature and flow rate, and the size and shape of the piece are specified to facilitate comparison of data among steels of different compositions. The strength along the rod, which is a function of the corresponding microstructure at that position, can be determined by hardness measurements.[4] Figure 8.3–15 compares the Jominy end-quench data for several steels. For all compositions, the

---

[4] Details of the hardness-testing technique are described in Chapter 9. At this point it suffices to say that hardness is well correlated with strength (i.e., high hardness implies high strength).

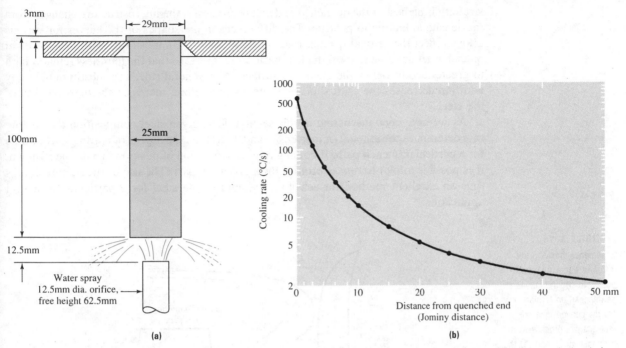

(a)   **(b)**

FIGURE 8.3–14   The Jominy end-quench test: **(a)** geometry of the test, and **(b)** cooling rates as a function of distance from the quenched end of the bar.   *(Source: Adapted from Albert G. Guy,* Elements of Physical Metallurgy, *2nd ed., Addison-Wesley, 1959, p. 484.)*

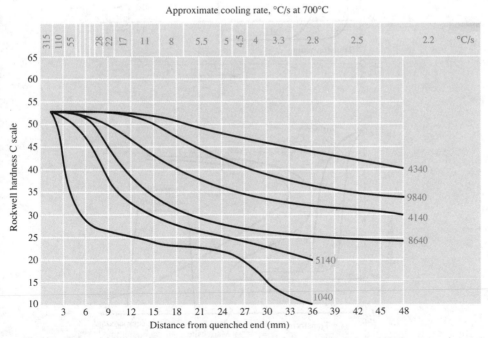

FIGURE 8.3–15   A comparison of the Jominy end-quench test data for a series of steels with the same carbon content (0.4 wt. %). Note the relationship between cooling rate (top) to distance from the quenched end. It has been assumed that the cooling behavior is independent of alloy composition.   *(Source:* The Making, Shaping and Treating of Steel, *10th ed. Copyright 1985 Association of Iron and Steel Engineers. Reprinted by permission of the publisher.)*

strength is highest at the quenched end and decreases as the microstructure changes from martensite to bainite to pearlite. The differences in the shapes of the curves for various alloys reflect the critical quench rates in those alloys. Compositions with lower critical quench rates are able to form thicker martensitic layers, so that the hardness remains high to greater depths below the quenched surface. As a general rule, any substitutional alloy addition decreases the critical quench rate and, therefore, increases the hardenability of the steel.

As we have been discussing in this section, for a given steel composition the cooling rate determines the amount of martensite and, therefore, the degree of hardness achievable for a particular quench path. If cooling rates as a function of position in a part are known, it is possible to plot hardness profiles throughout the part. The use of the Jominy data is thus an excellent method for selecting the appropriate steel for a particular structural application.

FIGURE 8.3–16

Schematics showing the different quenching methods, with cooling curves superimposed on IT diagrams: **(a)** the conventional, or direct quench, process (with a temper step), and **(b)** the martempering, or indirect quench, process (with a temper step).

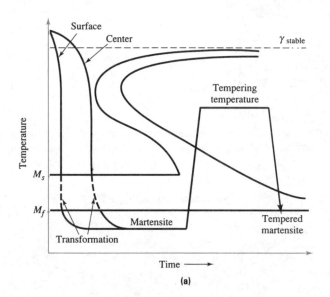

(a)

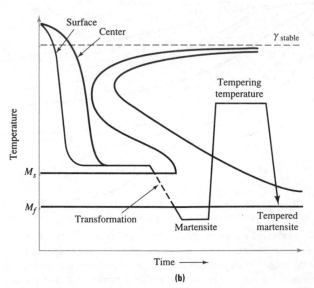

(b)

An important step in the production of a part fabricated from a quenched steel is tempering. A steel whose microstructure consists of as-quenched martensite, although quite strong, is also very brittle and thus such a material is useless for most engineering applications. To meet minimum ductility requirements, as-quenched steels are tempered. Tempering, as discussed previously, is accomplished by reheating a quenched steel to a temperature below the eutectoid temperature for a prescribed period of time; however, an unavoidable loss in strength accompanies the restoration of ductility. The change in mechanical properties is a direct result of changes occurring in the microstructure during tempering. The understanding we have derived in this chapter suggests that increasing temperature increases the rate of transformation from martensite to a tempered phase.

Two commercial quenching operations are schematically illustrated in Figure 8.3–16. The major difference between the conventional quench-and-temper process and the commercially important martempering process (also known as an indirect quench) is that in the latter case the steel is quenched rapidly enough to "miss" the start of the diffusion-based $\gamma \rightarrow \alpha + Fe_3C$ phase transformation, but the quench is then interrupted so that the part can stabilize at a temperature just above M_s. The second stage of the quench is then implemented (before the bainite transformation begins) in order to form the martensitic phase. The reason for holding the sample just above M_s is to minimize the thermal gradients through the thickness of the part that lead to the formation of surface cracks during the direct quench operation. The final stage of both the direct and indirect quench operations is the temper, which is designed to "trade" some strength for improved ductility.

EXAMPLE 8.3–4

Estimate the maximum amount of proeutectoid ferrite that can form in a plain carbon steel containing 0.5 wt. % carbon.

Solution

The maximum amount of proeutectoid ferrite is determined by applying the lever rule in the ferrite-austenite (α-γ) two-phase field just above the eutectoid isotherm. If the compositions just above 727°C can be approximated by the compositions at the eutectoid temperature, then the mass fraction of austenite is

$$f_\gamma = \frac{0.5 - 0.02}{0.77 - 0.02} = 0.64$$

and the mass fraction of the proeutectoid ferrite is

$$f_\alpha = \frac{0.77 - 0.5}{0.77 - 0.02} = 0.36$$

EXAMPLE 8.3–5

Is it possible by rapid quenching and isothermal holds to suppress the formation of proeutectoid ferrite?

Solution

Yes. According to the IT diagram in Figure 8.3–12, rapidly quenching to a temperature just below 520°C suppresses the formation of proeutectoid ferrite. An isothermal hold at 520°C for greater than 10 seconds results in a microstructure containing only bainite.

8.3.2 Precipitation from a Supersaturated Solid Solution

An important method for altering the properties of a material is to precipitate a second phase from a supersaturated solid solution (i.e., $\alpha \rightarrow \alpha + \beta$). Figure 8.3–17 illustrates this phenomenon. If an alloy of composition X_0 is held at temperature T_1, a process known as homogenization or solution heat treatment occurs, and all of the solute dissolves into the α phase. If the alloy is rapidly quenched to temperature T_2, the solute is not immediately able to diffuse out of the α phase, and the alloy is said to be **supersaturated.** Over time, however, diffusion of the solute out of the unstable α phase eventually results in a two-phase microstructure of $\alpha + \beta$. The relative amounts of α and β are determined by the phase diagram through the lever rule; however, it is the morphology of the alloy (e.g., the size and distribution of the second-phase particles) that determines many of the macroscopic properties of the alloy. Since the formation of a second-phase particle from a supersaturated solid solution is a nucleation and growth process, the size, number, and spatial distribution of precipitates are controlled by temperature and time.

The process of precipitation from a supersaturated solid solution is called **aging.** When aging occurs at room temperature, it is termed natural aging. Aging above room temperature is called artificial aging. The time-temperature sequence for the generic aging process is illustrated in Figure 8.3–17b. As mentioned above, the first step is a solution heat treatment followed by a rapid quench. The final step is to reheat the alloy, while remaining well below the solvus boundary, to accelerate the aging process (by increasing the diffusion rate).

Applications of Aging in the Aluminum-Copper System

The aluminum-copper system is well documented and is often used to introduce the concept of precipitation in supersaturated alloys. Figure 8.2–1 shows the aluminum-rich end of the A1-Cu phase diagram. Two necessary requirements for commercial application of precipitation are illustrated by this system:

1. There must be a temperature at which a significant amount of solute can be dissolved. The more solute that is dissolved, the greater the potential for property changes when precipitates form.

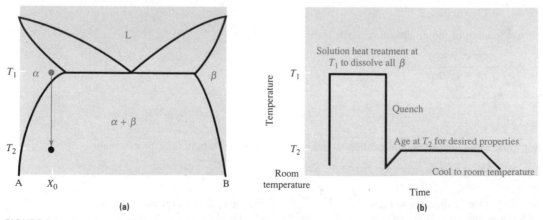

(a)

(b)

FIGURE 8.3–17 Schematic illustration of the age hardening process: **(a)** a phase diagram illustrating a binary system that can precipitate a second phase upon quenching from the single-phase field, and **(b)** The time-temperature process for creating an age-hardened structure.

2. There must be a significant decrease in solubility with decreasing temperature. The larger the difference between the solubility at the high and low temperatures, the larger is the volume fraction of the precipitating phase (a lever rule argument).

Suppose an alloy containing 4.5% Cu is solution-heat-treated at 820 K. At that temperature all the copper in the alloy is soluble in the α phase. A rapid quench to room temperature ensures two things. First, there is insufficient time at elevated temperatures to nucleate the second phase; consequently, the α solution is supersaturated. Second, there is a supersaturation of vacancies. These excess vacancies arise because the quench maintains the vacancy concentration developed at 820 K. These vacancies accelerate precipitation to the extent that it occurs at temperatures at which it would ordinarily have prohibitively low diffusion rates. Even with excess vacancies, however, precipitation at room temperature is too slow to be commercially acceptable. While artificial aging reduces the time required for the phase transformation, ensuring its commercial viability, it also results in a decrease in the amount of second phase that can precipitate. Consequently, the selection of an appropriate aging temperature requires a balance between the time to age and the maximum achievable properties.

The precipitation of copper-rich phases from a supersaturated matrix is a complex process. Thus far, all the diffusion-based phase transformations we have investigated have involved equilibrium phases. However, in many cases, including the Al-Cu system, when alloys are rapidly quenched from the single-phase field to a two-phase field, the phase that forms initially is a metastable, or nonequilibrium, phase. To reduce interfacial energy between the matrix and precipitate, the metastable phase tends to have a crystal structure similar to that of the matrix. The composition of the metastable phase, however, resembles that of the equilibrium phase. Consequently, the precipitation process in many alloy systems is determined by the most efficient way to decompose the unstable, supersaturated solid solution into the equilibrium two-phase mixture.

Detailed microstructural investigations of a variety of Al-Cu alloys aged at various times and temperatures have revealed several different metastable copper-rich phases. The size and structure of these phases are summarized below.

1. *GP zones.*[5] These are copper-rich clusters about 80 Å in diameter and 3 to 6 Å thick. The GP zones have the same structure as the matrix, but have a much higher copper content. Because Cu atoms are smaller than Al atoms, strain fields around the precipitate are created. Figure 8.3–18a shows a TEM micrograph of an Al–4.5% Cu alloy aged to produce GP zones.

2. θ'' *precipitates.* Upon heating, GP zones begin to dissolve, and a second precipitate appears. The precipitate maintains a platelike morphology about 300 Å in diameter and 20 Å thick. θ'' is tetragonal with two of its lattice parameters matching those of aluminum ($a = b = 4.04$ Å) and with the c axis ($c = 7.68$ Å) a bit less than twice as long as the Al lattice parameter. The θ'' precipitates are coherent on all faces, but because of the relatively poor fit in the c direction, large elastic strains are present in the matrix. The crystal structure for θ'' is shown in Figure 8.3–19, and Figure 8.3–18b is a TEM micrograph of an aged Al–4.5% Cu alloy containing θ''.

3. θ' *precipitates.* θ' precipitates are on the order of 1000 Å in diameter. θ' is tetragonal with $a = b = 4.04$ Å and $c = 5.8$ Å. Coherency is lost in the c

---

[5] GP zones are named for the two investigators (Guinier and Preston) who independently identified these structures in quenched Al-Cu alloys.

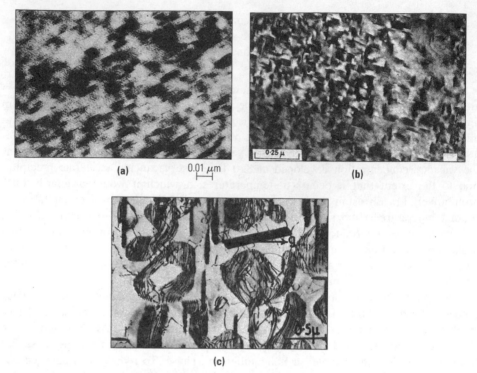

FIGURE 8.3–18 Transmission electron micrograph of an aged Al–4.5% Cu alloy to produce: **(a)** GP zones, **(b)** θ'', and **(c)** θ'. *(Sources:* **(a, b)** *G. Thomas and J. Nutting,* Journal of the Institute of Materials *87 (1958–59), p. 431, Institute of Materials. Reprinted by permission of the publisher.* **(c)** *G. C. Weatherly and R. B. Nicholson,* Philosophical Magazine, *Taylor and Francis, p. 813. Reprinted by permission of the publisher.)*

direction but is maintained in the $a = b$ plane, as shown in Figure 8.3–19. Figure 8.3–18c is a TEM micrograph of an A1–4.5% Cu alloy aged to produce θ'.

4. θ *precipitates.* θ is the equilibrium phase and is incoherent with the matrix. The structure is tetragonal with $a = b = 6.06$ Å and $c = 4.87$ Å, as shown in Figure 8.3–19.

There are several important observations. First, the GP zones are more like the matrix than any of the other phases, including the equilibrium θ phase. Second, the magnitude of the precipitate-matrix interfacial energy increases from GP zones $\rightarrow \theta'' \rightarrow \theta' \rightarrow \theta$. Third, the stability of the phases increases from GP zones $\rightarrow \theta'' \rightarrow \theta' \rightarrow \theta$ so that regardless of which phase forms first, eventually the equilibrium θ phase will be present.

Which phase nucleates first during a quench, and why? To answer this question, nucleation theory must be applied. The nucleation process can be either homogeneous or heterogeneous. Homogeneous nucleation provides the greatest number of nuclei; hence, it is a more efficient mechanism to relieve supersaturation than heterogeneous nucleation, provided sufficient energy is available to overcome the homogeneous nucleation barrier. The barrier to homogeneous nucleation, as derived in Section 8.2.3, has the form:

$$\Delta G_{\text{hom}}^* \propto \gamma_{\alpha\beta}^3 \left(\frac{1}{(\Delta T)^2} \right) \tag{8.3–1}$$

The larger the interfacial energy, the larger the barrier will be. Consequently, the probability of homogeneous nucleation decreases from GP zones $\rightarrow \theta'' \rightarrow \theta' \rightarrow \theta$. That is, GP zones are the most likely phase to nucleate homogeneously.

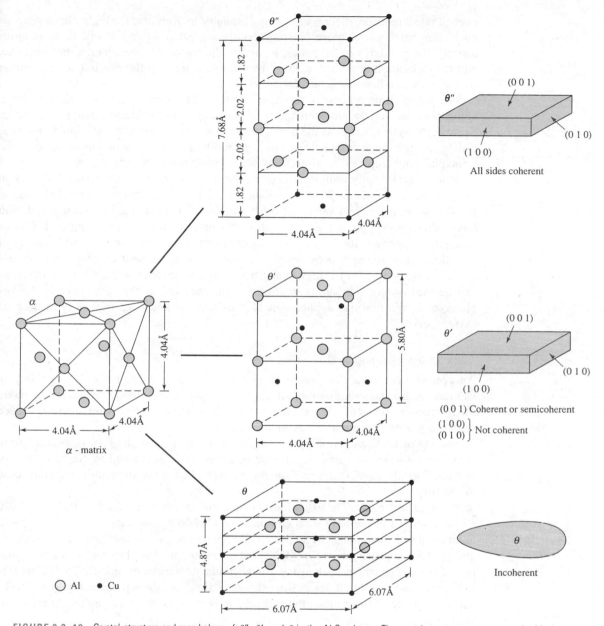

FIGURE 8.3–19   Crystal structure and morphology of θ'', θ', and θ in the Al-Cu phases. The crystal structures are compared with the crystal structure of the aluminum matrix.   *(Source: D. A. Porter and K. E. Easterline, Phase Transformations in Metals and Alloys, Chapman & Hall, p. 147. Reprinted by permission of the publisher.)*

Heterogeneous nucleation requires a nucleation site such as a grain boundary or dislocation. Near these sites (within distance $\sqrt{Dt}$) heterogeneous nucleation of θ occurs at grain boundaries and θ' grows from dislocations. The reason for the preferential nucleation sites is that the higher interfacial energy associated with the θ phase requires additional driving force be provided by the elimination of the comparatively high-energy grain boundaries.

Thus, during a rapid quench one might reasonably expect to find GP zones nucleating homogeneously in the grain interiors (where heterogeneous sites are absent) and θ and θ' nucleating heterogeneously at grain boundaries and dislocations, respectively. If the quench rate is reduced, there is less energy available to overcome the higher homogeneous nucleation barrier, and more heterogeneously nucleated θ' and θ will occur. Consequently, the initial mix of GP zones, θ', and θ depends on the quench rate, the grain size, and the dislocation density. The latter two factors determine the precipitate distribution within $\sqrt{Dt}$ of a heterogeneous nucleation site.

The precipitation process in the Al-Cu system was presented for three reasons. First, aging has been well documented and is often discussed in the literature on precipitation. Second, this system serves as the prototype for a number of commercial aluminum alloys. Finally, aging is quite complicated, but if the subtleties of this system are understood, the principles can be applied to almost any other precipitation system.

Many commercially important engineering materials derive their high strength from the presence of a fine, randomly distributed precipitate. In addition to the Al-Cu system just discussed, there are a variety of other aluminum alloys that are called **precipitation hardening** alloys. This name comes about because, as the size and number density of precipitates increase, there is a corresponding increase in the strength and hardness of the alloy. These issues will be developed more fully in the next chapter, on mechanical behavior. Because of the high strength and low density of these aluminum alloys, they can be used as structural units in airplanes and bicycles, as the skin covering the airplane, or in architectural applications such as window and door frames or facades on skyscrapers.

Applications of Aging in Other Systems

Another important class of precipitation hardening alloys is the nickel alloys. Alloying elements such as aluminum, titanium, chromium, cobalt, and niobium result in two-phase microstructures that can be heat-treated at a variety of temperatures and times to produce different second-phase particle morphologies.

In addition to Al and Ni alloys, precipitation strengthening plays an important role in some steel systems. In particular, maraging steels are strengthened by the formation of fine intermetallic precipitates and are among the highest-strength metallic materials used in industry.

Precipitation hardening also finds application in ceramic systems. Figure 8.3–20 shows a section of the ZrO_2-rich end of the ZrO_2-MgO phase diagram. For example, a composition of 8.64 mol % MgO permits fabrication in the single-phase field at temperatures above approximately 1750°C. Rapid quenching to avoid heterogeneous precipitation of the tetragonal phase along the grain boundary, followed by aging at 1400°C, results in the distribution of small, lense-shaped precipitates from the supersaturated cubic ZrO_2. The distribution of these precipitates is used to control the strength and toughness of the ceramic.

8.3.3 Solidification and Homogenization of an Alloy

In the previous sections we dealt with the process of heat treating. For most commercial materials, the heat-treating step is one of the last steps in the fabrication sequence. In general, alloys are cast, homogenized (heat treated to produce a uniform structure), and then worked into a product form by rolling, extruding, or forging.[6] It is not until after the forming operation that many materials are heat treated. We began the applications section

---

[6] Forming methods are discussed in Chapter 16.

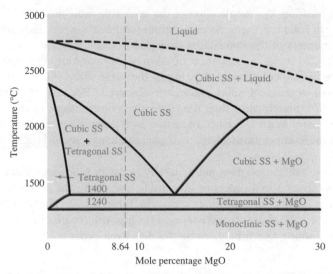

FIGURE 8.3–20   ZrO$_2$-rich end of the ZrO$_2$-MgO equilibrium phase diagram. Note that SS stands for solid solution.

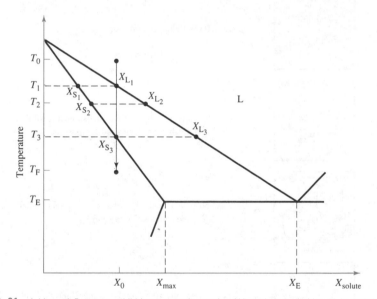

FIGURE 8.3–21   A binary A-B system exhibiting a eutectic reaction. When the composition X_0 is cooled under equilibrium conditions from T_0 to T_f the compositions of the phases present follow the equilibrium phase diagram.

of this chapter with the heat-treating step because we wanted to reinforce the principles of nucleation and growth presented at the beginning of the chapter. In this section we extend the application of kinetics to the solidification and homogenization steps in the production of an alloy.

The compositional changes of the phases present during equilibrium freezing can be obtained directly from a phase diagram. For example, Figure 8.3–21 represents a binary eutectic A-B system. An alloy of composition X_0 at a temperature T_0 is a single-phase liquid with a uniform composition X_0. If the temperature is lowered at an infinitesimally slow rate to T_1, the microstructure consists of two phases, a liquid of composition X_{L_1} and a solid of composition X_{S_1}. Application of the lever rule indicates that an infinitesimally small quantity of solid is present. As equilibrium solidification proceeds, the compositions and relative amounts of the two phases change.

Figure 8.3–22 shows a sequence of sketches illustrating the development of a microstructure during equilibrium freezing of the same alloy. At T_1, after nucleation is complete, small regions of solid phase are observed. Because the process is being conducted at an infinitely slow rate, the respective compositions in the solid and liquid phases are uniform. With a further reduction in temperature, the amount of solid increases while the compositions of the two phases change according to the equilibrium boundaries.

Equilibrium solidification does not generally occur in practice. Diffusion coefficients in the solid are about 3 to 4 orders of magnitude lower than the corresponding values in the liquid at the same temperature, as shown in Table 8.3–2. The slow diffusion in the solid prevents the solid phases from readjusting composition with changes in temperature. Consequently, cooling rate influences the solidification process.

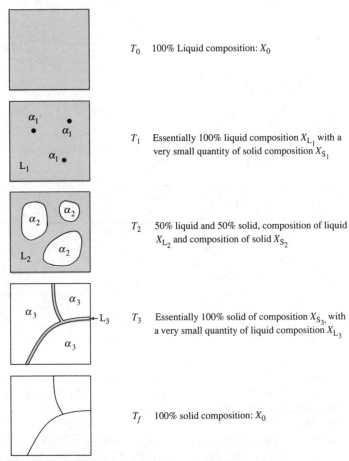

T_0 100% Liquid composition: X_0

T_1 Essentially 100% liquid composition X_{L_1} with a very small quantity of solid composition X_{S_1}

T_2 50% liquid and 50% solid, composition of liquid X_{L_2} and composition of solid X_{S_2}

T_3 Essentially 100% solid of composition X_{S_3}, with a very small quantity of liquid composition X_{L_3}

T_f 100% solid composition: X_0

FIGURE 8.3–22   The development of microstructure occurring during equilibrium solidification of the alloy X_0 shown in Figure 8.3–21.

TABLE 8.3–2   Comparison of approximate solute diffusion coefficients at the respective liquidus temperature.

Variable	Units	Al-Cu	Al-Si	γ Fe–Ni
C_0	wt. %	2	6	10
T_l	°C (K)	656 (929)	624 (897)	1503 (1776)
D_L	m²/s	3×10^{-9}	3×10^{-9}	7.5×10^{-9}
D_S	m²/s	3×10^{-13}	1×10^{-12}	3×10^{-13}

A series of schematics that illustrate the development of microstructures during non-equilibrium solidification are shown in Figure 8.3–23. The construction of the figure is based on the following assumptions:

1. Solid-state diffusion is negligible.
2. Complete mixing occurs in the liquid as the temperature changes.
3. At any temperature, the composition of the solid and liquid in contact with one another at the interface is given by the equilibrium phase diagram.

At the start of solidification $(T = T_1)$, the solid has composition X_{S_1} and the liquid has composition X_{L_1}. When the temperature is changed instantaneously to T_2, the composition of the next element to solidify is X_{S_2} and the composition of the liquid instantaneously changes to X_{L_2} (assumptions 2 and 3). Since solid-state diffusion is not permitted

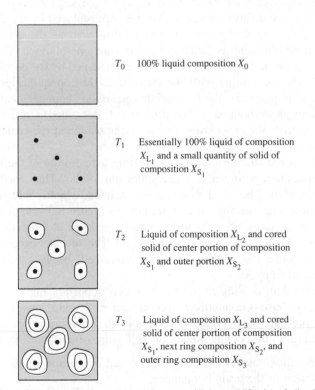

T_0 100% liquid composition X_0

T_1 Essentially 100% liquid of composition X_{L_1} and a small quantity of solid of composition X_{S_1}

T_2 Liquid of composition X_{L_2} and cored solid of center portion of composition X_{S_1} and outer portion X_{S_2}

T_3 Liquid of composition X_{L_3} and cored solid of center portion of composition X_{S_1}, next ring composition X_{S_2}, and outer ring composition X_{S_3}

FIGURE 8.3–23   The development of microstructure during nonequilibrium solidification of alloy X_0 in Figure 8.3–21.

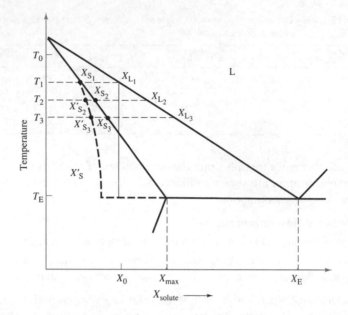

(assumption 1), the composition of the first element remains X_{S_1}. The average composition of the solid is represented by point X'_{S_2} on the diagram shown in Figure 8.3–24. When the temperature is instantaneously decreased to T_3, the composition of the new solid in contact with the liquid of composition X_{L_3} is X_{S_3} and the average solid composition is X'_{S_3}.

If the finite instantaneous temperature changes are replaced by the differential quantity dT, the variation of composition in the solid as well as the average composition of the solid can be obtained as functions of the fraction solidified. The details of this analysis are not presented here, but the result is represented by the dashed line shown on Figure 8.3–24. The dashed line is often referred to as the nonequilibrium solidus boundary. For calculation purposes, the nonequilibrium solidus boundary can be treated as a quasi-equilibrium phase boundary. The process just described is referred to as **coring.** This term comes from the observation that the composition at the center, or core, of the solid grain is different from that at the grain surfaces.

Note that if a cored structure forms, the melting temperature of the last solid to form is much lower than that predicted by the equilibrium solidus. Thus, melting would occur at the grain boundaries below the equilibrium solidus, which could lead to problems during fabrication. For example, with reference to Figure 8.3–24, if a severely cored structure of overall composition X_0 were to be hot-worked at temperature T_E, then the grain boundaries would have a continuous liquid film of eutectic composition. Upon application of pressure, the structure would be "squashed" because the continuous liquid film could not carry any load.

The nonequilibrium cooling model predicts two important microstructural features. First, the primary phase is nonuniform in composition. Second, low-melting temperature phases may be present, which are not predicted by the equilibrium phase diagram. When using the nonequilibrium cooling model, the following general statements can be made:

1. Increasing alloy content increases the amount of the nonequilibrium eutectic phase present at the grain boundaries.
2. In comparing alloy systems, the larger the freezing ranges of the alloy—that is, the larger the difference between the solidus and liquidus temperatures—the more extensive the coring and the more nonequilibrium eutectic material is present.

Figure 8.3–25 shows the microstructures of two aluminum-copper alloys cast under similar conditions. As expected, the alloy containing 2 wt. % Cu contains less eutectic than the alloy with 5 wt. % Cu. Both alloy compositions lie to the left of the maximum solid solubility of copper in α aluminum, so that we expect them to have single-phase microstructures if cooled under equilibrium conditions. However, because solid-state diffusion is so slow, cored microstructures containing nonequilibrium eutectic occur.

The preceding discussion addressed the phenomenon of coring, but not the effect of the solidification rate on the development of the microstructure. In fact, we anticipate some cooling-rate effects. For example, faster cooling rates lead to greater undercooling, which increases the nucleation rate markedly and leads to a finer-grained structure than expected. Figure 8.3–26 contains micrographs of an Al–5 wt. % Cu alloy cast at two solidification rates. All microstructural features appear finer in the sample solidified at the

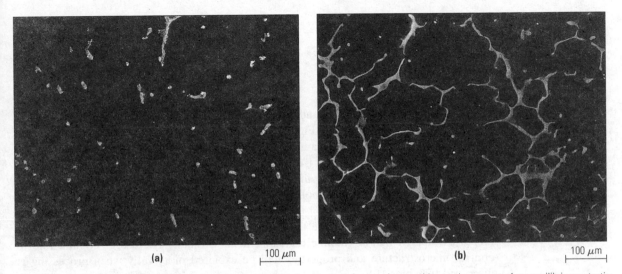

FIGURE 8.3–25   A pair of as-cast aluminum copper alloy samples showing the influence of composition on the amount of nonequilibrium eutectic: **(a)** Al–2 wt. % Cu, and **(b)** Al–5 wt. % Cu.

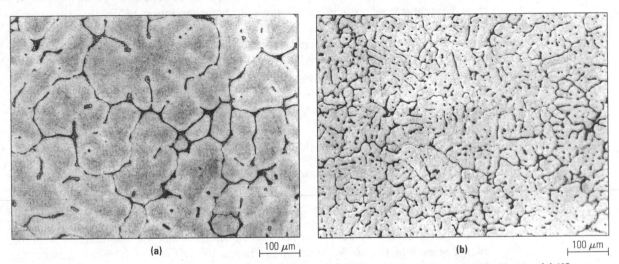

FIGURE 8.3–26   Photomicrographs of an Al–5% Cu alloy showing the influence of the cooling rate on the as-cast microstructure: **(a)** 1°C per second and **(b)** 10°C per second.

FIGURE 8.3–27 The effect of cooling rate on the size of the grain structure in a series of aluminum alloys. *(Source: R. E. Spear and G. R. Gardner, Transactions of the American Foundermen's Society, 71 (1963). Reprinted by permission of the publisher.)*

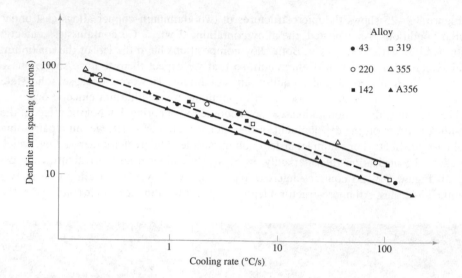

higher rate, including grain size and nonequilibrium eutectic size. However, both micrographs show evidence of coring. Figure 8.3–27 shows a plot of the effect of cooling rate on the microstructural scale for a number of aluminum alloys.

Not all materials are affected like aluminum alloys by cooling rate. For example, as discussed in Chapter 6, when certain oxides or polymers are cooled, instead of crystallizing, they remain amorphous at cooling rates that would stimulate crystallization in most metallic systems. The retention of the nonequilibrium amorphous structure is related to the viscosity of the melt. As the temperature drops, the viscosity of the liquid phase increases rapidly, making atomic rearrangements difficult. The use of cooling rate to control microstructure and properties will be explored more fully when processing is discussed (Chapter 16).

After solidification, ingot composition is nonuniform due to coring and the formation of nonequilibrium second-phase particles, often of near-eutectic composition. The ingots must be heat-treated in a process known as homogenization to dissolve the soluble second-phase particles and eliminate coring prior to shaping and finishing operations. Since homogenization is diffusion-controlled process, increasing temperature shortens the time it takes to reach a desired degree of homogenization. The homogenization time is also influenced by the distance over which diffusion must occur. Since the greatest compositional variation occurs from the center of a grain to the grain boundary, the grain size is the important spatial variable. Thus, decreasing the grain size of the alloy shortens the homogenization time. As will be discussed in Chapter 16, hot working, defined as plastically deforming the alloy at elevated temperatures, has several beneficial effects, including decreasing the grain size of the alloy and increasing the diffusion rate, through both an increase in temperature and the generation of additional vacancies.

8.3.4 Recovery and Recrystallization Processes

When a metal is cold-worked, that is, plastically deformed at low temperature, most of the energy goes into plastic deformation to change the shape and into heat generation. Depending upon the material, however, a small portion of the energy, up to ~5%, remains stored in the material. The stored energy is mainly in the form of elastic energy in the

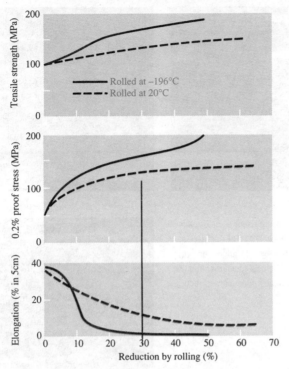

FIGURE 8.3–28   The influence of % cold working and cold-working temperature on the properties of an aluminum 1100 alloy: **(a)** tensile strength, and **(b)** elongation (ductility).   *(Source: K. van Horn, ed.,* Aluminum, *Vol. I, 1967, ASM International, Materials Park, OH. Reprinted by permission of the publisher.)*

strain fields surrounding dislocations and point defects generated during the cold-work process. The presence of these defects in concentrations far above those in the annealed state significantly affects the material properties.[7] These concepts were developed in Chapters 4 and 5. Some of the property changes resulting from the presence of these defects can be understood qualitatively.

Figure 8.3–28 shows the variation in strength and ductility of commercially pure aluminum as a function of percent cold work by rolling. The defect density increases with the amount of cold work. Increasing the amount of deformation increases tensile strength and decreases ductility as measured by elongation to failure. Furthermore, decreasing deformation temperature increases strength and lowers ductility. Thus, the mechanical properties plotted in Figure 8.3–28 are a function of the amount of deformation and the deformation temperature.

The effect of these two parameters on properties can be understood on the basis of knowledge of their contributions to the dislocation density in a rolled sheet. Figure 8.3–29 shows a series of optical micrographs illustrating the effect of cold rolling on the microstructure of aluminum alloy 1100 (this is the industrial designation for commercially pure aluminum). As the amount of deformation increases, grains are flattened in the transverse (thickness) direction and elongated in the direction parallel to rolling. To understand the influence of the rolling process on the mechanical properties of alloys, magnifications higher than those possible with an optical microscope are necessary.

---

[7] Specific property changes resulting from the change in defect concentrations are discussed in Chapters 9 through 13.

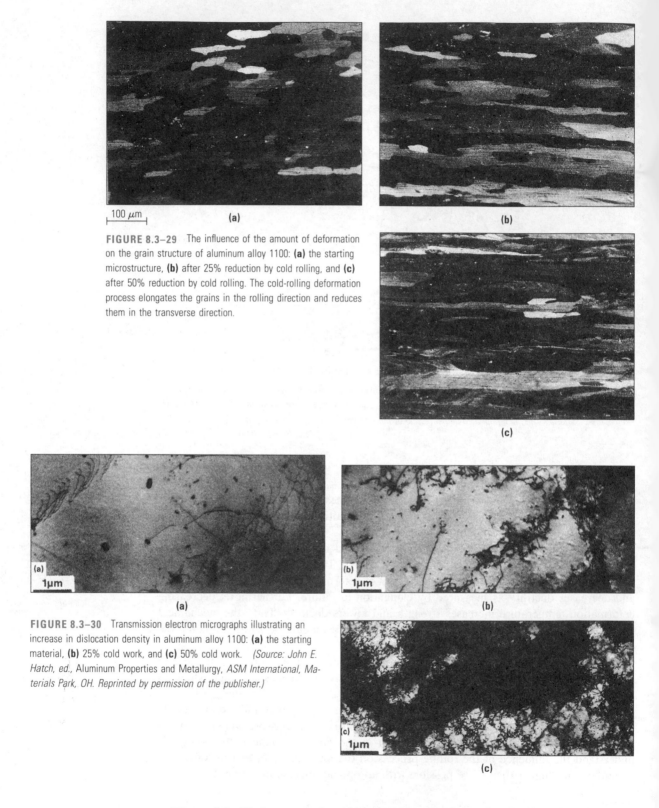

FIGURE 8.3–29   The influence of the amount of deformation on the grain structure of aluminum alloy 1100: **(a)** the starting microstructure, **(b)** after 25% reduction by cold rolling, and **(c)** after 50% reduction by cold rolling. The cold-rolling deformation process elongates the grains in the rolling direction and reduces them in the transverse direction.

FIGURE 8.3–30   Transmission electron micrographs illustrating an increase in dislocation density in aluminum alloy 1100: **(a)** the starting material, **(b)** 25% cold work, and **(c)** 50% cold work.   *(Source: John E. Hatch, ed.,* Aluminum Properties and Metallurgy, *ASM International, Materials Park, OH. Reprinted by permission of the publisher.)*

Figure 8.3–30 shows a series of transmission electron micrographs taken from aluminum alloy 1100 rolled to different levels of deformation. As the amount of deformation increases, the dislocation density increases. As derived in Section 5.5.2, the flow stress follows the relationship:

$$\tau_{flow} = \tau_0 + k\sqrt{\rho}$$

Thus, the increase in strength with increasing cold work is easily explained. What about the influence of deformation temperature? Since dislocations, like all other defects, possess excess energy, there is a thermodynamic driving force for the elimination of dislocations in order to reduce the total free energy of the system. Since many of the mechanisms by which dislocations can be eliminated involve diffusion, increasing the deformation temperature will enhance diffusion and lead to the elimination of a large number of dislocations. Hence, rolling at 20°C results in lower dislocation densities and lower strength than a similar deformation process conducted at −196°C.

Although some applications require a particular strength, ductility is also an important property. As a general rule, when the strength of a material is increased, the ductility is reduced. We already know that cold working increases strength, but how can we increase the ductility of an alloy? To regain some of the ductility without a significant loss in tensile strength, rolled alloys may be given a short heat treatment designed to reduce the defect concentration. Such a heat treatment is known as an **anneal,** and it takes place in two stages: recovery and recrystallization. The **recovery** process occurs at temperatures roughly equal to one-half the absolute melting temperature of the alloy. Thus, high-melting temperature alloys require high annealing temperatures. For aluminum alloy 1100, the appropriate temperature range for recovery is 150–250°C. During recovery, some dislocations are eliminated, and those remaining are rearranged into arrays that minimize the excess energy of the crystal. The influence of the recovery process on the tensile strength of aluminum alloy 1100 is illustrated in Figure 8.3–31a over the

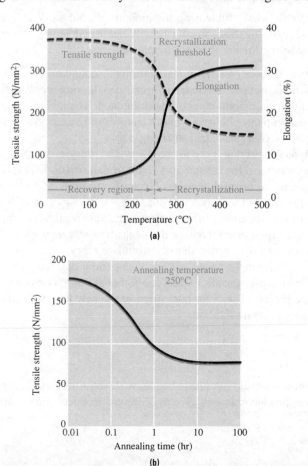

(a)

(b)

FIGURE 8.3–31 The change in properties resulting from annealing. **(a)** Influence of annealing temperature on the tensile strength and elongation to failure (ductility) of deformed aluminum alloy 1100. The annealing time was five minutes. **(b)** Influence of annealing time at 250°C on the tensile strength of aluminum alloy 1100. *(Source: D. Altenpohl, Aluminum Viewed from Within, Springer-Verlag, Düsseldorf, 1982, p. 134. Reprinted by permission of the publisher.)*

temperature range 150–250°C. During recovery, there is a gradual change in the magnitude of both ductility and tensile strength with increasing temperature.

When the annealing temperature is further increased, the rate of change of the particular property increases. A new mechanism of microstructural change, termed **recrystallization,** occurs in this region (250–350°C in Figure 8.3–31a). This process consists of the replacement of grains containing high concentrations of dislocations with new grains containing much lower dislocation densities.

Since recovery and recrystallization are both diffusion-controlled processes, increasing the time while holding the temperature constant has a similar but less powerful effect than increasing the temperature while holding the treatment time constant. (Recall that the effective diffusion distance is proportional to $\sqrt{Dt}$.) Thus, while Figure 8.3–31a shows the change in properties as a function of annealing temperature at constant time, Figure 8.3–31b shows similar data as a function of time at constant temperature.

The recrystallization temperature is influenced by the following factors:

1. *Alloy additions.* For a single-phase alloy, increasing the amount of solute increases the resistance to recrystallization. The same is true for two-phase alloys when the particles are small (diameters < 0.3 μm) and closely spaced (interparticle spacings < 1 μm).

2. *Annealing time.* The higher the recrystallization temperature, the shorter the annealing time for a given degree of recrystallization. For example, heavily deformed pure aluminum annealed at 500°C recrystallizes in a few seconds, that annealed at 380°C recrystallizes in a few minutes, and that annealed at 280°C recrystallizes in a few hours.

3. *Amount of cold work.* Increasing the amount of cold work increases the driving force for recrystallization, since a heavily deformed sample contains more stored energy necessary for the recrystallization process.

The processes of recovery and recrystallization of a cold-worked sheet represent *structural transformations,* not true phase transformations. During recovery and recrystallization, only dislocations and point defects are eliminated. The loss of these defects does not constitute a change in phase; however, many features of the recrystallization process can be described in terms of a phase transformation.

The thermodynamic driving force for recovery and recrystallization is associated with the excess energy stored in the crystal as a result of deformation. Therefore, unlike polymorphic transformations or solidification of a pure substance, ΔG^* (the barrier to the nucleation process) associated with recrystallization is not a function of temperature. Thus, rather than developing a maximum in the nucleation rate at a particular temperature, increasing the annealing temperature increases the nucleation rate as well as the growth rate.

Since regions of high dislocation density stimulate recrystallization, nucleation is heterogeneous. In rolled single-phase alloys, dislocation densities are higher at grain boundaries, while in two-phase alloys they are higher at interphase boundaries. Therefore, these sites are the preferred sites for nucleation of the new low-dislocation density grains that develop during recrystallization.

8.3.5 Sintering

Fabrication of most metallic materials begins with solidification. However, some metal and a few polymer products, as well as many ceramic products, are produced from powder form. **Sintering** is a process by which powder particles are densified, at elevated temperature, either with or without pressure, to form a nearly fully dense solid. For products that require strength and ductility, it is necessary to eliminate as much of the porosity as possible. If transparency is essential, the number of pores must be yet further reduced and their size controlled so they do not scatter light. Figure 8.3–32 shows the effect of the sintering process on the transparency of Al_2O_3.

While many products benefit from a minimization of porosity, there are some applications of powder products that require a structure with retained interconnected porosity.

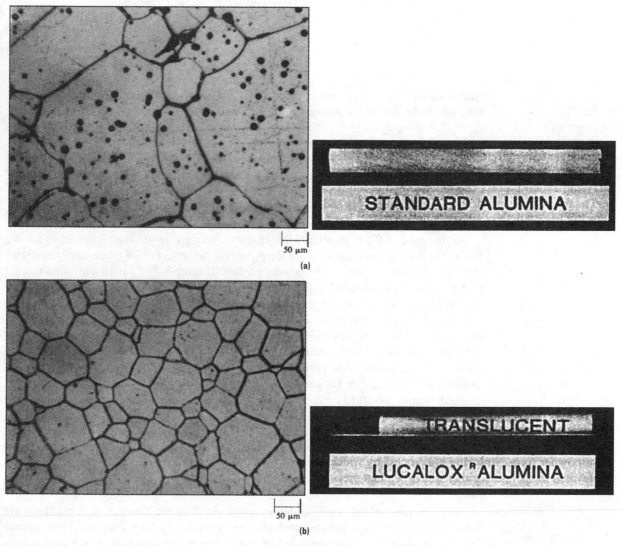

FIGURE 8.3–32 The presence of pores with dimensions in the range 30–70 nm renders a ceramic opaque: **(a)** a sample sintered in a manner yielding pores of the critical size, and **(b)** a sample containing additives sintered in a manner that accelerates the densification rate so that pores in the critical size range are eliminated. *(Source: Reprinted with permission from the General Electric Company. LUCALOX is a registered trademark of General Electric Company.)*

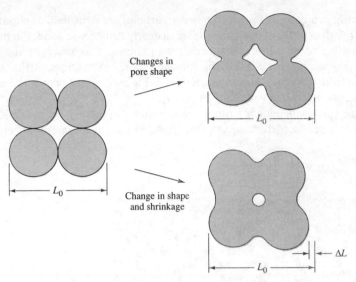

Changes in
pore shape

Change in shape
and shrinkage

L_0

L_0

L_0

ΔL

FIGURE 8.3–33 Changes in pore shape during sintering. Initially there may be only a change in the shape of the
pore. With time, not only does the pore size decrease, but there is shrinkage associated with sintering.

Consider, for example, the sintered aluminum alloy sheet developed by the Koreak
company and Chungnam National University in Korea. The material has structural
integrity, yet the pore morphology acts as a highly efficient sound-absorbing material.
Foams that are used as through-filters also require an interconnected pore structure.
Whatever the final structure, an understanding of the factors controlling the sintering
process is essential in order to create the desired properties of a solid.

As discussed at the beginning of this chapter, every phase transformation has a driving
force. The driving force for the sintering process is the reduction of free energy associated
with the high surface area of the powder particles. Figure 8.3–33 illustrates the sintering
process using a two-dimensional representation with four equal-diameter circles in contact with one another. During the initial stages of sintering there is a change in the pore
shape without any change in the area (volume) of the pore. Note that this process
decreases the free energy of the system by reducing the excess energy associated with the
solid/pore interface. During the second stage of sintering the pore volume is reduced and
there is an associated reduction in the macroscopic sample volume, since the average
center-to-center distance between adjacent particles decreases.

To develop a model to predict the effect of temperature on the sintering process, the
sintering mechanism must be understood. As shown in Figure 8.3–34, sintering requires
mass transport so that a "neck" or "bridge" can be formed between adjacent powder
particles. Mass transport can occur along several paths—along the surface of the particles, through the interior of the particles, or through the vapor phase. Since all of these
processes involve diffusion, increasing the sintering temperature increases the sintering
rate.

While increasing the sintering temperature shortens the fabrication time, there are
some negative effects associated with the use of extremely high sintering temperatures.
One disadvantage is an increase in the grain size of the material, which may have a
negative impact on the strength or optical properties of the part. Thus, it is important to
identify an appropriate sintering schedule to achieve rapid results at moderate

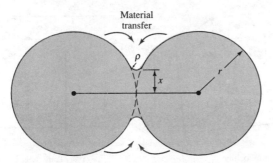

FIGURE 8.3–34   The formation of a "neck" between two adjacent particles during the sintering process requires the diffusion of atoms into the region of the neck.   *(Source: W. D. Kingery et al., Introduction to Ceramics. Copyright © 1960 by John Wiley & Sons, Inc. Reprinted by permission of John Wiley & Sons, Inc.)*

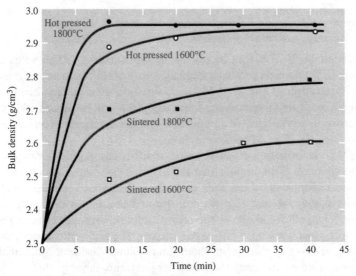

FIGURE 8.3–35   Densification of beryllia by sintering and hot pressing at 130 atmospheres.   *(Source: W. D. Kingery et al., Introduction to Ceramics, p. 502. Copyright © 1960 by John Wiley & Sons, Inc. Reprinted by permission of John Wiley & Sons, Inc.)*

temperatures. This can be accomplished by applying pressure to the part during sintering. In addition to increasing the sintering rate at a constant temperature, pressure aids in the densification process and reduces the amount of porosity in the final product. Figure 8.3–35 shows the influence of temperature, without external pressure, on sintering kinetics. We will return to the issues of sintering in Chapter 16, which deals with processing methods for engineering materials.

8.3.6 Martensitic (Displacive) Transformations in Zirconia

At normal pressures, zirconia (ZrO_2) exhibits three solid-state polymorphs: the monoclinic form stable up to 1170°C, the tetragonal form stable up to 2370°C, and the cubic form stable up to the melting point of 2680°C. The tetragonal $\rightarrow$ monoclinic (T $\rightarrow$ M) transformation is a transformation of particular commercial significance because of the

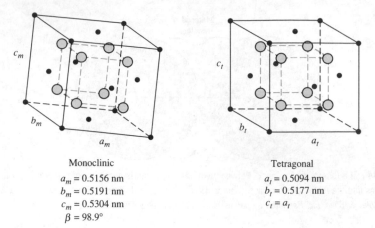

Monoclinic

$a_m = 0.5156$ nm
$b_m = 0.5191$ nm
$c_m = 0.5304$ nm
$\beta = 98.9°$

Tetragonal

$a_t = 0.5094$ nm
$b_t = 0.5177$ nm
$c_t = a_t$

FIGURE 8.3–36 Crystal structures of monoclinic and tetragonal ZrO_2 with the lattice parameters at 950°C for both phases. *(Source: R. Stevens, Zirconia and Zirconia Ceramics, MEL Chemicals (part of Alcan Chemicals, Ltd.) No. 113, 2nd ed., p. 12. Reprinted by permission of the publisher.)*

large volume change (3–5%) that occurs during transformation. This volume change can impose severe limitations on products fabricated from ZrO_2, such as crucibles and refractories. The implication of the T → M transformation as a mechanism to improve the fracture toughness of Al_2O_3 containing tetragonal ZrO_2 will be presented in Chapter 9. In order to prepare us to describe this important toughening mechanism, let us discuss the kinetic aspects of the T → M transformation.

The T → M transformation is a martensitic or diffusionless phase change similar to that which occurs in steels. This implies that the atoms do not move past their neighbors by diffusion but are displaced by less than one atomic distance to initiate the phase change. Consequently, it may be said that the transformed phase is a distorted version of the parent phase. Figure 8.3–36 shows the relative atomic arrangements in the two structures along with their associated lattice parameters. The figure illustrates that the T → M tranformation involves a change in shape and volume inside a solid matrix. As the fraction of transformed material increases, large strains develop that, for diffusionless transformations, are not relieved by atomic migration. Thus, the strain opposes the transformation. Because of this increasing strain energy, a reduction in temperature is required to advance the reaction. Figure 8.3–37 shows the fraction of tetragonal ZrO_2 as a function of cooling.

An important question arises as to whether the development of a polymorphic phase change over a temperature range violates the phase rule. Recall that the phase rule for conditions of constant pressure is $F = C - P + 1$. For a one-component system with two phases in equilibrium, $F = 0$, which implies that there is a fixed temperature at which this phase change can occur. However, the large strains developed by the T → M transition are accompanied by large local stresses, and hence, the local pressure is not fixed at one atmosphere. Instead, pressure is indeed a variable, and the form of the phase rule that must be used is $F = C - P + 2$. When this is done, $F = 1$ and the transformation may in fact occur over a temperature range. Thus, the consideration of pressure is the key to understanding why this and other similar transformations occur over a range of temperatures.

Another interesting observation regarding martensitic transformation in ZrO_2 is that it is reversible, although it occurs over a different temperature range depending on direction.

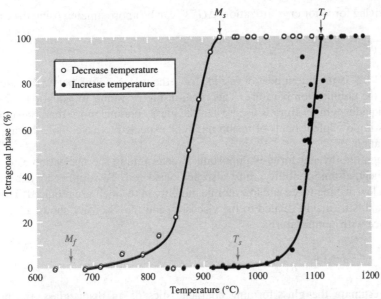

FIGURE 8.3-37 Tetragonal-monoclinic transformation in ZrO_2. *(Source: G. M. Wolten, Journal of the American Ceramic Society 46 (1963), p. 418. Reprinted by permission of the American Ceramic Society.)*

The reverse reaction is also martensitic (displacive) in nature. These properties are not unique to ZrO_2 but occur in other systems such as the indium-thallium system (18% Tl) and the iron-nickel system (29.5% Ni).

8.3.7 Devitrification of an Oxide Glass

Most inorganic glasses can be made to transform from a noncrystalline state to one that is crystalline by an appropriate high-temperature heat treatment. This process, termed **devitrification,** is ordinarily avoided because a devitrified glass is not transparent. However, there are numerous examples of uniformly crystallized glasses, or **glass-ceramics,** that were first developed at Corning Glass Works. Such materials are made by heat-treating particular glass compositions to give a fine, uniform dispersion of low–thermal properties such as high strength and impact resistance, low coefficient of thermal expansion, and a range of optical properties. Table 8.3–3 includes a number of commercially available products with their properties.

TABLE 8.3-3 Properties and uses of some commercial glass-ceramics.

Glass type and company	System	Properties	Uses
Corning 9606	$2MgO \cdot 2Al_2O_3$ (Cordierite)	Low expansion, transparent to radar	Radomes
Corning 9608	β-Spodumene	Low expansion, low chemical reactivity	Cookware
Owens-Illinois Cer-vit	β-Quartz	Very low expansion	Telescope mirrors
General Electric Re-X	Li_2O-$2SiO_2$	Low electrical conduction, high strength	Insulators

The driving force for crystallization, ΔG^{crys}, can be approximated using the expression:

$$\Delta G^{crys} = \left(\frac{\Delta H^{fusion}}{T_{liquidus}}\right) \Delta T \tag{8.3–2}$$

where ΔH^{fusion} is the latent heat of fusion, ΔT is the undercooling below the liquidus, and $T_{liquidus}$ is the liquidus temperature. This relationship is similar to Equation 8.2–4 except that the liquidus temperature is used when the glass contains more than one component. Systems with low latent heats of fusion might be expected to have a low driving force for crystallization.

Although the driving force is important in determining the nucleation rate, a second term, namely, atomic mobility, must also be considered. To suppress crystallization on cooling, a low driving force and low atomic mobility in the melt are required. The mobility in the liquid is inversely related to the viscosity, and the viscosity shows an exponential dependence with temperature:

$$\frac{1}{\eta} = C_0 \exp\left(-\frac{Q}{RT}\right) \tag{8.3–3}$$

Thus, to estimate the glass-forming characteristics of an oxide glass, the product of $(\Delta H^{fusion}/T_{liquidus})$ and $(1/\eta)$ has been used as an index for the tendency to form glasses on cooling. Note that a low index indicates a low driving force for crystallization and a high viscosity. Thus, a low index implies that it is comparatively easy to form a glass in the corresponding system. Table 8.3–4 summarizes the results for several oxide glasses and crystalline materials.

To produce a crystalline product, nucleating agents such as TiO_2 are often added to stimulate crystallization throughout the glass. The heat treatment of glasses containing nucleating agents is usually carried out in two stages. First, the glass is given a nucleation treatment in which a large number of nuclei per unit volume are produced. Once a suitable number density of nuclei are present, the glass is heated to a higher temperature to stimulate rapid growth of the crystalline phase. The appropriate heat-treating times and temperatures must be established for each glass composition.

8.3.8 Crystallization of Polymers

As discussed in Chapter 6, some polymers form extensive crystalline regions if cooled slowly enough that there is sufficient time for the relatively complex building blocks to arrange themselves into a regular repeating pattern. In fact, the kinetics of crystal

TABLE 8.3–4 Factors affecting glass-forming ability.

Composition	T_{mp} (°C)	$\Delta H_f/T_{mp}$ [(cal/mol)/K]	$(1/\eta)_{mp}$ (poise^{-1})	$(\Delta H_f/T_{mp}) \times (1/\eta)_{mp}$	Comments
B_2O_3	450	7.3	2×10^{-5}	1.5×10^{-4}	Good glass former
SiO_2	1713	1.1	1×10^{-6}	1.1×10^{-6}	Good glass former
$Na_2Si_2O_5$	874	7.4	5×10^{-4}	3.7×10^{-3}	Good glass former
Na_2SiO_3	1088	9.2	5×10^{-3}	4.5×10^{-2}	Poor glass former
$CaSiO_3$	1544	7.4	10^{-1}	0.74	Very difficult to form as glass
NaCl	800	6.9	50	345	Not a glass former

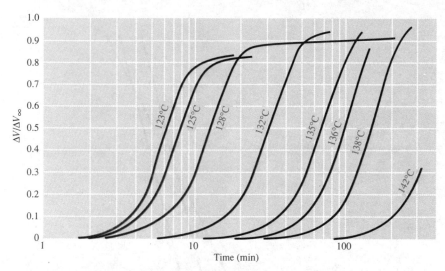

FIGURE 8.3–38 Volume changes during crystallization of isotactic polypropylene at various temperatures.

formation in polymers is very similar to that for metals and ceramics. One of the reasons we are interested in polymer crystallization kinetics is that the strength of polymers with a high degree of crystallinity is often limited by the size of the voids that develop between neighboring spherulites during the crystallization process. The size of these voids scales directly with the size of the spherulites. In turn, spherulite size is a function of the nucleation and growth rates, both of which vary with the transformation temperature. Thus, understanding the kinetics of polymer crystal formation is essential for selecting the processing temperature that will maximize the strength of a crystalline polymer.

As shown in Figure 8.3–38, by monitoring the change in volume of the polymer melt as it begins to crystallize, one can investigate the kinetics of spherulite formation. These data can be modeled using Equation 8.2–12, since the plot displays the characteristic sigmoidal shape. That is:

$$X = 1 - \exp[-(kt)^n]$$

where X is the fraction of polymer crystallized, k is an effective rate constant, t is time, and n is a constant characteristic of the crystallization mechanism. Taking the natural logarithm of both sides of this equation yields:

$$\ln(1 - X) = -(kt)^n \tag{8.3–4}$$

Taking the natural log of each side of Equation 8.3–4, after multiplying by negative 1, yields:

$$\ln[-\ln(1 - X)] = n \ln(k) + n \ln(t) \tag{8.3–5}$$

Thus, a plot of $\ln[-\ln(1 - X)]$ versus $\ln(t)$ should yield a straight line with slope n and intercept $n \ln(k)$. Figure 8.3–39 shows this type of plot for the crystallization of poly(ethylene terephthalate), which has a melting temperature of 265°C. Note that the slope changes from a value of $\sim$2 at low crystallization temperature to a value of $\sim$4 at higher transformation temperatures, indicating a change in the crystallization mechanism with temperature. In fact, the change in n indicates a change in the growth geometry.

FIGURE 8.3–39

Effect of transformation temperature on the mechanism of crystallization in poly(ethylene terephthalate).

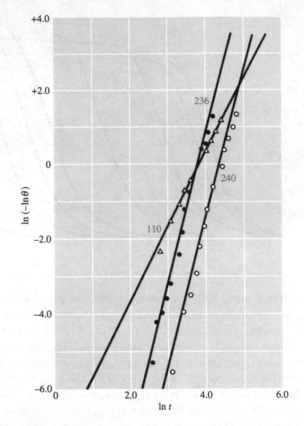

FIGURE 8.3–40

Growth rate of spherulites of poly(ethylene adipate) at various temperatures.

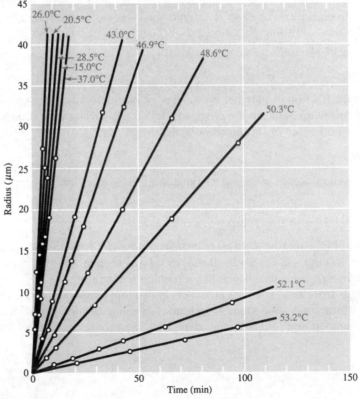

Temperature also has an effect on the growth rate of the spherulites, just as in the case of diffusion-based transformation in other classes of crystals. The maximum growth rate of spherulites occurs at intermediate temperatures (i.e., typical C-curve kinetics). Figure 8.3–40 shows the relationship between the radius of poly(ethylene adipate) spherulites and time for a series of transformation temperatures. The maximum growth rate occurs at 26°C. In summary, the nucleation and growth of polymer spherulites is governed by the same kinetic principles that apply to all other classes of diffusion-based phase transformations.

SUMMARY

This chapter introduced the basic concepts necessary to understand the development of microstructure through a variety of phase transformations and solid-state reactions occurring in some important engineering materials. We have seen that there are two general kinds of phase transformations: diffusional and martensitic. Diffusional phase transformations depend upon both time and temperature and generally involve two processes, nucleation and growth. Nucleation initiates a phase transformation and requires the formation of a stable particle called a critical nucleus. It can occur either homogeneously, throughout the material; or heterogeneously, at preferred locations such as mold walls, grain boundaries, dislocations, and precipitate particles. Heterogeneous nucleation lowers the free-energy barrier and results in more rapid nucleation just below the transformation temperature.

For liquid-to-solid transformations, nucleation is virtually always heterogeneous. On the other hand, for solid-state transformations, nucleation is usually heterogeneous but may, in some instances, be homogeneous. The nucleation rate depends on both the concentration of critical-size nuclei and atomic mobility, and thus exhibits a maximum at some temperature below the equilibrium temperature. This means that nucleation follows typical C-curve kinetics such that the minimum time required for a given amount of nucleation occurs at an intermediate temperature.

The growth rate depends on the temperature and the diffusion distance. At high temperatures, even though the diffusion rate is large, the growth rate is low because the diffusion distance is generally quite large. At low temperatures the growth rate is low because the diffusion rate is low. The growth rate, similar to the nucleation rate, exhibits a maximum for some intermediate temperature below the equilibrium temperature. Thus, C-curve kinetics are also observed for the growth portion of a phase transformation. The overall transformation rate is proportional to the nucleation and growth rates and exhibits C-curve behavior as well.

At a given temperature, the fraction transformed depends on time and follows a sigmoidal relationship in which the initial and final rates are low and the intermediate rate is a maximum. Nucleation and growth processes are involved in all commercial aging treatments in which small, second-phase particles form from a supersaturated solid solution. The precipitation of second-phase particles is responsible for the great variety of mechanical and physical properties that can be obtained in practical alloy systems.

Diffusionless transformations can also occur. They occur by processes involving both shear displacements and changes in volume. Perhaps the most important commercial example of such a transformation occurs in the formation of martensite, which is the starting phase for the majority of heat treatments in steel. Martensitic transformations also occur in various ceramic systems and can have a profound effect on mechanical properties. An example of such a system is the tetragonal-to-monoclinic transformation in ZrO_2.

Although a limited number of examples have been provided, our emphasis on fundamentals makes it a reasonably straightforward matter to extend these principles to other systems.

KEY TERMS

aging	devitrification	Jominy end-quench test	recrystallization
anneal	displacive	kinetics	semicoherent interface
bainite	glass-ceramics	martensite	sintering
C curve	hardenability	nucleation	spheroidite
coherent interface	heterogeneous nucleation	pearlite	supersaturation
coring	homogeneous nucleation	precipitation hardening	tempering
critical radius	incoherent interface	recovery	

HOMEWORK PROBLEMS

SECTION 8.2:
Fundamental Aspects of Structural Transformations

1. When a phase transformation occurs, such as a liquid phase transforming to a solid below its melting temperature, what are the two steps involved in the process? Briefly describe each.

2. We presented a derivation in Section 8.2.3 showing that the barrier for nucleation, ΔG^*, decreases with increasing undercooling following the proportionality

$$\Delta G^* \propto \frac{1}{(\Delta T)^2}$$

By starting with an expression for the free energy of a distribution of spherical particles of radius r, derive Equation 8.2–9a. Explain each step in the derivation. Explain any assumptions that are made.

3. Explain the simultaneous influence that undercooling has on the barrier to nucleation and the atomic rearrangements necessary to initiate the transformation. Show how these competing effects lead to classical C-curve behavior in the nucleation of diffusional transformations.

4. Explain how the value of interfacial energy between the parent phase and the transforming phase affects the critical radius and the barrier to nucleation.

5. Compare homogeneous nucleation and heterogeneous nucleation.

6. What is the difference between (a) coherent, (b) partially coherent (semicoherent), and (c) incoherent interfaces?

7. How does interfacial energy vary with coherency?

8. Based upon your answer to Problem 7, explain how the probability for heterogeneous nucleation changes as the type of interface changes from coherent to partially coherent to incoherent.

9. Figure 5.3–5 contains a schematic illustration of a twin boundary in a crystal. From the point of view of coherency, what is the nature of the type of twin boundary illustrated in the figure? Comment on the relative energy of a twin boundary compared with a random grain boundary in a polycrystalline material.

10. In certain nickel-base superalloys, a second phase can precipitate coherently from the matrix during aging because the lattice parameters of the two phases are very close and both

phases are cubic. For a coherent precipitate in this system, what is the most likely relationship between the crystallographic axes in the matrix phase and that of the precipitate? Explain, using sketches.

11. The transition precipitate $\gamma'(Al_2Ag)$ in the aluminum-silver system is a hexagonal close-packed structure. The a axis of the HCP phase is 0.2858 nm, and the c axis is 0.4607 nm. What crystallographic plane and direction in the aluminum matrix defines the coherent interface between matrix and precipitate?

12. The expression: $X = 1 - \exp[-(kt)^n]$, Equation 8.2–12 in the text, is a powerful empirical function that is useful in describing the kinetics of diffusional transformations. In the equation, X is the fraction transformed, k is a rate constant having units of reciprocal time, t is time, and n is a unitless constant. Sketch the behavior of this function over a range of times that demonstrate why this expression is useful for describing microstructural changes like recrystallization, or the decomposition of austenite to form pearlite.

13. The values of n and k for the decomposition of a particular steel have been investigated by following the fraction transformed versus time. From the experimental data, values for n and k were determined at two temperatures. Plot the fraction transformed as a function of time at 400 and 360°C using the data in the following table:

Decomposition temperature (°C)	n	k (s^{-1})
400	2.0	0.085
360	2.0	0.028

14. Explain how the data from Problem 13 could be used to determine the location of the start and finish times for the transformation.

15. When the fraction transformed as a function of the logarithm of time, plotted at two different temperatures, results in two curves that are parallel (same n but different k), the mechanisms for the transformation are the same. Sometimes we refer to the process as being *isokinetic*. For the kinetic data shown below, plot fraction transformed versus time for the two temperatures. Does the process appear to be isokinetic?

Isothermal transformation temperature 415°C		Isothermal transformation temperature 375°C	
Time (s)	Fraction transformed	Time (s)	Fraction transformed
2.0	0.032	4.0	0.025
2.5	0.048	5.0	0.037
3.0	0.072	7.5	0.084
4.0	0.129	9.0	0.124
4.5	0.157	10.0	0.151
5.0	0.189	13.5	0.262
6.0	0.272	16.5	0.369
9.0	0.518	20.0	0.502
11.0	0.669	24.5	0.653
13.5	0.813	30.0	0.799
16.5	0.921	36.5	0.912
		40.5	0.950

16. The value of k in Equation 8.2–12, for a set of kinetic data, can be determined by noting:

$$X = 1 - \exp[-(kt)^n]$$

When $kt = 1$,

$$X = 1 - \exp(-1) = 1 - \frac{1}{e} = 0.632$$

What are the values for the rate constants for the data in Problem 15?

17. When the mechanisms controlling a particular transformation are independent of temperature, we can define an empirical activation energy Q for the process by noting:

$$k = A \exp\left(-\frac{Q}{RT}\right)$$

Determine the empirical activation energy for the data given in Problem 15.

18. The value of n from a set of data can be found by noting:

$$X = 1 - \exp[-(kt)^n]$$
$$1 - X = \exp[-(kt)^n]$$
$$\ln(1 - X) = -(kt)^n$$
$$\ln\left[\frac{1}{(1-X)}\right] = (kt)^n$$
$$\ln\ln\left[\frac{1}{(1-X)}\right] = n \ln t + n \ln k$$

Determine n at the two temperatures from the data in Problem 15.

19. In a diffusion-controlled transformation, the time t at different temperatures to yield the same fraction transformed X is as follows:

T (°C)	t (h)
25	24
100	1
150	0.25

Calculate the empirical activation energy for the transformation process.

20. Using the data from Problem 19, calculate the time required to yield the same fraction transformed at 125°C.

21. The precipitation of carbides in certain steels can increase their strength. Shown below are data relating time to reach peak strength and the isothermal hold time. From the data, determine the activation energy for the precipitation process. Compare your results with the activation energy for diffusion of carbon in BCC iron. Explain why the similarities between the two activation energies are not too surprising.

T (°C)	Time to Peak Hardness (min.)
353	0.9
375	3.0
400	4.0
425	10.0
450	20.0

22. Sketch an isothermal transformation (IT) diagram for a plain carbon eutectoid steel. Label the various decomposition regions on the diagram.

23. Sketch a section of the Fe-Fe$_3$C diagram over the composition range 0 to 2 wt. % carbon and over the temperature range from 900°C to room temperature. Label the phase fields and compare the information that can be extracted from the phase diagram and the IT diagram.

24. Explain experimentally how you would determine an IT diagram for a particular steel.

25. Based upon the methods you outlined in Problem 24, explain the limitations of using such a diagram. In particular, explain why the application of an IT diagram is restricted and may not be applied directly to the production environment.

26. In our discussions in Chapter 7, we introduced the concept of a phase, and in this chapter we have been concerned with the microstructure of an alloy and ultimately how a particular microstructure affects properties. Explain the difference between the concept of a phase and the phases that are present, and the microstructure of the material.

27. Using the concepts associated with *phases* and *microstructure,* explain what is meant by the terms:
    a. Austenite
    b. Ferrite
    c. Pearlite
    d. Bainite
    e. Martensite
    f. Cementite
    g. Spheroidite

Problems 28–31 can be solved using Figures 8.3–6 and 8.3–11.

28. Thin specimens of a plain carbon steel having eutectoid composition are held at 800°C and have been at that temperature long enough to have achieved a complete and homogeneous austenitic structure. Describe the phases present and the microstructures that would occur using the quench paths given below.
    a. Instantaneous quench to 650°C; hold at that temperature for 200 seconds, and quench to room temperature.
    b. Instantaneous quench at 300°C; hold for 1000 seconds, and quench to room temperature.
    c. Instantaneous quench to room temperature.
    d. Instantaneous quench to 500°C; hold for 3 seconds, and quench to room temperature.

29. Using the same initial conditions outlined in Problem 28, describe the phases present and the microstructures that would occur using the quench paths given below.
    a. Instantaneous quench to 650°C; hold for 15 seconds, and quench to room temperature.
    b. Instantaneous quench to 500°C; hold for 60 seconds, and quench to room temperature.
    c. Instantaneous quench to 170°C; hold for 100 seconds, and quench to room temperature.
    d. Instantaneous quench to 170°C; hold at that temperature.

30. Briefly describe a heat treatment that would result in the following microstructure for a plain carbon eutectoid steel:
    a. Coarse pearlite
    b. Bainite
    c. A 50-50 mixture of coarse pearlite and bainite
    d. A 50-50 mixture of bainite and martensite

31. Describe the heat treatments that would be required to transform the following microstructure of a eutectoid steel from one microstructure to another:
    a. Pearlite to bainite
    b. Bainite to pearlite

SECTION 8.3:
Applications to
Engineering Materials

    c. Coarse pearlite to a mixture of pearlite and bainite

    d. Martensite to pearlite

32. Sketch an IT diagram for a hypoeutectoid, eutectoid, and hypereutectoid alloy. Compare the decomposition regions on the three diagrams with respect to proeutectoid phases and the martensite start and finish temperatures.

Shown in Figure HP8.1 are IT diagrams for a hypoeutectoid, eutectoid, and hypereutectoid steel. Questions 33–39 pertain to these diagrams.

33. Which diagram is associated with a composition that is hypoeutectoid, eutectoid, and hypereutectoid, and why?

34. Label the regions on each diagram.

35. Estimate the carbon content for the hypoeutectoid and hypereutectoid steels. Explain your reasoning.

36. If a thin specimen of alloy II is quenched instantaneously from 900°C to 550°C, held for 1 second, and quenched to room temperature, what phases will be present in the microstructure?

 37. Design two heat treatments that would result in the maximum difference in the amount of proeutectoid phase that would be present in the microstructure of the hypereutectoid steel after heat treating.

38. Calculate the maximum amount of proeutectoid phase that can form when heat-treating a plain carbon steel containing 0.4 wt. % C.

39. Calculate the maximum amount of proeutectoid phase that can form when heat-treating a plain carbon steel containing 1.2 wt. % C.

Problems 40–44 pertain to the IT diagram for two important steels, 4140 and 4340, shown in Figure HP8.2.

40. From the I-T diagrams shown in Figure HP8.2, are the two alloys hypo- or hypereutectoid? Explain.

41. Which of the two alloys, 4140 or 4340, has the greater hardenability?

42. What are the maximum isothermal hold temperatures that a thin piece of either 4140 or 4340 can be quenched to while avoiding the formation of proeutectoid phase? Explain by comparing the complex 4140 and 4340 diagrams with the off-eutectoid compositions in plain carbon steel.

43. What phases and microstructure are present after thin pieces of 4140 are homogenized in the austenite phase field and given the following quench paths?

    a. Instantaneous quench to 600°C; hold for 10^4 seconds, and quench to room temperature.

    b. Instantaneous quench to 600°C; hold for 10^3 seconds, and quench to room temperature.

    c. Instantaneous quench at 400°C; hold for 10^3 seconds, and quench to room temperature.

    d. Instantaneous quench to room temperature.

44. What phases and microstructures are present after thin pieces of 4340 are homogenized in the austenite phase field and subjected to the following quench paths?

    a. Instantaneous quench to 650°C; hold for 10^5 seconds, and quench to room temperature.

    b. Instantaneous quench at 350°C; hold for 10^4 seconds, and quench to room temperature.

    c. Instantaneous quench to room temperature.

45. Briefly describe a heat treatment that would result in the following microstructures for 4140:

    a. A 50-50 mixture of martensite and bainite (ferrite + carbide) where none of the ferrite is in the form of proeutectoid ferrite

    b. A mixture of proeutectoid ferrite and coarse pearlite

    c. A mixture of 50% bainite, 25% austenite, and 25% martensite

    d. A completely martensitic microstructure

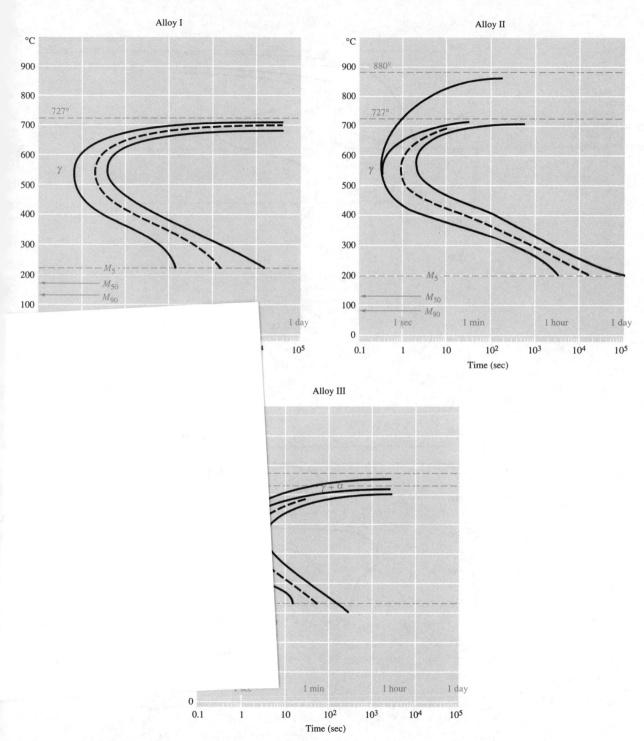

FIGURE HP8.1   IT diagrams for three plain carbon steels: a hypoeutectoid, a hypereutectoid, and a eutectoid.   *(Source:* Atlas of Isothermal Transformation and Cooling Transformation Diagrams, *ASM International, Metals Park, OH, 1977. Reprinted with permission.)*

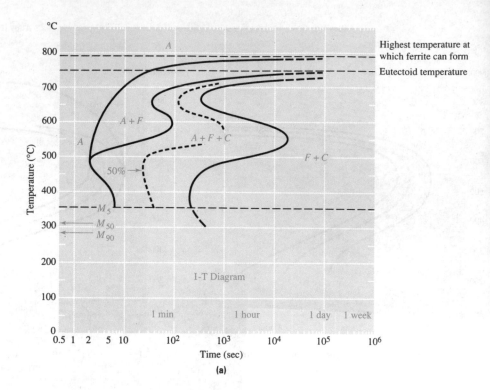

(a)

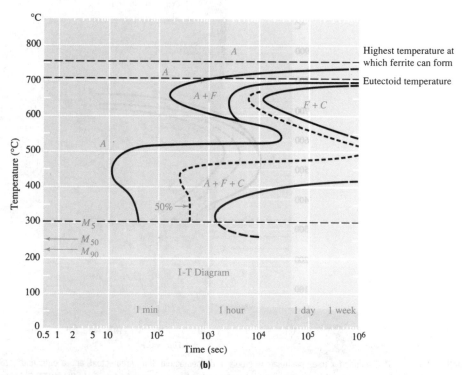

(b)

FIGURE HP8.2 IT diagrams for **(a)** 4140 and **(b)** 4340. Note that A = austenite, F = ferrite, and C = cementite *(Source: Atlas of Isothermal Transformation and Cooling Transformation Diagrams, ASM International, Metals Park, OH, 1977. Reprinted with permission.)*

46. Briefly describe a heat treatment that would result in the following microstructures for 4340:
    a. A mixture of proeutectoid ferrite and coarse pearlite
    b. A 50-50 mixture of bainite and martensite
    c. 100% bainite
    d. A mixture of 50% bainite, 25% austenite, and 25% martensite

47. When reporting isothermal transformation diagrams, the original austenite grain size is always reported. What effect, if any, would you expect the austenite grain size to have on the rate of transformation of pearlite?

48. Figure 8.3–15 shows Jominy end-quench test data for a series of steels. Explain why these hardness results are consistent with the IT diagrams shown in Figure HP8.2.

49. True or false; if you believe the statement is false, please explain the error and correct the statement:
    a. The general shape of a TTT curve describing the start of a diffusional phase transformation can be explained by noting that both nucleation rate and diffusion rate increase with increasing temperature.
    b. In a eutectoid steel, as the transformation temperature is lowered, the carbide phase becomes more coarsely distributed.
    c. For a plain carbon steel, increasing the carbon content increases both the martensite start and finish temperatures.
    d. IT diagrams, like phase diagrams, provide equilibrium information about the phases present in a microstructure.

50. Using the continuous cooling curve for a eutectoid steel shown in Figure 8.3–13, calculate the slowest cooling rate that would ensure a 100% martensitic structure at room temperature.

51. Suppose you have a thin sheet of eutectoid steel and you desire a microstructure with the finest pearlite spacing achievable, without any martensite. What cooling rate would you suggest?

52. The cooling rate along the radius of a bar fabricated from plain carbon eutectoid steel was determined to be 35°C/s at the center, 55°C/s at mid-radius, and 200°C/s at the surface. What microstructure would you expect at each of these locations?

53. Sketch a portion of a binary phase diagram that can be used to help define the following terms associated with alloys that are precipitation-hardenable:
    a. Solution heat treating
    b. Artificial aging

54. A simple binary eutectic diagram can be very useful in defining the conditions that are necessary for an alloy to precipitation-harden. Construct such a diagram paying careful attention to explain specific features that would tend to optimize a potential system for precipitation hardening. The factors that should be considered are:
    a. Maximum solid solubility of the solute in the matrix phase
    b. The solubility in the matrix at room temperature
    c. Change in solubility at temperatures between room temperature and the eutectic temperature

55. In alloys that can be strengthened by age hardening, why is a longer time required to reach maximum hardness at a lower temperature? Why would you expect the maximum hardness to be higher for lower aging temperatures?

56. Explain why you would expect the composition of a precipitation hardening alloy to affect the choice of:
    a. Artificial aging temperature
    b. Solution heat treatment temperature

57. Explain why alloy composition affects the strength of a precipitation-hardenable alloy.

58. Why is it important to rapidly quench a precipitation hardening alloy from the solution heat treatment temperature?

59. Compare the distribution of precipitates that would form in a precipitation hardening alloy that is slowly cooled to the distribution that would result if the alloy is rapidly cooled, then artificially aged.

60. It takes 200 hours for an aluminum-copper alloy to reach maximum hardness at 150°C, and 5 hours at 190°C. Calculate the temperature to reach the maximum hardness in 8 hours.

61. Calculate the activation energy for the precipitation process from the data given in the previous problem.

62. Figure 7.7–3 in Chapter 7 shows the Al-Ni phase diagram. The precipitation of the phase γ' [denoted as α(Ni-Al) on the figure] having composition near Ni_3Al is the basis of high-temperature nickel-base superalloys. Using the Al-Ni phase diagram, explain the factors that are necessary for nickel alloys containing aluminum to be candidates for precipitation hardening alloys.

Figure HP8.3 shows the yield strength dependence on artificial aging response for two precipitation hardening aluminum alloys. Although these alloys are complex in that they contain more than one solute element, the age hardening principles that apply to binary alloys like Al-Cu can be applied to these complex alloys. Questions 63–69 pertain to these diagrams.

63. Explain why the time to reach peak strength for 2014 and 6061 decreases with increasing aging temperature.

64. Why does the value of peak strength increase with decreasing artificial aging temperature?

65. Although the precipitation process is a very complex phenomenon, it is often useful to apply the concept of activation energy as determined through an Arrhenius plot to assist in interpolating data. Determine the activation energy for the artificial aging of 2014.

66. Using the approach developed in Problem 65, determine the activation energy for artificially aging 6061.

67. Determine the time to reach peak strength for alloy 2014 using an artificial aging temperature of 210°C.

68. Determine the time to reach peak strength for alloy 6061 using an artificial aging temperature of 210°C.

69. The aging response curves used in the preceding problems were constructed from numerous tensile tests taken at each of eight aging temperatures. Explain how you might apply the knowledge gained from these data to reduce the number of temperatures to be investigated if small changes to the alloy compositions were made to either 2014 or 6061.

70. Sketch a binary phase diagram containing a simple eutectic. Using this diagram, explain what is meant by coring and why it occurs.

71. If an Al-Cu alloy containing 5% by weight of copper is solidified to produce a cored microstructure, what is the minimum local copper concentration in the aluminum phase? What is the maximum local copper concentration in the aluminum phase?

72. Suppose the cored Al-Cu microstructure is rapidly heated to above the liquidus temperature. At what temperature would the first melting reaction occur, and why?

73. Figure HP8.4 shows a hypothetical binary phase diagram for a peritectic system. The dashed lines show the location of nonequilibrium solidus and have the same meaning as those shown in Figure 8.3–24. For the alloy of composition X, explain why you expect both the α and β phases to be cored.

74. Figure HP8.5 shows percent recrystallized as a function of time at several temperatures. Calculate the activation energy for the recrystallization process over the temperature range of 43 to 135°C.

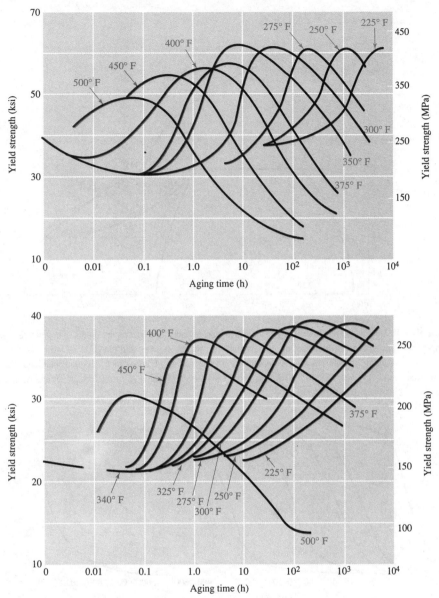

FIGURE HP8.3 Aging characteristics of two aluminum sheet alloys at elevated temperatures. The alloys were solution heat-treated, quenched to room temperature, and naturally aged at room temperature to a stable microstructure prior to artificial aging. The top set of data is for alloy 2014 and the other data set is for alloy 6061. *(Source: John E. Hatch, ed., Aluminum: Properties and Physical Metalling, p. 17, ASM International, Metals Park, OH. Reprinted with permission.)*

75. Figure HP8.6 shows the isothermal recrystallization behavior of zone-melted Fe–0.6% Mn alloy. Explain how you would determine the activation energy for the recrystallization process.

76. For the data in Figure HP8.6, determine the activation energy for the recrystallization process.

77. Using the data in Figure 8.3–38 for the crystallization of isotactic polypropylene at 123, 128, and 132°C, determine n and k in Equation 8.2–12.

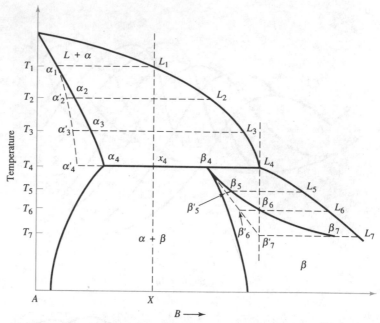

FIGURE HP8.4   This hypothetical binary peritectic phase diagram illustrates how coring occurs during natural freezing. Rapid cooling causes the development of the nonequilibrium solidus α_1 to α_4 and the β_4 to β_7', which lead to cored alpha phase and cored beta phase. The surrounding phenomenon also occurs during the rapid solidification of peritectic-type alloys.   *(Source: Frederick N. Rhines, Phase Diagrams in Metallurgy: Their Development and Application, McGraw-Hill, 1956. Reprinted with permission.)*

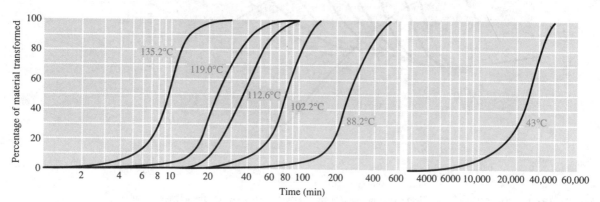

FIGURE HP8.5   Percent recrystallization as a function of time at several temperatures for pure copper.   *(Source: B. F. Decker and D. Harker, "Recrystallization in Cold Rolled Copper," Transactions AIME 188 (1950), Metallurgical Society of AIME, Reprinted with permission.)*

78. Using the results from Problem 77, compare the actual kinetic data with the calculated fraction transformed using Equation 8.2–12.

79. Show why spherulite growth is characterized by radius $\propto$ time.

80. Why is the growth rate in polymers so much slower than that in metals at a similar relative undercooling?

81. Phalocyanines, are complex aromatic molecules in which a metal is coordinated, are often added to polymers to pigment them. When phthalocyanine blue is added to PET, the crystallization rate is not affected, but when phthalocyanine green is added, the crystallization rate increases by about a factor of 4. How might this be possible?

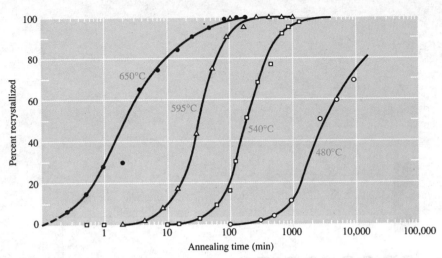

FIGURE HP8.6   Isothermal recrystallization of zone-melted Fe–0.60% Mn alloy.   *(Source: Reprinted with permission from Transactions of the Metallurgical Society 221 (1961), p. 691, a publication of The Minerals, Metals & Materials Society, 420 Commonwealth Drive, Warrendale, PA 15086.)*

PROPERTIES

We have established that the choice of materials in any application is based on the properties required of the material. In Part III, we will be concerned with defining, measuring, and manipulating the engineering properties of materials. These properties include mechanical, electrical, optical, dielectric, magnetic, and thermal properties, and susceptibility to environmental degradation. The discussion emphasizes the definition and measurement of material constants that collectively characterize material properties. These constants are reported in handbooks and are frequently used by engineers to compare candidate materials for a given application. An equally important emphasis will be on the relationships between the properties and the structure of the materials. The camcorder again offers examples of the need to choose materials to fulfill the property requirements of various components.

Mechanical properties are important in selecting housing materials for the camcorder because these materials must be able to resist impact during handling and adequately protect the fragile electronic circuitry and lenses. They must also be lightweight to reduce the camcorder's weight. Plastics, high-strength aluminum alloys, graphite-reinforced plastics (polymer composites), or SiC-reinforced aluminum or magnesium alloys (metal-matrix composites), which all possess high strength-to-density ratios, are good candidate materials for this application. The properties of composites are discussed in Chapter 14.

Also with regard to mechanical properties, all the structural components within a device must be sufficient in strength, stiffness, ductility, and other measures to perform their required functions. The mechanical properties of materials are discussed in Chapter 9.

Chapter 10 deals with the electrical properties of materials. Portions of the camcorder must be high-quality electrical conductors, while other components must be excellent electrical insulators. The camcorder will contain integrated circuits composed of semiconductor devices that serve many purposes, including the conversion of optical images and sound to electrical impulses. These electrical signals are then stored on magnetic tape so that the information can be retrieved at a later time. Chapter 12 describes magnetic information storage and several other magnetic properties of materials, including the operation of the motors within the camcorder.

The camcorder lenses must be able to form clear images that are free from distortion. The ability to manipulate the refractive index of the lens material is an important property in designing high-quality lenses. The refractive index determines the angle at which an incident optical ray is refracted, or bent, as it goes through the lens. Optical properties such as luminescence and phosphorescence are also important in designing display screens in camcorders. Electroluminescence is a phenomenon by which a stream of electrons is moved across a screen made from a phosphorescent material. In these materials (examples include GaAs and InP), the change in the energy levels occupied by electrons results in the production of light. This is the principle behind the television tube and the viewing screen on a camcorder. The optical and dielectric properties of materials are discussed in Chapter 11.

Whenever electrical current moves through a material, some of the energy is lost in the form of heat. If this heat is not transported away from the semiconducting devices, it can alter their electrical properties and degrade the performance of the camcorder. In Chapter 13, we will discuss the thermal properties of materials and describe the mechanisms and materials used to transport thermal energy.

If the camcorder housing is made from a polymer, exposure to ultraviolet light (sunlight) can degrade its mechanical properties. In Chapter 15, we will describe this environmental degradation mechanism as well as the methods used to minimize its impact on materials. Another mechanism, known as corrosion, can degrade the properties of metal components in the camcorder if it is inadvertently exposed to atmospheric moisture.

MECHANICAL PROPERTIES

MATERIALS IN ACTION Mechanical Behavior: Fracture and Economics

In the chapter we are about to study, we will learn about conditions (e.g., forces, temperatures, environments) to which materials are subjected in load-bearing applications. We will also learn that different characteristics are required for components to function properly under the conditions of use. These characteristics are called mechanical properties.

For example, in a general way, the strength of a material (we will define strength more precisely later) is one mechanical property of interest. At first it may seem as if the stronger the material, the better. However, this is not always true. Strong materials, as we shall see later in detail, may be brittle. That means that they may be sensitive to the existence of small flaws, which, when present, can cause fracture with little absorption of energy. The ability of a material to resist fracture is termed its toughness. When cracks become large enough, all materials will fracture under a given load. The tougher the material, the longer the crack will be before fracture.

Small cracks grow when loads are repeatedly applied through a process called fatigue. The cracks frequently form at imperfections on the surface of a component. For example, in Figure 9.5–8a later in this chapter, the fractured surface of an engine crankshaft fabricated from a high-strength steel is shown. In this figure two drilled holes are seen at approximately three o'clock and 12 o'clock. Around these holes, faint markings in the form of arcs can be seen. These markings are made by a propagating fatigue crack. In this example, the engine was being used in the extreme cold of the Arctic. Microscopic examination showed that the drilled surface was very rough because of poor machining practice. At the point along the drill surface where the stresses in the body reached a maximum, repeated loading associated with the rotation of the shaft caused the initiation and propagation of a fatigue crack, which eventually led to failure of the shaft and considerable downtime and economic loss. The engine manufacturer was under pressure to rectify the situation or lose customers. The solution in this case was rather simple: go to a slightly more expensive machining process to ensure good surface integrity and eliminate crack initiation. This was done, and no subsequent failures were observed. This example shows the interplay between design (applied loads), processing (in this case, the machining process), and performance. It also shows that it was a false economy to attempt to make a small savings on drilling while exposing the shaft to considerable risk of fracture, which had negative economic and reputation consequences.

On a larger scale, in most advanced industrial countries, the economic loss due to fracture, fatigue, and corrosion has been analyzed in detail and found to be about 10% of the cost of all goods and services produced in one year. In the United States and Japan, this is in excess of $200 billion per year for each country. Approximately half of this could be saved by application of known principles.

In this chapter, we will develop the tools necessary to minimize this severe economic loss.

9.1 INTRODUCTION

Mechanical properties describe the behavior of a material subjected to mechanical forces. Materials used in load-bearing applications are called structural materials and may be metals, ceramics, polymers, or composites. Selecting a material for a structural application is a difficult process; it typically involves consideration of several candidate materials whose mechanical properties under a given set of service conditions and cost constraints must be compared in order to make an optimum choice. Additional considerations may include processing options, available resources, or both as discussed in Chapters 16 and 17.

New technological developments often follow advancements in materials science. For example, the efficiency of converting thermal energy into mechanical and subsequently electrical energy is directly related to the inlet gas temperature in a turbine. Therefore, the maximum turbine efficiency is related to the high-temperature strength of the materials used in fabricating different turbine components. A revolution in the jet engine industry was brought about in the 1950s and 1960s by the introduction of nickel-based alloys, which can operate at temperatures up to 1200°C. In comparison, the maximum service temperature for steels is ~550°C. Use of ceramic materials is expected to further boost the efficiency of aircraft and automobile engines by permitting operating temperatures as high as 3000°C, provided the problems associated with brittleness can be solved or dealt with effectively.

Another example of the influence of new materials on technological advances is the introduction of lightweight composite and oriented polymer structural materials in aircraft. These materials have higher strength-to-weight ratios and allow for lighter aircraft. Such planes can carry more passengers, cargo, and fuel than conventional aluminum-based aircraft and can travel longer distances without stopping. Without these materials the 16-hour nonstop flight between New York and the Pacific Rim Countries would be impossible.

In this chapter we define the mechanical properties used by engineers to compare materials' suitability for structural applications. We will also review testing procedures for measuring these properties, and approaches for strengthening and toughening materials. The emphasis is on defining material properties that are independent of specimen size and geometry. These properties are typically cataloged in materials data handbooks that engineers use for material selection.

9.2 DEFORMATION AND FRACTURE OF ENGINEERING MATERIALS

All materials undergo changes in dimensions in response to mechanical forces. This phenomenon is called deformation. If the material reverts back to its original size and shape upon removal of the load, the deformation is said to be **elastic.** On the other hand, if application and removal of the load results in a permanent shape change, the specimen is said to have undergone **plastic deformation.**

Fracture (or rupture) occurs when a structural component or specimen separates into two or more pieces. While fracture clearly represents failure of a component, it should be noted that, depending on the design criteria, **failure** (an inability of a component to perform its desired function) may occur prior to fracture. For example, in many applications plastic deformation represents failure without fracture. A car axle that bends when you drive over a pothole, or a lawn chair that buckles and collapses, is an example of a component that has failed without fracture. In this section, we explain the phenomenon of deformation and consider test methods for measuring the resistance of materials to these processes.

9.2.1 Elastic Deformation

Figure 9.2–1a shows a cylindrical specimen with an original cross-sectional area A_0 and length l_0 subjected to a uniaxial force F. We define **engineering stress** σ and **engineering strain** ε as follows:

$$\sigma = \frac{F}{A_0} \tag{9.2-1}$$

$$\varepsilon = \frac{\Delta l}{l_0} = \frac{l - l_0}{l_0} \tag{9.2-2}$$

where l is the instantaneous length of the rod. Figure 9.2–1b shows the stress-strain relationship when a tensile specimen is subjected to a small load. For small strains, stress and strain are linearly related. This linear relationship between stress and strain is known as Hooke's law. Furthermore, the specimen is restored to its original condition when the force is removed (i.e., the strain is elastic). The ratio of stress to strain, σ/ε, in the linear elastic region is called **Young's modulus,** E. The physical significance of Young's modulus, also known as the elastic modulus, is that it measures the interatomic bonding forces and, therefore, the stiffness of the material. A material with a high elastic modulus is comparatively stiff, which means it exhibits a small amount of deformation under an applied load.

Recall from Section 2.5 that elastic modulus is related to the shape of the bond-energy curve. Materials with deep, sharp bond-energy curves have high elastic moduli. Examples of high-modulus materials include most ceramics with covalent or mixed ionic and covalent bonds such as diamond, graphite (in the directions of covalent bonding), and alumina (Al_2O_3). The bond energies and elastic moduli of metals are also relatively high but below those of most ceramics. In general, unoriented thermoplastic polymers display lower E values than ceramics and metals because of the comparatively weak secondary bonds between adjacent chains. When, however, polymer molecules are well aligned

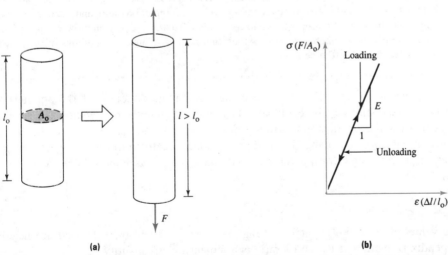

(a)	(b)

FIGURE 9.2–1   The response of a cylindrical specimen of original cross section A_0 and length l_0 to a tensile force F. **(a)** Under load the specimen elongates to length l. **(b)** The corresponding stress-strain behavior for materials at small strains. The elastic modulus E is defined as the ratio of stress to strain in the linear region. When the load is removed, the specimen returns to its original length.

along the direction of stress, polymers may also have high moduli. The moduli of amorphous materials are discussed in Section 9.2.2. When two materials with different modulus values are subjected to the same stress, the material with the higher modulus value experiences less deformation. Elastic modulus values for several materials are listed in Appendix D.

....

EXAMPLE 9.2–1

Consider three cylindrical specimens, each with a diameter of 10 mm and a length of 1 m. One specimen is aluminum ($E = 70$ GPa), the second is Al_2O_3 ($E = 380$ GPa), and the third is polystyrene ($E = 3.1$ GPa). A force of 2000 N is applied along the axis of each specimen. Assuming that the deformation is elastic, estimate the elongation in each specimen.

Solution

We must use the definitions of stress, strain, and modulus. The elongation can be obtained from Equation 9.2–2 if the strain in the sample is known. That is, $\Delta l = \varepsilon \times l_0$. The strain, in turn, can be determined using the modulus equation in the form $\varepsilon = \sigma/E$, and the stress can be calculated directly using Equation 9.2–1 ($\sigma = F/A_0$). Substituting the modulus equation and the definition of stress into the elongation equation yields:

$$\Delta l = \left(\frac{\sigma}{E}\right)l_0 = \left(\frac{F/A_0}{E}\right)l_0 = \frac{Fl_0/A_0}{E}$$

Since the sample dimensions and load are the same in all samples, the elongation equation reduces to:

$$\Delta l = \frac{(2000 \text{ N})(1.0 \text{ m})/[(\pi/4)(0.01 \text{ m})^2]}{E}$$

$$= \frac{25.5 \text{ MPa-m}}{E}$$

Finally, by substituting the appropriate moduli into this expression, we find that the elongations for the aluminum, Al_2O_3, and polystyrene are respectively 0.36 mm, 0.067 mm, and 8.2 mm. Note that the elongations are inversely proportional to the moduli. That is, the deformation in polystyrene is more than two orders of magnitude greater than that in Al_2O_3, since the modulus of the polymer is less than 1% of that of the ceramic.

....

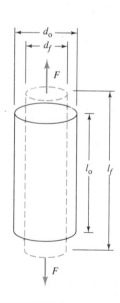

FIGURE 9.2–2

Elastic elongation in the direction of the applied load is accompanied by a contraction in the perpendicular direction.

As shown in Figure 9.2–2, elastic elongation in the direction of the applied load (known as axial strain ε_a) is accompanied by contraction in the perpendicular directions. The perpendicular or transverse strain is defined as $\varepsilon_t = \Delta d/d_0$, where d_0 is the original diameter and Δd is the change in the diameter. The negative ratio of transverse strain to axial strain is constant for a given material and is known as **Poisson's ratio** ν:

$$\nu = -\frac{\varepsilon_t}{\varepsilon_a} \tag{9.2–3}$$

The values of ν for most materials range between 0.25 and 0.35; several are listed in Appendix D. Note that Poisson's ratio is a dimensionless quantity.

The concepts of shear stress, shear strain, and shear modulus were introduced in Sections 5.2.1 and 6.3. Recall that the **shear modulus** G is defined as the ratio of the applied shear stress τ to the resultant shear strain γ, which makes G similar to E. Shear stress and strain are illustrated in Figure 9.2–3.

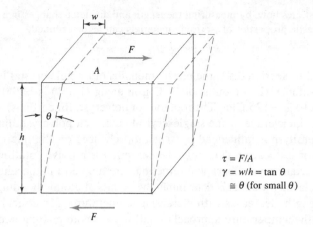

$$\tau = F/A$$
$$\gamma = w/h = \tan \theta$$
$$\cong \theta \text{ (for small } \theta)$$

FIGURE 9.2–3
A cube of edge length h and face area A is subjected to a shear force F. The solid lines show the cube in the unloaded state and the dotted lines show the cube shape after the application of the shear force. The shear stress τ and shear strain γ are defined in the figure.

The quantities E, G, and ν are called elastic constants. For isotropic materials the following relationship is valid:

$$G = \frac{E}{2(1 + \nu)} \tag{9.2–4}$$

Hence, only two of the three elastic constants are independent. For anisotropic materials such as composites, oriented polymers, and single crystals, the number of elastic constants varies with the degree of symmetry. A brief description of elastic constants for composites is contained in Chapter 14.

···

EXAMPLE 9.2–2

A cylindrical steel specimen is subjected to a stress of 100 MPa. The underformed specimen has a diameter of 10 mm and a length of 40 mm. The length and diameter of the deformed specimen are 40.019 mm and 9.9986 mm, respectively. Assuming the specimen remained elastic, calculate the elastic modulus, shear modulus, and Poisson's ratio for this steel.

Solution

Let the axis of the cylinder lie along the y axis and the x axis lie along a radial (transverse) direction in the cylinder. From the given information, we find that:

$$\varepsilon_y = \varepsilon_a = \frac{\Delta l}{l_0} = \frac{40.019 \text{ mm} - 40 \text{ mm}}{40 \text{ mm}} = 4.75 \times 10^{-4}$$

$$\varepsilon_x = \varepsilon_t = \frac{\Delta d}{d_0} = \frac{9.9986 \text{ mm} - 10 \text{ mm}}{10 \text{ mm}} = -1.4 \times 10^{-4}$$

Using the definition of Poisson's ratio (Equation 9.2–3):

$$\nu = -\frac{\varepsilon_x}{\varepsilon_y} = -\frac{-1.4 \times 10^{-4}}{4.75 \times 10^{-4}} = 0.295$$

The elastic modulus is defined as $E = \sigma/\varepsilon_y$, so that

$$E = \frac{100 \text{ MPa}}{4.75 \times 10^{-4}} = 210 \times 10^3 \text{ MPa} = 210 \text{ GPa}$$

Finally, using Equation 9.2–4, G is calculated as:

$$G = \frac{E}{2(1 + \nu)} = \frac{210 \text{ GPa}}{2(1 + 0.295)} = 81.1 \text{ GPa}$$

This problem illustrates how, by measuring the length and diameter changes in a cylindrical specimen, all three elastic properties of isotropic materials can be determined.

....

As mentioned in Section 3.11, the elastic modulus of a single crystal is an anisotropic property. For example, the modulus of BCC iron along $\langle 1\ 1\ 1 \rangle$ is ~280 GPa, while the value along $\langle 1\ 0\ 0 \rangle$ is ~125 GPa. The modulus of polycrystalline iron is ~200–205 GPa, which represents an average of the single-crystal values. In polycrystalline materials the modulus is generally not influenced by structural changes on the microstructural level (i.e., grain size or dislocation density). The exception is a polycrystalline material processed so that certain crystallographic directions are aligned in adjacent grains.

Because the elastic modulus is determined by details of atomic bonding, its value, like that of bond strength, decreases with increasing temperature. The effect becomes significant only when the temperature approaches half the absolute melting temperature of the crystal. The higher melting temperatures of ceramics (as compared with most metals) make them strong candidates of elevated-temperature applications such as heat exchangers, crucibles and molds for containing molten metals, and engine components. It must be remembered, however, that ceramics generally have limited ductility and are susceptible to brittle fracture.

9.2.2 Deformation of Polymers

This section expands our discussion of viscous deformation of polymers introduced in Section 6.3. In polymers the bonds between adjacent macromolecules are comparatively weak secondary bonds. In adddition, the tangled macromolecules can accommodate deformation by uncoiling. Thus, polymers have much lower elastic moduli than crystalline metals and ceramics. Thermoplastic polymers below their glass transition temperature T_g, and thermoset polymers resist sliding motion between macromolecules. Thus, deformation occurs only by stretching of secondary bonds and uncoiling of long-chain molecules. These deformation characteristics are not significantly influenced by temperature; therefore, the elastic moduli of thermoplastics below T_g and of crosslinked polymers do not change significantly with temperature. Also, such deformation is fully recoverable upon removal of stress, and it follows a stress-strain relationship similar to the one followed by metals and ceramics.

When the temperature in thermoplastics exceeds T_g, molecular motion can occur and can cause viscous deformation as described in Section 6.3. The extent of viscous deformation increases rapidly with temperature, causing the elastic modulus to decrease rapidly with increasing T. Since viscous deformation for $T > T_g$ occurs simultaneously with elastic deformation, the behavior in this regime is known as **viscoelasticity.** The viscous portion of the viscoelastic modulus is also time-dependent. In other words, if the loading rate is high and consequently less time is available for time-dependent deformation, the viscous behavior is suppressed and the elastic modulus is higher than for lower loading rates.

Viscoelastic deformation is often studied using either of two common mechanical tests. The first method involves a time-dependent reduction in stress under constant strain known as **stress relaxation,** and the second involves a time-dependent increase in strain under constant stress (**time-dependent deformation**). If you have ever placed a rubber band around a pack of papers and returned sometime later to find that the rubber band had "relaxed" and was no longer holding the pack together, you have witnessed stress relaxation. In contrast, a vinyl phonograph record left leaning against the seat of a hot car may warp under its own weight as a result of time-dependent deformation.

In a stress relaxation test, a specimen is quickly loaded (at $T > T_g$) to a constant strain level ε_0 that is maintained throughout the test. The stress measured as a function of time can be roughly described by the relationship

$$\sigma(t) = \sigma_0 \exp\left(-\frac{t}{\tau_0}\right) \tag{9.2-5}$$

where σ_0 is the initial stress, t is time, and τ_0 is the relaxation time constant (it is *not* a shear stress). The relationship between stress, strain, and time during stress relaxation is shown in Figure 9.2–4a. The relaxation time constant τ_0 is proportional to the viscosity of the polymer and, therefore, decreases exponentially with an increase in temperature (see Equation 6.3–5a or b). The corresponding expression for the relaxation modulus is

$$E_r(t) = \frac{\sigma(t)}{\varepsilon_0} \tag{9.2-6}$$

which shows that the modulus also decreases exponentially with time during a stress relaxation experiment.

As shown in Figure 9.2–4b, in a time-dependent deformation test a polymer is subjected to a constant stress σ_0 and the corresponding increase in strain with time, $\varepsilon(t)$, is monitored. The corresponding relaxation modulus expression is

$$E_r(t) = \frac{\sigma_0}{\varepsilon(t)} \tag{9.2-7}$$

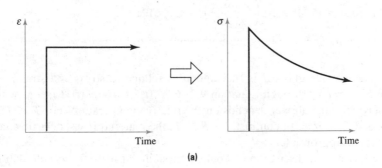

(a)

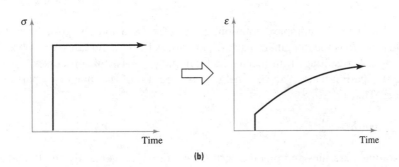

(b)

FIGURE 9.2–4

Time-dependent viscoelastic deformation in polymers. **(a)** During stress relaxation, a constant applied strain results in a decrease in stress over time. **(b)** During time-dependent deformation, a constant applied stress results in an increase in strain over time.

which shows that the modulus decreases with increasing strain during a time-dependent deformation test. The behavior of thermoplastic polymers is often more complex than described above. However, this treatment illustrates the main features of viscoelastic polymer behavior.

EXAMPLE 9.2–3

An instantaneous strain of 0.4 is applied to a polymer sample and the sample is maintained under strain. The initial stress is 5 MPa and decays to 2 MPa after 50 s. Estimate the stress in the polymer at $t = 10$ s.

Solution

This is a stress relaxation experiment with $\sigma_0 = 5$ MPa and $\sigma(50 \text{ s}) = 2$ MPa. Substituting $\sigma(50 \text{ s})$, σ_0, and t in Equation 9.2–5 yields:

$$2 \text{ MPa} = (5 \text{ MPa}) \exp\left(-\frac{50 \text{ s}}{\tau_0}\right)$$

Solving for τ_0 gives:

$$\tau_0 = -\frac{50 \text{ s}}{\ln(2 \text{ MPa})/(5 \text{ MPa})} = 54.6 \text{ s}$$

Substituting for τ_0 in Equation 9.2–5 yields:

$$\sigma(t) = (5 \text{ MPa}) \exp\left(-\frac{t}{54.6 \text{ s}}\right)$$

The stress after 10 s is:

$$\sigma(10 \text{ s}) = (5 \text{ MPa}) \exp\left(-\frac{10 \text{ s}}{54.6 \text{ s}}\right) = 4.16 \text{ MPa}$$

What about oxide glasses? Is their modulus related to stress and strain through the equation $E = \sigma/\varepsilon$ or through Equation 9.2–6 [$E_r(t) = \sigma(t)/\varepsilon(t)$]? Just as with thermoplastic polymers, the answer depends on the relative temperature. For $T < T_g$, the glass is elastic and $E = \sigma/\varepsilon$. In contrast, for $T > T_g$ the material is viscoelastic and Equation 9.2–6 is more appropriate.

Before leaving our discussion of Young's modulus, it is useful to note that the elastic modulus is in fact a complex quantity of the form:

$$E^* = E + iE' \tag{9.2–8}$$

The real part of the complex notation, E, is often used loosely to describe the elastic modulus, whereas the imaginary part, E', known as the loss modulus, describes the extent of energy loss resulting from mechanical damping, or viscous, processes.

The dissipation factor, $\tan \delta$, is defined as the ratio of the imaginary part of E^* to its real part. That is,

$$\tan \delta = \frac{E'}{E} \tag{9.2–9}$$

In crystalline ceramics and metals, both E' and $\tan \delta$ are generally small, and therefore, little mechanical damping occurs. As a result, the elastic modulus can be approximated by its real part: $E^* \approx E$. In contrast, E' and $\tan \delta$ are significant for viscoelastic materials,

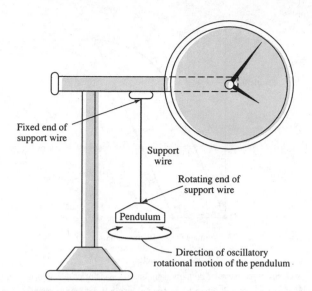

FIGURE 9.2–5

An illustration of a torsion
pendulum clock.

Fixed end of
support wire

Support
wire

Rotating end of
support wire

Pendulum

Direction of oscillatory
rotational motion of the pendulum

including most amorphous polymers above T_g. The energy loss mechanism is associated with molecular friction and results in heat generation.

Consider the operation of a clock that makes use of a torsion pendulum, as shown in Figure 9.2–5. Should the thin wire that supports the pendulum be fabricated from a ceramic, a metal, or a polymer? Such a pendulum oscillates back and forth several times a minute, and the clocks are often designed to be wound only once a year. Thus, the support wire will experience more than 10^6 cycles between windings, so energy losses must be minimized. Thus, a polymer is probably not ideal. Metals are usually favored over ceramics for this application, since they are easier to fabricate into the required shape and easier to attach to the other components in the clock.

The damping characteristics of polymers are highly desirable for other applications. For example, vibrating equipment such as pumps and motors is often mounted on pads designed to absorb the vibrations and isolate the equipment from the surroundings. In this case we would select a polymer with a high tan δ value, such as polychloroprene, to use in the fabrication of the mounting plate or engine mounts in cars.

9.2.3 Plastic Deformation

As shown in Figure 9.2–6, when the applied stress exceeds a critical value called the **elastic limit,** deformation becomes permanent. When a specimen is loaded beyond this limit, it no longer returns to its original length upon removal of the force. Such behavior is termed plastic or permanent deformation. The stress-strain behavior during plastic deformation becomes nonlinear and no longer obeys Hooke's law.

In most materials, elastic deformation is associated with bond stretching, as shown in Chapter 2. In crystals, plastic deformation is primarily associated with the movement of dislocations, as discussed in Chapter 5. In most thermoplastic polymers, plasticity is associated with sliding of entangled long-chain molecules past each other, an essentially irreversible process that also depends on time (recall our discussion of viscoelastic behavior in Section 9.2.2). Although the slope of the σ-ε curve in the plastic region decreases with increasing strain, continued plastic deformation requires a continuing increase in stress. That is, materials harden upon plastic straining. This phenomenon, known as **strain hardening,** is the result of dislocation-dislocation interactions in metallic crystals.

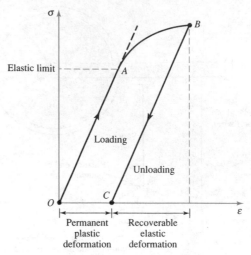

FIGURE 9.2–6   Stress versus strain behavior during loading and unloading for elastic and plastic loading conditions. In the elastic region (*OA*), the specimen will return to its original length if the load is removed prior to reaching the elastic limit (point *A*). If, on the other hand, the specimen is loaded to point *B* before the load is released, it will unload along *BC* (i.e., parallel to the elastic region) and will show a permanent deformation of *OC*.

These interactions either significantly reduce the dislocation mobility or stop dislocations from moving entirely. In the case of polymers, strain hardening is a result of chain alignment in the stress direction. In ceramics dislocation motion is difficult, as explained in Chapter 5. Therefore, plastic deformation in ceramics is quite restricted, so that these materials tend to be brittle. In other words, unlike polymers and metals, no mechanism of plastic deformation is available in ceramics.

Another way to understand the phenomenon of strain hardening is to do a thought experiment in which the specimen in Figure 9.2–6 is reloaded from point *C*. To promote dislocation motion upon reloading, a stress corresponding to point *B* will be required. Hence, the effective strength of the material, as measured by the stress necessary to cause dislocation motion, has increased as a result of plastic strain during the first loading. Strain hardening may occur when forming a component into a desired shape. The material may become so hard during forming that intermediate thermal treatments are necessary to soften the metal so that it can be formed into its final shape. A similar process, known as mechanical conditioning, can be used to improve the properties of polymer fiber by straining to align the molecules.

Another difference between elastic and plastic deformation is the magnitude of the volume and shape changes in the specimen associated with each type of strain. Elastic deformation (i.e., atomic bond stretching) changes the equilibrium separation distance between atoms and therefore changes the volume of the sample. Since atoms retain the same nearest neighbors during elastic deformation, however, there are no major changes in the shape of the specimen. In contrast, plastic deformation does not alter significantly either the bond length or crystal volume, but the slip process changes the shape of the material.

9.2.4 Tensile Testing

A tensile test is used to quantitatively measure some of the key mechanical properties of structural materials. Historically, this test was developed and standardized for metals, but the same principles apply to polymers, ceramics, and composites. The procedure, however, varies somewhat for the different classes of materials. We begin by discussing

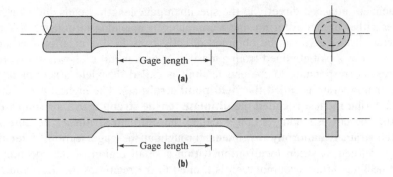

FIGURE 9.2–7

Specimen geometries used for tensile testing: **(a)** cylindrical and **(b)** flat specimens. *(Source: Copyright ASTM. Reprinted with permission.)*

tensile testing of metals, and then describe the corresponding procedures for ceramics and polymers (testing of composites is described in Chapter 14).

Figure 9.2–7 shows two specimen geometries recommended by the American Society for Testing and Materials (ASTM) for tensile testing of metals. The choice of specimen geometry and size often depends on the product form in which the material is to be used or the amount of material available for samples. A flat specimen geometry is preferred when the end product is a thin plate or sheet. Round–cross section specimens are preferred for products such as extruded bars, forgings, and castings.

As shown in Figure 9.2–8a, one end of the specimen is gripped in a fixture that is attached to the stationary end of the testing machine; the other end is gripped in a fixture attached to the actuator (moving portion) of the testing machine. The actuator usually moves at a fixed rate of displacement and thus applies load to the specimen. The test usually continues until the specimen fractures.

During the test, the load on the specimen is measured by a transducer called a load cell; the strain is measured by an extensometer (a device for measuring the change in length of

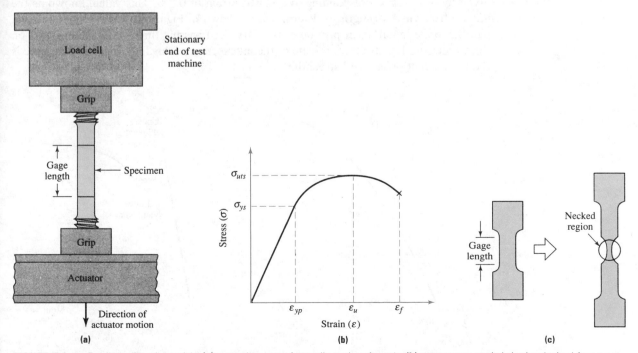

FIGURE 9.2–8 Tensile testing of materials: **(a)** a complete setup for tensile testing of metals, **(b)** stress versus strain behavior obtained from a tensile test, and **(c)** the formation of a "neck" within the gage length of the sample.

the specimen) attached directly to the specimen gage length. Loads and elongations are recorded either in digital form using a computer or in analog form using X-Y recorders. The stress-strain curve can be obtained directly from load-elongation measurements.

A typical σ-ε plot obtained from a tensile test on a metal is shown in Figure 9.2–8b. The stress corresponding to the elastic limit is called the **yield strength** σ_{ys}, and the corresponding strain is called the **yield point strain** ε_{yp}. The highest engineering stress reached during the test is called the **ultimate tensile strength** σ_{uts}, or simply the tensile strength. The corresponding strain is called the **uniform strain** ε_u, because up to this point the strain is uniformly distributed throughout the gage section. After this point, necking, defined as strain localization within a small region of the specimen, occurs. During necking, strain accumulation is limited to the region of the neck and is nonuniform, as shown in Figure 9.2–8c.

The **engineering strain at fracture,** ε_f, is usually reported as the percentage elongation (i.e., $\varepsilon_f \times 100$). This quantity is also referred to as the **ductility** of the sample. When reporting the percent elongation of a material, it is customary to specify the initial gage length of the specimen, since the value ε_f depends on the length-to-diameter ratio for the sample. The higher this ratio, the lower the engineering strain to fracture.

Percent reduction in area (%RA) is also commonly reported. %RA has the advantage of being independent of the length-to-diameter ratio. It is calculated as follows:

$$\%\text{RA} = \frac{A_0 - A_f}{A_0} \times 100 \tag{9.2–10}$$

where A_0 is the original cross-sectional area and A_f is the final area of the necked region. Values for the tensile properties of the common structural metals can be found in Appendix D.

In several FCC metals, such as copper and aluminum, the yield point is not well defined (see Figure 9.2–9a). The operational definition of yield strength for such materials is given by the stress corresponding to a plastic strain of 0.2%. This value, known as the **0.2% offset yield strength,** is determined as shown in Figure 9.2–9a. A line is drawn parallel to the initial linear portion of the curve and passing through the point 0.002 on the strain axis. The stress coordinate of the intersection of this line with the σ-ε curve is the 0.2 percent offset yield strength.

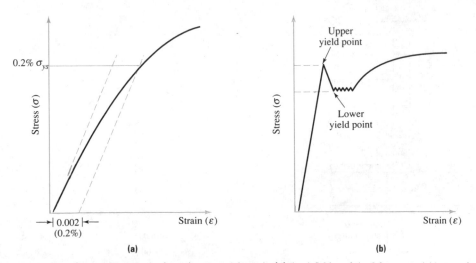

FIGURE 9.2–9   Stress-strain behavior for various types of metals: **(a)** the definition of the 0.2 percent yield stress for FCC metals, and **(b)** the upper and lower yield point phenomenon exhibited by some materials, such as carbon steels.

Some materials, including carbon steels, exhibit a complex yield behavior, as shown in Figure 9.2–9b. The transition from elastic to plastic deformation occurs abruptly and is accompanied by a reduction in stress. With continued deformation, the stress level remains constant, then begins to rise. The drop in stress is caused by the sudden mobility of dislocations as they are released from the strain fields associated with interstitial atoms (i.e., C in steel). The yield strength is defined by the lowest stress at which plastic deformation occurs and is identified as the lower yield point. The upper yield point characterizes the stress at which plastic deformation first begins.

The area under the σ-ε curve is a measure of the energy per unit volume required to cause the material to fracture. This quantity, given the symbol U, is one measure of **toughness** of a material and is calculated as:

$$U = \int_0^{\varepsilon_f} \sigma \, d\varepsilon \qquad (9.2\text{–}11)$$

The units for toughness are (force per unit area) $\times$ (length per unit length) = (force $\times$ length) per unit volume = energy per unit volume.

....

EXAMPLE 9.2–4

A metal tensile specimen has an initial diameter of 10 mm and is 50 mm long. The yield strength is 400 MPa, the elastic modulus is 70 GPa, and the ultimate tensile strength is 500 MPa. Calculate the yield point strain and the maximum load during the test.

Solution

During elastic deformation, stress is linearly related to strain through the equation $E = \sigma/\varepsilon$. Thus, at the yield point, $E = \sigma_{ys}/\varepsilon_{yp}$. Solving for ε_{yp} and substituting the values in the problem statement gives:

$$\varepsilon_{yp} = \frac{\sigma_{ys}}{E} = \frac{400 \text{ MPa}}{70 \times 10^3 \text{ MPa}} = 5.7 \times 10^{-3}$$

Since $\sigma = F/A_0$, and since the maximum load corresponds to the ultimate tensile stress, we find that $\sigma_{uts} = F_{max}/A_0$. Solving for F_{max} and substituting the appropriate values gives:

$$F_{max} = \sigma_{uts} \times A_0 = (500 \text{ MPa})(\pi/4)(0.01 \text{ m})^2$$
$$= 3.9 \times 10^{-2} \text{ MN} = 39.2 \times 10^3 \text{ N}$$

....

EXAMPLE 9.2–5

Figure 9.2–10 shows stress-strain curves for three materials.

    a. Which material has the highest modulus?
    b. Which material has the highest ductility?
    c. Which material has the highest toughness?
    d. Which material does not exhibit any significant plastic deformation prior to fracture?

Solution

    a. The material with the highest modulus is the one with the steepest slope in the initial region of the σ-ε curve. In this case material I has the highest modulus.
    b. Since ductility is defined as $\varepsilon_f \times 100$, material III displays the highest ductility.
    c. One measure of toughness is the area under the σ-ε curve. While material I has a high failure stress, it also has a low ductility, so it displays limited toughness. Material III has a high ductility but a low ultimate tensile strength, so it too displays limited toughness. Material II, with its moderate strength and ductility, is the toughest of the three materials.

FIGURE 9.2–10

A comparison of stress-strain curves for three different materials for use in Example 9.2–5.

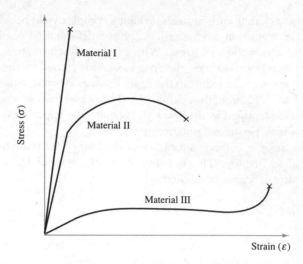

d. Only material I fractures without any significant amount of plastic deformation. The other two materials exhibit substantial plastic deformation, as indicated by the extensive nonlinear regions on their σ-ε curves.

We will show later in this chapter that many ceramics behave like material I, many metals like material II, and many rubbers like material III.

To this point we have been discussing engineering stress and strain. Both quantities are based on *original* specimen dimensions and do not take into account the fact that sample dimensions change during a tensile test. The corresponding quantities that reflect changing sample dimensions are known as true stress and true strain. For small deformations the differences between the engineering and true quantities can be safely neglected. As the deformation increases, particularly after the onset of necking, the use of the true stress and strain is recommended whenever precise quantitative relationships are required.

True stress σ_t is defined as the load F divided by the instantaneous area A_i:

$$\sigma_t = \frac{F}{A_i} \tag{9.2–12}$$

True strain ε_t is related to the differential change in length, dl, divided by the instantaneous length l and is calculated as:

$$\varepsilon_t = \int_{l_0}^{l} \frac{dl}{l} = \ln\left(\frac{l}{l_0}\right) \tag{9.2–13}$$

Prior to the onset of necking, the following relationships between the true and engineering stresses and strains are valid:

$$\varepsilon_t = \ln(1 + \varepsilon) \tag{9.2–14}$$

and

$$\sigma_t = \frac{\sigma}{1 + \varepsilon} \tag{9.2–15}$$

Another useful relationship is that between the true fracture strain ε_{tf} and the percent reduction in area, %RA, given by:

$$\varepsilon_{tf} = \ln\left(\frac{100}{100 - RA}\right) \tag{9.2–16}$$

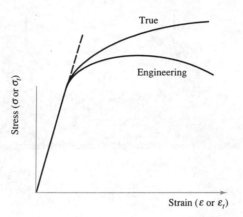

FIGURE 9.2–11

Engineering and true stress-strain diagrams obtained from the same tensile test.

Figure 9.2–11 shows a typical engineering stress-strain diagram and the corresponding true stress-strain diagram.

For many metals, the relationship between stress and plastic strain can be written as:

$$\sigma_t = K\,\varepsilon_{tp}^n \tag{9.2–17}$$

where K is the strength coefficient (in stress units), n is the strain hardening exponent, and ε_{tp} is the true plastic strain. This relationship may be used to compare the plastic behavior of metals.

EXAMPLE 9.2–6

A cylindrical metal tensile specimen has a diameter of 10 mm and a gage length of 50 mm. After the tensile test, the diameter in the necked region of the specimen was 6 mm. Calculate the %RA and the true strain at fracture.

Solution

To calculate the % RA we use Equation 9.2–10:

$$\%RA = \frac{A_0 - A_f}{A_0} \times 100$$

Recalling that the area is proportional to the square of the diameter, we may also write:

$$\%RA = \frac{d_0^2 - d^2}{d_0^2} \times 100$$

Substituting the values given in the problem statement yields:

$$\%RA = \frac{10^2 - 6^2}{10^2} \times 100 = 64$$

The true fracture strain is given by Equation 9.2–16 and requires knowledge of the percent reduction in area:

$$\varepsilon_{tf} = \ln\!\left(\frac{100}{100 - 64}\right) = 1.02$$

Testing of Ceramics

From a testing point of view, the major difference between metals and ceramics is the inherent brittleness of ceramics. Thus, it is difficult to machine ceramic samples into the shapes necessary for tensile testing, particularly the formation of the reduced cross-sectional area in the gage length and the threads used to attach the specimens to the testing machine.

FIGURE 9.2–12

The experimental setup for
four-point bend testing.

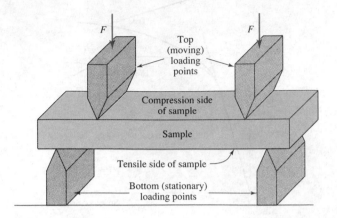

As shown in Figure 9.2–12, a common solution to this problem is to test ceramics in bending rather than in tension. The advantages of the bend test include simple sample geometry (rectangular or cylindrical specimens), a simplified testing procedure, and generally lower cost. The major disadvantage of the four-point bend test is that in contrast to a tensile test, which results in a nearly uniform stress throughout the gage length, the stress distribution in the sample is nonuniform. The maximum stress (i.e., the maximum force recorded by the instrument during the test divided by the cross-sectional area of the sample) is obtained only on the specimen surface in the region between the two central supports. The consequence of this nonuniform stress state is that under certain conditions, particularly when the largest flaws in the sample are located in the interior of the specimen, the strength of the ceramic will be overestimated during four-point bend testing.

Despite this disadvantage, the load-deflection data from a bend test can be used to construct stress-strain curves similar to those obtained from a tensile test. The σ-ε curves for a typical metal and a typical ceramic are shown in Figure 9.2–13a. The key features to note are: (1) the modulus for a ceramic is generally higher than that for a metal, (2) ceramics rarely exhibit significant plastic deformation, and (3) the fracture stress of a *flaw-free* ceramic is often higher than that of a metal. All three differences can be related to the different primary bond types of these materials.

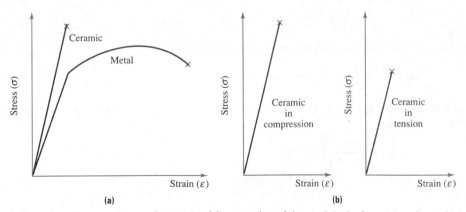

FIGURE 9.2–13   Stress-strain curves for ceramics: **(a)** a comparison of the σ-ε behavior for metals and ceramics, and **(b)** the influence of the state of stress (tension or compression) on the response of a ceramic.

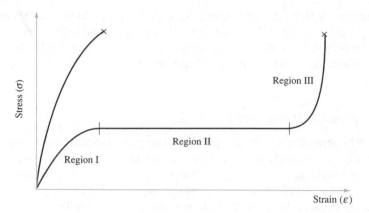

FIGURE 9.2–14
Stress-strain curves for polymers showing two types of general behavior. The highly nonlinear type of behavior can be subdivided into three regions. See text for a discussion of the types of polymers that exhibit each type of behavior.

Another difference between ceramics and metals is that while the σ-ε curves for metals tested in tension and compression are nearly identical, the curves for ceramics depend on the state of stress (compression or tension) during the test.[1] As shown in Figure 9.2–13b, while the slopes of the σ-ε curve are the same in both testing modes, a ceramic sample tested in tension generally fractures at a much lower stress than an identical specimen loaded in compression. The explanation for this phenomenon is related to preexisting flaws in the ceramic and is described in detail in Section 9.4.

Testing of Polymers

The inherent ductility of most polymers means that machining to obtain complex specimen geometries is not difficult. As such, polymer testing may be performed by either tension or bending. Typical stress-strain curves for two general types of polymer behavior are shown in Figure 9.2–14. Examples of classes of polymers that exhibit roughly linear behavior include thermosets, thermoplastics below T_g, and thermoplastic polymers composed of chains that have been aligned along the tensile axis prior to testing at any temperature.

In contrast, semicrystalline thermoplastic polymers composed of unaligned chains display a different type of σ-ε behavior. For these polymers the σ-ε curve can be divided into three regions. Initially, there is a nearly linear region characterized by a shallow slope (modulus) that reflects the increase in stress necessary to overcome intermolecular secondary bonds. As the deformation proceeds, the spherulites fragment and a neck region forms. At this point, the σ-ε curve remains approximately horizontal, indicating that a constant force is required to extend the necked region throughout the bulk of the polymer sample. Finally, once the majority of the spherulites have broken up and chains are partially aligned, additional strain causes homogeneous deformation, or further chain alignment, and the curve again displays a positive slope. The slope or modulus in the high-strain region is greater than that in the low-strain region, since it reflects the strength of the primary bonds within the aligned chains. You can perform this experiment yourself by pulling on the plastic (polyethylene) used to hold a six-pack of soda cans together. You should be able to see the formation and propagation of the neck region and then "feel" the increase in stress necessary to continue the deformation process to failure.

---

[1] In Figure 9.2–12 the bottom surface of the ceramic sample is in tension and the top surface is in compression. Failure will initiate in tension.

In comparison with metals and ceramics, polymers generally display lower modulus values (except in region III of the unaligned thermoplastics) and lower fracture strengths, but significantly higher ductilities as measured by strain to failure. Highly oriented polymers, such as industrial fibers for use in composites, on the other hand, can be as stiff and strong as metals or ceramics.

9.2.5 Strengthening Mechanisms

The strength of ceramics, metals, and polymers can be modified by changes in chemistry and morphology brought about by thermal and mechanical processing. Approaches for strengthening metals were discussed in Section 5.5, and we suggest that you review that material. In general, the idea is to restrict dislocation motion by adding point defects (solid solution strengthening), line defects (cold working or strain hardening), planar defects (grain size refinement), or volume defects (precipitation hardening). These strengthening mechanisms influence most of the mechanical properties of the metals, including yield strength, ductility, and toughness, as illustrated in the following example.

EXAMPLE 9.2–7

Explain the relationship between each of the following strengthening mechanism and mechanical property pairs in metals.

- *a.* Solid solution strengthening; yield strength
- *b.* Cold working; ductility
- *c.* Grain refinement; modulus of elasticity

Solution

- *a.* Incorporating solute atoms into the solvent crystal structure impedes the motion of dislocations through an interaction between their respective strain fields (see Section 5.5.1). Thus, solid solution strengthening raises the yield strength of a metal.
- *b.* Cold working increases the yield strength of a metal, since it increases the dislocation density with an associated decrease in dislocation mobility (see Section 5.5.2). Cold working also reduces the ductility of the alloy. This can be seen in Figure 9.2–6, which shows that the strain to failure, ε_f, is reduced by the deformation process.
- *c.* The modulus of elasticity is a function of the atomic bond characteristics of the metal. It is not influenced in any significant way by changes in microstructural features such as grain size; however, grain refinement generally does increase the yield strength of the metal (see Section 5.5.3).

Because of the inherent limited mobility of dislocations in ceramic crystals, inserting additional atomic scale defects does not increase the strength of ceramic crystals nearly as much as it does in metals. In structural ceramics, emphasis is placed on increasing toughness rather than increasing strength. This topic will be discussed in Section 9.4.

How can one increase the strength of an isotropic, or unoriented, polymer? If we interpret strength as the ability of the polymer to resist the relative motion of adjacent polymer chains, then the key to strengthening is to find ways to restrict such motion. We have already dealt with this issue in Section 6.4 when we noted that the following factors tend to inhibit molecular motion: longer chains (increased average molecular weight or crosslink formation) and increased crystallinity.

For example, the influence of degree of crystallinity on the strength of polyethylene is shown in Figure 9.2–15a. Similarly, the strength of polycarbonate, as a function of

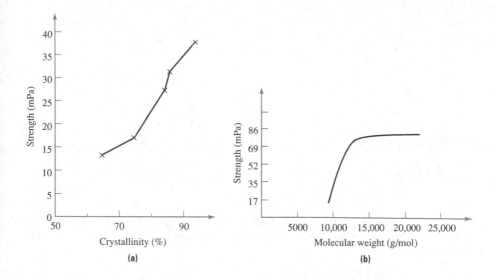

FIGURE 9.2–15

Factors affecting the strength of polymers: **(a)** the influence of crystallinity on the strength of polyethylene, and **(b)** the influence of molecular weight on the strength of polycarbonate. *(Source: Data for* **(a)**, *H. V. Boeing, Polyolefins: Structure and Properties,* Elsevier Press, *Lausanne, 1966.)*

average molecular weight, is shown in Figure 9.2–15b. Note that after some value of molecular weight the strength tends to level off. The reason is that once the chains become long enough, the stress required to overcome the entanglements is equal to that required to break primary bonds. At that point further increases in chain length will have no additional strengthening effect.

9.2.6 Ductile and Brittle Fracture

If the process of deformation is continued, fracture eventually occurs. Materials that sustain a large amount of plastic deformation before fracture are ductile, and those that fracture with little accompanying plastic deformation are brittle. Ductile and brittle are relative terms; metals and polymers are inherently more ductile than ceramics, but in a given class of materials, ductility can vary significantly.

As shown in Figure 9.2–16, ductile fracture in metals is usually nucleated at inhomogeneities such as inclusions. It is preceded by severe localized deformation in the necked

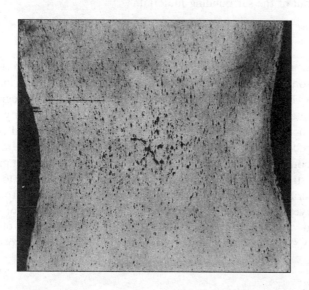

FIGURE 9.2–16

Initiation of ductile fracture around inclusions in the necked region of pure copper. *(Source: J. I. Bluhm and R. J. Morrissey,* International Conference on Fractures, 1965, *Sendai, Japan, Vol. D-II, p. 73.)*

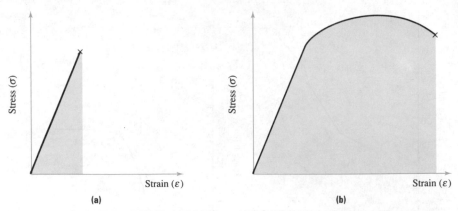

FIGURE 9.2–17   Stress-strain relationship for **(a)** brittle and **(b)** ductile materials. The colored areas represent (qualitatively) the energy absorbed per unit volume prior to fracture up to the point of homogeneous deformation in the specimen gage length.

region of the specimen. Ductile fracture requires a significant amount of energy, since work must be done to plastically deform the material in the necked region. In contrast, if no energy absorption mechanisms are available, then brittle fracture may occur.

Figure 9.2–17 shows the σ-ε behavior of brittle and ductile materials. The appearance of ductile and brittle fractures at the microscopic level is shown in Figure 9.2–18a and b; examples of ductile and brittle fractures obtained during tensile testing of metals are shown macroscopically in Figure 9.2–18c and d. Note that toughness of a ductile material, as measured by the area under the σ-ε curve, is much greater than that of a brittle material.

Why are metals generally more ductile than ceramics? The simplest answer is that metals display a natural energy absorption mechanism—the motion of dislocations on multiple slip systems. In ceramics, the restricted dislocation motion does not require significant energy absorption, so ceramics generally display brittle fracture. To improve a ceramic's resistance to brittle fracture (i.e., increase its toughness), some sort of energy absorption mechanism must be found. As will be discussed in more detail in Chapter 14, one reason for incorporating fibers into ceramic composites is that energy is required to pull the fibers out of the surrounding material.

9.2.7 Hardness Testing

Hardness is a measure of a material's resistance to plastic deformation (for materials that exhibit at least some ductility). In a hardness test a load is placed on an indenter (a pointed probe), which is driven into the surface of the test material. The degree to which the indenter penetrates the sample is a measure of the material's ability to resist plastic deformation. Since hardness testing is essentially nondestructive and no special specimens are required, it is a comparatively inexpensive quality assurance test and an indicator of material condition. Another advantage of hardness testing is that properties such as ultimate tensile strength, wear resistance due to friction, and resistance to fatigue (a failure mechanism described in Section 9.5) can all be accurately predicted from hardness data.

There are several ways to measure hardness. The shape and size of the indenter and the applied load vary with the type of material being tested. Because of the flexibility to choose a variety of loads during its measurement, the Brinell scale covers a wide range

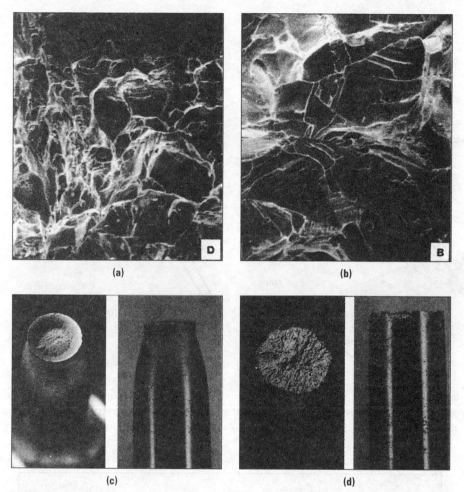

FIGURE 9.2–18 Photographs of the ductile and brittle fractures: **(a)** a microscopic view of a ductile fracture in low-carbon steel tested at high temperatures, **(b)** a microscopic view of a brittle fracture of low-carbon steel tested at low temperature, **(c)** a macroscopic view of a ductile tensile failure (note the characteristic "cup-and-cone" appearance), and **(d)** a macroscopic view of a brittle tensile failure. Note: D and B in parts a and b refer to the ductile and brittle fracture surfaces also shown in Figure 9.2–24. *(Source: Adapted from* Metals Handbook, *Vol. 11, "Failure Analysis and Prevention," 9th ed., 1986, ASM International, Materials Park, OH. Reprinted by permission of the publisher.)*

of hardnesses, from those of case-hardened steels to those of soft FCC metals such as annealed aluminum or copper. Figure 9.2–19a illustrates the geometry associated with a Brinell hardness measurement. The Brinell hardness number, BHN, is calculated using the equation:

$$\text{BHN} = \frac{2P}{\pi D(D - \sqrt{D^2 - d^2})} \qquad (9.2\text{--}18)$$

where P is the applied load in kg, D is the indenter diameter ($= 10$ mm), and d is the diameter in millimeters of the indentation measured on the sample surface. Standard loads vary between 500 and 3000 kg. Figure 9.2–19b shows the correlation between BHN and the tensile strength of carbon steels.

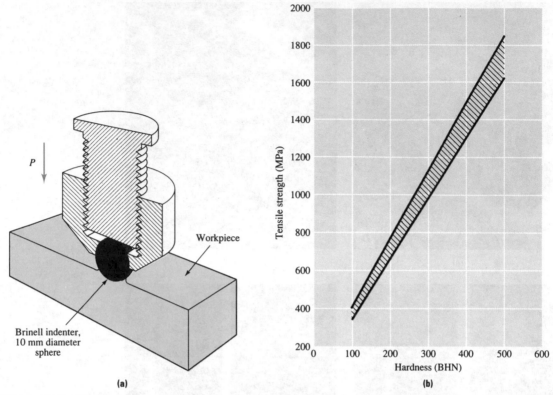

FIGURE 9.2–19 The Brinell hardness test: **(a)** the geometry of the test, and **(b)** correlations between the Brinell hardness number (BHN) and tensile strength of carbon steels. *(Source:* Metals Handbook, Desk Edition, *1985, p. 34.5, ASM International, Materials Park, OH. Reprinted by permission of the publisher.)*

Load levels and indenter sizes for Rockwell hardness tests.

Symbol, indenter	Minor (pre-) load (kg)	Major (total) load (kg)	Coefficients in $R = C_1 - C_2\,\Delta t$	
			C_1	$C_2(\mathrm{mm}^{-1})$
Normal scales				
R_B, 1/16 ball*	10	100	130	500
R_C, cont[†]	10	150	100	500
R_A, cone	10	60	100	500
R_D, cone	10	100	100	500
R_E, 1/8 ball	10	100	130	500
R_F, 1/16 ball	10	60	130	500
R_G, 1/16 ball	10	150	130	500
Superficial scales				
R_{15N}, cone[‡]	3	15	100	1000
R_{30N}, cone	3	30	100	1000
R_{45N}, cone	3	45	100	1000
R_{15T}, 1/16 ball	3	15	100	1000
R_{30T}, 1/16 ball	3	30	100	1000
R_{45T}, 1/16 ball	3	45	100	1000

*Ball is steel of the diameter shown in inches.
[†]Normal cone is diamond with 120° included angle and a spherical apex of 0.2 mm radius.
[‡]Superficial cone is similar to normal cone but not interchangeable.

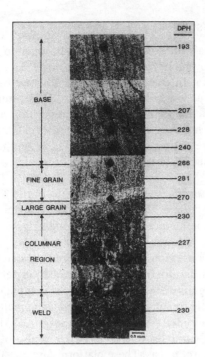

FIGURE 9.2–20
Microhardness variation in a 9 Cr–1 Mo steel weldment as a function of position. The dark diamond-shaped regions are the hardness indentations. Note the structural changes and the corresponding changes in hardness as the fusion line is traversed.   *(Source: Courtesy of Kelly Payne.)*

The Rockwell hardness value is based on the depth of the indentation, Δt, rather than the indentation diameter used in the Brinell method. Table 9.2–1 summarizes the load levels and indenter sizes used for the Rockwell scales. In general, different scales are used for different classes of materials. For example, R_C is used for high-strength steels and R_B for low-strength steels.

The Vickers microhardness test is used when a local hardness reading is desired within a microstructural entity on the size of a grain. Such measurements are useful in estimating the variability in mechanical properties between various regions of the test piece. For example, it is common for the hardness to change when going from the base metal to the fusion zone in weldments. This is shown in Figure 9.2–20, which represents a microhardness trace across a steel weldment.

EXAMPLE 9.2–8

The Brinell hardness of an alloy steel is 355. Compute the diameter of the indentation if a load of 2000 kg was used, and estimate the corresponding tensile strength of the material.

Solution

Equation 9.2–18 states that:

$$BHN = \frac{2P}{\pi D(D - \sqrt{D^2 - d^2})}$$

Substituting the values from the problem statement, and noting that $D = 10$ mm, yields:

$$355 = \frac{(2)(2000)}{10\pi(10 - \sqrt{10^2 - d^2})}$$

which after some algebra gives $d = 2.65$ mm. From Figure 9.2–19b, a BHN of 355 corresponds to a tensile strength of approximately 1200 MPa.

9.2.8 Charpy Impact Testing

Body-centered cubic metals, including ferritic steels, undergo a large change in energy absorbed during fracture over a limited temperature range. This phenomenon is called the ductile-to-brittle transition. In these materials the fracture behavior is ductile at high temperatures and brittle at low temperatures. The temperature range over which this transition occurs is a function of the chemical composition and microstructure of the metal. It should be noted, however, that not all materials exhibit a ductile-to-brittle transition. Such transitions are not generally observed in crystalline ceramics or polymers, composites, or FCC metals. Materials that show a pronounced glass transition temperature are usually brittle at $T < T_g$.

Figure 9.2–21 schematically shows the ductile-to-brittle transition phenomenon; it also illustrates the various regimes. The upper shelf is the region in which the temperature exceeds the upper transition temperature. Correspondingly, the lower shelf is the region in which the temperature is lower than the lower transition temperature. The middle region is called the transition region. The temperature at which the behavior is 50% brittle and 50% ductile is called the **ductile-to-brittle transition temperature (DBTT).** This temperature has great significance in the design of components for low-temperature service. For example, brittle fracture is of concern in equipment designed for application in cold climates. The winter of 1988–89 was very severe in Alaska, and in that year an unusually high number of car axles fractured. DBTT is also of concern in the design of offshore platforms in the Arctic seas. These platforms are constructed entirely from steel and have numerous weldments. Thus, it is desirable for the lowest service temperature to lie in the upper shelf region where significant energy must be supplied to cause ductile fracture. Similarly, ships that sail in cold waters are at risk unless they were made from steels with low transition temperatures. The Titanic may have sunk after impact with an iceberg due to brittle failure in the hull.

Experience gained through the analysis of past engineering failures has shown that toughness assessed using the area under a tensile σ-ε curve is not suitable for establishing the DBTT. The reason is that in tensile tests the loading rates are much lower than those encountered in service, and surface flaws are not present. In engineering practice, however, impact (rapid) loading is common, and most components contain notches. Both factors limit plasticity and lead to mechanical behavior that is more brittle than would be

FIGURE 9.2–21

Schematic of the ductile-to-brittle transition behavior in BCC metals.

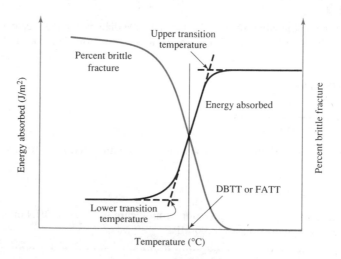

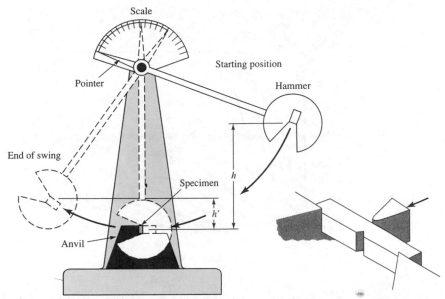

FIGURE 9.2–22   A Charpy impact test setup.   *(Source: Wayne Hayden, W. Moffatt, and J. Wulff, Mechanical Behavior, p. 13, Copyright © 1965 by John Wiley & Sons. Reprinted by permission of John Wiley & Sons, Inc.)*

apparent from a tensile test. Therefore, an alternative test method is required. The Charpy impact test meets these needs. It is simple, inexpensive, and widely used. Figure 9.2–22 shows the geometry of the Charpy impact testing device. The operating principle is that some of the energy initially present as potential energy in the pendulum is absorbed by the specimen during fracture. The difference between the initial pendulum height and the maximum height achieved during the follow-through (i.e., $h - h'$ in Figure 9.2–22) can be converted directly into the fracture energy of the specimen.

To determine the DBTT of steels, several Charpy specimens are tested over a wide temperature range. Specimens are immersed in liquid nitrogen, dry ice, or ice water before testing to obtain data in the low-temperature regime. Specimens are heated in boiling water or oil to obtain elevated temperature data.

Typical Charpy curves for steels with high and low ductile-to-brittle transition temperatures are shown in Figure 9.2–23. Note the influence of composition—as the Mn content increases or the carbon content decreases, the DBTT decreases. Figure 9.2–23c shows the lack of a DBTT in FCC metals, high-strength metal alloys, and some ceramics. The FCC alloys are ductile at all temperatures while the high-strength alloys and some ceramics are brittle at all temperatures.

Figure 9.2–24 shows Charpy specimens tested at various temperatures. Brittle fracture surfaces appear flat with no shear lips (ridges oriented at approximately 45° to the main fracture surface and located at the sides of the specimen). As the temperature and ductility increase, the area occupied by the shear lips increases, and the brittle fracture area decreases. As shown in Figure 9.2–21, the temperature at which the fracture surface is 50% flat is known as the fracture appearance transition temperature (FATT). In most cases, the DBTT and FATT are nearly the same.

As discussed in Section 6.2, amorphous solids are known to undergo severe reductions in ductility at low temperatures. In the glassy state (i.e., $T < T_g$), molecular motion is essentially frozen so the material behaves in a brittle fashion. For example, when natural rubber is cooled to liquid nitrogen temperatures ($\sim$77 K), it becomes brittle. The

FIGURE 9.2–23

Charpy impact test results for several types of materials: **(a)** the influence of carbon content on the DBTT of plain carbon steels, **(b)** the influence of manganese content on the DBTT of steels containing 0.05 percent carbon, and **(c)** a comparison of the data for FCC metal alloys, BCC steels, high-strength metal alloys, and ceramics. *(Source: Metals Handbook, Desk Edition, 1984, p. 4.85, ASM International, Materials Park, OH. Reprinted by permission of the publisher.)*

(a)

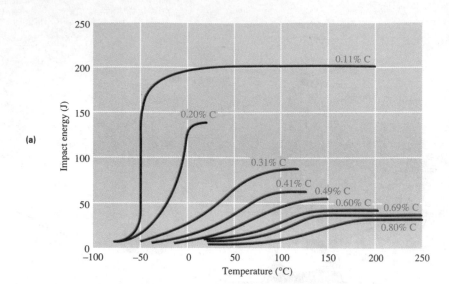

(b)

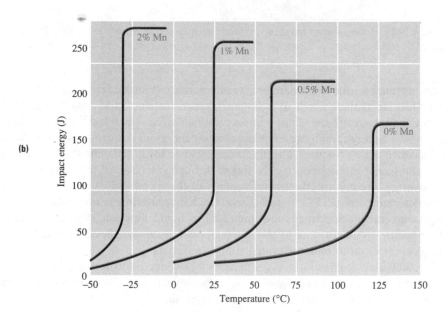

(c)

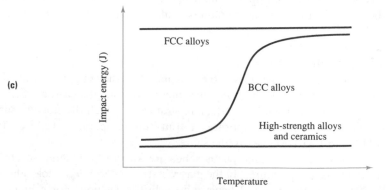

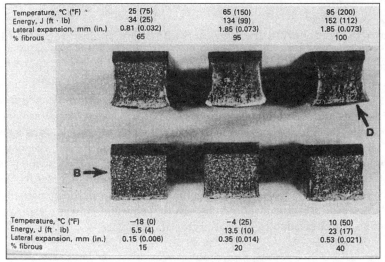

Temperature, °C (°F)	25 (75)	65 (150)	95 (200)
Energy, J (ft · lb)	34 (25)	134 (99)	152 (112)
Lateral expansion, mm (in.)	0.81 (0.032)	1.85 (0.073)	1.85 (0.073)
% fibrous	65	95	100

Temperature, °C (°F)	−18 (0)	−4 (25)	10 (50)
Energy, J (ft · lb)	5.5 (4)	13.5 (10)	23 (17)
Lateral expansion, mm (in.)	0.15 (0.006)	0.35 (0.014)	0.53 (0.021)
% fibrous	15	20	40

FIGURE 9.2–24 Photographs of fractured Charpy steel specimens tested at different temperatures. Note the increased area fraction associated with the (ductile) shear lips as the test temperature is increased. Note: D and B refer to ductile and brittle fracture surfaces. *(Source:* Metals Handbook: Failure Analysis and Prevention, *Vol. 11, 9th ed., 1986, ASM International, Materials Park, OH. Reprinted by permission of the publisher.)*

explosion of the space shuttle *Challenger* in 1986 was attributed to the embrittlement of rubber seals in the booster rocket because of cold weather.

Finally, it should be noted that some crystalline ceramics also display ductile-to-brittle transitions, but at comparatively high temperatures. The deformation of ceramics at elevated temperatures will be discussed in more detail in Section 9.6.2.

DESIGN EXAMPLE 9.2–9

Consider each application below and select the most appropriate material from the following list: an aluminum alloy, SiC (a structural ceramic), steel A with composition Fe–0.8% C–0% Mn, or steel B with composition Fe–0.05% C–2% Mn.

a. A material for construction of a vessel designed to contain liquid nitrogen at ~77 K

b. A steel to be used in the support structure of a snowmobile

c. A material that must retain its stiffness at elevated temperatures but will not experience impact loading

Solution

a. To minimize the potential for a brittle (low-energy) failure, a material that is ductile at 77 K is preferred. Figure 9.2–23 shows that of the candidate materials, only the FCC aluminum alloy is likely to retain its ductility at this service temperature.

b. The key requirement is that the BCC steel have a DBTT less than the service temperature so that it retains its ductility in use. From Figure 9.2–23, steel A has a DBTT of ~130°C, while steel B has a DBTT of ~−30°C. Therefore, steel B should be chosen for this application.

c. Impact resistance is not the problem. Instead, we require a material that retains its high modulus at elevated temperatures. Since ceramics generally have higher melting temperatures than metals, and since the ability to retain stiffness at high temperatures increases with melting temperature, the ceramic is the best choice for this application.

9.3 BRITTLE FRACTURE

As mentioned previously, brittle fracture occurs with little accompanying plastic deformation and comparatively little energy absorption. Such fracture occurs rapidly with little or no warning and can take place in all classes of materials. Hence, brittle fracture has received considerable attention among materials and mechanical engineers. In the previous section, we discussed the Charpy impact test, which is frequently used to assess a material's resistance to brittle fracture. However, this test has several drawbacks. First, the results of the Charpy test provide only a qualitative ranking of a material's resistance to brittle fracture and do not provide quantitative design data. Second, the loading used in these tests, while closer to reality than that in a tensile test, does not necessarily represent the loading in many typical applications. This section discusses the analytical approaches used to tackle the brittle fracture problem. Before proceeding with mathematical models, however, it is worthwhile to describe the sequence of events leading to brittle fracture and to consider some significant historical examples.

9.3.1 Examples and Sequence of Events Leading to Brittle Fracture

A classic example of brittle fracture occurred in World War II Liberty ships. The most spectacular failure was the USS *Schenectady,* whose hull completely fractured while it was docked in the Pacific Northwest. The fractured ship is shown in Figure 9.3–1. This problem was in part related to the welding methods used to construct the ships. When riveting was introduced to replace welding as a joining technique, the incidence of fracture was markedly reduced.

Brittle fracture has plagued the aviation industry. In the 1950s several Comet aircraft mysteriously exploded while in level flight. These problems were eventually traced to a design defect in which high stresses around the windows, caused by sharp corners, initiated small cracks from which the fractures emanated. In the late 1960s and early 1970s, the F-111 fighter experienced catastrophic failure of the wing thru-box (the structure at which the wings join to the fuselage). Failures of the F-111 were related to the poor

FIGURE 9.3–1   Fracture on the USS *Schenectady* at pier in the Pacific Northwest.   *(Source: Earl R. Parker,* Brittle Behavior of Engineering Structures, *National Academy of Science, National Research Council, copyright © 1957 by John Wiley & Sons, Inc.)*

choice of material (a high-strength tool steel) and a heat-treating procedure that produced nonuniform microstructures. In 1988, the canopy of a Boeing 737 operated by Aloha Airlines fractured without warning during level flight over the Pacific Ocean (Figure 9.3–2). The reasons for this were related to corrosion of the Al alloy used as a skin material. Additional examples of brittle fracture are seen in numerous bridges, train wheels, rolling mills, basketball backboards, hockey glass, and so on.

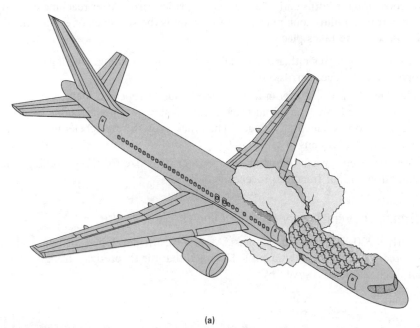

(a)

(b)

FIGURE 9.3–2

Boeing 737 Aloha Airlines incident: **(a)** schematic drawing of plane in flight; **(b)** Photograph of fuselage after landing. The canopy of the aircraft fractured in midflight over Hawaii. Subsequent investigation revealed that the canopy was weakened as a result of extensive corrosion and fatigue.   *(Source:* **(b)** © *Robert Nichols, Black Star.)*

There are some common factors in the examples above.[2] Brittle fracture generally occurs in high-strength materials (D6AC steel for the F-111 wing box, high-strength Al alloy for the Comets and 737), welded structures (Liberty ships, bridges), or cast structures (train wheels). It is significant that all failures started at small flaws that had escaped detection during inspection (in the case of the F-111, many inspections). Subsequent analysis showed that in most instances, small flaws slowly grew as a result of repeated loads or corrosion (or both) until they reached a critical size. After reaching critical size, rapid, catastrophic failure took place. The following is the sequence of events that occurs when brittle fracture takes place:

1. A small flaw forms either during fabrication (e.g., welding, riveting) or during operation (fatigue, corrosion).

2. The flaw then propagates in a stable mode due to repeated loads, corrosive environments, or both. The initial growth rate is slow and undetectable by all but the most sophisticated techniques. The crack growth rate accelerates with time, but the crack remains "stable."

3. Fracture occurs when the crack reaches a critical size for the prevailing load conditions. Final fracture proceeds rapidly.

9.3.2 Griffith-Orowan Theory for Predicting Brittle Fracture

In the early 1920s, the British physicist A. A. Griffith developed an approach to put fracture prediction on an analytic basis. He noted that the theoretical strength of a brittle material, such as glass, is given by:

$$\sigma_{th} = \sqrt{\frac{2E\gamma}{a_0}} \tag{9.3-1}$$

where E is Young's modulus, γ is the specific surface energy, a_0 is the lattice parameter, and σ_{th} is the theoretical strength of the material. Equation 9.3–1 reflects the strength of a defect-free material. If reasonable numbers are substituted into Equation 9.3–1 for the various material constants, strength is predicted to be about $E/10$, which is orders of magnitude higher than that usually observed. Griffith attributed this discrepancy to *preexisting* flaws, which greatly reduce fracture strength. Using this idea, he developed an approach for predicting conditions that facilitate rapid flaw propagation, leading to catastrophic facture.[3]

Griffith considered a semi-infinite panel of material containing a central crack of length $2a$ subjected to a remote stress σ. The load and crack geometry is shown in Figure 9.3–3. He recognized that as a crack extends, energy exchange occurs, which implies that cracks cannot grow unless the process is energetically favored. In other words, as the crack extends, energy is required to form new surfaces. This energy must be

---

[2] All these failures involve the brittle fracture of metals. This does not mean that nonmetals are immune to brittle fracture. Design engineers, however, have historically been more comfortable specifying metals for critical structural components. The reason is that they have more experience and data for metals than for any other class of materials. This trend is likely to change in the future as we gain more experience with structural ceramics, polymers, and composites.

[3] Note the similarity between Griffith's use of a defect (a crack) to explain the apparent discrepancy between the theoretical and experimental values for fracture stress and Taylor's use of a defect (a dislocation) to explain the apparent discrepancy between the theoretical and experimental values for the shear stress required for plastic deformation (see Section 5.2.2).

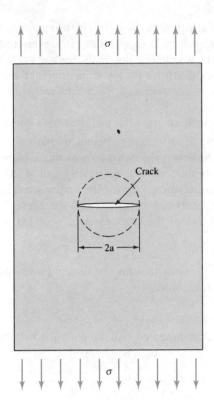

FIGURE 9.3–3

A semi-infinite panel of material containing a central crack of length 2*a* subjected to a remote stress σ.

supplied either by a corresponding reduction in the internal strain energy of the cracked body, or by the work done by the external forces, or by both. Griffith derived the critical stress for fracture for the case in which the energy for fracture is supplied totally by a reduction in stored elastic energy within the elastic body. He estimated the elastic energy reduction associated with a crack of length 2*a* (see Figure 9.3–3) as being equal to $\pi a^2(\sigma^2/E)$, where σ is the applied stress and the thickness of the body is assumed to be unity. The (surface) energy needed to form a crack of length 2*a* is given by $4a\gamma_s$, recognizing that two surfaces are created as the crack extends. Here, γ_s is the surface energy per unit area. For unstable fracture to occur, the rate of change of the release of strain energy with respect to crack size must at least equal the rate at which energy is consumed to create new surfaces. Thus, the critical condition is

$$\frac{d}{da}\left[\pi a^2\left(\frac{\sigma_G^2}{E}\right)\right] = \frac{d}{da}(4\gamma_s a) \tag{9.3–2a}$$

where σ_G is the critical stress (or the Griffith's stress) for unstable fracture. The results are expressed in the equation

$$\sigma_G = \sqrt{\frac{2E\gamma_s}{\pi a}} \tag{9.3–2b}$$

Equation 9.3–2b is known as the Griffith equation. Since it was developed on the basis of energy minimization, it is an energetically necessary condition for fracture. That is, the applied stress must equal or exceed σ_G for brittle fracture to occur. The converse, however, is not true. A stress of magnitude σ_G may not be sufficient for fracture, as explained below.

Equation 9.3–2b applies only to brittle materials that do not deform plastically—that is, all the work done on the material goes into forming new surfaces. Examples of materials that can be usefully treated using the Griffith analysis include oxide glasses and most crystalline ceramics, such as Al_2O_3 and SiO_2. However, many structural components are fabricated from metals, which, like polymers, undergo at least some plastic deformation prior to fracture. Thus, Griffith's theory in its original form does not apply to metals and cannot be used in many engineering applications without modifications proposed by Orowan two decades later.

Orowan noticed that for sharp cracks in metals, where ductility is significant, fracture occurs at a constant value of $\sigma\sqrt{\pi a}$ for specimens having a geometry similar to that shown in Figure 9.3–3; however, the constant was much higher than that predicted by Equation 9.3–2b. He hypothesized that a quantity called the effective surface energy should replace the true surface energy in Griffith's equation. The effective surface energy γ_e is the sum of the true surface energy γ_s and the energy dissipated during plastic deformation around the crack as the crack extends, γ_p. That is:

$$\gamma_e = \gamma_s + \gamma_p \tag{9.3–3}$$

Experimental measurements show that for metals and polymers, γ_p is much greater than γ_s, and to a good approximation:

$$\gamma_e \simeq \gamma_p \tag{9.3–4}$$

With this approximation, the modified Griffith equation becomes:

$$\sigma_f = \sqrt{\frac{2E\gamma_e}{\pi a}} \tag{9.3–5}$$

This equation shows that a concept initially developed for brittle materials can be modified and applied to materials with some ductility. While this equation is not based on first principles, as is Griffith's, it is a useful working empirical expression. Note that as the crack length increases, the fracture stress decreases.

In summary, the key result of the Griffith-Orowan theory is that for a crack to grow, energy must be supplied to the system to form new surfaces and to deform the material ahead of the advancing crack tip.

When glassy polymers are subjected to tensile stresses, cracklike structures called crazes appear prior to failure. The crazes are oriented normal to the stress direction and are regions in which molecules become highly oriented due to the high local deformation. A craze consists of ~50% void and 50% highly oriented polymer. Crazes can be seen most easily by allowing light to reflect off their surfaces. Crazes can often be seen in airline windows, which are made of polycarbonate glass. Crazing requires tremendous energy to orient molecules in, and just ahead of, the craze. Thus $\gamma_p \gg \gamma_s$ and $\gamma_e \approx \gamma_p$.

9.4 FRACTURE MECHANICS: A MODERN APPROACH

Unfortunately, the brittle fracture model described in Section 9.3.2 is limited to the precise geometry shown in Figure 9.3–3. Ideally, we seek a parameter that is an index of a material's toughness, independent of geometry, and can be used with a stress analysis to predict fracture loads and critical crack sizes. Once this toughness parameter is determined using a simple laboratory specimen, it can be used to predict the flaw size at which fracture will occur in components of arbitrary geometry and size subjected to specified loads. Conversely, given the flaw size, it should be possible to predict the maximum safe operating stress. This is what fracture mechanics is all about. The fracture mechanics framework described in this section is a result of the pioneering work of George Irwin done in the 1950s and 1960s.

One key concept is that the presence of a geometric discontinuity, such as a crack amplifies local stress in the vicinity of the flaw. Consider two glass rods of equal dimension, one of which has been scratched with a file. Which rod is easier to break? If you have ever tried this experiment, you know that the scratched rod can be broken with much less effort. The reason is that the applied stress is amplified by the crack. It is also the reason why modern airplanes have oval rather than rectangular windows (the windows are also made out of polycarbonate rather than oxide glass).

9.4.1 The Stress Intensity Parameter

Consider a structural component containing a sharp crack, subjected to a load applied in a direction normal to the crack surface, as shown in Figure 9.4–1. Any geometry for which the load is applied perpendicular to the crack surface is referred to as mode I loading. For isotropic and elastic materials, the normal stress in the y direction, σ_y, at a point located at the angle θ and distance r from the crack tip can be expressed as:

$$\sigma_y = \left(\frac{K}{\sqrt{2\pi r}} \right) f(\theta) \tag{9.4--1}$$

where $f(\theta)$ is trigonometric function of the angle θ. The most common units for K in the SI system are MPa-$\sqrt{m}$. Equation 9.4–1, along with the corresponding equations for the normal stress in the x direction and the shear stresses in the xy plane, are called the field equations. The parameter K is a measure of the magnitude of the stress field in the crack tip region and is called the **stress intensity parameter.** In effect, it describes the extent of stress amplification resulting from the flaw. The utility of this analysis is that as long as the direction of loading is perpendicular to the crack plane, the only term in Equation 9.4–1 that depends on the sample and loading geometry is the stress intensity factor. Thus, geometries having the same values of K have identical crack tip stress fields. This observation provides the geometric generality lacking in the Griffith-Orowan analysis.

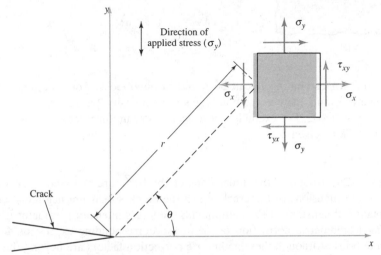

FIGURE 9.4–1   Mode I crack showing the coordinate system and stress components. A mode I crack opens so that all points on the crack surfaces are displaced parallel to the y axis.

FIGURE 9.4–2

Some typical load/crack geometries and their corresponding stress intensity parameters: **(a)** a tunnel crack, **(b)** a penny crack, **(c)** a wedge-opened crack, and **(d)** an eccentrically loaded crack.

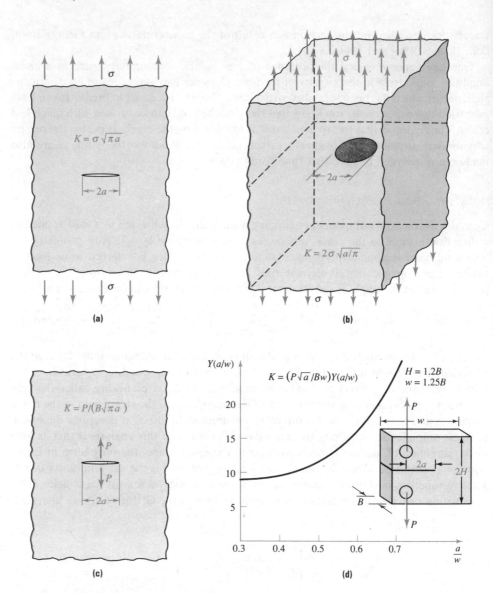

Expressions for estimating K can be found in handbooks. Some common load and crack geometries and their corresponding stress intensity parameter expressions are shown in Figure 9.4–2. For example, for a plate under uniform tensile stress containing a center crack, K is given by:

$$K = \sigma\sqrt{\pi a} \qquad\qquad (9.4\text{–}2)$$

In Figure 9.4–2a through c, the dimensions of the body are assumed to be very large relative to the dimensions of the crack. When the crack size is not negligible compared with the planar dimensions of the component, the stress intensity parameter is obtained by applying a geometric correction factor to the corresponding expression for K in a semi-infinite body. Although these geometric correction factors are beyond the scope of this text, we note that omitting their use, or, equivalently, assuming a geometric correction factor of 1, always leads to conservative design estimates.

The direct relationship between K and the stresses and strains at the crack tip led Irwin to propose the following hypothesis:

$$\text{At fracture, } K \rightarrow K_{critical} \tag{9.4-3}$$

This means that there is a critical value of the stress intensity parameter at which fracture will take place.

9.4.2 The Influence of Sample Thickness

The value of $K_{critical}$ depends not only on the material being considered but also on the thickness of the cracked body. The critical value of the stress intensity parameter in thick sections during mode I loading is denoted by the symbol K_{Ic}. K_{Ic} is a conservative measure of the material's toughness and is widely used for engineering calculations, as will be seen in subsequent sections.

For a given material, the toughness is higher in thin sections than it is in thick sections. The explanation, which we present without derivation, is that thin sections undergo comparatively more plastic deformation per unit volume than thicker sections. One practical application of the thickness dependence of toughness is that padlocks, which are designed to be as tough as possible to prevent opening when subjected to impact loading, are often fabricated out of many thin layers of metal. Experimental results of fracture toughness measurements have shown that the critical thickness B is related to both K_{Ic} and the yield stress of the material, σ_{ys}, through the equation:

$$B = 2.5\left(\frac{K_{Ic}}{\sigma_{ys}}\right)^2 \tag{9.4-4}$$

Thus, for a high-strength Al alloy for which K_{Ic} is about 28 MPa-$\sqrt{m}$ and the yield strength is about 520 MPa, the critical thickness is about 0.72 cm.

···

EXAMPLE 9.4–1

Suppose the fracture toughness of a titanium alloy has been determined to be 44 MPa-$\sqrt{m}$ and a penny-shaped crack of diameter 1.6 cm is present in a plate used in uniaxial tension. Calculate the maximum allowable stress that can be imposed without fracture. The yield stress of the material is 900 MPa, and the plate thickness is 5 cm.

Solution

The stress intensity parameter formula for a penny-shaped crack is given in Figure 9.4–2b as:

$$K = 2\sigma\sqrt{\frac{a}{\pi}}$$

where a is the crack length and σ is the applied stress. At fracture, the applied stress intensity is equal to the fracture toughness. In equation form, $K = K_{Ic}$. Thus, the failure condition becomes:

$$K_{Ic} = 2\sigma_f\sqrt{\frac{a}{\pi}}$$

Solving for σ_f and substituting the values in the problem statement gives:

$$\sigma_f = \frac{K_{Ic}}{2}\sqrt{\frac{\pi}{a}} = \left(\frac{44 \text{ MPa-}\sqrt{m}}{2}\right)\sqrt{\frac{\pi}{0.008 \text{ m}}}$$

$$= 436 \text{ MPa}$$

For this analysis to be valid, we must check to make sure that the sample thickness exceeds the critical thickness given in Equation 9.4–4 as:

$$B = 2.5\left(\frac{K_{Ic}}{\sigma_{ys}}\right)^2 = 2.5\left(\frac{44 \text{ MPa-}\sqrt{m}}{900 \text{ MPa}}\right)^2$$

$$= 0.006 \text{ m}$$

Since the plate thickness is greater than B, the use of K_{Ic} in the failure condition is appropriate. Note that brittle fracture will occur well below the material's yield stress. Thus, there is no guarantee that fracture will be prevented simply by specifying that the applied stress must be below the yield stress.

EXAMPLE 9.4–2

Maraging steel (300 grade) has a yield strength of approximately 2100 MPa and a toughness of 66 MPa-$\sqrt{m}$. A landing gear is to be fabricated from this material, and the maximum design stress is 70% of yield. If flaws must be 2.5 mm long to be detectable, is this a reasonable stress at which to operate? Assume that small edge cracks are present and the stress intensity parameter for this geometry is $K = 1.12 \, \sigma \sqrt{\pi a}$.

Solution

The flaw size at which fracture occurs is calculated by noting that at fracture, $K = K_{Ic}$. Thus, the failure condition $K_{Ic} = 1.12\sigma\sqrt{\pi a}$. Solving this expression for a and substituting the values in the problem statement gives:

$$a_f = \frac{1}{\pi}\left(\frac{K_{Ic}}{1.12\sigma}\right)^2 = \frac{1}{\pi}\left[\frac{K_{Ic}}{(1.12)(0.7\sigma_{ys})}\right]^2$$

$$= \frac{1}{\pi}\left[\frac{66 \text{ MPa-}\sqrt{m}}{1.12(0.7)(2100 \text{ MPa})}\right]^2$$

$$= 5.1 \times 10^{-4} \text{ m} = 0.51 \text{ mm}$$

Thus, critical flaws may escape detection even though the design stresses for the part are below the yield stress. Consequently, the stress is too high to ensure safe operation of the landing gear.

9.4.3 Relationship between Fracture Toughness and Tensile Properties

The fracture toughness of a material generally scales with the area under the stress-strain curve. This observation is useful in helping to explain why metals generally exhibit higher K_{Ic} values than either ceramics of polymers. Ceramics have high strength but low ductility, and unoriented polymers have high ductility but low strength. In either case, the area under the σ-ε curve is limited, and the corresponding K_{Ic} values are low. In contrast, metals have reasonably high strength and ductility. This combination gives them a large area under the σ-ε curve and the highest fracture toughness. In fact, this is why metals are favored in many critical structural applications.

What about the variation of fracture toughness values within any single class of materials? For metals, fracture toughness has been related to other mechanical properties through the expression:

$$K_{Ic} \propto n(E\sigma_{ys}\varepsilon_{ft})^{1/2} \tag{9.4–5}$$

where n is the strain hardening exponent, E is Young's modulus, σ_{ys} is the yield strength, and ε_{ft} is the true fracture strain. As a practical matter, K_{Ic} generally decreases with increasing strength, since both n and ε_f decrease with increasing strength. As a design rule, therefore, one is usually more concerned about brittle failure in high-strength metals (and in other high-strength materials) than in ductile metals.

9.4.4 Application of Fracture Mechanics to Various Classes of Materials

Fracture mechanics was initially developed so that high-strength materials of limited ductility could be safely used in engineering situations. This formulation is referred to as linear elastic fracture mechanics (LEFM). While a fracture mechanics approach has been developed for more ductile materials, its treatment is beyond the scope of this text. To apply the LEFM methodology correctly, fracture must occur under essentially elastic conditions. Practically, this means that there can be only limited plastic deformation at the crack tip at the time of fracture. Additionally, the materials must be homogenous and isotropic, and the failure must be the result of the growth of a well-defined single crack. In materials for which the above conditions are met, the LEFM approach works well.

One important class of materials for which fracture mechanics is frequently not applicable is composites. In composites, cracks tend to be spatially distributed, and the requirement of a single, well-defined crack is not met. Also, the homogenous and isotropic requirements are not satisfied for most composite or oriented polymer systems.

Application in Ceramics

Recall that one of the reasons ceramics are more brittle than metals is that dislocation motion is severely restricted in ceramics. How can this effect be explained within the LEFM model? The extent of stress amplification at the crack tip is related to the geometry of the crack through the equation:

$$\sigma_{max} \propto \sqrt{\frac{a}{\rho}} \tag{9.4-6}$$

where σ_{max} is the maximum stress at the tip of the crack, a is the crack length, and ρ is the crack tip radius (see Figure 9.4–3). Note that Equation 9.4–6 implies that long and sharp cracks are the most serious types of flaws. In metals, the motion of dislocations results in an increase in the crack tip radius known as crack blunting. This reduces the stress amplification and, therefore, the driving force for crack extension. Since this crack blunting mechanism is not available in ceramics, they generally display much lower K_{Ic} values.

Once we understand the crack blunting mechanism in metals, however, we can appreciate the ceramic toughening mechanism known as microcracking. As shown in Figure 9.4–4, if the microstructure of a ceramic can be altered so that it contains microscopic voids, then when an advancing crack enters one of these voids, its tip radius will increase significantly. Although the crack length also increases upon entering the

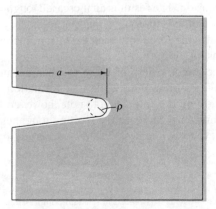

FIGURE 9.4–3

An illustration of the geometry of a surface crack showing the crack tip radius ρ.

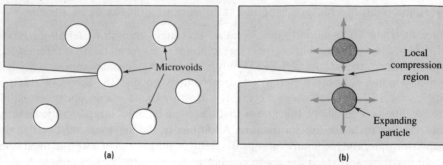

FIGURE 9.4–4 Toughening mechanisms for ceramics. **(a)** Microcracking—when the advancing crack enters the microvoid, its length increases slightly but its crack tip radius increases significantly, decreasing the stress amplification and correspondingly reducing the driving force for crack extension. **(b)** Residual compression at the crack tip—selected regions of a microstructure are induced to expand in the vicinity of the crack tip so that a local state of compression counteracts the externally applied tensile stress.

microvoid, the tip radius increases by a much larger factor so that the ratio a/ρ in Equation 9.4–6 decreases significantly. Thus, although the crack blunting mechanism is different in ceramics than in metals, the result is the same—a decrease in the driving force for crack extension.

How can these microvoids be inserted into the ceramic microstructure? One method is to obtain a microstructure that at elevated temperatures is composed of a roughly spherical phase surrounded by a second-phase matrix. If the coefficient of expansion for the spherical phase is greater than that of the matrix, then when the ceramic is cooled to room temperature, the spherical phase will contract more than the matrix. If the phase boundary is weak, this differential contraction results in the formation of a "gap" between the two phases that displays the desired characteristics.

The same mechanism is occasionally used to stop crack extension in large-scale metal structures. If the tip of an advancing crack can be located using a nondestructive testing method, then one can drill a hole in front of the crack so that when the crack enters the hole, its radius increases significantly. (This was used on the Liberty ships).

Let us return to the idea of a two-phase ceramic microstructure containing a spherical second phase. Suppose that the spherical phase has a lower expansion coefficient than the matrix. What happens when the ceramic is cooled from the fabrication temperature to room temperature? As shown in Figure 9.4–4b, the spherical phase contracts less than the matirx. The result is that the matrix material located between two nearby second-phase particles is placed in residual compression.[4] A crack attempting to enter this volume of the matrix phase will experience an effective reduction in the stress component responsible for crack extension. The result is an increased toughness value for the ceramic. Although there are other methods for toughening ceramics, they are all based on the same principles: decrease the driving force for crack extension, increase the amount of energy required for crack extension, or both.

As an example of the effectiveness of these toughening mechanisms, pure zirconia (ZrO_2) has a K_{Ic} value of ~2 MPa-$\sqrt{m}$, while transformation-toughened zirconia has a K_{Ic} value of ~9–13 MPa-$\sqrt{m}$. It is important to note, however, that while the mechanism described above can significantly increase the fracture toughness of ceramics, the data in Appendix D show that even the toughest ceramics generally have lower K_{Ic} values than most metals.

---

[4] There are other methods for creating a volume expansion in the spherical second phase, but the result is the same—the matrix between particles is placed in residual compression.

EXAMPLE 9.4–3

Consider a ceramic component containing a crack of initial length 0.5 mm with a crack tip radius of 0.5 nm. Estimate the reduction in the driving force for crack extension if this crack enters an adjacent roughly spherical microvoid with a diameter of 1 μm.

Solution

The driving force for crack growth scales with the magnitude of the stress amplification at the crack tip. Equation 9.4–6 states that the maximum stress at the crack tip is given by the expression:

$$\sigma_{max} \propto \sqrt{\frac{a}{\rho}}$$

Therefore, the ratio of the driving force before and after the crack enters the void ($D_{final}/D_{initial}$) is given by:

$$\frac{D_{final}}{D_{initial}} = \frac{\sqrt{a_f/\rho_f}}{\sqrt{a_i/\rho_i}}$$

The final crack length is equal to the initial crack length plus the diameter of the microvoid, and the final crack tip radius is equal to the radius of the microvoid. Substituting the appropriate values into the driving-force ratio yields:

$$\frac{D_f}{D_i} = \sqrt{\frac{(5 \times 10^{-4}\ m + 1 \times 10^{-6}\ m)/(5 \times 10^{-7}\ m)}{(5 \times 10^{-4}\ m)/(5 \times 10^{-10}\ m)}}$$

$$= 0.032$$

Thus, in this example the presence of the microvoids decreases the driving force for crack extension to roughly 3 percent of its initial value.

EXAMPLE 9.4–4

Compare the critical flaw sizes in a ductile aluminum sample ($K_{Ic} = 250$ MPa-$\sqrt{m}$), a high-strength steel ($K_{Ic} = 50$ MPa-$\sqrt{m}$), a pure zirconia sample ($K_{Ic} = 2$ MPa-$\sqrt{m}$), and a transformation-toughened zirconia sample ($K_{Ic} = 12$ MPa-$\sqrt{m}$). Each is subjected to a tensile stress of 1500 MPa. Assume that the stress intensity parameter for this geometry is $K = 1.12\sigma\sqrt{\pi a}$.

Solution

Solving the stress intensity parameter equation for the critical flaw size yields:

$$a_c = \frac{1}{\pi}\left(\frac{K_{Ic}}{1.12\sigma}\right)^2$$

Substituting the values in the problem statement gives critical flaw sizes of 7 mm (7000 μm), 280 μm, 0.45 μm, and 16 μm, respectively, for the aluminum, steel, pure zirconia, and transformation-toughened zirconia. This problem points out the extreme sensitivity of ceramics to the presence of even microscopic flaws.

Applications in Polymers

The combination of high ductility but low strength in polymers generally results in a limited area under the σ-ε curve and correspondingly low fracture toughness values. Typical values for polymers are in the 0.5 to 7 MPa-$\sqrt{m}$ range. This compares with a range of 0.5 to 13 MPa-$\sqrt{m}$ for ceramics and 20 to 150 MPa-$\sqrt{m}$ for most metals.

Although it is possible to find tabulated K_{Ic} values for most polymers, the use of these values in design calculations requires some caution. The reason is that some of the

TABLE 9.4–1   Calibration function for compact tension specimens.

a/W	$f(a/W)$	a/W	$f(a/W)$
0.450	8.34	0.505	9.81
0.455	8.46	0.510	9.96
0.460	8.58	0.515	10.12
0.465	8.70	0.520	10.29
0.470	8.83	0.525	10.45
0.475	8.96	0.530	10.63
0.480	9.09	0.535	10.80
0.485	9.23	0.540	10.98
0.490	9.37	0.545	11.17
0.495	9.51	0.550	11.36
0.500	9.66		

assumptions on which fracture mechanics (and, in particular, LEFM) are based are rather questionable for many polymers. Specifically, the assumptions of limited plastic deformation at the crack tip and constant modulus are not appropriate for thermoplastic polymers above their glass transition temperature or semicrystalline polymers. For thermoplastics below T_g and most thermoset polymers, however, one can use the concept of fracture toughness with some confidence.

Fracture mechanics can be used to explain the mechanism of energy absorption in bulletproof vests made from polymers. The fibers used (Kevlar and Spectra) are composed of highly oriented molecules. The impact of the projectile splits the fibers longitudinally into *many* fibrils, creating a substantial amount of new surface. The energy of the projectile is dissipated by the formation of the new surfaces.

9.4.5 Experimental Determination of Fracture Toughness

K_{Ic} is a materials parameter of considerable engineering significance. The American Society for Testing Materials (ASTM) has developed detailed procedures[5] for determining K_{Ic}. Frequently a standard compact-type specimen, shown in Figure 9.4–2, is used to experimentally determine the fracture toughness of materials. The critical K at fracture is calculated using the expression:

$$K_{Ic} = \left(\frac{P_f}{BW^{1/2}}\right) f\left(\frac{a}{W}\right) \tag{9.4–7}$$

where P_f is the fracture load, B is the specimen thickness, W is the specimen width, and $f(a/W)$ is a calibration function given in Table 9.4–1.

Variations of this procedure are recommended for polymers and ceramics where the considerations for loads, rates, and gripping are somewhat different. Obtaining K_{Ic} values for ceramics requires specialized techniques, since they are usually brittle, difficult to machine, and difficult to load into conventional test fixtures without breaking. To obtain toughness values for ceramics, a test very similar to a hardness test is frequently used. In this technique an indentation is made on the polished surface of a ceramic using a diamond indenter. Cracks form at the corners of the indent mark. The size of the cracks can be measured, and—from the knowledge of the stress intensity parameter for this geometry—

---

[5] Standard Test Method for Fracture Toughness of Metallic Materials, ASTM Standard E-399, *Annual Book of ASTM Standards*.

FIGURE 9.4–5
Vickers indentation in the surface of a single-crystal cubic zirconia (Y_2O_3-stabilized). Radial cracks are used to calculate the fracture toughness. *(Source: Courtesy of Joseph K. Cochran.)*

reasonable estimates of the fracture toughness can be obtained. An indentation on the surface of cubic zirconia is shown in Figure 9.4–5, with cracks emanating from the corners.

Typical fracture toughness values for some common metals, ceramics, and polymers are given in Appendix D.

9.5 FATIGUE FRACTURE

Fatigue is the most common mechanism of failure and is believed to be either fully or partially responsible for 90% of all structural failures. This failure mechanism is known to occur in metals, polymers, and ceramics. Of these three classes of structural materials, ceramics are least susceptible to fatigue fractures. The phenomenon of fatigue is best illustrated by a simple experiment. Take a metal paper clip and bend it in one direction until it forms a sharp kink. The clip undergoes plastic deformation in the region of the kink but does not fracture. If we now reverse the direction of bending and repeat this process a few times, the paper clip will fracture. Thus, under the action of cyclic loading, the paper clip breaks at a much lower load than would be required if it were pulled to fracture using a monotonically increasing load. While the initial loading causes the metal in the paper clip to strain-harden, repeated load application causes internal fatigue damage. In a simplified view of this process, the plastic deformation causes dislocations to move and to intersect one another. The intersections decrease the mobility of the dislocations, and continued deformation requires the nucleation of more dislocations. The increased dislocation density degrades the crystallographic perfection of the material, and eventually microcracks form and grow to a sufficiently large size that failure occurs.

9.5.1 Definitions Relating to Fatigue Fracture

Figure 9.5–1 shows a typical fatigue load cycle as characterized by a variation in stress as a function of time. The maximum and minimum levels of stress are denoted by S_{max} and S_{min}, respectively.[6] The range of stress, ΔS, is equal to $S_{max} - S_{min}$, and the stress amplitude, S_a, is $\Delta S/2$. A fatigue cycle is defined by successive maxima (or minima) in load or

---

[6] The symbol S is used to represent engineering stress by most specialists in the area of fatigue. We have therefore elected to employ this convention in our discussion of fatigue.

FIGURE 9.5–1

Typical fatigue loading cycles and associated
definitions (see text for discussion).

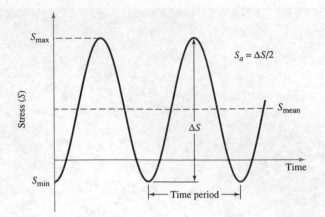

stress. The number of fatigue cycles to failure is designated by N_f. The number of fatigue
cycles per second is called the cyclic frequency. The average of the maximum and
minimum stress levels is called the mean stress, S_{mean}.

If we subject a well-polished test specimen of a structural material to fatigue stresses
of different amplitudes and keep other load variables the same, the number of fatigue
cycles to failure, N_f, is found to correlate uniquely with the stress amplitude. If the same
set of experiments is repeated at a different mean stress or stress ratio, the relationship
between N_f and S_a changes (see Figure 9.5–2). The curves shown in these figures are
commonly referred to as S-N curves (stress versus number of cycles).

Several classes of materials, including carbon steels and some polymers, exhibit a
limiting stress amplitude called the **endurance limit** S_e, below which fatigue failure does
not occur regardless of the number of cycles. The endurance limit is a function of the
applied mean stress and decreases with increasing mean stress (see Figure 9.5–2). This
is an important material property for cyclically loaded components designed for long life,
including rotating parts such as train wheels and axles and reciprocating parts like pistons
rods.

Some materials, such as nylon, aluminum, copper, and other FCC metals, do not
exhibit a well-defined endurance limit. The S-N curve continues to slope downward (see
Figure 9.5–3). An operational endurance limit is defined for such materials as the stress
amplitude corresponding to 10^7 cycles to failure. Loading frequency is important in determining the fatigue behavior only when time-dependent effects are significant.

FIGURE 9.5–2

The number of fatigue cy-
cles to failure, N_f, as related to the amplitude of
the fatigue stress, S_a, and
the mean stress, S_{mean}.

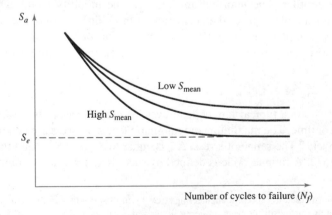

FIGURE 9.5–3

An operational definition of fatigue endurance limit based on a fatigue life of 10^7 cycles.

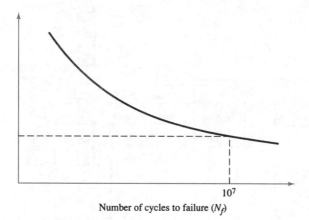

Number of cycles to failure (N_f)

When the amplitude of cyclic loading exceeds the yield strength of the material, plastic deformation occurs in the specimen during each cycle. Under these conditions, it is observed that N_f begins to decrease even more with increasing stress amplitude. In several materials this transition occurs at about 10^4 cycles to failure. For lives less than 10^4 cycles, the process is called low cycle fatigue (LFC), while for lives greater then 10^4 cycles, the process is called high cycle fatigue (HCF). Since the microscopic fatigue mechanism differs in the two regimes, it is important that test data obtained in one regime not be extrapolated to the other regime.

9.5.2 Fatigue Testing

The fatigue test most commonly used to determine the endurance limit is the rotating bending test. A simple setup for rotating bending tests is shown in Figure 9.5–4. While these tests are an inexpensive means of obtaining endurance limit data, the control of mean stress and other test parameters is not always easy. Furthermore, the volume of material at maximum stress is small and may not be representative of the microstructure being tested.

Several advances in fatigue-testing technology have occurred since the late 1960s. Fatigue machines with the ability to control the applied load at high frequencies have

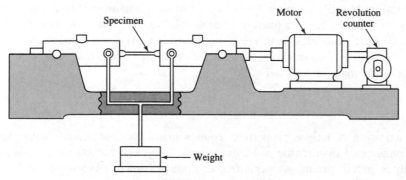

FIGURE 9.5–4 Schematic of the R. R. Moore reversed bending fatigue machine. *(Source: Wayne Hayden, W. Moffatt, and J. Wulff, Mechanical Behavior, "The Structure and Properties of Materials," Vol. III, p. 15. Copyright © 1965 by John Wiley & Sons. Reprinted by permission of John Wiley & Sons, Inc.)*

FIGURE 9.5–5

A modern setup used for fatigue testing. *(Source: Metals Handbook, "Mechanical Testing," Vol. 8, 9th ed., 1985, p. 395, ASM International, Materials Park, OH. Reprinted by permission of the publisher.)*

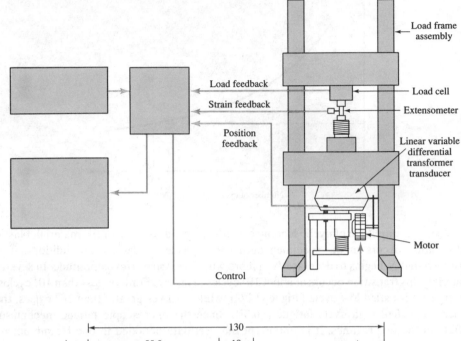

FIGURE 9.5–6

Geometry of a typical axial fatigue specimen (dimensions in mm).

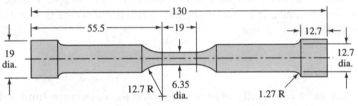

become available. Figure 9.5–5 shows a modern fatigue test setup. The loads are applied axially instead of in the bending mode used in earlier test methods. Thus, for specimens having uniform cross sections, the fatigue stress is constant throughout the specimen. A typical specimen geometry is shown in Figure 9.5–6. These advances have lead to a more accurate characterization of fatigue behavior. Also, tests can be conducted under more realistic service conditions, such as at elevated temperature or under corrosive environments.

A variation of fatigue testing, known as dynamic mechanical testing, can be used to measure the quantity tan δ defined in Section 9.2.2 as the ratio of the complex and real parts of the elastic modulus. In this test, the system input is $\varepsilon = \varepsilon_0 \sin(\omega t)$ and the output is $\sigma = \sigma_0 \sin(\omega t + \delta)$. For elastic materials, $\delta = 0$; for viscous materials, $\delta = 90°$; and for viscoelastic materials, $0 < \delta < 90°$.

9.5.3 Correlations between Fatigue Strength and Other Mechanical Properties

Characterization of fatigue properties requires many specimens, and fatigue tests are more complicated than tensile or hardness tests. They also require more elaborate and carefully prepared specimens, specialized equipment (which is expensive), and considerable time (ranging from several hours to weeks). It is therefore important to establish correlations between fatigue behavior and other more easily measurable mechanical

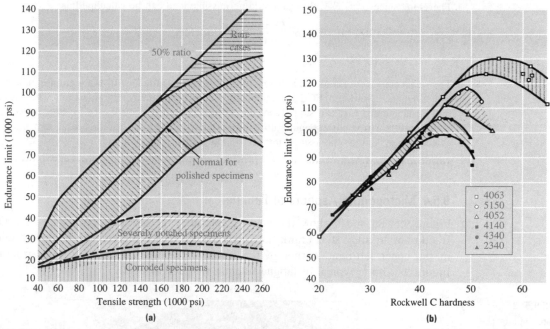

FIGURE 9.5–7   Correlation between fatigue and tensile properties: **(a)** endurance limit and tensile strength, and **(b)** endurance limit and hardness.   *(Source: **(a)** O. Horger, ASME Handbook of Metals Engineering Design, McGraw-Hill, 1953. Reprinted with permission of McGraw-Hill, Inc.* **(b)** *Interpretation of Tests and Correlation of Service, 1951, p. 12, ASM International, Materials Park, OH. Reprinted by permission of the publisher.)*

properties, such as tensile or yield strength or surface hardness. Figure 9.5–7a shows the correlation between the endurance limit and tensile strength for several materials; Figure 9.5–7b shows the correlation between endurance limit and Rockwell C hardness. As a rule of thumb, the following relationship between tensile strength and endurance limit is observed:

$$0.25 < \frac{S_e}{\sigma_{uts}} < 0.5 \tag{9.5–1}$$

EXAMPLE 9.5–1

A structural component of cross-sectional area 5 cm^2 is fabricated from a plain carbon steel with $\sigma_{uts} = 800$ MPa. Calculate the maximum permissible stress to which this component can be subjected if it must survive an infinite number of loading cycles. Repeat this calculation for a ductile aluminum alloy with $\sigma_{uts} = 280$ MPa.

Solution

If the component must survive an infinite number of loading cycles, then the applied stress must be below the endurance limit, S_e, for the alloy. Equation 9.5–1 gives the relationship between S_e and the ultimate tensile strength as:

$$0.25 < \frac{S_e}{\sigma_{uts}} < 0.5$$

To provide a conservative estimate, we will use the lower bound for S_e:

$$S_e \approx 0.25\sigma_{uts} = (0.25)(800 \text{ MPa}) = 200 \text{ MPa}$$

Therefore, since $\sigma = F/A$, the maximum permissible load can be calculated as:

$$F = (200 \text{ MPa})(5 \text{ cm}^2)\left(\frac{1 \text{ m}}{100 \text{ cm}}\right)^2 = 0.1 \text{ MN}$$

$$= 10^5 \text{ N}$$

The steel component should be able to withstand a cyclic load of $\sim 10^5$ N for an infinite number of cycles. On the other hand, FCC aluminum alloys do not exhibit an endurance limit. There is no load that an aluminum part can support for an "infinite" number of cycles. In practice, however, one could find the stress corresponding to failure after 10^7 cycles and use this value in a similar load calculation as long as the assumption of 10^7 was stated in the solution.

9.5.4 Microscopic Aspects of Fatigue

Figure 9.5–8a and b show the macroscopic and microscopic appearance of a fatigue fracture of an alloy steel crankshaft from a truck diesel engine. The fracture originated in the upper left-hand corner of the image in Figure 9.5–8a. The fine structure in Figure 9.5–8b is typical of fatigue fracture surfaces in steel.

FIGURE 9.5–8

Photographs of a fatigue fracture surface taken from a diesel engine component: **(a)** macroscopic appearance, and **(b)** microscopic view obtained from examination in an electron microscope in the cracked propagation area (200×).

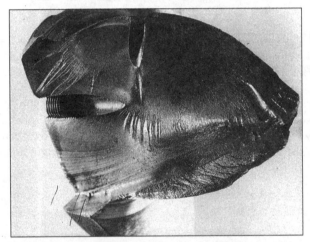

(a)

(b)

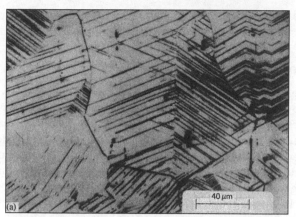

FIGURE 9.5–9   Slip band formation leading to localized deformation during fatigue loading in Waspaloy (a Ni-based alloy used in jet engines). The dark-appearing regions are slip bands that have formed during fatigue and are precursors to crack initiation.   *(Source: S. D. Antolovich and J.-P. Baïlon, "Standard Technical Publication (STP)," Reference 811, p. 340. Copyright ASTM. Reprinted with permission.)*

Fatigue damage in crystalline metals occurs by a localized slip process. Even though the bulk of the specimen may be undergoing elastic deformation, plastic deformation can occur in localized regions of the specimen surface, as shown by the presence of slip bands in Figure 9.5–9. These regions may have stress raisers, such as small surface imperfections, notches, holes, slightly different chemistry, or dislocation sources that can be more easily activated than in other regions of the specimen. The back-and-forth slip during cycling causes intrusions and extrusions that result in the formation of a notch within the slip band, as shown in Figure 9.5–10a and b. This notch is the nucleus of the fatigue crack, which grows during subsequent cycling and eventually causes catastrophic fracture.

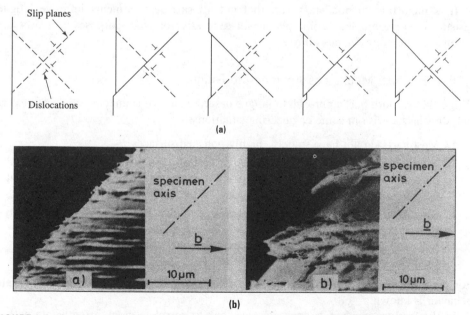

FIGURE 9.5–10   Fatigue crack initiation: **(a)** mechanism for the formation of slip band extrusions and intrusions, and **(b)** notches being formed at the surface of copper specimens as a result of fatigue loading and cyclic deformation. *(Source: **(a)** Cottrell and Hull, Processing 242A (1957), pp. 211–213, The Royal Society. **(b)** Copyright ASTM. Reprinted with permission.)*

Since plastic deformation is necessary to cause fatigue cracks, ceramic materials are highly resistant to fatigue fractures.

9.5.5 Prevention of Fatigue Fractures

One approach to prolonging fatigue life is to reduce the applied stress range. This can be accomplished by avoiding sharp corners and other stress raisers, such as bolt and rivet holes, during design. This is not always a viable solution, because stress raisers cannot be completely removed from the design of complicated components. Since fatigue fractures usually originate from the surface, surface treatments are the most common means for improving fatigue resistance. For example, a component with a smooth surface produced by fine grinding will have superior fatigue resistance to one with a rough surface. If tool or grinding marks cannot be avoided, they should run in a direction parallel to the primary loading direction.

Shot peening is another effective means of improving the fatigue resistance of some metal components. In this process small particles of hard material are shot at a high velocity onto the surface of the part. These particles produce plastic deformation at the surface, which has two beneficial effects. First, strain hardening occurs, thereby providing more resistance to plastic deformation, which is responsible for initiating fatigue cracks. The second effect is to develop compressive residual stresses on the surface that reduce the effective mean stress level during fatigue loading.

Case carburizing (introduced in Section 4.4.1) is a means for improving the fatigue resistance of automobile engine crankshafts made from carbon steel. After the crankshaft is formed, it is heated in a high-carbon-containing environment. This raises the carbon content of the steel near the surface through adsorption and subsequent diffusion. The increased carbon content results in surface hardening and provides superior fatigue resistance. In addition, it also causes compressive residual stresses, which enhance the fatigue performance.

It is important to note, however, that not all surface treatments improve fatigue resistance. For example, chrome plating for decorative or other purposes decreases the fatigue resistance of ferrous alloys.

9.5.6 A Fracture Mechanics Approach to Fatigue

While the "smooth bar" approach to fatigue described above is undoubtedly useful, certain drawbacks arise in some engineering situations:

1. Most parts do not have smooth, highly polished surfaces.
2. In general, the fatigue life is made up of an initiation phase and a propagation phase; however, results are usually given in terms of the total life to failure with no indication of the fraction of life spent in the each phase.
3. Structures usually contain small preexisting cracks. These cracks are frequently unavoidable and arise from fabrication procedures and material defects. They will propagate under repeated loads until they become critical, at which point fracture occurs. In such cases, the entire life is spent in the propagation phase.

For these reasons, overall fatigue life can be calculated only when crack growth rate behavior is known.

In the fracture mechanics approach, we seek to correlate crack growth rate with a parameter that is independent of geometry (i.e., the stress intensity parameter K). By crack growth rate we refer to the amount of crack extension per loading cycle, da/dN. It

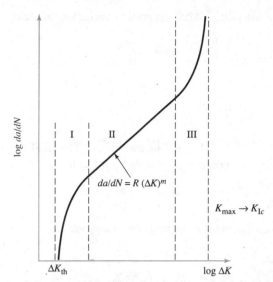

FIGURE 9.5–11   Schematic representation of fatigue crack propagation behavior. In regime I the crack growth rate is low, since the threshold for crack propagation is approached. In regime II the so-called Paris law is obeyed. In regime III the crack growth rate increases above that predicted by the Paris equation, since the fracture toughness of the material is approached and there is local tensile overload fracture.

was experimentally demonstrated by Paul Paris in the 1960s that the rate of fatigue crack growth is uniquely determined by the range of the cyclic stress intensity parameter, $\Delta K = K_{max} - K_{min}$. K_{max} and K_{min} are, respectively, the values of stress intensity corresponding to the maximum and minimum values of the fatigue load or stress. Thus,

$$\frac{da}{dN} = f(\Delta K) \tag{9.5–2}$$

The functional dependence of da/dN on ΔK is shown in Figure 9.5–11. Note that the trends are plotted on a log-log plot. There is a large range of ΔK in which the relationship between $\log(da/dN)$ and $\log(\Delta K)$ is linear. This region is called region II, or the Paris regime, in which the following relationship can be used:

$$\frac{da}{dN} = R(\Delta K)^m \tag{9.5–3}$$

where R and m are material constants. This is the most significant region of fatigue crack growth behavior in engineering applications.

We will illustrate the fracture mechanics approach to fatigue with an example. Suppose that the crack growth rate of an aluminum alloy in region II is given by $da/dN = (8.5 \times 10^{-12})(\Delta K)^4$ where ΔK has units of MPa-$\sqrt{m}$ and da/dN has units of m/cycle. A structural component contains a tunnel crack (Figure 9.4–2a) that is 5 mm long, and the applied stresses vary from 0 to 200 MPa. Given that the alloy has a fracture toughness of 27 MPa-$\sqrt{m}$, how would we estimate the fatigue life of the component?

For this geometry the stress intensity parameter is $K = \sigma\sqrt{\pi a}$. However, to use Equation 9.5–3, we require a formula for the stress intensity parameter range, ΔK, in terms of stress and crack length. The appropriate expression is $\Delta K = \Delta\sigma\sqrt{\pi a}$. Using the information given, the expression for the crack growth rate is:

$$\frac{da}{dN} = (8.5 \times 10^{-12})[(200)\sqrt{\pi a}]^4 = 0.134a^2$$

This equation can be integrated, after separating variables, to yield:

$$\int_{N_0}^{N_f} dN = \left(\frac{1}{0.134}\right) \int_{a_0}^{a_f} \left(\frac{1}{a^2}\right) da$$

or

$$N_f - N_0 = 7.46\left(\frac{1}{a_0} - \frac{1}{a_f}\right)$$

Now $N_0 = 0$ and $a_0 = 2.5 \times 10^{-3}$ m, but what is a_f? The final crack length is that for which fracture occurs and corresponds to the condition $K_{max} = K_{Ic}$. Thus, the failure condition is:

$$K_{Ic} = \sigma_{max}\sqrt{\pi a_f}$$

Solving for a_f and substituting the appropriate values yields:

$$a_f = \frac{1}{\pi}\left(\frac{K_{Ic}}{\sigma_{max}}\right)^2$$

$$= \frac{1}{\pi}\left(\frac{27 \text{ MPa-}\sqrt{m}}{200 \text{ MPa}}\right)^2 = 5.8 \times 10^{-3} \text{ m}$$

Substituting the values for N_0, a_0, and a_f into the expression for N_f gives:

$$N_f = 7.46\left(\frac{1}{2.5 \times 10^{-3}} - \frac{1}{5.8 \times 10^{-3}}\right) = 1698$$

The part is thus expected to last for ~1700 fatigue cycles.

To consider the fatigue crack growth rates in the full ΔK regime, additional factors must be considered at both low and high stress intensity parameter ranges. Although the details of this process are beyond the scope of this text, we note that the fatigue life may *always* be computed by integrating the actual crack growth curve (Figure 9.5–11).

DESIGN EXAMPLE 9.5–2

A steel component with $\sigma_{uts} = 800$ MPa and $K_{Ic} = 20$ MPa-$\sqrt{m}$ is known to contain a tunnel crack (Figure 9.4–2a) of length 1.4 mm. This alloy is being considered for use in a cyclic loading application for which the design stresses vary from 0 to 410 MPa. Would you recommend using this alloy in this application?

Solution

The brittle fracture failure condition for this crack geometry is $K_{Ic} = \sigma_f\sqrt{\pi a}$. Solving for the failure stress and substituting the given values yields:

$$\sigma_f = \frac{K_{Ic}}{\sqrt{\pi a}} = \frac{20 \text{ MPa-}\sqrt{m}}{\sqrt{\pi(7 \times 10^{-4} \text{ m})}} = 426 \text{ MPa}$$

Thus, the component can withstand the maximum static stress of 410 MPa, but the margin of safety is limited. Using a similar calculation, the crack length that will result in brittle failure under a stress of 410 MPa is:

$$a = \frac{1}{\pi}\left(\frac{K_{Ic}}{\sigma}\right)^2 = \frac{1}{\pi}\left(\frac{20 \text{ MPa-}\sqrt{m}}{410 \text{ MPa}}\right)^2$$

$$= 7.57 \times 10^{-4} \text{ m}$$

This represents a crack extension of less than 6 μm. Even without a crack growth rate integration based on the fracture mechanics approach, we can recognize that this material is unsuitable for the intended application. If, however, a more appropriate material is identified, the number of cycles to failure could be estimated using the procedure outlined in this section.

9.6 TIME-DEPENDENT BEHAVIOR

In aggressive environments or at elevated temperatures, both the stress-strain behavior and the fracture of materials become time-dependent. For example, during cyclic loading at elevated temperature or in a corrosive environment, the frequency of loading becomes important in determining the number of fatigue cycles to failure. Some polymers and low–melting-point metals, such as lead, can exhibit time-dependent deformation at room temperature. In such instances, the loading rate must be taken into account in order to accurately describe the mechanical response. For example, the influence of loading rate on the stress-strain behavior of polycarbonate is shown in Figure 9.6–1a.

9.6.1 Environmentally Induced Fracture

An aggressive environment may reduce the ability of a structural material to bear load. Environmentally induced failures occur over a period of time as the environment reacts with the material to degrade material properties. Depending on the applied load level, the time to failure may vary considerably. This is shown in Figure 9.6–2 for high-alloy stainless steels in magnesium chloride solution. This phenomenon is also called delayed

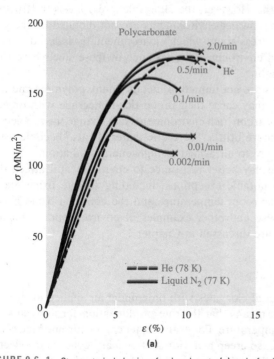
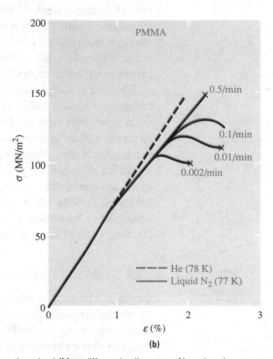

FIGURE 9.6–1   Stress-strain behavior of polycarbonate **(a)** and of poly(methyl methacrylate) **(b)** at different loading rates. Note that the stress response of the material depends strongly on the loading rate.   *(Source: T. Courtney,* Mechanical Behavior of Materials, *McGraw-Hill, 1986. Reprinted with permission of McGraw-Hill, Inc.)*

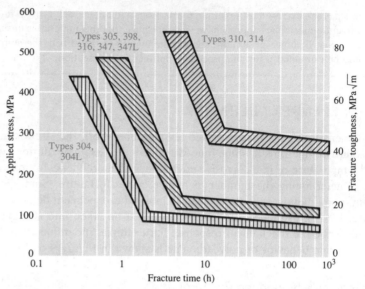

FIGURE 9.6–2   Applied stress versus life of tensile-type stainless steel specimens in a magnesium chloride solution. *(Source: Metals Handbook, "Corrosion," Vol. 13, 9th ed., p. 272, ASM International, Materials Park, OH. Reprinted by permission of the publisher.)*

failure and is particularly important when selecting steels for marine applications such as ship hulls and offshore platforms. For both applications, it is attractive to consider the use of high-strength steels to save weight. However, the allowable stress levels in structural members of high-strength steels in seawater environments are not dictated by the yield strength, but rather by the level of stress above which environmentally induced fracture occurs. Therefore, the strength and environmental resistance of these steels have to be carefully balanced to optimize weight.

Environmentally induced failures are not limited to metals. Many polymers and their composites absorb moisture, which may cause their properties to degrade with time. In other cases, polymers used in hydrocarbon-rich environments can change their molecular structure and in so doing become more brittle. Polymeric materials also become brittle when subjected to prolonged exposure to ultraviolet or high-energy radiation.

Although ceramics generally display better resistance to environmental attack than other materials, there are several notable exceptions, including silicate oxide glasses subjected to water or its vapor near room temperature and the common oxide glasses subjected to hydrofluoric acid. These and other examples of environmentally induced fracture in all classes of materials are discussed in Chapter 15.

9.6.2 Creep in Metals and Ceramics

Creep is a process in which a material elongates with time under an applied load. It is a thermally activated process, which means that the rate of elongation for a given stress level increases significantly with temperature. For example, jet engine turbine blades may reach local temperatures of 1200°C, so creep behavior is of primary concern in selecting suitable materials and processes for them. It should be noted, however, that the term "high temperature" is relative and depends on the material being considered. For jet engine

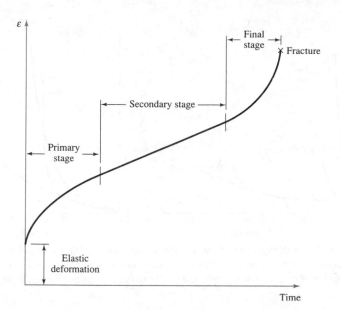

FIGURE 9.6–3

Schematic representation of creep strain as a function of time during a constant-stress creep test.

materials high temperature may be above 800°C, while for polymers and tin-lead solder alloys, high temperature may be 25°C! Creep behavior is extremely sensitive to the microstructure of the material, to its prior processing and mechanical history, and to composition. It is thus an important property that can be usefully manipulated through judicious choices of composition and processing history.

A schematic creep curve is shown in Figure 9.6–3. Initially there is an instantaneous elastic strain ε_0. After that, the strain begins to change with time. We see that although the creep strain *increases* with time, the creep rate in the initial region *decreases* with increasing time. In essence, the internal substructure (e.g., arrangement of dislocations) is changing to be in equilibrium with the applied load. This region is referred to as stage I, or primary, creep and is often described by the equation:

$$\varepsilon = At^{1/3}. \tag{9.6–1}$$

where ε is the creep strain, A is a material constant, and t is time.

At some point, dynamic equilibrium is established between the applied load and the microstructure of the material so that a minimum creep rate is attained. In this region there is a linear relationship between creep strain and time:

$$\varepsilon = \varepsilon_0 + \beta t \tag{9.6–2}$$

where β is a material constant.[7] This region is called stage II, or secondary, creep. For temperatures and loads typical of engineering applications, stage II lasts far longer than any other stage and is the most important engineering property of the stress/strain/time curve. The strain rate in this regime, termed the minimum creep rate β, is used in computations of the useful life of the component.

Finally, when considerable elongation has taken place, gross defects begin to appear inside the material, and its rate of elongation increases rapidly. This is called stage III, or

---

[7] Note the similarity between the equation for the steady-state creep rate and that for the phenomenon described as time-dependent deformation in Section 9.2.2.

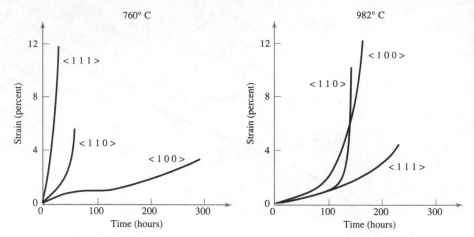

FIGURE 9.6–4 Typical creep curves for a single-crystal Ni-base superalloy PWA 1480 used in jet engines. Note that the behavior depends on temperature, stress level, and the crystallographic direction. *(Source:* Processing and Properties of Advanced High Temperature Alloys, *1986, p. 41, ASM International, Materials Park, OH. Reprinted by permission of the publisher.)*

tertiary, creep, and in this region the strain increases exponentially with time. The strain versus time relationship for constant stress in stage III is given by:

$$\varepsilon = B + C \exp(\gamma t) \tag{9.6–3}$$

where B, C, and γ are material constants. While most of the deformation in creep is accumulated during stage III, this strain is not useful in an engineering sense, since it is accumulated in such a short time. A typical set of creep curves is given in Figure 9.6–4 for a single-crystal Ni-base superalloy used for turbine blades in jet engines.

9.6.3 Mechanisms of Creep Deformation

In Chapter 4 we saw that both the equilibrium concentration of point defects and their mobility increase exponentially with temperature. What happens when a stress is applied to a crystal at elevated temperatures?

Diffusion at the atomic level is involved in a significant way in all the above creep processes; therefore, the increased atomic mobility at elevated temperature is quite important. Creep thus begins to be a concern at temperatures greater than $0.3T_m$, where T_m is the absolute melting temperature. At $0.5T_m$, creep is a serious concern and may be the dominating factor in design considerations.

Consider the geometry in Figure 9.6–5. Excess vacancies are created on those grain boundary facets that are normal to the applied load (facets AB and CD) and tend to migrate toward the facets of the grains that are parallel to the stress direction (facets AC and BD). The atomic movement is in the direction opposite to that of vacancy movement. At a constant stress, this results in a net elongation of the grain in the applied stress direction. This mechanism of deformation is called Nabarro-Herring creep. Coble creep is another mechanism of creep deformation in which vacancies migrate via diffusion along the grain boundaries rather than through the interior of the grain.

Figure 9.6–6 shows a third mechanism of creep deformation, called dislocation climb. In this mechanism the dislocation climbs, or moves up, one atomic distance by migration of an entire row of vacancies to the extra half plane of atoms associated with an edge dislocation. Grain boundary sliding is yet another mechanism of accommodating creep

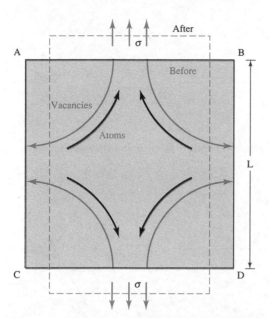

FIGURE 9.6–5 Nabarro-Herring creep mechanism based on migration of vacancies through the bulk of the grain.

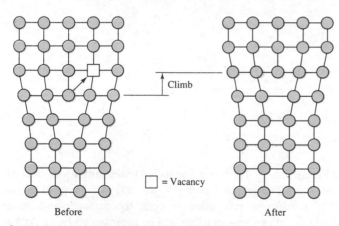

IGURE 9.6–6 Creep deformation mechanism involving dislocation climb.

train. In this mechanism, two adjoining grains slide along their common boundary under he action of shear stress, as shown in Figure 9.6–7.

Creep deformation occurs in ceramics, but it is considerably more difficult than in 1etals because vacancy migration can take place only under conditions of electroneutral-y. This means that there is a coupling between the migration of anions and cations. This oupling, as discussed in Chapter 4, puts an additional restriction on diffusion and thus mits the creep rate. As a result, ceramics are inherently more resistant to creep deforma-on than are metals. In addition, because of the stronger bonds of ceramics, the melting :mperature is higher and the point defect concentration lower than in metals. Both actors contribute to the low creep rates characteristic of ceramics; however, ceramics sed at very high temperatures have significant creep rates.

As discussed above, creep deformation may eventually lead to rupture. The rate of creep eformation and creep rupture times for a given material vary with temperature and the pplied stress level. Thus, test methods for creep focus on measuring the creep rate (strain er hour) and rupture time as a function of these variables. A typical setup for a creep

FIGURE 9.6–7
Grain boundary sliding and
the resultant cracks that
form at triple points.

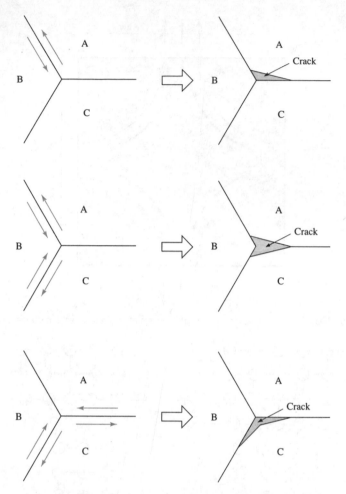

test is shown in Figure 9.6–8. The test is conducted under conditions of constant temperature and constant stress. The graphs shown in Figure 9.6–4 are typical results of such tests.

Since creep is a diffusion-controlled process, the primary and steady-state creep behavior for a wide variety of temperatures can be represented by an Arrhenius relationship. Dorn defined a time parameter θ as follows:

$$\theta = t \exp\left(-\frac{Q}{RT}\right) \tag{9.6–4}$$

where t is time, Q is the activation energy for self-diffusion, T is absolute temperature, and R is the universal gas constant. The creep strain in region II can be plotted as a function of the θ parameter to collapse all data onto a single curve, as shown at several temperatures for aluminum in Figure 9.6–9.

Larsen and Miller proposed a parameter P that is widely used to correlate stress versus time to rupture data over a range of temperatures. P is defined as:

$$P = T(\log t_r + C_l) \tag{9.6–5}$$

where t_r is the rupture time and C_l is a constant determined experimentally. Figure 9.6–10 is a plot of rupture stress versus the Larsen-Miller parameter for a Ni-base alloy known as Astroly. The curve can be used to determine the creep rupture life at a given temperature. It can also be used to determine the highest temperature at which the material can be used, given the operating stress and the required lifetime.

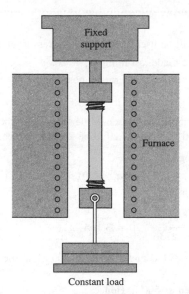

FIGURE 9.6–8

A typical setup for creep testing.

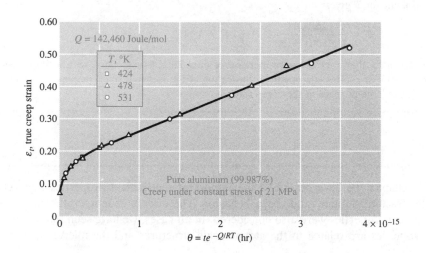

FIGURE 9.6–9

Representation of strain versus θ. θ is a parameter that incorporates both time and temperature and is defined by Equation 9.6–4. *(Source:* Creep and Recovery, *1957, ASM International, Materials Park, OH. Reprinted by permission of the publisher.)*

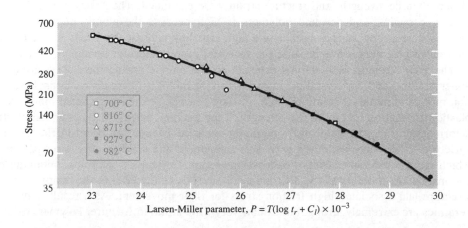

FIGURE 9.6–10

Master curve plotting creep rupture data at various temperatures for Astroly. Such a plot is called a Larsen-Miller diagram.

....

EXAMPLE 9.6–1

An Astroly jet engine blade will be used at 871°C at a stress level of 200 MPa.

    a. Determine the life of the blade, assuming $C_l = 20$.

    b. Estimate the maximum service temperature possible if a life of 500 hours is required.

Solution

    a. The first step is to determine the Larsen-Miller parameter for a stress level of 200 MPa. This value is found from Figure 9.6–10 to be 26,500. Hence,

$$P = T(\log t_r + C_l)$$
$$26{,}500 = (871 + 273)(\log t_r + 20)$$

Solving for $\log t_r$ gives:

$$\log t_r = \frac{26{,}500}{871 + 273} - 20 = 3.164$$

or $t_r = 1460$ h. Thus, at 871°C and a stress of 200 MPa, the turbine blade will last 1460 h.

    b. The rupture time is given as 500 h. Hence,

$$P = 26{,}500 = T[\log(500) + 20]$$

Solving for T gives:

$$T = \frac{26{,}500}{\log(500) + 20} = 1167 \text{ K} = 894°\text{C}$$

It is noteworthy that a temperature increase of only 23°C is associated with a life reduction of almost a factor of 3. This is a practical illustration of the fact that small increases in temperature can result in large changes in the creep rate.

....

SUMMARY

....

In this chapter we have described the important mechanical properties that relate to tensile, fracture, fatigue, and creep behavior. The effects of temperature and time on mechanical properties were also considered. In all cases, we have seen that the mechanical properties are related to the atomic scale structures and the microstructure of the materials.

In considering tensile behavior, properties such as Young's modulus, yield strength, ultimate tensile strength, and fracture strain were examined. The Young's modulus for metals, ceramics, and oriented polymers is high because the atomic bonds in these materials are strong. In the absence of a high crosslink density, the Young's modulus of unoriented polymers is low because the secondary intermolecular bonding is weak.

The yield strength of materials depends on the atomic arrangement (crystalline or noncrystalline) and, for crystalline materials, on the mobility of the dislocations. Dislocation motion in metals is relatively easy, so they generally exhibit significant amounts of plastic deformation and moderate strength. They become harder with continued plastic deformation, since the dislocations mutually interfere. Metals can be strengthened in a variety of other ways, including grain size refinement and solid solution strengthening, which are also ways of restricting dislocation motion. Crystalline ceramics contain dislocations, but the mobility of the dislocations is severely limited due to the relatively open structure and associated high friction stress for their movement. As a result, flaw-free ceramics are extremely strong, but exhibit very little strain to fracture. Polymers exhibit both brittle and nonbrittle behavior, depending on structure, temperature, and loading rate. At temperatures below the glass transition temperature, unoriented amorphous polymers display brittle behavior. Semicrystalline polymers can exhibit significant ductility because of the mobility of the long-chain molecules, and the fact that thermal energy assists in chains' sliding past one another.

Repeated application of loads is called fatigue, and we have seen that all materials are subject to fatigue failure. Fatigue is usually associated with permanent changes in the internal structure of a material brought about by deformation. For that reason, fatigue tends to be important in metals and polymers, but is much less of a consideration for ceramics. Some materials exhibit a plateau in the stress versus cycles to failure curves. The stress level associated with the plateau is called the endurance limit.

All materials are subject to deleterious interactions with some environments, and such interactions are usually strongly influenced by termperature.

All materials deform at high temperatures under sustained loads by a process called creep. In metals and ceramics, creep occurs because of processes such as dislocation climb, grain boundary sliding, and diffusion of point defects. In polymers, creep is associated with molecules sliding past one another. The specific mechanism depends on the temperature and stress level. Creep rate increases rapidly with temperature, since creep is a thermally activated process.

Many engineering structures contain preexisting cracks that are artifacts from fabrication and processing. They can have a significant influence on the service lives of the component. The methodology for dealing with the propagation of cracks and the resulting fractures is called fracture mechanics. In this approach, the crack tip stresses and strains are related to a quantity called the stress intensity parameter K. This parameter reaches a critical value at fracture called K_{Ic}. K_{Ic} is used to compute fracture loads (or safe operating loads) and critical crack lengths for a given service stress.

The fracture mechanics approach can also be used to calculate how long a component will last under fatigue loading conditions. During fatigue loading, the amount of crack extension per cycle uniquely correlates with the magnitude of the range of the stress intensity parameter, ΔK. The crack growth rate can be integrated to predict the lifetime of a component.

Use of fracture mechanics methodology provides the engineer with a method for making quantitative decisions about the probability of fracture and for avoiding severe economic loss and loss of life. However, it is important to remember that the concept of K is valid only when crack growth and fracture occur under dominantly elastic conditions, such as in high-strength materials.

KEY TERMS

creep	engineering stress σ	shear modulus G	true stress σ_t
ductile-to-brittle transition temperature (DBTT)	failure	strain hardening	ultimate tensile strength σ_{uts}
	fatigue	stress intensity parameter K	
ductility	fracture		uniform strain ε_u
elastic deformation	hardness	stress relaxation	viscoelasticity
elastic limit	percent reduction in area (%RA)	time-dependent deformation	yield point strain ε_{yp}
endurance limit S_e			yield strength σ_{ys}
engineering strain ε	plastic deformation	toughness	Young's modulus E
engineering strain at fracture ε_f	Poisson's ratio ν	true strain ε_t	0.2% offset yield strength

HOMEWORK PROBLEMS

1. What considerations and properties are required for a material to be used as a:
   a. Heart valve?
   b. Turbine blade?
   c. Leaf spring?
   d. Coffee mug?
   e. Golf club shaft?
   f. Suture?

2. Steel shows a very well defined yield point that is essentially associated with an avalanche of mobile dislocations. The upper yield point has been measured to be 207 MPa. The elastic strain at this point is 0.001. Compute Young's modulus.

3. The modulus of silicate glasses is on the order of 10^7 psi. The stress-strain curve is linear, indicating elastic behavior up to the point of catastrophic failure. The strength of ordinary window glass is typically about 5000 psi, that of tempered glass is about 50,000 psi, and that of optical waveguides can exceed 500,000 psi. Compute the strain to fail for each of these glasses. Sketch the stress-strain curves on one set of axes.

4. The yield stress for mild steel is 207 MPa. A specimen has a diameter of 0.01 m and a length of 0.10 m. It is loaded in tension to 1000 N and deflects 6.077×10^{-6} m.
   a. Compute whether the stress is above or below the yield stress.
   b. If the stress is below the yield, compute Young's modulus.

5. Young's moduli for Al, Cu, and W are 70,460 MPa, 122,500 MPa, and 388,080 MPa, respectively. Assuming that yielding does not occur, compute the deflections in specimens of each material when subjected to a load of 5000 N. The specimens are 1.00 m long with a cross section of 1 cm $\times$ 1 cm.

6. Young's modulus for a sample of nylon 6/6 is 2.83 GPa. If a load of 5000 N is applied to a 1-m-long specimen with a cross section of 1 cm $\times$ 1 cm, calculate the deflection. Compare your answer with the deflections for the metals in the previous problem.

7. A textile fiber has a circular cross section. The fiber radius is 10 μm. The fiber just fails when a load of 25 g is suspended from the fiber. Calculate the strength of the fiber.

8. Poisson's ratio for a particular steel is 0.295. Young's modulus for this steel is 205,000 MPa. The yield stress in tension is 300 MPa. If the shear yield stress is half of the tensile yield, compute the shear strain at yield.

9. Young's modulus of an Al alloy is 69×10^3 MPa. Since its structure is FCC, yielding is gradual. The stress-strain behavior is represented by the equation $\sigma = 295\varepsilon^{0.1}$ MPa. Compute the 0.2% offset yield stress. Compute the strain corresponding to the yield stress.

10. A cylindrical specimen of the alloy in the previous problem is loaded to half the yield stress. The longitudinal strain is 1.25×10^{-3} and the radial strain is 4.17×10^{-4}.
    a. Compute Poisson's ratio.
    b. If the radial strain is the same in all directions, compute the percentage change in volume at half the yield stress.
    c. Recompute the change in volume at the yield stress.

11. Give physical arguments for why steels have a definite yield point while Al and Cu do not. Based on these arguments, what type of yield behavior would you predict for amorphous polymers?

12. If the true stress–true strain behavior of a material is given by $\sigma = K\varepsilon^n$, find the ultimate tensile strength of the material. Hint: First use the above equation to find the relationship between engineering stress and engineering strain. Then differentiate it to find the strain at which the maximum engineering stress occurs. This value of strain can then be used to determine the ultimate tensile strength.

13. During plastic deformation of a crystalline material, it has been observed that the volume remains constant, as opposed to elastic deformation, where the volume increases.

a. Explain this effect based on physical reasoning.

b. As a result of the volume's being constant, show that Poisson's ratio for viscous or plastic deformation is 0.5.

14. Show that the relative load-bearing capacity of two materials is proportional to the ratios of the strengths divided by the densities of each material provided that the weights used of each material are the same. Use this concept to compute the relative load-bearing capacity of an aluminum alloy (strength = 400 MPa, specific gravity = 2.7) and polypropylene (strength = 40 MPa, specific gravity = 0.9).

15. The stress-strain relationship during plastic deformation of Cu may be described by the equation $\sigma = 310\varepsilon^{0.5}$ in MPa. Similarly, for a particular steel the equation $\sigma = 450\varepsilon^{0.3}$ is applicable. Calculate the energy required to deform each material to a strain of 0.01.

16. Construct a graph of the true fracture strain as a function of the percent reduction in area for a range of 0 to 70% reduction in area. Using the information provided in the preceding problem, place the steel and Cu on this graph.

17. Using the equation relating engineering strain to true strain, determine the value of the engineering strain at which the true strain value differs by 5%. Hint: Use a series expansion to write ε_t in terms of ε.

18. When reporting the ductility of polymers it is not common to report the gage length of the specimen as is required for metals. Can you give a reason for this practice?

19. Two amorphous polymers, A and B, have the same molecular weight but different glass transition temperatures. How will their elastic modulus values compare?

20. The stress relaxation characteristic of a viscoelastic polymer at room temperature is approximated by the following equation:

$$\sigma(t) = \sigma_0 \exp\left(-\frac{t}{\tau_0}\right)$$

where $\sigma(t)$ = stress at any given time t, σ_0 = initial stress, and τ_0 = time constant. A specimen of this polymer is subjected to a sudden strain of 0.2, at which time the stress rises to 2 MPa. The stress diminishes to 0.5 MPa after 50 seconds. Calculate the relaxation modulus of the polymer at 10 seconds.

21. A polymer sample is subjected to a constant stress by suspending a load from the sample, and the strain is monitored with time. The sample length varies as shown in Figure HP9.1. Sketch the relaxation modulus, $E(t) = \sigma_0/\varepsilon(t)$, with time.

FIGURE HP9.1

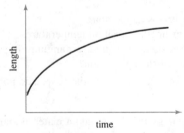

22. (a) Give reasons why a typical tensile test so commonly used for measuring strength of metals and polymers is not convenient for ceramic materials. (b) How does the three- or four-point bend specimen overcome these difficulties? (c) What limitations apply to the data obtained from bending tests?

23. The maximum tensile stress experienced by a cylindrical specimen loaded in three-point bending occurs at a point directly beneath the applied load and diametrically opposite the point of load application. If the distance between the support points is L, the maximum stress is given by:

$$\sigma = \frac{PLd}{4I}$$

where P = applied point load, d = diameter of the specimen, and I = cross-sectional moment of inertia = $\pi d^4/64$. The strength of ceramic materials is often determined in this way, and the resultant value is called the modulus of rupture. If the modulus of rupture of alumina is 3000 MPa, compute the load required to fracture a 5 mm-diameter specimen. The separation between support points is 25 mm.

24. The tensile strength of a metal is 800 MPa. Compute the diameter of an indentation made during Brinell hardness testing when using a 3000-kg load.

25. The data listed in the table were obtained for copper in the "1/4 hard" and "1/2 hard" conditions using an indenter of 10 mm diameter.

Load (kg)	Indent diameter (mm)	
	1/4 hard	1/2 hard
500	3.30	2.85
1000	4.00	3.75
1500	4.75	4.55
2000	5.40	5.25
2500	6.00	5.60

Calculate the hardness as a function of load.

26. Using the average hardness values as calculated in the previous problem, compute the expected indentation diameter if a load of 1500 kg were used with an indenter of 5 mm.

27. Explain why BCC metals exhibit a definite ductile-to-brittle transition while FCC metals do not.

28. You are given the following Charpy impact data for a low-carbon steel:

Temperature (°C)	Impact energy (J)
60	75
40	75
35	70
25	60
10	40
0	20
−20	5
−50	1

  a. Plot the impact energy versus temperature.
  b. Determine the ductile-to-brittle transition temperature.
  c. Selecting a steel for a car bumper requires a minimum impact energy of 10 J at −10°C. Is this steel appropriate for the application?

29. Cite factors that will cause brittle fracture in thermoplastic polymers.

SECTION 9.3
Brittle Fracture

30. tan δ is a measure of the energy that is put into a material and not returned, or the energy loss. This is an important consideration in many applications of polymers, such as running shoes, golf clubs and balls, and squash rackets and balls, to name just a few. Consider a squash ball. Before you play a game, you must warm up not only yourself but also the ball. The ball becomes somewhat bouncy as it heats up. Plot tan δ as a function of temperature for a squash ball.

31. Suppose you are given a sample of a polymer and asked to increase its strength to the highest value you can achieve. What would you do? (A huge amount of the polymer is available for you to work with.)

32. You need to determine the hardness of a series of rubber samples as a function of crosslink density. How might you proceed?

33. Most of the discussion in Section 9.3 focuses on brittle metals. Is brittle failure a problem with ceramics and oxide glasses? If so, then provide a few examples.

34. Differentiate between brittle and ductile fracture.

35. Explain why one expects to have significant scatter in fracture strength data of ceramic materials and why the fracture strength usually decreases with increasing specimen size.

36. Explain in your own words the influence of sample thickness on fracture toughness.

37. Fracture toughness values are shown for different materials in the following table:

Material	Fracture toughness (MPa-$\sqrt{m}$)
7075-T6 Al	28
300 maraging steel	66
Alumina	2.5

Assume that very wide panels are made from these materials. Assume also that the panels contain cracks of length $2a$ and are subjected to a stress of 350 MPa. Compute the maximum size of a crack that could be present in each of the panels.

38. The fracture toughness of a composite of Kevlar and epoxy has been measured to be 20 MPa-$\sqrt{m}$. The fracture toughness of a similar glass fiber–reinforced composite is 5 MPa-$\sqrt{m}$. Compute the relative fracture stresses for center-cracked panels with cracks that are 10 cm long.

39. Using the data in the preceding problem, calculate the crack lengths at fracture for applied stresses of 50 MPa.

40. The fracture toughness of Ca-stabilized ZrO_2 is 7.6 MPa-$\sqrt{m}$. The tensile strength is 140×10^3 MPa. If a center-cracked panel were loaded to this level without failure, compute the maximum size crack that could have been present. Would it be possible to detect such a crack? Is it meaningful to call a defect of this size a crack?

41. Using the data of the preceding problem and assuming that the smallest detectable crack is 1 mm, compute the maximum stress to which a part fabricated from this material could be loaded. Assume that during nondestructive inspection no indication of a crack was found.

42. In order to get a valid fracture toughness (K_{1c}) value for a material, the American Society for Testing and Materials (ASTM) standard E-399 requires that the following conditions be met:

$$B, W - a \geq 2.5 \left(\frac{K_Q}{\sigma_{ys}} \right)^2$$

where B = specimen thickness, W = specimen width, a = crack size, K_Q = candidate fracture toughness, and σ_{ys} = yield strength. Why are these requirements necessary?

43. The strength of aluminum oxide (alumina) can be as high as 4000 MPa and the fracture toughness can be as low as 2.5 MPa-$\sqrt{m}$. A sample of alumina contains flaws that are 100 μm or less in size. Estimate the stress at which fracture will occur in this specimen.

44. Estimate the size (width and thickness) of a specimen that will be needed to measure the fracture toughness of poly(methyl methacrylate) (PMMA) if the yield strength of the material is 7 MPa and its estimated fracture toughness is 1 MPa-$\sqrt{m}$. Do the same for an aluminum alloy for which the estimated fracture toughness is 36 MPa-$\sqrt{m}$ and the yield strength is 325 MPa, and for a stainless steel for which the yield strength is 200 MPa and fracture toughness is 300 MPa-$\sqrt{m}$.

45. If the surface energy of magnesium oxide is 1 J/m², its strength is 100 MPa, and the modulus of elasticity is 210,000 MPa, estimate the fracture toughness K_{1c}. What assumptions are implicit in your calculations?

46. Explain in your own words why the Griffith criterion for fracture, which worked exceedingly well for glass, had to be modified by Orowan to accommodate the behavior of metals.

 47. A component is designed to operate at a stress of 50 MPa. The component may contain a crack in the center of a wide panel.
   a. Construct a graph of critical crack length versus fracture toughness for this application.
   b. If the organization manufacturing the component is capable of reliably detecting cracks 2 mm long, refer to *Metals Handbook* to select a material with a minimum toughness that will meet the requirements.

48. In a shaft that is subjected to torsion, the maximum tensile stress is one-half of the shear stress and occurs at 45° to the axis of the shaft. For a solid shaft of radius r the relationship between torque T and shear stress τ is:

$$\tau = \frac{2T}{\pi r^3}$$

A shaft in a truck will be constructed from a cylindrical bar of steel that is 10 cm in diameter. The material is 4340 steel hardened to 42 R_C. In this condition the fracture toughness is 100 MPa-$\sqrt{m}$. If cracks 3 mm long may be present, compute the maximum torque that the shaft could sustain without failing.

49. The yield strength for 7075-T6 Al is 525 MPa and for 300 grade maraging steel it is 2100 MPa. The fracture toughness values are given in Problem 37. Both of these materials find application in aircraft. The aluminum alloy has been used for airframe applications, while the maraging steel has been used for landing gear and for shafts in jet engines. Frequently designers specify that a material can be stressed to one-half of the yield strength. Assume that cracks may be present as edge cracks of length a. Assume also that the formula $K_{Ic} = \sigma\sqrt{\pi a}$ applies to this situation and that the smallest crack that can be reliably detected is 1.4 mm. If components are inspected and then put into service at one-half the yield strength, can you guarantee that no parts will be put into service that have a crack length greater than the critical crack size?

50. A pressure vessel is made in the form of a cylinder 0.5 in thick, 8 inches in diameter, and 3 ft long. The ends are spherical. The operating pressure of the tank is 2250 psi. The material has a yield strength of 100,000 psi and measured toughness of 150 ksi-$\sqrt{in}$. Assuming semicircular-shaped flaws can be present on the interior, compute the maximum crack size before failure. What does your answer tell you about the likelihood of the component to experience abrupt, brittle fracture?

SECTION 9.5
Fatigue Fracture

51. You are flying on an airplane and notice the wings moving up and down several times with each bump of turbulence. It reminds you of this course and your brief study of fatigue of materials. Thinking further that the plane is structured chiefly of aluminum, do you rest comfortably?

52. The mean stress during a fatigue test was 100 MPa and the stress range was 50 MPa. Compute (*a*) maximum stress, (*b*) minimum stress, and (*c*) stress ratio.

53. The following stress amplitude and cycles to failure (N_f) data are given for SAE 1045 steel:

Stress amplitude (MPa)	Cycles to failure
500	10^4
400	7×10^4
350	10^5
300	10^6
275	$> 10^7$
250	$> 10^7$

   a. Construct an *S-N* curve and determine the endurance limit.

   b. What is the maximum allowable stress amplitude if a design application calls for a minimum fatigue life of 10^5 cycles?

54. List some of the factors you would consider in designing a crankshaft for an automobile engine. What would you suggest as steps to improve fatigue life of the crankshaft.

55. Consider a wide, flat panel in an aircraft application which experiences fatigue stresses. The maximum stress is 150 MPa and the minimum stress in a cycle is 0. During inspection, a 2-mm-deep flaw has been found on the edge of the panel. The flaw is normal to the direction of applied stresses. Estimate the remaining fatigue life of the panel given the following material data:

$$K_{Ic} = 30 \text{ MPa-}\sqrt{m} \qquad \frac{da}{dN} = (2.5 \times 10^{-12})(\Delta K)^{2.5}$$

where da/dN is in m/cycle and ΔK in MPa-$\sqrt{m}$.

56. You are given the following fatigue data for nylon:

Stress amplitude (MPa)	Cycles to failure
35	5×10^4
30	1.5×10^5
25	6×10^5
20	10^6
15	7×10^6
13	10^7

   a. Does this material have an endurance limit?

   b. Write a mathematical relationship which expresses the number of cycles to failure as a function of applied stress amplitude.

57. Many metallic materials obey an equation of the type:

$$\frac{da}{dN} = R(\Delta K)^4$$

If the initial crack size is a_0 and the final crack size is a_f, show that the total fatigue life may be increased much more by decreasing a_0 than by increasing the fracture toughness K_{Ic}. Hint: Integrate the crack growth rate between a_0 and a_f and express a_f in terms of K_{Ic}.

58. In a smooth-bar rotating-beam fatigue test, under fully reversed loading, it is found that failure of a mild steel occurs on loading (i.e., at 1/4 cycle) at a stress of 420 MPa. At a stress amplitude of 210 MPa the number of cycles to failure is 10^6. Just slightly below 210 MPa, fatigue failure never occurs. Give a mathematical description of the equation(s) that govern fatigue behavior of this material. Note that between 1/4 cycle and 10^6 cycles, the stress amplitude is linearly related to the $\log_{10}$ of the number of cycles.

59. Using the data of the previous problem, calculate the fatigue life at a stress amplitude of 315 MPa.

60. Why do fatigue failures often originate from the surface? Under what conditions would you expect the fatigue failures to initiate from the interior of the component?

61. In this chapter we discussed the *S-N* approach for predicting fatigue behavior of components and materials and also the fracture mechanics approach for predicting fatigue behavior. (*a*) Are these competing approaches for addressing fatigue problems? (*b*) Make a list of material constants you need to use both approaches. (*c*) Briefly compare the types of tests and specimens required to determine the material constants in these approaches.

62. Are ceramic materials more or less susceptible to fatigue than metals? Explain your answer.

63. In many situations the amplitude of the applied cyclic stress changes during use. For example, the stresses applied to a driveshaft in a truck change with operating conditions. Under such conditions, one theory states that when the sum of the fractions of life spent at each stress amplitude total unity, failure will occur. This theory is known as Miner's law.
   a. Express Miner's law in mathematical terms.
   b. Use the data in Problem 58 to compute how long a part will last at a stress amplitude of 280 MPa if it is first subjected to a stress amplitude of 315 MPa for 1000 cycles.

64. The same component mentioned in problem 50 is repressurized after each use. How many pressurization cycles can the part undergo before a crack breaks through to the outside surface if there is a small semicircular crack of 0.1 in radius on the interior of the wall? The crack growth rate has been found to obey an equation of the form:

$$\frac{da}{dN} = 10^{-10}(\Delta K)^2$$

where the units for ΔK are ksi-$\sqrt{in}$ and the units for da/dN are in/cycle.

SECTION 9.6
Time-Dependent
Behavior

65. Figure 9.6–1, which opened Section 9.6, showed that the stress-strain behavior of materials can depend on the temperature or rate of testing or use. Suppose you are selecting a fiber for use in a bulletproof vest. How do you screen the potential materials?

66. What are the different stages of creep deformation in metals? Draw a schematic diagram showing how creep strain varies with time for a constant stress. How does the strain-time behavior change if the stress increases?

67. The secondary-stage creep behavior of a high-temperature steel at 538°C is given by

$$\dot{\varepsilon} = 1.16 \times 10^{-24}\sigma^8$$

where $\dot{\varepsilon}$ = strain rate in h^{-1} and σ = stress in MPa. Predict the creep rate of this steel at a temperature of 500°C at a stress level of 150 MPa. The activation energy for creep is given as 100 kcal/mol.

68. The service temperature in the above problem is changed to 550°C and it is required that the creep strain during service not exceed 0.01. Estimate the maximum allowable stress if the service life is (a) 100,000 h and (b) 500,000 h.

69. For the Ni-base superalloy M252, the constant in the Larsen-Miller parameter is 20. At a stress of 10 ksi, the value of the parameter is 45.5×10^3. Compute the stress rupture life at 1500°F.

70. Using the information in the previous problem, determine the temperature that will give a rupture life of 10,000 h.

71. Which of the three creep deformation stages is most important, and why?

72. Wide panels of brass containing edge cracks of different sizes were subjected to the same sustained tensile stress in an ammonium sulfate solution. The tensile stress was 200 MPa. It was observed that all specimens that had cracks shorter than 12 cm did not fail in one year of exposure. Specimens with crack sizes longer than 12 cm failed at different times. If K_{ISCC} is defined as the critical stress intensity parameter below which environment-assisted failure will not occur, use the data provided to estimate a K_{ISCC} for brass in ammonium sulfate.

73. Compare the characteristics of creep in polymers with creep in metals and ceramics.

74. (a) What is the relaxation modulus of polymers? (b) How do crystallinity and crosslinking influence the relaxation modulus of polymers?

75. Creep or stress relaxation can become significant when the temperature approaches $0.4T_m$, where T is expressed in kelvins. What temperature minimum does this represent for Al_2O_3, PET, Al, and Ni alloys?

76. In a stress relaxation test, a sample is instantaneously strained and then the stress in the sample is measured with time under constant strain. Plot $\sigma(t)$ for a sample that shows relaxation, but not to a value of zero stress.

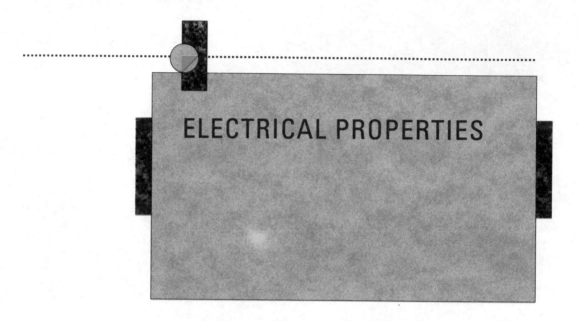

ELECTRICAL PROPERTIES

MATERIALS IN ACTION   Electrical Properties

A report of the U.S. National Academy of Sciences stated that in 1985, the worldwide electronics market was $2.5 billion and accounted for $400 billion of equipment sales. It is no exaggeration to state that materials and materials processing lie at the heart of this huge economic activity. The whole revolution in computing technology was enabled by the miniaturization associated with the advent of solid-state electronics. The first digital computer, the Univac, was produced at the University of Pennsylvania and was based on vacuum tube technology. The electronic components filled a good-sized laboratory. Furthermore, computations were very slow and there were frequent failures of the vacuum tubes. Today, because of the discovery of semiconductors and transistors and great advances in processing, faster and more reliable computing power is available in a small personal computer than was available in the Univac.

Electrical properties may be tailored to a given application by precise composition control. For example, Si is a poor conductor of electricity in the pure form. This is related to the fact that Si is a covalently bonded crystal in which all bonds consist of shared electron pairs. There simply are no free electrons to contribute to electrical conduction. If a small amount of P is added to Si, the P substitutes for Si in the crystalline lattice. However, as opposed to Si, which has four valence electrons, P has five valence electrons. Thus, one extra electron is available for conduction of electricity. It was pointed out in Chapter 1 that the addition of two P atoms to 1 million Si atoms increases the conductivity by a factor of 5 million. This type of conductivity is called n-type, since the current is carried by an electron, which is negative. In a similar way it is possible to add small amounts of a material whose valence is 1 less than that of Si. In this case there is a deficiency of electrons. This electron "hole," which carries a relative positive charge, migrates from bond to bond under the action of an electric field and results in current. This kind of conductivity is termed p-type, since it depends on the migration of a positive charge. When n-type and p-type semiconductors are joined together in a circuit, current can flow in only one direction, and the pn junction acts as a rectifier. Now, instead of a large vacuum tube, a small "chip" serves as a rectifier. This process of miniaturization has proceeded to a remarkable degree, and through process control many different types of circuit elements can be produced. It is currently possible to place well over 1 million circuit components on a small chip (~ 1 cm^2), and further advances are occurring on almost a daily basis. Miniaturization in microelectronic circuits is shown in the electron microscope photographs of Figure 1.4–5.

In this chapter we will study the relationship between the atomic-level structure of a material and its electrical properties. We will also study how various materials are used to produce electronic devices.

10.1 INTRODUCTION

In the previous chapter we demonstrated that an appreciation of bond types, crystalline and amorphous structures, defects and impurities, microstructure, and kinetics is necessary for an understanding of the mechanical properties of materials. In this chapter we show that the electrical properties of solids also depend on their atomic and microscopic structure. While mechanical properties of a material, such as elastic modulus, yield strength, and fracture toughness, describe a material's response to the application of external loads, the electrical properties determine a material's response to external electric fields.

Most of you have probably seen the inside of a compact disk player. Along with many other modern electronic devices, a CD player contains parts from all classes of materials. The active electronic components are semiconductors; electrical connections are made with polymer-coated metal wire; some of the capacitors are ceramic; and the printed circuit boards on which the devices are mounted are likely to be composites. Why are there so many different types of materials? Clearly the answer is that each offers certain advantages over the others for specific electronic applications. In this chapter we will explain why different classes of materials respond so differently to electric fields.

We begin with a physics refresher — that the response of a conducting material to an applied voltage is a flow of current through the material. The magnitude of the current is proportional to the voltage, and the proportionality constant is known as resistance. In turn, the resistance of a material is a complex function of several factors, including the type of atoms present, the atomic bonding type, the number of valence electrons per atom, the amount of crystallinity, defect density, microstructure, temperature, and macroscopic sample dimensions. After investigating the physical bases for the influence of each of these factors on the macroscopic material properties, we will describe a number of engineering devices developed from these principles, including superconductors, thermocouples, spark plug insulators, oxygen detection devices, and materials designed to prevent the accumulation of static electric charge.

We close the chapter with a discussion of the unique electrical properties of semiconductors, a technologically important group of materials. After describing the basic electrical conduction mechanisms in these materials, we investigate the influence of temperature and impurities on their properties. Even the most basic characteristics of semiconductors can be exploited to create useful engineering devices. Some of these include thermistors (used for temperature measurement), infrared detection devices, light-emitting diodes, and solid-state transistors. The section concludes with a discussion of fundamental materials issues in the microelectronics industry.

10.2 ELECTRICAL CONDUCTION

Consider an experiment in which a bar of material, of length L and cross-sectional area A, is subjected to an applied voltage V, as shown in Figure 10.2–1. For many materials the response of the system, that is, the current flow I, is a linear function of the driving voltage. Expressed mathematically, this relationship is known as **Ohm's law:**

$$V = IR \tag{10.2–1}$$

In this expression voltage V has units of volts (V), current has units of amperes (A), and the constant of proportionality R, known as the bar's **resistance** to current flow, has units of ohms (Ω). By convention, the direction of current flow is downhill, from the high

The primary theme presented in this book is the relationship between microstructure and properties. To optimize properties, the materials specialist must understand how microstructure evolves during processing. Consequently, the materials engineers focus on the interrelationship between processing, microstructure, and properties.

There are a variety of tools that enable the researcher to examine the details of the microstructure from 1X to 1,000,000X. However, depending upon the issue that is to be resolved, the appropriate level of examination must be chosen in order to identify the microstructural feature that is controlling the process. The features included in this insert present a variety of techniques that have been used to image a particular microstructural feature. What is special about these photomicrographs is the application of color to emphasize features in the microstructure. In some of the examples presented, the images are formed using light optical microscopes equipped with special lenses to separate the light into its basic colors. Because of the length scale of the feature of interest, magnifications of approximately 100X are sufficient. In other photographs, digitized images were formed and intensities were correlated to color to enhance variations in the microstructure.

FIGURE 1 Photomicrograph of an alloy of aluminum containing 1 weight percent magnesium after hot rolling. The alloy was partially annealed and the resulting microstructure was a mixture of small, equiaxed, recrystallized grains that formed at the original grain boundries.

Rolling is the most common metal-forming operation. To facilitate rolling, the starting ingots are hot-rolled since the force required to deform them decreases with increasing tempurature (Section 16.4.1 Metal Forming). Depending upon the rolling tempurature and the amount of deformation in the rolled product, recrystallization (the process by which grains containing high densities of dislocations are replaced by new strain-free grains) can occur in the product during post-rolling heat treatments such as annealing. The grain structure can affect the evolution of microstructure during heat treatment and hence the properties.

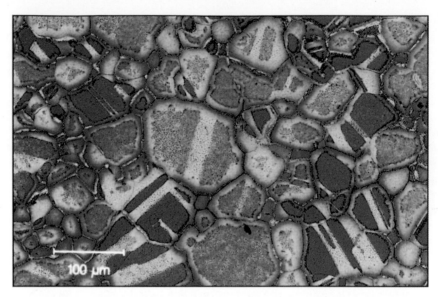

FIGURE 2 A Ni–20 Cr alloy, which has been oxidized for a short period of time. Nickel is an excellent structural metal for high-temperature applications. To improve the resistance to high-temperature oxidation, chromium is added to nickel alloys (see Section 15.3.6 Methods of Improving Resistance to Atmospheric Attack). The different colors exhibited are a result of different thicknesses of oxides grown and indicate that grains with different crystallographic orientations oxidize at different rates. Readily visible are twinned regions within grains. Segregation of Cr to grain boundries allows Cr_2O_3, a more protective oxide than NiO, to more readily form at grain boundries. (Source: Micrograph recorded by J. M. Hampikian, School of Materials Science and Engineering, Georgia Institute of Technology, 1993.)

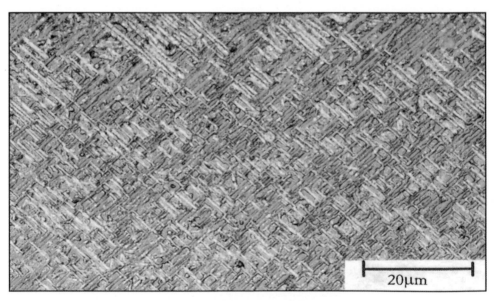

FIGURE 3 Nomarski interference photomicrograph of a Y_2BaCuO_5 coating on a (1 0 0) single-crystal substrate of magnesium oxide. Materials engineers often coat the surfaces of materials to improve surface-related properties. When crystalline films are deposited on crystalline substrate, there often exists a specific orientational relationship between the grains of the film and the surface of the part. This phenomenon is called epitaxy. Here, the elongated Y_2BaCuO_5 grains are predominantly aligned along two mutually perpendicular directions or epitaxially related. The Nomarski technique uses two cross-polarized light rays to contrast regions of the specimen with different orientations. (Source: A. T. Hunt and W. B. Carter, Georgia Institute of Technology.)

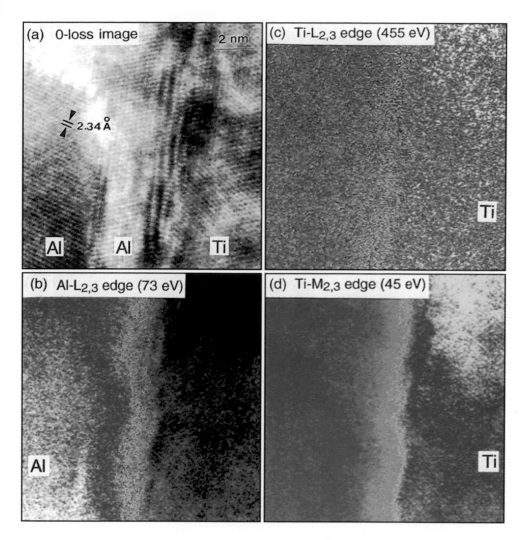

FIGURE 4 When surfaces are coated we are not only interested in the crystallographic relationships between the substrate and the coated films but the variation in the composition in the vicinity of the interface. Shown are transmission electron microscope images of an Al/Ti interface grown by the deposition process of sputtering. Images are formed by selecting electrons with specific energy losses that are characteristic of the elements of interest. (a) An image formed by the transmission of electrons without experiencing energy loss (0-loss image), showing the crystal lattices near the interface. (b, c, and d) Images formed by the transmitted electrons that have excited 2p core-shell electrons of Al and Ti, and 3d core-shell electrons of Ti, respectively. Color coding was used to indicate concentration. Yellow and red are highest concentrations and dark and blue are lowest concentrations. The chemical sharpness between the Al and Ti layers is approximately 1nm. (Source: Z. L. Wang, Georgia Institute of Technology.)

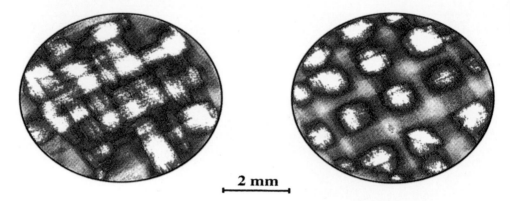

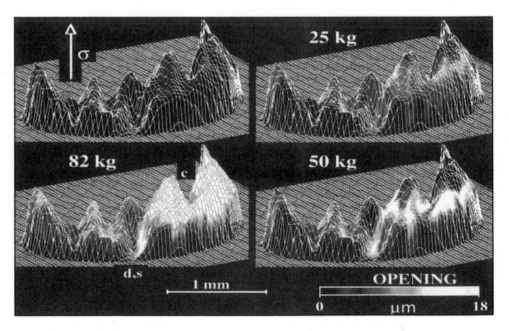

FIGURE 5 Properties of ceramic matrix composite (CMC) materials are highly dependent on the fabrication process. High-resolution X-ray tomography (similar to medical CAT scan) can be used to study the chemical vapor infiltration (CVI) process (see Section 16.11.3 CMCs), which is used to fabricate composites of silicon carbide reinforced with layers of woven silicon carbide cloth. These images show the pore structure of such a composite at an intermediate stage of processing. Red and black show densified regions of the composite. White and blue show open porosity. The pores between poorly aligned (left) and well-aligned (right) adjacent layers have different geometries. The open, connected pore structure created by the well-aligned layers produces more complete densification and less residual porosity in the final composite. (Source: Printed with permission from S. R. Stock, S.-B. Lee and T. L. Starr.)

FIGURE 6 X-ray computed tomography is a technique that allows materials engineers and scientists to obtain images of the interior of a part. This technique is particularly useful when defects such as cracks are present in a material. Predicting growth behavior of cracks is essential to producing safe structures such as airplanes. As shown in the figure, mating surfaces of fatigue cracks in a cylindrical sample of Al-Li alloy 2090 open and close in response to applied stress σ. The separation distance between the mating surfaces, or the crack opening, can vary considerably with the instantaneous stress during a fatigue cycle and the location on the crack surface. It is of considerable interest to researchers to be able to determine the opening between the mating crack surfaces at various stress levels throughout the cracked region including the interior of the specimen. Images (clockwise from upper left) show (1) the basic crack surface topography at zero load when the two crack surfaces are more or less completely closed and in (2), (3), and (4) where the surfaces are partially opened with the amount of opening increasing with the load. The color code at the bottom of the figure is provided to determine the precise crack opening at different points. (Source: Published with permission of A. Guvenilir, T. M. Breunig, J. H. Kinney, and S. R. Stock.)

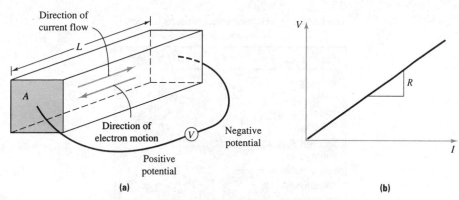

FIGURE 10.2–1   A simple electrical conduction experiment. **(a)** A bar of material of length L and cross-sectional area A is subjected to a voltage V. **(b)** The response of the system, current flow I, is a linear function of the magnitude of the driving force, voltage V. The constant of proportionality, or slope, is the resistance R.

voltage or potential to the lower potential: current flows opposite to the direction of electron flow.

The resistance of the bar is a property that depends on the geometry of the sample—it is an extrinsic property. The corresponding intrinsic or geometry-independent property is known as **resistivity** ρ and is defined as:

$$\rho = \left(\frac{A}{L}\right)R \tag{10.2–2}$$

where ρ has units of ohm-meter (Ω-m). It is often more convenient to use the inverse of resistivity, **conductivity** σ, defined as:

$$\sigma = \frac{1}{\rho} \tag{10.2–3}$$

where σ has units of $(\Omega$-m$)^{-1}$ or mho/m, where 1 mho $\equiv$ 1 ohm^{-1}. Like ρ, σ is an intrinsic property. Table 10.2–1 lists the conductivities of a variety of materials. Interestingly, the values of σ span more than 23 orders of magnitude—from 6×10^5 $(\Omega$-cm$)^{-1}$ for silver to $\sim 10^{-18}$ $(\Omega$-cm$)^{-1}$ for polytetrafluoroethylene. Few other material properties show such a wide variation.

....

EXAMPLE 10.2–1

The resistance of a metal rod with conductivity 3.5×10^8 $(\Omega$-cm$)^{-1}$ and length 10 m is measured to be 0.08 Ω. Calculate the cross-sectional area of the rod. What voltage is required to produce a current of 3.2 A in this rod?

Solution

Substitution of Equation 10.2–3 into Equation 10.2–2 yields $1/\sigma = (A/L)R$. Solving this expression for A gives $A = L/(\sigma R)$. Substituting the appropriate values from the problem yields:

$$A = \frac{10 \text{ m}}{[3.5 \times 10^8 \ (\Omega\text{-m})^{-1}](0.08 \ \Omega)} = 3.57 \times 10^{-7} \text{ m}^2 = 0.357 \text{ mm}^2$$

For the second part of the problem, Equation 10.2–1 states:

$$V = IR = 3.2 \text{ A} \times 0.08 \ \Omega = 0.256 \text{ V}$$

....

TABLE 10.2–1   Electrical conductivities for a variety of materials at room temperature.

Class of materials	$\sigma \, [(\Omega\text{-cm})^{-1}]$
Polymer	
Nylon	10^{-12}–10^{-15}
Polycarbonate	5×10^{-17}
Polyethylene	$< 10^{-16}$
Polypropylene	$< 10^{-15}$
Polystyrene	$< 10^{-16}$
Polytetrafluoroethylene	10^{-18}
Polyvinylchloride	10^{-12}–10^{-16}
Phenolformaldehyde	10^{-13}
Polyesters	10^{-11}
Silicones	$< 10^{-12}$
Acetal	10^{-15}
Metals and alloys	
Al	3.8×10^5
Ag	6.3×10^5
Au	4.3×10^5
Co	1.6×10^5
Cr	7.8×10^4
Cu	6.0×10^5
Fe	1.0×10^5
Mg	2.2×10^5
Ni	1.5×10^5
Pd	9.2×10^4
Pb	4.8×10^4
Pt	9.4×10^4
Sn	9.1×10^4
Ta	8.0×10^4
Zn	1.7×10^5
Zr	2.5×10^4
Plain carbon steel (1020)	1.0×10^5
Stainless steel (304)	1.4×10^4
Gray cast iron	1.5×10^4
Ceramics	
ReO_3	5.0×10^5
CrO_2	3.3×10^4
SiC	1.0×10^{-1}
Fe_3O_4	1.0×10^2
SiO_2	$< 10^{-14}$
Al_2O_3	$< 10^{-14}$
Si_3N_4	$< 10^{-14}$
MgO	$< 10^{-14}$
Si	1.0×10^{-4}
Ge	2.3×10^{-2}

Sources: Adapted from Richerson, *Modern Ceramic Engineering,* Marcel Dekker, Inc., NY. Reprinted by courtesy of Marcel Dekker, Inc. Also *Handbook of Materials Science,* Vol. I. Copyright CRC Press, Boca Raton, FL.

The remainder of this section focuses on the atomic, microstructural, and external factors, including temperature and electric field strength, that influence the magnitude of σ. Since conduction occurs by the motion of electrical charge through a solid, we anticipate the magnitude of σ to depend on three factors:

1. The number of mobile charge carriers per unit volume, N, with units of carriers per m³, or simply m^{-3}.
2. The charge per carrier, q, with units of coulombs (C)
3. The **mobility** of the charge carriers, μ, with units $m^2/(V\text{-}s)$

Recall that the concept of mobility was introduced in Chapter 2 and that electron mobility can be interpreted as the ease with which charge carriers can move through the atomic scale structure of a material in response to an applied electric field.

Consider the form of the relationship between N, q, μ, and σ. The product Nq represents the mobile charge density within the material. Therefore, we expect a linear relationship between σ and Nq. Increasing either N or q or both results in a proportional increase in conductivity. Anticipating a similar direct relationship between conductivity and mobility, we obtain the important relationship:

$$\sigma = Nq\mu \tag{10.2-4}$$

EXAMPLE 10.2-2

Consider a conductor with the following properties: $q = 1.6 \times 10^{-19}$ C, $\rho = 5 \times 10^{-8}$ Ω-m, and $N = 1 \times 10^{29}$ m^{-3}. Estimate the charge-carrier mobility μ in this material.

Solution

Solving Equation 10.2-4 for μ and using $\sigma = 1/\rho$, we find $\mu = 1/(Nq\rho)$. Substituting the appropriate values yields:

$$\mu = \frac{1}{(1 \times 10^{29}\ m^{-3})(1.6 \times 10^{-19}\ C)(5 \times 10^{-8}\ \Omega\text{-m})}$$
$$= 1.25 \times 10^{-3}\ m^2/(V\text{-}s)$$

EXAMPLE 10.2-3

The resistance of an Al wire ($L = 1$ m and $A = 1$ mm²) is 0.0283 Ω. Calculate the number of mobile electrons contributed by each Al atom. Assume current is carried entirely by electrons and that the electron mobility is 1.22×10^{-3} $m^2/(V\text{-}s)$.

Solution

We must determine both the density of mobile electrons and the density of aluminum atoms. The first calculation involves the equation $\sigma = Nq\mu$, and the second involves the concept of volume density of atoms introduced in Section 3.5.3. The conductivity can be obtained by combining Equations 10.2-2 and 10.2-3 to yield:

$$\sigma = \frac{1}{\rho} = \frac{L}{RA}$$

Substituting the quantities in the problem statement gives:

$$\sigma = \frac{1\ m}{(0.0283\ \Omega)(1\ mm^2)(1\ m/1000\ mm)^2} = 3.53 \times 10^7\ (\Omega\text{-m})^{-1}$$

The mobile electron density can be calculated:

$$N = \frac{\sigma}{q\mu} = \frac{3.53 \times 10^7 \ (\Omega\text{-m})^{-1}}{(1.6 \times 10^{-19} \ \text{C})[1.22 \times 10^{-3} \ \text{m}^2/(\text{V-s})]}$$

$$= 1.81 \times 10^{29} \ \text{mobile electrons/m}^3$$

Since Al is an FCC metal, we find from Section 3.5.3 that the atomic density is $\rho_V = 1/(4\sqrt{2}r^3)$. From Appendix C, $r(\text{Al}) = 0.143$ nm. Therefore,

$$\rho_V = \frac{1}{(4\sqrt{2})(0.143 \ \text{nm})^3(1 \ \text{m}/10^9 \ \text{nm})^3} = 6.05 \times 10^{28} \ \text{atoms/m}^3$$

The number of mobile electrons per atom is:

$$\frac{N}{\rho_V} = \frac{1.81 \times 10^{29} \ \text{electrons/m}^3}{6.05 \times 10^{28} \ \text{atoms/m}^3}$$

$$= 2.99 \ \text{mobile electrons per atom}$$

Note that this value is approximately equal to the number of valence electrons per aluminum atom, 3.

Although the electrical conductivity of all materials can be understood in terms of Equation 10.2–4, the critical term in the equation depends on the material. For example, in metals, N and q are essentially constant, so the influence of both internal and external variables on the conductivity of a metal can be understood by considering their influence on μ. In contrast, the conductivity of semiconductors is dominated by factors that alter N. In the next few subsections we investigate in greater detail the internal and external factors that influence N, q, μ, and, therefore, σ.

10.2.1 Charge per Carrier

In many materials, such as metals and some covalent solids, the electrical charge is carried by electrons. For these materials q is the charge on an electron, q_e, with value 1.6×10^{-19} C. In ionic solids there is another charged species that can contribute to electrical conduction—the ions themselves. In this case, $q = (q_e \times Z)$, where Z is the valence of the ion involved in electrical conduction. Ionic solids contain more than one type of ion, and each type of ion may contribute to the overall conductivity of the solid. Equation 10.2–4 can be readily modified to allow this possibility:

$$\sigma = \Sigma_i \ (N_i q_i \mu_i) \tag{10.2–5}$$

where the summation includes all types of charge carriers.

EXAMPLE 10.2–4

Write the form of the conductivity equation for the ionic solid Li_2O by first assuming that charge is transported by the motion of ions only, and then assuming that charge is transported by both ions and electrons.

Solution

Nothing that $q_{Li} = q_e$ and $q_O = 2q_e$, the appropriate form of Equation 10.2–5 under the original assumption is:

$$\sigma = N_{Li}q_{Li}\mu_{Li} + N_O q_O \mu_O = q_e(N_{Li}\mu_{Li} + 2N_O \mu_O)$$

When electrons also transport charge, the equation becomes:

$$\sigma = q_e(N_{Li}\,\mu_{Li} + 2N_O\,\mu_O + N_e\,\mu_e)$$

10.2.2 Charge Mobility

What factors influence the mobility of electrical charge carriers? One way to approach this question is to draw an analogy between the mobility μ and the diffusion coefficient D introduced in Chapter 4. While D represents the ease with which atoms move when driven by a concentration gradient, μ represents the ease with which charge carriers move through a solid in response to an electric field gradient. Since the magnitude of D is determined by several factors, including the nature of the diffusing species, the defect concentration, and temperature, it is not surprising to find that μ depends on all the same factors.

The two major types of charge carriers are electrons and ions. Ions move through the solid by diffusion, as described in Chapter 4. In contrast, the much smaller size of electrons permits them to move through the solid relatively unimpeded. There are fundamental differences between the conduction mechanisms of these two types of carriers; hence, they will be discussed independently. We first deal with the motion of electrons, postponing our discussion of ionic conduction until Section 10.2.6.

A complete description of electron motion requires quantum mechanics; however, we can model many important characteristics of this motion using classical mechanics and treating the electrons as rigid particles. As shown in Figure 10.2–2a, when an electric field is applied, mobile electrons are accelerated toward the positive pole. The electrons are subjected to a constant acceleration until they collide with an ion core. Assuming the collision resets the electron velocity to zero and electrons continuously undergo

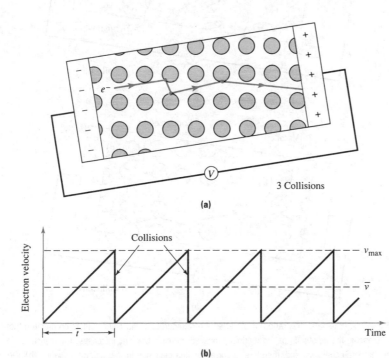

(a)

(b)

FIGURE 10.2–2

Particle model of an electron, e^-, moving through a crystal lattice. **(a)** Under an applied electric field, the mobile electrons are accelerated toward the positive potential and occasionally suffer collisions with the surrounding ion cores. **(b)** A plot of electron velocity versus time for an electron assumed to continuously undergo the acceleration-collision cycle ($\bar{t}$ is the mean time between collisions and $\bar{v}$ is the average velocity).

acceleration-collision cycles as shown in Figure 10.2–2b, then from classical mechanics:

$$\bar{v} = a\bar{t} \tag{10.2–6}$$

where $\bar{v}$ is the average or **drift velocity,** a is the acceleration due to the applied field, and $\bar{t}$ is the **mean time between collisions.**

Since the magnitude of the acceleration is proportional to the electric field strength, Equation 10.2–6 can be recast in the form:

$$\bar{v} \propto E \tag{10.2–7}$$

where E is the field strength in V/m. The proportionality constant is the electron mobility:

$$\bar{v} = \mu E \tag{10.2–8}$$

In fact, this equation may be used to define mobility as the ratio of the average drift velocity to the applied electric field strength. When the field strength is constant, Equations 10.2–6 and 10.2–8 can be combined to show that mobility is proportional to the mean time between collisions. This result is used to explain the effect of temperature and atomic scale defects on the mobility of electrons.

As temperature is increased, atoms gain thermal and kinetic energy and begin to vibrate about their equilibrium positions (as discussed in Section 2.5 dealing with thermal expansion). The magnitude of the atomic vibrations increases with temperature. This,

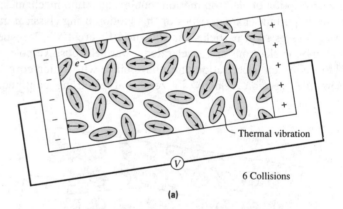

(a)

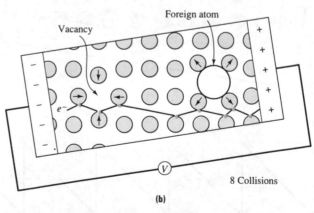

(b)

FIGURE 10.2–3   A schematic illustration of the effects of temperature and point defects on electron motion. **(a)** As temperature increases, the amplitude of the thermally induced atomic vibrations increases. This causes a decrease in the mean time between collisions and, therefore, a corresponding decrease in the electron mobility. **(b)** The mean time between collisions and, therefore, the electron mobility decrease with increasing defect concentration.

in turn, results in an enhanced perturbation of the crystal lattice, a corresponding decrease in $\bar{t}$, and ultimately a decrease in the electron mobility. The process is depicted schematically in Figure 10.2–3a. The influence of temperature on electron mobility can be expressed mathematically as:

$$\frac{\partial \mu}{\partial T} < 0 \qquad (10.2-9)$$

indicating that mobility decreases as temperature increases.

Now consider the influence of lattice defects on electron mobility. Referring to the schematic illustration shown in Figure 10.2–3b, the mean time between collisions will be decreased by atomic scale defects that perturb the perfection of the crystalline structure. Thus, all types of lattice defects—point, line, planar, and volume—will decrease the mobility of electrons. In practice, however, it is usually point and line defects that have the most significant influence on electron mobility. In analogy with Equation 10.2–9, we can summarize the influence of lattice defects (of concentration N_d) on electron mobility:

$$\frac{\partial \mu}{\partial N_d} < 0 \qquad (10.2-10)$$

The arguments above have treated an electron as a particle. An alternative approach is to consider the wave model of an electron. Any factor that disturbs the periodicity of the crystal lattice will interfere with the motion of the electron wave. Specifically, both lattice defects and vibrating ion cores act as local wave-scattering centers. Both the wave and particle models lead to the prediction that an increase in either temperature or defect density decreases electron mobility by decreasing the time between electron-scattering events.

EXAMPLE 10.2–5

Estimate the drift velocity of an electron through the Al wire described in Example 10.2–3 [$R = 0.0283\ \Omega$, $L = 1$ m, and $\mu_e = 1.22 \times 10^{-3}$ m^2/(V-s)] when the current flowing through the wire is 3 A.

Solution

Equation 10.2–8, $\bar{v} = \mu E$, may be used to determine $\bar{v}$ if E can be determined. The first step in calculating E is to use Ohm's law, $V = IR$, to find V. Substitution yields:

$$V = IR = (3\ \text{A})(0.0283\ \Omega) = 0.085\ \text{V}$$

The electric field strength is

$$E = \frac{V}{L} = \frac{0.085\ \text{V}}{1\ \text{m}} = 0.085\ \text{V/m}$$

Finally, the drift velocity is

$$\bar{v} = \mu E = [1.22 \times 10^{-3}\ \text{m}^2/\text{(V-s)}](0.085\ \text{V/m}) \approx 0.1\ \text{mm/s}$$

EXAMPLE 10.2–6

Consider two wires of equal dimensions made from the same metal. If the only significant difference between the wires is that one of them has a higher defect density than the other, predict which wire has lower resistance. Assume that conduction occurs by the motion of electrons and that the density of mobile charge carriers is the same in both wires.

Solution

Combining Equations 10.2–2 and 10.2–3 yields:

$$R\frac{A}{L} = \rho = \frac{1}{\sigma}$$

which for constant wire dimensions reduces to $R \propto (1/\sigma)$. From Equation 10.2–4, $\sigma = Nq\mu$. Since conduction in both wires occurs via the motion of electrons and the density of the charge carriers is constant, we find $\sigma \propto \mu$. Combining this with the previous result yields: $R \propto (1/\mu)$. Thus, the wire with the higher electron mobility has the lower resistance. Since defects decrease the mobility, the wire with the lower defect concentration has the higher mobility and, therefore, the lower resistance.

10.2.3 Energy Band Diagrams and Number of Charge Carriers

In Section 2.2 we noted that quantum mechanics places two key restrictions on electron energy levels. Within a single isolated atom, only certain discrete electron energy levels are allowed, and as the Pauli exclusion principle states, no two electrons with the same spin may occupy the same energy level. When a large number of atoms are brought together to form a solid, the permissible electron energy levels change in several ways, including:

1. The permissible energy levels are a function of the separation distance between the atoms within the solid.
2. The discrete energy levels in the isolated atoms spread into energy bands in the solid.
3. The outermost, or valence, electrons are no longer spatially localized at a particular atom.

The corresponding energy band diagrams are shown schematically in Figure 10.2–4. Note that for large separations the electrons associated with any individual atom are independent of those of other atoms. For separation distances less than x_2, however, the original isolated energy levels in a single atom spread into a band of discrete levels separated by small energy differences (adjacent energy levels are $\sim 10^{-23}$ eV apart), as shown in Figure 10.2–4c. Three important features of Figure 10.2–4 require emphasis:

1. The energy bands become wider as the amount of overlap increases (i.e., as x decreases).
2. The higher the energy level, the wider the corresponding energy band.
3. In metallic and ionic solids the number of energy levels in an energy band equals the number of atoms in the solid multiplied by the number of discrete energy states in an isolated atom.

A comparison of the energy level diagrams in Figure 10.2–4c and d shows an important distinction. At separation distance x_3, the energy band corresponding to the 2s level does not overlap the energy band corresponding to the 2p level. There is an **energy gap**, also known as a **band gap**, separating the two energy bands. The magnitude of the gap is labeled E_g. In contrast, for separation distance x_4, the 2s and 2p energy bands overlap to form an extended energy band. To determine which energy band diagram to use in our calculations, we must determine the equilibrium separation distance between atoms, x_0, and use the corresponding diagram.

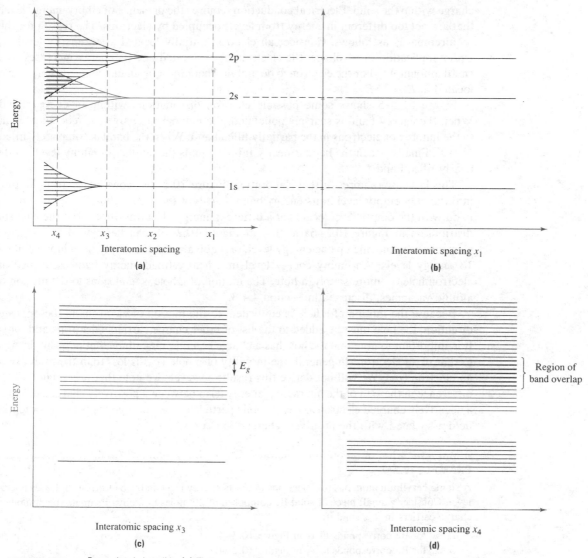

FIGURE 10.2–4 Energy bands in solids. **(a)** The energy band structure as a function of interatomic separation distance. **(b)** For large separation distances, the electrons associated with any atom are independent of those of all other atoms. **(c)** For separation distance x_3, the isolated energy levels in a single atom split into a band of discrete levels separated by small energy differences. **(d)** For separation distance x_4, the 2s and 2p energy bands overlap to form an extended energy band.

The energy bands in a solid can be classified as follows:

1. The highest energy band that is at least partially occupied is known as the **valence band.**
2. All the bands below the valence band are **core bands.**
3. The energy band above the valence band is the **conduction band.**
4. The term band gap refers to the magnitude of the forbidden energy range between the valence and conduction bands.

Another important consideration is the relationship between the energy bands and the number of electrons sufficiently mobile that they are capable of transporting electric charge within the solid. Electrical conduction requires the presence of empty energy levels that are not too different in energy from levels occupied by electrons. The reason for this requirement is as follows. Consider an electron initially located at an energy level E_0. Upon application of a voltage, the electron is accelerated and its energy increases by a small amount ΔE. For the electron to be mobile, there must be an unoccupied energy level located at $E_0 + \Delta E$.

Figure 10.2–5 shows some possible electron distributions within an energy band. When the energy band is sparsely populated, the number of charge carriers, N, is equal to the number of electrons in the partially filled band. When the band is completely filled, $N = 0$. Finally, when the band is nearly full, N equals the number of empty levels in the nearly filled band.

This last case requires further elaboration. Figure 10.2–6a shows an electron jumping into the only empty level in its energy band. Part b of this figure shows a similar energy band with the empty level located at a different energy. The transition from the electron distribution in Figure 10.2–6a to that in Figure 10.2–6b can be viewed as either 13 electrons each moving up one energy level, or, equivalently, the empty level moving down 13 energy levels. An empty energy level in a nearly filled energy band is termed an electron hole or, more simply, a **hole.** The motion of a hole is analogous to the motion of atomic vacancies, discussed in Section 4.4.3.

Because the motion of a hole is equivalent to the motion of electrons in the opposite direction, the hole must be added to the list of other charge carriers (electrons and ions). It is important to note that the hole has a charge equal in magnitude, but opposite in sign, to that of an electron. In general, the mobility of a hole (μ_h) is less than the mobility of an electron (μ_e). We will not derive this result; however, we note that it is related to the difference in the sign of the interaction energy term between a positively charged hole and a negatively charged electron as each mobile particle moves through the periodic potential field associated with the positively charged ion cores.

....

EXAMPLE 10.2–7

A single beryllium atom has two electrons in the 1s energy level and two electrons in the 2s energy level. Consider a small piece of solid Be composed of 20 moles of atoms. Estimate the number of charge carriers in this solid if:

   a. x_0 for Be corresponds to x_3 in Figure 10.2–4.
   b. x_0 for Be corresponds to x_4 in Figure 10.2–4.

Solution

   a. For this separation distance the 2s and 2p bands do not overlap. Since the 2s level is filled in an isolated Be atom, the 2s band will be filled in the solid and the material will not be a conductor.

   b. For this band structure, the 2s and 2p bands overlap. In this case, there will be two electrons per atom multiplied by 20 moles of atoms, or 40 moles of electrons (charge carriers), in a partially filled valence band. Since Be is a metallic conductor, the latter model is a better representation than the former.

....

10.2.4 The Influence of Temperature on Electrical Conductivity and the Fermi-Dirac Distribution Function

We have demonstrated that conductivity depends on temperature, since charge-carrier mobility is a function of temperature. In this subsection we will show that the number of charge carriers is also a function of temperature in band gap materials.

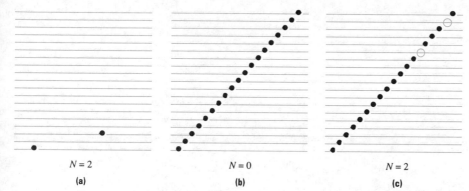

$N = 2$ $N = 0$ $N = 2$

(a)   **(b)**   **(c)**

FIGURE 10.2–5 Electron distributions within an energy band. **(a)** For a sparsely filled band, N equals the number of electrons in the band. **(b)** For a completely filled band, $N = 0$. **(c)** For a nearly filled band, N equals the number of empty levels in the band.

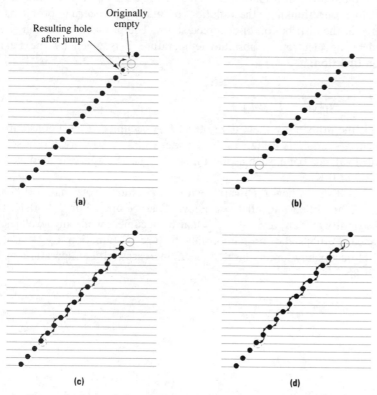

Originally empty

Resulting hole after jump

(a)   **(b)**

(c)   **(d)**

FIGURE 10.2–6 Electron distributions in a nearly filled energy band. **(a)** An electron "jumping" from a filled level into a nearby empty level. **(b)** An empty level is located near the bottom of the band. The transition from a to b may be thought of as 13 electrons each jumping up one level **(c)** or as the empty level moving down 13 energy levels **(d)**.

We begin our discussion of the relationship between N and T by considering the electron distribution in a solid with a partially filled valence band. As shown in Figure 10.2–7a, at 0 K the electrons are distributed in the lowest possible energy levels consistent with the Pauli exclusion principle. As the temperature increases, electrons gain energy and some of them are promoted to higher energy levels. The change in the occupancy of the energy levels with increasing temperature is shown in Figure 10.2–7b and c.

FIGURE 10.2–7
Electron distributions within
a partially filled valence
band as a function of temperature: **(a)** at 0 K, **(b)** at
$T_1 > 0$ K, and **(c)** at
$T_2 > T_1$.

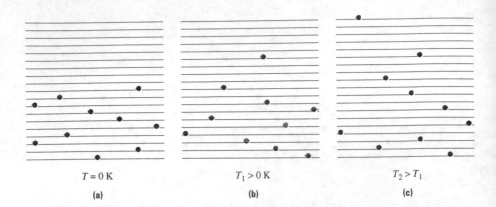

$T = 0\,\text{K}$

(a)

$T_1 > 0\,\text{K}$

(b)

$T_2 > T_1$

(c)

While the qualitative information depicted in Figure 10.2–7 is an extremely useful tool for guiding our thinking, the solution of many engineering problems requires quantification of the distribution of occupied states. The probability that an energy level is occupied by an electron at absolute temperature T is given by the **Fermi-Dirac distribution function:**

$$f(E) = \frac{1}{\exp[(E - E_f)/kT] + 1} \tag{10.2–11}$$

where $f(E)$ is the probability that energy level E is occupied, E_f is a constant known as the **Fermi energy,** and k is Boltzmann's constant (8.62×10^{-5} eV/K). In solids with a partially filled valence band, such as most metals, E_f can be approximated as the highest occupied level at 0 K.

Figure 10.2–8 shows how $f(E)$ varies with temperature. Note that at 0 K all energy levels below E_f are filled and all those above E_f are empty. For $T > 0$ K, there is a nonzero probability that any level above E_f may be occupied, while any level below E_f may be unoccupied. Perhaps the most important feature of Equation 10.2–11 is that the probability of occupancy for any level above E_f increases with temperature, as shown in Figure 10.2–8b.

FIGURE 10.2–8
(a) The change in the probability of occupation, as described by the Fermi-Dirac
distribution function $f(E)$,
as a function of tempera-
ture; and **(b)** the probability
that energy level E^* in part
a is occupied, given by
$f(E^*)$, as a function of temperature.

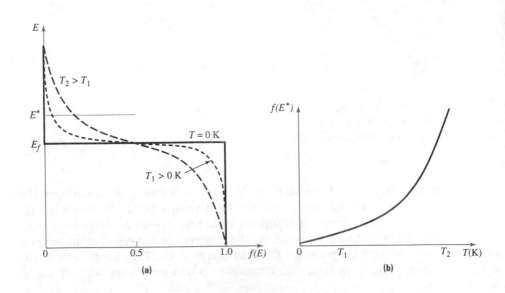

(a)

(b)

EXAMPLE 10.2–8

This example deals with the Fermi-Dirac distribution function.

  a. Show that $f(E_f) = 0.5$ at any temperature.
  b. Show that at 0 K, $f(E > E_f) = 0$ and $f(E < E_f) = 1.0$.
  c. Calculate the values of $f(E_f + 0.5 \text{ eV})$ at 0 K, 300 K, and 600 K.
  d. Repeat part c for $f(E_f + 2 \text{ eV})$.

Solution

  a. When $E = E_f$, Equation 10.2–11 reduces to

$$f(E_f) = \frac{1}{\exp(0) + 1} = 0.5$$

  b. When $T = 0$ K and $E > E_f$, Equation 10.2–11 reduces to

$$f(E > E_f) = \frac{1}{\exp(+\infty) + 1} = \frac{1}{+\infty} = 0$$

Similarly, when $T = 0$ K and $E < E_f$, we find

$$f(E < E_f) = \frac{1}{\exp(-\infty) + 1} = \frac{1}{0 + 1} = 1$$

  c, d. See the table:

Temperature	$f(E_f + 0.5 \text{ eV})$	$f(E_f + 2.0 \text{ eV})$
0 K	0	0
300 K	4.0×10^{-9}	2.6×10^{-34}
600 K	6.3×10^{-5}	1.6×10^{-17}

Figure 10.2–9 compares the qualitative and quantitative representations of the electron distribution within a partially filled valence band at several different temperatures. Notice that electrons are shown only in levels for which $f(E) > 0$.

Next, we address the issue we set out to examine in this section, the temperature dependence of the electron distribution in band gap materials. For an electron to occupy an energy level, two criteria must be satisfied:

1. Evaluated at the temperature of interest, the Fermi-Dirac distribution function must yield a reasonable probability of occupation.

2. The energy level of interest must be located within an allowed energy range (i.e., not located within a forbidden energy gap).

The occupied energy levels were not limited by the second constraint in solids with a partially filled energy band. In solids with completely filled energy bands at 0 K, however, the second constraint is important.

Figure 10.2–10 shows the relationship between the occupied energy levels, the Fermi-Dirac distribution function, and temperature for solids with an energy gap. In such solids the Fermi level is located in the center of the energy gap. In part a of the figure, corresponding to $T = 0$ K, the Fermi-Dirac function predicts that levels in the lower half of the band gap are occupied. Since this energy region is "forbidden" to electrons, however, these levels remain unoccupied. As temperature increases, the Fermi-Dirac distribution function assumes a sigmoidal shape. For the relatively low temperature corresponding to Figure 10.2–10b, all levels above E_f with a nonzero probability of occupancy, as given by the Fermi-Dirac function, are still in the band gap and, therefore, remain unoccupied. As the temperature continues to increase, the lowest levels in the

FIGURE 10.2–9

The relationship between the occupied energy levels and $f(E)$ as a function of temperature for solids with a partially filled energy band: (a) at 0 K, (b) at $T_1 > 0$ K, and (c) at $T_2 > T_1$.

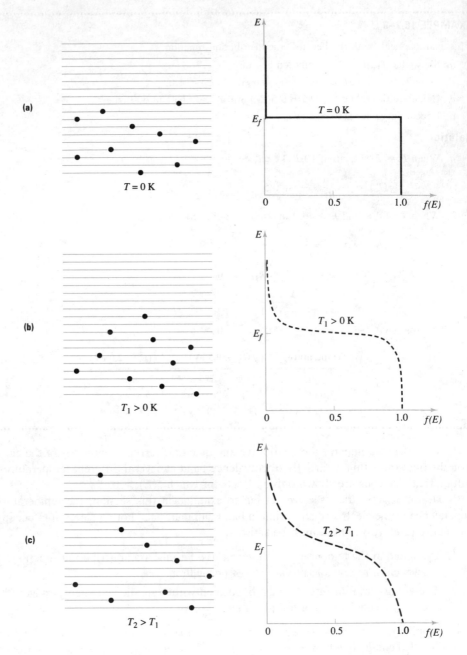

conduction band have a small probability of occupancy and the highest levels in the valence band have an identical probability of being unoccupied. This situation, illustrated in Figure 10.2–10c, corresponds to the temperature at which there is a real probability that electrons can be thermally promoted from the valence band to the conduction band.

How can we determine the probability that an electron is located in the conduction band at a specific temperature T? The solution is to determine the area of the (dark) shaded region in Figure 10.2–10d. This area can be expressed:

$$F(E > E_c) = \int_{E_c}^{\infty} f(E)\, dE, \qquad (10.2-12)$$

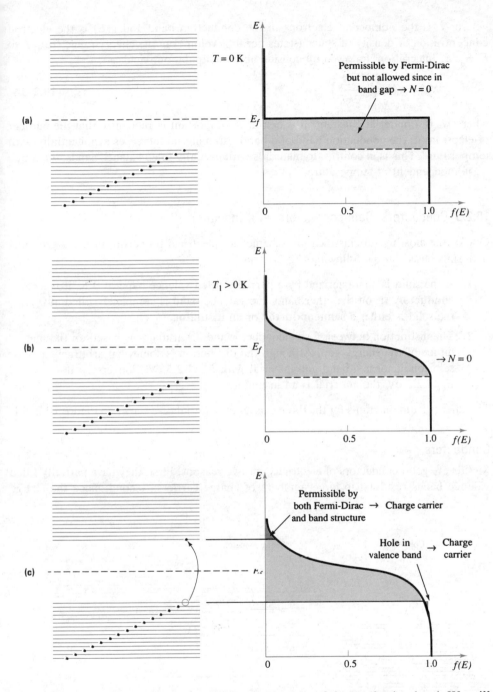

FIGURE 10.2–10

The relationship between
the occupied energy levels
and $f(E)$ as a function of
temperataure for solids
with a band gap: **(a)** at
0 K, **(b)** at $T_1 > 0$ K, and
(c) at $T_2 > T_1$.

where E_c corresponds to the energy level at the bottom of the conduction band. We will
not evaluate this integral. However, its solution shows that the probability of an electron
existing in the conduction band increases exponentially with temperature.

The final step in the process is to determine the number of electrons in the conduction
band of a band gap material as a function of temperature. Mathematically, this is accomplished by rewriting Equation 10.2–12 in the form

$$N_e = \int_{E_c}^{\infty} f(E) g(E) \, dE \qquad (10.2\text{–}13)$$

where N_e is the number of electrons in the conduction band and $g(E)$ is the electron concentration or density of states (states per unit volume) in the energy range from E to $E + dE$. Integration, followed by considerable manipulation, yields

$$N_e = N_0 \exp\left(\frac{-E_g}{2kT}\right) \tag{10.2–14}$$

where N_0 is a material constant. It is extremely important to recognize that the number of electrons in the conduction band of a band gap material increases exponentially with temperature. This is in contrast to materials with partially filled valence bands, for which N_e is independent of temperature.

10.2.5 Conductors, Semiconductors, and Insulators

One of the most useful classification schemes for electrical properties involves grouping materials based on the following two criteria:

1. If the solid is characterized by a partially filled valence band at 0 K, it is a **conductor.** If, on the other hand, the valence band is completely filled at 0 K, the solid is either a **semiconductor** or an **insulator.**

2. The distinction between a semiconductor and an insulator is based on the magnitude of the energy gap. Although the distinction is somewhat arbitrary, a semiconductor is a band gap material with $E_g \leq 2.5$ eV. Conversely, if $E_g > 2.5$ eV, the material is an insulator.

The energy band diagrams for the three classes of materials are shown in Figure 10.2–11.

Conductors

Metals are good conductors of electricity for two reasons. First, they have partially filled valence bands resulting in large numbers of charge carriers. Second, since the charge

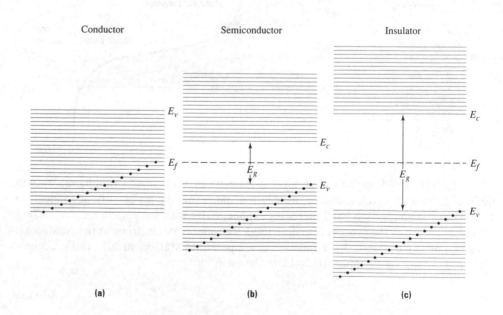

FIGURE 10.2–11

Energy band diagrams at 0 K for the three classes of electrical materials: **(a)** conductors, **(b)** semiconductors, and **(c)** insulators. E_v represents the top of the valence band, E_c represents the bottom of the conduction band, and E_f is the Fermi level.

carriers are electrons that are not localized to specific nuclei, the mobility term is reasonably high.

The energy band diagrams for metals fall into two general classes. As examples, we consider the first two elements in the third row of the periodic table, Na and Mg. The electronic structure of Na ($1s^2 2s^2 2p^6 3s^1$) shows that the 3s level in an isolated Na atom is half full. In bulk solid Na the 3s energy band contains twice as many levels as there are Na atoms; however, only half of these levels are occupied (Figure 10.2–12a through c).

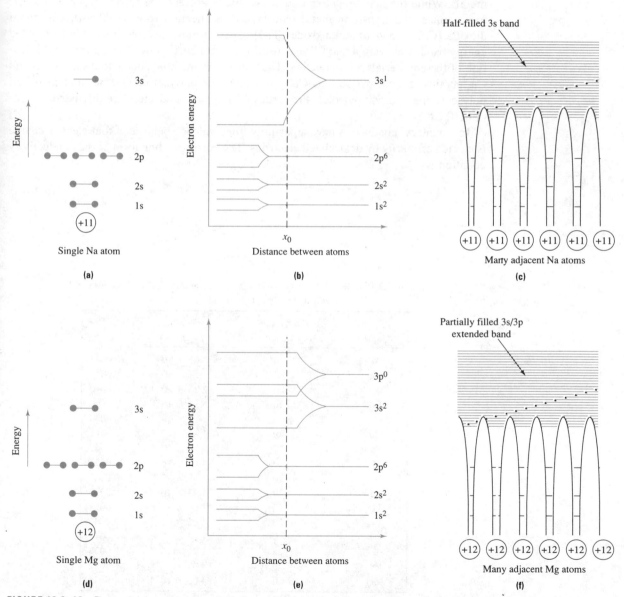

FIGURE 10.2–12 Electron band structures for metallic Na and Mg: **(a)** electron energy levels for an isolated Na atom, **(b)** splitting of the energy levels as a function of distance between two Na atoms, **(c)** the energy band diagram for Na showing a half-filled 3s band, **(d)** electron energy levels for an isolated Mg atom, **(e)** splitting of the energy levels as a function of distance between two Mg atoms, and **(f)** the energy band diagram for metallic Mg showing a partially filled 3s/3p extended band.

By comparison, the electronic structure of Mg ($1s^2 2s^2 2p^6 3s^2$) suggests that the 3s band should be completely filled in the solid. Hence, Mg would not be a conductor, a prediction that is inconsistent with reality. The solution to this apparent dilemma is that in Mg the 3s and 3p energy bands overlap to form an extended energy band. As shown in Figure 10.2–12d through f, the extended band is only partially filled, and hence, Mg satisfies the definition of a conductor. Similar arguments can be made for transition metals, although their band structures are more complex, involving overlaps among s, p, and d energy bands.

It is tempting to conclude that all metals are conductors and that all conductors are metals. While the first generalization is essentially correct, the second is not. Consider, for example, the transition metal oxides such as rhenium trioxide (ReO_3), chromium dioxide (CrO_2), and titanium oxide (TiO). These solids have partially filled extended energy bands involving d and f levels, resulting in conduction by electrons and classification of these materials as conductors. Thus, there are metallic conductors such as Al and Cu, ceramic conductors such as CrO_2 and TiO, and polymer conductors such as doped polyacetylene and polypyrrole. (Properties of polymer conductors are discussed in Section 10.2.8.)

In summary, conductors have a partially filled valence band such that electric charge is carried primarily by delocalized electrons. The corresponding form of the conductivity equation is

$$\sigma = N_e q_e \mu_e$$

(10.2–15)

TABLE 10.2–2   Electrical resistivities ρ_0 and temperature coefficients of resistivity α_e for selected conductors.

Conductor	Reference temperature (°C)	$\rho_0 (\Omega\text{-cm})$	α_e (°C^{-1})
Al	20	2.65×10^{-6}	0.0043
Ag	20	1.59×10^{-6}	0.0041
Au	20	2.35×10^{-6}	0.0040
Co	20	6.24×10^{-6}	0.0060
Cr	0	1.29×10^{-5}	0.0030
Cu	20	1.67×10^{-6}	0.0068
Fe	20	9.71×10^{-6}	0.0065
Mg	20	4.45×10^{-6}	0.0065
Ni	20	6.84×10^{-6}	0.0069
Pd	20	1.08×10^{-5}	0.0038
Pb	20	2.06×10^{-5}	0.0034
Pt	20	1.06×10^{-5}	0.0039
Sn	0	1.10×10^{-5}	0.0047
Ta	25	1.25×10^{-5}	0.0038
Zn	20	5.92×10^{-6}	0.0042
Zr	20	4.00×10^{-5}	0.0044

Source: Adapted from *Handbook of Materials Science*, Vol. I. Copyright CRC Press, Boca Raton, FL.

Since the product $N_e q_e$ is effectively independent of temperature and microstructure, the influence of internal and external variables on the conductivity of a conductor can be largely understood by determining the influence of these variables on electron mobility. For example, since we know that $\partial \mu / \partial T < 0$, we can now conclude that $\partial \sigma / \partial T < 0$ for an electrical conductor.

Since resistivity is the inverse of conductivity, it must increase as temperature increases. In fact, for many conductors, including most metals, the temperature dependence of resistivity, $\rho(T)$, can be expressed as

$$\rho(T) = \rho_0[1 + \alpha_e \, \Delta T] \tag{10.2–16}$$

where ρ_0 is the resistivity at a reference temperature, ΔT is the difference between the temperature of interest and the reference temperature, and α_e is a positive constant known as the **temperature coefficient of resistivity.** Values for ρ_0 and α_e for several conductors are given in Table 10.2–2.

....

EXAMPLE 10.2–9

Given that $\mu_e = 1.22 \times 10^{-3}$ m^2/(V-s) at 25°C, estimate the mobility of an electron in Al at 150°C.

Solution

Using Equation 10.2–16 and selecting the reference temperature as room temperature, we find the ratio of the resistivity at 150°C to that at 25°C:

$$\frac{\rho(150°C)}{\rho(25°C)} = \frac{\rho_0[1 + \alpha_e \, \Delta T]}{\rho_0} = 1 + \alpha_e(125°C),$$

or equivalently:

$$\frac{\sigma(150°C)}{\sigma(25°C)} = \frac{1}{1 + \alpha_e(125°C)}$$

Since for conductors $\sigma \propto \mu_e$, we have

$$\frac{\mu_e(150°C)}{\mu_e(25°C)} = \frac{\sigma(150°C)}{\sigma(25°C)} = \frac{1}{1 + \alpha_e(125°C)}$$

Solving for $\mu_e(150°C)$ and substituting the values given in the problem statement and in Table 10.2–2 yields:

$$\mu_e(150°C) = \frac{\mu_e(25°C)}{1 + \alpha_e(125°C)}$$

$$= \frac{1.22 \times 10^{-3} \text{ m}^2/(\text{V-s})}{1 + (0.00429°C^{-1})(125°C)}$$

$$= 7.94 \times 10^{-4} \text{ m}^2/(\text{V-s})$$

....

Semiconductors

The most common semiconducting materials are the covalently bonded solids, such as Si, Ge, and GaAs. There are, however, many other band gap solids with $E_g \leq 2.5$ eV.

Examples include ceramics such as SiC and Cu_2O and a few organic materials such as anthracene (coal) and naphthalene (mothballs). Table 10.2–3 lists the band gaps and charge-carrier mobilities for a number of semiconductors.

Consider the $1s^22s^22p^63s^23p^2$ electronic structure of Si. On the basis of previous arguments, we expect the partially filled 3p band to lead to extensive electron conduction in solid Si. This is not the case. The difference between Si and metals is that Si is covalently bonded. The formation of covalent bonds in Si involves a complex interaction, known as hybridization, among the four electrons in the 3s and 3p levels. Two electron energy bands, each containing four levels for each atom in the solid, are produced. At

TABLE 10.2–3   Band gaps and charge-carrier mobilities for a variety of semiconductors.

Material	Band gap (eV)	μ_e (cm²/V-s)	μ_h (cm²/V-s)
C*	5.4	1800	1400
Si	1.107	1900	500
Ge	0.67	3800	1820
III-V compounds with the zinc blende crystal structure			
BN	~4	—	—
AlP	2.5	—	—
AlAs	2.16	1,200	420
AlSb	1.60	200–400	550
GaP	2.24	300	100
GaAs	1.35	8,800	400
GaSb	0.67	4,000	1400
InP	1.27	4,600	150
InAs	0.36	33,000	460
InSb	0.165	78,000	740
II-VI compounds with the zinc blende crystal structure			
ZnS	3.54	180	5
ZnSe	2.58	540	28
ZnTe	2.26	340	100
CdTe	1.44	1200	50
Other crystalline semiconductors			
β-SiC	2.3	4000	—
α-SiC	2.9	4000	~2000
ZnO	3.2	180	—
CdS	2.42	400	—
CdSe	1.74	650	—
$CuAlS_2$	2.5	—	—
$CuFeSe_2$	0.16	—	—
$AgInSe_2$	1.18	—	—
$ZnSiAs_2$	1.7	—	50
PbS	0.37	600	600
CdO	2.5	100	—
$BaTiO_3$	2.8	—	—
Polymers			
Polyacetylene	1.4	—	—
Poly(p-phenylene sulfide)	4.0	—	—
Polypyrrole	3.0	—	—

*Although carbon (diamond) is an insulator by our definition, it is included here for comparison.

Source: Adapted from *Handbook of Chemistry and Physics*, 61st ed. Copyright CRC Press, Boca Raton, FL.

0 K the lower, or valence, band is completely filled and the upper, or conduction, band is completely empty. The result is a material with a band gap. Most, but not all, covalent solids exhibit this type of band structure and, therefore, are either semiconductors or insulators.

Although the details of the evolution of energy bands are somewhat different in ionic solids, the result is the same—most ionic solids are band gap materials. Since the individual ions have filled valence shells, the corresponding energy bands in ionic solids will be completely filled unless two bands overlap. If the bands overlap, the ionic solid is a conductor; if not, it is either an insulator or a semiconductor.

The technological impact of semiconductors has grown exponentially in the past few decades. The details of the conduction mechanisms in these materials will be discussed in Section 10.3. At this point it suffices to say that their behavior is complex for several reasons including:

1. Both electrons and holes can transport charge.
2. Changes in temperature and defect density influence both the number of charge carriers and their mobility.

Insulators

Examples of materials that are generally considered insulators include most pure metal oxides (e.g., Al_2O_3, MgO, SiO_2), the silicate ceramics, and the common organic polymers (e.g., polyolefins, vinyl polymers, polyamides, polytetrafluoroethylene, and polyesters).

By our present classification scheme the only difference between a semiconductor and an insulator is the size of the energy gap. This implies that at significantly high temperatures an insulator begins to conduct electricity. Although the statement is theoretically correct, the temperature required for conduction in most insulators is generally so high as to be of little practical importance or is above the temperature at which the material degrades. For example, a comparison of the occupational probabilities for diamond and silicon at 25°C shows that diamond is 36 orders of magnitude less likely to have an electron in the conduction band. This suggests that diamond will be an insulator at room temperature. At 1200°C, however, diamond has approximately the same chance of electron promotion as does Si at 25°C. Note, however, that diamond is one of the few solids that can exist at 1200°C. Most other materials would have degraded long before an equivalent conduction temperature was reached.

10.2.6 Ionic Conduction Mechanisms

For the most part, our discussion of electrical conduction has centered on the motion of electrons and holes. It is important to recognize, however, that in ionic solids and polymers charge transport may also occur by the motion of ions. Since electrical conduction via ions occurs by one of the diffusion mechanisms described in Chapter 4, we anticipate a direct relationship between the ionic mobility μ_{ion} and ionic diffusion coefficient D_{ion}. In fact, it is found that

$$\mu_{ion} = \left(\frac{q}{kT} D_{ion} \right) \tag{10.2-17}$$

where q, k, and T have their usual meanings. Equation 10.2–17 is known as the Einstein relationship. Since we already know that D_{ion} is a strong function of both temperature and defect density, we should expect the same factors to influence ionic mobility and, therefore, ionic conductivity.

The form of the conductivity equation for an ionic solid is

$$\sigma = q_e \left[N_e \mu_e + N_h \mu_h + \sum_i (N_{ion} Z_{ion} \mu_{ion})_i \right] \tag{10.2-18}$$

where Z_{ion} is the valence of the ion and the summation includes all types of mobile ions. The relative contribution of the carriers—electrons, holes, and ions—depends on several factors, including the band structure of the solid, temperature, and defect density.

When an ionic solid has a partially filled valence band, the contribution from electrons is significant. If the solid has a small band gap ($E_g < 2.5$ eV), then electrons and holes may both contribute to the overall conductivity. If, however, the band gap is large, as is often the case in ionic solids, then the conductivity will be dominated by the motion of ions.

The relative contributions from various charge carriers can be quantified using **transference numbers.** The transference number t_i of any charge carrier is the ratio of the conductivity due to that charge carrier to the total conductivity of the solid. For example, the transference number for cations, t_{cat}, is

$$t_{cat} = \frac{\sigma_{cat}}{\sigma_{total}} = \frac{N_{cat} q_{cat} \mu_{cat}}{\sigma_{total}} \tag{10.2-19}$$

Similar quantities can be defined for anions, electrons, and holes. As shown in Table 10.2–4, the conductivity in most ionic solids is dominated by a single type of charge carrier. The dominant carrier, however, may change with temperature. Consider, for example, CuCl. At low temperatures charge is transported by electrons, but at higher temperatures mobile cations dominate.

How does temperature influence the magnitude of ionic conduction? Recall from Chapter 4 that small ions diffuse by an interstitial mechanism while larger ions diffuse via a vacancy mechanism. Since the concentration of both vacancies and interstitials

TABLE 10.2–4 Transference numbers for selected ionic solids.

Compound	Temperature (°C)	t_{cation}	t_{anion}	$t_{electron/hole}$
NaCl	400	1.0	0	0
	600	0.95	0.05	0
KCl	435	0.96	0.04	0
	600	0.88	0.12	0
KCl + 0.02% $CaCl_2$	430	0.99	0.01	0
	600	0.99	0.01	0
AgCl	20–350	1.0	0	0
AgBr	20–350	1.0	0	0
BaF_2	500	0	1.0	0
PbF_2	200	0	1.0	0
CuCl	20	0	0	1.0
	366	1.0	0	0
ZrO_2 + 7% CaO	> 700	0	1.0	10^{-4}
$Na_2O \cdot 11Al_2O_3$	< 800	1.0 (Na^+)	0	$< 10^{-6}$
FeO	800	10^{-4}	0	1.0
ZrO_2 + 18% CeO_2	1500	0	0.52	0.48
ZrO_2 + 50% CeO_2	1500	0	0.15	0.85
$Na_2O \cdot CaO \cdot SiO_2$ glass	—	1.0 (Na^+)	0	0

Source: L. L. Hench and J. K. West, *Principles of Electronic Ceramics.* Copyright © 1990 by John Wiley & Sons. Reprinted by permission of John Wiley & Sons, Inc.

TABLE 10.2–5   A comparison of the activation energies for ionic conduction and the diffusion of Na^{+1} in a series of silicate glasses.

Composition (Mol %)					Activation energy (kJ/mol)	
Na_2O	CaO	Al_2O_3	SiO_2	GeO_2	Diffusion (Na^+)	Conduction
33.3	—	—	66.7	—	54–59	59–67
25.0	—	—	—	75.0	71–75	67–75
15.7	—	12.1	72.2	—	68.6	65.3
11.0	—	16.1	72.9	—	65.3	63.2
15.9	11.9	—	72.2	—	92.0	87.0
14.5	12.3	5.8	67.4	—	84.5	81.6

Source: L. L. Hench and J. K. West, *Principles of Electronic Ceramics.* Copyright © 1990 by John Wiley & Sons. Reprinted by permission of John Wiley & Sons, Inc.

increases exponentially with temperature, the value of N_{ion} will also increase with temperature. The influence of temperature on ionic mobility can be obtained by examining Equation 10.2–17. As the temperature increases, the exponential increase in the diffusion coefficient (numerator) overshadows the linear function of temperature (denominator). Since both N_{ion} and μ_{ion} increase exponentially with temperature, the ionic conductivity at temperature T, $\sigma(T)$, is given by

$$\sigma(T) = N_{ion} q_{ion} \mu_{ion} = \sigma_0 \exp\left(-\frac{Q}{kT}\right) \tag{10.2–20}$$

where σ_0 is a material-specific constant, Q is the activation energy for charge transfer, and k is Boltzmann's constant.

It is useful to compare the activation energies for diffusion with those for electrical conduction. Consider, for example, the case of ionic conduction in silicate glasses (Table 10.2–5). In these and many other ionic solids the most mobile ions are often the monovalent network modifier cations, such as Na^{+1}. The similarity of the activation energies suggests that both processes are controlled by the same mechanisms. An important distinction between ionic conduction, σ_{ion}, and electron conduction, σ_e, is that the former increases with increasing temperature ($\partial\sigma_{ion}/\partial T > 0$) while the latter decreases with increasing temperature ($\partial\sigma_e/\partial T < 0$).

An interesting class of ionic conductors is the so-called solid electrolytes, or fast ionic conductors. These compounds have conductivities of $\sim 10^{-2}(\Omega\text{-}cm)^{-1}$, which is at the lower end of the range associated with electron conductors. One subclass of these solid electrolytes is based on zirconia, ZrO_2. The addition of an oxide with a lower-valence cation, such as Na_2O, CaO, or Y_2O_3, results in a large number of point defects, which are believed to be responsible for the high mobilities (since they increase the diffusion rate) and high conductivities of these compounds. Thus, in addition to its excellent mechanical properties, partially stabilized zirconia also has interesting electrical properties.

10.2.7 Effects of Defects and Impurities

The influence of defects and impurities on conductivity can be understood in terms of their effects on the number of charge carriers and on mobility. Because they perturb the periodicity of the crystal lattice, both impurities and defects decrease μ in all classes of materials dominated by electron/hole conduction. For equal defect concentrations, however, a uniform distribution of defects decreases μ more than a clustered distribution, since the former represents a shorter distance between electron-scattering centers. Thus, an impurity in solid solution decreases the conductivity of an alloy more than it does if the same weight percent impurity were distributed as a coarse precipitate.

The influence of impurities and defects on the number of charge carriers in conductors is usually rather small. The reason is that the number of mobile electrons in the partially filled valence band is large in comparison with typical defect densities.

Recalling that for conductors $\sigma = N_e q_e \mu_e$, and recognizing that only μ_e is a strong function of defect density, one can conclude that the conductivity of a conductor decreases as the defect density increases. Combining this observation with our previous knowledge of the temperature dependence of ρ yields the general expression for the total resistivity of a conductor, ρ:

$$\rho = \rho(T) + \Sigma_i(\rho_d)_i \tag{10.2-21}$$

where $\rho(T)$ is as defined in Equation 10.2–16, $(\rho_d)_i$ is the resistivity increase due to the ith defect type, and the summation is carried out over all types of defects in the solid.

The form of the expression for $(\rho_d)_i$ depends on the defect type. For example, in dilute solid solutions, ρ_d takes the form

$$\rho_d = b_{AB}[x(1 - x)] \tag{10.2-22}$$

where b_{AB} is a positive constant describing the effect of additions of solution B into solvent A and x is the atom fraction of the solute. (Note that $b_{AB} \neq b_{BA}$, so that one must be sure to select the appropriate constant when solving quantitative problems.) Figure 10.2–13a shows the influence of alloy composition on the resistivity of binary Cu-Ni alloys. Often, the effect of defects on ρ, or equivalently σ, is presented graphically rather than as an analytical expression. For example, Figure 10.2–13b shows the effect of cold working on the conductivity of a series of binary Cu-Zn alloys. The decrease in σ due to cold working is related to the increase in the dislocation density that occurs during the plastic

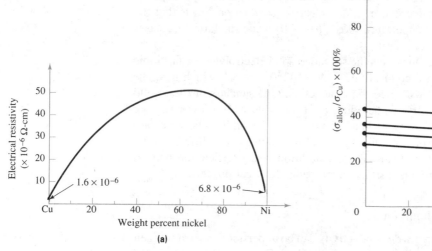

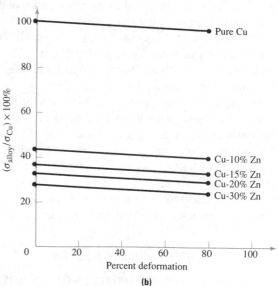

(a)

(b)

FIGURE 10.2–13 The influence of alloy additions and cold working on the resistivity of metals: **(a)** the variation of resistivity with composition in binary Cu-Ni alloys, and **(b)** the influence of cold working on the electrical conductivity of binary Cu-Zn alloys.   *(Source:* **(a)** *Askeland,* The Science and Engineering of Materials, *2nd. ed. Boston: PWS-Kent Publishing Company, 1989.* **(b)** *Askeland,* The Science and Engineering of Materials, *3rd ed. Boston: PWS Publishing Company, 1994.)*

deformation process. Note that the decrease in σ with increasing solution concentration is also apparent in this figure.

The effect of defects and impurities on the conductivity of semiconductors is more complex. In small-band gap materials at modest temperatures, the density of mobile charge carriers is approximately the same as the typical densities of defects and impurities. Thus, in contrast to conductors, the number of charge carriers can be substantially altered by the defects. This point and its consequences are discussed in more detail in Section 10.3.

In insulators the conductivity is low primarily because of the low density of mobile charge carriers. Since defects and impurities can increase N, they may substantially alter the conductivity of the solid. If the material was selected for its insulating properties, the defects may cause it to fail to serve its desired function. On the other hand, if an application requires a modest increase in conductivity, an insulator can be doped with "impurities" to increase σ intentionally. A good example is the incorporation of small amounts of arsenic pentafluoride into poly (p-phenylene sulfide), which increases the conductivity of the polymer from 10^{-16} to 10^1 $(\Omega\text{-cm})^{-1}$.

The influence of defects and impurities on solids that exhibit ionic conduction is complex. Although defects may increase N, their greatest influence is often on the ionic mobility. Since the rate-limiting step in ionic conduction is usually ionic diffusion, any impurity that introduces additional point defects or open volume into the lattice may increase ionic diffusion, ionic mobility, and ultimately ionic conductivity.

In summary, increases in the concentration of point defects or impurities tend to decrease the conductivity of electron conductors and increase the conductivity of most other types of materials.

DESIGN EXAMPLE 10.2–10

Many engineering applications call for a metal wire with a combination of high strength and high electrical conductivity. Assume that sufficient strength can be obtained by either cold working or solid solution strengthening. In view of the electrical conductivity requirement, which strengthening mechanism would you recommend? Why?

Solution

An increase in the defect concentration, either in the form of a solid solution or cold working, decreases the mobility and, therefore, the conductivity of the metal. The magnitude of the conductivity reduction, however, depends on the spatial distribution of the defects. The solute atoms will be arranged uniformly throughout the alloy. As such, they will have a small separation distance, act as effective electron-scattering centers, and significantly reduce the conductivity of the metal. In contrast, the more coarse defect distribution associated with the dislocations produced by cold working represents a less effective electron-scattering network and results in a less substantial decrease in σ. (See Figure 10.2–13b.) Thus, to increase strength with the minimal reduction in electrical conductivity, the cold-working process will be preferred.

10.2.8 Conducting Polymers

The basic principles of electron conduction are the same in polymers as they are in other classes of materials: there must be a partially filled valence band. However, the details of the mechanisms are sufficiently different that a separate treatment is helpful. Most polymers have filled valence bands and large band gaps. That is, they are insulators. There are, however, examples of conducting polymers.

A conducting polymer is one that has a conductivity at room temperature similar to that of metals and other materials with partially filled valence bands. By this definition there are no pure polymers that conduct. Several polymers, however, including poly-acetylene, polyparaphenylene, polypyrrole, polyaniline, and polythiophene can be made conductive by doping with controlled amounts of specific impurities. These polymers are all based on a conjugated carbon ring structure loosely similar to that found in graphite. It is believed that the conducting electrons are able to move along the carbon backbone of these polymers.

The addition of arsenic pentafluoride to polyacetylene can increase its conductivity to ~25% of that for copper. This number is even more impressive when it is recognized that on a per-unit-weight basis the specific conductivity of the doped polyacetylene is twice that of copper. Unfortunately, this polymer is unstable in air and atmospheric moisture. Additional research is required to solve the plethora of problems associated with conduct-ing polymers. The motivation for this research, however, is supplied by the potential to design materials with excellent specific conductivity. Potential applications include lightweight storage batteries, flexible and lightweight electrical cables in aerospace appli-cations, and radar-absorbing coatings on aircraft.

A second method for increasing the electrical conductivity of polymers is to form a composite by introducing a conducting filler material into the inherently insulating poly-mer. Consider, for example, silicone rubber with a conductivity at room temperature of 10^{-14} $(\Omega\text{-cm})^{-1}$. The addition of 15 wt. % carbon black into the rubber increases the conductivity of the composite to ~10^{-4} $(\Omega\text{-cm})^{-1}$. An additional eight orders of magni-tude increase to ~10^{4} $(\Omega\text{-cm})^{-1}$ can be achieved by replacing the carbon black with silver-coated metal particles. This class of conducting polymers has several applications, including battery cables in automobile engines and as a possible replacement for metal solders used in printed circuit boards.

10.2.9 Superconductivity

In Section 10.2.5 we defined an electrical conductor as a material with a partially filled valence band. In some conductive materials, the resistivity changes abruptly and ap-proaches zero as the temperature decreases below a critical value, T_c (see Figure 10.2–14). This phenomenon, known as **superconductivity,** was first demonstrated for Hg ($T_c = 4.12$ K) by H. Kamerlingh Onnes in 1911. This section provides a qualitative understanding of the atomic scale mechanisms responsible for superconductivity, de-scribes the crystal structure of one of the "high T_c" materials, and lists a number of applications for these materials. The interesting magnetic properties of these materials are discussed in Chapter 12.

FIGURE 10.2–14

Resistivity versus tempera-ture for Hg. Note the sub-stantial drop to a resistivity very close to zero at the superconducting transition temperature, T_c.

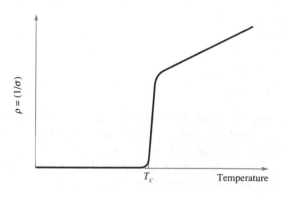

We begin our description of the mechanism of superconductivity with an experimental observation. Most pure metals that exhibit superconductivity also exhibit comparatively poor room-temperature conductivity. Since the conductivity of a defect-free metal is controlled by scattering of electrons due to the vibration of the atoms in the crystal lattice, metals with poor room-temperature conductivities have comparatively intense electron-lattice interactions. Thus, the experimental link between relatively poor room temperature conductivity and superconductivity suggests that electron-lattice interactions may be a key to understanding superconductivity.

Electron mobility in a defect-free crystal is limited by the time between collisions of electrons with the atomic cores. If a mechanism that prohibits these collisions were put in place, then the mobility, and hence conductivity, might increase without limit. While a complete explanation of this mechanism requires a quantum mechanical treatment, the essence of the idea can be expressed as follows. At low temperature, electrons with opposite spins are paired because of mutual attraction to a local area of positive charge within the lattice. At the critical temperature, the vibrational frequency of the electron pairs corresponds to the vibrational frequency of the atomic nuclei. The result of this correlation between vibrational frequencies is a synchronized motion of the electron pairs through the vibrating crystal lattice. Because the motion is synchronized, the time between collisions, and hence the conductivity, increases without limit. The model described above, known as the BCS theory of superconductivity, was developed by John Bardeen, Leon H. Cooper, and J. Robert Schrieffer in 1957. The electron pairs are known as Cooper pairs.

Until the mid-1980s, the highest observed value of T_c was less than 25 K. In 1987, however, a new class of superconducting oxides with superconducting transition temperatures near 100 K was discovered. One example of this class of superconductors, known as the 1-2-3 compounds because of the subscripts on the cations, is $YBa_2Cu_3O_{7-x}$. As shown in Figure 10.2–15, the orthorhombic unit cell for this material can be thought of as three adjacent pervoskite unit cells with an ordered array of cations and oxygen vacancies. (See Section 3.7.3 for a discussion of the perovskite structure.) The superconducting properties of the 1-2-3 compounds are believed to be related to (electron-pair)–(oxygen-vacancy) interactions.

A great many technologically significant developments may result from these high-T_c compounds. Examples of projects either in place or under development include more

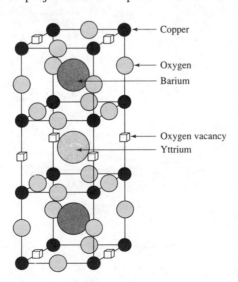

Copper

Oxygen
Barium

Oxygen vacancy
Yttrium

FIGURE 10.2–15

The orthorhombic unit cell for the 1-2-3 superconducting compound $YBa_2Cu_3O_{7-x}$. The unit cell can be thought of as three adjacent perovskite unit cells with an ordered array of cations and oxygen vacancies.

efficient ways to transmit electrical power at low voltages, high-speed electrical switching capabilities with applications in computers, large-scale electrical power generation by fusion made possible by high-field superconducting magnets, battery-powered pollution-free automobiles, and magnetically levitated (i.e., low-friction) trains.

10.2.10 Devices and Applications

In this section we describe some of the engineering devices and applications that have been developed based on the principles described so far in this chapter.

Thermocouples

A thermocouple is a device used to measure temperature. The underlying principle is that as a conductor is heated, some of its electrons are excited to higher energies. As shown in Figure 10.2–16a, the hotter end of the wire has a great number of excited electrons than does the cooler end. Some of the "extra" electrons flow from the hot end toward the cold end, resulting in a voltage difference V_a across the wire. Another wire, made from a different material, will develop a voltage V_b between its ends. In general, $V_a \neq V_b$, since the energy band diagrams of the two conductors differ. If the hot ends of the wires are placed in electrical contact, as shown in Figure 10.2–16b, a voltmeter connected to the opposite ends of the wires can be used to measure the voltage difference, $V_{ab} = (V_a - V_b)$. V_{ab} is a function of the temperature at the junction of the wires. The thermal generation of a voltage between the junction of two dissimilar wires and their free ends is known as the Seebeck effect.

FIGURE 10.2–16

A thermocouple. **(a)** A voltage gradient V_a is established between the hot and cold ends of a conductor. **(b)** If the hot ends of the two wires are placed in electrical contact, a voltmeter connected to the opposite ends of the wires measures the voltage difference V_{ab}, which depends directly on the temperature at the joined end of the wires.

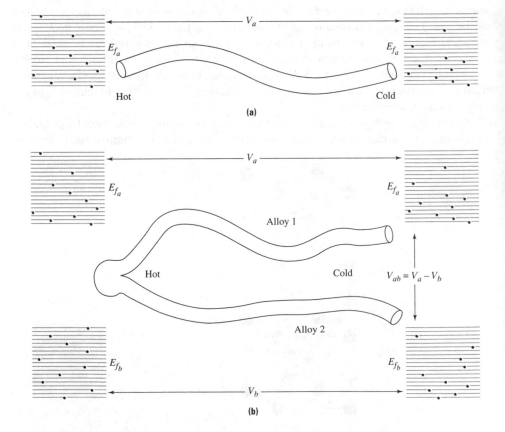

Spark Plug Insulators

The insulating portion of a spark plug is designed to survive the application of several thousand volts at high temperatures. The high resistivity requirement suggests the use of either an ionic material or a polymer. The high-temperature requirement, however, leads to the choice of a ceramic material. In fact, most common spark plug insulators are made from an Al_2O_3-based ceramic (steatite).

Elemental Sensing Devices

The form of the conductivity equation for ionic solids in which charge is carried by a single dominant mobile ion is $\sigma = N_{ion}q_{ion}\mu_{ion}$. So far, we have considered an applied voltage as the driving force and the motion of charge as the response. If, however, no voltage were applied, yet a flow of charge carries were driven across a conducting membrane by a concentration gradient, the moving charge would induce a voltage difference across the membrane. This phenomenon is the basis of a family of devices used to determine the concentration of specific elements. For example, zirconia (ZrO_2) is used as an oxygen sensor to monitor air-fuel ratios in internal-combustion engines.

Static Discharge

The problem of static discharge in synthetic polymers is well known. Shuffle across a nylon carpet in winter and touch another person or a metal object and there is a good chance that an arc will jump from your fingertip. Similarly, static discharge may be heard when an acrylic pullover sweater is removed. These phenomena are associated with the high insulating properties of polymeric materials. Commerical solutions to the problems associated with static discharge include the addition of a conductive fiber to each yarn or the application of an antistatic finish to the fibers. For example, the incorporation of a hygroscopic material, that is, one that attracts water, can result in the development of a wet surface layer that will dissipate the electric charge. Important applications of antistatic materials are in trolley wheels used in hospital operating rooms (where a spark could ignite oxygen gas) and in parachutes (where static could prevent opening of the parachute).

10.3 SEMICONDUCTORS

We begin our discussion of the electrical properties of semiconductors by refocusing our attention on the basic conduction mechanisms and the influence of temperature and impurities on σ. Next, after examining some semiconducting devices, we investigate materials-related issues in the microelectronics industry.

10.3.1 Intrinsic and Extrinsic Conduction

Intrinsic conduction is defined as semiconducting behavior resulting from the band structure of a pure element or compound. In contrast, **extrinsic conduction** is faciliated by the presence of impurities.

When ionic conduction is negligible, the conductivity equation is:

$$\sigma = q_e(N_e\mu_e + N_h\mu_h) \tag{10.3-1}$$

where the subscripts e and h refer to electrons and holes. As temperature increases, electrons are promoted from the valence band to the conduction band. Figure 10.3–1 shows that both electrons in the conduction band and holes in the valence band are charge carriers. (Recall the discussion of charge carriers in Section 10.2.3.) In fact, in intrinsic

FIGURE 10.3–1

Thermal generation of electrical charge carriers in an
intrinsic semiconductor.
Both the electron in the
nearly empty conduction
band and the hole in the
nearly filled valence band
are mobile charge carriers.

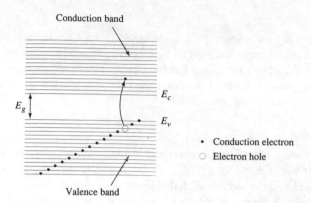

semiconductors there is a one-to-one correspondence between conduction electrons and valence holes—that is, $N_e = N_h$—and the conductivity equation reduces to:

$$\sigma = N_e q_e (\mu_e + \mu_h) \tag{10.3–2}$$

The important issue is how temperature influences the terms in this equation. We already know the answer:

1. q_e is not a function of T;
2. The mobility of both electrons and holes decreases linearly with an increase in temperature (i.e., $\partial\mu/\partial T < 0$); and
3. The number of mobile charge carriers increases exponentially with an increase in temperature (i.e., $\partial N/\partial T > 0$).

The exponential effect dominates and $\partial\sigma/\partial T > 0$ for an intrinsic semiconductor.

As shown previously (Equation 10.2–14), the number of conduction electrons in a band gap material varies with temperature as:

$$N_e \approx N_0 \exp\left(-\frac{E_g}{2kT}\right)$$

By combining Equations 10.3–2 and 10.2–14 we find:

$$\sigma = N_0 q_e (\mu_e + \mu_h) \exp\left(-\frac{E_g}{2kT}\right) \tag{10.3–3}$$

Since the effect of temperature on mobility over a limited temperature range is small, the conductivity of an intrinsic semiconductor becomes:

$$\sigma = \sigma_0 \exp\left(-\frac{E_g}{2kT}\right) \tag{10.3–4}$$

where $\sigma_0 = N_0 q_e (\mu_e + \mu_h)$.

Taking the logarithm of both sides of Equation 10.3–4 yields:

$$\ln \sigma = \ln \sigma_0 - \left(\frac{E_g}{2k}\right)\left(\frac{1}{T}\right) \tag{10.3–5}$$

Thus, as shown in Figure 10.3–2, the slope of a plot of $\ln \sigma$ versus $(1/T)$ is $-E_g/2k$. This observation provides investigators with an experimental method for determining the band gap of a semiconductor and, hence, a means of identifying an unknown material.

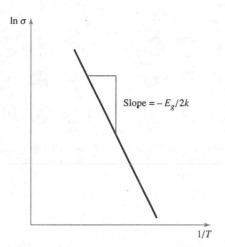

FIGURE 10.3–2

A plot of $\ln \sigma$ versus $1/T$ for an intrinsic semiconductor. The slope of the curve is $-E_g/2k$.

EXAMPLE 10.3–1

Consider an intrinsic semiconductor with a conductivity that triples between 25 and 53°C. Calculate the band gap of this material.

Solution

The conductivity of an intrinsic semiconductor is given by Equation 10.3–4 as $\sigma = \sigma_0 \exp(-E_g/2kT)$. Taking the ratio of the conductivities at the two temperatures yields

$$\frac{\sigma(326 \text{ K})}{\sigma(298 \text{ K})} = 3 = \frac{\sigma_0 \exp[-E_g/2k(326 \text{ K})]}{\sigma_0 \exp[-E_g/2k(298 \text{ K})]}$$

Taking the natural logarithm of both sides of the second equality yields:

$$\ln 3 = \left(-\frac{E_g}{2k}\right)\left(\frac{1}{326 \text{ K}} - \frac{1}{298 \text{ K}}\right)$$

Solving for E_g gives:

$$E_g = -\frac{2k(\ln 3)}{(1/326 \text{ K}) - (1/298 \text{ K})}$$

$$= -\frac{2(8.62 \times 10^{-5} \text{ eV/K})(\ln 3)}{(1/326 \text{ K}) - (1/298 \text{ K})} = 0.66 \text{ eV}$$

So far we have discussed only pure intrinsic semiconductors such as Si and Ge. How does the conductivity change if the semiconductor is impure or doped? To answer this question, let us reinvestigate the energy band structures in solids. Recall that energy band formation resulted in part from the Pauli exclusion principle, which requires electrons in neighboring atoms to "move" to different energy levels. The electrons associated with a foreign atom, however, are not restricted to the same energy levels as those of the host atoms. In fact, the impurity atom's electrons may reside in energy levels forbidden to the host element's electrons. This phenomenon is the heart of extrinsic conduction.

Consider the addition of a Group V atom, such as phosphorus (P) into a Si crystal. As shown in Figure 10.3–3, four of the five valence electrons from the P atom will be involved in covalent bonds with the neighboring Si atoms. The fifth valence electron, however, is only loosely bonded to the P atom. In effect, this is an "extra" electron. Because the extra electron is less tightly bound to the P atom, it requires less thermal energy to promote it into the conduction band. As shown in Figure 10.3–4a, the extra

FIGURE 10.3–3 A comparison of the bond structure of: **(a)** pure Si, and **(b)** Si doped with a Group V element such as P. Note the "extra" loosely bound electron associated with the dopant.

Shared electrons
in a covalent bond

(a)

Extra electron from
Group V dopant atom

(b)

FIGURE 10.3–4

(a) The energy band diagram for a semiconductor doped with a Group V element (i.e., an n-type semiconductor). E_d represents the energy level associated with the extra electrons donated by the dopant atoms. **(b)** A plot of ln σ versus $1/T$ for an extrinsic n-type semiconductor.

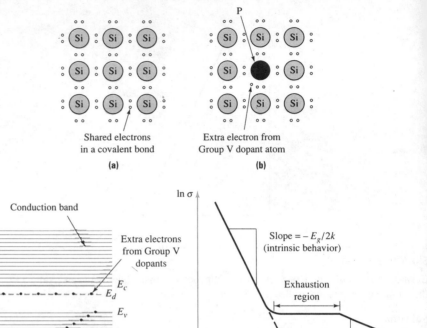

Conduction band

Extra electrons
from Group V
dopants

E_c
E_d

E_g

E_v

Valence band

(a)

ln σ

Slope $= -E_g/2k$
(intrinsic behavior)

Exhaustion
region

Slope $= -(E_c - E_d)/k$
(extrinsic behavior)

$1/T$

(b)

P electron resides in an energy level in the upper half of the forbidden gap of Si. Since the P atom "donated" an electron to the structure, the corresponding energy level is called a **donor level** and is labeled E_d. Extrinsic semiconductors of this genre are called **n-type** because of their excess negative charge.

How does the presence of the donor level change the conductivity of the semiconductor? Rather than having to supply thermal energy equivalent to the magnitude of the band gap (i.e., $E_g = E_c - E_v$) to get conduction, it is necessary to supply only enough thermal energy to promote an electron from E_d to E_c. It is therefore tempting to propose an equation for an n-type extrinsic semiconductor that is similar to Equation 10.3–5 with a simple substitution of $(E_c - E_d)$ for E_g. While such an equation would be nearly correct, it is deficient on one essential point: in intrinsic conduction, each electron in the conduction band is accompanied by a hole in the valence band. This is not the case for an n-type extrinsic material. Since the electron originally located at E_d was an "extra" electron, its promotion does not result in the formation of a hole. A more detailed treatment of this problem shows that Equation 10.3–5 must be modified by removing the factor of 2 from the denominator of the exponent. The proper form of the conductivity equation for an n-type extrinsic semiconductor at low temperatures is:

$$\sigma = \sigma_0 \exp\left(-\frac{E_c - E_d}{kT}\right) \tag{10.3–6}$$

where $\sigma_0 = N_0 q_e \mu_e$.

Figure 10.3–4b shows a plot of ln σ versus $1/T$ for an n-type semiconductor. From Equation 10.3–6, we see that the slope in the low T (high $1/T$) regime is $-(E_c - E_d)/k$.

As the temperature increases, the figure shows a flat, or exhaustion, region, followed by another region with a slope of $-E_g/2k$. The exhaustion region occurs because there are a finite number of electrons in the donor level. After all donor electrons have been promoted, σ remains constant until intrinsic conduction becomes possible. After the onset of intrinsic conduction, the conductivity change with temperature is described by Equation 10.3–5.

In the extrinsic conduction region (low-temperature region), the conductivity can be represented by the equation $\sigma = N_e q_e \mu_e$. At the temperature corresponding to the onset of the exhaustion region, every Group V dopant atom has contributed one electron to the conduction band. Therefore, the value of N_e at the temperature corresponding to the onset of exhaustion is a good estimate of the dopant concentration.

....

EXAMPLE 10.3–2

You are given two n-type semiconductors.

    a. How would you determine if they are the same host material?

    b. How would you determine if they contain the same dopant?

Solution

Both questions can be answered by examining the form of ln σ versus $1/T$ plots for the two extrinsic semiconductors. (See Figure 10.3–4b.)

    a. The slope in the intrinsic, or high-temperature, region identifies the host material.

    b. The slope in the extrinsic, or low-temperature, region identifies the E_d level and, hence, the dopant.

....

Next, we consider the addition of a Group III atom, such as boron (B), into a Si crystal. As shown in Figure 10.3–5a, the three valence electrons associated with a B atom will be involved in covalent bonds with three of the four neighboring Si atoms. The covalent bond between the B atom and the fourth Si neighbor will, however, be "missing" an electron. In effect, this is an extra hole. The hole associated with the B atom will reside in an energy level in the lower half of the forbidden gap of Si. Since the hole associated with the B atom can "accept" an electron from the valence band, the corresponding energy level is called an **acceptor level** and is labeled E_a. Extrinsic semiconductors of this kind are called **p-type** due to their excess positive charge.

As shown in Figure 10.3–5b, the significance of the acceptor level is that it requires less thermal energy to promote an electron from the valence band into the acceptor level than to promote the same electron into the conduction band. When an electron is promoted to E_a, the hole left behind in the valence band is a valid charge carrier. The electron located in E_a, however, is localized and is not able to conduct charge. Therefore, only one charge carrier per promotion event is created, a hole.

The proper form of the conductivity equation for a p-type extrinsic semiconductor is:

$$\sigma = \sigma_0 \exp\left(-\frac{E_a}{kT}\right) \qquad (10.3–7)$$

where $\sigma_0 = N_0 q_e \mu_h$. As shown in Figure 10.3–5c, a plot of ln σ versus $1/T$ for a p-type semiconductor is similar to that for an n-type semiconductor except that the slope in the extrinsic region is $-E_a/k$. In addition, the horizontal region of the curve is referred to as the saturation region rather than the exhaustion region, since it occurs when all of the holes in the acceptor level become filled, or saturated, with electrons from the valence band.

Table 10.3–1 shows the location of the impurity energy levels (E_d or E_a) for some common dopant materials in Si and Ge.

FIGURE 10.3–5

(a) An illustration of the bond structure of Si doped with a Group III element such as B. Note the location of the "missing" electron, or hole, associated with the dopant. (b) The energy band diagram for a semiconductor doped with a Group III element (i.e., a p-type semiconductor). E_a represents the energy level associated with the electron holes resulting from dopant atoms. (c) A plot of $\ln \sigma$ versus $1/T$ for an extrinsic p-type semiconductor.

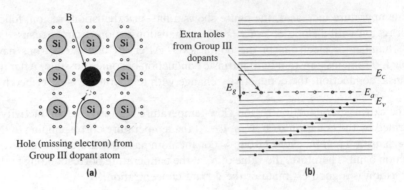

(a)

(b)

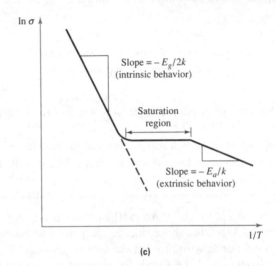

(c)

Impurity energy levels (E_a or E_d) in Si and Ge.

Host	Dopant	Energy level
Silicon	Sb	$E_c - E_d = 0.039$ eV
	P	$E_c - E_d = 0.044$ eV
	As	$E_c - E_d = 0.049$ eV
	Bi	$E_c - E_d = 0.069$ eV
	B	$E_a = 0.045$ eV
	Al	$E_a = 0.057$ eV
	Ga	$E_a = 0.065$ eV
	In	$E_a = 0.160$ eV
	Tl	$E_a = 0.260$ eV
Germanium	P	$E_c - E_d = 0.012$ eV
	As	$E_c - E_d = 0.013$ eV
	B	$E_a = 0.010$ eV
	Al	$E_a = 0.010$ eV

EXAMPLE 10.3–3

Consider a plot of $\ln \sigma$ versus $1/T$ for a sample of silicon doped with 10^{18} Al atoms/cm³. [This doping concentration corresponds to about 20 Al per 10^6 Si atoms, so the common nomenclature is 20 parts per million (ppm).]

a. Estimate the slope of the curve in the intrinsic (high-temperature) regime.

b. Estimate the slope of the curve in the extrinsic regime.

c. Estimate the conductivity of this material in the saturation (horizontal) region of the curve.

d. If $\sigma_0 = 600$ $(\Omega\text{-cm})^{-1}$ for this material, estimate the temperature at which the onset of saturation occurs.

Solution

a. Since the band gap of Si is 1.107 eV (Table 10.2–3), the slope in the intrinsic region is:

$$\text{Slope} = -\frac{E_g}{2k} = -\frac{1.107 \text{ eV}}{2(8.62 \times 10^{-5} \text{ eV/K})} = -6421 \text{ K}$$

b. From Table 10.3–1 we see that Al is a p-type dopant in Si and that $E_a = 0.057$ eV. Therefore, the slope in the extrinsic region is:

$$\text{Slope} = -\frac{E_a}{k} = -\frac{0.057 \text{ eV}}{(8.62 \times 10^{-5} \text{ eV/K})} = -661 \text{ K}$$

c. Since this is a p-type semiconductor, the appropriate equation for the conductivity is $\sigma = N_h q_h \mu_h$. Substituting the appropriate values (q is a constant, μ_h is found in Table 10.2–3, and N_h is equal to the doping concentration at the onset of saturation) yields the value:

$$\sigma = (10^{18} \text{ cm}^{-3})(1.6 \times 10^{-19} \text{ C})(500 \text{ cm}^2/\text{V-s}) = 80 \ (\Omega\text{-cm})^{-1}$$

d. Since this is a p-type semiconductor, the appropriate equation for the conductivity is $\sigma = \sigma_0 \exp(-E_a/kT)$. Solving for T:

$$T = -\frac{E_a/k}{\ln(\sigma/\sigma_0)}$$

Substituting the appropriate values (the value for σ was calculated in part c) yields:

$$T = \frac{(-0.057 \text{ eV})/(8.62 \times 10^{-5} \text{ eV/K})}{\ln\{[80 \ (\Omega\text{-cm})^{-1}]/[600(\Omega\text{-cm})^{-1}]\}}$$

$$= 328.2 \text{ K} = 55.2°\text{C}$$

Much of the previous discussion can be summarized by noting that there are three specific forms of the conductivity equation for a semiconductor, all with the general form $\sigma = \sigma_0 \exp(-E^*/nkT)$. The differences in the equations are related to the values of the constants σ_0, E^*, and n. E^* represents the magnitude of the effective energy gap, and n represents the number of charge carriers created per thermal excitation event. Table 10.3–2 lists the values for these constants for intrinsic, n-type, and p-type semiconductors.

TABLE 10.3–2   Values for the constants σ_0, E^*, and n in the equation $\sigma = \sigma_0 \exp(-E^*/nkT)$ for the various types of semiconductors.

Semiconductor type	σ_0	E^*	n
Intrinsic	$N_0 q_e(\mu_e + \mu_h)$	E_g	2
n-type	$N_0 q_e \mu_e$	$(E_c - E_d)$	1
p-type	$N_0 q_e \mu_h$	E_a	1

10.3.2 Compound Semiconductors

Si and Ge are the only pure elements that display semiconducting behavior at room temperature. As shown in Table 10.2–3, however, a large number of compounds display semiconducting behavior. The band gaps in these materials range from a few 10ths of an eV to several eV. Some of these compounds are similar in structure to Si. Others, such as those based on polymers, have very different structures.

Si and Ge have the diamond cubic crystal structure with atoms located at the FCC positions and half of the tetrahedral positions. (See Section 3.7.1.) The closely related zinc blende structure is obtained by placing one atom type on the FCC positions and a second atom type on half of the tetrahedral positions. For each atom in the zinc blende structure to have a filled valence shell, the average number of electrons per atom must be 4. This is true if both elements are from Group IV, as in SiC, or by combining 50 at. % (atomic percent) of a Group III element with 50 at. % of a group V element. Examples of these so-called III-V compound semiconductors include GaAs, GaP, and InP. Similarly, one can combine equal atomic fractions of Groups II and VI elements to form II-VI compound semiconductors such as CdTe and ZnTe.

Pure semiconductors exhibit intrinsic behavior regardless of whether they are single-element or compound materials. When dopants are added, an extrinsic compound semiconductor may be formed. Although the details of the conduction mechanisms in compound semiconductors are more complicated than those in elemental semiconductors, the relevant conductivity equations are identical. Examples of extrinsic compound semiconductors include GaAs doped with Si or Te (n-type) and GaAs doped with Zn (p-type).

As mentioned above, not all compound semiconductors are based on the zinc blende crystal structure. Some of the metal oxides, such as NiO (NaCl structure) and Fe_2O_3, exhibit intrinsic semiconducting properties near room temperature. $BaTiO_3$ (perovskite structure), Al_2O_3 (corundum structure), and others can be doped to exhibit extrinsic semiconducting properties.

It is also possible to obtain semiconductorlike properties from organic materials, including polymers. Anthracene ($C_{14}H_{10}$) and naphthalene ($C_{10}H_8$) are examples of organic semiconductors. In molecular crystals of these materials an increase in energy of approximately 1 eV is sufficient to free some electrons from their covalent bonds. Thus, these materials behave like semiconductors with a band gap of ~1 eV.

10.3.3 Role of Defects

Atomic scale defects can adversely affect the electrical properties of semiconductors in several ways. An important feature of semiconductors is the reproducibility of electrical properties among nominally equivalent specimens. This implies that an Al-doped Si wafer purchased from one vendor exhibits properties essentially the same as those of a similar wafer, doped with the same concentration of Al, purchased from a second vendor. The reproducibility is obtained as a result of minimizing the number of possible variations in the atomic structure of the material.

Consider, for example, the advantages of fabricating single crystals as opposed to polycrystals. Polycrystals contain grain boundaries. These planar defects influence the electrical properties of a semiconductor in the following ways:

1. They have a small negative impact on charge-carrier mobility.
2. They introduce additional energy levels into the "forbidden" energy gap.
3. They may act as charge-carrier recombination sites.

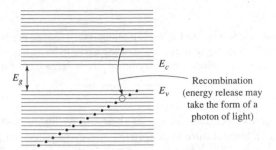

FIGURE 10.3–6
An energy band diagram showing recombination. The recombination process can be thought of as a conduction electron simply "filling a hole."

Since the significance of the first effect has already been considered in Section 10.2.7, only the last two effects will be discussed here. Recall that additional levels in the energy gap introduced by the grain boundaries will have a significant impact on the number of charge carriers. Therefore, the conductivity of a semiconductor will be a function of the number, type, and spatial distribution of grain boundaries in the polycrystal. Since it is extremely difficult to control precisely the morphology of the grain boundaries, their existence increases the variability among nominally equivalent polycrystals.

In our previous discussions we described the creation of charge carriers. Recall that the promotion of an electron from the filled valence band to the empty conduction band results in the formation of two charge carriers. We have not yet considered the possibility of an electron and a hole interacting so as to eliminate each other. The elimination process, known as **recombination,** can be envisioned as a conduction electron simply "filling a hole," as shown in Figure 10.3–6. Since recombination reduces the number of charge carriers, any factor that increases the recombination rate will decrease the electrical response of the material. Many defects, including grain boundaries, may act as recombination centers, that is, places at which recombination is comparatively easy.

Single crystals, therefore, offer the following advantages over polycrystals: lower variability between nominally similar samples, better charge-carrier mobility, and a lower charge-carrier recombination rate. Although we have cast our discussion in terms of a specific type of defect, the general argument is the same for a wide variety of defects, including point defects, impurity complexes, dislocations, grain and twin boundaries, and voids.

Because of the importance of minimizing defect concentrations in semiconductors, the purification and fabrication techniques for these materials are highly developed. Specific processing methods associated with these materials are discussed in Chapter 16.

10.3.4 Simple Devices

Even the most basic characteristics of semiconductors can be exploited to create useful engineering devices. Recall that a plot of $\ln \sigma$ versus $1/T$ for an intrinsic semiconductor has a slope of $-E_g/2k$. Therefore, if the material is well characterized and the slope known, a change in conductivity can be used to measure temperature changes. Devices based on this principle are known as thermistors. It is possible to detect temperature changes of $\sim 10^{-4}$ K by using appropriate materials and sensitive instruments for detecting conductivity changes. One practical application of a thermistor is as a fire alarm trigger.

....

DESIGN EXAMPLE 10.3–4

You have been asked to select a thermistor material. Would you pick a large or a small band gap material? Why?

Solution

The sensitivity of the device is increased if a small temperature change produces a relatively larg change in conductivity. That is, we want the slope of the curve to be as large as possible. Since th slope is $-E_g/2k$ select a semiconductor with a comparatively large band gap.

....

Another application of semiconductors is their use as light-sensing devices. To thi point we have relied on temperature to supply the energy required to promote an electro from the valence band to the conduction band (or similar transitions in extrinsic materi als). Another means of supplying the necessary energy is with a photon of light, which ha an energy

$$E = \frac{hc}{\lambda}$$

(10.3–8

where h is Planck's constant (4.135×10^{-15} eV-s), λ is the wavelength of the photon, an c is the speed of light (2.998×10^8 m/s). A photon with sufficient energy to creat additional charge carriers ($E > E_g$) will cause a concomitant increase in the conductivit of the semiconductor. By monitoring the conductivity, the intensity of the light can b determined. The band gap of the detector may be selected to produce a device that i sensitive to wavelengths ranging from ultraviolet to visible to infrared. Examples includ photosensors used to turn on streetlights at dusk and a variety of infrared detection device used to sense and locate people based on their body temperature.

It is possible to construct another class of devices by using a combination of two c more extrinsic semiconductors. Perhaps the simplest example is the **pn junction diod** (see Figure 10.3–7). The properties of a pn junction are dominated by extrinsic charg carriers: electrons on the n side and holes on the p side. If an external voltage is applie to the device as shown in Figure 10.3–7b, the loosely bound electrons on the n side of th junction will flow toward and recombine with the loosely bound holes on the p side. Nc all of the recombination occurs exactly at the junction. Some of the carriers recombin in thin regions on either side of the junction. The combined motion of electrons to the le and holes to the right results in a flow of current. When the junction is biased as describe above—that is, the negative potential is applied to the n-type material—it is known a a forward bias.

If, however, the external voltage is applied as shown in Figure 10.3–7c (i.e., a revers bias), the electrons on the n side flow to the right, and the holes on the p side flow to th left. This motion of electrons and holes away from the junction creates a volume c material that is essentially void of mobile extrinsic charge carriers, known as a deple tion zone. One might expect that the lack of charge-carrier recombination would indi cate that no current flows. This statement is true with respect to the extrinsic carrier: however, there is a small amount of external current resulting from the presence c intrinsic charge carriers on both sides of the junction. Since the number of intrin sic carriers is much smaller than the number of extrinsic carriers, the magnitude of th current flow under reverse bias is much less than under forward bias. Thus, as shown i Figure 10.3–7d, the pn junction acts as a rectifier, allowing significant current to pass i only one direction.

We have seen that a photon of light with sufficient energy can create electrical charg carriers in a semiconductor. A variation of this principle, applied to a simple pn junctior is the basis for construction of light-emitting diodes (LEDs). When a pn diode fabricate from the III-V compound GaAs is forward-biased, the electrons and holes recombine i

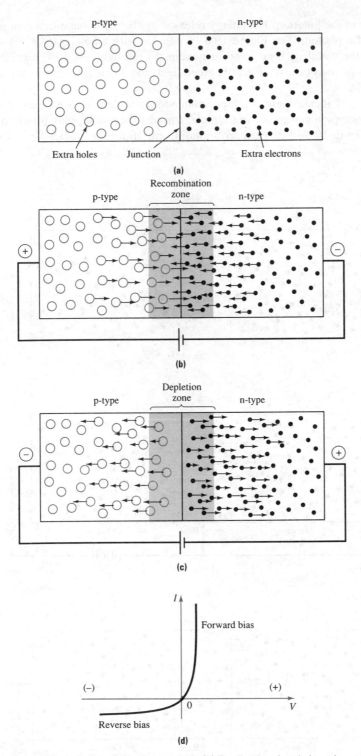

p-type   n-type

Extra holes   Junction   Extra electrons

(a)

Recombination
zone

p-type   n-type

⊕   ⊖

(b)

Depletion
zone

p-type   n-type

⊖   ⊕

(c)

I

Forward bias

(−)   (+)

0   *V*

Reverse bias

(d)

FIGURE 10.3–7   **(a)** The construction of a pn junction diode. **(b)** The direction of extrinsic carrier motion under forward bias. **(c)** The direction of extrinsic carrier motion under reverse bias. **(d)** A plot of the magnitude of the current flowing through the diode as a function of the applied bias.

the vicinity of the junction. The energy released by the recombination event is usually in the form of a photon. (See Figure 10.3–6.) GaAs, with a band gap of ~1.4 eV, emits photons in the wavelength range corresponding to red light. An LED fabricated from a ternary compound of Ga(As, P) emits green light, and a blue LED can be fabricated from Al-doped SiC.

A variety of increasingly complex semiconducting devices can be created by using more than one pn junction. As an example, consider the npn **bipolar junction transistor** shown in Figure 10.3–8 (or the corresponding pnp transistor, not shown). Note that this

FIGURE 10.3–8

An npn bipolar junction transistor: **(a)** the extrinsic charge carriers in each region; **(b)** the forward bias on the emitter-base junction, the reverse bias on the base-collector junction, and the mechanism or current flow through the external load; and **(c)** the amplification mechanism for the device.

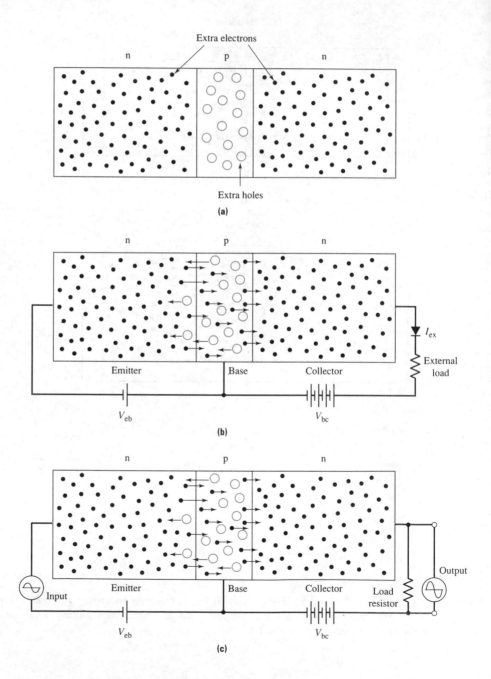

device consists of two pn junctions in series. The n-type region on the left side of the device is called the emitter, the thin p-type region in the center is the base, and the n-type region on the right is known as the collector. By appropriately biasing the two junctions, this device becomes an amplifier. That is, a small change in the magnitude of the bias on the emitter-base junction (V_{eb}) results in a significant change in the amount of current flowing through the external load (I_{ex}) on the base-collector side of the transistor.

The principle of operation of the transistor is as follows. The base-collector junction has a large reverse bias (V_{bc}). Therefore, considering the base-collector junction only, little if any current is flowing through the external load. The emitter-base junction has a small forward bias (V_{eb}). Thus, in the vicinity of the emitter-base junction, holes are flowing from right to left, and electrons from left to right, as shown in Figure 10.3–8b. In our discussion of a forward-biased junction we mentioned that not all recombination takes place exactly at the junction. In fact, some of the electrons from the emitter will cross the emitter-base junction and travel into the base prior to recombination. (See Figure 10.3–8b.) When the base is sufficiently thin, a fraction of the electrons from the emitter escape recombination in the base and travel across the base-collector junction. Once an electron has made it across the base-collector junction, it is quickly attracted to the positive potential at the right of the collector. That is, current can be made to flow through the external load if a sufficient number of electrons can be made to move from the emitter through the base and into the collector.

The two parameters controlling the number of electrons that cross the base-collector junction are the width of the base and the magnitude of the forward bias on the emitter-base junction (V_{eb}). Since the number of holes crossing the second junction is exponentially related to V_{eb}, the relationship between the current flowing through the external load (I_{ex}) and V_{eb} is of the form:

$$I_{ex} = I_0 \exp\left(\frac{qV_{eb}}{kT}\right) \tag{10.3–9}$$

where I_0 is a constant for a given transistor. As shown in Figure 10.3–8c by superimposing a weak signal on top of V_{eb}, one can amplify the superimposed signal. For example, transistors can be used to amplify a relatively weak radio signal and to drive an external speaker.

EXAMPLE 10.3–5

Consider an npn transistor with a characteristic value of $I_0 = 10^{-15}$ A.

a. What current flows through the load at room temperature when the voltage applied at the emitter-base junction is 0.75 V?

b. If an external signal is applied to the transistor as shown in Figure 10.3–8c, what magnitude of signal voltage results in a doubling of the current passing through the load?

Solution

a. Substituting the appropriate values into Equation 10.3–9, and recognizing that 1 V = 1 J/C, yields:

$$I_{ex} = I_0 \exp\left(\frac{qV_{eb}}{kT}\right)$$

$$= (10^{-15}\ \text{A}) \exp\left[\frac{(1.6 \times 10^{-19}\ \text{C})(0.75\ \text{V})}{(1.38 \times 10^{-23}\ \text{J/K})(300\ \text{K})}\right]$$

$$= 3.875 \times 10^{-3}\ \text{A}$$

b. Solving Equation 10.3–9 for V_{eb} yields:

$$V_{eb} = \left(\frac{kT}{q}\right) \ln\left(\frac{I_{ex}}{I_0}\right)$$

doubling the current flow means that

$$I_{ex} = 2(3.875 \times 10^{-3}\,\text{A}) = 7.75 \times 10^{-3}\,\text{A}$$

Substituting the appropriate values gives:

$$V_{eb} = \frac{(1.38 \times 10^{-23}\,\text{J/K})(300\,\text{K})}{(1.6 \times 10^{-19}\,\text{C})\,\ln[(7.75 \times 10^{-3}\,\text{A})/(10^{-15}\,\text{A})]}$$

$$= 0.7679\,\text{V}$$

Therefore, a signal of only (0.7679 V − 0.75 V) = 0.0179 V will double the current through the load.

10.3.5 Microelectronics

Electronics is the branch of science involving the design, production, and application of engineering devices based on the electrical properties of semiconductors. Over the past few decades the size of these devices has decreased to the extent that it is now possible to place several million of them in an area of less than 1 cm². The prefix *micro*- in microelectronics reflects the fact that a typical device dimension is now on the order of a micron or less. Examples of microelectronic devices include the pn junction diode and npn bipolar junction transistor described in the previous section. These and other devices can be linked together to fabricate microelectronic circuits, known as **integrated circuits (ICs),** ranging in complexity from a simple amplifier to a complete miniature digital computer known as a microprocessor.

Many of the key issues in the microelectronics industry are related to the materials, properties, and processing of semiconductors. The first semiconducting devices were fabricated from Ge. It was soon realized, however, that Si offered several advantages, including its lower raw material cost; its larger band gap, which permits operation at higher temperatures; and the superior characteristics of its natural surface oxide, SiO_2. To understand the significance of the oxide layer we must describe some of the processing operations used to fabricate integrated circuits. These processing issues are elaborated in Chapter 16.

From a materials point of view, an IC is a thin piece of intrinsic semiconductor, often Si, with well-defined surface regions processed to contain known amounts of electrically active dopants. Figure 10.3–9 shows three methods for fabricating a pn junction diode: (1) a controlled amount of an n-type dopant can be diffused into a p-type substrate, (2) a p-type dopant can be diffused into an n-type substrate, or (3) if the substrate is intrinsic, two sequential diffusion operations are required. One of the key steps in the process is the definition of the spatial region over which the dopant is introduced. As shown in Figure 10.3–10, this is accomplished by selectively removing the SiO_2 surface layer above the region to be doped. The dopant is then introduced onto the surface and thermally diffused into the substrate. Since the rate of diffusion is substantially slower through the SiO_2 than through the Si, the shape of the window in the oxide effectively defines the shape of the doped region.

In addition, SiO_2 is a high-quality electrical insulator that can become an integral part of an active device. For example, consider the metal-oxide-semiconductor (MOS) capacitor shown in Figure 10.3–11. As described in detail in the next chapter, a capacitor is

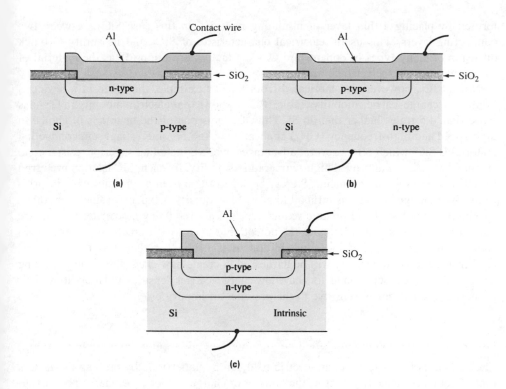

FIGURE 10.3–9
Three methods for fabricating a pn junction diode: **(a)** diffusion of an n-type dopant into a p-type substrate, **(b)** diffusion of a p-type dopant into an n-type substrate, or **(c)** sequential diffusion of n-type and p-type dopants into an intrinsic substrate.

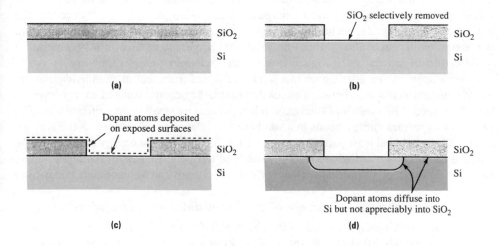

FIGURE 10.3–10
The method for introducing a controlled amount of a dopant into a silicon substrate: **(a)** the oxidized wafer, **(b)** the SiO$_2$ surface oxide layer is removed from the region to be doped, **(c)** dopant atoms are introduced onto the surface, and **(d)** the dopant is thermally diffused into the underlying silicon.

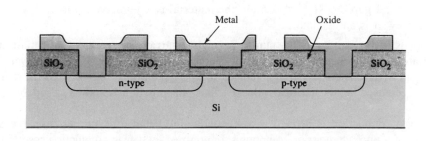

FIGURE 10.3–11
A metal-oxide-semiconductor (MOS) capacitor.

formed by placing a thin layer of insulating material, in this case SiO_2, between two conducting layers. This useful electrical characteristic of SiO_2 and its ability to block diffusion of many types of dopants are two key factors promoting Si as the material of choice in most microelectronic applications.

Not all ICs, however, are fabricated from Si. For example, the speed of a device is limited by charge-carrier mobility. Table 10.2–3 shows that electron mobility in GaAs is more than 4.5 times higher than in Si. Thus, GaAs is one of the materials of choice in high-speed and high-frequency (GHz) applications. Another interesting, semiconducting material is SiC, which offers several advantages over Si, including superior resistance to chemical attack and improved high-temperature stability. As such, SiC is often preferred for use in aggressive environments. SiC can be doped either n- or p-type; however, its principal disadvantage is that it is difficult to grow high-quality inexpensive single crystals.

The polymer-based and organic semiconductors are relatively new materials and are not yet fully understood. They have the potential to become extremely important, especially in biological applications. In addition, it is likely that their low mass density, anisotropy, and good ductility will be harnessed to create new classes of semiconducting devices. Since these semiconductors have low charge-carrier densities and mobilities, their properties are extremely sensitive to the presence of impurities.

SUMMARY

The response of a material to an electric field can be understood by probing the factors that affect its conductivity, that is, the number of charge carriers, the charge per carrier, and charge-carrier mobility. Mobility describes the ease of carrier motion through a solid. The total conductivity is the sum of the contributions from all carriers in the material— electrons, holes, and ions. When the charge carriers are electrons, factors that reduce the periodicity of the lattice, including an increase in temperature or defect concentration, decrease mobility. In contrast, when ions carry the charge, increases in either temperature or defect density increase the ionic diffusion rate and charge-carrier mobility.

The electron energy distribution in a solid is different from that in an individual atom. As the separation distance between atoms decreases, the original isolated energy levels in the atoms spread into bands of discrete levels separated by small energy differences. The two most important energy bands in a solid are the valence band, defined as the highest-energy band that is at least partially occupied at 0 K, and the first energy band above the valence band, known as the conduction band. The term energy gap, or band gap, refers to the magnitude of the forbidden energy range between the valence and conduction bands.

For electrical characterization, one of the most useful classification schemes is:

1. A solid with a partially filled valence band at 0 K is a conductor. If the valence band is completely filled at 0 K, the solid is either a semiconductor or an insulator.
2. The distinction between semiconductors and insulators is based on the size of the band gap, E_g. If $E_g \leq 2.5$ eV, the material is a semiconductor; if $E_g >$ 2.5 eV, it is an insulator.

In solids with a band gap, the number of charge carriers increases exponentially with temperature. In conductors the number of charge carriers is independent of temperature.

In some materials the resistivity changes abruptly and approaches zero as the temperature is decreased below a critical value T_c. This phenomenon is known as superconductivity.

Intrinsic conduction is defined as semiconducting behavior resulting from the band structure of a pure element or compound. In contrast, extrinsic conduction results from the interaction of impurities with the host material. Si and Ge are the only pure elements that exhibit semiconducting behavior at room temperature. A large number of compounds, however, display semiconducting behavior.

Modern electronic instruments or products rely on a number of devices based on conductor, insulator, and semiconductor materials.

KEY TERMS

acceptor level

band gap

bipolar junction transistor

conduction band

conductivity σ

conductor

core band

donor level

drift velocity

energy gap

extrinsic conduction

Fermi-Dirac distribution function

Fermi energy E_f

hole

insulator

integrated circuits (ICs)

intrinsic conduction

mean time between collisions

mobility μ

n-type

Ohm's law

p-type

pn junction diode

recombination

resistance R

resistivity ρ

semiconductor

superconductivity

temperature coefficient of resistivity

transference numbers

valence band

HOMEWORK PROBLEMS

SECTION 10.2
Electrical Conduction

1. Once again, consider the components of a CD player. For what specific purposes do you need insulators, conductors, and semiconductors?

2. The resistance of a polymer rod with a conductivity of 3.5×10^{-8} $(\Omega\text{-m})^{-1}$ and length of 10 m is measured to be 0.08 Ω. Calculate the cross-sectional area of this rod. What voltage is required to obtain a current of 3.2 A flowing through this rod? Compare your solutions with those of Example 10.2–1.

3. Table 10.2–1 shows the electrical conductivity of various materials set into each of three classes of materials—metals, ceramics, and polymers. What material class shows the greatest range? If the "polymers" section were changed to "organics" so as to include both short and long molecules, then how might the range change?

4. Use the fact that an ampere is defined to be a coulomb per second (i.e., 1 A = 1 C/s) to verify that the units of the $Nq\mu$ product are consistent with those of σ.

5. Predict whether the conductivity of a metallic glass is higher or lower than the conductivity of its crystalline counterpart. Assume conduction occurs by the motion of electrons and the number of mobile charge carriers per atom is the same for the two structures.

6. Suppose you have a material in which electrical charge is carried by both electrons and holes. What is the proper form of Equation 10.2–5 to use in a calculation of the conductivity of this material?

7. Assume you have a material in which an equal number of electrons and holes are available to transport charge. If the resistivity of this material is 3.15×10^{-1} Ω-m and the number of each type of charge carrier is 1×10^{13} cm^{-3}, estimate the mobility of the electrons. (Assume $\mu_e \approx 3\mu_h$.)

8. In terms of their energy band diagrams, describe the similarities and differences among conductors, semiconductors, and insulators.

9. The resistance of a rod of material 1 m long and 50 μm radius is 20 Ω. What is the resistance of the same material 1 m long and 500 μm in diameter?

10. Show that the probability of finding an electron in the conduction band increases exponentially with temperature for a band gap material with E_g greater than 0.2 eV.

11. Explain why the resistivity of a metal increases while the resistivity of a semiconductor decreases with an increase in temperature.

12. The fact that C as graphite does not appear in Table 10.2–1 may reflect the notion that many scientists do not know to which class of materials it properly belongs. Go to the library or the World Wide Web and ascertain the electrical conductivity of carbon (pyrolytic carbon if available). This value is similar to that of a carbon fiber. If the fiber is 10 μm in diameter and it is used in lengths of 1 m, what is its resistance? How much current will flow under a potential of 12 V?

13. Suppose a metal, a semiconductor, and an insulator all have the same resistance at 20°C. Which material will have the highest resistance at 50°C? Why?

14. Table 10.2–1 states that the conductivity of ReO_3 is 5×10^5 $(\Omega\text{-cm})^{-1}$. How can a material with ionic bonding have such a high conductivity?

15. PbS has a band gap of 0.37 eV. Calculate the fraction of electrons in the conduction band at 100°C compared with 25°C.

16. Calculate the resistivity of copper at 200°C and compare it with the value at 20°C.

17. Suppose you have a stoichiometric ionic solid that contains twice as many cations as anions. Also assume that in this hypothetical material electrical charge can be transported by electrons, holes, cations, and anions. Determine an appropriate expression for the conductivity of the material.

18. How does grain size affect the conductivity of a material in which the dominant charge carriers are electrons, holes, or both?

19. Give an example of a material in which an increase in the defect density results in:
    a. A decrease in σ
    b. An increase in σ

20. The curing of epoxy can be monitored by following the mobility of the ionic impurities present in the resin. Sketch a plot of the electrical conductivity of a room-temperature-curing epoxy as the material transforms from fluid to rubber to glass.

21. In Section 10.2.9 we listed five potential applications for high-temperature superconducting materials. Select any three of these and explain how superconductors will aid the development of each engineering application.

22. Electrical conductors can be metals, ceramics, or polymers.
    a. Give two examples of electrical applications that favor the use of conducting polymers.
    b. Give two examples of applications that favor the use of metals.
    c. Give two examples of applications that favor the use of conducting ceramics.

23. Explain why fine distributions of second-phase particles increase the resistance of an alloy more than an equivalent volume fraction of coarse precipitates.

24. Individual titanium and germanium atoms each have four electrons in their valence shells. Why is solid titanium a conductor while solid germanium is a semiconductor?

25. Why is it that an electron in the conduction band is an effective charge carrier while an electron in a donor level, E_d, is not a valid charge carrier?

26. a. What is meant by the term 1-2-3 superconductor?
    b. Give an example of a 1-2-3 superconductor.
    c. What is a Cooper pair, and how is it related to superconductivity?

27. a. How does a thermocouple work?
    b. How does the ZrO_2 oxygen sensor used to control the air-fuel mixture in an automobile engine work?

28. Generally, we think of metals as having higher electrical conductivities than ceramics. Show by example that this generalization is not always correct. Are polymers always less conductive than metals?

29. Fill in the table by indicating whether each of the terms in the column headings is less than or greater than zero for each class of materials. $\partial N/\partial T$ is the change in the number of a charge carriers with temperature, $\partial N/\partial \rho_d$ is the change in N with defect concentration, $\partial \mu/\partial T$ is the change in mobility with T, and $\partial \mu/\partial \rho_d$ is the change in μ with defect concentration.

Material class	$\partial N/\partial T$	$\partial N/\partial \rho_d$	$\partial \mu/\partial T$	$\partial \mu/\partial \rho_d$
Electron conductors				
Ionic conductors				
Semiconductors				

30. Determine the dominant charge carrier in each of the following materials:
    a. NaCl
    b. FeO
    c. CuCl
    d. ZrO_2 + 18% CeO_2

31. Is the electrical resistivity of the glass in a Coke bottle higher or lower than that of the glass in a crucible used for high-temperature experiments? Why?

32. Plot the variation in the resistivities of Mg and Ta over the temperature range $-200°C$ to $+200°C$. Which material has the higher conductivity?

33. In a sample of Ge at room temperature, calculate the percentage of electrical charge transported by holes.

34. Calculate the drift velocity in a material with a charge-carrier mobility of 0.43 $m^2/(V\text{-}s)$ if a sample of thickness 1 mm is subjected to a voltage of 110 mV.

35. a. What change in temperature is necessary to increase the conductivity of Cu by 10%?
    b. What change in temperature is necessary to increase the conductivity of Si by 10%?
    c. Which material is more sensitive to temperature fluctuations? Why?

36. Suppose you have two semiconducting rods made from the same material. If the diameter of the first rod is three times that of the second ($d_1 = 3.0d_2$) and the length of the first rod is half that of the second ($L_1 = 0.5L_2$), which rod has the higher resistivity?

37. For the conditions listed below, determine whether each of the indicated variables increases, decreases, or remains the same.
    a. As the temperature increases:
        i. Time between electron-scattering events (collisions):
        ii. Charge-carrier mobility in a metal:
        iii. Conductivity of a metal:
        iv. Number of charge carriers in a semiconductor:
        v. Charge-carrier mobility in a semiconductor:
        vi. Resistivity of a semiconductor:
    b. As the defect density increases:
        i. Time between electron-scattering events (collisions):
        ii. Resistance of a metal:
    c. As the cross-sectional area of a sample increases:
        i. Conductivity of the sample:
        ii. Resistance of the sample:

38. Explain how an increase in temperature influences each of the following factors in the conductivity equation:
    a. The number of mobile electrons in a semiconductor
    b. The number of mobile electrons in a solid with a partially filled valence band (i.e., a conductor)
    c. The number of mobile ions in an ionic solid
    d. The mobility of an electron
    e. The mobility of an ion

39. Consider an intrinsic semiconductor with a conductivity that doubles between 50 and 100°C.

    a. Calculate the band gap of this material.

    b. If the resistance of a piece of this material is 25 Ω at 50°C, calculate the resistance of the same piece of material at 100°C.

40. The text notes that anthracene and naphthalene, both fused aromatic ring organic compounds containing a continuous path of π electrons, have a band gap of about 1 eV. As certain donor or acceptor impurities are added, the conductivity increases. How do you know when the band gap has reached 0.0 eV? Will this material look like a metal (bright gray)?

41. Most audiophiles advise spending the bulk of your sound system budget on good speakers. You do this; however, when you crank up your stereo amplifier, the music eventually distorts. What is "distorting"? For the moment, assume that the distortion is not caused by the speakers.

 42. Suppose you have two semiconducting wafers: one is Si and the other is Ge, but you do not know which is which. Describe a conductivity experiment that would allow you to identify the wafers.

43. Demonstrate that the slope of the ln σ versus $1/T$ plot for an n-type semiconductor will always be steeper in the intrinsic region than in the extrinsic region. Assume that donor levels are located in the upper half of the band gap.

44. Suppose you are investigating the band structure of a semiconductor. You make a plot of ln σ versus $1/T$ and find the slopes of the two nonhorizontal regions of the curve to be -3480 K and -6960 K. What is the magnitude of the intrinsic band gap for this semiconductor?

45. Suppose you have an unknown material for which you measure the electrical conductivity versus temperature and find a linear relationship between ln σ and $(1/T)$.

    a. Assuming the material is an n-type semiconductor, calculate the energy difference (in eV) from the donor band to the bottom of the conduction band (i.e., find $E_c - E_d$) if the slope of the curve is -1200 K.

    b. A colleague suggests that this material may be an intrinsic semiconductor. How would you determine whether the material is intrinsic or extrinsic?

46. Intrinsic silicon [$\sigma_0 = 9950$ (Ω-cm)$^{-1}$] has a conductivity of 2.623 × 10^{-5} (Ω-cm)$^{-1}$ at 50°C. Silicon doped with Group III boron [$\sigma_0 = 4.9 \times 10^3$ (Ω-cm)$^{-1}$] has a conductivity of 858 (Ω-cm)$^{-1}$ at 60°C.

    a. Calculate the magnitude of the band gap for intrinsic Si.

    b. Sketch the energy band diagram for the p-type Si.

    c. Calculate the magnitude of the energy difference between the conduction band edge and the energy level introduced by the B atoms.

47. Sketch the band energy diagram for an n-type semiconductor. Given: (1) the conduction band is 1.2 eV above the top of the valence band (i.e., $E_g = 1.2$ eV), (2) the energy level introduced by the n-type dopant is 0.3 eV below the conduction band, and (3) the conductivity of this material is 5 × 10^{-6} (Ω-cm)$^{-1}$ at 25°C. Calculate the conductivity of this material at 150°C.

48. The band gap of intrinsic germanium is 0.66 eV. Suppose a Group V element is intentionally added as a dopant.

    a. Calculate the value of $E_g - E_d$ (in units of eV) if the slope of the ln σ versus $1/T$ curve is -1700 K (in the extrinsic region).

    b. What is the conductivity of this material at 200°C if $\sigma_0 = 4.0 \times 10^3$ (Ω-cm)$^{-1}$ (assume extrinsic behavior)?

49. Consider the stoichiometric compound GaP.

    a. Based on the average number of electrons per atom, do you think this material could be a semiconductor?

    b. What crystal structure do you think GaP displays?

   c. What phase diagram information would you require to determine if GaP is a likely candidate to exhibit semiconducting behavior?

50. In the context of semiconductors, what is a recombination center? How do recombination centers affect electrical conductivity?

51. With reference to an extrinsic semiconductor, what is the difference between exhaustion and saturation?

52. List the dominant charge carrier(s) in each of the following semiconductors:
   a. Intrinsic Si
   b. Intrinsic GaAs
   c. Si doped with P (at low temperatures)
   d. Ge doped with P (at high temperatures)
   e. GaAs doped with Zn (at low temperatures)
   f. CdTe doped with anything (at high temperatures)

53. *a.* Sketch a pnp bipolar junction transistor and label the emitter, base, and collector regions.
   b. Show the polarity of voltages applied to the emitter-base and base-collector junctions during normal operation of the device.

54. Suppose an application requires a semiconductor that does not change its resistivity over a limited temperature range. What type of semiconductor do you suggest? Why?

55. Consider the following list of semiconductors: Si, GaAs, SiC. In a given application how would you decide which of these three materials is the best choice?

56. What is the difference between an intrinsic compound semiconductor and an extrinsic elemental semiconductor?

57. What is the minimum number of elements necessary to fabricate an extrinsic compound semiconductor?

58. Give examples of applications that favor the use of a larger–band gap semiconductor. Repeat for a smaller–band gap material.

59. Sketch the structure of a blue LED and show the events that lead to blue light.

60. What advantages do single-crystal semiconductors offer over similar polycrystals?

61. What is a thermistor, and how does it work?

62. Discuss the characteristics of SiO_2 that help to make silicon the most commonly used semiconducting material.

63. At what temperature is the conductivity of silicon equal to that of germanium?

64. At what temperature is the conductivity of Bi-doped Si equal to 90% of its room-temperature value (assume extrinsic conduction)?

65. A sample of Ge is known to be doped with either P or B. If the conductivity increases by 10% between 25°C and 76.4°C, determine whether the material is an n- or a p-type semiconductor.

66. A silicon (Si) wafer is intentionally doped with arsenic (As).
   a. Sketch the band diagram for this material at 0 K.
   b. If the intrinsic band gap for Si is 1.017 eV and the level introduced by the As atoms is located 0.3 eV from the nearest band edge, calculate the conductivity of this material at 50°C. Note: The conductivity of this material at 25°C is 5×10^{-6} $(\Omega\text{-cm})^{-1}$
   c. Sketch the relationship between ln σ and $1/T$ for this material.

67. Consider the extrinsic semiconductor obtained by adding Group III Al to Ge. The energy level introduced in the band gap by the dopant atoms is 0.01 eV from the nearest band edge.
   a. Calculate the temperature at which this extrinsic semiconductor will have a conductivity equal to 3 times its room-temperature value.
   b. Sketch ln σ versus $1/T$ for this material and calculate the nonzero slopes.

68. Explain why the conductivity of wool, which is an insulator, increases as much as 8 orders of magnitude at room temperature as the relative humidity is increased from 0 to 99%.

69. How can you very simply rectify AC? Is the rectification perfect?

C H A P T E R 11

OPTICAL AND
DIELECTRIC PROPERTIES

MATERIALS IN ACTION   Optical and Dielectric Properties

The optical and dielectric properties of materials play an important role in many industries, but perhaps none more than the communications industry, which is one of the dominant industries in modern economies. Telecommunications in the United States alone accounts for approximately $200 billion in sales and employs in excess of 1 million people.

An example of the impact of materials in the communications industry may be gained by considering the size of radios. In tuning a radio, the receiver frequency is adjusted to match that of the desired signal. The frequency of an oscillating circuit may be changed by adjusting the capacitance of the circuit. In the early 1940s, the frequency of radios was tuned by changing the effective area of a variable capacitor. The dielectric material, the insulating material between the plates of the capacitor which determines the capacitance, was rutile (TiO_2), which has a relative permittivity directly related to the capacitance of about 80. This would require a capacitor about 9 cm in thickness if modern radios used the same capacitance. Huge increases in the capacitance of radio tuning circuits, with associated improvements in signal quality, were achieved by the introduction of $BaTiO_3$-type ceramic materials (see Figure 11.2–2). The relative permittivity of such materials can exceed 5000 over a fairly broad temperature range. This huge increase in capacitance has allowed tuning capacitors to decrease in size to about 1 cm. Such advances in materials have driven the miniaturization of radios and just about all other electronic devices.

Advances in materials have been no less spectacular in the field of optics. From the first known occurrence of glass to about 1980, the percentage of light transmitted per kilometer increased from essentially zero to 10^{-100}, a small number indeed! However, with increased understanding of the structure of glasses and modern improvements in processing, the percent transmission of light has increased to near 100%. Such increases have spurred the use of optical fibers for telecommunications. It is now possible for a fiber of 0.001 cm diameter to transmit thousands of telephone conversations simultaneously!

In this chapter we will study the fundamental relationships between the structure of materials and their dielectric and optical properties, as well as how these advances are used in engineering.

11.1 INTRODUCTION

The study of the optical properties of materials is concerned with the response of materials to electromagnetic radiation. Similarly, the study of dielectric properties deals with the response to electric fields of materials characterized as poor electrical conductors. Although these two classes of properties may seem unrelated, both optical and dielectric properties (and the magnetic properties discussed in the next chapter) depend to a large extent on the polarization of the material.

Most materials used for their special optical properties in visible light, such as transparency, are also good dielectric materials. They are not good electrical conductors: rather, they are typically insulators with large band gaps. In fact, the difficulty of promoting an electron to the conduction band provides the materials with many of their dielectric and optical properties. Therefore, most dielectric and optical materials are either ceramics or polymers (i.e., nonmetals).

As shown in Figure 11.1–1, the visible spectrum is a small but important part of the entire electromagnetic spectrum. In general we will not confine our treatment of electromagnetic radiation to any particular part of the full spectrum. When discussing optical properties, however, one often tacitly assumes that the relevant behavior of the material occurs upon exposure to visible light.

While the mechanical properties of materials have been exploited for over a thousand years and their electrical properties have been utilized for more than a century, the optical properties of materials have been developed extensively only during the past few decades. It is likely, however, that we are on the brink of an explosion of activity in optical

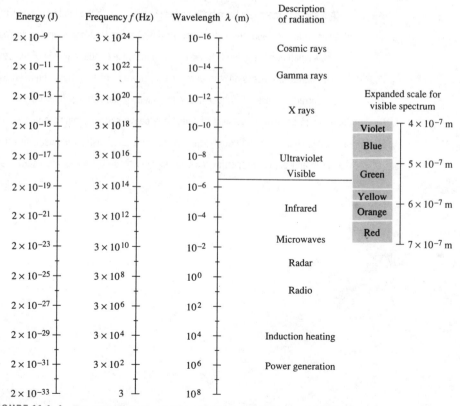

FIGURE 11.1–1   The electromagnetic spectrum showing the narrow range of visible light.

echnology similar to that experienced in the electronics industry during the 1950s. Thus, while the immediate goal of this chapter is to help you understand existing devices, including lasers, fiber optics, and solar cells, it is also designed to prepare you to appreciate the new optical devices that will undoubtedly be developed in the near future.

11.2 POLARIZATION

Dipoles are atoms or molecules that have a positive and a negative pole. That is, although the total charge on the atom or molecule is zero, the centers of negative and positive charge do not coincide. As discussed in Chapter 2, dipoles can be either permanent or induced. In the presence of an external electric field the dipoles align with the field causing **polarization** P of magnitude:

$$P = Zqd \qquad\qquad (11.2\text{--}1)$$

where Z is the number of charge centers displaced per unit volume, q is the electronic charge, and d is the displacement between the poles. The units of polarization are C/m^2. Of immediate concern to us are:

1. The origin of polarization
2. The magnitude of polarization in various solids
3. The speed with which dipoles can align

11.2.1 Electronic Polarization

The mechanisms of polarization in a dielectric material are shown in Figure 11.2–1. One process that occurs frequently is **electronic polarization,** shown in Figure 11.2–1a. When an external electric field E (in volts) is applied to a material, the electron cloud of an atom shifts off the center of the positive nuclear charge. The atom then acts as a temporary or induced dipole. This sort of polarization can occur in all materials, regardless of the type of bonding in the solid. When the field is removed, the polarization vanishes.

....

EXAMPLE 11.2–1

In an electric field of 5000 V, the center of the electron cloud in a Ni atom is 10^{-9} nm from that of the nuclear charge. Given that Ni has an FCC structure with a lattice parameter of 0.351 nm, and assuming that all the electrons in the metal contribute to the electronic polarization, calculate the polarization.

Solution

Since each electron represents a displaced charge center, the value of Z can be found by multiplying the number of atoms per unit volume by the number of electrons per atom:

$$Z = \frac{4 \text{ atoms/unit cell}}{(0.351 \times 10^{-9})^3 (m^3/\text{unit cell})} \times (28\ e/\text{atom})$$

$$= 2.68 \times 10^{30}\ e/m^3$$

Then Equation 11.2–1 can be used to find the polarization:

$$P = Zqd = (2.68 \times 10^{30}\ e/m^3)(1.6 \times 10^{-19}\ C/e)(10^{-18}\ m)$$

$$= 4.29 \times 10^{-7}\ C/m^2$$

....

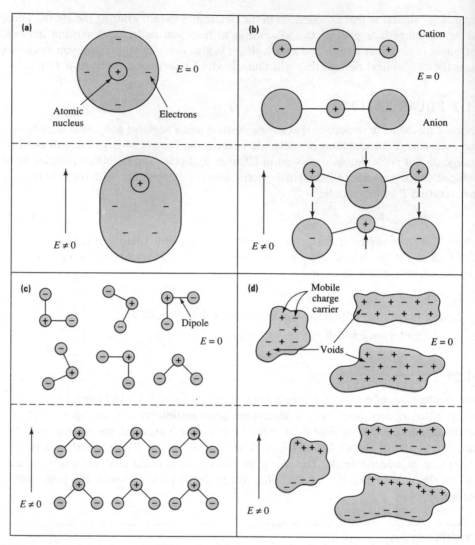

FIGURE 11.2–1 Polarization mechanisms in solids: **(a)** electronic—center of electron cloud differs from center of positive charge; **(b)** ionic—cations and anions shift relative positions; **(c)** molecular—permanent dipoles align with external field; and **(d)** interfacial—mobile charges responding to an external field are blocked at interfaces.

11.2.2 Ionic Polarization

Another mechanism of polarization is the displacement of anions and cations in crystals relative to their normal positions. As shown in Figure 11.2–1b, the applied field causes separation of charge—the cations are attracted toward the negative electrode and the anions move toward the positive electrode. This type of polarization is called **ionic** or **atomic polarization.** It is extremely important in ionic materials.

EXAMPLE 11.2–2

Calculate the increase in separation of Cs^+ and Cl^- in a CsCl crystal when an ionic polarization of 4×10^{-8} C/m^2 is achieved by the application of an electric field.

Solution

The lattice parameter of CsCl can be calculated from the ionic radii (0.165 and 0.181 nm) using Equation 3.7–1, and is found to be $a_0 = 0.402$ nm. Since there is one pair of ions per cell,

$$Z = \frac{(1 \text{ dipole/cell})(1 \text{ charge/dipole})}{(0.402 \times 10^{-9})^3 (\text{m}^3/\text{cell})}$$

$$= 1.54 \times 10^{28} \text{ charges/m}^3$$

Using Equation 11.2–1, the increase in separation distance is given by:

$$d = \frac{P}{Zq} = \frac{(4 \times 10^{-8} \text{ C/m}^2)}{(1.54 \times 10^{28} \text{ charges/m}^3)(1.6 \times 10^{-19} \text{ C/charge})}$$

$$= 1.62 \times 10^{-17} \text{ m} = 1.62 \times 10^{-8} \text{ nm}$$

Since the separation distance in the absence of an applied field is $0.167 + 0.181 = 0.348$ nm, the increase of 1.62×10^{-8} nm represents a change of less than $5 \times 10^{-6}\%$. Despite the small magnitude of this displacement, the corresponding polarization influences the macroscopic dielectric properties of the crystal.

11.2.3 Molecular Polarization

Yet a third mechanism of polarization, **molecular** or **dipole polarization,** is associated with the alignment of permanent dipole molecules. As shown in Figure 11.2–1c, the external field causes the dipoles to orient parallel to the field. Dipole polarization is common in ionic ceramics, including silicates, and in polar polymers.

In contrast to electronic polarization, which involves temporary dipoles, molecular polarization can be retained after removal of the external field because of the permanent nature of the dipoles involved. For example, an amorphous polar polymer can be subjected to an electric field at a temperature above its glass transition, where significant molecular motion can occur, and then cooled to below its glass transition in the presence of the field to "freeze in" the aligned structure. Polymers in this metastable state are commonly referred to as poled. Polymers that retain this poled structure for long times are called electrets and are used in a number of areas, including air filtration. In this application charged dust particles are removed from an airstream by passing them through a web of fine charged electret fibers.

DESIGN EXAMPLE 11.2–3

Which of the following polymers would you select for use in an application requiring an electret: $(C_2H_4)_n$, $(C_2H_2F_2)_n$, or $(C_2F_4)_n$?

Solution

The question is equivalent to asking which polymer is most likely to exhibit molecular polarization. Both the C_2H_4 and C_2F_4 groups are symmetric, so their net dipole moment is zero. The asymmetry of the $C_2H_2F_2$ group, combined with the highly polar nature of the C—F bond (as demonstrated by the difference in the atoms' electronegativities), suggests that $(C_2H_2F_2)_n$ will be susceptible to molecular polarization. In fact, this polymer, known as poly (vinylidene fluoride), is one of the most common commercial electrets. It is used as a transducer in telephone receivers.

11.2.4 Interfacial Polarization

A final source of polarization is **interfacial polarization,** as illustrated in Figure 11.2–1d. This type of polarization occurs when mobile charge carriers accelerated by an external electric field are impeded by, and pile up at, a physical barrier. The most common physical barriers are grain or phase boundaries and free surfaces. Interfacial polarization is an important concept in ceramic defect structures and an important consideration in the dielectric behavior of ionic solids.

11.2.5 Net Polarization

We have just learned about the four mechanisms of polarization in materials: electronic, ionic, molecular, and interfacial. The net polarization of a dielectric is the sum of the contributions from these four mechanisms. The level of response from each mechanism is what differentiates materials. Both the atomic scale structure and the microstructure are important variables in determining the level of response from each of these mechanisms. For example, the extent of ionic and molecular polarization depends strongly on the type of atoms and type of bonding present in the sample, while the extent of interfacial polarization is related to the planar defect density.

Another important factor is the frequency of the applied electromagnetic field. Each of the four polarization mechanisms occurs on a different time scale and, therefore, has a limited range of frequencies over which it can contribute significantly to the net polarization of the sample. In general, the larger the masses involved, the slower the response upon application of a field and the slower the relaxation upon removal of the alternating field. Since electrons are much lighter than protons, they can react to much higher-frequency oscillations than can ions. In fact, electronic polarization is the only one of the four processes that is rapid enough to respond to radiation in the visible part of the spectrum (i.e., $\sim 10^{15}$ Hz). Ionic polarization is important in the infrared range ($\sim 10^{12}$ to $\sim 10^{13}$ Hz), molecular polarization in the subinfrared range ($\sim 10^{11}$ to $\sim 10^{12}$ Hz), and interfacial polarization in the lower-frequency range ($\sim 10^{-3}$ to $\sim 10^{3}$ Hz).

The high-frequency polarization mechanisms, electronic and ionic, dominate the optical properties of materials, while the lower-frequency mechanisms, molecular and interfacial, generally control the dielectric properties.

EXAMPLE 11.2–4

Describe what effect, if any, each of the following processes would have on the net polarization of a solid:

    a. Increasing the grain size of a ceramic polycrystal

    b. Increasing the concentration of network modifiers in a silicate glass

    c. Changing the frequency of the excitation field from the ultraviolet range to the microwave range

Solution

    a. Since the mobile charge carriers require more time to shuttle back and forth across the larger grains, the interfacial polarization occurs at lower frequencies.

    b. Increasing the network modifier concentration decreases the primary bond density and increases the number of secondary bonds (dipoles) in the silicate. The enhanced potential for molecular polarization increases the net polarization in the subinfrared range.

    c. Figure 11.1–1 shows that ultraviolet radiation has a frequency of $\sim 10^{16}$ Hz while microwaves have frequencies of $\sim 10^{11}$ Hz. Therefore, changing the excitation from ultraviolet to microwave changes the dominant mechanism from electronic to molecular polarization (assuming the latter is viable in the solid of interest).

11.2.6 Applications

A number of crystalline ceramics that assume the perovskite structure (see Section 3.7.3), such as barium titanate, have an interesting and useful property associated with charge distribution. Above 120°C, $BaTiO_3$ is a cubic crystal; however, below 120°C the central Ti^{4+} ion cannot fit into its octahedral lattice position. It is displaced 0.006 nm to one side of the cell, and the O^{2-} ions are displaced 0.008 nm in the opposite direction, as shown in Figure 11.2–2. The unit cell is tetragonal below 120°C. The temperature below which the unit cell exhibits asymmetric charge distribution (i.e., permanent polarization) is known as the ferroelectric Curie temperature.

Mechanical deformation, usually applied in the form of uniaxial compression, may decrease the separation of cation and anion. The motion of ions produces a current and induces an internal field (voltage). This is the **piezoelectric effect.** Conversely, application of an electric field along the correct crystallographic direction will cause changes in the anion-cation spacing. This effect, the inverse of the piezoelectric effect, is called **electrostriction.** Here, an applied voltage produces changes in the dimensions of a material. Mathematically, these relations may be expressed:

$$\text{Field produced by stress} = \xi = g\sigma \tag{11.2–2}$$

and

$$\text{Strain produced by field} = \varepsilon = d\xi \tag{11.2–3}$$

where ξ is the electric field, σ is the applied stress, ε is the strain, and g and d are constants related to the modulus of elasticity E through the equation:

$$E = \frac{1}{gd} \tag{11.2–4}$$

Typical values for d are given in Table 11.2–1. Examples of materials that show these effects are quartz, various titanates, lead zirconate, cadmium sulfide, and zinc oxide.

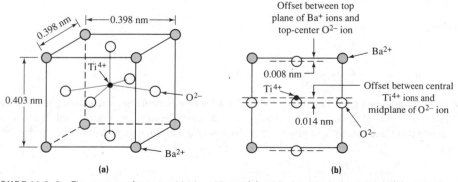

FIGURE 11.2–2   The structure of tetragonal barium titanate: **(a)** a 3-D view of the unit cell, and **(b)** a 2-D representation of the structure. Note the asymmetry of charge, and hence, the presence of a permanent dipole in each unit cell.

TABLE 11.2–1   Piezoelectric constants of materials.

Material	Piezoelectric constant d (10^{12} m/V)
Quartz	2.3
Li_2SO_4	16
$BaTiO_3$	149
$PbZrO_6$	250
$PbNb_2O_6$	85
PZTs ($PbZrO_3$ and $PbTiO_3$)	320–365

Source: Adapted from *CRC Handbook of Tables for Applied Engineering Science.*
Copyright CRC Press, Boca Raton, FL.

The piezoelectric effect is used in transducers such as phonograph cartridges. In this and other similar applications the role of the transducer is to convert an electrical signal into an acoustical wave or vice versa. Piezoelectric transducers have also been used historically in telephones; however, electrets, which show the same behavior, now capture a major share of the market. Recall that an electret is formed by aligning dipoles and then "freezing" the molecular conformation by decreasing the temperature below the glass transition temperature. Application of stress will cause deformation and charge motion, as in the case of true piezoelectric crystals. Poly(vinylidene fluoride), the polymer described in Design Example 11.2–3, is the electret polymer used most often in telephone sets.

Other applications for piezoelectric materials include pressure gages and high-frequency sound generators. Quartz crystals, historically one of the first piezoelectric materials in common use, are still used to provide precise timing of electrical circuits and watches using vibrational frequency control.

Barium titanate, with a ferroelectric Curie temperature of 120°C, is one of the most common low-temperature piezoelectric materials. For high-temperature applications, however, materials with a higher Curie temperature are required. Solid solutions of $PbZrO_3$ and $PbTiO_3$, known as PZTs, with Curie temperatures up to 490°C, have become the most widely used piezoelectric ceramics.

In some dielectric materials, such as perovskites, one dipole may influence adjacent dipoles in such a way that they align spontaneously even in the absence of an external field. This process is called ferroelectricity.

....

EXAMPLE 11.2–5

A compressive force of 5 N (~1 lb) is applied to a slice of quartz with dimensions 3 mm $\times$ 3 mm $\times$ 0.25 mm. If the elasticity modulus for quartz is 70 GPa, calculate the voltage created by the stress.

Solution

The stress on the quartz is

$$\sigma = \frac{F}{A} = \frac{5 \text{ N}}{(3 \times 10^{-3} \text{ m})^2} = 5.56 \times 10^5 \text{ Pa}$$

The field can be found from Equation 11.2–2 if a value of g is available. Recognizing that d values are given in Table 11.2–1, Equation 11.2–4 can be rearranged as $g = 1/Ed$. Substituting this expression into Equation 11.2–2 gives $\xi = \sigma/Ed$. Using the calculated stress and the d value from Table 11.2–1:

$$\xi = \frac{5.56 \times 10^5 \text{ Pa}}{(70 \times 10^9 \text{ Pa})(2.3 \times 10^{-12} \text{ m/V})}$$

$$= 3.5 \times 10^6 \text{ V/m}$$

Finally, the voltage can be calculated as:

$$V = \xi \times \text{separation} = (3.5 \times 10^6 \text{ V/m})(0.25 \times 10^{-3} \text{ m}) = 875 \text{ V}$$

The voltage developed is quite large. Consequently, even small loads can be easily detected using this technique. A phonograph cartridge operates on this principle.

11.3 DIELECTRIC CONSTANT AND CAPACITANCE

In this section we will investigate the ability of a material to polarize and hold charge. Since polarization is most easily illustrated by using a parallel-plate capacitor, we will first discuss the properties of capacitors.

11.3.1 Capacitance

A basic capacitor can be envisioned as two parallel plates of conducting material separated by an insulator, as shown in Figure 11.3–1. The electrical charge Q (in coulombs) stored on either plate is directly proportional to the applied voltage V (in volts):

$$Q = C \times V \tag{11.3–1}$$

where the proportionality constant, C, is defined as the capacitance, with units of farads, F (1 farad = 1 coulomb per volt).

11.3.2 Permittivity and Dielectric Constant

The capacitance contains both a geometrical and a material factor. The material factor, ε, called the **permittivity,** reflects the ability of the material to be polarized by the electric field. The geometric factor is the ratio of the capacitor area, A, to the separation distance, d. Capacitance is then defined as,

$$C = \varepsilon\left(\frac{A}{d}\right) \tag{11.3–2}$$

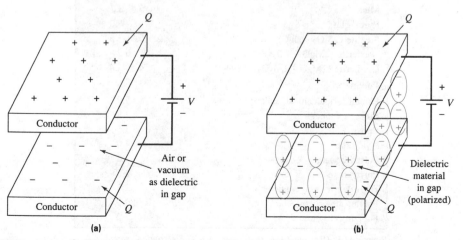

FIGURE 11.3–1   Structure of simple capacitor: **(a)** the dielectric is either air or vacuum; **(b)** the dielectric is a material more polarizable than air or vacuum.

When the dielectric is vacuum, then

$$C_0 = \varepsilon_0 \left(\frac{A}{d} \right) \qquad (11.3\text{-}3)$$

The permittivity of a vacuum, ε_0, is 8.85×10^{-12} F/m. Since in a capacitor any dielectric material is able to polarize more than does a vacuum, the permittivity of a material, ε, is greater than ε_0. In fact, it can be shown that the relationship between ε and the polarization P is

$$\varepsilon = \varepsilon_0 + \frac{P}{\xi} \qquad (11.3\text{-}4)$$

where ξ is the electric field strength. The derivation of this equation is left as a homework exercise.

The permittivity of a material is customarily described as its permittivity normalized to that of vacuum. Hence, the relative permittivity, or **dielectric constant,** κ is defined as

$$\kappa = \frac{\varepsilon}{\varepsilon_0} \qquad (11.3\text{-}5)$$

Values for the dielectric constant of several materials are given in Table 11.3–1. Be aware that the dielectric constant is a function of frequency in some materials, especially

TABLE 11.3–1   Dielectric constants of materials.

Material	Dielectric constant κ
Air (or vacuum)	1.0
Water	80.4
Ceramics	
Diamond	5.5–6.6
Al_2O_3 (polycrystals)	~9.0
SiO_2	3.7–3.8
MgO	9.6
NaCl	5.9
$BaTiO_3$	3000
Mica	5.4–8.7
Pyrex glass	4.0–6.0
Steatite ($2SiO_2 \cdot MgO$)	5.5–7.5
Forsterite ($2MgO \cdot SiO_2$)	6.2
Cordierite ($2MgO \cdot 2Al_2O_3 \cdot 5SiO_2$)	4.5–5.4
Polymers	
Phenol formaldehyde (Bakelite)	5.0
Silicone rubber	2.8
Epoxy	3.5
Nylon 6,6	4.0
Polycarbonate	3.0
Polystyrene	2.5
High-density polyethylene	2.3
Polytetrafluoroethylene	2.0
Polyvinylchloride	3.2

Source: L. L. Hench and J. K. West, *Principles of Electronic Ceramics.* Copyright © 1990 by John Wiley & Sons. Reprinted by permission of John Wiley & Sons, Inc.

polymers. This is to be expected, since it was shown in Section 11.2.5 that polarization is a function of frequency.

EXAMPLE 11.3–1

Polyethylene has a polarization of 5.75×10^{-8} C/m^2 in a field of 5000 V/m. Calculate the dielectric constant of this polymer.

Solution

An expression for κ can be found by dividing both sides of Equation 11.3–4 by ε_0:

$$\kappa = 1 + \frac{P}{\varepsilon_0 \xi}$$

Substituting the appropriate values:

$$\kappa = 1 + \frac{5.75 \times 10^{-8} \text{ C/m}^2}{(8.85 \times 10^{-12} \text{ F/m})(5000 \text{ V/m})}$$

$$= 2.3$$

Note the units of the dielectric constant. It is dimensionless.

EXAMPLE 11.3–2

A simple parallel-plate capacitor that can store a charge of 10^{-4} C at 3 kV is required for a particular application. The thickness of the dielectric is 0.02 cm. Calculate the area of the plates if the dielectric is: vacuum, PTFE (Teflon®), BaTiO$_3$, mica.

Solution

The required capacitance can be found from Equation 11.3–1 as

$$C = \frac{Q}{V} = \frac{10^{-4} \text{ C}}{3000 \text{ V}} = 3.33 \times 10^{-8} \text{ F}$$

The area can then be found by rearranging Equation 11.3–2 to yield

$$A = \frac{Cd}{\varepsilon} = \frac{Cd}{\kappa \varepsilon_0}$$

$$= \frac{(3.33 \times 10^{-8} \text{ F})(0.02 \text{ cm})}{(8.85 \times 10^{-14} \text{ F/cm})\kappa} = \frac{7525 \text{ cm}^2}{\kappa}$$

Using the values for κ in Table 11.3–1, we find

For a vacuum:	$\kappa = 1$	$\Rightarrow$	$A = 7525$ cm^2 = 0.75 m^2
For PTFE (Teflon®):	$\kappa = 2$	$\Rightarrow$	$A = 3763$ cm^2 = 0.38 m^2
For BaTiO$_3$:	$\kappa = 3000$	$\Rightarrow$	$A = 2.5$ cm^2 = 0.00025 m^2
For mica:	$\kappa = 7$	$\Rightarrow$	$A = 1075$ cm^2 = 0.11 m^2

As electronic devices and components become smaller and smaller, it is important to use new materials that enable the component to be constructed more efficiently. The use of either BaTiO$_3$ or mica in the above example allows the capacitor to be smaller. In addition, practical capacitors may not conform to the simple two-plate parallel array described above; rather, the dielectric may be vacuum metallized (i.e., coated with a conductive layer) and the two-layer material spiral-wound on itself to create a large surface area in a small volume.

11.3.3 Dielectric Strength and Breakdown

An important related material property is **dielectric strength.** This is the maximum electric field to which a dielectric material can be subjected without breaking down or discharging, and is defined as

$$\xi_{max} = \left(\frac{V}{d}\right)_{max} \tag{11.3--6}$$

where the subscript max indicates the onset of breakdown.

Dielectric breakdown is the electrical equivalent to mechanical failure. Once dielectric breakdown has occurred, the device will not perform acceptably. Breakdown occurs when the field strength is sufficient to produce a burst of current that creates pits, holes, and channels bridging the conductors. Similar to mechanical strength in brittle solids, dielectric strength is dependent upon the thickness of the material. Both dielectric and mechanical strength are determined by the probability of finding a critical flaw in the test region. Hence, strength increases with decreasing thickness.

To achieve maximum charge storage in an intense field, a material with both high dielectric constant and high dielectric strength is required. Table 11.3–2 provides the

TABLE 11.3–2   Dielectric strength of materials.

Material	Dielectric strength (10^6 V/cm)
Ceramics	
Al_2O_3 (0.03 μm)	7.0
Al_2O_3 (0.6 μm)	1.5
Al_2O_3 (0.63 cm)	0.18
SiO_2 (quartz, 0.005 cm)	6.0
NaCl (0.002 cm)	2.0
NaCl (0.014 cm)	1.3
$BaTiO_3$ (0.02 cm, single crystal)	0.04
$BaTiO_3$ (0.02 cm, polycrystal)	0.12
$PbZrO_3$ (poly., 0% porosity, 0.016 cm)	0.08
$PbZrO_3$ (poly., 10% porosity, 0.16 cm)	0.03
$PbZrO_3$ (poly., 22% porosity, 0.16 cm)	0.02
Mica (0.002 cm)	10.1
Mica (0.006 cm)	9.7
Pyrex glass (0.003 cm)	5.8
Pyrex glass (0.0005 cm)	6.5
Steatite (SiO_2 + MgO + Al_2O_3, 0.63 cm)	0.1
Forsterite ($2MgO \cdot SiO_2$, 0.63 cm)	0.15
Polymers	
Phenol formaldehyde (Bakelite)	120–160
Silicone rubber	220
Epoxy	160–200
Nylon 6,6	240
Polycarbonate	160
Polystyrene	200–280
High-density polyethylene	190–200
Polytetrafluoroethylene	160–200
Polyvinylchloride	160–590

Source: L. L. Hench and J. K. West, *Principles of Electronic Ceramics.* Copyright © 1990 by John Wiley & Sons. Reprinted by permission of John Wiley & Sons, Inc.

ielectric strength of a number of materials and shows why mica has been used historically
n applications where high dielectric strength, high dielectric constant, and good thermal
tability are required.

DESIGN EXAMPLE 11.3–3

Reconsider Example 11.3–2. Which materials may be used for the dielectric application described?

Solution

The applied field strength is:

$$\xi = \frac{3 \text{ kV}}{0.02 \text{ cm}} = 150 \text{ kV/cm}$$

Comparing this value with the dielectric strength values in Table 11.3–2 shows that the viable
materials for this application are Teflon and mica, although the area required in either case is
significant.

DESIGN EXAMPLE 11.3–4

Suggest an alternative design for the capacitor needed to fulfill the requirements of storing a charge
of 10^{-4} C at 3 kV, assuming that the raw material for the dielectric is 0.02-cm-thick sheets of
$BaTiO_3$.

Solution

To use $BaTiO_3$ the applied field strength must be reduced below the dielectric strength of 120 kV/cm
(obtained from Table 11.3–2). This corresponds to a minimum dielectric thickness of

$$d = \frac{3 \text{ kV}}{120 \text{ kV/cm}} = 0.025 \text{ cm}$$

Thus, two sheets of $BaTiO_3$, each 0.02 cm in thickness, must be stacked together to achieve a total
dielectric thickness of 0.04 cm. The required area for the capacitor is calculated using Equation 11.3–2 in the form

$$A = \frac{Cd}{\varepsilon} = \frac{Cd}{\kappa \varepsilon_0}$$

For $BaTiO_3$, $\kappa = 3000$ and $C = Q/V = 3.33 \times 10^{-8}$ F, so that the area of the $BaTiO_3$ capacitor
must be:

$$A = \frac{(3.33 \times 10^{-8} \text{ F})(0.04 \text{ cm})}{(3000)(8.84 \times 10^{-14} \text{ F/cm})}$$

$$= 5.02 \text{ cm}^2$$

Breakdown of a material subject to high fields may not occur through the bulk of the
material as described above. Rather, breakdown may occur on the surface of the material.
This sort of breakdown can be aggravated by surface moisture or contamination. A
familiar application where this behavior cannot be tolerated is spark plugs. To withstand
high temperatures and high fields, spark plug insulators are ceramic, usually steatite (a
mixture of SiO_2, MgO, and Al_2O_3). This and other common ceramics are typically porous
and hygroscopic (water absorbing), a combination that offers the potential for surface
breakdown. The most common solution to this problem is to coat the steatite with a
water-impervious ceramic glaze, often a glassy silicate.

11.4 DISSIPATION AND DIELECTRIC LOSS

Chapter 9 discussed the effect of reversing the direction of shear stress on a polymer solid or melt. It was shown that the molecules interact, producing friction, which is converted to heat. The amount of heat produced is a function of the frequency of stress reversal. Similarly, when an alternating electric field is applied to a dielectric, some of the power is converted to heat. The dipoles effectively are not able to respond to the field, but rather, experience drag or friction from neighbors.

The amount of power lost as heat is a function of the frequency f of the applied field, the field strength ξ, the permittivity ε of the material, and a measure of the extent of molecular friction in the dielectric. This last factor is known as the dissipation factor, or loss tangent, and is represented by the symbol $\tan \delta$. The power loss per unit volume, W, is given by:

$$W \approx \pi \varepsilon f \xi^2 \tan \delta \tag{11.4--1}$$

with the units for W being watts per cubic meter (W/m^3). Table 11.4–1 lists values of $\tan \delta$ for a number of materials. While these values are expressed as constants so that you can develop an appreciation for the magnitude of losses in various materials, $\tan \delta$ is itself a complex function of frequency, temperature, and the atomic scale structure of the material.

The loss factor is a major consideration for the usefulness of a dielectric material in a particular insulating application. It is often necessary to select a material with a low dielectric constant and a very low dielectric loss. When high capacitance is required in a minimal amount of space, a material with a high dielectric constant is called for; however, it is also necessary to have low loss to prevent heat buildup.

TABLE 11.4–1 Dielectric loss of materials.

Material	$\tan \delta$
Ceramics	
Al_2O_3	0.0002–0.01
SiO_2	0.00038
$BaTiO_3$	0.0001–0.02
Mica	0.0016
Pyrex glass	0.006–0.025
Steatite ($2SiO_2 \cdot MgO$)	0.0002–0.004
Forsterite ($2MgO \cdot SiO_2$)	0.0004
Cordierite ($2MgO \cdot 2Al_2O_3 \cdot 5SiO_2$)	0.004–0.012
Polymers	
Phenol formaldehyde (Bakelite)	0.06–0.10
Silicone rubber	0.001–0.025
Epoxy	0.002–0.010
Nylon 6,6	0.01
Polycarbonate	0.0009
Polystyrene	0.0001–0.0006
High-density polyethylene	< 0.0001
Polytetrafluoroethylene	0.0002
Polyvinylchloride	0.007–0.020

Source : Adapted from J. Brandrup and E. H. Immergut, eds., *Polymer Handbook*, 2nd ed. Copyright © 1975 by John Wiley & Sons. Reprinted by permission of John Wiley & Sons, Inc. Data also from CRC *Handbook of Chemistry and Physics*, copyright CRC Press, Boca Raton, FL.

Losses from frequent charge reversals are often unwelcome. Such is not the case in dielectric heating. In this technique materials with high loss are subject to alternating electric fields for the purpose of increasing the temperature of the material. Materials with high losses can be rapidly and efficiently heated. This is the principle of the microwave oven, in which the O-H dipole in the water molecule is excited. This is also the principle of ultrasonic welding, used routinely to melt, fuse, and bond thermoplastic polymers.

The concept of dielectric loss can be used to explain why Tupperware, designed for use in a microwave oven, is fabricated from high-density polyethylene (HDPE). This application requires a low tan δ value so that the container is not influenced significantly by the microwave field. Table 11.4–1 shows that HDPE has a very low tan δ value of < 0.0001. Thus, HDPF is an ideal material for this application. The reason that HDPE has such a low loss factor is that the repeat unit in PE, C_2H_4, is completely symmetric. Therefore, there are no significant molecular dipoles that can respond to the applied alternating field in the microwave.

EXAMPLE 11.4–1

A simple parallel-plate capacitor, with mica of $d = 0.02$ cm and $A = 1$ cm^2 as the dielectric, is subjected to an alternating electric field of magnitude 120 V and frequency 60 Hz. Calculate the power loss for this capacitor.

Solution

The power loss per m^3 is found from Equation 11.4–1:

$$W \approx \pi \varepsilon f \xi^2 \tan \delta$$

From Table 11.3–1 we find that κ for mica is ~ 7, so we can use Equation 11.3–5 to find:

$$\varepsilon = \kappa \varepsilon_0 = 7 \times 8.85 \times 10^{-12} \text{ F/m} = 6.2 \times 10^{-11} \text{ F/m}$$

Next, we can calculate the field strength as

$$\xi = \frac{V}{d} = \frac{120 \text{ V}}{(0.02 \text{ cm})(1 \text{ m}/100 \text{ cm})} = 6 \times 10^5 \text{ V/m}$$

The tan δ value for mica, from Table 11.4–1, is 0.0016. Substituting these values into the power loss equation yields:

$$W \approx (3.14)(6.2 \times 10^{-11} \text{ F/m})(60 \text{ s}^{-1})(6 \times 10^5 \text{ V/m})^2(0.0016)$$
$$\approx 1.1 \times 10^{10}(\text{F-V}^2)/(\text{s-m}^3) = 6.7 \text{ W/m}^3$$

The power loss is found by multiplying by the capacitor volume:

$$\text{Power loss} = (6.7 \text{ W/m}^3)(0.02 \text{ cm})(1 \text{ cm}^2)(1 \text{ m}/100 \text{ cm})^3$$
$$= 1.34 \times 10^{-7} \text{ W} = 0.134 \text{ } \mu\text{W}$$

Before leaving our discussion of dielectric losses, it is helpful to point out that permittivity is in fact a complex quantity of the form:

$$\varepsilon^* = \varepsilon - i\varepsilon' \tag{11.4–2}$$

The real part of the complex notation is commonly used loosely to describe the named property, whereas the imaginary part describes loss processes.

This is not the first time that we have used a complex number to describe a material property. Recall that elastic modulus, E, is the real part of the complex modulus E^* (see Chapter 9). Often E', the imaginary part of E^* (which represents the extent of mechanical damping in the system), is small, leading to the good approximation that $E^* \approx E$. E' is significant for viscous or viscoelastic materials.

The dielectric dissipation factor, $\tan \delta$, is defined as the ratio of the imaginary part of ε^* to its real part. That is,

$$\tan \delta = \frac{\varepsilon'}{\varepsilon} \qquad (11.4\text{--}3)$$

When $\tan \delta$ is small, the electrical damping (power loss) is small, and the permittivity can be approximated by its real part: $\varepsilon^* \approx \varepsilon$. This simplification has been used throughout this chapter.

If the simplification leads to useful results, why have we bothered to introduce the idea of a complex dielectric permittivity? First, in more advanced treatments of the subject the complex notation is necessary to describe completely the dielectric properties of materials. Second, the similarity between the mechanical and electrical loss processes is easily recognized as being associated with the imaginary part of the complex elastic modulus and dielectric permittivity. Third, the complex notation will be important for understanding the absorption of light in Section 11.6.

11.5 REFRACTION AND REFLECTION

We are all familiar with reflection from a surface such as a mirror. We are equally familiar with refraction. A common example of refraction is the apparent bending of light rays or images as they pass from air into water—a spoon partially immersed in a glass of water or a branch protruding from the water in a pond. In this section we will discuss refractive index, a material property, in detail. We will then discuss and calculate the amount of light that is reflected from surfaces. Throughout this section we assume that the materials under discussion are transparent to electromagnetic radiation. Absorption of light is treated in Section 11.6.

FIGURE 11.5–1

Light reflection and refraction. At each interface a portion of light is reflected and a portion is refracted.

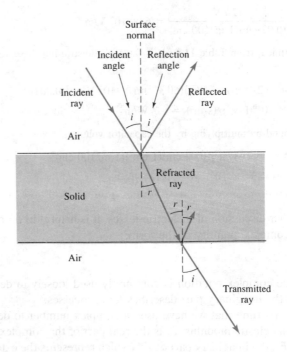

11.5.1 Refraction

Refractive index n can be described in a number of ways. Perhaps the simplest is that it is the ratio of the speed of light in a vacuum to that in the material:

$$n = \frac{v_{vac}}{v_{mat}}$$

(11.5–1)

where v is the speed of light. Refractive index is a fundamental property of a material. Typical values are shown in Table 11.5–1. All the values of refractive index are greater than 1.0, indicating that the velocity of light is less in the material than in a vacuum.

The behavior of light at boundaries or surfaces is determined chiefly by the relative refractive indices and the angle of incidence. Not only do the speed and wavelength change abruptly when light encounters a medium of different refractive index, but so also does the direction of the light. The phenomenon is called **refraction.** If the angle of incidence is i and the angle of refraction is r, as shown in Figure 11.5–1, then the

TABLE 11.5–1   Refractive indices of materials.

Material	Average refractive index
Air	1.00
Water	1.33
Ice	1.31
Ceramics	
Diamond	2.43
Al_2O_3	1.76
SiO_2	1.544, 1.553
MgO	1.74
NaCl	1.55
$BaTiO_3$	2.40
TiO_2	2.71
Pyrex glass	1.47
Soda-lime-silicate glass	1.51
Lead-silicate glass	2.50
Calcite	1.658, 1.486
Semiconductors	
Ge	4.00
Si	3.49
GaAs	3.63
Polymers	
Epoxy	1.58
Nylon 6,6	1.53
Polycarbonate	1.60
Polystyrene	1.59
High-density polyethylene	1.54
Polypropylene	1.49
Polytetrafluoroethylene	1.30–1.40
Polyvinylchloride	1.54
Poly(ethylene terephthalate)	1.57

Source : Adapted from *Handbook of Tables for Applied Engineering Science.* Copyright CRC Press, Boca Raton, FL.

relationship of these angles to the speed of light can be expressed using the refractive index:

$$n = \frac{v_{vac}}{v_{mat}} = \frac{\lambda_{vac}}{\lambda_{mat}} = \frac{\sin i}{\sin r} \tag{11.5-2}$$

This equation assumes that the light was traveling in vacuum (or air) prior to entering the material. When the light is traveling in material 1 and enters material 2, Equation 11.5–2 becomes:

$$\frac{v_1}{v_2} = \frac{n_2}{n_1} = \frac{\sin i}{\sin r} \tag{11.5-3}$$

EXAMPLE 11.5–1

Consider a beam of light traveling through soda-lime-silica glass and impinging upon a glass-air interface.

    a. Derive an expression that can be used to calculate the refraction angle as a function of the incident angle.

    b. Is this expression valid for all incidence angles?

Solution

    a. At low angles of incidence the angle of refraction can be calculated by solving Equation 11.5–3 for the angle *r*:

$$\sin r = (\sin i)\left(\frac{n_1}{n_2}\right)$$

        The refractive indices for glass and air are 1.51 and 1.0, respectively (Table 11.5–1). Substituting these values into the previous equation and solving for *r*,

$$r = \sin^{-1}(1.51 \sin i)$$

    b. This equation shows that the right-hand side can be evaluated only if

$$1.51 \sin i \leq 1.0$$

        Solving for the critical angle, i_{cr}, gives:

$$i_{cr} = \sin^{-1}\left(\frac{1.0}{1.51}\right) = 41.47°$$

    The previous example points out an extremely important result. As shown in Figure 11.5–2, for $i < i_{cr}$, *r* is given by the equation developed in Example 11.5–1 and $r < 90°$. For $i \geq i_{cr}$, $r = 90°$. That is, none of the light escapes from the glass. This phenomenon is called total internal reflection and is one of the key principles in the fiber optics industry (discussed in Section 11.5.5). Note that the concept of a critical angle has meaning only when light moves from a medium of high index to one of low index.

11.5.2 Specular Reflection

When a beam of light strikes an interface, not all of it is refracted. A portion of the incident light is typically reflected from the interface at an angle equal to that of the incident light. This is specular reflection. For a beam with normal incidence ($i = 90°$) moving from

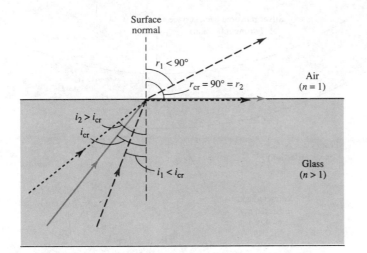

FIGURE 11.5–2

Concept of critical angle. Above the critical angle, all light is refracted parallel to the interface.

material 1 into material 2, the fraction of light reflected, R, is given by:

$$R \approx \left(\frac{n_2 - n_1}{n_2 + n_1}\right)^2 \qquad\qquad (11.5\text{–}4)$$

Equation 11.5–4 is known as Fresnel's formula, and the version we have presented assumes that both materials are transparent to visible light.

EXAMPLE 11.5–2

Camera and microscope lenses are several glasses cemented together. Consider reflection from one of these lenses. Light travels through silica glass into a high-lead silica glass at normal incidence. How much light is reflected at the internal surface?

Solution

Using the values of n found in Table 11.5–1 and substituting these values into Equation 11.5–4 yields:

$$R \approx \left(\frac{n_2 - n_1}{n_2 + n_1}\right)^2 = \left(\frac{2.5 - 1.5}{2.5 + 1.5}\right)^2 = 6\%$$

Surface reflection can be a desirable property in some applications, for example, when a reflected image is to be observed off the front surface of a glass. Consider the rearview mirror in an automobile, shown in Figure 11.5–3. Note that two beams of light are emerging from the mirror. Some of the incident light, in this example ~8%, is reflected from the front surface of the glass. Most of the remaining incident light is refracted at the air-glass interface, fully reflected from the silver coating on the back of the mirror, and then refracted again at the glass-air interface. Since the mirror is curved, the two beams are not parallel. The day-night switch on the mirror allows the driver to select which of the two beams he or she will see. In the "day" setting, the driver sees the comparatively high intensity beam reflected from the mirrored back surface. In the "night" setting, the driver selects the low-intensity beam reflected from the front surface to minimize the glare from the headlights of trailing cars.

For light of normal incidence the intensity of reflected light increases from 4% to 10% when changing from $n = 1.5$ to $n = 1.9$. Lead crystal, which is a silicate glass containing

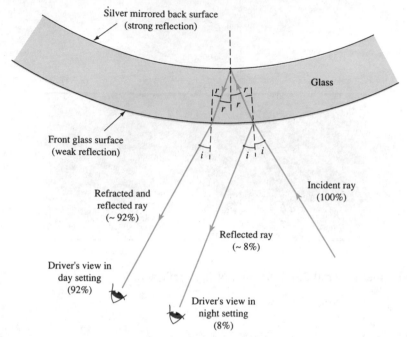

FIGURE 11.5–3 The operation of a day-night mirror shows that the intensity of the beam reflected from the glass front surface is less than that reflected from the silvered back surface. The driver selects the set of rays he or she "sees" by tilting the mirror.

a large fraction of lead oxide, has a high refractive index. Thus, lead crystal glistens because of the relatively large amount of light reflected from its surfaces.

In other applications it is desirable to have R as low as possible, for example, in microscopes, telescopes, and binoculars. In these applications the lenses are usually supplied with antireflective coatings. The operative principle of antireflective coatings is that the light reflected from the front and back surfaces of the coating is equally intense, as shown in Figure 11.5–4. If the coating is 1/4 of a wavelength thick, then the total

FIGURE 11.5–4
The principle of an antireflective coating. The light reflected from the back surface of the coating is 180° or $\lambda/2$ out of phase with that reflected from the front surface because it passes through the $\lambda/4$-thick antireflective coating twice.

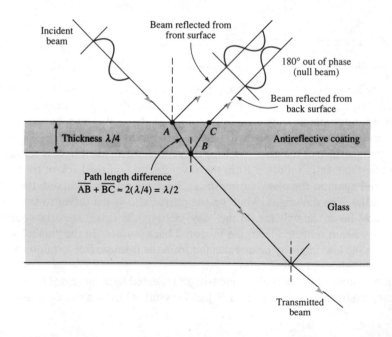

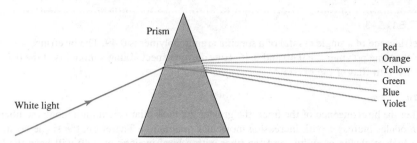

FIGURE 11.5–5 Light refraction in a prism. Dispersion produces the familiar colors of the rainbow.

path difference is $\lambda/2$, the two reflected beams are out of phase, and a null beam is reflected.

11.5.3 Dispersion

Refractive index varies with frequency, normally decreasing with increasing wavelength. This is because the refractive index is related to the extent of polarization, which in turn is a function of frequency. The change in refractive index with wavelength is called **dispersion.** The effect of dispersion on white light shining on a prism is shown in Figure 11.5–5. The longer the wavelength of the radiation, the lower the refractive index and the less it is refracted. Hence, white light separates into the well-known colors.

Dispersion is an important effect that must be dealt with in designing and making optical instruments with focal planes, where the response of the light to a lens varies with wavelength. Consider, for example, a lens system in an optical microscope. The lenses in the microscope can be designed to focus a beam of light at a particular point only if the light is monochromatic (i.e., of a single wavelength). In polychromatic light, each wavelength has a different focal point, leading to a blurry image with coloration at the edges. To overcome this problem, the highest-quality microscopes have complex lenses that have the same focal length for three wavelengths (red, blue, and yellow-green). They are called apochromatic lenses.

11.5.4 Birefringence

The refractive index has a single value in isotropic materials, such as ceramic glasses or polycrystalline materials. Single crystals and oriented polymers are anisotropic in electronic polarizability and have refractive indices that vary with direction. Consider, for example, an oriented polymer fiber. Assuming the fiber is orthotropic (i.e., all directions within the cross section parallel to the fiber axis are equivalent), the refractive indices encountered by a light ray traveling through a fiber with the light polarized parallel and normal to the fiber axis, $n_\parallel$ and $n_\perp$, respectively, characterize the molecular orientation of the material fully. The difference in refractive index is called birefringence Δn:

$$\Delta n = n_\parallel - n_\perp \qquad (11.5\text{–}5)$$

Birefringence measurement is an important tool for characterizing and identifying materials. It is used in forensics to help identify polymer fibers, since each class of fiber has a distinct range of birefringence. It is used by fiber scientists to understand elastic modulus, since modulus correlates with molecular orientation. Birefringence is also used by mineralogists to identify materials, since each crystal substance has characteristic and cataloged refractive indices, as indicated by the entries for quartz and calcite in Table 11.5–1.

EXAMPLE 11.5–3

The birefringence of a single crystal of a specific aramid polymer is 0.49. The birefringences of two fibers of the same polymer are 0.40 and 0.46. How do you expect Young's modulus of the two fibers to compare with that of the crystal?

Solution

The higher the birefringence of the fiber, the greater the molecular orientation along the fiber axis. Elastic modulus increases with increasing molecular orientation. Therefore, the single crystal will have the highest elastic modulus, and the fiber with a birefringence of 0.40 will have the lowest elastic modulus.

11.5.5 Application: Optical Waveguides

One particularly interesting application of a number of the principles presented above is optical waveguides. There are two classes of applications based on optical fibers. In the more straightforward technique, often called **fiber optics,** optical information is directly transmitted along the fiber. One property of materials that fiber-optic devices use is total internal reflection. An example of this sort of application is the laparoscope, which is a fiber-optic device inserted into the human body through a small incision in the navel. The surgeon can then move the end of the device to examine the body internally with minimal disturbance.

The other class of applications deals with the transmission of electrical signals over long distances. Copper wires are being supplanted with optical cables. Optical waveguides offer a number of advantages: they weigh less (mass density is lower), more information can be transmitted per unit time, fewer signal-boosting stations are required, and less information is lost. Several long-distance telephone companies boast of these advantages.

The process of telecommunication by optical waveguides consists of several steps:

1. The electrical signal is coded digitally and converted to an optical signal.
2. The digitized optical signal, consisting of high-frequency laser pulses, is sent along the waveguide.
3. The signals are reamplified periodically.
4. The signal is received and decoded.

A number of technical obstacles stood in the way of commercialization of optical waveguides. We will explore the nature of these obstacles by examining the properties of light traveling down a waveguide.

Consider the introduction of a well-collimated beam of light into a waveguide. While most of the light travels parallel to the fiber axis, the beam has some divergence. Rays that impinge upon the surface of the fiber may be scattered by surface defects. To prevent surface scattering, optical fibers are encased in protective polymers. Rays that travel toward the fiber surface must intersect the interface at an angle greater than the critical angle if they are to remain within the waveguide. The combination of the ray divergence and fiber diameter leads to a minimally acceptable bending radius for the waveguide.

Figure 11.5–6 shows that rays moving down the center of the waveguide have a much shorter path length than those that reflect from wall to wall. Since both beams travel at the same velocity, they require different amounts of time to traverse the waveguide. This results in a broadening of the original signal, which is undesirable. Consequently, waveguides are often constructed with a graded index. The composition in the center of the waveguide is silica doped with germania (SiO_2 doped with GeO_2), and the amount of

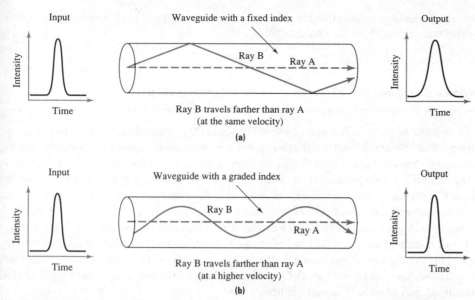

FIGURE 11.5–6   A comparison of two types of waveguides: **(a)** a single-index waveguide—the beam traveling through the center of the waveguide will arrive before others that travel longer distances, causing a broadening of the signal; **(b)** a graded-index waveguide—variation in refractive index yields higher velocity for waves traveling longer distance to eliminate undesirable signal broadening.

germania decreases with radial position in the fiber. Since GeO_2 has a higher refractive index than SiO_2, the index of refraction of the waveguide decreases with increasing distance from the center of the fiber. Thus, the light waves that travel off center travel at a faster rate, decreasing the undesirable broadening.

In some cases, such as on ships, the light signals need to travel only short distances. In other cases the light signals need to travel much further. For short distances, roughly one kilometer or less, polymer waveguides can be used. For long distances the signal in the fiber needs to travel virtually unattenuated (i.e., without loss). The amount of attenuation, α, is given by

$$\alpha = 10 \log_{10}\left(\frac{I}{I_0}\right) \tag{11.5–6}$$

where I_0 is the original intensity of the light beam and I is the final intensity. The units for attenuation are decibels per kilometer (dB/km).

··

EXAMPLE 11.5–4

The attenuation of an ordinary sample of soda-lime-silicate glass is about 3000 dB/km. What fraction of a light signal will be lost in 1 m?

Solution

If the attenuation is 3000 dB/km and the length of the glass is 1 m, then the total attenuation is

$$(3000 \text{ dB/km})(1 \text{ m})(1 \text{ km}/1000 \text{ m}) = 3 \text{ dB}$$

Substituting this value into Equation 11.5–6 yields:

$$-3.0 \text{ dB} = 10 \log_{10}\left(\frac{I}{I_0}\right)$$

where the minus sign signifies that the attenuation decreases the intensity of the transmitted signal. Solving this expression for the ratio I/I_0 yields:

$$\frac{I}{I_0} = 10^{-0.3} = 0.5$$

Ordinary glass allows only 50 percent of the initial light to travel through a 1-meter section.

A number of phenomena contribute to the scattering, absorption, and overall deterioration of an optical signal as it travels down a waveguide. Producing low-loss fibers requires systematic elimination of various large defects, such as air bubbles, grain boundaries, impurities, compositional fluctuations, and so on. To reach the targeted value of about 0.2 dB/km, even the smallest defects and impurities need to be eliminated.

Why are graded optical fibers fabricated by adding controlled amounts of GeO_2 to SiO_2 rather than doping with either Na_2O or CaO? As discussed in Example 11.2–4, adding network modifiers such as Na_2O or CaO to SiO_2 increases the potential for ionic polarization. This in turn would increase the attenuation in the fiber. In contrast, adding another network-forming oxide, such as GeO_2 or P_2O_5, changes the refractive index without the associated degradation in signal quality.

In practice, polychromatic, or white, light is not used in transmission of optical data. Rather, laser output, an intense monochromatic coherent beam, is used. Low absorption by the glass at the lasing frequency is necessary. Largely because of their organic structure, polymer glasses absorb more light at typical laser frequencies than do inorganic glasses.

11.6 ABSORPTION, TRANSMISSION, AND SCATTERING

Like the elastic modulus and dielectric permittivity, index of refraction is a simplification of a complex variable, the coefficient of refraction, n^*:

$$n^* = n - ik \tag{11.6–1}$$

The real part of n^* is simply refractive index. The imaginary part of n^* deals with absorption phenomena or losses and is called **index of absorption** k. When absorption is minimal, k is small, and the refractive index n can be used to describe optical phenomena. When absorption is significant, however, the magnitude of k increases, and the complex nature of the refractive index becomes important.

So far we have considered only properties of transparent materials. In this section we will consider why materials are transparent to electromagnetic radiation over only a relatively small range of frequencies. We will calculate the wavelength at which a material ceases to be transparent.

11.6.1 Absorption

The spectra in Figure 11.6–1 show the typical light transmission behavior of metals, semiconductors, and dielectrics as a function of frequency. One of the primary absorption mechanisms is the promotion of an electron from a filled level to a higher unfilled level within an energy band. Recall that the energy associated with a photon of wavelength λ was given in Equation 10.3–8 as $E = hc/\lambda$.

First consider the band structure of a metal. The partially filled valence band characteristic of conductors suggests that all wavelengths could be absorbed by electron promotion. This prediction, however, is only partially correct. Metals do in fact absorb photons in the low-frequency range, including those in the visible spectrum. For this reason, most

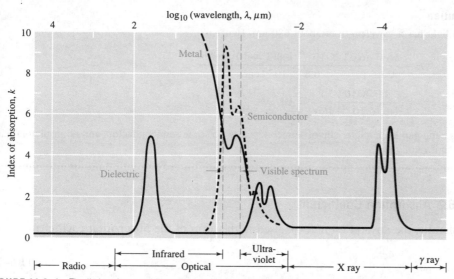

FIGURE 11.6–1 The limits of transparency. Transparency is limited at high frequencies by electronic transitions and at low frequencies by ionic transitions. Metals are opaque to visible light because electronic transitions are easy, whereas band gap materials may be transparent to visible light. *(Source: W. D. Kingery et al.,* Introduction to Ceramics, *2nd ed. Copyright © 1976 by John Wiley & Sons. Reprinted by permission of John Wiley & Sons, Inc.)*

metals are opaque to light in the visible range. High-frequency photons (e.g., X rays) are transmitted, however, because not even the electrons are able to respond rapidly enough to absorb the incident energy.

In contrast to conductors, semiconductors are transparent at low frequencies. They are unable to absorb light via electron promotion unless the photon energy is greater than the band gap energy (i.e., absorption occurs only if $E \geq E_g$). Coincidentally, the energy of light in the visible spectrum is approximately equal to the band gap energy of most semiconductors. Like metals, semiconductors are unable to absorb the high frequencies. Thus they typically display absorption only in the region of visible light.

Dielectrics, with their typically higher band gaps, exhibit an absorption peak at correspondingly higher frequencies than those for the semiconductors. In addition, ionic dielectrics also exhibit an absorption spike in the low-frequency range. The location of an absorption peak corresponds to the frequency at which the incident radiation is in resonance with that of the oscillatory motion of the anions and cations (i.e., ionic polarization). Multiple absorption peaks are possible if more than one type of anion or cation is present in the dielectric.

For a variety of applications it is desirable to have as wide a range of transparency as possible. Consider, for example, the requirements of CO and CO_2 lasers. These lasers may be high-powered and used to cut through materials, or they may be used in a number of important military applications, such as to detect motion at night. The wavelength of the radiation produced by these lasers is 5 μm and 10.6 μm, respectively. Window materials for these and other lasers require transmission in the infrared. The high–atomic weight monovalent alkali halide glasses generally provide the largest window. Also, CaF_2 can be used with the CO laser, and ZnSe or CdTe for the CO_2 laser.

EXAMPLE 11.6–1

What band gap is required to allow transmission of red light? What class of materials have such band gaps?

Solution

Red light has a wavelength of about 680 nm. Hence,

$$E = \frac{hc}{\lambda} = \frac{(6.62 \times 10^{-34} \text{ J-s})(3 \times 10^{10} \text{ cm/s})}{6.8 \times 10^{-5} \text{ cm}}$$

$$= \frac{2.9 \times 10^{-19} \text{ J}}{1.6 \times 10^{-19} \text{ J/eV}} = 1.8 \text{ eV}$$

This band gap is characteristic of semiconductors, so some semiconductors and all insulators may transmit red light.

11.6.2 Absorption Coefficient

Generally, the absorption index is used in the form of the **absorption coefficient** β:

$$\beta = \frac{4\pi k}{\lambda} \tag{11.6–2}$$

where k is the index of absorption and λ is the wavelength of the radiation. The fraction of light transmitted through a homogeneous sample, I/I_0, is a function of the absorption coefficient and the thickness x. Mathematically, the relationship is expressed by Lambert's law,

$$\frac{dI}{I_0} = -\beta \, dx \tag{11.6–3}$$

Upon integration, the equation appears similar to that for all decay functions:

$$\frac{I}{I_0} = \exp(-\beta x) \tag{11.6–4}$$

or

$$\ln\left(\frac{I}{I_0}\right) = -\beta x \tag{11.6–5}$$

The ratio of transmitted intensity to initial intensity, I/I_0, is the transmission. The sum of the fraction of light absorbed (A), transmitted (T), and reflected (R) for materials must be 1.0. Stated mathematically,

$$A + T + R = 1 \tag{11.6–6}$$

After considerable manipulation, transmission at normal incidence is:

$$T = \frac{I_{\text{out}}}{I_{\text{in}}} = (1 - R)^2 \exp(-\beta x) \tag{11.6–7}$$

where R is the reflectivity given in Equation 11.5–4.

EXAMPLE 11.6–2

Eighty percent of the light incident onto a 1.0-mm-thick plate of polyester-fiberglass composite is transmitted. The refractive index of both glass and polyester is 1.53. Calculate the reflectivity from the surface at normal incidence and the absorption coefficient.

Solution

The relationship among reflectance, transmission, and absorption coefficient is given in Equation 11.6–7 as:

$$T = (1 - R)^2 \exp(-\beta x)$$

Solving this expression for β yields

$$\beta = \frac{-\ln[T/(1 - R)^2]}{x}$$

To use this equation we must first solve for R using Equation 11.5–4:

$$R \approx \left(\frac{n - 1}{n + 1}\right)^2 = \left(\frac{1.53 - 1}{1.53 + 1}\right)^2 = 0.044$$

Substituting this R value, along with the T and x values given in the problem statement, in the expression for β yields:

$$\beta = \frac{-\ln[0.8/(1 - 0.044)^2]}{1 \text{ mm}} = 0.133 \text{ mm}^{-1}$$

This lightweight, nonbrittle material is used as an alternative to oxide glasses to glaze greenhouses. By matching the refractive index of the glass and the polyester, no refraction occurs within the composite. The fibers are invisible.

11.6.3 Absorption by Chromophores

Absorption, reflection, and transmission each vary with wavelength. As an example, the optical characteristics of a blue silicate glass are shown in Figure 11.6–2. The absorption cutoffs occur at wavelengths of about 0.3 and 3.0 microns. The light absorption in the blue region is the result of the presence of an impurity or additive that selectively absorbs blue light. A specific chemical species in a material that selectively absorbs is called a chromophore. Examples of additives used to provide color include Mn^{2+} in oxide glass to

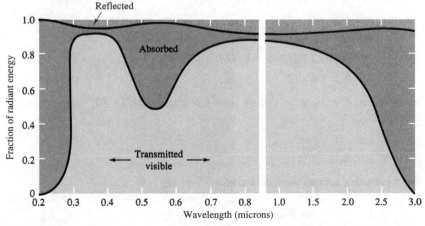

FIGURE 11.6–2 Absorption, transmission, and reflection of a deep-blue silicate glass. Color in glass is the result of absorption at a specific frequency. *(Source: W. D. Kingery et al.: Introduction to Ceramics, 2nd ed. Copyright © 1976 by John Wiley & Sons. Reprinted by permission of John Wiley & Sons, Inc.)*

give yellow, Co^{2+} to give blue, dyes in natural fibers, and various pigments added to polymer melts. Thus, the color of a transparent material is determined by the wavelength of absorption.

For the material shown in Figure 11.6–2, the value of the absorption coefficient varies directly with the concentration of the chromophore. Thus, the Beer-Lambert equation, a modification of Equation 11.6–4, is applicable:

$$\frac{I}{I_0} = \exp(-\varepsilon c x) \tag{11.6-8}$$

where ε is called the extinction coefficient, x is the thickness, and c is the concentration of the chromophore.

....

EXAMPLE 11.6–3

The concentration of a light-absorbing ion in solution is doubled.

    a. How does the absorption change?

    b. How must the thickness of the sample be altered to keep the transmission invariant?

Solution

Using Equation 11.6–8, let the transmission through the original and modified materials be represented by

$$\frac{I_1}{I_0} = \exp(-\varepsilon c_1 x_1) \quad\text{and}\quad \frac{I_2}{I_0} = \exp(-\varepsilon c_2 x_2)$$

respectively. Taking natural logarithms gives

$$\ln\left(\frac{I_1}{I_0}\right) = -\varepsilon c_1 x_1 \quad\text{and}\quad \ln\left(\frac{I_2}{I_0}\right) = -\varepsilon c_2 x_2$$

Dividing the second equation by the first gives

$$\frac{\ln(I_2/I_0)}{\ln(I_1/I_0)} = \frac{-\varepsilon c_2 x_2}{-\varepsilon c_1 x_1} = \frac{c_2 x_2}{c_1 x_1}$$

    a. In the first case, $x_2 = x_1$ and $c_2 = 2c_1$, so that

$$\frac{\ln(I_2/I_0)}{\ln(I_1/I_0)} = 2$$

$$\ln(I_2/I_0) = 2\ln(I_1/I_0) = \ln[(I_1/I_0)^2]$$

$$(I_2/I_0) = (I_1/I_0)^2$$

    b. In the second case, $c_2 = 2c_1$ and we must find x_2 such that $I_2 = I_1$. Beginning with the expression

$$\frac{\ln(I_2/I_0)}{\ln(I_1/I_0)} = \frac{c_2 x_2}{c_1 x_1}$$

and substituting appropriate values gives

$$1 = \frac{2x_2}{x_1} \quad\text{or equivalently}\quad x_2 = \frac{x_1}{2}$$

Hence, the thickness must be halved for the absorbed intensity to remain unchanged.

....

A clever application of absorption is photochromic glasses, in which exposure to light causes a glass to change color. The most common photochromic materials are photogray eyeglasses. Optical-quality glass is doped with small silver halide crystals and copper cations. Upon exposure to light, the silver halide crystals dissociate. The copper cations trap the halide anions. The silver ion is a chromophore, absorbing gray light. As the intensity of the light increases, more silver ions are released and the glasses take on a darker shade. As the light intensity decreases, the silver and halide recombine, the silver chromophore is no longer effective, and the gray color disappears.

11.6.4 Scattering and Opacity

We have largely avoided the question of opacity in materials. Materials are transparent if they do not absorb heavily and if they do not scatter light. Conversely, materials are opaque if they absorb heavily or scatter light. Light scattering was briefly touched on in the section on waveguides. The cause of scattering is perhaps most easily described as heterogeneities in refractive index on a scale comparable to that of the wavelength of light. The larger the refractive index difference between the medium and the heterogeneity, the more the scattering that results. This makes perfectly good sense, since scattering is largely reflection from internal surfaces. The effect of an isolated spherical void on the transmission of light through a homogeneous glass is shown in Figure 11.6–3.

Let us consider several materials. A single crystal of quartz (silica) is transparent, since it has a band gap and there are no intrinsic heterogeneities. The same is true for quartz glass. In fact, virtually all oxide and polymer glasses are transparent (in the absence of a chromophore). Now consider polycrystalline oxides and semicrystalline polymers. The refractive index difference between crystal and amorphous regions (grain boundaries in polycrystals) usually leads to heavy scattering in these materials and, hence, opacity. In addition, ceramics may contain significant amounts of porosity at their grain boundaries as a result of powder processing methods (to be discussed in Chapter 16).

An understanding of light-scattering mechanisms led to the development of Lucalox® alumina, which is a polycrystalline material whose grain size and extent of grain boundary porosity are minimized by inclusions of magnesia. Since the concentration of

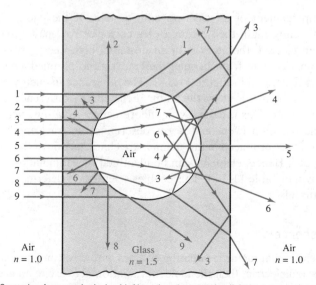

FIGURE 11.6–3   Scattering from a spherical void. Note that the emerging light rays are no longer parallel.

light-scattering defects (grain boundaries and pores) is minimized, the material is essentially transparent. Because of its translucency and high-temperature capability, Lucalox paved the way for high-intensity streetlights.

Let us now turn our attention to polymers, specifically textile fibers. One of the problems with bathing suits first made from synthetic fibers (nylon) was that they became nearly transparent when wet. The fibers lacked sufficient opacity themselves, and the water reduced the air-fiber scattering. To overcome this problem, TiO_2 is now commonly added to textile fibers. Titania has a very high refractive index.

A final example in the polymer area is Ivory® soap. The color of soap is normally amber. Ivory soap is white because of scattering. Ivory soap contains many small voids or pores of air. The refractive index of air is much less than that of the base polymer. Ivory was greeted favorably by the consumer from its initial introduction. White symbolizes purity, and better yet, Ivory floats, whereas other soaps do not.

The same principle can be used to explain why when clear or colored transparent materials are ground to a fine powder, the powder is usually white. Before grinding, the light can enter the material and interact with the atoms of the solid. After grinding, however, the bulk of the light is scattered from the surface of the fine particles.

11.7 ELECTRONIC PROCESSES

Electromagnetic radiation causes a variety of electronic processes in solids. These processes can be loosely characterized as either emission or absorption processes. Many are the basis of important technological innovations such as optical pyrometers and lasers.

We begin by discussing the similarities and differences among luminescence, fluorescence, and phosphorescence. **Luminescence** is defined as light emission in the visible spectrum that results for reasons other than the temperature of the solid. **Fluorescence** is the emission of electromagnetic radiation that occurs within $\sim 10^{-8}$ s of an excitation event. In contrast, **phosphorescence** is the emission of electromagnetic radiation over an extended period of time after the excitation event is over.

11.7.1 X-Ray Fluorescence

Consider the impingement of an electron onto a solid. If the energy of the incident radiation is sufficiently high, then ejection of a core electron may ensue. After a core electron has been ejected, the atom is in an unstable, high-energy state. To lower the energy of the atom, electrons from an outer shell move almost immediately into the vacant core positions, as illustrated in Figure 11.7–1. The difference in energy between the two electron states must be manifest by the atom. If the energy difference is large, then an X ray may be given off. This is called X-ray fluorescence—*fluorescence* referring to the time scale of the emission event and *X ray* describing the wavelength of the emitted photons. X rays that are produced in this manner are characteristic of the atom from which they were produced. Hence, elements can be identified by examination of characteristic radiation, as shown in Table 11.7–1. This enables scientists to determine the composition of unknown materials.

11.7.2 Luminescence

Suppose, once again, that incident radiation collides with electrons surrounding an atom. If the material is a dielectric, then the incident radiation may excite an electron from the valence band to the conduction band. If the magnitude of the band gap is in the appropriate

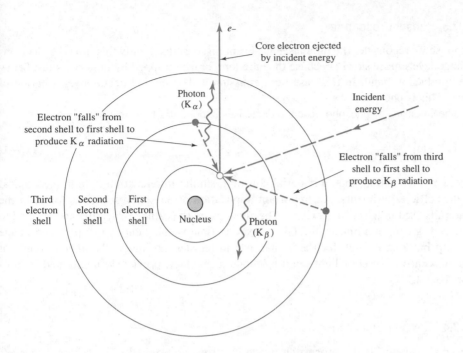

FIGURE 11.7–1

Schematic energy structure
of X-ray fluorescence. Incident radiation knocks an
electron from the core. An
electron from an outer shell
falls into the vacant core
position and releases an X
ray of characteristic wavelength.

TABLE 11.7–1 Characteristic emission
lines.

Element	Wavelength of emission (nm)	
	K_α	K_β
C	4.47	—
N	3.16	—
O	2.36	—
F	1.83	—
Na	1.19	—
Mg	0.99	–
Al	0.834	0.796
Si	0.713	0.675
Cr	0.229	0.209
Fe	0.194	0.176
Co	0.179	0.162
Ni	0.166	0.150
Cu	0.154	0.139

Source: Courtesy of American Physical
Society.

range, then a photon with a wavelength in the visible spectrum may be produced when
electrons fall back from the conduction to the valence band. This effect is called luminescence. If the luminescence occurs rapidly—within 10^{-8} s—the process is still fluorescence. As an example, a fluorescent light exhibits both luminescence and fluorescence,
since the emitted radiation is in the visible spectrum and the emission occurs soon after
the excitation radiation (provided by mercury gas contained within the fluorescent light
tube) is absorbed. The glow of a firefly is an example of bioluminescence.

11.7.3 Phosphorescence

Suppose we reconsider the luminescence event just described, only this time the electrons change levels more slowly because they are temporarily trapped by impurities just below the conduction band. In this case light emission is delayed and occurs over a period of time. This is phosphorescence.

The intensity of phosphorescence decreases exponentially with time:

$$\ln\left(\frac{I}{I_0}\right) = -\frac{t}{\tau} \tag{11.7-1}$$

where τ is relaxation time, a material constant, similar to relaxation time in mechanical or dielectric experiments. An important application of phosphorescence is found in the materials used in television screens. The coatings on the cathode-ray tube are selected to give red, green, and blue light. The relaxation time is short enough to preclude image overlap but long enough for the human eye to register the image. Another example of luminescence is LED or light-emitting diode technology, previously mentioned in Section 10.3.4.

11.7.4 Thermal Emission

Thermal emission is the process in which electrons are excited to higher energy levels using high temperature. Some of the electrons drop back into core positions, releasing photons in the process. Consequently, a continuous spectrum of photons is emitted as the temperature of a material is increased. The minimum wavelength and intensity of the radiation are temperature-dependent. Hence, the color of a material changes with temperature, independent of the composition of the material. The lower the temperature, the longer the wavelength of the radiation. At about 700°C, a material is tinted red; at 1500°C a material appears orange; and finally, at very high temperatures, a sample, like the sun, is "white hot." Ceramists routinely use an optical pyrometer to assess the temperature of ceramics as they are heated in furnaces. This instrument facilitates temperature measurement by matching the color of the sample to that of standards.

11.7.5 Photoconductivity

Photon-induced electrical conductivity, or **photoconductivity,** can result when photons strike the surface of a semiconductor. If the incident light has sufficient energy, then electrons may be promoted (excited) into the conduction band. This results in an increase in the number of charge carriers and a concomitant increase in the conductivity of the material. (Recall from Chapter 10 that $\sigma = Nq\mu$.) The minimum energy required for photoconduction is related to the band gap:

$$E_{\text{electron}} = \frac{hc}{\lambda} > E_g \tag{11.7-2}$$

Note that in photoconductivity a photon generates an electron-hole pair, while in luminescence the recombination of a hole and an electron produces a photon. Thus, photoconductivity is the inverse of luminescence. In older elevators, the doors were triggered by the interruption of a beam of light. The sensor was likely a photoconductive material, perhaps CdS or CdSe, since these materials respond to visible radiation. Another familiar application of photoconductors is as light meters in cameras.

EXAMPLE 11.7–1

Calculate the wavelength of light required to stimulate Ge ($E_g = 0.67$ eV) and ZnO ($E_g = 3.2$ eV) into photoconduction.

Solution

This problem is solved using Equation 10.3–8:

$$\lambda = \frac{(6.62 \times 10^{-34} \text{ J-s})(3 \times 10^8 \text{ m/s})}{(1.6 \times 10^{-19} \text{ J/eV})(E_g \text{ eV})}.$$

$\lambda = 388$ nm for $E_g = 3.2$ eV (ZnO), which is ultraviolet radiation; and $\lambda = 1852$ nm for $E_g = 0.67$ eV (Ge), which is infrared radiation.

A similar phenomenon is the **photovoltaic effect,** which is the key principle behind the operation of a solar cell. In its simplest form a solar cell is just a pn junction connected to an external circuit. The absorbance of photons of a specific wavelength leads not only to the generation of additional charge carriers, as in the case of a photoconductive material, but also to the generation of a voltage across the pn junction. The photo-induced voltage can drive the photo-induced carriers through an external load to perform electrical work.

11.7.6 Application: Lasers

LASER is the acronym for *light amplification by stimulated emission of radiation.* Lasers have become an extremely important tool for a host of applications. Lasers can be used to produce and direct high-intensity heat to specific areas, such as in surgical devices; they produce light that can be used as the carrier for optical signals in waveguides; they can be used in optical instruments such as in a laser Doppler anemometer (used, for example, to measure the speed of dust particles, automobiles, or baseballs) or inter-ferometer (for measuring thickness); and they can also be used for a variety of other purposes.

The solid-state laser is an example of luminescent material in which the light emitted by the fluorescence of an excited atomic cluster stimulates other clusters to emit light in phase. Let us consider the ruby laser, shown in Figure 11.7–2, as a typical example. The heart of the laser is a single crystal of alumina doped with about 0.05 wt. % Cr. The ends of the crystal are polished flat. One end is coated with silver, so that all photons are reflected. The other end is only lightly silvered, so that some of the light is transmitted and some is reflected. Around the laser body is a xenon flashlamp.

To display lasing characteristics, at least three energy levels are required, as shown in Figure 11.7–3. The highest energy level, E_2, and the ground state, E_0, are intrinsic levels associated with the Al_2O_3 single crystal. The intermediate level, E_1, is an extrinsic level

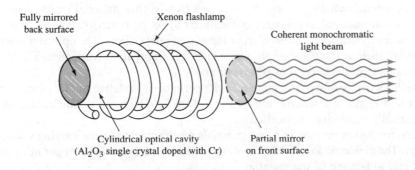

Fully mirrored back surface

Xenon flashlamp

Coherent monochromatic light beam

Cylindrical optical cavity (Al_2O_3 single crystal doped with Cr)

Partial mirror on front surface

FIGURE 11.7–2

Structure of a ruby laser.

FIGURE 11.7–3

Schematic representation of the energy levels of a three-level laser. Electrons are promoted from the ground level E_0 to the excited level E_2 by the action of the flash. The electrons immediately decay to the metastable extrinsic level E_1. Some of the electrons in E_1 fluoresce immediately (back to the ground state), stimulating others to do the same.

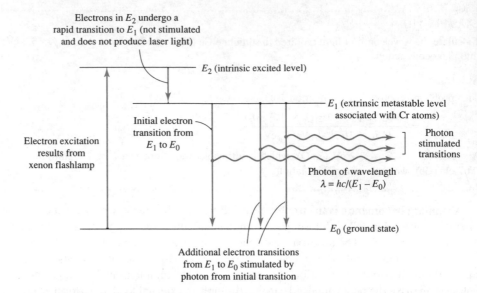

Electrons in E_2 undergo a rapid transition to E_1 (not stimulated and does not produce laser light)

E_2 (intrinsic excited level)

E_1 (extrinsic metastable level associated with Cr atoms)

Photon stimulated transitions

Initial electron transition from E_1 to E_0

Electron excitation results from xenon flashlamp

Photon of wavelength $\lambda = hc/(E_1 - E_0)$

E_0 (ground state)

Additional electron transitions from E_1 to E_0 stimulated by photon from initial transition

associated with the Cr dopant atoms. The lasing action is developed when the xenon lamp is flashed. The energy from the lamp excites electrons from E_0 to E_2. These electrons fall almost immediately into the metastable level E_1, where they are able to remain for an extended period of time. If the electrons fell from the E_1 level to the ground level E_0 in a random fashion, there would be no lasing action, because the light would not be coherent (in phase). Under certain conditions, however, when one electron undergoes the transition from E_1 to E_0, the corresponding photon, of wavelength $\lambda = hc/(E_1 - E_0)$, triggers another electron in the metastable state to make the same transition. That is, one photon "stimulates the emission" of a second photon with the same wavelength and phase. In turn, the second photon stimulates the emission of a third photon, and the chain reaction continues to produce an intense, coherent, monochromatic light beam. The partially silvered mirror permits some of the light to leave the optical cavity while the rest of the light is trapped inside the alumina crystal to sustain the process.

Since the frequency of laser light is a function of the energy difference $\Delta E = E_1 - E_0$, and since ΔE depends on both the host material and the dopant, different materials lase at different frequencies. Lasing materials can be glasses, crystals, semiconductors, or gases. The most inexpensive lasers are the He-Ne (red) lasers used as optical scanners in grocery stores. The most powerful lasers are the CO and CO_2 (infrared) lasers used for heating and a plethora of military applications.

SUMMARY

Electromagnetic radiation of energy hc/λ interacts with a material on the basis of the ability of that material to polarize, as measured by its permittivity. Polarization may be based on electronic, ionic, molecular, or interfacial interactions. The optical properties of a material, such as the wavelength over which it is transparent, depend to a large extent on the dominant mechanism of polarization.

The dielectric constant of a material is its permittivity relative to that of vacuum. The dielectric strength of a material is a measure of the field strength at which a material fails in electrically insulating applications.

Dielectric losses develop in some materials when subject to time-varying voltages and current. The dielectric loss factor, tan δ, is a measure of the power loss per m³. Dielectric losses lead to heating of the material.

The complex coefficient of refraction is an important characteristic of a material. It contains information about the refractive index of a material, which is a measure of the speed of light through the material. It also contains information on the index of absorption. Dispersion is the change in refractive index with wavelength. Dispersion accounts for the splitting of white light into colors as it passes through prisms and other lenses.

The sum of the absorption, reflection, and transmission of a material is 1.0. The amount of light reflected at a surface of a material can be determined by application of Fresnel's formula. Total reflection occurs at all angles of incidence above a certain critical angle, which is determined by the refractive index of the materials on both sides of the interface. Absorption losses can be calculated by using a form of Lambert's law and the absorption coefficient. Transmission through a transparent material is then simply the amount of light that is neither reflected nor absorbed. Scattering is the result of heterogeneities in refractive index in a material. Scattering and absorption must be minimized if light is to travel large distances in materials.

Transducers are often made of piezoelectric materials, in which force on the material produces an electrical voltage. Luminescent materials are materials in which light is produced upon stimulation with electromagnetic radiation. Photoconductivity is the phenomenon occurring when certain materials are exposed to electromagnetic radiation and their electrical resistance decreases substantially. The development of lasers and optical waveguides has revitalized the telecommunications industry.

KEY TERMS

absorption coefficient β	fiber optics	molecular (dipole) polarization	photovoltaic effect
dielectric constant κ	fluorescence	permittivity	piezoelectric effect
dielectric strength	index of absorption k	phosphorescence	polarization P
dispersion	interfacial polarization	photoconductivity	refraction
electronic polarization	ionic (atomic) polarization		refractive index n
electrostriction	luminescence		

HOMEWORK PROBLEMS

1. Some polymers are transparent or translucent. Some ceramics and oxide glasses are transparent or translucent. Why are no metals transparent or translucent?

2. Give an example of a transparent material with high electrical conductivity. Polymer scientists can now synthesize polymers with high electrical conductivity. What color does a polymer become as its conductivity is increased to that of a good conductor?

3. The following data were found in various tables in this text:

SECTION 11.1
Introduction

Material	Resistivity (Ω-cm)
Silicon carbide	10^{-1}
Calcium	2.5×10^5
Boron nitride	10^{+13}
Epoxy	10^{+15}
	Band gap (eV)
Diamond	5.4
Cadmium sulfide	2.42
Tin	0.08

Which of these materials may be considered "dielectrics"?

4. Identify the types of polarization that can occur in the following materials: $-[-CH_2-CF_2-]_n-$, NaCl, Na, single-crystal diamond, PE, and Al_2O_3.

5. Use your knowledge of electronegativity to show which dipole has a larger moment, the CF groups in PVDF, $-[-CHF-CHF-]-$, or the CN group in $-[-CH-CH_2-]-$
$$C\equiv N$$

6. A field is applied to tin and the electronic displacement is 2×10^{-10} nm from the center of the nuclear charge. Calculate the polarization.

7. Copper is placed in a strong electric field and the polarization is measured to be 8×10^{-8} C/m^2. Determine the average separation of charge.

8. Calculate the ionic polarization of MgO, which has the NaCl crystal structure, when an applied electric field causes a displacement of 5×10^{-9} nm between ions.

9. Calculate the percentage increase in separation of ions in ZnS when an applied field causes an ionic polarization of 7×10^{-8} C/m^2.

10. A force of 10^{-2} N is applied to a crystal of $PbZrO_6$ that is $0.1 \times 0.1 \times 0.1$ mm on a side. What voltage results?

11. Suppose positive charge is produced in an ion gun and accelerated toward a molten stream of polypropylene (a good insulator). The ions penetrate the PP and are trapped as the PP cools. Describe the properties of the resulting fiber.

12. A capacitor is made using polycrystalline alumina 0.1 mm thick. What voltage is required to produce a polarization of 10^{-7} C/m^2? Will the dielectric tolerate the voltage?

13. What is the chief source of polarization when polyethylene is placed in an electric field? Calculate the ratio of thickness of mica to polyethylene required to achieve similar polarization in an electric field.

14. Polyacrylonitrile has a higher dipole strength than polyvinylchloride. Which material will have a higher dielectric constant? Why?

15. Derive Equation 11.3–4 using the procedure described below.
   a. Write an expression for the polarization of the dielectric by noting that the net polarization represents the increase in charge per unit area that can be stored because of the presence of the dielectric material.
   b. Use Equations 11.3–1 and 11.3–2 to solve for the quantity Q/A in terms of ε and ξ.
   c. Combine the expressions obtained in parts a and b to obtain the desired result.

16. A simple capacitor with a capacitance of 0.05 μF is required. You must decide between cordierite and polyvinylchloride of equal thickness for the dielectric. The capacitor will not be subject to high fields, but space is limited. What material do you suggest, and how large must the capacitor be?

17. A capacitor is usually not simply two parallel plates with a dielectric material between them, but rather a stack of parallel plates, each of the same area, with the dielectric between successive plates. Show that the capacitance of such an assembly is:
$$C = \frac{\kappa \varepsilon_0 A(n-1)}{d}$$
where n is the number of conducting plates.

18. Table 11.3–1 gives a range of relative permittivity values for mica. For our application we need an accurate value, so a test is conducted. The sample to be used is placed in a 1 kV/m field. The measured polarization is 5.3×10^{-8} C/m^2. Determine κ.

19. Calculate the field strength required to obtain a polarization of 10^{-7} C/m^2 in polyethylene. Will the polyethylene survive the field?

20. The dielectric constant of a new polycrystalline ceramic is measured to be 6.5, 5.5, and 4.5 at 10^2, 10^{11}, and 10^{16} Hz. Justify the variation.

21. What voltage is required to develop a charge of 2.5×10^{-10} C on two capacitor plates 20×20 mm separated by 0.01 mm of (a) vacuum and (b) polytetrafluoroethylene?

22. A spark plug wire carries a voltage upwards of 15,000 V. With what should the conductor be insulated? How is the thickness of the insulator determined?

23. Explain why the dielectric constant of a polar polymer might be a function of temperature. (Hint: Consider the effect of temperature on molecular polarization).

24. Explain why the dielectric constant of polyethylene is insensitive to frequency, whereas that of polyvinylchloride is not.

25. What is the approximate value of κ for Al?

26. Is dielectric strength an instrinsic or extrinsic property? What about relative permittivity? Polarization? What factors reduce the dielectric strength from one sample to another, supposedly identical sample?

27. Which material will generate more heat in a rapidly fluctuating electric field—Bakelite, or nylon? These two materials share the market for distributor caps and rotors. (Note: $W \propto \varepsilon \times \tan \delta$.)

SECTION 11.4
Dissipation and Dielectric
Loss

28. Polystyrene film $25 \times 25 \times 0.01$ mm is the dielectric in a capacitor. Calculate the maximum voltage that can be safely used if the power loss is not to exceed 0.1 W. Assume the field vibrates at 10^6 Hz.

29. Tupperware is not the perfect material for use in a microwave oven if the contents are in whole or part oil. The oil may get hot and melt the polyethylene. What polymer is yet a better choice? To brown the top of a casserole in a microwave, you might place a high absorbing material above the casserole and the radiated heat will brown the top of the casserole. What material might be a good choice for browning?

30. Calculate ε', the imaginary part of the dielectric constant, for polycarbonate (PC), using the data from Tables 11.3–1 and 11.4–1.

31. Determine the refractive index of polypropylene from the observation that the critical angle in air is 42.86°.

SECTION 11.5
Refraction and Reflection

32. Plates of window and silica glass are transparent when viewed through the thickness; however, when viewed on edge, a plate of silica is clear and window glass is green. Explain this observation.

33. Calculate the reflectivity from acrylic plastic (cheap camera lenses) at normal incidence. The refractive index is 1.50.

34. A layered composite is made using polyethylene and a soda-lime-silicate glass. The polyethylene film top layer is 0.2 mm thick and the glass is 1 mm thick. At what angle does the light pass through the polyethylene and glass if the incident beam is (a) 20° or (b) 70° from the normal?

35. The refractive index of poly(ethylene terephthalate) fiber parallel to the fiber axis is 1.660, and the birefringence is 0.090. Calculate the refractive index normal to the fiber axis.

36. The surface of a glass plate is rough on the scale of the incident light wavelength. Use a sketch to show what happens when the beam strikes the surface at glancing incidence. Show what happens when the surface is wet with a liquid of equal refractive index. (Why do car headlights appear brighter when the road is wet?)

37. What property is required of a transparent object such that when you immerse it in water it disappears?

38. Explain why optical waveguides often have a refractive index gradient.

39. We have discussed how wave velocity depends on n; however, not only does v change as a wave enters a new material, but so also does λ. How does the wavelength of the radiation vary with n?

40. Compare the critical angle when looking upward from underwater to that looking into the water from above.

41. Can a fish swimming in the water look up and see you fishing from a boat or from the shore?

42. How much light at normal incidence is reflected from the surface of water? of ice?

43. Show how refractive index can be calculated from a knowledge of the critical angle. This is the operating principle of a refractometer.

44. Optical waveguides must be able to transmit light with minimal losses and be able to resist mechanical failure due to tensile loads. Design a waveguide assembly, containing seven optical fibers, for this application.

45. In optical microscopy, oil immersion lenses are used at very high magnification ($1000\times$). In this technique the region between the objective lens and the sample is filled with oil (rather than air). Show with a sketch why this technique is useful.

46. A camera may consist of upwards of five sets of compound lenses. If the lenses were not made antireflective, estimate the fraction of available light that would not reach the film.

47. You are asked by a friend to help select glass for use in a beveled glass window. Not being familiar with beveled glass windows, you ask your friend why he wants one. He explains that sunlight passing through the window creates beautiful rainbows. What characteristics of the glass are important, and which glass would you select—soda-lime-silicate, high-lead silicate, or polymethylmethacrylate?

48. Why can people who wear glasses see better under water without corrective lenses?

49. How far can a laser beam travel along a waveguide that has a loss of 2 dB/km? The signal is considered lost when the intensity falls to $0.2I_0$.

SECTION 11.6
Absorption, Transmission, and Scattering

50. Show why only electronic polarization determines refractive index in solids.

51. What wavelength light will stimulate photoconduction in Sn, which has a band gap of 0.08 eV?

52. Which of the following materials do you anticipate may be transparent to visible light:

Material	E_g (eV)
Diamond	5.4
ZnS	3.54
CdS	2.42
GaAs	1.35
PbTe	0.25

53. Might any of the materials in Problem 52, with minor modification, be able to lase in the visible? If so, what is the modification, and what will the lasing wavelength be?

54. Given that the linear extinction coefficient is 573 mm^{-1}, estimate the concentration of Fe^{+2} in a Coke bottle.

55. Beer degrades in ultraviolet sunlight. Only one or two brands are sold in clear bottles. Why is beer normally sold in brown or green bottles?

56. Photosensitive sunglasses are made using oxide glass. Might polymers also be made photosensitive? If so, describe the characteristics of the best candidates. What might be the problems with such materials?

57. Can you get sunburned through a car window? Explain your answer.

58. A glass is to be used as a window for a very high power density CO (infrared) laser. What are the required characteristics of the glass?

9. Plot optical transmission as a function of grain size for a silica sample 10 cm on a side. Let the crystal size range from 10 nm to 10 cm in the sample.

60. Glassy silica prepared using the "sol-gel" processing method is only 10% of the theoretical mass density of glass. The material is highly porous, yet it is highly transparent. Characterize the pore structure.

61. When used at a level of 0.001 g/cm^3 in a host material, a chromophore absorbs 10% of incident light. How much light will be absorbed when the concentration of the chromophore is doubled?

62. Why is snow white?

63. Why are there no K$_\beta$ values for the first few entries in Table 11.7–1?

64. Suggest reasons that a black (UV) light makes some white fabrics and paper glow as long as the stimulus is present.

65. Most household mirrors are glass silvered on the back; however, most laser applications use front-surface mirrors. Explain this observation.

66. A laser shines on a prism. What happens? An LED shines on a prism. What happens?

67. The intensity of a phosphor decreases 50% in 10 s. How long before the intensity has decreased to ~0.37 of its original value?

68. Describe the properties of a semiconductor necessary to make a green LED. Also, a red LED, and a blue LED.

69. Describe a suitable detector for the receiving end of a laser signal traveling along a waveguide.

70. What properties of a laser are utilized in the scanning units in the grocery store?

71. When ZnS is doped with Cl$^-$, the emitted radiation is violet. When ZnS is doped with Cu^{2+}, the radiation is red. Use a band diagram for the doped and undoped ZnS to explain the observations.

72. Optical pyrometers can be used to measure high temperature noninvasively. Thermocouples can be used to measure moderate temperature, say, 200°C, easily. When the 200°C environment is in an atmosphere of rapidly moving air, then the errors induced by convective heat loss at the thermocouple tip preclude simple use of a thermocouple. In this application, UV sensors can be used. What is their operating principle?

SECTION 11.7
Electronic Processes

MAGNETIC PROPERTIES

MATERIALS IN ACTION   Magnetic Properties

Magnets are used in many applications involving motors, power generation, tape recording, and data storage. Strong magnets are also the basis of transportation systems in which the vehicles are levitated above a track. The earliest known magnets were probably naturally occurring permanent magnets of the Fe_3O_4 type and were probably more of a curiosity.

There has been an accelerating increase in available magnetic field strength since the beginning of the 20th century. In the early part of the century, most magnets were made from steels and were used in power generation and transformers. Only small improvements were made up to the early 1950s, at which time Alnico (AlNiCo) alloys were discovered. Advances continued to be made through the discovery of complex compounds, especially ceramic compounds. Field strengths well over 100 times those obtainable in steels are now commonly available. Hard ferrites (i.e., compounds such as $PbFe_{12}O_{10}$ with strong permanent magnetism) are used in small, battery-driven motors which have replaced copper-wound DC motors. Motors employing magnets made from these ceramics are used in electric toothbrushes, windshield wipers, power windows, and a myriad of other such devices.

Another important use of magnetic materials is for data storage (i.e., memory) in computers. As you know, computers operate on a binary system of zeros and ones. These two values may be very conveniently represented by two distinct magnetic states that may easily be changed by changes in an applied magnetic field. Such readily reversible magnetic characteristics are available in the so-called square-loop magnetic ceramics.

Closely allied to magnetic properties, at least in terms of applications, is superconductivity. Superconductivity refers to the flow of electricity in a body without electrical resistance. It was discovered by the Dutch scientist Heike Kamerlingh Onnes in 1911; he achieved it by cooling Hg to 4.2 K. Through the mid-1980s, progress in developing superconducting materials was slow; the highest temperature at which superconductivity was documented was 23 K. It is difficult and costly to cool to this temperature, and so superconductivity had very limited applications. In 1987 the situation changed dramatically when Paul Chu and co-workers at the University of Houston announced that they observed superconductivity at 92 K in a compound of $YBa_2Cu_3O_{7-x}$. This was especially significant because liquid nitrogen could be used to cool a material to the superconducting state, as opposed to liquid helium, which is very expensive and impractical in an engineering sense.

One potential application of superconductivity (among many) would be efficient levitation of ground transportation vehicles such as trains. When a superconductor is cooled into the superconducting regime, it has the property of excluding a magnetic field. This is called the Meissner effect and is very useful, since any magnet in the immediate region is repelled from the superconductor. Thus, if suitable magnets are attached to train cars and the track is superconducting, the cars can travel suspended above the track. Such applications require strong permanent magnets.

In this chapter we will study the physical basis of magnetism, relate it to the structure of a wide variety of materials, and discuss some important applications.

12.1 INTRODUCTION

The previous chapter showed that the optical and dielectric properties of materials result from the interaction between electromagnetic radiation and atomic scale dipoles. This chapter extends the concept of polarization to describe the magnetic properties of materials. We will show that atomic scale magnetic dipoles result from the motion of electrons. The details of the interaction between these internal magnetic dipoles and external magnetic fields determine the magnetic characteristics of a material.

Long before reading this book most of you knew something about mechanical properties (e.g., it is easier to break window glass than metals or plastics), electrical properties (e.g., metals conduct while plastics insulate), and perhaps optical properties (e.g., a prism converts white light into colors). For many of us, however, magnetism is more mysterious. Can you list three engineering applications that depend on the magnetic properties of materials? It is not uncommon for students to answer no to this question. Therefore, we begin our discussion of the magnetic properties by describing why engineers need to learn about magnetism.

We first offer a qualitative appreciation for magnetic behavior. We then lay the quantitative groundwork for the chapter by reviewing the physical basis for magnetism. We will find that the equations necessary to describe the response of a material to an external magnetic field are similar to those developed previously for an electric field. For example, the magnetic analog to the dielectric constant κ is the magnetic susceptibility χ. Based on the magnitude and sign of the susceptibility, materials can be classified as diamagnetic, paramagnetic, ferromagnetic, antiferromagnetic, or ferrimagnetic. After a brief description of each class of magnetic materials, we discuss some of the engineering devices and applications resulting from the magnetic properties of materials. Examples include permanent magnets, electrical transformer cores, computer memory devices, and superconducting magnets.

12.2 MATERIALS AND MAGNETISM

For many of us the word *magnet* brings to mind the image of a bar or horseshoe-shaped piece of metal with the ability to attract small objects over limited distances. The ability to attract and hold certain objects is one of the properties of magnetic materials exploited by engineers to create useful devices. Perhaps you have seen old cars being moved around a junkyard by a crane suspending a large magnet from a cable, or maybe you have witnessed the use of magnets to separate "tin" cans from waste streams containing paper, plastic, and other solids at a recycling facility. What materials are used to fabricate magnets? Why is it that only certain materials are affected significantly by magnetic fields?

Recall that electric dipoles have two poles—a positive and a negative pole. Similarly, magnets have two poles—a north and a south pole. In both cases, like poles repel and unlike poles attract. Since the earth is itself a huge magnet, its magnetic north pole attracts the south pole of, for example, a small needle-shaped magnet floating in a dense liquid. This is, of course, the principle on which a compass operates. What are the atomic mechanisms associated with magnetic dipole formation?

The mutual repulsion of like magnetic poles is also one of the principles of electric motors. Understanding the conversion of electricity to mechanical energy (a motor) or the conversion of mechanical energy to electricity (a generator) requires familiarity with Maxwell's equations. Although we will not discuss Maxwell's equations in this text, it is important for you to recognize the intimate relationship between electric and magnetic fields. To help you understand this relationship, and to prepare you to answer the questions posed above, we will develop a limited quantitative model for magnetism in the next section.

12.3 PHYSICAL BASIS OF MAGNETISM

The motion of an electric charge results in the production of a magnetic field. For example, an electric current loop of area A and current I produces a magnetic dipole with strength

$$\mathbf{m} = IA \tag{12.3-1}$$

where $\mathbf{m}$ is a vector perpendicular to the loop. The direction of $\mathbf{m}$ can be determined by what is known as the "right hand rule." If the fingers of the right hand point in the direction of current flow, then the thumb of the right hand points in the direction of $\mathbf{m}$. By convention the direction of current flow is opposite to that of the motion of electrons.

The physical basis of the magnetic properties of engineering materials is the current loops produced by the motion of electrons. As shown in Figure 12.3–1, the two relevant types of electron motion are: (1) the orbital motion of electrons around the atomic nucleus, and (2) the spin of an electron about its own axis of rotation. The interaction of the electron-induced magnetic moments with an externally applied magnetic field results in the macroscopic magnetic properties of a material. Thus, like all the other material properties discussed in this book, macroscopic magnetic properties depend on the atomic and microstructural characteristics of a material.

An external magnetic field can be described by either of two parallel vectors, the **magnetic induction B**, or the **magnetic field strength H**. In a vacuum, the quantities **B** and **H** are related through the equation:

$$\mathbf{B} = \mu_0 \mathbf{H} \tag{12.3-2}$$

where μ_0 is the permeability of a vacuum. The SI unit for magnetic induction $\mathbf{B}$ is the tesla (T). The units for $\mathbf{H}$ are A/m, and those for permeability are T-m/A. The magnitude of μ_0 is $4\pi \times 10^{-7}$ T-m/A. [Note: The tesla can be defined in terms of units you are more familiar with, as $1\ T = 1\ (V\text{-}s)/m^2$. To give you a feeling for the magnitude of a tesla, the magnetic induction of the earth is $\sim 6 \times 10^{-5}$ T and a typical bar magnet has an induction of ~ 1 T.]

A material becomes magnetized when it is placed in an external magnetic field of strength $\mathbf{H}$. The extent of this magnetization can be quantified by the **magnetic vector M**, which represents the induced dipole moment per unit volume within the material. The total magnetic induction is the sum of the contributions from the external magnetic field ($\mu_0\mathbf{H}$) and the internal response of the material ($\mu_0\mathbf{M}$). That is:

$$\mathbf{B} = \mu_0\mathbf{H} + \mu_0\mathbf{M} \tag{12.3-3}$$

Figure 12.3–2 illustrates this summation concept schematically.

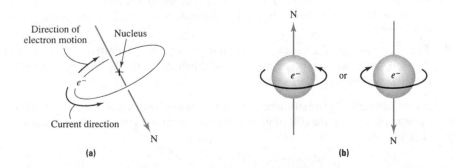

(a)   (b)

FIGURE 12.3–1

An electric current loop produces a magnetic dipole. Two types of current loops exist in atoms: **(a)** the orbital motion of an electron around its nucleus, and **(b)** the spin of an electron around its axis of rotation.

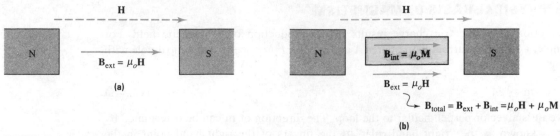

FIGURE 12.3-2   A comparison of total magnetic induction resulting from a field of strength **H**: **(a)** when a vacuum exists between the poles of the magnet, and **(b)** when a material is placed in the field. In the latter case the total induction is the sum of the contributions from the external and internal fields.

Since the magnetic vector **M**, also known as the magnetization, is induced by the external magnetic field **H**, it is reasonable to assume that **M** is proportional to **H**. That is:

$$\mathbf{M} = \chi \mathbf{H} \qquad\qquad (12.3\text{--}4)$$

where χ is known as the **magnetic susceptibility.** By combining the previous two equations, one obtains an alternative expression for **B** of the form:

$$\mathbf{B} = \mu_0(1 + \chi)\mathbf{H} = \mu\mathbf{H} \qquad \text{with} \qquad \mu = \mu_0(1 + \chi) \qquad (12.3\text{--}5)$$

where μ is the **permeability** of the material. The ratio

$$\mu_r = \frac{\mu}{\mu_0} = 1 + \chi \qquad\qquad (12.3\text{--}6)$$

is known as the **relative permeability** of the material.

At this point it is useful to recognize the similarities between the magnetic parameters **B**, μ_0, χ, and **H** and the corresponding dielectric parameters P, ε_0, κ, and ξ. Equations 11.3–4 and 11.3–5 gave the relationship among the latter group of parameters. Solving these expressions for P yields:

$$P = \varepsilon_0(\kappa - 1)\xi \qquad\qquad (12.3\text{--}7)$$

Comparison of Equations 12.3–5 and 12.3–7 shows that:

1. The magnetic field strength **H**, with units of A/m, is comparable to the electric field strength ξ, with units of V/m. Both are measures of the external driving force for polarization.

2. The magnetic induction **B**, with units of T $= $ (V-s)/m^2, is analogous to the dielectric polarization P, with units of C/m^2. Both are measures of the response of the material to the external field.

3. The magnetic susceptibility χ is similar to the dielectric constant κ. Both are dimensionless material properties that describe a material's response to the external field.

4. The permeability of a vacuum μ_0 is analogous to the permittivity of a vacuum ε_0. Both are reference points used to establish the scale for the corresponding material parameters (i.e., χ and κ).

The striking similarity between these parameters is not surprising, since both dielectric and magnetic properties are determined by the interaction between internal dipoles and external fields.

In fact, the magnetic, dielectric, and optical properties of materials are more than similar—they are mathematically related through the Maxwell equations. The refractive index n is related to the relative magnetic permeability μ_r and the dielectric constant κ through the equation

$$n = \sqrt{\mu_r \kappa} \tag{12.3–8}$$

While this equation is important theoretically, it is difficult to use in practice, since all the quantities are functions of the frequency of the external field.

EXAMPLE 12.3–1

Calculate the magnetic field strength required to create an induction equal to that of the earth in an aluminum sample. The magnetic susceptibility of Al is 16.5×10^{-6}.

Solution

Solving Equation 12.3–5 for field strength yields:

$$\mathbf{H} = \frac{\mathbf{B}}{\mu_0(1 + \chi)}$$

Substituting the value for χ given in the problem statement and using the values for μ_0 and the earth's induction given in the text yields:

$$\mathbf{H} = \frac{6 \times 10^{-5}\ \text{T}}{[4\pi \times 10^{-7}\ \text{(T-m)/A}][1 + (16.5 \times 10^{-6})]}$$

$$= 50\ \text{A/m}$$

12.4 CLASSIFICATION OF MAGNETIC MATERIALS

Based on the extent and nature of the interaction between electrons in the solid and an external magnetic field, it is possible to group materials into five classes. Three of the classes of magnetic materials—paramagnetic, diamagnetic, and antiferromagnetic solids—show almost no response to external magnetic fields. In contrast, ferromagnetic and ferrimagnetic materials interact strongly with external magnetic fields and are used in a variety of magnetic applications, including electrical transformers, information storage devices (magnetic tapes and computer disks), motors and generators, and loudspeakers.

The property that quantitatively describes the material's response to an external field is its magnetic susceptibility. If $|\chi| \ll 1$, the material has a weak response to the external field; if $|\chi| \gg 1$, the material has a strong magnetic response.

12.5 DIAMAGNETISM AND PARAMAGNETISM

Diamagnetism and **paramagnetism** are both weak forms of interaction between solids and external magnetic fields. In diamagnetic solids the internal magnetic field is antiparallel to the external field, while in paramagnetic solids the internal and external fields point in the same direction. The orbital motion of an electron around its nucleus *always* results in a contribution to the diamagnetic response of a material. The electron's spin, however, *may* lead to a paramagnetic contribution.

What factors determine whether a material will be dia- or paramagnetic? Diamagnetism is usually observed in solids composed of atoms with completely filled electron shells. The reason is that in a filled shell there are equal numbers of electrons with positive

FIGURE 12.5–1

Distribution of electron
spins in the 3d shells of
some of the transition
metals.

Magnetic moment	Element	Number of electrons	Electronic structure 3d shell					4s electrons
1	Sc	21	↑					2
2	Ti	22	↑	↑				2
3	V	23	↑	↑	↑			2
5	Cr	24	↑	↑	↑	↑	↑	1
5	Mn	25	↑	↑	↑	↑	↑	2
4	Fe	26	↑ ↓	↑	↑	↑	↑	2
3	Co	27	↑ ↓	↑ ↓	↑	↑	↑	2
2	Ni	28	↑ ↓	↑ ↓	↑ ↓	↑	↑	2
0	Cu	29	↑ ↓	↑ ↓	↑ ↓	↑ ↓	↑ ↓	1

and negative spins, so that the total magnetic moment (from spin) is zero. Examples of diamagnetic materials include most ionic and covalent crystals, almost all organic compounds including polymers, and some metals, notably Cu, Ag, and Au. The interaction between a diamagnetic solid and an external field is very weak. In fact, the magnitude of χ for most diamagnetic solids is on the order of 10^{-4} to 10^{-5}. As such, diamagnetic materials find few magnetic applications.

Paramagnetism is related to the magnetic moment resulting from unpaired electrons in unfilled inner electron shells (unpaired electrons in the valence shell do not contribute to paramagnetic behavior). Thus, most transition metals are paramagnetic. Figure 12.5–1 shows the distribution of electron spins in the 3d shells of some transition metals. The number of unpaired spins per atom can be found using Hund's rule, which states that in an unfilled shell the number of unpaired spins will be as large as possible within the constraints of the Pauli exclusion principle. As with diamagnetic solids, the strength of the interaction between a paramagnetic material and an external magnetic field is limited (χ is on the order of 10^{-2} to 10^{-3}). As a result paramagnetic materials find few magnetic applications. A comparison of Figure 12.5–1 and Table 12.5–1 shows the relationship between electronic structure and magnetic behavior. For example, paramagnetic Cr has five unpaired inner shell electrons, while diamagnetic Cu has none.

EXAMPLE 12.5–1

Compare the magnetic field strength required to produce an induction of 2 T in a vacuum, in Al_2O_3, and in Al.

Solution

Solving Equation 12.3–5 for magnetic field strength gives: $\mathbf{H} = \mathbf{B}/[\mu_0(1 + \chi)]$. From Table 12.5–1, $\chi(Al_2O_3) = -37 \times 10^{-6}$ and $\chi(Al) = 16.5 \times 10^{-6}$. Therefore,

$$\mathbf{H}(Al_2O_3) = \frac{2\ T}{[4\pi \times 10^{-7}\ (\text{T-m})/\text{A}][1 - (37 \times 10^{-6})]}$$

$$\mathbf{H}(Al_2O_3) = 1.59161 \times 10^6\ \text{A/m}$$

Similarly $\mathbf{H}(Al) = 1.59152 \times 10^6$ A/m and $\mathbf{H}(\text{vac}) = 1.59155 \times 10^6$ A/m. The values for Al_2O_3 and Al differ from that for the vacuum by less than 0.006%. From an engineering point of view, the presence of either Al_2O_3 or Al makes no difference in the calculation (i.e., their contribution to magnetic induction is negligible). This is why neither dia- nor paramagnetic materials have important magnetic applications.

TABLE 12.5–1 Magnetic susceptibilities χ for several paramagnetic and diamagnetic materials.

Material	$\chi (\times 10^{-6})$	Class	Comment
Al	+16.5	para-	An example of a paramagnetic metal.
Al_2O_3	−37.0	dia-	Oxides generally have ionic and/or covalent bonds.
Be	−9.0	dia-	An example of a diamagnetic metal.
BeO	−11.9	dia-	
Bi	−280.1	dia-	
B	−6.7	dia-	
Ca	+40.0	para-	
CaO	−15.0	dia-	
CaF_2	−28.0	dia-	
C (diamond)	−5.9	dia-	
C (graphite)	−6.0	dia-	Note the influence of crystal structure.
Ce	+2450	para-	Ce is a transition metal.
CeO_2	+26.0	para-	Not all oxides are diamagnetic.
Cr	+180	para-	
Cr_2O_3	+1965	para-	
Cu	−5.5	dia-	
CuCl	−40.0	dia-	
$CuCl_2$	+1080	para-	Note the influence of the valence of Cu.
Ge	−76.8	dia-	
Au	−28.0	dia-	
Pb	−23.0	dia-	
Li	+14.2	para-	
LiF	−10.1	dia-	
Mg	+13.1	para-	
MgO	−10.2	dia-	
Mn (α-phase)	+529	para-	
Mn (β-phase)	+483	para-	Note influence of crystal structure.
Si	−3.9	dia-	
Ag	−19.5	dia-	
Na	+16.0	para-	
NaCl	−30.3	dia-	
Sn (gray)	−37.0	dia-	
Sn (white)	+3.1	para-	Influence of crystal structure and bond type.
Ti	+153.0	para-	
TiO_2	+5.9	para-	
H_2O (gas)	−13.1	dia-	
H_2O (liquid)	−12.9	dia-	Note influence of phase.
H_2O (solid)	−12.7	dia-	

Source: Adopted from *Handbook of Chemistry and Physics, 61st ed.* Copyright CRC Press, Boca Raton, FL.

12.6 FERROMAGNETISM

Ferromagnetic solids display magnetic susceptibility values much greater than 1. The primary difference between para- and ferromagnetic materials is in the strength of the interaction between adjacent atomic magnetic dipoles. While the dipoles are essentially independent in paramagnetic materials, they interact strongly in ferromagnetic materials. At low temperatures the exchange interaction between adjacent atomic magnetic dipoles (associated with unpaired spins) in ferromagnetic solids is strong enough to overcome the thermal fluctuations attempting to randomize the orientation of the magnetic dipoles. The

TABLE 12.6–1 Curie temperatures for selected ferromagnetic elements and alloys.

Material	Curie temperature (K)	Comments
Fe	1063	Fe, Co, and Ni are the classic examples with partially filled 3d shells.
Co	1390	
Ni	627	
Dy	85	
Gd	294	Dy, Gd, Tb, and Ho are the materials with partially filled 4f shells.
Tb	210	
Ho	20	

result is that even without an external field, neighboring dipoles align with each other. (Note the similarity between ferromagnetism and ferroelectricity, as discussed in Section 11.2.6.)

Why do some materials with partially filled inner electron shells exhibit weak dipole interactions (paramagnetism) while others show strong dipole interactions (ferromagnetism)? Detailed calculations show that the requirements for ferromagnetism are:

1. The inner electron shell is unfilled.
2. The unfilled shell must have a small radius.
3. The electron energy band associated with the unfilled shell must be narrow.

These criteria are satisfied by the 3d band in Fe, Ni, and Co and the 4f band in several rare earth metals. In addition, the hybrid band structures in certain compounds can occasionally satisfy the requirements. Table 12.6–1 lists some ferromagnetic materials.

12.6.1 Magnetic Domains

The alignment of neighboring dipoles described above does not extend throughout the sample. Instead, it is limited to finite microscopic volumes of material known as **magnetic domains.** The domains are separated by boundaries known as Bloch walls. Bloch walls are analogous to grain boundaries in that they represent a disordered transition region between two highly ordered regions. All the dipoles within a domain are aligned. In the absence of an external field, the direction of alignment varies from domain to domain, resulting in zero net magnetization for the macroscopic sample.

One might reasonably wonder why magnetic domains exist. We need only consider the alternative: all dipoles in the sample aligned in the same direction as shown in Figure 12.6–1a. In this case, since the lines of magnetic field force must close, an external magnetic field develops to balance the internal field. A significant amount of energy is associated with the external field. As shown in Figure 12.6–1b through e, the extent of the external field, and hence the total energy of the system, can be reduced by the development of multiple magnetic domains within the sample.

There is, however, a price to pay for the formation of magnetic domains. When a ferromagnetic crystal is magnetized, its length increases in the magnetization direction; that is, there is a magnetically induced strain. If two adjacent domains have different dipole orientation, then a mismatch in strain across the domain wall develops. The elastic deformation necessary to accommodate this strain mismatch produces an increase in the energy of the system. The balance between this energy increase—known as the magnetostrictive energy—and the energy decrease associated with the minimization of the external magnetic field determines the lowest-energy domain structure of a material.

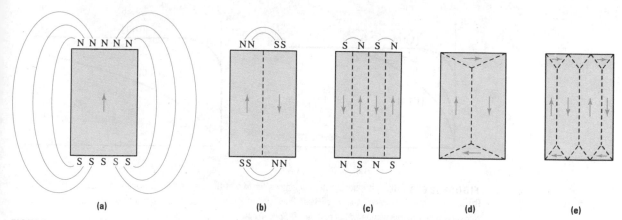

(a)   (b)   (c)   (d)   (e)

FIGURE 12.6–1   Magnetic domain structures: **(a)** all the dipoles in the sample are aligned in the same direction, resulting in the generation of external lines of force necessary for closure; **(b–e)** the extent of the external field can be reduced by the development of multiple magnetic domains within the sample.   (*Source: L. L. Hench and J. K. West*, Principles of Electronic Ceramics, *Copyright © 1990 by John Wiley & Sons. Reprinted by permission of John Wiley & Sons, Inc.*)

Figure 12.6–1 is a schematic illustration of magnetic domains and external field lines. The same concepts can be illustrated with photographs using a technique developed by Francis Bitter. As shown in Figure 12.6–2, when magnetic filings are sprinkled in the region between two magnetic poles, the filings become aligned with the external magnetic force lines.

As discussed in Section 3.11, most crystals are anisotropic. One manifestation of crystalline anisotropy is that the ease of magnetization is a function of crystallographic direction. Directions in which magnetization is comparatively easy are called magnetically soft directions, while those for which magnetization is more difficult are termed hard directions. The soft magnetization directions for the common ferromagnetic metals are shown in Figure 12.6–3. The magnetic dipole orientation within a domain usually corresponds to one of the crystallographic soft magnetization directions in the absence of an external magnetic field.

FIGURE 12.6–2   Bitter patterns in the region between two magnetic poles.   (*Source: © Richard Megna, Fundamental Photographs.*)

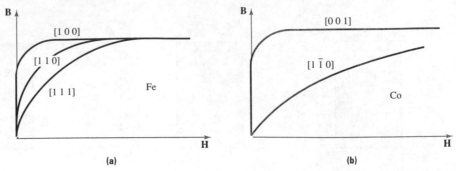

FIGURE 12.6–3 The soft magnetization directions for two of the common ferromagnetic elements: **(a)** Fe and **(b)** Co. (*Source: L. L. Hench and J. K. West,* Principles of Electronic Ceramics, *Copyright © 1990 by John Wiley & Sons. Reprinted by permission of John Wiley & Sons, Inc.*)

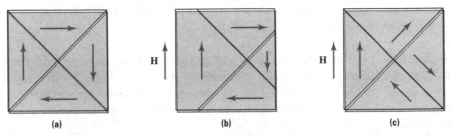

FIGURE 12.6–4 Several aspects of the magnetic domain structure in ferromagnetic materials. **(a)** All the dipoles within a domain are aligned, and when no external field is present the direction of alignment varies from domain to domain, resulting in no net magnetization for the macroscopic sample. **(b)** When an external field is applied the favorably oriented domains can "grow" at the expense of those that are less favorably oriented. At higher external field strengths **(c)**, the aligned dipoles within a domain can rotate in such a way as to align their moments more closely with the applied field.

An applied external field increases the net magnetic moment in the direction of the field. This can occur in one of two ways. Either the favorably oriented domains can "grow" at the expense of those that are less favorably oriented, or the dipoles within a domain can rotate to become more closely aligned with the applied field. These two mechanisms are shown schematically in Figure 12.6–4. The ratio of the energy required for magnetization in the hard and soft directions is termed the magnetic anisotropy ratio. The higher the magnitude of this ratio, the more difficult it is to rotate dipoles within a domain.

12.6.2 Response of Ferromagnetic Materials to External Fields

The response of a ferromagnetic material to an external field is more complex than that of either dia- or paramagnetic materials. Figure 12.6–5 shows the change in the magnetic induction **B** as a function of the applied magnetic field strength **H** for a sample that has not been magnetized previously. At point 1, the sample is in its virgin state with its domains randomly oriented to produce no net magnetization. From point 1 to point 3 the favorably oriented domains grow at the expense of the unaligned domains. The driving force for the Bloch wall motion is the reduction in the free energy of the system. The aligned domains have a lower free energy than misaligned domains. The domain growth stage can be divided into two regions. From point 1 to point 2 the growth is reversible,while from point 2 to point 3 the Bloch wall motion is irreversible.

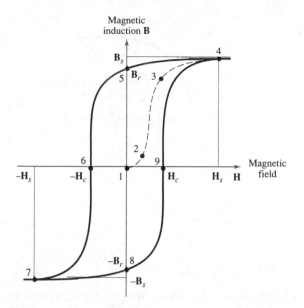

FIGURE 12.6–5

The change in the magnetic induction **B** as a function of the applied magnetic field strength **H** for a sample that has not been magnetized previously. The solid curve is known as the ferromagnetic hysteresis loop.

Once the unfavorably oriented domains are consumed, additional increases in **B** occur by dipole rotation within domains away from the directions of easy magnetization toward the direction of the external field (point 3 to point 4). Eventually, a saturation induction, B_s, is reached. At this point the dipoles in all domains are aligned with the field. The minimum field required to obtain saturation induction is termed the saturation field, H_s.

When the field is gradually reduced to zero, the induction does not return to zero along the path 4–3–2–1. Such a path would require significant amounts of dipole rotation and irreversible Bloch wall motion. Instead, the induction decreases only slightly, along the path 4–5. The decrease in **B** occurs because of the recovery of the reversible Bloch wall motion and dipole rotation away from the direction of magnetization back to a direction of easy magnetization. Even after the external field has been reduced to zero, some induction remains. This residual induction, at **H** = 0, is known as the remanent induction, or simply the remanence, B_r.

When the magnitude of the applied field in the reverse direction increases, the processes of favored domain growth and dipole rotation begin again. Note that an external field of strength $-H_c$, called the coercive field, is needed to obtain a microstructure with an equal volume fraction of domains aligned parallel and antiparallel to the external field (i.e., **B** = 0). Further increases in the field strength result in movement along path 6–7 until a saturation induction, $-B_s$, is obtained for a saturation field strength of $-H_s$. The path 7–8–9–4 is the complementary path of 4–5–6–7.

At this point the careful reader may have noted an apparent contradiction between the information in Figure 12.6–5 and Equation 12.3–5. The equation states that for a material with constant χ, **B** must be directly (i.e., linearly) related to **H**. Examination of Figure 12.6–5, however, shows that for a ferromagnetic material, **B** is not proportional to **H**. We must therefore conclude either that Equation 12.3–5 does not apply to ferromagnetic materials or that χ is not constant for these materials. The latter conclusion is correct: for ferromagnetic solids, χ is a function of **H**. Thus, you will not find a table of χ values for ferromagnetic materials in this text. The statement that $\chi \gg 1$ for ferromagnetic materials describes the initial slope of the **B**-**H** curve (i.e., the slope of the tangent to the curve at point 1 in Figure 12.6–5).

FIGURE 12.6-6

The differences in the shape and size of the hysteresis loops for "hard" and "soft" magnets.

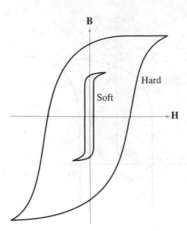

12.6.3 The Shape of the Hysteresis Loop

The solid curve in Figure 12.6–5 is referred to as the ferromagnetic hysteresis loop. The shape of this hysteresis loop, and more specifically the magnitude of the area enclosed within the loop, is an important characteristic of ferromagnetic materials. Figure 9.2–17 showed that the area under the stress-strain curve represents the work done during deformation. Similarly, the area within the ferromagnetic hysteresis loop represents the irreversible work lost during one cycle of the applied external magnetic field.

Materials with high values of $\mathbf{B}_s$, $\mathbf{B}_r$, $\mathbf{H}_s$, and $\mathbf{H}_c$ have a significant area enclosed within their hysteresis loops. Such materials are classified as **hard magnets.** Conversely, materials with low values of these four parameters are **soft magnets.** The hysteresis loops for hard and soft magnets are compared qualitatively in Figure 12.6–6.

The shape of the hysteresis loop is often the most important factor in the selection of a ferromagnetic material for a particular application. As will be discussed in more detail in Section 12.8, soft magnets are used when alternating electromagnetic fields are present and energy losses must be minimized (e.g., in electrical transformer cores), while hard magnets are employed as permanent magnets.

....

EXAMPLE 12.6–1

Demonstrate that the area enclosed within the magnetic hysteresis loop represents the work per unit volume done by the external field on the ferromagnetic material during one magnetization cycle.

Solution

The area within the hysteresis loop can be expressed as

$$\text{Area} = \int \mathbf{B} \, d\mathbf{H}$$

where the integration is carried out over the closed loop. Since the units for $\mathbf{B}$ are T and for $\mathbf{H}$ are A/m, the units for the integral are (T-A)/m. Using the definition of the tesla given after Equation 12.3–2, we see that (T-A)/m corresponds to $[(\text{V-s})/\text{m}^2](\text{A/m})$. Next, we note that a volt-ampere is a watt and a watt-second is a joule. Therefore, $[(\text{V-s})/\text{m}^2](\text{A/m}) = (\text{W-s})/\text{m}^3 = \text{J/m}^3$. The area within the magnetic hysteresis loop does indeed have units of energy per unit volume.

....

12.6.4 Microstructural Effects

The magnitudes of the magnetic remanence $\mathbf{B}_r$ and coercive field $\mathbf{H}_c$ are inversely related to the ease with which domain boundaries are able to move through the crystal lattice. Thus, it is possible to alter the magnitudes of $\mathbf{B}_r$ and $\mathbf{H}_c$, and hence the relative "hardness" of a ferromagnetic material, by tailoring the microstructure to influence the mobility of domain walls.

Some of the microstructural features that restrict domain wall motion are grain or phase boundaries and dislocations. Therefore, for a given alloy, the "softest" magnetic structure is obtained from a single-phase, fully annealed microstructure. Conversely, promoting a fine dispersion of a second phase, reducing the grain size, and cold-working the material increases the magnetic "hardness" of a crystal.

Is the electrical conductivity of a crystalline solid more likely to increase or decrease as its microstructure is modified to maximize its magnetic "hardness"? Magnetic hardness is maximized by tailoring the microstructure to minimize domain wall motion. This is accomplished by inserting obstacles such as dislocations and grain or phase boundaries. Since these factors also impede the motion of electrical charge carriers, magnetically hard microstructures are likely to display relatively low electron mobilities and, therefore, relatively low electrical conductivities.

12.6.5 Temperature Effects

Since ferromagnetism is related to the ability of interacting atomic magnetic dipoles to overcome thermal fluctuation, we expect an increase in temperature to decrease the relative ability of the dipoles to align. Thus, the saturation magnetization is a decreasing function of temperature. In fact, above a critical temperature, known as the **Curie temperature,** ferromagnetic materials become paramagnetic.[1] (See Table 12.6–1.) The Curie temperature is an important material property, since it effectively defines the maximum service temperature for a ferromagnetic material.

12.6.6 Estimating the Magnitude of M

Recall that M represents the induced dipole moment per unit volume within the solid. In ferromagnetic materials the maximum saturation magnetization M_s occurs when all the atoms have their magnetic moments, associated with the spin of unpaired inner shell electrons, aligned. Therefore,

$$M_s = N_v N_s \mu_B \tag{12.6--1}$$

where N_v is the number of atoms per unit volume, N_s is the number of unpaired spins per atom, and μ_B is the magnetic moment associated with the spin of an electron. The magnitude of μ_B is given by:

$$|\mu_B| = \frac{qh}{4\pi m_e} = 9.27 \times 10^{-24} \text{ A-m}^2 \tag{12.6--2}$$

where q is the charge of an electron, h is Planck's constant, and m_e is the mass of an electron. Just as q is the unit of electrical charge, the quantity above, known as the Bohr magneton (μ_B), is the unit of magnetic moment. N_s is found using Hund's rule, described in Section 12.5.

---

[1] Note the similarity between the ferromagnetic Curie temperature and the ferroelectric Curie temperature defined in Section 11.2.6.

EXAMPLE 12.6–2

Estimate the maximum saturation magnetization and saturation induction for Fe.

Solution

M_s can be estimated using Equation 12.6–1:

$$M_s = N_c N_s \mu_B$$

where $\mu_B = 9.27 \times 10^{-24}$ A-m^2. Using Figure 12.5–1, the number of unpaired spins per iron atom, N_s, is 4. N_c can be calculated on a macroscopic basis using the data in Appendix B:

$$N_c = \left(\frac{7.87 \text{ g/cm}^3}{55.85 \text{ g/mol}} \right)(6.02 \times 10^{23} \text{ atoms/mol})$$

$$= 8.48 \times 10^{22} \text{ atoms/cm}^3 = 8.48 \times 10^{28} \text{ atoms/m}^3$$

Substituting values into the expression for M_s gives:

$$M_s = (8.48 \times 10^{28} \text{ atoms/m}^3)(4 \text{ spins/atom})[9.27 \times 10^{-24} \text{ (A-m}^2)/\text{spin}]$$

$$M_s = 3.14 \times 10^6 \text{ A/m}$$

Equation 12.3–5 states that $\mathbf{B} = \mu_0(1 + \chi)\mathbf{H}$. Since for ferromagnetic material $\chi \gg 1$, we find $\mathbf{B} \approx \mu_0\chi\mathbf{H} = \mu_0 M$. Therefore, the saturation induction for Fe is:

$$\mathbf{B}_s \approx \mu_0 M_s = [4\pi \times 10^{-7} \text{ (T-m)/A}](3.14 \times 10^6 \text{ A/m}) \approx 3.95 \text{ T}$$

12.7 ANTIFERROMAGNETISM AND FERRIMAGNETISM

In the previous section we showed that adjacent atomic dipoles can interact to form an ordered array of magnetic dipoles in some materials. If, as shown schematically in Figure 12.7–1a, all the dipoles point in the same direction, the material is said to be ferromagnetic. However, other orientational relationships between neighboring dipoles are possible. For example, the neighboring dipoles can align themselves in an antiparallel configuration. If a material is characterized by a dipole arrangement, as shown Figure 12.7–1b, in which the alternating dipoles have equal strength, it is **antiferromagnetic.** If the alternating dipoles have different strengths, as shown in part c of the figure, the material is **ferrimagnetic.**

Most common antiferromagnetic materials are compounds with the chemical formula MX_n, where M is a transition metal atom, X is an electronegative atom (O, S, Te, or F), and n is either 1 or 2. The magnetic dipoles occur within the unfilled inner shell of the metal atoms. Since only the metal atoms contribute magnetic dipoles, all the dipoles have equal strength.

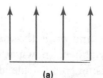 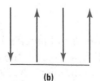 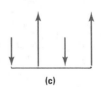

(a)   (b)   (c)

FIGURE 12.7–1 Three variations of orientational relationships between neighboring magnetic dipoles. **(a)** When all of the dipoles point in the same direction, the material is ferromagnetic. **(b)** When the neighboring dipoles are of equal strength but aligned in an antiparallel configuration, the material is antiferromagnetic. **(c)** When alternating dipoles have different strengths, the material is ferrimagnetic.

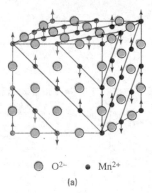
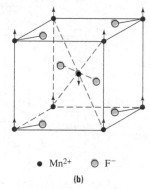

O^{2-} • Mn^{2+} • Mn^{2+} O F^{-}

(a)   (b)

FIGURE 12.7–2   The crystal structures and dipole arrangements for two antiferromagnetic materials: **(a)** MnO and **(b)** MnF$_2$.

The dipole arrangements in two antiferromagnetic materials, MnO and MnF$_2$, are shown in Figure 12.7–2. The common feature in these two structures is that the magnetic moments of the metal atoms are arranged so as to oppose those of their neighbors. The result is a material with a near zero net magnetic moment and a correspondingly small value of magnetic susceptibility. Typically, antiferromagnetic materials have $10^{-4} <$ $\chi < 10^{-2}$. Therefore, like diamagnetic and paramagnetic materials, antiferromagnetic materials find few magnetic applications.

Ferrimagnetism occurs when more than one type of metal atom is in the structure. If the two metal atoms have unequal dipole strengths, then even if their dipoles are aligned so as to oppose one another, there will be a nonzero net magnetic dipole. Thus, ferrimagnetic materials have magnetic susceptibilities between those of the antiferromagnetic and the ferromagnetic materials. Their properties and applications, however, are similar to those of the ferromagnetic materials.

The three most common types of ferrimagnetic materials, all of which are ceramic oxides, are summarized in Table 12.7–1. The spinels, or cubic ferrites, are used in applications requiring soft magnets, including transformer cores, inductors, and memory devices. The garnets, or rare earth ferrites, are favored in high-frequency applications such as in microwave devices. The magnetoplumbites, or hexagonal ferrites, are preferred in hard magnet applications (i.e., they are used as permanent magnets).

The cubic ferrites have the chemical formula MFe$_2$O$_4$, where M represents a divalent metallic atom (Mg, Fe, Ni, Mn, or Zn). Lodestone, or Fe$_3$O$_4$ (FeFe$_2$O$_4$), is an example of a cubic ferrite and was the original material used in magnetic compasses. The dipoles associated with the Fe^{3+} ions have a different strength from those of the Fe^{2+} ions. These

TABLE 12.7–1   Classes of magnetic ceramics (ferrites).

Class	Composition	Comments
Cubic ferrites	MFe$_2$O$_4$ (or MO·Fe$_2$O$_3$) (where M is a divalent transition metal, e.g., Ni, Co, Fe, Mn, Zn)	Soft magnets used in transformer cores, inductors, memory devices, etc.
Rare earth ferrites (garnets)	M$_3$Fe$_5$O$_{12}$ (or 3M$_2$O$_3$·5Fe$_2$O$_3$) (where M is a trivalent rare earth metal, e.g., Y or Gd)	Soft magnets capable of high-frequency operation (microwave applications)
Hexagonal ferrites (magnetoplumbites)	MFe$_{12}$O$_{19}$ (or MO·6Fe$_2$O$_3$) (where M is a divalent metal from Group IIA, e.g., Ba, Ca, Sr)	Hard, or permanent, magnet applications

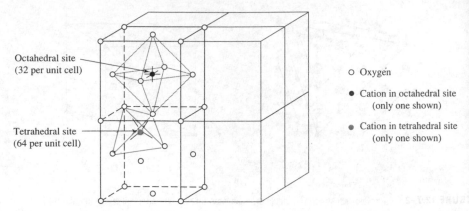

FIGURE 12.7–3   The spinel crystal structure.

compounds are based on the spinel crystal structure shown in Figure 12.7–3. The spinel structure consists of eight FCC unit cells of oxygen atoms with the cations located in the tetrahedral and octahedral interstitial positions. The tetrahedral cations have their moments aligned opposite to the external field, while those in the octahedral sites reinforce the external field. All the trivalent ions, located in octahedral sites, have their moments aligned with the external field, while all the divalent ions, located in the tetrahedral sites, have their moments opposing the external field.

In another variation of this structure, known as inverse spinel, the divalent cations are located in the octahedral positions, and the trivalent cations are equally split between the octahedral and tetrahedral interstitial positions. As with the spinel structure, the unequal dipole strength in the tetrahedral and octahedral positions produces ferrimagnetic behavior. The inverse spinel structure is more common, since the divalent cations, which in general are a bit larger than the trivalent cations, tend to prefer the larger octahedral positions.

The cubic ferrites have relatively high crystallographic symmetry and comparatively low values of magnetic anisotropy. This combination suggests that the magnetic dipoles within a domain can be rotated away from the soft magnetization directions to become aligned with a relatively weak external field. Thus, cubic ferrites tend to have low saturation field values H_s and are classified as soft magnets.

One of the principal advantages of ferrimagnetic compounds over their ferromagnetic counterparts is the ability to tailor their magnetic properties through compositional modifications. For example, since $NiFe_2O_4$ and $ZnFe_2O_4$ are soluble in one another, it is possible to design solid solutions with properties intermediate to those of the parent compounds. In this way, specific ferrimagnetic compounds with specific magnetic properties can be obtained for individual applications.

The second class of ferrimagnetic materials, the rare earth ferrites, have the generic formula $3M_2O_3 : 5Fe_2O_3$ ($M_3Fe_5O_{12}$), where M is a trivalent rare earth metal. The most common example is yttrium-iron-garnet (YIG). The unit cell for the rare earth ferrites is cubic, but the basis is complex. It suffices for our purposes to note that the cations can reside in three types of interstitial positions. The "extra" type of interstitial position gives the materials engineer added flexibility for tailoring the properties of these magnetic compounds to meet unique design specifications. As will be discussed in Section 12.8, the second principal competitive advantage of the rare earth ferrites is their comparatively high electrical resistivity. This characteristic makes them attractive for high-frequency applications.

The third class of ferrimagnetic compounds is the magnetoplumbites, or hexagonal ferrites. The two most common examples are $BaO:6Fe_2O_3$ and $SrO:6Fe_2O_3$. These materials display "hard" magnetic properties. Since hexagonal ferrites have three types of interstitial positions, they offer the potential for extensive property tailoring through compositional control.

EXAMPLE 12.7-1

Determine whether each of the following materials is likely to be antiferromagnetic or ferrimagnetic. If you believe it will be ferrimagnetic, predict whether it will be a cubic, rare earth, or hexagonal ferrite: CrF_2, $MnFe_2O_4$, $CaFe_{12}O_{19}$, CuO, and $Gd_3Fe_5O_{12}$.

Solution

Antiferromagnetic materials have the formula MX_n, where M is a transition metal atom, X is O, S, Te, or F, and n is either 1 or 2. The two materials on the list that satisfy these requirements are CrF_2 and CuO. The cubic ferrites have formula MFe_2O_4, where M is a divalent metal atom. The only appropriate compound on the list is $MnFe_2O_4$. The rare earth ferrites have the formula $M_3Fe_5O_{12}$, where M is a trivalent rare earth metal. $Gd_3Fe_5O_{12}$ is a rare earth ferrite. The hexagonal ferrites have formula $MO:6Fe_2O_3$. Thus, $CaFe_{12}O_{19}$, which is chemically equivalent to $CaO:6Fe_2O_3$, is a hexagonal ferrite.

DESIGN EXAMPLE 12.7-2

Consider the compounds CoO, $ZnFe_2O_4$, and $Y_3Fe_5O_{12}$. Which material would you select for each of the following applications:

 a. A component that must not respond to an external magnetic field
 b. A component that must respond to high-frequency magnetic fields
 c. A component to be used as a hard magnet

Solution

Using the method outlined in the previous example, we suspect that CoO is antiferromagnetic, $ZnFe_2O_4$ is a cubic ferrite, and $Y_3Fe_5O_{12}$ is a rare earth ferrite. If the component must not respond to an applied magnetic field, then antiferromagnetic CoO is the best choice. As shown in Table 12.7-1, the rare earth ferrite, in this case $Y_3Fe_5O_{12}$, should be used in the high-frequency application. Since none of the three candidate materials is a hexagonal ferrite, none is a good choice for the hard magnet application.

12.8 DEVICES AND APPLICATIONS

In this section we will briefly describe some important magnetic devices: (1) permanent magnets, (2) transformer cores, and (3) magnetic storage devices.

12.8.1 Permanent Magnets

A permanent magnet must retain its magnetization in the absence of an applied field and resist demagnetization in the presence of modest magnetic fields in the service environment. These two constraints point to materials with high values of remanence $\mathbf{B}_r$ and high values of coercive field strength $\mathbf{H}_c$—that is, hard magnets. To maximize $\mathbf{B}_r$ and $\mathbf{H}_c$, the microstructure of the material should be tailored to minimize Bloch wall mobility. In addition, these applications favor materials in which dipole rotation is difficult. The

TABLE 12.8–1   Properties of hard magnets.

Name (composition)	Class	Remanence B_r (T)	Coercive field H_c (A/m)	Resistivity (Ω-m)	Density (g/cm^3)	Comments
3.5% Cr–Steel Fe–3.5 Cr–1 C	Metal (ferro)	0.95	5,280	290×10^{-9}	7.3	Low-cost alternative; not often specified in new designs; application: telephone bell ringer.
Cunife Cu–20 Fe–20 Ni	Metal (ferro)	0.54	44,000	180×10^{-9}	8.6	Used in tachometers and speedometers; one of the early replacements for Fe-base alloys.
Cunico Cu–29 Co–21 Ni	Metal (ferro)	0.34	54,400	240×10^{-9}	8.3	Substituted for Cunife if higher ductility or higher H_c is required.
Cast Alnico 5 Fe–8 Al–14 Ni–24 Co–3 Cu	Metal (ferro)	1.25	49,600	470×10^{-9}	7.3	One of the most popular modern metal magnets; used in relays, rotors in small motors, loudspeakers, magnetos, and amp/volt meters.
Co–rare earth, Co$_5$M where M is Sm, Y, Ce, La, Nd, or Pr	Metal (ferro)	0.86	640,000	500×10^{-9}	8.2	High H_c and B_r values; better thermal stability than previous classes; along with Alnico and ferrites (below), one of the three most common classes of magnets.
BaO · 6Fe$_2$O$_3$ (sintered)	Hexagonal (ferrite) (ferri)	0.2–0.4	1,800–3,000	~10^4	4.5–5.0	Ferrites offer lower cost (Co is expensive), lower mass density, and higher H_c values than most alternatives; their market share is steadily increasing.
SrO · 6Fe$_2$O$_3$	Hexagonal ferrite (ferri)	0.35–0.4	2,200–3,200	~10^4	4.5–4.8	

Source: Adapted from *Metals Handbook*, 9th ed., Vol. 3, ASM International, Materials Park, OH, 1980, p. 625, Table 3, and p. 632, Table 6. Reprinted with permission of ASM International.

magnetic properties of several hard ferro- and ferrimagnetic materials are given in Table 12.8–1.

There are many different types of permanent magnetic materials. In general, these materials are classified according to the processing method used to decrease Bloch wall mobility. The oldest class of permanent magnets is the cold-worked steels. For these materials the ease with which domain growth occurs is limited by the high dislocation densities. The amount of strain in the lattice can also be increased by quenching the Fe-C alloys to produce a martensitic structure. These alloys are not often specified in new designs, although they are occasionally considered in less demanding applications, such as part of a telephone bell ringer, as a low-cost alternative.

It is also possible to restrict Bloch wall mobility using a heat treatment that produces a finely dispersed second phase. This method is extremely successful if the size of the second phase is comparable to the domain size. In such cases, domain growth is essentially eliminated, and magnetization can occur via only the less efficient method of dipole rotation. If, in addition, the second phase has high magnetic anisotropy, the magnitudes of B_r and H_c are further increased. Three examples of alloy classes that can be processed in this way are the Cunife (Cu-Ni-Fe), Cunico (Cu-Ni-Co), and Alnico (Fe-Al-Ni-Co) families.

Because of their improved magnetic properties, Alnico magnets are gradually replacing the older steel- and copper-based alloys in applications such as rotors in small DC motors, loudspeakers, relays, and amp/volt meters. The superior ductility of the copper-based alloys, however, has helped them to retain a reasonable but decreasing share of the market. The speedometer or tachometer in your car may well contain a Cunife alloy magnet.

The concept of creating a microstructure with a length scale similar to that of the domain size can be extended to single-phase materials by using powder processing techniques. If the size of the raw powder particles is approximately that of the magnetic domains and if sintering is controlled to limit grain growth, then the resulting microstructure displays an extremely fine grain size. Since grain boundaries are effective obstacles to Bloch wall motion, domain growth is severely restricted.

This process can be used successfully with ceramic materials as well as with metals. The most common family of metallic magnets produced in this way is the cobalt–rare earth alloys. These alloys have the general formula Co_5RE where the rare earth (RE) element can be Y, La, Ce, Nd, Sm, or Pr. The superior magnetic properties of the Co_5RE alloys have given them an increasing market share in recent years. Cobalt–rare earth alloys show better high-temperature stability than any of the other metal alloys.

As suggested previously, the preferred ceramic compounds for hard magnet applications are the hexagonal ferrites. These compounds can be processed so that the only magnetically soft direction in the hexagonal crystal, the c direction, is perpendicular to the applied field direction. Thus, it is possible to produce a magnet with both a small grain size (limited domain growth) and substantial resistance to dipole rotation. Another advantage of the hexagonal ferrites—and for that matter ceramic magnets in general—is that they are generally less dense than their metallic counterparts. Their low density, high coercive field values, and generally lower cost (which is related to the absence of Co) allow them to compete effectively with the other two dominant families of magnets—Alnico and Co_5RE—in most hard magnet applications.

Cobalt is considered to be a strategically important material in the United States. This means, among other things, that the use of Co should be minimized or eliminated in products associated with national defense. What do you think are the favorite classes of hard magnets in the strategic U.S. industries? Since the Alnico, Cunico, and Co_5RE families all contain Co, they can be eliminated from further consideration. That leaves the steel, Cunife, and hexagonal ferrite families. Steel is undesirable because of its inferior magnetic properties. Except in certain applications requiring high ductility, the low cost, low density, and higher H_c values of the hexagonal ferrites make them a likely choice for many critical U.S. industries.

12.8.2 Transformer Cores

The materials used in transformer cores are continually subjected to alternating electrical and magnetic fields. The selection of a material for this application is based on minimizing both magnetic hysteresis and eddy current power losses. A fluctuating magnetic field induces a voltage within the transformer core. This voltage causes a current flow known as an eddy current. Since the power losses (resulting from resistance heating) are described by the expression V^2/R, maximizing the resistance of the transformer core minimizes eddy current losses. To minimize magnetic hysteresis losses, the transformer cores must be fabricated from soft magnetic materials. The magnetic properties of several soft ferro- and ferrimagnetic materials are given in Table 12.8–2.

The requirements of high Bloch wall mobility and high electrical resistivity are difficult to satisfy simultaneously, since most microstructural characteristics that increase one of these factors decrease the other. One of the most commonly used materials for this application is an Fe-Si alloy. This alloy is magnetically soft because in its annealed condition it is relatively strain-free, coarse-grained, and single-phase. The addition of silicon in solid solution significantly increases the resistivity of the alloy with only minor decreases in the Bloch wall mobility. In contrast, small amounts of the impurities C, S, N, or O can degrade the properties of an iron-based soft magnet by restricting the mobility of Bloch walls.

Further improvements in the magnetic response of the alloy can be obtained using one additional processing step. The magnitude of $\mathbf{H}_c$ can be decreased by cold working the microstructure prior to the final annealing treatment so as to align the magnetically soft crystallographic directions, $\langle 1\ 0\ 0 \rangle$, with the direction of the external magnetic field. This process essentially eliminates the need for dipole rotation to achieve saturation magnetization.

TABLE 12.8–2   Properties of soft magnets.

Name (composition)	Class	Remanence B_r (T)	Saturation induction B_s (T)	Coercive field H_c (A/m)	Hysteresis loss/cycle (J/m³)	Resistivity (Ω-m)	Density (g/cm³)
Commercial Fe 99.95% Fe	Metal (ferro)	0.77	2.14	80	270	100×10^{-9}	7.87
3% Si–steel (low carbon)	Metal (ferro)	—	2.01	56	160	470×10^{-9}	—
Oriented 3% Si–steel (low carbon)	Metal (ferro)	1.2	2.01	7.2	40	500×10^{-9}	—
45 Permalloy Fe–45% Ni	Metal (ferro)	0.9	~1.58	4.0–20.0	25	$500–600 \times 10^{-9}$	8.17
Supermalloy Ni–16% Fe–5% Mo	Metal (ferro)	~0.5	0.68–0.78	0.48	2	650×10^{-9}	8.77
Metglass (see text)	Metal (ferro)	—	1.61	6	—	1400×10^{-9}	7.05
$(MnO \cdot Fe_2O_3)$ + $(ZnO \cdot Fe_2O_3)$	Cubic ferrite (ferri)	—	~0.3	16–56	~40	~2000	—
$(NiO \cdot Fe_2O_3)$ + $(ZnO \cdot Fe_2O_3)$	Cubic ferrite (ferri)	—	~0.4	320	~35	$~10^7–10^8$	—

Source: Adapted from *Metals Handbook*, 9th ed., Vol. 3, ASM International, Materials Park, OH, 1980, p. 602, Table 3, and p. 603, Table 5. Reprinted with permission of ASM International.

In some applications, particularly high-frequency applications, the eddy current losses and the associated heating effects of the magnetic material are the design-limiting constraint. In these applications ceramic magnets, with their substantially higher electrical resistivities, become the material of choice. The most common soft ceramic magnets are the cubic and rare earth ferrites described in Section 12.7. Although the cubic ferrites in which the M^{2+} sites are occupied by Zn^{2+} ions together with Mn^{2+}, Ni^{2+}, or $Ni^{2+} + Cu^{2+}$ ions have the smallest hysteresis loops, the rare earth ferrites have somewhat higher resistivities. Since the latter factor becomes dominant at higher frequencies, the rare earth ferrites are preferred in applications such as microwave devices. Disadvantages of the ceramic magnets include lower ductility, lower magnetic saturation, and lower maximum service temperatures (Curie temperatures) than their metallic counterparts.

Another option for obtaining high Bloch wall mobility and a high electrical resistivity is provided by metallic glasses. These materials have high electrical resistivities because the nonperiodic nature of their amorphous structure results in extremely low electron mobilities. The lack of extended defects such as dislocations and grain and phase boundaries, however, results in relatively easy domain boundary motion. The most common metallic glasses are formed from complex Fe-based alloys containing B, C, Cr, Ni, P, and Si. Refer to Section 6.5.3 for a review of the structure and composition of metallic glasses.

DESIGN EXAMPLE 12.8–1

Suppose you work for a large company that manufactures transformer cores. An inventor comes to you with a new alloy she has just developed. This alloy has an $\mathbf{H}_c$ value of 700,000 A/m, a $\mathbf{B}_r$ value of 1.3 T, a density of 4.0 g/cm³, and a resistivity of 200×10^{-9} Ω-m. Would you like to purchase the patent rights for this alloy?

Solution

Transformer cores require materials with low $\mathbf{H}_c$ and low $\mathbf{B}_r$ values and high electrical resistivities. A comparison of the properties of the new alloy with those of the soft magnet materials in Table 12.8–2 shows that the new alloy is a terrible soft magnet. If you are smart, however, you will negotiate for the rights to this alloy as soon as possible. Comparing its properties with those of the alloys listed in Table 12.8–1 shows that it is an excellent hard magnet. Its $\mathbf{H}_c$ and $\mathbf{B}_r$ values are higher than those of any existing alloy, and its low density is a substantial bonus.

12.8.3 Magnetic Storage Devices

The two major requirements for a material intended for use as a magnetic storage device are that it be easily magnetized (low $\mathbf{H}_s$) and that it retain most of its magnetization after the external field is removed (high $\mathbf{B}_r$). As shown in Figure 12.8–1, another desirable material characteristic for this application is a hysteresis loop that has nearly square corners in the second and fourth quadrants. The advantage of such a shape is that the magnetic device can be easily and cleanly "erased" by application of a reversed field with magnitude $\mathbf{H}_c$. The most common magnetic materials for this application are the cubic ferrites in which the M^{2+} ions are either Co^{2+} or a combination of Mn^{2+} and Mg^{2+}. These ceramic magnets are used in applications ranging from computer disks (floppy and hard) to audio- and videotape.

One of the most common systems for storing information is magnetic tape recording. A typical magnetic tape consists of a thin polymer substrate coated with a monolayer of usually either iron oxide or chromium dioxide particles. The information storage occurs by moving the tape past a recording head as shown in Figure 12.8–2. A wire carrying a

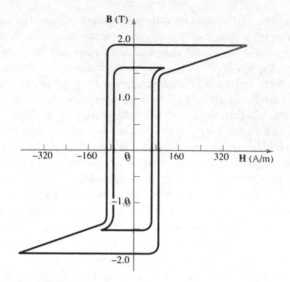

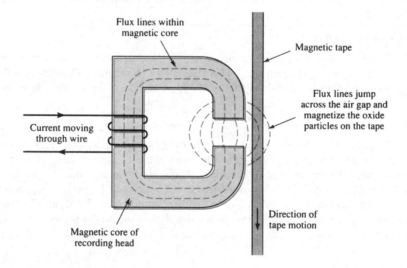

pulsating electric current is wrapped around the magnetic core of the recording head. The current induces a pulsating magnetic field through the core and across the air gap in the vicinity of the moving tape. As the magnetic flux lines impinge upon the tape, they magnetize the oxide particles on the moving tape. Information retrieval is accomplished using the inverse procedure. That is, as the magnetized tape moves past the playback head it induces an electric current in the wire.

DESIGN EXAMPLE 12.8–2

Explain why people who work with magnetic fields often have problems with their credit cards. How would you change the characteristics of the magnetic material to minimize this problem?

Solution

The materials used to store magnetic information are selected to have low values of coercive fields so that they can be easily and cleanly erased. If these materials are unintentionally subjected to large magnetic fields, the stored information can be lost. To minimize the problem one should select one of the cubic ferrites with a relatively large value for $\mathbf{H}_c$ (i.e., the material with the largest hysteresis loop in Figure 12.8–1).

12.9 SUPERCONDUCTING MAGNETS

In Section 10.2.9 we introduced the concept of superconductors and briefly described the electrical properties of these materials. Recall that their resistivity drops to zero as the temperature decreases below a critical value T_c. Zero resistivity is a necessary but not sufficient condition for a material to become a superconductor. In addition, the magnetic properties of the material must also change. Specifically, as shown in Figure 12.9–1, the superconducting transition results in a material that expels all magnetic fields.

Recall that the relationship between the total magnetic induction in a material and the external field strength is given by Equation 12.3–5 as $\mathbf{B} = \mu_0(1 + \chi)\mathbf{H}$. The expulsion of the magnetic field means that the total induction inside the superconductor is zero. That is:

$$\mathbf{B} = \mu_0(1 + \chi)\mathbf{H} = 0 \tag{12.9–1}$$

which implies that $\chi = -1$ for a superconductor. The only class of materials with magnetic susceptibilities less than zero is the diamagnetic materials. Note that a value of $\chi = -1$ implies that the internal magnetic field is of the same magnitude as the external field and antiparallel to it. For this reason, superconductors are classified as perfect diamagnetic materials.

In most materials the mechanism of diamagnetism is related to the orbital motion of electrons; however, in superconductors the mechanism is different. As depicted schematically in Figure 12.9–1, diamagnetism in superconductors results from the circular motion of electrons near the surface of the material. This electron motion produces an internal field that exactly balances the external field. This type of electron motion occurs

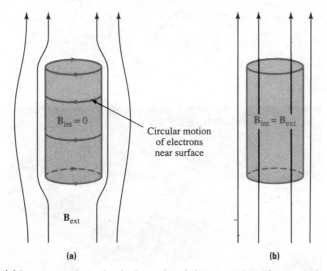

FIGURE 12.9–1 **(a)** In a superconductor the circular motion of electrons at the surface results in perfect diamagnetic behavior and the exclusion of the external magnetic field from the interior of the superconducting solid. **(b)** The external field penetrates a nonsuperconducting solid.   (*Source: L. L. Hench and J. K. West, Principles of Electronic Ceramics. Copyright © 1990 by John Wiley & Sons. Reprinted by permission of John Wiley & Sons, Inc.*)

on a much larger scale than the atomic motion, orbital and spin, discussed previously. It is possible to maintain this macroscopic electron motion only when there is no resistance to electron flow—that is, in a superconductor.

The magnitude of the internal magnetic field that can be supported by the macroscopic electron motion is limited. If the external field exceeds this limiting value, then the material is no longer able to maintain its perfect diamagnetic behavior and the superconducting condition is lost. Thus, there is a critical external field strength, $\mathbf{B}_c$, as well as a critical temperature, T_c, that limits superconducting behavior. In fact, the two limiting quantities are related through the expression:

$$\mathbf{B}_c = \mathbf{B}_0\left[1 - \left(\frac{T}{T_c}\right)^2\right]$$
(12.9–2)

where T is the temperature of the system and $\mathbf{B}_0$ is a material-specific constant.

The relationship between $\mathbf{B}_c$ and T_c in three low-temperature metallic superconductors is shown graphically in Figure 12.9–2. It is important to note that as the external field strength increases, the effective critical temperature decreases. The significance of this

FIGURE 12.9–2

The relationship between $\mathbf{B}_c$ and T_c for Hg, Sn, and Tl.

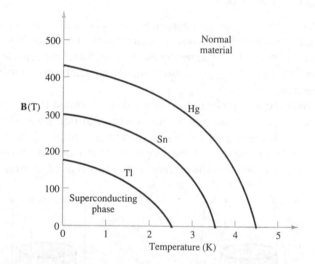

FIGURE 12.9–3

A photograph showing the Meissner effect.

observation is that in almost any conceivable application a superconductor carries a current and this current produces a magnetic field. The inevitable production of the magnetic field serves to lower the maximum operating temperature of the superconductor.

The total expulsion of a magnetic field by a superconductor, known as the Meissner effect, is the basis of the classical levitation experiment where a magnet "floats" above a superconductor lying in a pool of liquid nitrogen. A photograph of this demonstration is shown in Figure 12.9–3. On the more practical side, the Meissner effect offers the possibility of producing magnetically levitated low-friction trains as well as other useful engineering devices (see Section 10.2.9).

SUMMARY

The physical bases for the magnetic properties of solids are the current loops produced by the orbital and rotational motions of electrons. The interaction of the electron-induced magnetic moments with an external magnetic field results in the macroscopic magnetic properties of the material.

Diamagnetic materials are composed of atoms that have completely filled electron shells and have magnetic susceptibilities in the range $0 > \chi \gg -1$. Paramagnetic behavior is related to the nonzero net magnetic moment resulting from unpaired electrons in inner electron shells. These materials have susceptibilities in the range $0 < \chi \ll 1$. Because the magnitude of χ is small, neither dia- nor paramagnetic materials have many significant engineering applications.

If χ is positive and much greater than 1, the material is either ferri- or ferromagnetic. The three most common ferromagnetic elements are Fe, Ni, and Co.

Magnetic domains are microscopic volumes of material in which the dipoles are aligned. When an external field is applied, it increases the net magnetic moment in the direction of the field. This process can occur either by the growth of favorably oriented domains or by dipole rotation. Because of the irreversible work required to move domain walls, a hysteresis loop is associated with a ferri- or ferromagnetic material's response to alternating external magnetic fields. The microstructure of a material controls the Bloch wall mobility and, hence, the shape of the hysteresis loop. Some of the microstructural features that restrict domain wall motion are grain boundaries and dislocations.

In ferromagnetic materials all the dipoles within a domain point in the same direction; in other materials the neighboring dipoles can align themselves in an antiparallel configuration. Alternating dipoles of equal strength cause a material to be antiferromagnetic. These materials have a near zero net magnetic moment and a correspondingly small value of χ. If, however, the alternating dipoles have different strengths, the material is ferrimagnetic. Ferrimagnetic materials have properties similar to those of ferromagnetic materials and, therefore, have many engineering applications, including high-frequency microwave devices and magnetic storage of information.

KEY TERMS

antiferromagnetic	ferromagnetic	magnetic induction B	permeability μ
Curie temperature	hard magnets	magnetic susceptibility χ	relative permeability μ_r
diamagnetic	magnetic domains	magnetic vector M	soft magnets
ferrimagnetic	magnetic field strength H	paramagnetic	

HOMEWORK PROBLEMS

SECTION 12.2
Materials and Magnetism

1. Give two examples of engineering applications that are related to a magnet's ability to attract and hold certain objects.

2. Give two examples of engineering applications that are related to the mutual repulsion of like magnetic poles.

3. Since magnetic dipoles result from the motion of electrons which classes of materials do you expect to show magnetic behavior? Don't forget to discuss ceramics, which are combinations of metals and nonmetals.

4. Does a compass needle point precisely to the North Pole?

SECTION 12.3
Physical Basis of
Magnetism

5. Use Equations 12.3–2 and 10.2–1 through 10.2–3 to relate the quantities $\mathbf{B}$, $\mathbf{H}$, and μ to their electrical counterparts I, V, and σ.

6. As discussed in the text, the motion of electrons results in the atomic scale magnetic properties of materials.
   a. Which type of electron motion always results in a diamagnetic contribution?
   b. Which type of electron motion may result in a paramagnetic contribution?

7. Describe in your own words the difference between the meanings of the terms *magnetic induction* and *magnetic field strength*.

8. What are the units for each of the following magnetic parameters?
   a. Magnetic permeability
   b. Relative permeability
   c. Susceptibility
   Which set(s) of values do you anticipate are more often tabulated in handbooks? Why?

9. Estimate μ_r from the data provided in Tables 11.3–1 and 11.5–1 for air, water, diamond, alumina, Pyrex glass, and epoxy. Compare these estimates with accurate values computed using Equation 12.3–6 and data from Table 12.5–1.

10. Calculate the magnetic field strength in magnesium required to create an induction equal to that of the earth. The magnetic susceptibility of Mg is 13.1×10^{-6}.

11. Using the data given in the previous problem, estimate each of the following quantities:
    a. The relative permeability of Mg
    b. The permeability of Mg

12. Calculate the magnetization and inductance of a Mg sample (with $\chi = 13.1 \times 10^{-6}$) when it is placed in a magnetic field of strength 10^5 A/m.

SECTION 12.4
Classification of
Magnetic Materials

13. Some stainless steels have $\chi < 1$. How can you easily pick them out of a pile of scrap metal that contains other stainless steels, ordinary carbon steels, brass, aluminum, and so on?

14. If you placed a giant ferromagnetic single crystal in a magnetic field, would it actually measurably grow in length and then shrink when removed from the magnetic field?

15. Describe the relationship among the directions of the vectors $\mathbf{B}$, $\mathbf{H}$, and $\mathbf{M}$ in both diamagnetic and paramagnetic materials.

16. Explain why diamagnetic and paramagnetic materials find few applications based on their magnetic properties.

SECTION 12.5
Diamagnetism and
Paramagnetism

17. NaCl, LiF, Si, Ge, Cu, and Ag are all diamagnetic materials. Explain this observation.

18. Show the predicted distribution of electron spins in the 4d and 5s shells for the elements from Sr to Cd. What type of magnetic behavior do you predict for each of these elements?

19. Section 12.2 mentioned that "tin" cans are often separated from a solid waste stream using a magnet. Why do you think the word "tin" was placed in quotation marks in this discussion?

20. Explain why unpaired electrons in the valence shell of an atom do not yield paramagnetic behavior.

21. Explain why most oxides are diamagnetic. Examine Table 12.5–1 and find exceptions to this trend.

22. Describe the two mechanisms by which an applied external magnetic field can increase the magnetization of a ferromagnetic material.

SECTION 12.6
Ferromagnetism

23. Define each of the following terms as they relate to a ferromagnetic material:
    a. Saturation field
    b. Remanence
    c. Coercive field
    d. Hysteresis loop

24. Sketch the **B-H** loading curve for a previously unmagnetized material that is initially exposed to a negative **H**.

25. Consider a pair of polyphase, polycrystalline metallic ferromagnets with similar elemental compositions.
    a. Suggest several processing steps that could be used to convert one of the samples into a hard magnet and the other into a soft magnet.
    b. Which microstructure would have the higher electrical resistivity?

26. How does the magnitude of the magnetic anisotropy ratio influence the shape of the hysteresis loop for a ferromagnetic material?

27. A magnetic material is surrounded by insulation, so that no heat can escape from the magnetic material. The material is subjected to a rapidly fluctuating magnetic field. Sketch the temperature of the material as a function of time in the field. Be sure to label important temperatures on the graph. If your graph is for a hard magnet, make a similar graph for a soft magnet, or vice versa.

28. Sketch the **B-H** curve for a steel (1) after annealing and (2) after heavy cold working.

29. Estimate the saturation magnetization and induction for Co.

30. Why do ferromagnetic materials revert to paramagnetic behavior at high temperatures?

31. Why are ferromagnetic materials composed of multiple magnetic domains rather than one single domain?

32. Sketch a **B** versus **H** curve for a ferromagnetic crystal that is originally in the virgin state (i.e., **B** = **H** = 0) and is then subjected to an external field that increases to a value above H_s and then decreases to **H** = 0. Label the regions of this curve that are associated with reversible domain growth, irreversible domain growth, and dipole rotation.

33. Use the data in Table 12.8–2 to estimate the number of Bohr magnetons associated with each Fe atom. Compare your result with the value obtained from Figure 12.5–1.

34. Consider the **B-H** curve shown in Figure HP12.1. Calculate the initial and maximum permeability of this material. What is the coercive field strength? What is the remanence?

35. A magnetic field of 2×10^5 A/m is applied to the ferromagnetic material whose **B-H** curve is given in Figure HP12.1. Determine the magnetic induction and permeability of the material.

36. Reconsider the material whose **B-H** curve is given in Figure HP12.1. Calculate the work lost per cycle and temperature rise for the material when it is cycled between a large positive and negative magnetic field.

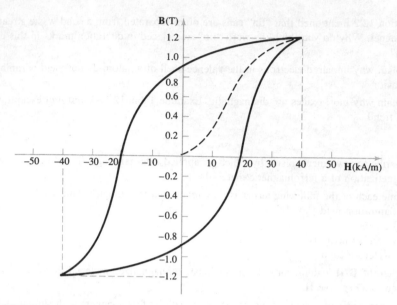

FIGURE HP12.1

37. An inductor consisting of a 45 Permalloy core and a 10-m-long, 20-turn coil of copper is activated with 3 A of current. Calculate **H**, **M**, and **B**. [Hint: **H** in units of A/m is given by the expression $\mathbf{H} = 0.4(\pi nI/L)$.]

SECTION 12.7
Antiferromagnetism and
Ferrimagnetism

38. Discuss the similarities and differences among ferro-, ferri-, and antiferromagnetic materials.

39. Estimate the saturation magnetization for each of the following materials:
    a. Ni
    b. MnO
    c. NiFe$_2$O$_4$ (inverse spinel structure)

40. Explain why the crystal structure of the common ferrimagnetic solids permits them to be tailored to specific applications more easily than their ferromagnetic counterparts.

41. Explain why the compound Fe$_3$O$_4$ can be ferrimagnetic even though only one type of metal atom is in this structure.

42. Why is the inverse spinel structure more common than the spinel crystal structure?

43. Explain how the hexagonal crystal structure of BaO · 6Fe$_2$O$_3$ helps to make this material well suited for hard magnet applications.

44. Determine whether each of the following materials is likely to be antiferromagnetic or ferrimagnetic. If you believe it will be ferrimagnetic, predict whether it will be a cubic, rare earth, or hexagonal ferrite: 3Sm$_2$O$_3$ · 5Fe$_2$O$_3$, NiF$_2$, ZnFe$_2$O$_4$, FeO, and SrO · 6Fe$_2$O$_3$.

45. List the three classes of ceramic magnets and describe the type of application for which each class is best suited.

SECTION 12.8
Devices and Applications

46. Discuss the magnetic properties of metallic glasses.

47. Examine Table 12.8–1. One of the most popular magnets is cast Alnico. Why are there so many elements—Al, Ni, Co, Cu, and Fe—in Alnico? Is Alnico single-phase?

48. The text states that many recording tapes consist of metal oxide particles attached to polymer film. Oxides are generally excellent abrasives. Won't they wear out the tape heads and anything they touch? Also, the tape itself, being a polymer (polyester), will not be dimensionally stable for many years. Is there then a potential archive problem?

49. Suppose the same inventor you were introduced to in Design Example 12.8–1 shows up at your office again with a second alloy. This material has an H_c value of 5 A/m, a B_r value of 0.5 T, a density of 5 g/cm^3, and a resistivity of 5×10^7 Ω-m. Would you like to purchase the patent rights for this alloy? Why or why not?

50. Consider the following four methods for increasing the "hardness" of a magnetic alloy: (a) cold working, (b) powder processing, (c) incorporating a finely distributed second phase, and (d) magnetic anisotropy. Explain how each method influences the magnetic properties of the alloy, and list at least one class of alloys associated with each processing method.

51. Suppose you must select a solid for use in a magnetic application at ~100°C. Which material(s) would you select? Why?

52. What type of application would favor the use of either an Alnico or a Cunife alloy?

53. Discuss the relative advantages and disadvantages of Co–rare earth magnetic alloys.

54. Soft magnetic crystals generally contain relatively low concentrations of structural defects. Why are Fe-Si alloys heavily cold-worked prior to their final annealing step?

55. In general, one typically uses ceramic materials in high-temperature applications. Is this reasonable in magnetic applications? Why or why not?

56. Why is it a bad idea to store credit cards in your wallet with their magnetic strips adjacent to one another?

57. A magnet of volume 1 cm^3 fabricated from Supermalloy is subjected to an alternating field of frequency 60 Hz. Assuming that the maximum field strength exceeds the saturation field strength for this alloy, calculate the power loss in this magnet. Report your answer in watts.

58. Repeat the previous problem for a similar commercial iron magnet and comment on the significance of the difference in the answers.

59. Will there ever be a room-temperature superconductor?

60. Is the Meissner effect the only way to levitate materials.

61. Previously we stated that diamagnetic materials have no important engineering applications. Explain why this statement does not apply to superconductors.

62. Are superconductors ever operated at T_c in an engineering application? Why or why not?

63. Estimate the value of the constant B_0 for Sn and then use that estimate to calculate the critical field strength, B_c, at 2 K.

64. A certain 1-2-3 superconductor is characterized by $T_c = 90$ K and $H_0 = 10^4$ A/m.
    a. Estimate the temperature at which the critical field strength is 5×10^3 A/m.
    b. Estimate the magnetic susceptibility of this material when $H = 5 \times 10^3$ A/m and $T = 2$ K.

65. Describe the Meissner effect and give a commercial application based on it.

SECTION 12.9
Superconducting
Magnets

THERMAL PROPERTIES

MATERIALS IN ACTION   Thermal Properties

In many engineering applications, the effects of temperature and thermal energy must be managed to achieve a given set of objectives. Thus, the thermal properties of materials often play a key role in practical applications.

In Chapter 1 we saw that the space shuttle experiences temperatures perhaps as high as 1600°C upon reentry into the earth's atmosphere. The shuttle must be protected by materials that have good insulating properties. Specifically, such materials should have:

Low thermal conductivity, so that thermal energy is transmitted slowly

High heat capacity, so that a large amount of thermal energy is required to raise the temperature

High density, so that much thermal energy can be stored in a relatively small volume

Materials with these properties are said to have a low thermal diffusivity. We saw that materials such as silica tiles, borosilicate glasses, and carbon/carbon composites were used to meet this challenge. Materials used in applications such as the space shuttle must also not expand or contract very much as the temperature changes. If they do change their dimensions significantly, high stresses can build up and cracking might occur. In engineering terms, they must have a low coefficient of thermal expansion.

Another example of the need for thermal protection occurs in the turbine section of jet engines. Turbine blades operate in an extremely aggressive environment at temperatures up to 1200°C. In such cases, thermal barrier coatings of ZrO_2 may be used to protect the underlying metal superalloys. These coatings, in addition to having a very low thermal diffusivity, have low coefficients of thermal expansion to prevent the buildup of stresses at the interface and subsequent spalling.

Control of thermal properties is important in common household applications as well. For example, cookware fabricated from polycrystalline lithium aluminum silicate (known as LAS) has an extremely low coefficient of thermal expansion. Consequently, it is possible to take a hot container directly from the stove top and immerse it in cold water without cracking. Similarly, stove tops with embedded heating elements are made from LAS and are able to withstand extreme temperature gradients.

Some applications require materials with *high* thermal diffusivities. An example is found in computers. The computational speed of a computer is determined by the time it takes electrical signals to travel. The speed of computers has been increased dramatically by increasing the density of circuit elements on a chip. However, as the circuit density has increased, the thermal energy density increased as well. Eventually, the point is reached where further miniaturization (and hence increased speed) is limited by the rate at which heat can be extracted. Massive efforts are under way to develop methods to extract heat efficiently, including the development of substrate materials with greater thermal diffusivities.

In this chapter we will study the relationship between atomic structures, microstructures, and thermal properties. We shall see that thermal properties may be engineered to meet application objectives.

13.1 INTRODUCTION

The inclusion of a separate chapter on thermal properties may seem unusual, since throughout the text a recurring theme has been the influence of temperature on the structure and properties of materials. Recall our previous discussions of: (1) the difference in the response of thermoplastic and thermoset polymers to changes in temperature; (2) the Arrhenius nature of diffusion, creep, stress relaxation, and so on; (3) the change from ductile to brittle behavior in BCC metals at the nil-ductility temperature and in amorphous materials at the glass transition temperature; and (4) the reason the electrical conductivity of a metal decreases while that of a semiconductor increases with an increase in temperature. This list reminds us of the importance of temperature in all aspects of materials design and selection. The list, however, is not complete. The purpose of this chapter is to describe the important thermal properties of materials that have not been touched upon previously.

In particular, we discuss the coefficient of thermal expansion, heat capacity, and thermal conductivity. As always, we stress the relationship between each of these properties and the atomic structure of materials. Some of the engineering applications related to the thermal properties of materials include bimetallic strips used as switches in thermostats, insulating materials used on the space shuttle and in protective clothing, thermal shock–resistant cookware, and tempered glass.

When heat is supplied to a solid, the average thermal energy of the atoms increases. Approximately half of the thermal energy increases the potential energy of the atoms while the other half increases their kinetic, or vibrational, energy. This atomic vibration, or oscillation, generates vibrational wave packets, known as *phonons*,[1] that propagate through the solid. Since energy is quantized, atoms can change their energy only by discrete amounts. Thus, the energy of the phonon, E_{ph}, is quantized and related to the wavelength of the atomic vibrations, λ, through Equation 10.3–8 as $E_{ph} = hc/\lambda = h\nu$, where ν is the frequency of the vibrations. We will see that these atomic vibrations are important for understanding the physical basis of the thermal properties of materials.

13.2 COEFFICIENT OF THERMAL EXPANSION

Most solids expand when they are heated and contract when they are cooled (the major exception being oriented polymers, as discussed in Section 6.6.5). The material property that describes the extent of expansion and contraction with changing temperature is the linear coefficient of thermal expansion, α_{th}. The defining equation is:

$$\alpha_{th} = \frac{\varepsilon_{th}}{\Delta T} \tag{13.2–1}$$

where ε_{th} is the thermal strain ($\Delta L/L_0$) resulting from a temperature change ΔT. Recall from Section 2.5 that as atoms gain thermal energy, the magnitude of their oscillations increases. Since the bond-energy curve is asymmetric (see Figure 2.5–2),

---

[1] Phonons, much like photons and electrons, display both wave and particle characteristics.

he average separation distance between neighboring atoms increases with increasing emperature.

In addition to length changes, one may also be interested in how the volume of a solid changes with temperature. By defining a volume coefficient of thermal expansion α_v, an equation similar to 13.2–1 can be obtained:

$$\frac{\Delta V}{V_0} = \alpha_v \, \Delta T \qquad (13.2–2)$$

where V_0 is the original volume and ΔV is the change in volume resulting from a temperature change ΔT. For isotropic solids the relationship between the volume and linear coefficients of thermal expansion is:

$$\alpha_v \approx 3\alpha_{th} \qquad (13.2–3)$$

Recognize, however, that many but not all materials are isotropic (see Section 3.11).

From the discussion so far you might erroneously conclude that a plot of ε_{th} as a function of temperature should yield a single straight line with a slope given by α_{th}. Some deviations from this behavior are shown in Figure 13.2–1. Part a shows several abrupt changes in ε_{th} with temperature corresponding to changes in the crystalline forms of SiO_2. The critical temperatures are those at which the materials undergo polymorphic phase transformations (see Section 3.10). Even for a single-phase material, however, deviations from linear ε_{th}-ΔT behavior are often observed. As shown in Figure 13.2–1b, many ceramics exhibit nonconstant values of α_{th}. Also note that fused silica has a near zero value of α_{th} over the entire temperature range indicated in the figure. Other materials— for example, graphite—display near zero expansion over limited temperature ranges. Thus, Equation 13.2–1 should be considered valid over only relatively small temperature ranges, and one must use caution when selecting handbook values for α_{th} to be used in calculations.

Based on atomic bonding characteristics, it is possible to make some reasonable generalizations about the magnitudes of α_{th} for several classes of materials. Table 13.2–1 shows that because of the lack of primary bonds between the chains of thermoplastic polymers, these materials exhibit comparatively high values of α_{th}. In contrast, the network polymers show somewhat lower values for α_{th} in the unstretched state.

Of the crystalline materials, metals usually show the highest values of expansion coefficients. These values are, however, generally lower than those of most polymers. Many of the ceramic crystals are more "open" (i.e., have lower atomic or ionic packing factors) than their metallic counterparts. Thus, some of the atomic vibration can be accommodated without expansion. Therefore, ceramic crystals, in general, have lower values of α_{th} than do metals. As shown in the table, the α_{th} values for noncrystalline ceramic materials vary significantly. The expansion behavior of a glass is a complex function of its composition and thermal history.

··

DESIGN EXAMPLE 13.2–1

Consider a basketball arena in which the basket is suspended from the ceiling by a metal support system so that at 21°C the rim is 3.048 m (10 ft) above the floor. The ceiling is 15.250 m above the floor. During a game the temperature can increase by as much as 15°C. If the thermal properties

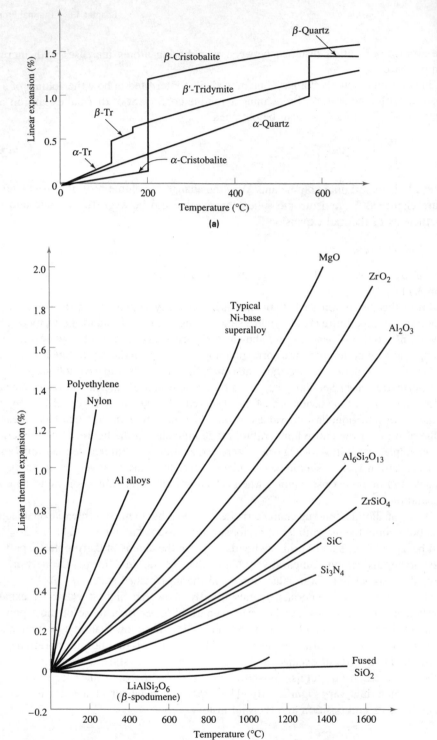

FIGURE 13.2–1 The variation in linear thermal expansion, in percent, as a function of temperature for: **(a)** several crystalline forms of SiO₂, and **(b)** some representative metals, polymers, and ceramics. The slope of each curve is the magnitude of α_{th} for each material at a given temperature. *(Source:* **(a)** *Zbigniew D. Jastrzebski, The Nature and Properties of Engineering Materials, 3rd ed. Copyright © 1987, John Wiley & Sons, Inc. Reprinted by permission of John Wiley & Sons, Inc.* **(b)** *Reprinted from David W. Richerson, Modern Ceramic Engineering, 2nd ed., Marcel Dekker, Inc., New York, 1992, p. 147, by courtesy of Marcel Dekker, Inc.).*

TABLE 13.2-1 Coefficient of (linear) thermal expansion α_{th} for selected materials.

Material	$\alpha_{th}(\times\ 10^{-6}\ °C^{-1})$	Material	$\alpha_{th}(\times\ 10^{-6}\ °C^{-1})$
Metals		**Ceramics**	
Al	25	Al_2O_3	6.5–8.8
Cr	6	BeO	9
Co	12	MgO	13.5
Cu	17	SiC	4.8
Au	14	Si	2.6
Fe	12	Si_3N_4 (α phase)	2.9
Pb	29	Si_3N_4 (β phase)	2.3
Mg	25	Spinel ($MgAl_2O_4$)	7.6
Mo	5	Soda-lime-silicate glass	9.2 (used in light bulbs)
Ni	13	Borosilicate glass	4.6 (used with Kovar)
Pt	9	Silica (96% pure)	0.8
K	83	Silica (99.9% pure)	0.55
Ag	19	**Polymers (unoriented)**	
Na	70	Polyethylene	100–200
Ta	7	Polypropylene	58–100
Sn	20	Polystyrene	60–80
Ti	9	Polytetrafluoroethylene	100
W	5	Polycarbonate	66
Zn	35	Nylon (6/6)	80
1020 steel	12	Cellulose acetate	80–160
Stainless steel	17	Polymethylmethacrylate	50–90
3003 aluminum alloy	23.2	Epoxy	45–90
2017 aluminum alloy	22.9	Phenolformaldehyde	60–80
ASTM B 152 copper alloy	17	Silicones	20–40
Brass	18		
Pb-Sn solder (50-50)	24		
AZ31B magnesium alloy	26		
ASTM B160 nickel alloy	12		
Commercial titanium	8.8		
Kovar (Fe-Ni-Co)	5		

of the suspension system can be modeled as if it were a single cable, would you recommend aluminum or tungsten for this application?

Solution

The length of the support "cable" changes with temperature, causing a change in the rim-to-floor distance. Solving Equation 13.2–1 for the final cable length yields:

$$L_f = L_0(1 + \alpha_{th}\ \Delta T)$$

Table 13.2–1 gives $\alpha_{th}(Al) = 25 \times 10^{-6}\ °C^{-1}$, and $\alpha_{th}(W) = 4.5 \times 10^{-6}\ °C^{-1}$. Noting that $\Delta T = 15°C$ and $L_0 = (15.250 - 3.084) = 12.166$ m, and substituting the appropriate α_{th} values into the equation, gives: $L_f(Al) = (12.166\ \text{m})[1 + (25 \times 10^{-6}\ °C^{-1})(15°C)] = 12.171$ m and $L_f(W) = 12.167$ m. For Al this represents a change of ~5 mm, and for W the change is ~1 mm. Even though the W expands less, either material can be used to meet the thermal expansion requirement, since neither change is significant. Since Al is less expensive and easier to fabricate, it is the preferred material for this application. (Note: In practice the structure would probably be steel, but that was not an option in this problem.)

13.3 HEAT CAPACITY

The **heat capacity,** c, of a material is the amount of heat energy required to raise the temperature of one mole of the material by one degree. Mathematically,

$$c = \frac{dQ}{dT} \tag{13.3-1}$$

where dQ is the amount of heat added to produce a temperature change dT. To obtain a theoretical estimate for the magnitude of the heat capacity of a material, one must obtain an expression for dQ/dT.

Classical thermodynamic calculations show that the heat energy of an atom oscillating in three dimensions in a solid at temperature T is given by the expression:

$$Q(\text{atom}) = 3kT \quad \text{or} \quad Q(\text{mole}) = 3N_a kT \tag{13.3-2}$$

where k is Boltzmann's constant and N_a is Avogadro's number. By differentiating with respect to temperature, we find

$$c = 3N_a k = 3R = 24.9 \text{ J/(mol-K)} \tag{13.3-3}$$

where R $(= N_a k)$ is the gas constant, with a value of 8.314 J/(mol-K). Equation 13.3–3 is known as the Dulong-Petit law.[2]

This simplified expression assumes that heat is added at constant volume. Thus, the heat capacity in Equation 13.3–3 is referred to as c_v. Alternatively, the heat energy could have been added to the solid at constant pressure. The corresponding heat capacity, c_p, is greater than c_v because of the extra energy required for the expansion of the solid at constant pressure. The difference between c_v and c_p increases with temperature and is significant only near the melting point of a crystalline solid or the glass transition temperature of an amorphous solid.

The heat capacity per unit mass is known as the **specific heat** of the material. We will use the capital letter C to represent specific heat [with units of J/(kg-K)] and the lowercase letter c to represent heat capacity [with units J/(mol-K)]. Values for specific heat at constant pressure, C_p, are given in Table 13.3–1 for several materials. As with the corresponding heat capacities, the difference between C_v and C_p can usually be safely neglected. Table 13.3–1 shows the agreement between the measured specific heat C_p and the theoretical estimate for C_v calculated as:

$$C_v = \frac{c_v}{\text{Atomic weight}} \tag{13.3-4}$$

Note that both heat capacity and specific heat are intrinsic properties. The corresponding extrinsic property, known as the **thermal capacitance** C_{th}, is calculated by multiplying the intrinsic property, C_v, by the mass of the material, M:

$$C_{\text{th}} = MC_v = \rho V C_v \tag{13.3-5}$$

where ρ is the density of the material and V is its volume. Thermal capacitance rep-

---

[2] The estimate for c given in Equation 13.3–3 is valid at relatively high temperature—for metals, above room temperature, and for ceramics, above ~1000°C. A more complete (quantum mechanical) treatment of the problem shows that at low temperatures, c gradually approaches zero.

TABLE 13.3–1   Specific heats for selected materials. C_p is the experimental value at constant pressure and C_v/(atomic weight) is the theoretical estimate (see text for discussion).

Material	C_p (J/kg-K)	C_v/(at. wt.) (J/kg-K)
Metals		
Al	900	923
Cr	448	479
Co	456	423
Cu	386	392
Au	130	126
Fe	448	446
Pb	159	120
Mg	1017	1024
Mo	251	260
Ni	443	424
Pt	133	128
K	753	637
Ag	235	231
Na	1226	1083
Ta	140	138
Sn	218	210
Ti	523	520
W	142	135
Zn	388	381
1020 steel	486	—
Stainless steel	502	—
Brass	375	—
Kovar	460	—
Ceramics		
Al_2O_3	775	244
BeO	1050	996
MgO	940	618
Soda-lime-silicate glass	840	—
Silica	740	—
Polymers		
Polyethylene	2100	—
Polypropylene	1880	—
Polystyrene	1360	—
Polytetrafluoroethylene	1050	—
Nylon (6/6)	1670	—
Phenolformaldehyde	1650	—
Semiconductors		
Si	702	886
Ge	322	343
GaAs	350	172

resents the amount of energy necessary to change the temperature of a block of material by 1°C.

Thermal capacitance is one of the design parameters in applications where the time rate of temperature change, $\partial T/\partial t$, is important. If all other variables are constant, the maximum rate at which a system can change its temperature is inversely related to its thermal capacitance. Thus, low-C_{th} materials are preferred when rapid changes in temperature are required. For example, porous ceramic firebricks are preferred over solid bricks of the same material for use in lining furnaces, since their lower thermal capacitance permits more rapid heating and cooling rates. In a later section of this chapter we will see that the porous material also offers superior thermal insulating characteristics.

One of the material issues in the design of integrated circuits is the problem of temperature increases resulting from resistive heating in active electrical components. When current passes through a device, the heat dissipated, Q, can be calculated as:

$$Q = I^2 Rt \qquad (13.3-6)$$

where Q has units of joules, I is current in amperes, R is resistance in ohms, and t is time in seconds. The heat generated by this process can be substantial. If the heat is not removed from the vicinity of the active devices by thermal conduction through the substrate, it can result in an increase in the temperature of the device and a corresponding degradation in performance. The topic of thermal conductivity is discussed in the next section.

....

EXAMPLE 13.3–1

Suppose a doped-Si resistor of length 10 μm, cross-sectional area 5 μm^2, and resistivity 0.001 Ω-m carries a current of 10^{-5} A for 0.1 s. Assuming that all the heat generated goes into increasing the temperature of the resistor, calculate the resulting temperature rise.

Solution

Equation 13.3–1 gives the relationship among change in heat dQ, changes in temperature dT, and heat capacity c. Dividing both sides of this equation by the mass of the system M and substituting ΔQ and ΔT for their differential counterparts yields an expression for the specific heat of the material:

$$C \approx \frac{\Delta Q}{M \, \Delta T}$$

Solving this expression for the temperature rise ΔT and substituting the quantity $I^2 Rt$ for the heat increase ΔQ gives

$$\Delta T \approx \frac{I^2 Rt}{MC}$$

From Table 13.3–1, $C_p(\text{Si}) = 700$ J/(kg-K). M is calculated by multiplying the density of Si by the volume of the resistor:

$$M = (2.33 \times 10^{-3} \text{ kg/cm}^3)(10 \times 10^{-4} \text{ cm})(5 \times 10^{-8} \text{ cm}^2)$$
$$= 1.16 \times 10^{-13} \text{ kg}$$

The resistance is calculated using Equation 10.2–2:

$$R = \rho\left(\frac{L}{A}\right) = \frac{(0.001 \ \Omega\text{-m})(10 \times 10^{-6} \text{ m})}{5 \times 10^{-12} \text{ m}^2}$$
$$= 2 \times 10^3 \ \Omega$$

Substituting these values into the equation for ΔT gives:

$$\Delta T = \frac{(10^{-5} \text{ A})^2 (2 \times 10^3 \ \Omega)(0.1 \text{ s})}{(1.16 \times 10^{-13} \text{ kg})[700 \text{ J/(kg-K)}]}$$
$$= 246°C$$

A temperature increase of this magnitude will degrade device performance. Consequently, most devices such as this have conductive paths or forced convection to remove heat generated.

....

13.4 THERMAL CONDUCTION MECHANISMS

While cooking pots and pans may be fabricated entirely from metal, many metal pans have wood or plastic handles. Why was this design modification implemented? If you have ever picked up a heavy pan by its thick metal handle after it has been over a fire for some time, you know the answer—the metal handle gets very hot. In contrast, the plastic handle does not transport (conduct) the heat from the heat source to your hand, so it remains cool even though the pan gets hot. In this section we will explain the link between atomic structure and the ability of a material to conduct heat.

The conduction of thermal energy through a solid as a result of a temperature gradient is analogous to the diffusion of an atomic species as a result of a concentration gradient. The direction of heat transfer is from high-temperature regions toward low-temperature regions (i.e., "down" the temperature gradient), and the property that describes a material's ability to transport heat is the **thermal conductivity.** The relationship between temperature gradient $\partial T/\partial x$ and thermal conductivity K has the same form as the diffusion equation introduced in Section 4.4.2. With respect to a 1-D heat flow problem as shown in Figure 13.4–1, the relevant equation is

$$\left(\frac{\partial Q}{\partial t}\right)\left(\frac{1}{A}\right) = K\left(\frac{\partial T}{\partial x}\right) \tag{13.4-1}$$

where $\partial Q/\partial t$ is the heat transferred per unit time across a plane of area A. The units for K are J/(s-m-K) or, equivalently, W/(m-K). Equation 13.4–1 states that the heat flow per unit time per unit area (i.e., the heat flux) is proportional to the temperature gradient, and that the constant of proportionality is the thermal conductivity. Values for the thermal conductivities of several materials are given in Table 13.4–1.

Thermal energy can be conducted through a material by two mechanisms: lattice vibrations (phonons), or the motion of free electrons. Thus, in the most simplistic model

$$K = K_p + K_e \tag{13.4-2}$$

where K_p represents the contribution from phonons and K_e represents the electron contribution. The relative importance of these two mechanisms depends primarily on the electronic band structure of the material. Materials with a partially filled valence band (e.g., metals) have thermal conductivities dominated by the motion of free electrons. Small–band gap materials (e.g., semiconductors) can have significant contributions from

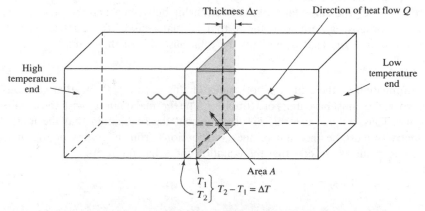

FIGURE 13.4–1   Rate of heat flow $\partial Q/\partial t$ through an area A as a result of a thermal gradient $\partial T/\partial x$.

TABLE 13.4–1 Thermal conductivities K for selected materials.

Material	K(W/m-K)	Material	K(W/m-K)
Metals		**Ceramics (approximate values)**	
Al	300	Al_2O_3	34
Cr	158	BeO	216
Cu	483	MgO	37
Au	345	SiC	93
Fe	132	SiO_2	1.4
Pb	40	Spinel ($MgAl_2O_4$)	12
Mg	169	Soda-lime-silicate glass	1.7
Mo	179	Silica glass	2
Ni	158	**Polymers (unoriented)**	
Pt	79	Polyethylene	0.38
Ag	450	Polypropylene	0.12
Ta	59	Polystyrene	0.13
Sn	85	Polytetrafluoroethylene	0.25
Ti	31	Polyisoprene	0.14
W	235	Nylon	0.24
Zn	132	Phenolformaldehyde	0.15
1020 steel	52	**Semiconductors**	
Stainless steel	16	Si	148
Brass	120	Ge	60
		GaAs	46

both mechanisms, while the thermal conductivities of large–band gap materials (e.g., diamond) are dominated by the phonon mechanism.

The influence of temperature on K is complex. The magnitude of the thermal conductivity of a solid is proportional to several factors, including:

1. The number N of thermal energy carriers (phonons, electrons, or both)
2. The mean velocity $\bar{v}$ of the carriers
3. The average distance λ traveled by a carrier before being scattered by the lattice

This last factor is analogous to the mean time between collisions discussed in Section 10.2.2 on electrical conduction. In fact, the $\bar{v}\lambda$ product can be envisioned as a quasi-mobility term for thermal conductivity.

How can we predict the form of the relationship between K and the variables N, $\bar{v}$, and λ? Let us propose an equation for K that is analogous to the electrical conductivity equation, $\sigma = Nq\mu$. That is, perhaps the relationship is of the form:

$$K \propto N\bar{v}\lambda \tag{13.4–3}$$

An examination of the units for both sides of this equation shows that the constant of proportionality must have units of J/(mol-K). The thermal quantity with these units is heat capacity. Thus, we can use this observation and the realization that the total thermal conductivity must be the sum of the contributions from all types of thermal energy carriers to predict correctly that the equation for K has the form:

$$K \propto \sum_i c(N_i \bar{v}_i \lambda_i) \tag{13.4–4}$$

where the summation is carried out for the contributions of both phonons and electrons.

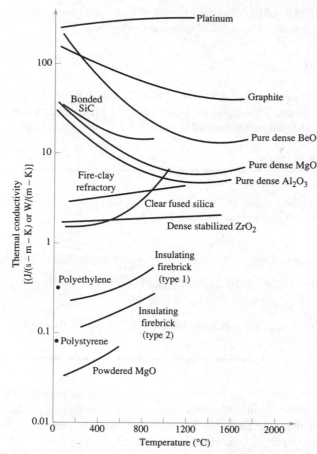

FIGURE 13.4–2 Variation in the thermal conductivity K as a function of temperature for a variety of materials. *(Source: Reprinted from David W. Richerson, Modern Ceramic Engineering, 2nd ed., Marcel Dekker, Inc., New York, 1992, p. 137, by courtesy of Marcel Dekker, Inc.)*

We expect an increase in temperature to influence all four terms on the right hand side of the previous equation. An increase in T results in either no change or an increase in c, an increase in N, a slight increase in $\bar{v}$, but a decrease in λ. The relative magnitude of the changes in these four quantities depends on the particular material of interest.

The temperature dependence of thermal conductivity is shown in Figure 13.4–2 for a variety of materials. There are two general types of behavior. Some materials exhibit a conductivity that continually increases with temperature. Examples include most of the glasses, insulating firebrick, nylon, and platinum. Most other materials, including iron, alumina, fused silica, and graphite, display a minimum value of conductivity at an intermediate temperature. That is, as temperature increases, the conductivity of these materials initially decreases, reaches a minimum, and then increases again.

Despite the complexity of the temperature dependence of K, it is possible to form some reasonable generalizations about the relative magnitude of K for the different classes of materials. Because of their large number of mobile free carriers, metals tend to have comparatively high thermal conductivities. Since both the electrical and thermal conductivities are dominated by the motion of free electrons, there is a direct relationship between the magnitude of the electrical conductivity, σ, and the electron contribution thermal conductivity, K_e. The form of the relationship is

$$\frac{K_e}{\sigma} = L_0 T \qquad (13.4\text{--}5$$

where L_0 is the Lorenz factor (2.45×10^{-8} W-Ω/K^2). Equation 13.4–5 is known as the Wiedemann-Franz law. This equation has significant practical importance, since it much easier to find tabulated values of σ than it is to find values of K_e.

....

EXAMPLE 13.4–1

Estimate the thermal conductivity of Cu at 25°C using the data in Table 10.2–1, and compare your estimate with the experimental value listed in Table 13.4–1.

Solution

Substituting the value of $\sigma(\text{Cu}) = 6 \times 10^7$ $(\Omega\text{-m})^{-1}$ found in the table into Equation 13.4–5:

$$K = L_0 T\sigma = [2.45 \times 10^{-8} \text{ (W-}\Omega/\text{K}^2)][300 \text{ K}][5.88 \times 10^7 \text{ }(\Omega\text{-m})^{-1}]$$

$$= 432 \text{ W/(m-K)}$$

Since the experimental value listed in Table 13.4–1 is 398 W/(m-K), the error associated with this estimate is ~8.5%.

....

The thermal conductivities of crystalline ceramics, which are dominated by the phonon mechanism, can vary significantly depending on the complexity of the structure. The structural factors that minimize phonon scattering and, therefore, favor high thermal conductivity values in ionic or covalent crystals include:

1. Open crystal structures (i.e., structures with low packing factors)
2. Simple crystal structures (i.e., structures with a basis of only one or two atoms)
3. Atoms or ions of similar size and weight

As an extreme example, the ordered and open structure of diamond displays one of the highest values of K of any material known. The crystals of BeO and SiC, in which the ions have similar sizes and atomic weights, show intermediate values of K. The more complex structures, such as the spinels, and those with significant size and atomic weight differences between the ions, such as UO_2, have the lowest relative conductivities.

The influence of atomic scale defects on thermal conductivity is similar to that for electrical conductivity. That is, defects tend to decrease the mobility of thermal carriers. Thus, the thermal conductivity of most crystalline solids decreases as the defect (and impurity) density increases.

The thermal conductivities of amorphous ceramics tend to be lower than those of their corresponding crystals. The lack of long-range order increases the likelihood of phonon scattering, reduces λ, and results in low values of K.

Most polymers have rather low values of K. The electron contribution to thermal conduction is generally small. In addition, the lack of periodicity in the solid results in a large amount of scattering and, hence, small values of λ. The conductivity of materials can be decreased by foaming with air. This technique is common in the formation of polymer-based thermal insulators, such as Styrofoam. Similarly, porous ceramic materials can be fabricated for use as high-temperature thermal insulators. The conductivity of oriented

~~highly~~ ighly crystalline polymers (such as PE in filament form) can be very high, as much as ~~two~~ wo orders of magnitude greater than that of unoriented polymers.

EXAMPLE 13.4–2

Consider the three ceramics $MgAl_2O_4$ (spinel crystal structure), MgO (NaCl crystal structure), and amorphous Na_2O-CaO-SiO_2 (window glass). Predict which material will have the highest and lowest value of thermal conductivity and then compare your predictions with the data in Table 13.4–1.

Solution

Since the NaCl crystal structure is much less complex than the spinel structure, we should predict that MgO will have a higher thermal conductivity than $MgAl_2O_3$. The lack of periodicity in the glass should result in limited phonon mobility and a correspondingly low value of thermal conductivity. These predictions are consistent with the experimental values in Table 13.4–1: K(MgO) = 37 W/(m-K), K($MgAl_2O_3$) = 12 W/(m-K), and K(glass) = 1.7 W/(m-K).

Equation 13.4–1 shows that under steady-state conditions the only materials parameter that influences heat conduction is thermal conductivity. Under non-steady-state conditions, however, the situation becomes a bit more complicated. The governing equation, which we present without derivation, is:

$$\frac{\partial T}{\partial t} = D_{th}\left(\frac{\partial^2 T}{\partial x^2}\right) \tag{13.4–6}$$

where D_{th} is a materials property known as the **thermal diffusivity.** Not surprisingly, this equation for time-dependent heat transfer is virtually identical to the time-dependent diffusion equation (i.e., Fick's second law) introduced in Chapter 4. D_{th} can be defined in terms of the more fundamental thermal parameters K and C_p as:

$$D_{th} = \frac{K}{\rho C_p} \tag{13.4–7}$$

where ρ is the density of the material. Using either of the previous two equations, you can convince yourself that the units for D_{th} are m^2/s. Note that these units are identical to those for the diffusion coefficient D.

Since a high value of D_{th} means that a material is capable of responding rapidly to temperature changes, thermal diffusivity is perhaps the most critical design parameter in engineering applications where $\partial T/\partial t$ is important. What about the thermal capacitance introduced in the previous section? Equation 13.4–7 shows that D_{th} actually includes the concept of C_{th}, since the denominator is in effect the thermal capacitance on a per unit mass basis. The important point is that rapid heat transfer requires not only low thermal capacitance but also high thermal conductivity.

EXAMPLE 13.4–3

Which family of alloys, aluminum or ferrous, allows the achievement of higher quench rates?

Solution

Since quenching is a non-steady-state heat transfer situation, the important materials property is thermal diffusivity. Equation 13.4–7 defines D_{th} as

$$D_{th} = \frac{K}{\rho C_p}$$

Using the data in Tables 13.3–1 and 13.4–1 along with the density values in Appendix B, we find

$$D_{th}(Al) = \frac{247 \text{ J/(s-m-K)}}{(2.7 \times 10^3 \text{ kg/m}^3)[900 \text{ J/(kg-K)}]} \approx 1.0 \times 10^{-4} \text{ m}^2/\text{s}$$

$$D_{th}(Fe) = \frac{80.4 \text{ J/(s-m-K)}}{(7.87 \times 10^3 \text{ kg/m}^3)[448 \text{ J/(kg-K)}]} \approx 0.2 \times 10^{-4} \text{ m}^2/\text{s}$$

Therefore, since Al has a higher thermal diffusivity, we can correctly predict that Al-based alloys can be more rapidly quenched than ferrous alloys.

13.5 THERMAL STRESSES

Have you ever taken a warm glass out of a dishwasher and broken it by filling it with a cold beverage? Have you ever stored leftovers in the refrigerator in a glass pan and cracked the pan by placing it directly on a hot stove burner? These are two examples of material failures resulting from temperature-induced stresses. A typical industrial example of a thermal stress–induced failure might occur when a cooling water pipe near a furnace develops a leak and sprays cold water on the furnace lining. In this section we describe the mechanism by which thermal stresses are developed and discuss ways to select materials to minimize the possibility of thermal stress–induced failures.

For a material with a nonzero value of α_{th}, a temperature change results in a dimensional change known as a thermal strain, ε_{th}. Under certain conditions a thermal strain can in turn generate a thermal stress of sufficient magnitude to cause material failure. Consider the situation shown in Figure 13.5–1. In part a of this figure an unconstrained bar of original length L_0 expands by an amount $\Delta L = L_0 \alpha_{th} \Delta T$. In part b of the figure, an identical bar is constrained (restricted from movement) so that it cannot expand when the temperature is increased by an amount ΔT. In order for the total strain ε_t in the bar to be zero, the thermal strain must be balanced by an induced mechanical strain ε_m that has an equal magnitude but opposite sign. Mathematically,

$$\varepsilon_t = \varepsilon_{th} + \varepsilon_m = 0 \quad \text{or} \quad \varepsilon_m = -\varepsilon_{th} \tag{13.5–1}$$

In this particular situation we have

$$\varepsilon_m = -\alpha_{th} \Delta T \tag{13.5–2}$$

and if the strains are small enough that the deformation is elastic, we find

$$\sigma_{th} = \varepsilon_m E = -\alpha_{th} E \Delta T \tag{13.5–3}$$

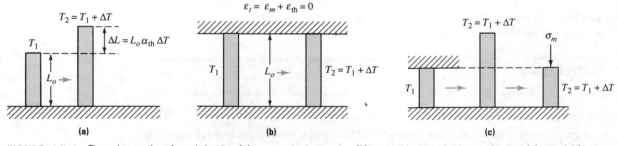

FIGURE 13.5–1   Thermal expansion of a rod showing: **(a)** unconstrained expansion, **(b)** completely constrained expansion, and **(c)** a model for calculating the resulting thermal stress by allowing unconstrained expansion and then applying a compressive stress in a way that returns the bar to its original length.

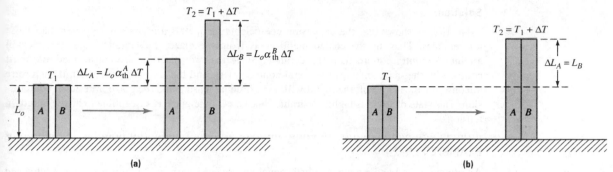

FIGURE 13.5–2   Thermal expansion of two rods showing: **(a)** different amounts of thermal expansion when the rods are isolated, and **(b)** an intermediate amount of thermal expansion when the rods are rigidly connected.

That is, a compressive stress of magnitude $\alpha_{th}E\,\Delta T$ has developed in the constrained bar as a result of the temperature increase. Alternatively, one can consider the origin of the compressive stress to be as shown in Figure 13.5–1c. In this case the bar is imagined to expand as if it were unconstrained and then a compressive stress is applied such that the bar returns to its original length.

Another type of thermally generated stress occurs when two materials with different coefficients of thermal expansion are rigidly connected and subjected to a change in temperature. This situation is shown in Figure 13.5–2. Isolated rods A and B would experience thermal strains of magnitude $\alpha_{th}^{A}\,\Delta T$ and $\alpha_{th}^{B}\,\Delta T$, respectively. If rigidly connected, however, the total strain in rod A must equal that in rod B. Thus, there must be some additional mechanical strains in the two rods such that

$$\varepsilon_t^A = \alpha_{th}^A\,\Delta T + \varepsilon_m^A = \alpha_{th}^B\,\Delta T + \varepsilon_m^B = \varepsilon_t^B \qquad (13.5\text{–}4)$$

By assuming elastic deformation and using a force balance at the interface, this equation can be used to determine the thermal stresses in each of the rods.

This type of thermally generated stress is common in composite materials. For example, layered electronic devices may fail because of thermal expansion mismatch. Printed circuit boards, as well as integrated circuits, are composed of layers of conductive metal and dielectric ceramic or polymer. The difference in thermal expansion coefficient of the various materials, $\Delta\alpha$, leads to stresses that begin during fabrication and may develop further during use. The magnitude of the strain can be approximated by

$$\varepsilon = \Delta\alpha\,\Delta T \qquad (13.5\text{–}5)$$

The stresses that result from mismatch can lead to circuit board warping or loss of adhesion between layers. Similar effects can occur in other composite structures. Note that no stress develops when the expansion coefficients match. Microstresses can even develop between grains in single-phase polycrystalline solids because of crystalline anisotropy. This problem can be especially serious in ceramics.

....

EXAMPLE 13.5–1

Metal components are often coated with a ceramic glaze for decorative purposes or to protect the metal from aggressive environments. Recognizing that glazes are applied to the metal at elevated temperatures (where the viscosity of the glaze is low), do you think the glaze is in residual compression or residual tension at room temperature?

Solution

Table 13.2–1 shows that the expansion coefficients for metals are generally greater than those for ceramics. Thus, as the coated metal cools from the glaze application temperature, it will attempt to contract more than the ceramic. Since the two materials are constrained and must contract by the same amount, the metal contracts less, and the ceramic more, than if each were unconstrained. As a result the metal will be in residual tension and the glaze in residual compression. This state of stress is highly desirable, since a ceramic glaze is susceptible to brittle failure in tension.

EXAMPLE 13.5–2

An aluminum rod and a nylon rod have equal unstressed lengths at 20°C. If each rod is subjected to a tensile stress of 5×10^6 Pa, find the temperature at which the two stressed rods will have equal lengths. [$E(\text{aluminum}) = E^{Al} = 70 \times 10^9$ Pa, $E(\text{nylon}) = E^n = 2.8 \times 10^9$ Pa, $\alpha_{th}^{Al} = 25 \times 10^{-6}$ °C^{-1}, $\alpha_{th}^n = 80 \times 10^{-6}$ °C^{-1}.]

Solution

Since the two rods have equal length at 20°C, the problem reduces to finding the temperature at which the two stressed rods have the same total strain, ε_t. The relevant equation is

$$\varepsilon_t^{Al} = \varepsilon_t^n$$

or, equivalently,

$$\varepsilon_{th}^{Al} + \varepsilon_m^{Al} = \varepsilon_{th}^n + \varepsilon_m^n$$

where the total strain is the sum of the thermal (ε_{th}) and mechanical (ε_m) strains in each material. Using the definitions $\varepsilon_{th} = \alpha_{th}\,\Delta T$ and $\varepsilon_m = \sigma/E$ yields

$$\alpha_{th}^{Al}\,\Delta T + \frac{\sigma}{E^{Al}} = \alpha_{th}^n\,\Delta T + \frac{\sigma}{E^n}$$

Solving for ΔT and substituting the given data yields:

$$\Delta T = \frac{\sigma(1/E^n - 1/E^{Al})}{\alpha_{th}^{Al} - \alpha_{th}^n}$$

$$= \frac{(5 \times 10^6 \text{ Pa})[1/(2.8 \times 10^9 \text{ Pa}) - 1/(70 \times 10^9 \text{ Pa})]}{25 \times 10^{-6} \text{ °C}^{-1} - 80 \times 10^{-6} \text{ °C}^{-1}}$$

$$= T - T_0 = -31.2\text{°C}$$

Solving for T and noting that $T_0 = 20$°C gives:

$$T = 20\text{°C} - 31.2\text{°C} = -11.2\text{°C}$$

As a final example of a mechanism for generating thermal strains, consider the situation shown in Figure 13.5–3. In this case a hot solid is being sprayed with cold water. If surface flaws are present, the potential for brittle failure exists. This phenomenon is known as **thermal shock**. (Recall the three examples of thermally induced failure that opened our discussion of thermal stresses.)

Several material properties can contribute to resistance to failure by thermal shock. First, since the magnitude of the thermal stress is proportional to $E\alpha_{th}\,\Delta T$, a small value of the $E\alpha_{th}$ product will help to minimize the problem. Second, the source of the thermal stress is a temperature gradient across the thickness of the sample. Since the magnitude of the gradient will be inversely related to the thermal conductivity of the sample,

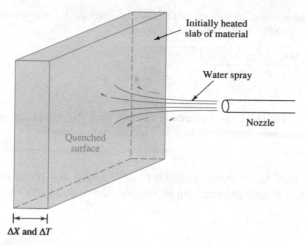

FIGURE 13.5–3   Generation of thermal stress because of differential cooling rates resulting in a temperature gradient across a slab of material.

large values of K are desirable. Third, the material will fail when the fracture stress, defined by

$$\sigma_f = \frac{K_{Ic}}{\sqrt{\pi a}} \tag{13.5–6}$$

is exceeded (see Chapter 9). Therefore, to choose among candidate materials for an application in which thermal shock resistance is important, one would select the material with the highest ratio $K\sigma_f/E\alpha_{th}$.

....

EXAMPLE 13.5–3

A ceramic component with $E = 300 \times 10^3$ MPa, $K_{Ic} = 2$ MPa-$\sqrt{m}$, and a maximum flaw size of 2.5 mm must be subjected to a cold water quench from 1000°C to 25°C. What value of α_{th} must this ceramic have in order to survive this quench? How does the required α_{th} value compare with typical α_{th} values for ceramics (see Table 13.2–1)?

Solution

The thermal stress generated by the quench is given by $\sigma_{th} = E\alpha_{th}\,\Delta T$. Failure will occur if this stress exceeds the fracture stress of the material given in Equation 13.5–7 as $\sigma_f = K_{Ic}/\sqrt{\pi a}$. Thus, the failure condition is:

$$\sigma_{th} > \sigma_f$$

or, equivalently,

$$E\alpha_{th}\,\Delta T > \frac{K_{Ic}}{\sqrt{\pi a}}$$

Solving the second inequality for α_{th} yields:

$$\alpha_{th} > \frac{K_{Ic}/\sqrt{\pi a}}{E\,\Delta T}$$

Using the data in the problem statement, and noting that since the material is effectively infinitely thick so that $\Delta T = 1000°C - 25°C$, failure occurs if:

$$\alpha_{th} > \frac{(2 \text{ MPa-}\sqrt{m})/\sqrt{\pi(2.5 \times 10^{-3} \text{ m})}}{(300 \times 10^3 \text{ MPa})(1000°C - 25°C)}$$

$$\alpha_{th} > 7.7 \times 10^{-8} \text{ °C}^{-1}$$

Since this value is two orders of magnitude smaller than the values for ceramics listed in Table 13.2–1, it is unlikely that the ceramic can survive such a quench.

Note that our analysis of thermal stresses has assumed that the materials are brittle elastic solids. When plastic deformation is possible, thermal stresses rarely cause failure.

13.6 APPLICATIONS

The scientific principles described in this chapter may be used to develop many useful engineering devices. These devices include bimetallic strips, thermal insulation, thermal shock–resistant cookware, tempered glass, support structures for orbiting telescopes, ceramic-to-metal joints, and cryogenic materials.

13.6.1 Bimetallic Strip

The basis for the operation of bimetallic strips used as switches in home thermostats is the differential strain generated when two materials are rigidly connected and subjected to a change in temperature. A schematic of this device is shown in Figure 13.6–1. If $\alpha_{th}^A > \alpha_{th}^B$,

FIGURE 13.6–1

A schematic illustration of a bimetallic strip used as a switch in a thermostat: **(a)** the differential expansion induces stresses in the two metals, **(b)** the induced stresses cause the strip to bend, and **(c)** an illustration of the device.

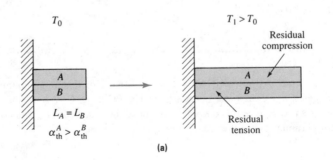

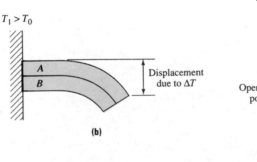

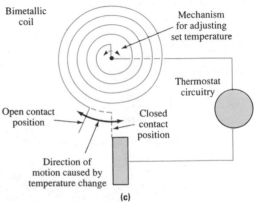

then when the temperature increases, materials A and B will be in residual compression and tension, respectively. If the metal strips are thin, then the stresses can be relaxed by bending, as shown in Figure 13.6–1b. As shown in Figure 13.6–1c, the total displacement of one end of the strip can be magnified by the use of a coil of the two materials rather than a straight strip. The expansion of the coil breaks the circuit when the desired temperature is reached. The operational action at the critical temperature is adjusted by moving, or twisting, the opposite end of the bimetallic coil.

13.6.2 Thermal Insulation

The applications of thermal insulating materials range from reducing residential energy consumption to protecting the space shuttle during reentry into the earth's atmosphere. In straightforward applications the objective is to fabricate a material with a low thermal conductivity. One of the more common examples is the low-density fiberglass-based insulation found in most homes. Although fiberglass has a low intrinsic value of K, its conductivity has been decreased substantially by trapping air in the space between fibers. Conduction along the thin entangled fibers contributes little to heat transfer. On the other hand, the conductivity of air is so low that the less efficient heat transfer mechanisms, convection and radiation, may become significant.

The ceramic tiles on the space shuttle contain a controlled amount of internal porosity that improves their insulating characteristics. In this application, however, it is important to control not only the amount of porosity, but also the pore morphology. Specifically, the pores form a continuous interconnected network to accommodate the pressure changes associated with space travel. The tiles on the shuttle are 15 cm × 15 cm and are composed of ~80 weight percent (wt. %) silica (SiO_2) and 20 wt. % of an alumina-borosilicate fiber.

13.6.3 Thermal Shock–Resistant Cookware

Failure resulting from thermal shock occurs as a result of expansion differences caused by a thermal gradient across a material. As mentioned in Section 13.5, materials with good thermal shock resistance have a high ratio $K\sigma_f/E\alpha_{th}$. One way to obtain a high ratio is to design a material with a near zero value for α_{th}. Certain anisotropic materials have positive α_{th} values in some crystallographic directions but negative values in other crystallographic directions. An example of a material of this type is lithium aluminosilicate, $LiAlSi_2O_6$, which is the composition of the familiar Corning Ware®. By forming a polycrystalline structure in which the grains are properly oriented with respect to each other, it is possible to create a material with a macroscopic coefficient of thermal expansion near zero. Since $\alpha_{th} \approx 0$, thermal shock is not a problem and Corning Ware can be taken directly from the freezer to a preheated oven.

A related application is the development of ceramic stove tops containing embedded electrical heating elements. The thermal shock resistance allows you to safely place a chilled pan directly on a heating element. In addition, the anisotropic thermal conductivity (high K in the thickness direction with low K in the plane of the stove top) allows you to place your hand adjacent to the heating element without fear of injury.

13.6.4 Tempered Glass

As discussed in Chapter 6, amorphous materials are brittle solids when the service temperature is below the glass transition temperature T_g. Below T_g, oxide glasses are extremely sensitive to surface flaws, exhibit low fracture toughness values, and, therefore,

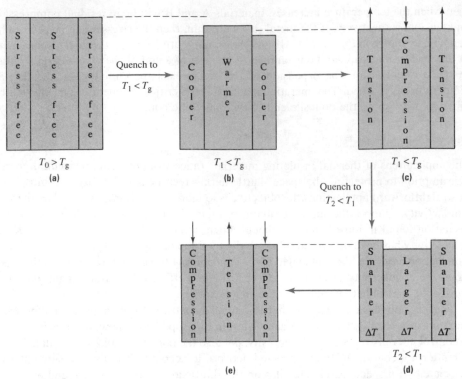

FIGURE 13.6–2 The thermal process for producing tempered glass is a two-step quenching operation: **(a)** initially the material is stress free; **(b)** during the first quench the exterior experiences a larger temperature change than the interior so the exterior attempts to contract more; **(c)** as a result, the first quench places the interior in compression and the exterior in tension; **(d)** during the second quenching operation the interior experiences a larger total temperature change than the exterior; **(e)** as a result, the stress profile is inverted and the surface is placed in residual compression.

can fail under even modest tensile stresses. One method used to increase the tensile fracture stress of a glass is to thermally treat the material so that the surface is placed in residual compression.

The appropriate thermal process, illustrated in Figure 13.6–2, is a two-step quenching operation beginning at a temperature slightly above T_g. During the first processing step the surface of the glass is quenched to a temperature just below T_g. As in the case of thermal shock, the temperature gradient across the thickness of the sample results in differential thermal strains, which in turn generate a tensile stress at the surface. If the original temperature is too high, the glass will fail during the initial quenching operation. If, however, the initial temperature is selected judiciously, the glass will survive the first quenching step. The result is a material with a warm viscous interior ($T_{\text{interior}} > T_g$) and a relatively cool surface layer ($T_{\text{surface}} < T_g$) in residual tension. The stress profile in the glass plate after the first quench is shown in Figure 13.6–2c. Since the interior is viscous, the internal stresses relax.

During the second quenching operation the entire glass sample is cooled to room temperature. During this stage the total temperature drop of the interior is larger than that of the exterior. Thus, the interior contracts more than the already rigid surface layer, resulting in the development of a residual compressive stress on the surface. (See Figure 13.6–2e.)

Tempered glass is routinely used for the side and rear windows of automobiles. Although the tempering process raises the tensile fracture stress of the glass, it also generates a significant amount of stored energy within the material. When an advancing crack passes through the compressive surface layer and enters the interior tensile region, it releases this stored energy. The result is the generation of stress waves that propagate through the strained material, shattering the glass into a great many small fragments. Car windshields are not fabricated from tempered glass alone, because a small stone that impacts the windshield at high velocity could shatter and scatter it. Instead of being made from only tempered glass, by law windshields are composed of a "sandwich" of two sheets of glass held together by a layer of elastomeric polymer between them. If the windshield shatters, the polymer sheet anchors the fragments of glass together so they do not become projectiles.

13.6.5 Support Structure for Orbiting Telescopes

Telescopes are instruments composed of mirrors, lenses, and electrooptical devices used to collect electromagnetic radiation from distant objects in space. The resolution of these instruments is sensitive to the optical alignment of the components. As such, the support structure for the active elements must be rigid and stable. Some of the more demanding applications from a materials point of view are in orbiting space stations or satellites. In this environment the support structure material must possess several characteristics, including:

1. A low coefficient of thermal expansion ($< 1 \times 10^{-6}\ °C^{-1}$) to minimize thermal strains over the service temperature range (typically -150 to $100°C$)
2. A high specific elastic modulus to minimize elastic strains and the mass of material that must be placed in orbit
3. Reasonable impact resistance to minimize the potential for catastrophic failure from a collision with a small meteorite
4. The ability to resist thermal and radiation damage (to be discussed in Chapter 15)

It is difficult, if not impossible, to find any single material that simultaneously satisfies all these requirements. Most common materials cannot satisfy the low α_{th} requirement. Of those that can, most ceramics are susceptible to impact damage, most polymers are subject to radiation damage, and most metals are too dense to satisfy the specific elastic modulus requirements. Thus, as is often the case in demanding applications with apparently competing property requirements, the answer is a composite material. Since composites are described in detail in the next chapter, we simply state here that the material of choice in this application is a fiber-reinforced composite consisting of carbon fibers in a stabilized epoxy matrix.

13.6.6 Ceramic-to-Metal Joints

Many devices require that a ceramic or oxide glass be attached to a metal. Examples include the glass globe of a light bulb attached to its metallic base and a silicon integrated circuit chip attached to a metal lead frame (substrate). If a ceramic-to-metal joint is subjected to changes in temperature, the difference in the expansion coefficients of the materials can lead to thermally induced stresses via the mechanism discussed in

Section 13.5. This problem is amplified if the joint is poorly designed so that the ceramic is placed in tension.

How can the problem be minimized? Consider the attachment of a silicon chip to a metal lead frame. One might consider using either Al or Au because of their high electrical conductivities. Unfortunately, their thermal expansion coefficients, $\alpha_{th}(Al) = 23.6 \times 10^{-6}\ °C^{-1}$ and $\alpha_{th}(Au) = 14.2 \times 10^{-6}\ °C^{-1}$, are much higher than the value for silicon, $\alpha_{th}(Si) = 2.6 \times 10^{-6}\ °C^{-1}$. One relatively straightforward approach to minimizing the problem is to use a different metal with a lower α_{th} value. Kovar, which is a ternary alloy of Fe-Ni-Co, is a good alternative, since it has $\alpha_{th} \approx 5 \times 10^{-6}\ °C^{-1}$. Other ceramic-to-metal joining options will be discussed in Chapter 17.

13.6.7 Cryogenic Materials

Cryogenic engineering deals with the storage and use of materials at very low temperatures. Several specific applications are summarized in Table 13.6–1. Cryogenic engineering places severe demands on the materials used at these temperatures. Specifically, high-quality insulating materials are needed to maintain the operating temperature, and structural integrity demands that the containment materials maintain strength and ductility at low temperatures.

The principles of insulation were discussed previously. The key property requirement is low thermal conductivity K. Typical cryogenic insulators include fiber blankets, foams, vacuum flasks, and gas-filled powders. A vacuum flask is a container in which the inside and outside walls are separated by an evacuated region. Heat transfer across two noncontacting materials is not possible by conduction, but rather, heat transfer may occur by radiation or convection. By evacuating the space between the layers the convective heat loss is minimized. By metallizing one surface, radiation losses are kept to a minimum. This is the idea behind the familiar thermos bottle. The concept can be extended by fabricating hollow powder spheres of a low-K material and then backfilling these spheres with an inert gas characterized by a K lower than that of air. This design modification eliminates the difficulties of generating and maintaining a vacuum.

Low-temperature structural metals are most frequently FCC metal alloys, usually aluminum- or nickel-based. BCC alloys are generally unsuitable for these applications, since they usually undergo a ductile-to-brittle transformation as the service temperature decreases. In some less mechanically demanding applications, a polymer, such as nylon or Teflon, can serve as both the insulator and the structural support for the system.

TABLE 13.6–1   Examples of cryogenic engineering applications.

1. *Gas liquefication separation.* Each of the constituent gases in air liquefies at a different temperature (O_2 at 90 K, N_2 at 77 K, H_2 at 20 K, and He at 4 K). This principle permits their separation in a distillation column. It is also used in the production and transportation of natural gas.
2. *Food freezing.* Food is often frozen by spraying the package with liquid N_2 (77 K).
3. *Superconductivity.* Superconducting devices must operate below the critical temperature of the superconducting material. Before the advent of high-T_c materials, this meant that all superconducting devices operated in the cryogenic regime.
4. *Rocket propulsion.* Liquid hydrogen and liquid oxygen are important lightweight fuels for aerospace applications.
5. *Cryobiology.* Long-term preservation of biological samples and some applications in surgery.

Source: Adapted from Table 4–46 in *CRC Handbook of Materials Science,* Vol. I, CRC Press, Boca Raton, FL.

SUMMARY

In addition to the temperatures that characterize large-scale changes in atomic structure, such as the melting temperature or the glass transition temperature, the three most important thermal properties of a material are coefficient of thermal expansion, heat capacity, and thermal conductivity.

The heat capacity of a material represents the amount of energy required to raise the temperature of one mole of the material by one degree. Specific heat is the analogous quantity for a kilogram rather than a mole of material.

The conduction of thermal energy through a solid as a result of a temperature gradient is analogous to the diffusion of an atomic species as a result of a concentration gradient. In steady-state heat flow, the relevant materials property is thermal conductivity. In time-dependent heat flow the important materials property is thermal diffusivity.

Temperature changes result in dimensional changes known as thermal strains. If the object is unconstrained and the temperature is uniform throughout a material, then thermal strain can occur without the generation of thermal stress. If, however, the object is constrained so that thermal strains cannot be accommodated, a thermal stress will develop. In turn, the thermal stress can lead to temperature-induced failure through one of several mechanisms, including thermal shock in ceramics, or delamination or warping in composites.

KEY TERMS

heat capacity c

specific heat C

thermal capacitance C_{th}

thermal conductivity K

thermal diffusivity D_{th}

thermal shock

HOMEWORK PROBLEMS

1. Provide five examples of applications of materials in which thermal properties are of utmost importance.

2. Calculate the energy of a photon with a wavelength of 1 μm.

SECTION 13.1
Introduction

3. Verify that for an isotropic solid the relationship between α_v and α_{th} is as given in Equation 13.2–3.

4. The bond-energy curves for two materials are sketched in Figure HP13.1.
   a. Which material exhibits a larger ε_{th} for a given ΔT?
   b. Which material has the higher melting temperature?

5. The text suggests that the thermal expansion of ceramics is low in part because they have more open structures than do most metals. Cite data to support or refute this contention.

6. All highly oriented polymers have a negative thermal expansion coefficient along the orientation axis. Suggest a mechanism for this observation. (You might want to consider the thermal expansion behavior of a graphite crystal.)

7. Problem 3 showed that for an isotropic solid the relationship between α_v and α_{th} is $\alpha_v \approx 3\alpha_{th}$. For copper, estimate the maximum change in temperature for which this estimate results in less than 1% error.

8. A 4-cm-diameter metal bar is 2.000 m long at 25°C. If this bar is 2.002 m long at 1000°C, calculate α_{th}.

SECTION 13.2
Coefficient of Thermal Expansion

FIGURE HP13.1

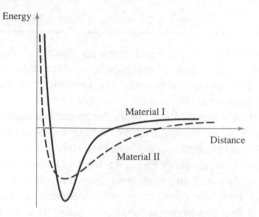

9. Recently, Japanese scientists developed polyimide polymers that show a near zero thermal expansion coefficient in a plane. We suspect the chemical structure is less important than the arrangement of the molecules. How might the molecules be arranged?

10. When you drive over a bridge, you probably notice the "clunks" that the car makes as you cross. The clunks are expansion gates. Describe the reason for their necessary presence. If they occur every 100 ft, about how much expansion/contraction must they permit?

11. Suppose you have an Al, a polycarbonate, and a SiC rod each of which is 1 m long at 25°C. Calculate the temperature change necessary to cause each bar to increase its length by 1 mm.

12. Rank the following classes of materials in order of increasing magnitude of coefficient of thermal expansion: metals, ceramics, thermoplastic polymers, thermoset polymers. Justify your answer.

13. Based on the data in Table 13.2–1, do you think Fe or W has a higher melting temperature? Which material has a higher modulus of elasticity? Why?

14. A wire spans two aluminum supports 1000 ft apart. The wire is attached 75 ft off the ground at the poles, but sags to 55 ft at the midpoint on a −20°F Maine winter night. How high will the wire be from the ground on a 95°F sunny Maine August day? State your assumptions, but include two possible cases: (a) the wire is a high-voltage wire with a copper core, and (b) the "wire" is an optical waveguide reinforced with Kevlar®.

15. Give an example of a material that contracts when heated.

16. How does the coefficient of thermal expansion for an amorphous solid compare with the corresponding value for its crystalline counterpart? Does your answer depend on whether the temperature range of interest is above or below the glass transition temperature?

17. Is it possible to increase the volume of a sample of silicon, initially at room temperature, by 1% through a change in temperature? Is it possible to decrease the volume of the same sample to 99% of its room temperature value by a change in temperature?

18. If you want to remove a metal lid that is "stuck" on a glass jar, should you hold the container under hot or cold water? Explain why.

19. Suppose you are trying to set a world record in the long jump on a hot day. Would you rather have the judges use a metal or a plastic measuring tape, assuming each has been calibrated at room temperature? Explain why.

SECTION 13.3
Heat Capacity

20. Why is the heat capacity of polymers so high? How does it compare with that of other materials on a per unit volume basis?

21. Sketch a plot of the heat capacity of a metal as it is heated through its melting temperature.

22. Suppose an aluminum pathway that forms part of an integrated circuit has a length 100 μm and a cross-sectional area 2 μm^2 and carries a current of 10^{-4} A for 2 s. Assuming that all the heat generated goes into increasing the temperature of the Al strip, calculate the resulting temperature rise.

23. How much heat energy is necessary to raise the temperature of each of the following amounts of material by 5°C?
    a. One mole of aluminum.
    b. One kilogram of beryllia.
    c. One cubic meter of polystyrene.

24. A fellow student suggests that for equal volumes, materials with higher values of specific heat generally have high values of thermal capacitance. Comment on the validity of this statement after examining the characteristics of Al, Cu, and Fe.

25. Which would cause a more severe burn on your skin: a drop of molten polyethylene, or a drop of molten polytetrafluoroethylene? Assume that both drops have the same initial temperature and mass and both stick to your skin.

26. Give an example of an engineering application in which you would use a material with:
    a. A low value of specific heat
    b. A high value of specific heat

27. From the perspective of energy cost per kilogram, is it more expensive to melt Al, Fe, MgO, or PET?

28. a. Estimate the thermal conductivity of Al at 100°C.
    b. Estimate the thermal conductivity of Si at 25°C.

29. a. Consider a plate-glass window with a thickness of 0.5 cm and an area of 1000 cm^2. If the thermal conductivity for soda-lime-silica glass is 1.7 W/(m-K), estimate the heat loss through this window if the inside temperature is 25°C and the outside temperature is 0°C.
    b. Some insulating windows are composed of two parallel plates of glass with an air gap between them. Explain why this is an effective method for reducing heat loss.

30. Explain why many cooking pots have a copper bottom and a plastic handle. Do you think the handle is more likely to be a thermoplastic or a thermoset material?

31. In general, which class of materials—ceramics, metals, or polymers—can be subjected to higher quench rates? Why?

32. Use the Wiedemann-Franz law and the data in Table 10.2–1 to estimate the thermal conductivities of gold, alumina, and polytetrafluoroethylene at room temperature. Compare these estimates with the experimental values given in Table 13.4–1 and comment on their accuracy.

33. Explain why it is difficult to change the electrical conductivity of a metal without altering its thermal conductivity.

34. A typical porous ceramic brick used to line furnaces has a thermal conductivity of ~1.3 W/(m-K). Estimate the heat flux through this material if the furnace temperature is 1200°C and the ambient temperature is 25°C.

35. Why does a metal ice cube tray feel colder than a plastic ice cube tray even though both are at the same temperature?

36. The thermal conductivity of air is roughly 0.026 W/(m-K). Examine the data for the thermal conductivity of oxide glass and the polymers, shown in Table 13.4–1. Why is melt-blown oxide glass fiber bat the preferred material for home insulation?

37. As stated in the previous problem, the thermal conductivity of air is roughly 0.026 W/(m-K). The thermal conductivity of polymers, shown in Table 13.4–1, is much higher than that of air and varies only by a factor of about 3 from highest to lowest. Does

SECTION 13.4
Thermal Conduction
Mechanisms

it really matter what fiber is used to insulate sleeping bags or coats, so long as plenty of trapped air is included?

38. Pyrex glass can go from boiling water into ice water. Ordinary window or container glass will fail under this test. What is the difference between the two glasses? (We realize that the thermal expansion coefficient of Pyrex is about 1/2 that of container glass, but how is this achieved?)

39. In Chapter 10 we learned that scientists are working to increase the electrical conductivity of polymers. Great strides have been made. Predict the thermal conductivity of a polymer that has metal-like electrical conductivity.

SECTION 13.5
Thermal Stresses

40. A rod is 0.5100 m long at 25°C. At 225°C the rod is loaded with a stress of 276 MPa and its length is found to be 0.5358 m (under load at 225°C). If the elastic modulus of this material is 5.52 GPa, calculate the coefficient of thermal expansion for this material. State any assumptions.

41. A wire is 2 m long at 50°C. It has a coefficient of thermal expansion of $\alpha = 3 \times 10^{-6}\ °C^{-1}$. At 100°C a stress of 50,000 psi is required to stretch the wire (elastically) to a length of 2.002 m. What is the modulus of elasticity of this material?

 42. Suppose you need to fabricate a thick ceramic component that will experience large temperature fluctuations while in service. Which of the materials below would you select?

Material	Fracture stress	Thermal conductivity	Coefficient of thermal expansion	Modulus
A	700 MPa	290 W/(m-K)	$9 \times 10^{-6}\ °C^{-1}$	210 GPa
B	800 MPa	120 W/(m-K)	$3 \times 10^{-6}\ °C^{-1}$	140 GPa
C	900 MPa	300 W/(m-K)	$7 \times 10^{-6}\ °C^{-1}$	70 GPa
D	1000 MPa	180 W/(m-K)	$4 \times 10^{-6}\ °C^{-1}$	140 GPa

43. Is thermal shock more likely to result from rapid heating or from rapid cooling of a material? Why?

44. Explain how the use of a ceramic glaze on a ceramic component (of different composition than the glaze) can influence the effective fracture toughness of the component. Would you suggest using a glaze with a higher or a lower coefficient of thermal expansion if you are trying to maximize the effective fracture toughness of the component? Why?

45. Give an example of a situation in which a temperature change induces each of the following types of behavior:
    a. A thermal strain but no thermal stress
    b. A thermal stress but no thermal strain
    c. Both a thermal strain and a thermal stress

46. A nylon rod and an alumina rod are each inserted through two identical holes drilled through a brass plate. The outside diameter of the rods and inside diameter of the holes are identical at 25°C.
    a. Determine the state of stress (compression or tension) in each rod and in the plate when the temperature is raised above 25°C.
    b. How does your answer change if the temperature is decreased below 25°C?

47. An iron rod is constrained in such a way that at 0°C it is stress-free. Calculate the maximum temperature to which this rod can be raised without the thermally induced stress exceeding its yield strength.

48. A glass fiber with a thermal expansion coefficient of $5 \times 10^{-6}\ °C^{-1}$, an elastic modulus of 72 GPa, and a fracture toughness of 0.75 MPa-$\sqrt{m}$ is rigidly supported between two posts so that it is stress-free at 25°C. Calculate the maximum flaw size that this fiber can tolerate if it must survive a temperature increase of 200°C.

9. When washing a drinking glass, a person wearing a diamond ring can occasionally scratch the surface of the glass. If the glass has the same properties as the fiber described in Problem 48, estimate the depth of such a scratch necessary to cause the glass to fracture when it is taken from a dishwasher ($T = 50°C$) and immediately submerged in ice water.

0. A soda-lime-silicate glass is sealed to Kovar at 500°C. Will the seal crack when cooled to room temperature? If the seal is made between borosilicate glass and Kovar, will the problem develop? (Assume oxide glass has a modulus of 10^7 psi and a strength of 6000 psi.)

51. Sketch the stress through the thickness of a plate of glass (from surface to surface) that has been tempered. Repeat for a glazed ceramic object, assuming the glaze is a glass in compression.

52. A Rupert's drop is a drop of glass that looks like a drop of water that has just released from a dripping faucet. In fact, it is formed by allowing a drop of oxide glass to fall into water. The surface of the cooled drop has enormous surface residual compressive stresses and, hence, is so strong it can be hit by a hammer without shattering. The tail of the drop, however, is so thin that is can be broken by hand. When the tail is snapped, what is the fate of the glass?

53. Suppose you take a handful of aligned fibers, say, glass fibers, and put them into a mold to make a fiber-reinforced rod. You mix some resin, say, epoxy, and pour it into the mold over the fibers. After the epoxy cures, you take the composite out of the mold and note that all the fibers are toward one side. You decide to used the "unbalanced" composite as a flag holder for ice fishing. When you retrieve the composite from your garage that winter, when it is $-10°F$, it is no longer straight. Why?

54. What would you get if you combined carbon fiber (which has a near zero thermal expansion coefficient) with β spodumene (Figure 13.2–1) to make a fiber-reinforced glass-ceramic?

COMPOSITE MATERIALS

MATERIALS IN ACTION   Composites

Composites are a class of materials that incorporate two or more different materials. Composites are designed to take advantage of the most desirable characteristics of each material while eliminating or at least reducing the negative properties of the constituents.

A highly publicized use of composite materials occurs in sporting equipment such as fishing rods, tennis rackets, golf clubs, and skis, to name but a few. These items were traditionally made from wood, which in itself is a natural composite. While acceptable equipment can be made using wood, high performance man-made fibrous composites offer superior performance. With the space program as a technology driver, many new materials and processes were developed. For example, carbon fibers, composed of graphite-like crystals oriented along the fiber axis, can be very strong and lightweight. Epoxy glue, on the other hand, is not very strong but can serve as a glue to bind carbon fibers. By incorporating carbon fibers in epoxy glue it is possible to obtain a material that is light, tough, and strong! By changing the size, volume fraction, and geometric arrangement of the fibers, elastic and strength properties can be tailored to meet specific demands. Suddenly new possibilities are available for creative design that were previously unavailable. Nowhere have these opportunities been more vigorously pursued than in the sporting equipment field. New equipment appeared that made extensive use of composites to achieve new possibilities. Suddenly club players had the power and control of champions of the past. The technological revolution in sports was thrust to new heights—bobsleds, skis, vaulting poles, running shoes, tennis rackets, golf clubs, and so on. Companies who failed to grasp this new technology faded from the scene, while new ones with a strong technology and marketing bases appeared and became household words—Nike, Calloway, and so on.

In this chapter we will develop the fundamentals of composite materials.

14.1 INTRODUCTION

In the last few chapters we learned that some properties of materials are directly related to one another. Examples include: (1) the link among elastic modulus, bond strength, and thermal expansion coefficient (all characteristics of the bond-energy curve); (2) the inverse relationship between strength and ductility (or toughness) within a class of materials; and (3) the proportionality of electrical and thermal conductivity in metals (the Wiedemann-Franz law). Often these correlated properties cause problems for design engineers. In particular, the inverse relationship between strength and toughness is unfortunate, since in many applications one would like to maximize both properties. When no single conventional material is able to satisfy the competing design specifications for a given application, one solution may be a composite material.

Composites are combinations of two materials in which one of the materials, called the **reinforcing phase,** is in the form of fibers, sheets, or particles and is embedded in the other material, called the **matrix phase.** The reinforcing material and the matrix material can be metal, ceramic, or polymer. Typically reinforcing materials are strong with low densities while the matrix is usually a ductile, or tough, material. If the composite is designed and fabricated correctly, it combines the strength of the reinforcement with the toughness of the matrix to achieve a combination of desirable properties not available in any single conventional material. The downside is that man-made composites are generally more expensive than conventional materials. Examples of some current applications of composites include the diesel piston, brake shoes and pads, tires, and the Beechcraft aircraft in which 100% of the structural components are composites.

Because of the variety of available reinforcement and matrix materials, as well as the ability to combine them in a wide range of volume fractions, composites can be produced with a broad range of properties, notably elastic modulus, strength, and toughness combinations. The ease of tailoring properties to specific needs is one of the most important attributes of composites. Another advantage over conventional materials is that composites can be designed to exhibit specific properties in specific directions (i.e., their anisotropy can be beneficial).

So far, the bulk of our discussion has tacitly assumed that a composite is designed and selected for its mechanical properties. This is a reasonable assumption, since the majority of modern composites are designed as structural materials. Thus, the term *reinforcing phase* is appropriate. Other important classes of composites, however, are used for their interesting electrical, thermal, or magnetic properties. For these composites the terms *matrix* and *reinforcing phase* may not be suitable. For example, research is now in progress for producing superconducting wires. These composites will likely consist of a flexible core to provide mechanical support, a thin concentric layer of superconducting material to carry the current, and a protective and insulating surface layer. While the bulk of this chapter discusses composites designed for structural applications, we will investigate composites designed for other properties in Section 14.6.

14.2 HISTORY AND CLASSIFICATION OF COMPOSITES

A composite is a combination of a reinforcing phase present in the form of particles, whiskers, or fibers in a matrix that holds the discrete reinforcement pieces together and provides them with lateral support. Many composites occur naturally. Examples include wood, which essentially consists of cellulose fibers in a matrix of hemicellulose and lignin; and bones, which are composites of soft protein collagen and hard minerals called apatites. However, the majority of advanced composite materials with unique properties are synthetic. These are the focus of discussion in this chapter.

Composite materials have been developed in response to the need for man-made materials with unusual combinations of properties such as high **specific strength** (i.e., high strength-to-density ratio) and stiffness in one direction, or a combination of high strength and toughness not available from any single material. Straw-reinforced mud bricks for construction of huts, and laminated bows made from wood, animal tendons, and silk, are examples of composites that were in use as early as several centuries B.C. A few thousand years later, the Wright brothers used "fibers and schellac" to strengthen the structure of their airplane.

The history of modern composites, however, begins with the introduction of fabric-reinforced phenolic resins in the 1930s and glass fiber–reinforced plastics in the 1940s. Since then, a steady stream of new composite materials has evolved.

One of the most common classification schemes for composites is based on the geometry of the reinforcing phase. **Unidirectional fiber–reinforced composites** (i.e., composites in which all of the fibers are aligned parallel to one another) are strong in the direction of fiber alignment but are weak in the transverse direction. That is, most fiber-reinforced composites are anisotropic. On the other hand, particle-reinforced composites such as concrete are nearly isotropic in behavior. As shown in Figure 14.2–1, **concrete** contains a mixture of (small) sand particles and (large) gravel stones in a cement matrix. Since the particles are nearly spherical and are mixed randomly, all directions within the composite are equivalent.

The aerospace industry is weight-conscious, and therefore requires materials with high specific strength and stiffness. Figure 14.2–2 shows a picture of the vertical takeoff and landing AV-88 Harrier aircraft in which composites account for 28% of the structural weight. Composites save weight, increase maneuverability, and may provide the capability to avoid radar detection. Figure 14.2–3 shows the composite parts in a commercial Boeing 757 aircraft.

Figure 14.2–4 shows specific strength versus specific modulus for a number of composite and conventional materials. The materials in the top right hand corner of the plot are desirable because of their high specific stiffness and specific strength. Fibrous composite materials score consistently higher than conventional materials. As discussed in Chapter 9, the elastic modulus is fixed for a given class of metals, while strength can be varied by control of the microstructure. In contrast, both the modulus and strength of composites can be varied considerably.

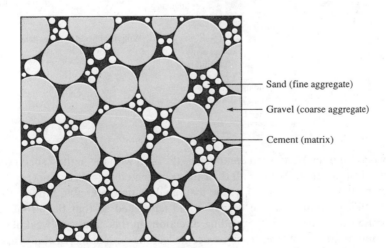

Sand (fine aggregate)

Gravel (coarse aggregate)

Cement (matrix)

FIGURE 14.2–1

A schematic illustration of the idealized structure of concrete containing large gravel stones and small sand particles in a matrix of cement. The wide particle size distribution aids in achieving a higher volume fraction of the reinforcing phase.

FIGURE 14.2–2   A picture of the AV-88 Harrier aircraft, in which composites account for 28% of structural weight. *(Source:* Journal of Metals *43, no. 12 (December 1991), p. 16, a publication of the Minerals, Metals & Materials Society, Warrendale, PA.)*

FIGURE 14.2–3

List of composite parts in the main structure of the Boeing 757-200 aircraft. •
(Source: Boeing Commercial Airplane Company.)

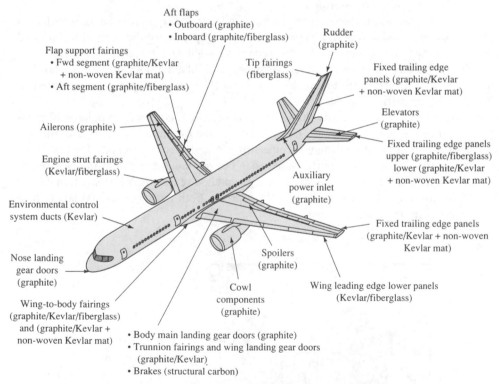

Aft flaps
• Outboard (graphite)
• Inboard (graphite/fiberglass)

Rudder (graphite)

Flap support fairings
• Fwd segment (graphite/Kevlar + non-woven Kevlar mat)
• Aft segment (graphite/fiberglass)

Tip fairings (fiberglass)

Fixed trailing edge panels (graphite/Kevlar + non-woven Kevlar mat)

Ailerons (graphite)

Elevators (graphite)

Engine strut fairings (Kevlar/fiberglass)

Fixed trailing edge panels upper (graphite/fiberglass) lower (graphite/Kevlar + non-woven Kevlar mat)

Auxiliary power inlet (graphite)

Environmental control system ducts (Kevlar)

Fixed trailing edge panels (graphite/Kevlar + non-woven Kevlar mat)

Nose landing gear doors (graphite)

Spoilers (graphite)

Wing leading edge lower panels (Kevlar/fiberglass)

Wing-to-body fairings (graphite/Kevlar/fiberglass) and (graphite/Kevlar + non-woven Kevlar mat)

Cowl components (graphite)

• Body main landing gear doors (graphite)
• Trunnion fairings and wing landing gear doors (graphite/Kevlar)
• Brakes (structural carbon)

For any class of fiber-reinforced composites, the ones with the highest specific strength and modulus values generally have all their fibers aligned in one direction (i.e., unidirectional fiber-reinforced composites). If the loading direction is known (and is always the same), then the composite can be designed and fabricated so that the strong and stiff (fiber) direction coincides with the loading direction. In this case the "weakness" in the other directions is not a problem. If, however, the loading direction is not known, or varies

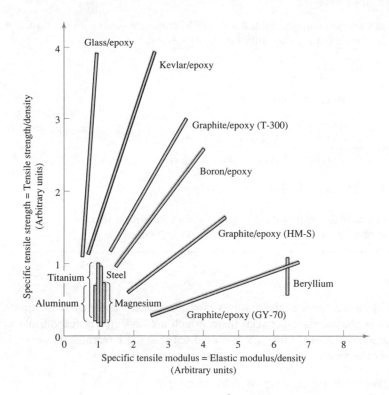

FIGURE 14.2–4

A plot of specific strength versus specific elastic modulus for several monolithic structural materials and polymer-based composite materials. The data in the figure demonstrate the superiority of composites as structural materials.   ·
(Source: Adapted from C. Zweben.)

with time, then a nearly isotropic composite is required. The fibers must be arranged so that a portion of them is oriented in each of several directions within the material. This type of fiber architecture yields properties that are between those of the "strong" and "weak" directions in aligned fiber composites. That data at the lower left end of the ranges in Figure 14.2–4 correspond to nearly isotropic composites.

···

EXAMPLE 14.2–1

A robotic arm with a circular cross section used in a space vehicle is subjected to a constant tensile force along its axis. The length of the arm is fixed by other design considerations. Derive an expression for use in selecting a material that will minimize the weight of the arm. Assume that the applied stress cannot exceed 50% of the ultimate strength. How would the material choice change if the design requirement called for a fixed stiffness value? [Hint: From mechanics, the bending stiffness S of the arm is given by $S = EI$, where E is the elastic modulus and I is the bending moment of inertia (I for a circular rod is $\pi d^4/64$).]

Solution

Since this is unidirectional loading, we can use the properties of unidirectional composites when comparing materials. If the applied force is F, the stress is $F/A = F/(\pi d^2/4)$ where d is the arm diameter. We can thus write,

$$\frac{4F}{\pi d^2} = 0.5\sigma_u$$

where σ_u = ultimate tensile strength. The mass M of an arm with length l and density ρ is $M = \rho(\pi d^2/4)l$. If we solve the stress equation for πd^2 and substitute this expression into the mass equation, we find:

$$M = 2Fl\left(\frac{\rho}{\sigma_u}\right)$$

Since F and l are constant, the lightest arm will be from a material with the highest σ_u/ρ value. What changes if we require fixed stiffness? If the stiffness is S, then,

$$I = \frac{\pi d^4}{64} = \frac{S}{E}$$

Solving this expression for d yields:

$$d = 2\sqrt{2}\left(\frac{S}{E\pi}\right)^{0.25}$$

Substituting this value into the equation for mass gives:

$$M = 2l(S\pi)^{0.5}\left(\frac{\rho}{E^{0.5}}\right)$$

The lightest arm with a specified stiffness is one with the highest value of $E^{0.5}/\rho$. It is customary, however, to provide the data in the form of E/ρ, as in Figure 14.2–4.

14.3 GENERAL CONCEPTS

In this section we consider the features of reinforcement and matrix materials, and the properties of the interface between them, which are collectively responsible for the unusual characteristics of composites. Before discussing these features, however, it is useful to consider the mechanics of fiber strengthening in composites.

14.3.1 Strengthening by Fiber Reinforcement

In most unidirectional fiber-reinforced composites, the fibers do not run continuously from one end of the component to the other. If the fiber length is significantly less than the component dimensions, then the material is known as a discontinuous fiber-reinforced composite. When a discontinuous fiber with a high elastic modulus is embedded in a low-modulus material and the resulting composite is loaded in the fiber direction, the fibers carry a higher load than does the matrix. This is the principle of fiber strengthening. There are, however, several conditions that must be met in order to achieve maximum strengthening of the composite from the fibers. One requirement is that the fiber length exceed some critical (minimum) value.

Figure 14.3–1a shows an isolated fiber embedded in a matrix in the unloaded state. Let the modulus of the fiber be E_f and that of the matrix be E_m. If a normal stress σ is applied to the composite, the matrix and fiber will change shape as shown in Figure 14.3–1b. The applied stress generates both normal and shear stresses in the fiber as shown in Figure 14.3–1c. Note that the types of stress present (shear and/or normal stress) and the stress magnitude vary along the length of the fiber. The tensile stress in the fiber increases from zero at the ends to a maximum value in the central region. The shear stress distribution is nearly the mirror image of the tensile stress state. The fiber length required for the normal stress to reach its peak value is determined by a number of factors including the ratio $(E_m/E_f)^{0.5}$ and the spatial arrangement of fibers within the matrix. Optimum strengthening requires that the fibers be long enough that the stress developed within them is significantly larger than the nominal stress on the composite. If the fibers are too short, then the normal stress that they carry at the time the composite is carrying its capacity load will be well below their ultimate tensile strength and the advantages of strengthening by the fibers will be less than fully realized.

The minimum or **critical fiber length** l_c required for effective strengthening is a function of several variables including the diameter of the fiber d, the ultimate tensile

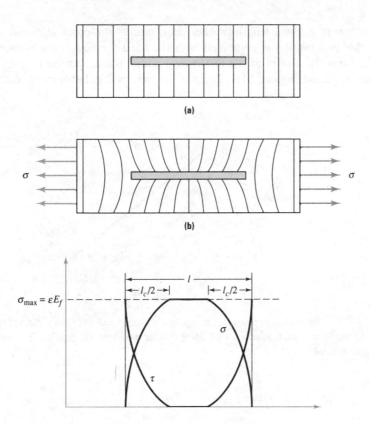

FIGURE 14.3–1

(a) Isolated fiber embedded in a matrix, **(b)** the distortions associated with tensile loading of the composite in part a, and **(c)** the distribution of shear and normal stresses on the outer fiber surface (also the interface region) along the fiber length in response to the loading.

strength of the fibers σ_{fu}, and the matrix shear yield strength τ_{my}. The second factor sets the limit on the load carried by the fibers, and the third factor describes the ability of the matrix to transmit the load from one fiber to the next. The relationship among these variables is found using a force balance. The force carried by the fibers is equal to the normal stress in the fibers multiplied by their cross-sectional area. This force is transferred to the fibers via a shear stress acting on the fiber surface. Mathematically, the force balance is:

$$\pi d \left(\frac{l_c}{2}\right)\tau_{my} = \left(\frac{\pi d^2}{4}\right)\sigma_{fu} \tag{14.3–1}$$

or, equivalently,

$$\frac{l_c}{d} = \frac{\sigma_{fu}}{2\tau_{my}} \tag{14.3–2}$$

The quantity l_c/d is known as the **critical aspect ratio.** Typically, the critical aspect ratio ranges from 20 to 150 for most fiber and matrix materials. Since a typical fiber diameter is between 10 and 30 μm, critical fiber lengths are on the order of 0.2–4.5 mm. In most practical instances, the fiber length l is much larger than l_c.

14.3.2 Characteristics of Fiber Materials

The most common geometrical shape for the reinforcing phase in a high-performance structural composite is a fiber. The reason for this is that the strength of a brittle material is inversely related to the square root of its maximum flaw size and the chance of having a large flaw in a given length of fiber decreases as the cross-sectional area decreases. When this observation is combined with the need for significant surface area for load transfer from the matrix to the reinforcing phase, the advantage of long, slender fibers becomes apparent. Figure 14.3–2 shows the relationship between strength and diameter

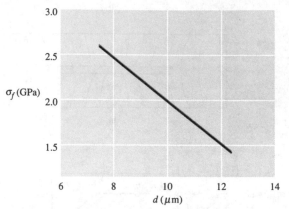

FIGURE 14.3–2 Relationship between fiber strength and diameter for carbon fibers. *(Source: K. K. Chawla, Composites Material Science and Engineering, 1987, Springer-Verlag, New York. Reprinted with permission of Springer-Verlag, New York Publishers.)*

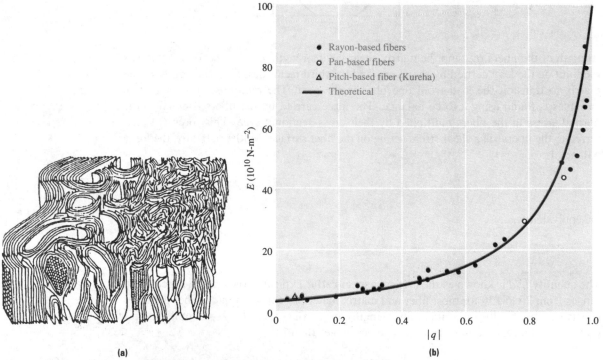

(a)

(b)

FIGURE 14.3–3   **(a)** A three-dimensional schematic of a carbon fiber showing how the graphitic-like plane can be oriented so that the fiber axis lies in this plane. Since all graphitic-like planes do not lie along the fiber axis, variability in fiber properties depends on the degree of orientation **(b)** Variation in elastic modulus of carbon fibers as a function of the degree of orientation represented by the parameter $|q|$. A value of $|q|$ equal to zero represents random orientation, and 1 represents perfect orientation. *(Source: K. K. Chawla, Composites Material Science and Engineering, 1987, Springer-Verlag, New York. Reprinted with permission of Springer-Verlag, New York Publishers.)*

for carbon fibers. Thus, a small diameter is an advantage. In fact, the strongest materials on earth are fibers.

In Chapters 2 and 9 we learned that the elastic modulus of materials increases with bond energy. Thus, for a fiber with mixed covalent/secondary bonding (a common characteristic of many reinforcing fibers) the modulus can be increased by orienting the covalent bonds along the fiber axis. For example, the elastic modulus of carbon fibers can be increased significantly by orienting the graphitic-like planes to coincide with the fiber axis, as shown in Figure 14.3–3a. The relationship between the elastic modulus and degree of orientation is shown in Figure 14.3–3b.

A small diameter is also important in providing a fiber with much needed flexibility. (See Example 14.2–1, in which we showed that fiber bending stiffness is proportional to the fourth power of the diameter.) Thus, minor decreases in fiber diameter result in an enormous decrease in bending stiffness or increase in flexibility. Flexible fibers are much better suited for complex fiber weaves during manufacturing of composites. In sum, as the diameter of a fiber decreases, its elastic modulus, strength, elongation-to-break, and flexibility all increase.

EXAMPLE 14.3–1

Flexible nylon fibers, with $E = 3$ GPa and diameter 25 μm, can be easily wound on spools. What must be the diameter of the following fibers (which have circular cross section) to show the same flexibility?

   a. SiC
   b. Pitch-based C fibers
   c. Kevlar-49®
   d. Boron

Solution

In Example 14.2–1, we used the result that the bending stiffness of a circular fiber is equal to EI with $I = \pi d^4/64$. Since flexibility is the inverse of bending stiffness, the values of $1/EI$, or, equivalently, the EI product itself, for the various fibers must be equal. The elastic moduli for the fibers are listed in Table 14.3–1. Using these data we find:

$$EI(\text{nylon}) = 3 \text{ GPa} \times \frac{\pi(25\ \mu\text{m})^4}{64}$$

and

$$EI(\text{SiC}) = 430 \text{ GPa} \times \frac{\pi(d^4)}{64}$$

Hence,

$$d(\text{SiC}) = (3/430)^{1/4}(25\ \mu\text{m}) = 7.2\ \mu\text{m}$$

Similarly,

$$d(\text{pitch-based C}) = (3/140)^{1/4}(25\ \mu\text{m}) = 9.5\ \mu\text{m}$$
$$d(\text{Kevlar-49}) = (3/131)^{1/4}(25\ \mu\text{m}) = 9.7\ \mu\text{m}$$

and

$$d(\text{boron}) = (3/400)^{1/4}(25\ \mu\text{m}) = 7.3\ \mu\text{m}$$

TABLE 14.3–1   Properties of typical fibers used in composite materials.

Fibers	Density (g/cm^3)	Elastic modulus (GPa)	Tensile strength (MPa)	Axial CTE (°C^{-1})
E-glass	2.6	72	1.7×10^3	5×10^{-6}
S-glass	2.5	87	2.5×10^3	5.6×10^{-6}
PAN-based C-fiber	1.7–1.9	230–370	1.8×10^3	-0.5×10^{-6}
Pitch-based C-fiber	1.6–1.8	41–140	1.4×10^3	-0.9×10^{-6}
Single-crystal graphite	2.25	1000	20.6×10^3	—
Kevlar-49	1.44	131	3.8×10^3	—
Kevlar-149	1.47	186	3.4×10^3	—
Spectra (polyethylene)	0.97	117	2.6×10^3	—
Boron	2.5	400	2.8×10^3	4.9×10^{-6}
FP (alumina)	3.9	379	1.38×10^3	6.7×10^{-6}
SiC particles	3.3	430	3.5×10^3	4.9×10^{-6}
SiC whiskers	3.5	580	8×10^3	4.9×10^{-6}
SiC fibers	2.6–3.3	180–430	2–3.5×10^3	4.9×10^{-6}
Stainless steel	8.0	198	0.7–1.0×10^3	18×10^{-6}
Tungsten	19.3	360	3.8×10^3	11.6×10^{-6}
Molybdenum	10.2	310	2.45×10^3	6.0×10^{-6}

Note: CTE = coefficient of thermal expansion.

Attractive fibers for composite reinforcement must have high strength and high elastic modulus. They must also be suitable for production in small diameters. Fibers have been successfully fabricated from metals, ceramics, and polymers. Properties of some common fiber materials are discussed in this section; the methods used for producing fibers are discussed in Chapter 16. The fiber properties tabulated include tensile strength, elastic modulus, density, and coefficient of thermal expansion.

Polymers are frequently used to bind glass fibers together in composites. The fibers are made from oxide glasses containing various fractions of oxides of silicon, sodium, aluminum, calcium, magnesium, potassium, and boron. Glass fibers are categorized on the basis of their compositions and corresponding properties as E-glass, C-glass, or S-glass. E-glass fibers are very good electrical insulators, C-glass fibers have high chemical corrosion resistance, and S-glass fibers have high strength and can withstand high temperatures. Table 14.3–1 lists typical properties of glass fibers.

Boron fibers are used for stiffening aluminum matrices. Since boron is inherently brittle, it is chemically deposited on a tungsten (W) wire or a carbon-coated glass fiber. The fibers containing the W filament are expensive, but have superior properties compared with glass-filament fibers. Since the chemical vapor deposition process (discussed in Chapter 16) often results in surface defects, the surface of boron fibers is polished to remove defects and is also coated with a thin layer of SiC to produce compressive surface stresses and pacify the surface, as shown in Figure 14.3–4. The surface treatments considerably enhance the fracture strength of B fibers; however, they also add to their cost. Properties of boron fibers are given in Table 14.3–1.

Carbon fibers have a highly distorted and defective graphitic-like structure. The atoms within the basal planes are covalently bonded. Parallel basal planes are joined together by

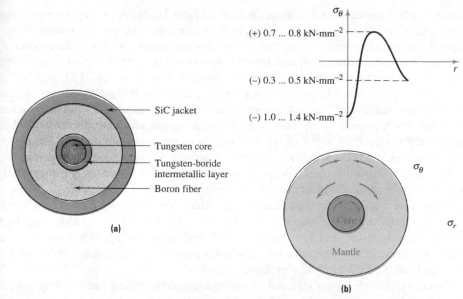

FIGURE 14.3–4 Schematic cross sections of a boron fiber. **(a)** The fiber is a composite consisting of a series of concentric layers. **(b)** The residual stress pattern across a section of the boron fiber. In this view the intermetallic layer and SiC jacket are not shown. *(Source: K. K. Chawla,* Composites Material Science and Engineering, *1987, Springer-Verlag, New York, Reprinted with permission of Springer-Verlag, New York Publishers.)*

weak van der Waals bonds. The bonds perpendicular to the basal planes are secondary bonds. Carbon fibers are produced so that the basal plane lies along the fiber axis. Therefore, the elastic modulus along the fiber axis can be as high as 1000 GPa, as shown in Figure 14.3–3b. In the transverse direction (i.e., the direction perpendicular to the fiber axis), the modulus can be as low as 35 GPa.

The mechanical, chemical, electrical, and other physical properties of carbon fibers vary widely depending on the raw material and the processing treatment. The raw material for carbon fiber is called a *precursor* and consists of a fiber itself. While practically any fiber can be transformed with heat into carbon fibers, only a few precursors are commercially viable. Polyacrylonitrile may be used to make high-modulus, high-strength fibers using pyrolysis, which is described in Chapter 16. In addition, carbon fibers are made on a commercial scale using pitch. Pitch is a product of oil refining (asphalt pitch) or of combining coal with appropriate fluids (coal tar pitch). The pitch is converted into a liquid crystalline material, melt-spun into fibers, and then pyrolyzed.

The first synthetic polymer reinforcing fibers were nylon and polyester. These fibers have reasonable strength and modulus, giving good toughness. They are generally used in soft matrices, such as reinforcement of rubbers in tires, belts, and hoses. About two decades ago polymer fibers with much higher molecular orientation and without chain folds were developed. These materials represented a significant increase in strength and modulus over conventional nylons and polyesters, revolutionizing the development of stiff, strong, lightweight (all polymer) composites. Examples of these polymers are Kevlar and Spectra. The high molecular orientation of Kevlar and other similar aramids (short for aromatic polyamides) is achieved by dissolving the inherently stiff polymer in a solvent so that a liquid crystalline solution is obtained. The solution is spun into a fiber with extremely high axial molecular alignment. Spectra is formed by dissolving ultrahigh-molecular weight flexible polyethylene chains in a solvent to form a gel. The gel is spun into a fiber and collapsed by multiple drawings at high temperatures. In both of these processes, which are described in more detail in Chapter 16, chain folding is almost completely absent and molecular orientation is high. While these fibers have excellent tensile properties at room temperature, their use temperature is limited to a few hundred degrees C or less. In addition, their axial and lateral compressive properties are extremely poor.

The most common ceramic reinforcing materials include fibers of Al_2O_3 and SiC as well as SiC whiskers and particles. Other ceramic fibers with good properties include silicon nitride, boron carbide, and boron nitride. SiC fibers are made using a chemical vapor deposition process or by controlled pyrolysis of organosilane precursors, while whiskers are obtained by vapor phase growth. These processes are discussed in Chapter 16. Much like carbon fibers, the properties of ceramic fibers depend on the processing method used to fabricate them. Because ceramic fibers are brittle, it is important to produce them in small diameters to achieve flexibility and good strength.

A variety of metals can be used to draw high-strength wires, which can serve as metal fibers. The most prominent metal fibers include beryllium alloys—which have high strength, high modulus, and low density—steel, and tungsten. The strengths of metal fibers are consistent and reproducible. The properties of typical carbon, polymer, ceramic, and metal fibers are given in Table 14.3–1.

In summary, the ideal fiber material for strengthing and stiffening a matrix requires the following attributes: low density, high tensile strength, high modulus of elasticity, high flexibility, and the ability to form a strong interface with the matrix.

14.3.3 Characteristics of Matrix Materials

Like fibers, matrix materials can be polymers, ceramics, or metals. Carbon is also used as a matrix material with carbon fibers in a class of composites known as carbon-carbon composites. The primary purposes of the matrix materials are to provide lateral support to the fibers and transfer loads. They also are a source of toughness in the composite, since the majority of fiber materials are brittle. Cracks that have propagated through a brittle fiber are stopped when their tips encounter relatively tougher matrix materials. An exception to ductile matrix materials is ceramic matrix materials, which are inherently brittle. Composites using ceramic matrices, such as reinforced concrete, are used in compressive load applications, or the brittle behavior is countered by carefully tailoring the interface properties. This approach is discussed in Section 14.3.4. Typical matrix materials and their properties are given in Table 14.3–2.

TABLE 14.3–2 Typical properties of matrix materials.

Material	Density (g/cm^3)	Elastic modulus (GPa)	Tensile strength (MPa)	CTE (°C^{-1})
Epoxy	1.05–1.35	2.8–4.5	55–130	30–45×10^{-6}
Polyester	1.12–1.46	2–4.4	30–70	40–60×10^{-6}
Copper	8.9	120	400	16.5×10^{-6}
Ti–6Al–4V	4.5	110	1000	
Stainless steel	8.0	198	700–1000	18×10^{-6}
High-strength aluminum alloys	2.7	70	250–480	23.6×10^{-6}
Magnesia (MgO)	3.6	210–310	97–130	13.8×10^{-6}
Lithium-alumino-silicate (glass ceramic)	2.0	100	100–150	1.5×10^{-6}
Silicon carbide	3.2	400–440	310	4.8×10^{-6}

Note: CTE = coefficient of thermal expansion.

4.3.4 Role of Interfaces

Interfaces play an important role in determining the properties of composites. Consider 1 cm^3 of a unidirectional composite made from 25-μm-diameter continuous fibers embedded in a matrix. If the fibers are arranged in a square array and are on average 50 μm apart, the volume fraction of fibers is approximately 20%. The total fiber-matrix interface area is approximately 314 cm^2, compared with the 6 cm^2 external surface area. The typical fiber volume fraction in composites is 2 to 3 times the above amount, and the interfacial area increases proportionally. The large interfacial area can significantly affects the properties of composites, in particular the crucial properties of toughness and ductility.

If the matrix and the fibers have different coefficients of thermal expansion, then cooling from a high fabrication temperature causes differential thermal contraction between the fiber and the matrix and results in thermal stresses at the interfaces (see Section 13.5). This problem can be minimized by matching the expansion coefficients of the fibers and matrix. However, some differential expansion will always be present, and the fiber-matrix interface must have sufficient strength to survive temperature changes. Interfaces also provide a preferential path for oxygen diffusion or moisture uptake that may facilitate composite property degradation. For example, if moisture diffuses along the fiber-matrix interface while an airplane is on the ground, the moisture may freeze upon flight, causing further degradation of the interface and cracking. Thus, the nature of bonding at the interfaces is important.

The fiber-matrix interfacial bond strength can be characterized by the shear stress required to cause sliding between the fiber and the matrix. The interface bonding can be mechanical, chemical, or both.

Under certain conditions the differential thermal expansion described above can actually aid in the formation of a mechanical bond between the matrix and the fiber. Since most matrix materials have a higher coefficient of thermal expansion than the fibers, cooling from elevated processing temperatures results in compression, or frictional, bonding across the fiber-matrix interface. Curing temperatures for thermosetting resins are in the 100–200°C range, and the expansion coefficients of these materials are high compared with glass and carbon (see Tables 14.3–1 and 14.3–2), which are the most common reinforcing materials. Thus, this type of bonding can be easily accomplished in these composites. Mechanical bond strength can be increased by making the fiber surface rougher, which must be done carefully to avoid compromising fiber strength. However, in most instances, mechanical bonding alone is not sufficient; it is used to supplement strength derived from chemical interactions.

Chemical bonding can be in the form of a wettability or secondary bond, which results from electron interactions between the fiber and matrix. This type of bonding occurs when, during fabrication, the molten matrix "wets" the fiber. As discussed in Section 8.2.4, the important variables are the surface energies of the matrix and fibers, which in turn determine the contact angle, the area of contact, and, ultimately, the bond strength. Because the electronic interactions associated with this type of bond occur over a distance of only 2–4 angstroms, it is important that the two bonding surfaces come into intimate contact with each other. Cleanliness of the fiber surface is necessary for such contact to occur. Layers of dust or oil can prevent these interactions, thereby reducing the bond strength.

A second type of chemical bonding, referred to as reaction or primary bonding, can occur as a result of mass transfer through solid-state diffusion between the fiber surface and the surrounding matrix. This can result in the formation of a compound at the interface. In such cases, the interface becomes less sharply defined and is more appropriately

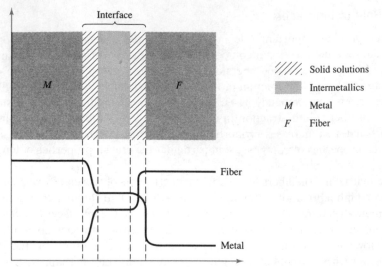

FIGURE 14.3–5   Schematic of the interface region in a metal-matrix composite showing the metal (M) and fiber (F) concentrations and the metal-fiber reaction compounds that may form in the interface region.   *(Source: K. K. Chawla, Composites Material Science and Engineering, 1987, Springer-Verlag, New York. Reprinted with permission of Springer-Verlag, New York Publishers.)*

referred to as an interface region (or interphase region). Figure 14.3–5 shows the various regions of such an interface in a metal-matrix composite. Since the bonds are primary and a result of extensive chemical interactions between fiber and matrix, the interfacial strength is much greater than with wettability bonds alone. Coatings and coupling agents are frequently used to promote chemical reactions and increase the bond strength.

EXAMPLE 14.3–2

Many carbon fibers have an imperfect graphitic structure often with the basal planes forming the surface of the fiber. These surfaces are difficult to wet and are essentially incapable of forming primary bonds with epoxies or other matrices. Fibers may be treated with oxidizing reagents such as nitric acid to change the surface properties. The acid attacks the imperfections in the fiber surface. This results in blunting of surface cracks, an increase in surface area, and creation of a "rough" surface. In addition, the acid alters the fiber surface chemistry by creating hydroxyl (—OH) and other reactive groups that are capable of forming primary bonds. Describe how the acid treatment changes the fiber and composite properties.

Solution

The blunting of surface cracks increases the strength of the fiber with a corresponding increase in the strength of the composite. The rough surface improves the mechanical bonding, the increased surface area permits a higher density of secondary bonds across the fiber-matrix interface, and the generation of reactive surface groups increases the density of primary bonds across the interface. Hence, the interfacial strength increases, with a corresponding improvement in composite properties. It is interesting to note that epoxy composites made using untreated carbon fibres are virtually useless.

14.3.5 Fiber Architecture

In previous sections we discussed how fiber, matrix, and interface characteristics influence the properties of composites. One additional factor in determining composite properties is the fiber arrangement, also known as the fiber architecture.

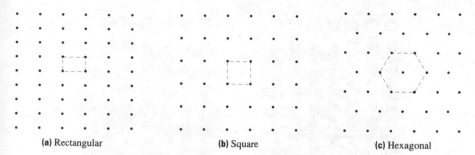

(a) Rectangular   **(b)** Square   **(c)** Hexagonal

FIGURE 14.3–6 Rectangular, square, and hexagonal array of fiber arrangements in unidirectional composites. The hexagonal array results in a transversely isotropic material.

As mentioned previously, unidirectional composites have all their fibers oriented in a single direction. Figure 14.3–6 shows the various geometric arrangements possible in the plane perpendicular to the fiber direction, called the transverse plane. These arrangements can be rectangular, square, or hexagonal. A hexagonal fiber array has the same nearest-neighbor distance between fiber centers along three directions in the transverse plane. Because of this symmetry, these composites behave roughly like isotropic materials in the transverse plane and are termed transversely isotropic. They are preferred by many design engineers because their analysis is considerably simplified. Regardless of the fiber architecture, however, the properties of unidirectional composites in the transverse plane are vastly inferior to those in the fiber direction. Therefore, these composites are unsuitable for applications involving multiaxial (multidirectional) loading.

Laminate composites can be designed to have fibers arranged along different directions. Figure 14.3–7 shows thin sheets of unidirectional composites stacked in an arrangement so that fibers are oriented along 0°, 90°, and ±45° directions. The thin sheets, called the laminae, can be stacked together in a regular arrangement to make up the laminate composite. If the fibers are stacked in the 0° and 90° directions only, the strength of the composite is high along these directions, but such composites have poor shear resistance. To obtain good shear resistance, the fibers must also be stacked along the ±45° orientations. Such laminates are strong in all directions within the plane containing the fibers but are weak in the direction normal to the fiber planes.

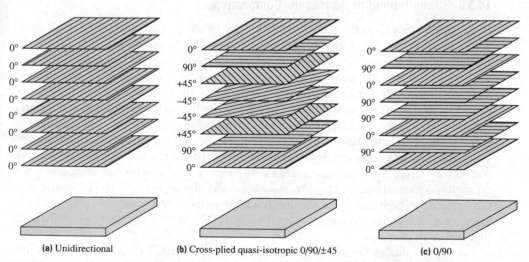

(a) Unidirectional   **(b)** Cross-plied quasi-isotropic 0/90/±45   **(c)** 0/90

FIGURE 14.3–7 Arrangement of fibers in laminate composites: **(a)** unidirectional, **(b)** cross-plied, transversely isotropic containing fibers in 0°, 90°, and ±45° directions, and **(c)** two-directional composite with fibers in 0° and 90° directions.

FIGURE 14.3–8

Two-dimensional fabric weave architecture: **(a)** plain weave, and **(b)** five-harness satin weave.   *(Source: K. K. Chawla, Composites Material Science and Engineering, 1987, Springer-Verlag, New York. Reprinted with permission of Springer-Verlag, New York Publishers.)*

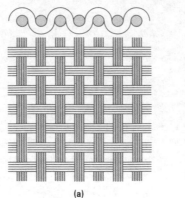

(a)   (b)

DESIGN EXAMPLE 14.3–3

The lid of an electronic package is made from a glass-reinforced polymer composite. The corners of the lid must be able to withstand minor impact loads that may be applied in any direction during service. What fiber architecture would you recommend for this application?

Solution

The corner regions will be subjected to multiaxial loads. Hence, the fiber architecture should consist of plies in the $0°$, $\pm 45°$, and $90°$ directions. This will ensure nearly uniform high strength in all directions in the plane of the composite (not in the thickness direction). Another common arrangement in multiaxial applications is the 0, $\pm 30°$, $\pm 60°$, $90°$ stacking arrangement.

Another type of fiber arrangement is a two-dimensional fabric, as shown in Figure 14.3–8. It is possible to have three-dimensional weaves that result in composites with nearly isotropic properties. A three-dimensional network of fibers can also be obtained by random orientation of discontinuous fibers. These composites also display isotropic properties and are less expensive than their woven counterparts. Their properties, however, are inferior to those of woven composites because of the short fiber length and inherent low fiber volume fraction.

14.3.6 Strengthening in Aggregate Composites

If, in contrast to a fiber, the reinforcing phase has roughly equal dimensions in all directions, the material is known as an **aggregate composite.** As shown in Figure 14.2–1, an example of such a composite is concrete, which is a mixture of fine aggregate in the form of sand and coarse aggregate in the form of gravel, in a matrix of cement (calcium aluminosilicate). By mixing large and small particles, the volume fraction of reinforcement can be increased because the small particles can fit into the cavities among the large particles. Typical total aggregate (fine plus coarse) volume fractions in concrete are 0.60 to 0.75. The aggregate strengthens the composite in two ways. First, the inherent compressive strength of the aggregate is higher than that of the matrix. Second, the high volume fraction of irregularly shaped aggregate particles provides strength in compression through mechanical interlocking. As an added bonus, the aggregate is less expensive than the cement, so high aggregate volume fractions reduce the composite cost. Similarly, asphalt on the highway, which is about 90% aggregate and 10% tar, is a low cost material that serves its purpose.

Concrete is one of the most common structural materials. Its use, however, is somewhat limited, since its tensile strength is much lower than its compressive strength. A composite's tensile strength can be increased by "reinforcing" the aggregate composite with steel rods. These reinforcing rods are placed in the composites in the locations that will bear the largest tensile loads. The interfacial strength between the "fibers" and the matrix is increased by having ridges on the steel rods, which form a mechanical bond with the surrounding matrix. The addition of the steel rods transforms the composite from nearly isotropic to highly anisotropic.

Another type of aggregate composite is a **particulate composite.** Examples include SiC or Al_2O_3 particles embedded in an aluminum alloy matrix, and rubber particles in epoxy. In these composites the volume fraction of the reinforcing phase is much lower than that in cement. Typically, reinforcing-phase volume fractions are about 0.15. Since the aspect ratio of the reinforcing particles is approximately 1, particulate composites are essentially isotropic.

The strengthening mechanism in particulate composites is different from that in either concrete or fiber-reinforced composites. In particulate composites the particle-matrix thermal expansion coefficient mismatch results in a high dislocation density in the metal matrix during cooling from the fabrication temperature. Thus, matrix strain hardening contributes significantly to the strength of the composite.

14.4 PRACTICAL COMPOSITE SYSTEMS

So far composites have been classified as unidirectional fiber composites, continuous and discontinuous fiber composites, laminates, two- and three-dimensional woven-fabric composites, particle-reinforced composites, and so on. These classifications are based on the architecture and characteristics of the reinforcing phase. Composites are also frequently classified by their matrix types, including metal-matrix, polymer-matrix, ceramic-matrix, and carbon-carbon composites. Some important commercial composites in each of these classes are described below.

14.4.1 Metal-Matrix Composites

Metal-matrix composites (MMCs) are fabricated by plasma spray of the matrix material over properly laid fibers, liquid infiltration methods, physical vapor deposition, hot pressing, squeeze casting, and other methods. These methods are discussed in Chapter 16.

MMCs have good high-temperature capability, good transverse properties, and reasonably high compressive and shear strength, in part because of a combination of good strength and toughness of the metal matrix and good interface bonding. Common MMC systems include boron fibers in an aluminum matrix, SiC fibers or whiskers in an aluminum matrix, SiC in a titanium matrix, and carbon fibers in magnesium, copper, or aluminum matrices. Applications include piston heads for automotive engines, connecting rods in engine crankshafts, gas turbine blades, radar domes, and electronic packages. Typical properties of some MMCs are given in Table 14.4–1.

14.4.2 Polymer-Matrix Composites

Polymer-matrix composites (PMCs) are the workhorse of the composites industry. They can have excellent room-temperature properties at a comparatively low cost. They have been in use for about four decades. Two of the best known early uses in the 1950s are in body panels for the Chevy Corvette and in pleasure boat hulls. The matrix consists of various thermosetting resins and, more recently, thermoplastic polymers reinforced by glass, carbon, boron, or organic fibers. The mechanical and physical properties of some typical PMCs are listed in Table 14.4–1.

TABLE 14.4–1   Properties of some typical MMC and PMC composites in the longitudinal and transverse directions.

Matrix	Fiber	Volume fraction	Density (g/cm³)	Elastic modulus (GPa) Long.	Elastic modulus (GPa) Trans.	Tensile strength (MPa) Long.	Tensile strength (MPa) Trans.
Al	B	0.50	2.65	210	150	1500	140
Ti–6Al–4V	SiC	0.35	3.86	300	150	1750	410
Al–Li	Al₂O₃	0.60	3.45	262	152	690	180
Epoxy	E-glass (unidirectional)	0.60	2.0	40	10	780	28
Epoxy	2-D glass cloth	0.35	1.7	16.5	16.5	280	280
Epoxy	Boron (unidirectional)	0.60	2.1	215	9.3	1400	63
Epoxy	Carbon	0.60	1.9	145	9.4	1860	65
Polyester	Chopped glass	0.70	1.8	55–138	—	103–206	—

The traditional application of PMCs was in secondary load-bearing aerospace structures. Now they are used in I-beams in civil structures, various automotive parts, steel-belted tires, and in sports goods. The cost of PMCs has been steadily decreasing, which makes them attractive enough for use in consumer goods. Another burgeoning application of PMCs is in architectural structures such as the Georgia Dome: the roof is composed of seven acres of woven glass fiber cloth in a PTFE matrix.

14.4.3 Ceramic-Matrix Composites

Ceramic-matrix composites (CMCs) have been used in the form of reinforced concrete for a long time. Because of their low tensile strength, their use has been limited to structures subject to compressive loading. Ceramics are well known for their elevated-temperature oxidation and creep resistance. Therefore, if their brittle behavior could be controlled, these materials would be excellent candidates for use in automotive and aircraft gas turbine hot-section components such as blades, disks, pistons, and rotors. The fracture toughness of conventional ceramics is on the order of 1 to 5 MPa-$\sqrt{m}$. The toughness of CMCs has been boosted to levels of 15 to 20 MPa-$\sqrt{m}$ by embedding ceramic fibers in ceramic-matrix materials. Examples include SiC-reinforced SiC, SiC-reinforced silicon nitride (Si_3N_4), and carbon fiber-reinforced glass. Table 14.4–2 lists properties of typical CMCs.

TABLE 14.4–2   Properties of some representative ceramics and CMC composites.

Matrix	Fiber	Flexure strength (MPa)*	Fracture toughness (MPa-$\sqrt{m}$)
Al₂O₃	—	350–700	2–5
MgO	—	200–500	1–3
SiC	—	500–800	3–6
SiO₂ glass	—	70–150	1
Al₂O₃	SiC whiskers	800	10
SiO₂ glass	SiC fibers	1000	~20
Al₂O₃	BN particulates	350	7

*Flexural strength of brittle materials is determined by using a four-point bend test instead of tensile specimens used for ductile materials to prevent failure from occurring in the grips. It represents an estimation of the tensile strength of the material.

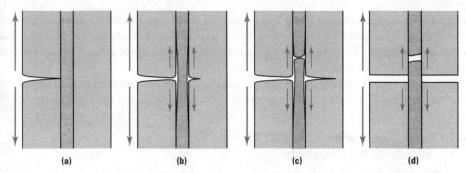

FIGURE 14.4–1   A schematic illustration of how the presence of fibers can be used for toughening of brittle ceramic matrices: **(a)** matrix crack approaching a fiber, **(b)** crack developing on the other side of the fiber while the fiber remains intact, **(c)** fiber break occurring away from the matrix crack plane, and **(d)** fiber pullout accompanied by energy absorption.

Figure 14.4–1 shows the principle behind fiber toughening in CMCs. The interface in these materials consists of mechanical bonding with some strength derived from interdiffusion between the fiber and the matrix. Cracking begins in the brittlest material, often the matrix. When a crack tip encounters a fiber, two things can happen. If the fiber-matrix bond is strong, the crack will continue to grow across the fiber. In this instance, having fibers offers no apparent advantage. However, if the bond is relatively weak, as in CMCs, delamination occurs at the interface, and the crack is temporarily stopped, with some of the energy going into the formation of new surfaces (Figure 14.4–1a). When the load is raised further, the matrix on the other side of the fiber begins to crack (Figure 14.4–1b), but the fiber is still capable of transferring load across the crack faces. This phenomenon is called fiber bridging. When the fiber breaks, usually at a location away from the matrix crack plane (Figure 14.4–1c) in a section containing a random flaw, the matrix crack advances further. However, some load is still carried by the fiber (Figure 14.4–1d) across the crack plane because of continued fiber bridging. During subsequent loading, the fiber is pulled out of the matrix near the short end. This last step requires energy to overcome the frictional forces between the fiber and the matrix and raises the toughness level of CMCs. CMCs represent a relatively new class of materials that have caused considerable excitement in the composite and ceramic industries.

14.4.4 Carbon-Carbon Composites

As the name implies, in this class of fiber-reinforced composites both the matrix and the fibers are fabricated from carbon. These materials offer a unique combination of properties, including the ability to withstand extremely high service temperatures ($>3000°C$), high specific strength, excellent resistance to wear (they can be self-lubricating), good resistance to thermal shock, and reasonable machinability. The maximum use temperature for these composites is limited by oxidation problems. Typical applications include brake components, heat shields, and rocket nozzles—not your household composites because of high costs. Carbon-carbon composites can be fabricated using the CVD methods described in Chapter 16, by impregnating graphite fibers with a carbon-based polymer that is then pyrolyzed to "burn off" the noncarbon atoms in the polymer, or combinations of the two.

14.5 PREDICTION OF COMPOSITE PROPERTIES

Testing of materials is both expensive and time-consuming. It also causes delays in the time required for bringing a new material into the marketplace. Therefore, materials scientists and engineers have spent a good part of their time predicting material properties. This section presents some elementary approaches for estimating properties of composite materials. For simplicity, we limit our discussion to properties of unidirectional fiber–reinforced composites.

Important factors in any model for predicting composite properties are the volume fraction of fibers (V_f), the volume fraction of the matrix (V_m), and the fiber diameter (d).

14.5.1 Estimation of Fiber Diameter, Volume Fraction, and Density of the Composite

In an ideal two-component composite, the fiber and matrix volume fractions are related by the equation:

$$V_f + V_m = 1 \tag{14.5-1}$$

This equation assumes no porosity. In a unidirectional composite the fiber volume fraction is equal to the area fraction occupied by fibers in the transverse plane. Figure 14.5–1 shows a picture of a transverse plane obtained by sectioning a small sample of a composite. V_f can be determined by using a PC with a simple image analysis program such as Adobe Photoworkshop. Notice in Figure 14.5–1 that the fiber diameter and spacing vary significantly as a result of the manufacturing process.

The density of the composite, ρ_c, can be determined by a simple **rule of mixtures** as follows:

$$\rho_c = V_f \rho_f + V_m \rho_m \tag{14.5-2}$$

where ρ_f is the fiber density and ρ_m is the matrix density. The rule of mixtures can be used to calculate any composite property that does not depend on the spatial relationship of its components.

14.5.2 Estimation of Elastic Modulus and Strength

In this section, we derive simple relationships for estimating the elastic modulus and strength of uniaxial composites using the properties of the constituent materials. We

FIGURE 14.5–1

Photomicrograph of the transverse plane of a composite containing alumina fibers (dark circles) in a matrix of Al-Li alloy. The picture shows the distribution of fiber diameters as well as the distances between fibers. *(Source: Reprinted with permission from A. M. Gokhale.)*

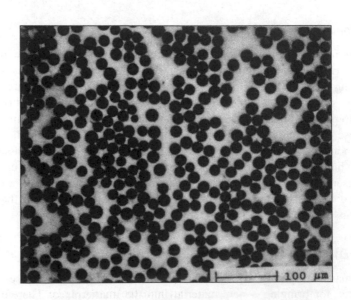

100 μm

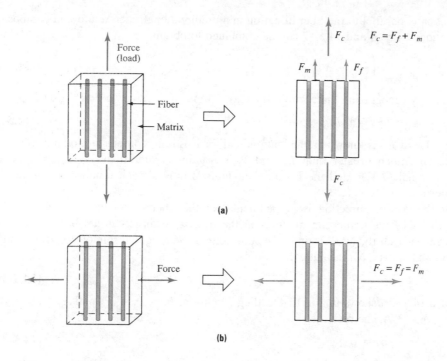

FIGURE 14.5–2

Schematic of a unidirectional composite loaded by a force F: **(a)** in the axial direction, and **(b)** in the transverse direction.

assume for simplicity that the constituent materials are isotropic. This assumption is reasonable except when organic fibers are used.

If the composite is subjected to an axial force F_c, as shown in Figure 14.5–2a, the total force is partitioned into two parts, F_f (force on the fibers) and F_m (force on the matrix). We may write,

$$F_c = F_f + F_m \tag{14.5–3}$$

Noting that $F = \sigma A$, Equation 14.5–3 can be written as:

$$\sigma_c A_c = \sigma_f A_f + \sigma_m A_m \tag{14.5–4}$$

where σ_c, σ_f, and σ_m are the average normal stresses on the composite, fibers, and matrix and A_c, A_f, and A_m are the corresponding cross-sectional areas. Dividing both sides of Equation 14.5–4 by A_c yields:

$$\sigma_c = \sigma_f\left(\frac{A_f}{A_c}\right) + \sigma_m\left(\frac{A_m}{A_c}\right) \tag{14.5–5}$$

Since for continuous-fiber composites the fiber and matrix area fractions are the same as the corresponding volume fractions (i.e., $A_f/A_c = V_f$ and $A_m/A_c = V_m$), the previous equation can be rewritten as:

$$\sigma_c = \sigma_f V_f + \sigma_m V_m \tag{14.5–6}$$

If we assume the interface is "perfect" in that no sliding occurs between the fiber and matrix, then

$$\varepsilon_c = \varepsilon_f = \varepsilon_m \tag{14.5–7}$$

where ε_c, ε_f, and ε_m are the strains in the composite, fiber, and matrix. Equation 14.5–7, known as the **isostrain** condition, is usually assumed to be valid when the loading direction is parallel to the fiber direction in a unidirectional fiber-reinforced composite. Equations 14.5–6 and 14.5–7 can be combined to obtain:

$$\frac{\sigma_c}{\varepsilon_c} = V_f\left(\frac{\sigma_f}{\varepsilon_f}\right) + V_m\left(\frac{\sigma_m}{\varepsilon_m}\right) \tag{14.5–8}$$

Finally, by recognizing that, within the elastic limit, $E = \sigma/\varepsilon$, we find

$$E_{cl} = E_f V_f + E_m V_m \tag{14.5–9}$$

The value of E_{cl} is known as the longitudinal, or isostrain, composite modulus. Since E_f is usually much greater than E_m and V_f is usually greater than V_m, E_{cl} is approximately equal to $V_f E_f$ and the isostrain modulus is only a weak function of the matrix properties.

If the loading direction is perpendicular to the fiber direction, as shown in Figure 14.5–2b, the composite modulus in the transverse direction can be estimated by recognizing that the total strain in the composite is the weighted sum of the strains in the fibers and matrix. Mathematically,

$$\varepsilon_c = \varepsilon_f V_f + \varepsilon_m V_m \tag{14.5–10}$$

Figure 14.5–2b shows that in this loading condition, $F_c = F_f = F_m$ and $A_c = A_f = A_m$. Since $\sigma = F/A$, the result is

$$\sigma_c = \sigma_f = \sigma_m \tag{14.5–11}$$

Therefore, when the load is perpendicular to the fibers, we have what are known as **isostress** conditions. Combining the previous two equations gives:

$$\frac{\varepsilon_c}{\sigma_c} = V_f\left(\frac{\varepsilon_f}{\sigma_f}\right) + V_m\left(\frac{\varepsilon_m}{\sigma_m}\right) \tag{14.5–12}$$

Noting that $\varepsilon/\sigma = 1/E$ and substituting into the previous equation gives:

$$\frac{1}{E_{ct}} = \frac{V_f}{E_f} + \frac{V_m}{E_m} \tag{14.5–13}$$

E_{ct} is known as the transverse, or **isostress,** composite modulus. Since $E_f \gg E_m$, and $V_f \sim V_m$, we find $V_f/E_{ff} \ll V_m/E_m$. Hence, fiber modulus has relatively little influence on E_{ct}.

If, as shown in Figure 14.5–3, the load is neither parallel nor perpendicular to the fiber direction but instead forms an angle θ with the fibers, the composite's elastic modulus, $E_{c\theta}$, varies with θ, as shown in Figure 14.5–4.

....

EXAMPLE 14.5–1

Estimate the elastic modulus of a uniaxial epoxy-glass fiber composite in the longitudinal and transverse directions given the following data: $E_m = 3.0$ GPa, $E_f = 70$ GPa, and $V_f = 0.60$. Compare your result with the data in Figure 14.5–4.

Solution

From Equation 14.5–9, the longitudinal elastic modulus is given by:

$$E_{cl} = (70 \text{ GPa} \times 0.6) + (3 \text{ GPa} \times 0.4) = 43.2 \text{ GPa}$$

From Equation 14.5–13, the transverse modulus is given by:

$$\frac{1}{E_{ct}} = \frac{0.6}{70 \text{ GPa}} + \frac{0.4}{3 \text{ GPa}} \qquad \text{or} \qquad E_{ct} = 7.05 \text{ GPa}$$

....

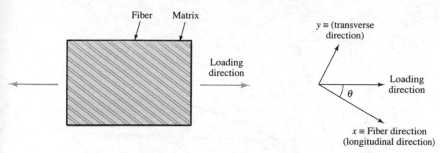

FIGURE 14.5-3   A unidirectional composite system with the x axis oriented along the fiber axis, y axis in the transverse direction, and loading direction at an angle θ to the x axis.

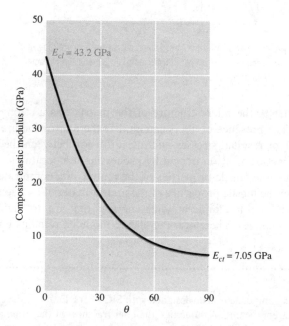

FIGURE 14.5–4

The variation of elastic modulus as a function of orientation with respect to the fiber axis for epoxy-glass composite ($V_f = 0.6$).

The achievable tensile strength of composites loaded under isostrain conditions can be estimated using a variation of equation 14.5–6 as follows:

$$\sigma_{cu} = \sigma_{fu} V_f + \sigma'_m V_m \tag{14.5–14}$$

where σ_{cu} is the ultimate strength of the composite, σ_{fu} is the ultimate strength of the fiber, and σ'_m is the flow stress of the matrix at a strain corresponding to the failure strain of the fiber. It is assumed that once widespread failure occurs in the fibers, the matrix will be unable to withstand the increased stress and the composite will fail. This requires that the volume fraction of fiber be larger than the following critical value V_{crit}:

$$V_{crit} = \frac{\sigma_{mu} - \sigma'_m}{\sigma_{fu} - \sigma'_m} \tag{14.5–15}$$

where σ_{mu} is the ultimate strength of the matrix. We can see from this equation that V_{crit} is higher if the strain hardening capability of the matrix material is high and the strain to fracture in the fiber is low. The important quantities are shown graphically in Figure 14.5–5.

Equation 14.5–14 yields composite strength values that are often not as accurate as predictions of the modulus. A number of factors are responsible for this behavior. As

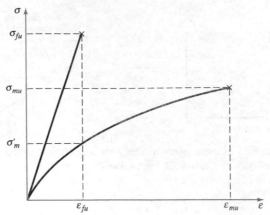

FIGURE 14.5–5 A schematic illustration of the stress-strain curves for a typical fiber and matrix showing the ultimate strength of the fiber (σ_{fu}), the ultimate strength of the matrix (σ_{mu}), and the strength of the matrix at the strain corresponding to the failure strain of the fibers (σ'_m).

discussed in Chapter 9, the microstructure of the matrix has a strong influence on its tensile properties. The presence of fibers can modify the microstructure of the surrounding matrix because of residual stresses and interdiffusion between the matrix and fiber materials. Microstructure is also affected by processing temperatures and conditions of the composite. Thus, the tensile properties of the matrix in a composite can be substantially different from the tensile properties of the matrix material by itself. In contrast, the elastic modulus is related to atomic bonding energies that are not significantly affected by microstructural changes. Therefore, the elastic modulus remains unaffected by processing.

....

EXAMPLE 14.5–2

An aluminum-matrix composite is to be designed with SiC fibers. Estimate the critical fiber volume fraction needed for strengthening. Assume that the fiber fractures at the strain at which the matrix begins to yield. The following data are provided: the yield strength of the matrix is 400 MPa, the ultimate strength of the matrix is 482 MPa, and the fiber strength is 2000 MPa.

Solution

Since the matrix yields at the strain when fiber fracture occurs, we can take σ'_m = yield strength of the matrix = 400 MPa. Thus, from Equation 14.5–15, we find:

$$V_{\text{crit}} = \frac{482 - 400}{2000 - 400} = 0.051$$

Hence, fiber strengthening should occur in this system for a fiber volume fraction as little as 5.1%.

....

14.5.3 Estimation of the Coefficient of Thermal Expansion

The coefficient of thermal expansion is an important design parameter for all structural materials subjected to even modest variations in temperature during service. As a first estimate, the expansion coefficient of the composite, α_c, can be obtained by the rule of mixtures:

$$\alpha_c = V_m \alpha_m + V_f \alpha_f \tag{14.5–16}$$

where α_m and α_f are the expansion coefficients of the matrix and fiber materials. In most common composites, $\alpha_m \gg \alpha_f$, so in a uniaxial composite, the fibers constrain the expansion of the matrix in the longitudinal direction. This causes the composite to expand less in the longitudinal direction and more in the transverse direction than predicted by Equation 14.5–16. More accurate expressions for estimating the expansion coefficients in the longitudinal and transverse directions have been derived, but are outside the scope of this introductory chapter.

14.5.4 Fracture Behavior of Composites

Fracture in composites usually begins with cracking of the most brittle phase. In metal-matrix and polymer-matrix composites this usually means cracking begins in the brittle fibers; in ceramic-matrix composites this usually means cracking begins in the matrix. The manner in which this initial fracture progresses determines the toughness of the composite. To illustrate this point, consider a system of brittle fibers in a ductile matrix. When fracture occurs in an isolated fiber at any point along its length, the stresses carried by the fiber in the vicinity of the crack must be transferred to the surrounding matrix and the other fibers. If the surrounding matrix and fibers are able to withstand the stresses, the fracture will stabilize at that location.

Interfaces play a major role in stabilizing fractures. If delamination occurs at the interface, effective blunting of the crack occurs and the fracture is stabilized. Crack blunting also occurs because of plastic deformation in the ductile matrix. If the fracture is stabilized at one location, it will begin at other locations if the deformation is continued. This process will continue until the damage is so widely spread that the stress originally carried by the fractured fibers can no longer be carried by the uncracked matrix. At this point, ultimate fracture of the composite occurs. This process is schematically illustrated in Figure 14.5–6a along with an accompanying stress-strain diagram. Such a fracture mode is preferable because substantial energy is absorbed both during plastic deformation of the matrix and during delamination. These processes contribute to the overall toughness of the composite materials.

In the previous example widespread damage develops in the composite prior to fracture, leading to a tough material. On the contrary, if the interfaces are so strong that no

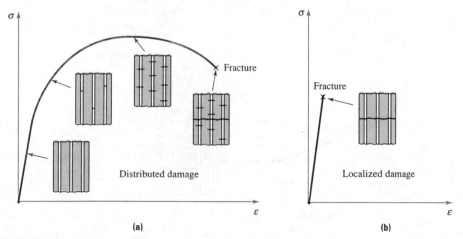

FIGURE 14.5–6   (a) Schematic of a tensile stress-strain diagram and the progress of damage in ductile composites, and (b) the same as part a but in a brittle system.

delamination is possible and the matrix is not ductile enough to effectively blunt the fiber cracks, then the progression of damage remains localized and a low-energy fracture results. This process is shown schematically in Figure 14.5–6b.

The same principles apply to ceramic-matrix composites. Typically the cracking begins in the matrix and the toughness is enhanced by fiber bridging and fiber pullout, as described in Section 14.4.

14.5.5 Fatigue Behavior of Composites

As discussed in Chapter 9, fatigue failures in metals generally initiate at the specimen surface because of microplasticity, which leads to crack formation. These cracks propagate and become larger, causing the final fracture. Fatigue mechanisms in composites are considerably different. The different stages of fatigue of laminates consist of ply cracking, delamination, and ultimately fiber fatigue. This process is illustrated in Figure 14.5–7 in a laminate with fibers in the 0° and 90° orientations. The number of applied fatigue cycles is divided by the number of cycles of failure to derive a cycle ratio. The microstructural changes that result from fatigue damage are then correlated with the cycle ratio. The extent of ply cracking is best represented by the crack density in the laminae oriented at 90° to the loading axis. As shown in Figure 14.5–7a, the crack density increases rapidly at first and then reaches a constant value. Delamination does not occur initially, but occurs rapidly after saturation of ply cracking (Figure 14.5–7b). Delamination saturates when the last stage of composite fatigue, fiber fatigue, begins. As shown in Figure 14.5–7c, when enough fibers have fractured because of fatigue, the composite

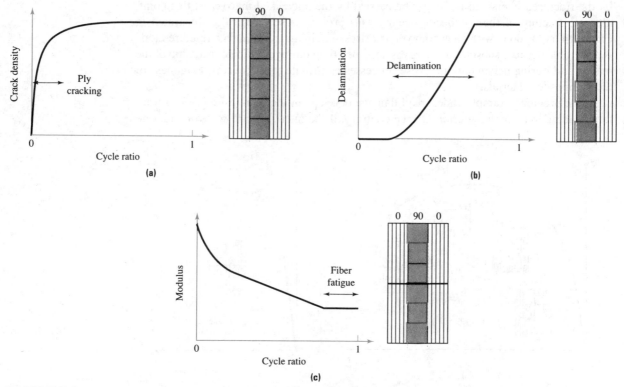

FIGURE 14.5–7   Schematic representation of stages of fatigue damage accumulation: **(a)** ply cracking, **(b)** delamination, and **(c)** fiber cracking. *(Source: H. T. Hahn and L. Lorenzo,* Advances in Fracture Research, *Pergamon Press, 1984.)*

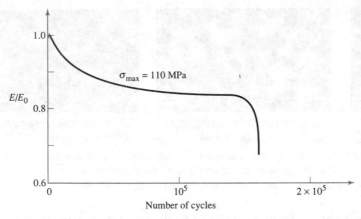

FIGURE 14.5–8   Normalized elastic modulus as a function of fatigue cycles in a glass fiber–polymer matrix composite.   *(Source: K. K. Chawla, Composites Material Science and Engineering, 1987, Springer-Verlag, New York. Reprinted with permission of Springer-Verlag, New York Publishers.)*

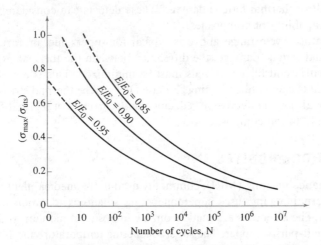

FIGURE 14.5–9   Number of fatigue cycles required to achieve different fractions of stiffness reduction as a function of applied maximum fatigue stress.   *(Source: K. K. Chawla, Composites Material Science and Engineering, 1987, Springer-Verlag, New York. Reprinted with permission of Springer-Verlag, New York Publishers.)*

fails. The elastic modulus of the composite decreases continuously as fatigue damage accumulates and is frequently used as an indicator of the progression of fatigue damage. The normalized modulus value (E/E_0) as a function of cycle ratio is shown schematically in Figure 14.5–8, where E is the elastic modulus and E_0 is the modulus of the undamaged composite.

Figure 14.5–9 shows the relationships between the ratio of the applied maximum stress during a fatigue cycle to the ultimate strength and the number of fatigue cycles to failure. It is common to define fatigue life in terms of the modulus ratio, E/E_0, because loss of stiffness in many composite structural applications constitutes functional failure.

As discussed earlier, high stresses are generated at the interfaces every time a composite containing components with significantly different coefficients of thermal expansion is subjected to a temperature change. If repeated heating and cooling are part of a component's service duty cycle, **thermal fatigue** can occur. This phenomenon affects the interfaces most significantly because the thermal fatigue shear stresses obtain their

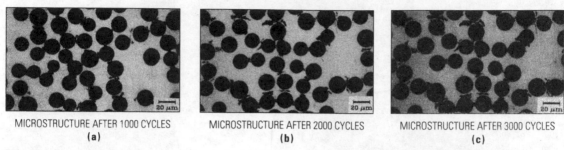

MICROSTRUCTURE AFTER 1000 CYCLES   MICROSTRUCTURE AFTER 2000 CYCLES   MICROSTRUCTURE AFTER 3000 CYCLES
(a)   **(b)**   **(c)**

FIGURE 14.5–10   Damage accumulation in the matrix of an alumina fiber-reinforced Al-Li matrix composite as a result of thermal cycling: **(a)** after 1000 thermal cycles, **(b)** after 2000 thermal cycles, and **(c)** after 3000 thermal cycles. The damaged regions appear as noncircular dark patches within the lighter-colored matrix.   *(Source: Reprinted with permission from A. M. Gokhale.)*

maximum value at that location. Figure 14.5–10 shows the progression of thermal fatigue damage in a metal-matrix composite containing alumina (Al_2O_3) fibers in a matrix of an aluminum-lithium alloy. The specimens were thermally cycled between 300°C and room temperature with no mechanical loading. Cracks and voids appear in the matrix between fibers as a result of thermal fatigue damage. These defects can considerably degrade the load-carrying capability of composites.

The temperature cycle range above is typical for supersonic aircraft structures. To minimize thermal fatigue damage, the difference between the thermal expansion coefficients of the matrix and fiber materials must be minimized. This is not always possible; hence, compliant (flexible) fiber coatings are used to ease the thermal stresses at the interfaces. This step, although effective in reducing the occurrence of thermal fatigue failures, adds considerably to the composite cost.

14.6 OTHER APPLICATIONS OF COMPOSITES

Magnetic resonance imaging (MRI), commonly used in the medical field for noninvasive imaging of internal organs, uses superconducting solenoids to produce high-intensity magnetic fields. The solenoids are made from a composite of niobium-titanium filament in a matrix of high-purity copper. The superconducting temperature for this alloy is about 23 K. Thus, the superconducting cable is immersed in liquid helium at all times. Any change in the applied field leads to heating of the filament, which in turn may cause the filament to lose its superconducting characteristic. This raises its resistance, leading to additional heat generation. To contain runaway behavior, several small-diameter superconducting filaments are embedded in a copper matrix. When the resistance of the filament changes, the current can be conducted by the surrounding copper, which also effectively removes the heat. Figure 14.6–1 shows a close-up of a multifilament superconducting cable.

Recently, superconducting materials containing copper, yttrium, barium, and oxygen have been developed with critical temperatures greater than 100 K (see Sections 10.2.9 and 12.9). This will allow superconducting behavior at liquid nitrogen temperatures. The wires made from these materials will also be composites. An unrefined example of such a wire is shown in Figure 14.6–2.

Electronic package casings are another example of composites used in the electronics industry. With the revolution in integrated chip manufacturing, more-active devices are packaged in smaller and lighter packages (see Section 10.3.5). One factor that limits the density of active devices is the ability to rapidly remove heat. A 10°C rise in temperature can reduce a chip's lifetime by a factor of 2.5 to 3. Thus, thermal conductivity of the casing

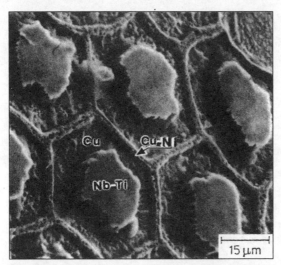

FIGURE 14.6-1   Close-up view of Nb-Ti multifilamentary superconducting cable. The superconducting Nb-Ti filaments are surrounded by a copper matrix.   *(Source: K. K. Chawla, Composites Material Science and Engineering, 1987, Springer-Verlag, New York. Reprinted with permission of Springer-Verlag, New York Publishers.)*

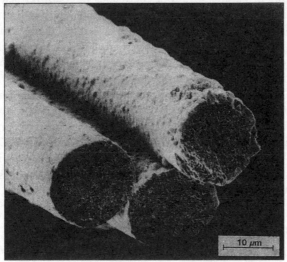

FIGURE 14.6-2   A picture of a Y-Ba-Cu-O superconducting composite wire fabricated by chemical vapor deposition of $YBa_2Cu_3O_{7-x}$ (light surface layer) onto a multifilament Al_2O_3 fiber (darker core).   *(Source: Reprinted with permission from W. J. Lackey).*

material (without the associated high electrical conductivity of metals) becomes critical. Carbon fiber composites can be used to increase the thermal conductivity of package materials. An additional design option to reduce thermal fatigue problems is tailoring the coefficient of thermal expansion of the composite casing to match that of the silicon. Thus, when silicon chips are mounted on the casing, the thermal stresses are low, extending the life of the package. Package casings are commonly made from either an iron-copper alloy called Kovar or a silicon carbide particulate-reinforced aluminum alloy. The composite casing has superior heat transfer characteristics and represents a significant weight reduction.

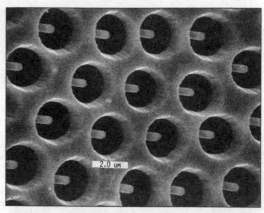

FIGURE 14.6–3   A picture of a eutectic in situ composite used as a low-voltage field emitter array cathode.   *(Source: Reprinted with permission from D. N. Hill.)*

Figure 14.6–3 shows a low-voltage field emitter array cathode for potential use in microwave amplifiers. It is made from a directionally solidified oxide-metal eutectic composite consisting of parallel arrays of continuous fibers of refractory metals such as W and Mo in an oxide matrix. The fibers range from 0.3 to 1 μm in diameter and are about 2 to 5 μm apart. Typical systems include a UO_2 matrix with W fibers, a ZrO_2 matrix with W fibers, and a Gd_2O_3 matrix with Mo fibers. The composites are grown from a melt produced by induction heating, which allows the center of the ingot to melt while maintaining a solid skin of unmelted but compacted powder on the outside. The unmelted skin serves as a crucible to contain the melt. Since the melt temperature ranges from 2200 to 2800°C, any other type of crucible would contaminate the melt. The oxide matrix can be partially etched away and other elements can be vapor-deposited on the surface. The primary application of such composites is a high–current density electron source for high-power, high-frequency microwave devices. When a sufficiently high electric field is applied to the tips of the exposed fibers, electrons are drawn from the surface of the fiber in a process known as field emission.

14.6.1 Estimation of Nonmechanical Properties of Composites

We have learned that several properties of composites can be estimated using the equation

$$P_c^n = \sum_i \left(V_i P_i^n \right) \tag{14.6–1}$$

where P_c is the estimate for the property of the composite, the P_i values are the properties of components of the composite, the V_i values are the corresponding volume fractions, and the exponent n takes on values of $+1$ or -1 depending on the direction for which the property is being estimated.

This equation can be used to estimate several other properties of composites, including density and, under certain conditions, the electrical and thermal conductivities of the material. In the case of density, the value of n is $+1$ regardless of the spatial distribution of the components of the composite. For both thermal and electrical conductivities, the equation is only useful if the second phase is in the form of fibers or sheets. For such composites, n has the value of -1 in a direction perpendicular to the fibers (or sheets) and the value $+1$ in the direction parallel to the fibers (or sheets).

SUMMARY

Composites are an important class of materials used in high-performance structural, thermal, and electrical applications. Structural composites often combine the properties of low density, high strength, high elastic modulus, and good fracture and fatigue resistance. They provide an opportunity for tailoring the properties to meet the needs in any given direction or combination of directions.

Most composites consist of a reinforcement component in the form of small-diameter fibers, whiskers, particles, or flakes. Aside from particles in concrete, fibers are the most common reinforcing agents. The use of small-diameter fibers made from various polymers, metals, and ceramics facilitates the achievement of both high strength and flexibility, while retaining high modulus. Matrix materials may be polymers, metals, or ceramics. The matrix provides lateral support to the fibers in continuous-fiber composites and also transfers the load to the fibers through the fiber-matrix interface in short-fiber composites. The fiber architecture determines the strength and modulus of the composite in various directions. The optimal fiber arrangement depends primarily on the intended application.

The properties of composites also depend on the interfacial characteristics. The interfacial strength determines how efficiently the stress is transferred to the fibers, which is especially critical in short-fiber composites. It controls the fracture toughness and the fatigue resistance of all composites. Interfacial strength is derived from mechanical bonding, chemical bonding, or both.

Important properties such as elastic modulus in various directions, tensile strength, coefficient of thermal expansion, and thermal and electrical conductivity of composites can be estimated from the fiber arrangement and the properties and volume fractions of the matrix and reinforcing materials. Properties that do not depend on the spatial arrangement can be determined using the rule of mixtures.

The mechanisms of fracture and fatigue of composites are different from those in conventional materials. Composite fracture originates in the brittle phase and is slowed from spreading by either the ductile phase or interfacial failure. The high fracture toughness in these systems is derived from spreading the damage over a large area in front of the main crack as opposed to relying solely on plastic deformation of the ductile phase. Composite laminate fatigue failure occurs in three stages: ply cracking, delamination, and fiber fatigue. Composites are also particularly susceptible to thermal fatigue failures.

Although most applications for composites are in high-performance structures, their use is growing rapidly in the electrical and electronic industries. Examples include superconducting cables and electronic package casings.

KEY TERMS

aggregate composites	isostrain	reinforcing phase	unidirectional fiber-reinforced composites
composites	isostress	rule of mixtures	
concrete	laminate composites	specific modulus	
critical aspect ratio	matrix phase	specific strength	
critical fiber length l_0	particulate composites	thermal fatigue	

HOMEWORK PROBLEMS

SECTION 14.2
History and Classification
of Composites

1. Most fiber-reinforced composites consist of three "parts": the fibers, the matrix, and the interfaces. Describe the major functions of each of these "parts."

2. Briefly describe each of the following composite classes:
   a. Unidirectional fiber-reinforced composite
   b. Pseudo-isotropic composite
   c. Particulate composite
   d. Transversely isotropic composite

3. Explain why composites are common in the aerospace industry.

4. Define the terms *specific strength* and *specific modulus*.

5. Give examples of composite materials in the home. Provide a reason why these products contain composite materials.

SECTION 14.3
General Concepts

6. Compare the following strengthening mechanisms: oxide strengthening of nickel alloys, precipitation hardening in aluminum alloys, and dispersion strengthening in concrete.

7. Discuss the advantage of reinforcing mud bricks with straw.

8. Using the data in Tables 14.3–1 and 14.3–2, estimate the percentage weight savings in the robotic arm in Example 14.2–1 if it were to be made from 304 stainless steel, epoxy resin, Ti–6Al–4V, (Al/B)–0.50V_f composite, and (Ti–6Al–4V/SiC)–0.35V_f composite. Use 304 stainless steel as your reference material.

9. What are the advantages of producing fibers with small diameters?

10. Equation 14.3–2 shows the effect of the relationship between fiber properties and matrix properties on the critical length required to achieve maximum effective reinforcement.
    a. What occurs when $1/d$ does not equal or exceed $\sigma_{fu}/2\tau_{my}$?
    b. If a particularly good fiber is available in two diameters, large and small, and two corresponding lengths, long and short, respectively, which fiber do you select for use in a composite?
    c. If the fiber is inadvertently sprayed with a very thin layer of lubricant, what will be the effect on composite properties?

11. Calculate the specific strength (strength/density) and specific modulus, (elastic modulus/density) of stainless steel, Kevlar, and Spectra fibers. Which fiber is stronger, pound for pound?

12. Give an example of fibrous material for which a decrease in diameter will result in an increase in strength. Give an example of a fiber for which you would expect a correlation between modulus and diameter.

13. Explain in your own words the principle behind mechanical bonding. Suggest a means by which you can improve the interfacial strength in steel-reinforced concrete.

14. Explain what is meant by the term *chemical bonding* as it relates to a fiber-matrix interface.

15. Table 14.3–1 does not give the cost of fibers, which is important in manufacturing of composites. If boron fiber cost \$400/lb, carbon fiber costs \$50/lb, and oxide glass fiber costs \$2/lb, which is the best choice for:
    a. High specific strength in a cost-critical application?
    b. High specific modulus in a cost-critical application?

16. What is a lamina? Describe the stacking sequence of laminae in a laminated composite that has good tensile properties in 0° and 90° orientations and also has good shear stress resistance. What design modification would you suggest if shear resistance is not required.

17. Suggest an appropriate matrix to be used with each of the following fiber types:
    a. E-glass

    b. Boron

    c. Oriented polyethylene

    d. SiC

18. Suggest an appropriate application for each of the following fiber types:

    a. S-glass

    b. Kevlar

    c. Al_2O_3

19. Suppose you need to achieve certain mechanical properties in a composite. What is the principal basis for selecting the matrix material? For example, why choose epoxy rather than aluminum?

SECTION 14.4
Practical Composite
Systems

20. Why is it necessary to thoroughly clean the surfaces of glass fibers to achieve a strong bond with epoxy?

21. Explain why a strong interface is a detriment to the properties of ceramic-matrix composites.

22. List two applications that favor the use of

    a. Metal-matrix composites

    b. Ceramic-matrix composites

    c. Polymer-matrix composites

23. Which class of composites (CMCs, MMCs, or PMCs) is generally the least expensive class? Why?

24. What is generally considered to be the most serious problem associated with the use of CMCs? How is this problem addressed?

25. Discuss the advantages and disadvantages of carbon-carbon composites over traditional PMCs.

SECTION 14.5
Prediction of Composite
Properties

26. Using the data in Tables 14.3–1 and 14.3–2 calculate the following properties of composite materials:

    a. The isostrain modulus of epoxy reinforced with 60 volume % for

        (i) B fiber

        (ii) C fiber

    b. The isostress modulus of the composites in part *a*

    c. The isostrain modulus of Al reinforced with 60 volume % for

        (i) B fiber

        (ii) C fiber

    d. The isostress modulus of the composites in part *c*

    e. Estimate the maximum achievable strength of the composites described in parts *a* and *c*.

    Note that the composite may not reach this strength due to the presence of defects.

27. The elastic modulus of magnesium is about 50 GPa, and its density is 1.74 g/cm^3. Find the volume fraction of SiC whiskers all aligned in one direction that must be added to a magnesium-based alloy to increase its specific elastic modulus by 50%.

28. The shear strength of an aluminum alloy is 150 MPa. This alloy is under consideration for being reinforced by SiC whiskers, which have a tensile strength of 8 GPa. Estimate the critical aspect ratio to achieve fiber strengthening assuming *(a)* the interface strength is larger than the matrix shear strength and *(b)* the interface strength is only half of the matrix shear strength.

29. Compare the advantage of 2024-T4 Al (tensile strength = 400 MPa), Ti–6Al–4V (tensile strength = 1100 MPa), and 300 grade maraging steel (tensile strength = 2000 MPa) for use as matrix materials for SiC whisker-reinforced composites in which the whiskers are randomly oriented.

30. How does Equation 14.5–9 depend on fiber length?

31. Show that in 1 cm^3 of a composite containing 50-μm-diameter fibers in a square array with a center-to-center spacing of 100 μm, the volume fraction of fibers is 0.2 and the interface area is approximately 157 cm^2. Assume that the composite is a continuous fiber composite. Calculate the volume fraction of fibers and the interface area if the fibers are arranged in a hexagonal array and the nearest-neighbor distance is 50 μm.

32. Explain how delamination at the fiber-matrix interface can enhance the fracture toughness of composites.

33. Ceramic matrix and fiber materials often contain porosity levels on the order of 10%. How will porosity influence the elastic modulus and the tensile strength of these materials?

34. What is fiber bridging, and how does it enhance the fracture toughness of ceramic-matrix composites? Give an example of a composite system where fiber bridging is the dominant mechanism for toughness.

35. In an epoxy/S-glass composite system, it is desired that the longitudinal elastic modulus be at least twice the transverse elastic modulus. What volume fraction of fibers is necessary to achieve this condition?

36. Show that the critical volume fraction of fibers necessary for fiber strengthening is given by Equation 14.5–15.

37. Explain why the rule of mixtures does not yield a reasonable estimate for the coefficient of thermal expansion.

38. E-glass and epoxy both exhibit isotropic thermal expansion. Do you expect that the coefficient of thermal expansion of an E-glass/epoxy composite will also be isotropic? Explain your answer.

39. Explain why carbon fibers have a modulus of elasticity that is so different along the fiber axis compared with the value in the transverse direction. Do you expect the elastic modulus of a diamond fiber to be different in different directions? Explain.

40. List and briefly describe the three stages of fatigue failure in laminate composites.

41. Explain why the fracture toughness is high in composites with distributed damage.

42. A unidirectional fiber–reinforced composite has a modulus of 20 GPa in the longitudinal direction and 3 GPa in the transverse direction. If the volume fraction of fibers is 0.25, what are the elastic moduli of the matrix and fibers?

43. The pictures shown in Figure 14.5–1 and Figure 14.5–10a, b, and c belong to the same Al-Li/alumina composite system. Construct a grid of test points and estimate the volume fraction of the fibers using all four pictures. Select one of the pictures and measure the area of the composite and fibers directly. Then compute the volume fraction. Compare the two estimates and comment. Now double the test point spacing in the grid and make the same measurements. Compare your answers.

44. Consider a unidirectional fiber-reinforced composite loaded perpendicular to the fibers. What is the relationship between the stress in the composite (σ_c), the stress in the fibers (σ_f), and the stress in the matrix (σ_m)? What is the stress relationship if the load is parallel to the fiber axis?

45. Consider a unidirectional fiber-reinforced composite with the following properties: $E_f = 70$ GPa, $E_m = 6$ GPa, and the isostress composite modulus $E_c = 9$ GPa.
    a. Calculate the isostrain modulus of this composite.
    b. If this composite is loaded in a direction parallel to the fibers and the stress in the matrix is found to be 3 MPa ($= 3 \times 10^{-3}$ GPa), calculate the stress in the fibers under the same load.
    c. Which of the two constituents of this composite, the fibers or the matrix, is more likely to be a ceramic material? Why?

6. A typical textile fiber is composed of crystallites embedded in a matrix of noncrystalline polymer. Can composite theory be used to determine properties? How could you calculate the density of the fiber?

7. Why is it necessary to embed Nb-Ti superconductor filaments in a copper matrix in super-conducting cables?

8. Why are boron fibers coated with SiC?

9. A silver-tungsten composite for an electrical contact is produced by first making a porous W powder compact, then infiltrating pure Ag into the pores. The density of the W compact before infiltration is 13.3 g/cm^3. Calculate the volume fraction of porosity (before infiltration) and the final weight percent of Ag (after infiltration).

50. The density of a composite made from boron fibers in an epoxy matrix is 1.8 g/cm^3. The density of boron is 2.36 g/cm^3 and that of the epoxy is 1.38 g/cm^3. Calculate the volume fraction of boron fibers in the composite.

51. Estimate the thermal conductivity in the longitudinal direction of a unidirectional fiber−reinforced composite with 30 volume percent SiC fibers in an aluminum matrix.

MATERIALS-ENVIRONMENT INTERACTIONS

MATERIALS IN ACTION   Materials-Environment Interactions

Practically all materials can undergo reactions with their environment. Such reactions can occur in many ways, including direct dissolution, absorption of foreign species into the base material, or formation of new phases. While such reactions are usually undesirable, this is not always the case. For example, a very adherent, protective oxide forms on aluminum alloys. Thus, in many applications (but not all, as we shall see), aluminum is impervious to further attack from the environment. A common example occurs with cookware fabricated from aluminum, which is very durable. Another example is found in copper, which has been used for centuries for roofing. Copper soon forms a greenish patina, composed of copper sulfates. The copper sulfates are formed through reactions between atmospherically borne sulfur compounds and the copper. The patina, once formed, acts as a corrosion barrier, and the rate of copper loss is less than 0.1 mm per century.

Most frequently, however, material-environment interactions are undesirable and result in economic loss. In most advanced industrial societies, corrosion represents an annual economic loss of approximately 5% of the gross domestic product. In the United States, that amounts to well over $150 billion every year! A spectacular example of corrosion-induced failure occurred in 1988. The canopy of an Aloha Airlines 737 jet aircraft operating on a routine flight in the Hawaiian Islands suddenly ripped off, and one flight attendant lost her life. The cause of this was related to extensive corrosion of the aluminum alloy canopy material in a salt air environment.

(Source: Robert Nichols, Black Star.)

In this chapter we will study mechanisms of environment-material interactions and ways to eliminate or reduce their most deleterious effects.

15.1 INTRODUCTION

An engineering design is not complete until the interaction of materials with the environment has been evaluated. We must consider both the environment's influence on materials and the materials' influence on the environment. The latter is more true today than ever before, largely because manufacturers are being forced to plan for the disposal of used products. Also, an estimated $150 billion is lost annually to corrosion of products and equipment in the United States alone. Thus, the materials-environment interaction is an important design consideration in a wide range of industries, such as aerospace, power generation, chemical and petroleum, automotive, construction, marine, and electronics.

The environmental degradation of materials and the degradation of the environment by discarded products are intimately related issues. For example, corrosion frequently limits the life of products such as appliances and automobiles. Doubling the service life of these products decreases both waste and pollution from the manufacturing process. Thus, there are financial and ecological incentives for judiciously choosing materials that can resist environmental attack.

Our discussion focuses on designing to minimize the degradation of material properties resulting from interactions with the service environment. The chapter is arranged as follows: liquid-solid reactions (corrosion), gas-solid reactions (oxidation), solid-solid interactions (friction and wear), thermal degradation, and radiation damage.

15.2 LIQUID-SOLID REACTIONS

Many different types of liquid-solid reactions degrade the physical or mechanical properties of engineering materials. Familiar examples include rusting of ferrous alloys in (for example) bridge support structures or car fenders, and cracking of concrete because of expansion of trapped water as it freezes. We divide our discussion of the diverse range of liquid-solid degradation mechanisms into three parts. The key factor used to classify the reactions is whether or not electrical charge transfer plays an important role in the degradation mechanism. If so, the reaction will be classified as **electrochemical corrosion.** The discussion of electrochemical corrosion reactions forms the bulk of Section 15.2 and begins in Section 15.2.2. Many of the remaining degradation mechanisms can be described as **direct dissolution mechanisms** and involve solid dissolving in the surrounding liquid without substantial charge transfer. These reactions are the subject of Section 15.2.1. We begin by investigating two other types of liquid-solid degradation mechanisms: the swelling of polymers in solvents, and the degradation of the strength of silicates with time in the presence of water.

Solvents interact with polymers much as they do with other materials. If the molecules in a polymer can form secondary bonds with a solvent, then interaction occurs. Polymers that are not crosslinked (thermoplastics) can dissolve by creating new secondary bonds between the solvent and the polymer. If the polymer is thermoset, then the solvent cannot completely separate the covalently bonded polymer chains. Consequently, crosslinked polymers may only swell in the presence of a solvent. Examples include the expansion of a rubber band when placed in toluene and the swelling of wool and cotton fibers in water or humid air. In contrast, polyethylene, other hydrocarbon polymers, and polytetrafluoroethylene absorb very little water.

The silicate materials—that is, ceramics based on Si and O—can be degraded by water at or near room temperature. The mechanism is shown in Figure 15.2–1a. The H_2O

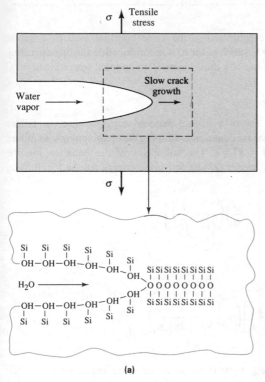

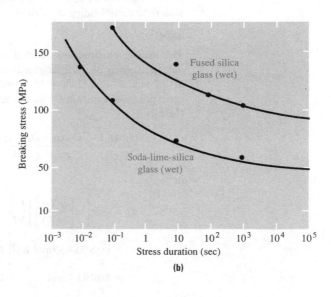

FIGURE 15.2–1   The mechanism of static fatigue. **(a)** An H_2O molecule reacts with one of the Si—O—Si groups to produce two Si—OH groups. Since the Si—OH groups are not joined by a primary bond, the effective length of the crack is increased. **(b)** During static fatigue the stress causing failure decreases with time due to the increase in the effective length of the flaw.

molecule reacts with one of the Si—O—Si groups to produce two Si—OH groups. This reaction is assisted by tensile stress. Since the two Si—OH groups are not joined by a primary bond, the effective length of the preexisting flaw increases. As discussed in Chapter 9, ceramics loaded in tension are extremely sensitive to flaws. Thus, extending the flaw reduces the failure stress of the material.

Figure 15.2–1b shows the relationship between the silicate failure stress and time of exposure to the water. The similarity of the shape of this curve to the fatigue curves discussed in Section 9.5 is one reason why this degradation mechanism is known as **static fatigue.** Note, however, that this is chemical rather than mechanical degradation and that it occurs under sustained rather than cyclic loading conditions.

For static fatigue the rate of attack, as represented by the inverse of the time to failure under constant stress, increases with temperature according to the familiar Arrhenius relationship:

$$\frac{1}{t_f} = C \exp\left(-\frac{Q}{RT}\right) \tag{15.2–1}$$

where t_f is the time to failure, C is a constant for a given stress (but its value changes with changes in stress), Q is the activation energy for the process, and R and T have their usual meaning.

....

EXAMPLE 15.2–1

Two identical silicate glass components are subjected to the same constant load in a humid environment. The component operating at 10°C survives for 83.4 seconds, while the one operating at 25°C survives for 15 seconds. Estimate the life of a third identical component operating at 40°C under the same loading conditions.

Solution

The loading environment coupled with the use of a silicate material suggests that the failure mechanism may be static fatigue. If so, the relevant equation is 15.2–1. We are given time to failure at 25 and 10°C, $t_f(25°) = 15$ s and $t_f(10°) = 83.4$ s. The unknowns in the equation are the preexponential constant C and the activation energy Q. The activation energy can be found using the ratio:

$$\frac{1/t_f(T_1)}{1/t_f(T_2)} = \frac{C \exp(-Q/RT_1)}{C \exp(-Q/RT_2)} = \exp\left[\left(-\frac{Q}{R}\right)\left(\frac{1}{T_1} - \frac{1}{T_2}\right)\right]$$

or

$$Q = -R\left\{\ln\left[\frac{t_f(T_2)}{t_f(T_1)}\right]\right\}\left(\frac{1}{T_1} - \frac{1}{T_2}\right)^{-1}$$

$$= [-8.314 \text{ J/(mol-K)}]\left[\ln\left(\frac{83.4 \text{ s}}{15 \text{ s}}\right)\right]\left(\frac{1}{298 \text{ K}} - \frac{1}{283 \text{ K}}\right)^{-1}$$

$$= 80,193 \text{ J/mol}$$

Once the value of Q has been determined, $t_f(40°)$ can be found using the same technique:

$$\frac{1/t_f(25°)}{1/t_f(40°)} = \exp\left[\left(-\frac{Q}{R}\right)\left(\frac{1}{T_1} - \frac{1}{T_2}\right)\right]$$

$$t_f(40°) = (15 \text{ s}) \exp\left[-\left(\frac{80,193 \text{ J/mol}}{8.314 \text{ J/mol-K}}\right)\left(\frac{1}{298 \text{ K}} - \frac{1}{313 \text{ K}}\right)\right]$$

$$= 3.18 \text{ s}$$

The predicted time is 3.18 s on the basis that static fatigue is the operative mechanism leading to failure.

....

15.2.1 Direct Dissolution Mechanisms

The mechanisms discussed in this section involve the dissolution of a solid directly into the surrounding liquid without the aid of electrical charge transfer. This type of degradation occurs in virtually all types of materials. Once the reaction begins, it will continue until either the material is consumed or the liquid is saturated. A familiar example is the dissolution of salt in water.

One of the general design rules for direct dissolution of polymers and organic materials is that "like dissolves like." For example, polar solvents, such as acetone, alcohol, and water, dissolve polymers containing polar groups such as —OH, —NH, and —COOH. Similarly, nonpolar solvents, such as carbon tetrachloride, benzene, and gasoline, dissolve (or at least swell) nonpolar polymers like polyethylene and polypropylene (i.e., those with nonpolar groups such as —CH and —CH$_3$). In addition, straight-chain saturated hydrocarbon polymers such as polyethylene and polypropylene are resistant to most acidic and basic liquids, but dissolve in straight-chain hydrocarbon solvents such as hot xylene.

TABLE 15.2–1 Chemical resistance of common polymers to aggressive liquids.

	Poly-propylene; Polyethylene	CAB*	ABS†	PVC‡	Saran§	Polyester glass‖	Epoxy glass	Phenolic asbestos	Fluoro-carbons	Chlorinated polyether (penton)	Poly-carbonate
10% H_2SO_4	Excel.	Good	Excel.	Excel.	Excel.	Excel.	Excel.	Excel.	Excel.	Excel.	Excel.
50% H_2SO_4	Excel.	Poor	Excel.	Excel.	Excel.	Good	Excel.	Excel.	Excel.	Excel.	Excel.
10% HCl	Excel.	Excel.	Excel.	Excel.	Excel.	Excel.	Excel.	Excel.	Excel.	Excel.	Excel.
10% HNO_3	Excel.	Poor	Good	Excel.	Excel.	Good	Good	Fair	Excel.	Excel.	Excel.
10% acetic acid	Excel.	Good	Excel.	Excel.	Excel.	Excel.	Excel.	Excel.	Excel.	Excel.	Excel.
10% NaOH	Excel.	Fair	Excel.	Good	Fair	Fair	Excel.	Poor	Excel.	Excel.	Excel.
50% NaOH	Excel.	Poor	Excel.	Excel.	Fair	Poor	Good	Poor	Excel.	Excel.	Excel.
NH_4OH	Excel.	Poor	Excel.	Excel.	Poor	Fair	Excel.	Poor	Excel.	Excel.	Excel.
NaCl	Excel.	Excel.	Excel.	Excel.	Excel.	Excel.	Excel.	Excel.	Excel.	Excel.	Excel.
$FeCl_3$	Excel.	Excel.	Excel.	Excel.	Excel.	Excel.	Excl.	Excel.	Excel.	Excel.	Excel.
$CuSO_4$	Excel.	Excel.	Excel.	Excel.	Excel.	Excel.	Excel.	Excel.	Excel.	Excel.	Excel.
NH_4NO_3	Excel.	Excel.	Excel.	Excel.	Excel.	Excel.	Excel.	Good	Excel.	Excel.	Excel.
Wet H_2S	Excel.	Excel.	Excel.	Excel.	Excel.	Excel.	Excel.	Excel.	Excel.	Excel.	
Wet Cl_2	Poor	Poor	Excel.	Good	Poor	Poor	Poor	Excel.	Excel.	Excel.	
Wet SO_2	Excel.	Poor	Excel.	Excel.	Good	Excel.	Excel.	Excel.	Excel.	Excel.	
Gasoline	Poor	Excel.	Excel.	Excel.	Excel.	Excel.	Excel.	Excel.	Excel.	Excel.	Excel.
Benzene	Poor	Poor	Poor	Poor	Fair	Good	Excel.	Excel.	Excel.	Fair	Fair
CCl_4	Poor	Poor	Poor	Fair	Fair	Excel.	Good	Excel.	Excel.	Fair	Poor
Acetone	Poor	Poor	Poor	Poor	Fair	Poor	Good	Poor	Excel.	Good	Good
Alcohol	Poor	Poor	Excel.	Excel.	Excel.	Excel.	Excel.	Excel.	Excel.	Excel.	Excel.

Note: Ratings are for long-term exposures at ambient temperatures [less than 38°C (100°F)].
* Cellulose acetate butyrate.
† Acrylonitrile butadiene styrene polymer.
‡ Polyvinylchloride.
§ Chemical resistance of Saran-lined pipe is superior to extruded Saran in some environments.
‖ Refers to general-purpose polyesters. Special polyesters have superior resistance, particularly in alkalies.
Source: Mars G. Fontana and Norbert D. Greene, *Corrosion Engineering,* 2nd ed. Copyright 1978, McGraw-Hill, Inc. Reprinted with permission of McGraw-Hill, Inc.

In general, an increase in molecular weight, percent crystallinity, or crosslink density improves a polymer's resistance to solvents. Most polymers resist most weak acids and alkalis. Silicones and fully fluorinated hydrocarbons [e.g., $(C_2F_4)_n$] are quite resistant to chemical degradation. Silicone polymers are typically crosslinked rubbers that resist swelling in most solvents (but not gasoline). The polar C—F bonds make polytetrafluoroethylene, a highly crystalline thermoplastic polymer, insensitive to most organic liquids. Table 15.2–1 shows the resistance of a variety of polymers to a representative list of liquid service environments.

It is often useful for an engineer to know what liquid will dissolve a polymer. Consider the following three examples:

1. The knowledge that polymethylmethacrylate (PMMA) dissolves in methylene chloride allows one to develop the right kind of adhesive for joining Plexiglas®.

2. Since polyethylene does not dissolve in acids, it is safe to store or use sulfuric acid in PE containers, such as battery cases.

3. The dry-cleaning business is based on finding a solvent that dissolves dirt but not the cotton, PET, wool, or nylon fibers of the underlying material. Therefore, the most commonly used solutions are partially chlorinated hydrocarbons.

DESIGN EXAMPLE 15.2–2

Gasoline is to be stored in a container made from either synthetic rubber (polychloroprene) or natural rubber (polyisoprene). Which material would you recommend?

Solution

Natural rubber is based on the nonpolar monomer C_5H_8. In contrast, the polychloroprene mer (C_4ClH_5) is more polar because of the Cl atom. Since gasoline is a nonpolar solvent, the "like-dissolves-like" rule correctly predicts that the polar polychloroprene structure is more resistant to gasoline. Use the polychloroprene container.

The resistance of a particular ceramic to an aggressive environment scales with the strength of its bonds. Therefore, we expect the ionic Group I–Group VII compounds (NaCl, KCl, LiF) to have a lower resistance to corrosion than mixed ionic-covalent compounds (SiO_2, Al_2O_3, Si_3N_4) or the pure covalent materials (diamond and graphite). This general concept can be extended: most borides, carbides, nitrides, oxides, and silicates resist attack by organic solvents and most acids. In contrast, many chlorides, nitrates, and sulfates, while able to resist organic solvents, are attacked by most acids. An important exception is that many of even the strongly bonded ceramics are attacked by hydrofluoric acid.

Vitreous silica (SiO_2 glass) and borosilicate glasses (B_2O_3-SiO_2) are resistant to alkali solutions, while soda-lime-silicate glass (NaO-CaO-SiO_2) is rapidly attacked by these solutions. The decreased corrosion resistance is a result of the disruption of the strong 3-D primary bond structure in the network-forming oxides (B_2O_3 and SiO_2) caused by the network modifier oxides (NaO and CaO).

Other general rules for predicting the rate of degradation of a ceramic (or any other material, for that matter) in an aggressive liquid environment include the following:

1. Increasing the temperature generally increases the corrosion rate, since it is a thermally activated (Arrhenius) process.

2. The degradation rate increases as the grain size decreases. This is because the energy of an ion at the grain boundary is higher than that of an ion in the bulk, so the removal of a grain boundary ion requires a smaller energy input.

3. The corrosion rate increases as the amount of relative motion between the ceramic and the liquid increases. In the absence of relative motion the liquid becomes locally saturated with degradation by-products. In contrast, relative motion continually brings "fresh" liquid into contact with the ceramic surface.

DESIGN EXAMPLE 15.2–3

Offer a suggestion to eliminate or minimize each of these problems:

a. A ceramic component formed from a very fine powder is rapidly attacked in an aggressive liquid environment. The component experiences minimal stresses, but it is possible neither to change the composition of the ceramic nor to alter the service environment.

b. A small industrial laboratory is having problems with both hydrofluoric acid (HF) and alkali solution leakage from their soda-lime-silicate glass containers.

Solution

a. Fabrication of a component from a fine powder will result in a small grain size. While this is often advantageous in applications where maximum strength and toughness are required, it can be detrimental in a low-stress aggressive liquid environment. The problem might be minimized by increasing the raw material powder size used to fabricate the component (although it is difficult to sinter coarse ceramic powders).

b. Soda-lime-silicate glass is susceptible to attack by alkali solutions. The alkali leakage problem can be eliminated by using either borosilicate glass or vitreous silica. This substitution, however, will not solve the HF problem. Proper HF storage is in a nonpolar polymer container, such as one made from polyethylene. In fact, both solutions could be stored in PE containers.

··

When the corrosive liquid is a molten metal, the degradation phenomenon is known as **liquid metal embrittlement.** Examples include the degradation of ceramic refractories in the steelmaking industry because of contact with molten metal slags, and the degradation of metal pipes filled with liquid sodium used to cool nuclear (breeder) reactors. In general, the liquid metal preferentially attacks the grain boundaries in the solid. One of the driving forces for this reaction is that the energy of the system is reduced by the penetration of the liquid along the prior solid grain boundaries. (See Figure 15.2–2.) The requirement for liquid penetration is $\gamma_{ls}/\gamma_{ss} < 0.5$, where γ_{ls} and γ_{ss} are the surface energies of the liquid-solid and solid-solid interfaces. Liquid metal embrittlement can also occur during brazing, soldering, or coating operations involving molten Zn, Sn, Pb, or Cd, or exposure to Hg.

Similar problems occur in oxide glassmaking where molten glass is continually in contact with Al_2O_3 refractory bricks used in furnace linings. As suggested previously, the problem can be minimized by increasing the grain size of the Al_2O_3 and by minimizing the relative motion between the molten glass and the refractory linings.

15.2.2 Electrochemical Corrosion—Half-Cell Potentials

In contrast to direct dissolution mechanisms, electrochemical corrosion requires electric charge transfer. Therefore, it is a phenomenon essentially confined to metals. Like direct dissolution, electrochemical corrosion results in material loss and a corresponding

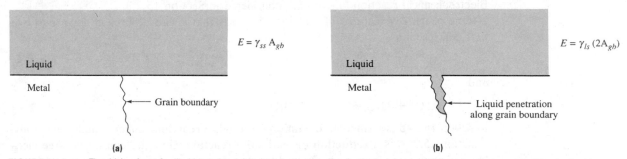

FIGURE 15.2–2   The driving force for liquid metal embrittlement is the reduction in the free energy of the system that occurs when the liquid penetrates along and wets grain boundaries. For this process to reduce the system energy the magnitude of the liquid-solid surface tension (γ_{ls}) must be less than half of the corresponding solid-solid value (γ_{ss}).

degradation of properties. We begin our discussion by investigating the driving force for electrochemical reactions—the reduction of the free energy of the system.

Consider the reaction:

$$M(OH)_2 \rightarrow M + 0.5O_2 + H_2O \tag{15.2-2}$$

and the reverse reaction:

$$M + 0.5O_2 + H_2O \rightarrow M(OH)_2 \tag{15.2-3}$$

where M represents a divalent metal atom. Since many metals exist in nature as oxides, hydroxides, carbonates, sulfides, sulfates, or silicates, reaction 15.2–2 represents the extraction of metals from their ores. Conversely, reaction 15.2–3 represents one of many possible forms of electrochemical corrosion.

Which reaction is thermodynamically favored? Theoretically, the answer is obtained by examining the sign of the Gibbs free energy change, ΔG, for each reaction. In practice, it is difficult to measure ΔG directly. Instead, a related quantity, the electrochemical cell potential E, is monitored. For an electrochemical reaction, the important equation, which we present without derivation, is the **Nernst equation:**

$$E = E° + \frac{RT}{nF} \ln\left(\frac{a_r}{a_p}\right) \tag{15.2-4a}$$

where $E°$ is the cell potential under standard conditions, a_r and a_p are respectively the activities[1] of the reactants and products, n is the number of moles of electrons involved in the reaction, and F is Faraday's constant ($= 96{,}500$ coulombs/mole). At room temperature, and noting that $\ln x = 2.3 \log_{10} x$, the equation reduces to:

$$E = E° + \frac{0.0592}{n} \log\left(\frac{a_r}{a_p}\right) \tag{15.2-4b}$$

If the cell potential E for a reaction is greater than zero, then the reaction is energetically favorable. For example, if, as is the case with many metals, $E < 0$ for reaction 15.2–2 and $E > 0$ for reaction 15.2–3, then the electrochemical corrosion of the metal is a spontaneous process. This means that metal oxides are more stable than pure metal. (Gold is an exception, which is in part why it is the international currency standard.)

It is important to note that satisfying the thermodynamic requirement is a necessary, but not sufficient, requirement for determining whether a reaction will occur. The reaction kinetics must also be considered; that is, the reaction may proceed imperceptibly slowly. Electrochemical reaction kinetics are considered in Section 15.2.3.

One of the key features of reaction 15.2–3 is that it can be broken down into two half-cell reactions:

$$M \rightarrow M^{+2} + 2e^- \tag{15.2-5}$$

and

$$0.5O_2 + H_2O + 2e^- \rightarrow 2(OH)^- \tag{15.2-6}$$

Reaction 15.2–5 is termed an **oxidation** (or **anodic**) **reaction,** since it "adds" electrons. Reaction 15.2–6 is a **reduction** (or **cathodic**) **reaction,** since it "consumes" free electrons. Any reaction that can be divided into two or more oxidation-reduction reactions is by definition an electrochemical reaction.

---

[1] Thermodynamic activity can be loosely defined as a concentration-like term that has been corrected to account for the nonideal behavior of the substance.

In electrochemical corrosion, the oxidation reaction is always the oxidation of a metal to its ion. The reduction reaction can take several forms depending on the corrosive environment. Reaction 15.2–6, known as hydroxyl formation, is common in aerated water. In oxygen-free liquids such as hydrochloric acid and stagnant water, the reduction reaction is often hydrogen evolution:

$$2H^+ + 2e^- \rightarrow H_2(gas) \tag{15.2–7}$$

In oxidizing acids the reaction usually involves water formation:

$$O_2 + 4H^+ + 4e \rightarrow 2H_2O \tag{15.2–8}$$

Finally, in certain situations the reaction may involve the reduction of a metal ion:

$$M^{n+} + e^- \rightarrow M^{(n-1)+} \quad \text{or} \quad M^+ + e^- \rightarrow M \tag{15.2–9}$$

EXAMPLE 15.2–4

Assuming that the oxidation reaction is as given in Equation 15.2–5, give the appropriate reduction half-cell reaction and the net reaction for each electrochemical cell:

  a. A divalent metal in oxygen-free hydrochloric acid
  b. A divalent metal in aerated hydrochloric acid

Solution

The appropriate reduction half-cell reaction depends on the solution. In oxygen-free hydrochloric acid the reduction reaction will be hydrogen evolution (Equation 15.2–7). Therefore, the net reaction is:

$$M + 2HCl \rightarrow MCl_2 + H_2$$

In aerated hydrochloric acid the reduction reaction involves water formation (Equation 15.2–8). Thus, the net reaction is:

$$M + 2HCl + 0.5O_2 \rightarrow MCl_2 + H_2O$$

One important characteristic of these reduction and oxidation reactions is that neither reaction can occur alone. Stopping or slowing either reaction stops or slows the entire electrochemical process. This has important implications for corrosion minimization.

One of the engineering goals of the study of electrochemical corrosion is to be able to predict the severity of the corrosion problem for any specific metal-environment couple. Consider the application of Equation 15.2–4 to the metal reduction reaction ($M^+ + e^- \rightarrow M$). The activity of a solid metal is approximately 1, and the activity of ions in a dilute solution can be approximated as the concentration of the ions, C_{M^+}. Therefore,

$$E_M = E_M^\circ + \frac{0.0592}{n} \log C_{M^+} \tag{15.2–10}$$

Note that the sign of E (and hence the thermodynamically favored direction for the reaction) depends on the ion concentration. Similarly, for the case of the reduction form of the hydrogen reaction (Equation 15.2–7), we find:

$$E_H = E_H^\circ + 0.0592 \log C_{H^+} \tag{15.2–11}$$

where C_{H^+} is the concentration of the hydrogen ions.

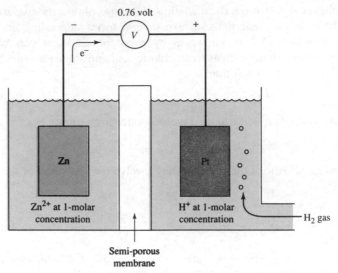

FIGURE 15.2–3   The experimental system for determining the reduction potential of zinc consists of two half cells, one containing the metal in a 1-molar concentration of its ions and the other containing a Pt electrode in a 1-molar concentration of H^+ ions and bubbling hydrogen gas. A voltage meter is used to measure the total cell potential.

A ranking of the corrosion potentials of metals can be performed by comparing cell potentials E for each metal reduction reaction with the corresponding value of a standard reference reaction. The most common choice for the reference reaction is the hydrogen reduction reaction. In order to compare "apples to apples," all of the reaction must be written as reduction reactions even though each electrochemical cell contains one reduction and one oxidation reaction.

As shown in Figure 15.2–3, the experiment is performed by placing the metal in a 1-molar solution of its ions, and in electrical contact with a Pt electrode in a 1-molar solution of H^+ ions and bubbling hydrogen gas. Thus, both the metal and hydrogen reactions can proceed in either the oxidation or the reduction direction. The Pt electrode does not take part in the reaction. It serves only as a substrate for the hydrogen reaction. The sign and magnitude of the open-circuit potential can be used to evaluate the tendency of each metal to corrode.

Two possible sets of half-cell reactions may occur in this experiment. When the metal is more reactive, or anodic, than the reference hydrogen reaction, then the metal goes into solution and the reduction reaction will be hydrogen evolution. The half-cell reactions are

$$M \rightarrow M^{2+} + 2e^-$$ (15.2–12a)

and

$$2H^+ + 2e^- \rightarrow H_2$$ (15.2–12b)

The magnitude of the total reaction potential, as measured by the potential meter in the circuit, is $|E_{H^+} - E_{M^+}|$. The half-cell potential for the metal reaction is subtracted rather than added, since the oxidation reaction is proceeding in the reverse direction from that used to define Equation 15.2–10.

Alternatively, when the metal is less reactive, or cathodic, than the hydrogen reaction, the half-cell reactions are

$$M^{2+} + 2e^- \rightarrow M$$ (15.2–13a)

and

$$H_2 \rightarrow 2H^+ + 2e^- \tag{15.2-13b}$$

The magnitude of the total reaction potential is $|E_{M^+} - E_{H^+}|$. As before, the potential for the oxidation reaction is subtracted.

The judicious choice of 1-molar solutions in this experiment simplifies the form of Equations 15.2–10 and 15.2–11 in that the log C terms go to zero. Therefore, $E_{M^+} = E^\circ_{M^+}$, and $E_{H^+} = E^\circ_{H^+}$. In addition, since the hydrogen reaction serves only as a reference reaction, we are free to assign an arbitrary value (usually zero) to the reference potential $E^\circ_{H^+}$. Under these assumptions we find that the magnitude of the measured cell potential is $|E^\circ_{M^+}|$, regardless of the direction of the reaction.

The sign of the measured potential is extremely important, since it describes the direction of the reactions. We must, however, adopt yet another convention, since the measured potential depends on which pole of the meter is attached to the metal electrode. By specifying that the meter be connected so that it shows a positive potential when the metal is the cathode, we find that the measured potential corresponds directly to the reduction potential of the metal.

The results from a series of experiments for a variety of metals are given in Table 15.2–2, which shows the standard **reduction potentials** for these reactions. The

TABLE 15.2–2 Electrode reduction potentials (compiled from a variety of sources).

Reduction reaction	Potential (V)	
$Au^{3+} + 3e^- \rightarrow Au$	+1.50	↑
$O_2 + 4H^+ + 4e^- \rightarrow 2H_2O$	+1.23	Cathodic
$Pt^{4+} + 4e^- \rightarrow Pt$	+1.20	
$Pd^{2+} + 2e^- \rightarrow Pd$	+0.99	
$Ag^+ + e^- \rightarrow Ag$	+0.80	
$Hg^{2+} + 2e^- \rightarrow Hg$	+0.79	
$Fe^{3+} + e^- \rightarrow Fe^{2+}$	+0.77	
$O_2 + H_2O + 4e^- \rightarrow 4(OH)^-$	+0.40	
$Cu^{2+} + 2e^- \rightarrow Cu$	+0.34	
$2H^+ + 2e^- \rightarrow H_2$	0.00	
$Pb^{2+} + 2e^- \rightarrow Pb$	−0.13	
$Sn^{2+} + 2e^- \rightarrow Sn$	−0.14	
$Ni^{2+} + 2e^- \rightarrow Ni$	−0.25	
$Fe^{2+} + 2e^- \rightarrow Fe$	−0.44	
$Cr^{2+} + 2e^- \rightarrow Cr$	−0.56	
$Zn^{2+} + 2e^- \rightarrow Zn$	−0.76	
$Mn^{2+} + 2e^- \rightarrow Mn$	−1.63	
$Ti^{2+} + 2e^- \rightarrow Ti$	−1.63	
$Al^{3+} + 3e^- \rightarrow Al$	−1.66	
$Be^{2+} + 2e^- \rightarrow Be$	−1.85	
$Mg^{2+} + 2e^- \rightarrow Mg$	−2.36	
$Na^+ + e^- \rightarrow Na$	−2.71	
$Ca^+ + e^- \rightarrow Ca$	−2.76	
$K^+ + e^- \rightarrow K$	−2.92	
$Rb^+ + e^- \rightarrow Rb$	−2.93	Anodic
$Li^+ + e^- \rightarrow Li$	−2.96	↓

key concept is that metals with reduction potentials less than zero are anodic with respect to the hydrogen reaction and, hence, corrode in acid.[2]

The next question to be addressed is, What happens when the Pt/H half cell is replaced with a different half cell? Consider the case of Fe and Mg, each in 1-molar solutions of their ions. Two possibilities exist. If Fe is anodic with respect to Mg, then the half-cell reactions are:

$$Fe \rightarrow Fe^{2+} + 2e^- \tag{15.2-14a}$$

and

$$Mg^{2+} + 2e^- \rightarrow Mg \tag{15.2-14b}$$

The total reaction potential will be $E^\circ_{Mg} - E^\circ_{Fe} = [-2.36 \text{ V} - (-0.44 \text{ V})] = -1.92 \text{ V}$. Since $E_{total} < 0$, the reactions will not proceed in this direction. If, however, the reactions are

$$Mg \rightarrow Mg^{2+} + 2e^- \tag{15.2-15a}$$

and

$$Fe^{2+} + 2e^- \rightarrow Fe \tag{15.2-15b}$$

then the total reaction potential is $[-0.44 \text{ V} - (-2.36 \text{ V})] = 1.92 \text{ V}$. The positive potential means this reaction is thermodynamically favored, and Mg will corrode preferentially when in electrical contact with Fe.

The generalization of the previous result is that when two half-cell reactions are combined, *the metal with the more negative reduction potential will become the anode*. In addition, the greater the algebraic difference between the potentials of the two metals in contact with one another, the greater the initial thermodynamic driving force for the corrosion of the more anodic metal.

....

EXAMPLE 15.2–5

Determine which metal in each pair is the anode when the two metals form an electrochemical cell:

   a. Cu and Al
   b. Fe and Al
   c. Ag and Cu

Solution

The metal with the more negative reduction potential will be the anode. Using Table 15.2–2, we find $E^\circ_{Cu} = +0.34 \text{ V}$, $E^\circ_{Al} = -1.66 \text{ V}$, $E^\circ_{Fe} = -0.44 \text{ V}$, and $E^\circ_{Ag} = +0.80 \text{ V}$.

   a. In the Cu/Al cell, since E°_{Al} is more negative than E°_{Cu}, the Al is more anodic and corrodes preferentially.
   b. In the Fe/Al cell, since E°_{Al} is more negative than E°_{Fe}, the Al is again the anode and corrodes preferentially. By placing the Fe in electrical contact with the Al, the corrosion of the Fe is prevented even though it has a negative reduction potential.
   c. In the Ag/Cu cell, since E°_{Cu} is more negative than E°_{Ag}, the Cu is more anodic and corrodes even though it has a positive reduction potential.

....

---

[2] The sign convention described above is based on reduction potentials. It is, however, equally valid to evaluate corrosion potentials in terms of the oxidation reaction potentials. Under this convention a negative potential implies that the metal is cathodic with respect to the reference hydrogen reaction. Since some handbooks employ the oxidation potential convention, one must be sure to note which convention is being used before predicting anodic/cathodic behavior.

Although the ranking of the reduction potentials in Table 15.2–2 is a helpful guide, its use in a quantitative fashion is limited. In practice one rarely encounters pure metals in 1-molar concentration solutions of their ions. Fortunately, some of the deviations can be treated in a straightforward manner. The influence of ion concentration can be handled through the Nernst equation. The influence of solution changes (e.g., freshwater, seawater, industrial chemical environments) is more difficult to deal with. Perhaps the most useful general guide is the galvanic series (Table 15.2–3), which ranks the corrosion potentials of common alloys in seawater.

There are some significant differences between the reduction potential series and the galvanic series. As an example, note the relative positions of Al and Zn in the two tables. Further examination of the galvanic series shows that alloy additions can strongly influence the corrosion potential. (See the relative positions of Fe, steel, and stainless steel.) In the absence of data for a "new" corrosion environment, it is better to use the galvanic series than the standard reduction potential series as a guide for predicting corrosion behavior.

TABLE 15.2–3   The galvanic series in seawater.

Platinum	↑
Gold	Cathodic
Graphite	
Titanium	
Silver	
Stainless steel (passive)	
Nickle-base alloys (passive)	
Cu–30% Ni alloy	
Copper	
Aluminum bronze	
Cu–35% Zn brass	
Nickle-base alloys (active)	
Manganese bronze	
Cu–40% Zn brass	
Tin	
Lead	
316 stainless steel (active)	
50% Pb–50% Sn solder	
410 stainless steel (active)	
Cast iron	
Low-carbon steel	
2024 aluminum	
2017 aluminum	
Cadmium	
Alclad	
1100 aluminum	
5052 aluminum	
Galvanized steel	
Zinc	
Magnesium alloys	Anodic
Magnesium	↓

Source: Adapted from *Metals Handbook,* 8th ed.,
Vol. 1, "Properties and Selection of Metals," 1961,
ASM International, Materials Park, OH. Reprinted by
permission of the publisher.

EXAMPLE 15.2–6

Consider an electrochemical cell of the type shown in Figure 15.2–3 with one half cell containing a standard Pt/H electrode and the other containing a Pb electrode in a 1-molar solution of its ions. What concentration of H^+ ions is necessary for the reaction to be in equilibrium with no transfer of electrons through the external circuit?

Solution

No electron transfer means the electrical potentials must be the same in each half cell. Since the Pb half cell is a standard half cell, we obtain the value of $E_{Pb} = E_{Pb}^{\circ}$ directly from Table 15.2–2. The H^+ concentration can then be adjusted so that $E_H = E_{Pb}^{\circ}$. The relationship between concentration and potential is given by the Nernst equation (15.2–4). E_H° is defined as zero in the Pt/H half cell, and H^+ is monovalent ($n = 1$). Therefore, the Nernst equation reduces to:

$$E_H = 0.0592 \log C_{H^+}$$

Since we require $E_H = E_{Pb}^{\circ}$,

$$E_{Pb}^{\circ} = 0.0592 \log C_{H^+}$$

Substituting the value for E_{Pb}° from Table 15.2–2 yields:

$$-0.126/0.0592 = -2.13 = \log C_{H^+}$$

$$C_{H^+} = 0.0074 \text{ molar concentration}$$

Noting that $pH \equiv -\log C_{H^+}$, this corresponds to a pH of 2.13.

15.2.3 Kinetics of Corrosion Reactions

The appropriate use of engineering materials requires knowledge not only of thermodynamics and mechanisms of corrosion but also knowledge of the kinetics of corrosion. The importance of kinetics can be illustrated with an example. Use of the galvanic series alone might suggest that Fe is a better material than Al to use in a marine environment. Experiment, however, shows that the steady-state corrosion rate for Al is orders of magnitude less than that of Fe.

The corrosion rate is often determined indirectly by measuring the corrosion current. The relationship between these two quantities is given by **Faraday's equation:**

$$w = \frac{IM}{nF} \tag{15.2–16}$$

where w is the weight loss (g/s), I is the corrosion current (A), M is the atomic mass of the metal (g/mol), n is the valence of the metal ion, and F is Faraday's constant, 96,500 C/mol. This expression has limited appeal, since the surface area undergoing corrosion is not specified. It is often more useful to divide both sides of the equation by the surface area A, to obtain:

$$W = \left(\frac{I}{A}\right)\left(\frac{M}{nF}\right) \tag{15.2–17}$$

where W is the weight loss per unit time per unit area of exposed surface [g/(s-m^2)], M/nF is a constant for a given metal, and I/A is the readily measured current density (A/m^2). An even more useful quantity is the thickness of metal lost per unit time of exposure (y')

as given by:

$$y' = \left(\frac{I}{A}\right)\left(\frac{M}{n\rho F}\right)$$

(15.2–18)

where y' has units of m/s. Note that Faraday's equation, in which the corrosion current is normalized by the surface area, is meaningful only for uniform corrosion.

....

EXAMPLE 15.2–7

During corrosion tests a current density of 0.86 A/m^2 was measured for a steel sample immersed in seawater. Calculate the weight loss per second and the thickness loss per year for this steel.

Solution

The weight loss is calculated using Equation 15.2–17. For steel, which is essentially iron, the atomic weight is 55.85 g/mol and the valence of the metal ion is 2. Substituting these values into the expression along with the value of Faraday's constant and the measured current density, 0.86 A/m^2, yields:

$$W = \left(\frac{I}{m^2}\right)\left(\frac{M}{nF}\right) = \frac{(0.86 \text{ A/m}^2)(55.85 \text{ g/mol})}{(2)(96{,}500 \text{ C/mol})}$$

$$= 2.5 \times 10^{-4} \text{ g/(s-m}^2)$$

The thickness of metal lost per year is calculated using Equation 15.2–18. Using $\rho(\text{Fe}) = 7.87 \times 10^6$ g/m^3, we find:

$$y' = \left(\frac{I}{m^2}\right)\left(\frac{M}{n\rho F}\right)$$

$$= \frac{(0.86 \text{ A/m}^2)(55.85 \text{ g/mol})}{(2)(96{,}500 \text{ C/mol})(7.87 \times 10^6 \text{ g/m}^3)}$$

$$= 3.2 \times 10^{-11} \text{ m/s} \approx 1 \text{ mm/year}$$

Thus, the structural steel components in a bridge over a saltwater bay would lose 1 cm of thickness per decade.

....

Polarization

Measurements of current density show that for many metals the corrosion rate decreases with time. This phenomenon, known as polarization, results from irreversible changes in the system that occur when current begins to flow.

Concentration polarization occurs when the rate of corrosion is limited by the diffusion of ions within the electrolyte. If the corrosion reaction proceeds more rapidly than the rate at which ions can be supplied to the cathode (or removed from the anode), then the local ion concentration changes with time. The change in ion concentration alters the half-cell potentials, as suggested by the Nernst equation. Either a decrease in ion concentration at the cathode, or an increase in metal ion concentration at the anode, decreases the cell potential and, hence, the corrosion rate.

Why do some metals experience higher corrosion rates when immersed in a moving fluid than in a stagnant fluid of the same composition? Since a moving fluid increases the ion removal rate, concentration polarization is reduced and corrosion rate increases.

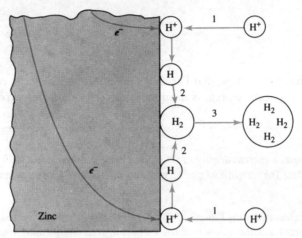

FIGURE 15.2–4 The hydrogen evolution reaction involves a sequence of steps including: (1) the combination of a hydrogen ion with an electron to form a neutral hydrogen atom, (2) combination of two hydrogen atoms to form diatomic hydrogen, and (3) combination of multiple H_2 molecules to form a hydrogen bubble.

Activation polarization occurs at the solid–electrolyte interface and is related to the energy barriers associated with the corrosion process. Consider the hydrogen evolution reaction occurring on the cathode surface. As shown in Figure 15.2–4, this process involves a sequence of steps, including combination of a hydrogen ion with an electron to form a neutral hydrogen atom, combination of two hydrogen atoms to form diatomic hydrogen, and combination of multiple H_2 molecules to form a hydrogen bubble.

The reaction rate for each step is related to temperature through the Arrhenius equation:

$$\text{Reaction rate} \propto \exp\left(-\frac{Q}{RT}\right) \tag{15.2–19}$$

where Q is the activation energy barrier for the process. As with any sequential reaction, the slowest step controls the net reaction rate. Since the reaction rate decreases as the activation energy increases, the process with the highest activation barrier controls the rate of the reaction. In the hydrogen evolution reaction the rate-controlling step depends on the metal substrate. For nickel, the rate-controlling step is the combination of a hydrogen ion with an electron, while for platinum and niobium it is formation of diatomic hydrogen. The Arrhenius nature of activation polarization emphasizes the design rule that corrosion rate usually, but not always, increases with temperature.

Passivation

In Table 15.2–3, some materials were listed twice—once in the active state and once in the passive state. The passive state refers to a surface condition under which a material displays a significant positive shift in its reduction potential. This phenomenon, known as **passivation,** results in substantial decreases (often three to six orders of magnitude) in the corrosion rate of many common engineering alloys, including those based on Fe, Ni, Cr, and Ti. In most cases, passivation is related to the formation of a protective surface film. The film is usually a nonporous, adherent, insulating oxide that protects the metal either by breaking the electrical circuit or substantially limiting diffusion of the required ionic species. Passivation, therefore, can be envisioned as an extreme case of concentration polarization.

Passivation is not a general phenomenon. It occurs only for certain alloy-environment pairs. For example, Al is rapidly attacked by dilute nitric acid but is nearly inert (fully passivated) in concentrated nitric acid. Austenitic (FCC) stainless steels are passive in aerated dilute sulfuric acid but attacked in air-free acid.

While an understanding of passivation allows us to use certain alloys under aggressive environmental conditions, an incomplete understanding of the phenomenon may lead to serious errors. For example, someone without the proper knowledge might attempt to reduce the corrosion rate of an aluminum vessel designed to contain concentrated nitric acid by diluting the acid. This would lead to catastrophic failure, since the passive oxide film would not be sustained in the dilute acid.

15.2.4 Specific Types of Corrosion

Electrochemical corrosion occurs under a wide variety of circumstances and involves several different mechanisms. This section describes some of the more common mechanisms.

Uniform Corrosion

Even on a nominally homogeneous metal surface there are local anodic and cathodic regions. These local regions can arise either from compositional variations in the alloy or from chemical variations in the electrolyte. With time, the locations of the anodic and cathodic regions change randomly so that the result is a uniform loss of metal over the entire exposed surface. While **uniform corrosion** is common and results in the loss of a tremendous amount of structural metal each year, it is relatively easy to design around. By measuring the current density in a test sample, the average thickness loss with time can be calculated. The thickness of the component can then be increased to the point where it can perform its intended function over its design life.

Galvanic Corrosion (Composition Cells)

In contrast to uniform corrosion, there are specific areas in many alloys that are always anodic relative to the bulk of the structure. These regions corrode preferentially. When the anodic regions are a result of compositional variation, the problem is referred to as **galvanic corrosion.** An example of a galvanic corrosion cell, or galvanic couple, occurs when two dissimilar metals are placed in electrical contact and immersed in an electrolyte. As described previously, the alloy with the more negative corrosion potential will corrode preferentially. The severity of the problem is determined by several factors, including:

1. The potential difference between the half-cell reactions
2. Kinetic factors, such as polarization and passivation
3. The ratio of the anodic surface area to the cathodic surface area

To understand the importance of the anode-to-cathode area ratio, consider two galvanic couples composed of the same pair of metals. Suppose the cathode in each couple has an area of 1 m^2 but the anode in couple 1 has an area of 0.1 m^2 while the anode in couple 2 has an area of 10 m^2. (See Figure 15.2–5.) If the system is under cathodic activation polarization, then both cathodes need the same number of electrons from their respective anodes. Thus, the current supplied from each anode will be the same. The current density at the smaller anode, however, will be two orders of magnitude higher than at the larger anode. Since corrosion rate scales with current density, the rate of metal loss

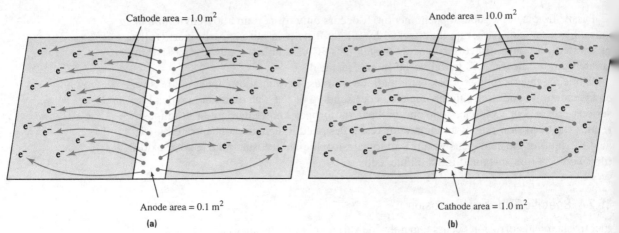

Cathode area = 1.0 m²

Anode area = 10.0 m²

Anode area = 0.1 m²

Cathode area = 1.0 m²

(a)

(b)

FIGURE 15.2–5 The figure shows the importance of the anode-to-cathode area ratio. **(a)** A large cathode in contact with a small anode results in a high current density at the anode. **(b)** A much lower current density is developed in the anode when its area greatly exceeds that of the cathode.

will also be two orders of magnitude higher for the couple with the smaller anode-to-cathode area ratio.

Based on the above example, two general rules for corrosion protection can be developed. First, avoid the unnecessary use of dissimilar metals in electrical contact. Second, when dissimilar metals must be used together, keep the anode-to-cathode area ratio as large as possible.

Consider the relative advantages and disadvantages of protecting steel using Zn and Cr coatings. As long as the coating is continuous and defect-free, either metal protects the underlying steel. A difference in behavior occurs when the inevitable scratch or flaw in the coating exposes the underlying metal. The galvanic series (Table 15.2–3) show that Zn is anodic to steel and, therefore, will continue to protect the steel. In contrast, Cr is cathodic to steel, so once the coating is scratched, the steel will begin to corrode preferentially. The problem is amplified by the small exposed area of the steel, which represents a highly unfavorable anode-to-cathode area ratio. Thus, the aesthetic quality of chrome-plated components must be balanced against the potential for serious corrosion problems. Car bumpers are chrome-plated, and hence, when rust first appears, its volume increases rapidly. Zinc-plated steel, known as galvanized steel, is a better choice if corrosion resistance is the primary objective. Steel trash cans are galvanized, not chrome-plated. An even better solution, however, is to use lightweight corrosion-proof plastic trash containers.

Galvanic couples are not always as easy to detect and design against as in the example above. For example, if the protective film on a passivated metal is scratched it is possible to create an extremely unfavorable anode-to-cathode area ratio. The entire passivated surface will become cathodic with respect to the small region of exposed metal.

Galvanic couples may develop on the microscopic rather than the macroscopic level. One example of a microscopic galvanic cell occurs in some grades of welded stainless steel. The corrosion resistance of stainless steel is the result of a passive Cr_2O_3 oxide film. Unfortunately, during welding of these alloys some of the Cr combines with C to produce chromium carbides. The process, known as sensitization, occurs preferentially at grain boundaries in regions of the weldment that have been heated to temperatures in the range of 400 to 900°C. As shown in Figure 15.2–6, the areas adjacent to the grain boundaries where carbides precipitate are depleted of Cr. When the local free Cr content drops below the 10–13% range required for the formation of a passive film, the metal in this area is

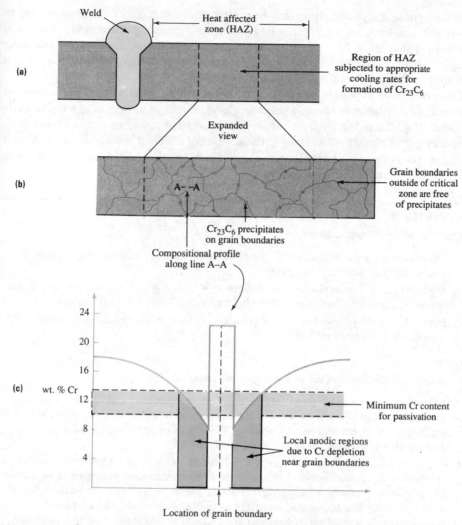

(a)

(b)

(c)

FIGURE 15.2–6 During the welding of certain stainless steels, some of the Cr can combine with C to form chromium carbide particles. **(a)** The sensitized volume of metal is located some distance from the actual weld. **(b)** The chromium carbide particles form at grain boundaries. **(c)** The metal adjacent to the grain boundary is depleted of chromium and becomes the anode of the microscopic galvanic cell.

not stainless steel and is anodic with respect to the bulk of the structure. The resulting galvanic corrosion is further aggravated by the low anode-to-cathode area ratio.

An understanding of corrosion meachanisms allows engineers to modify alloys to eliminate problems. In the case of welding stainless steels, the problem may be approached in either of two ways: use a grade of stainless steel with a very low C content, or add Ti or Nb to the alloy, since these metals have a stronger affinity for C than does Cr. Also, a postwelding heat treatment can dissolve the chromium carbides and diffuse the chromium back into solution.

Grain boundaries and heavily cold-worked areas act as microscopic anodic regions relative to defect-free regions. This is because atoms near crystalline defects are at a higher energy level than those in equilibrium lattice positions. The higher energy state of these atoms permits them to go into solution more easily than lattice atoms; thus, the defected regions corrode preferentially.

In multiphase alloys one phase may be preferentially attacked. Thus, heat-treating to age-harden an alloy from a supersaturated single phase generally decreases the corrosion resistance of that alloy. Even single-phase alloys can contain microscopic galvanic cells brought about by alloy segregation during cooling. This is particularly common in cast structures in the vicinity of the interdendritic channels.

These observations lead to several general design rules that minimize microcorrosion problems. First, whenever possible select single-phase alloys for use in aggressive environments. If additional strength is required, solid solution strengthening is a more judicious choice than either cold working or the addition of a second phase. Second, it is good practice to subject single-phase alloys to an additional solution heat treatment to promote diffusion and chemical homogenization.

EXAMPLE 15.2–8

Explain each of these observations:

  a. Crime laboratories can "magically" reveal the serial numbers on guns after these numbers have been "removed" by grinding.
  b. While the pearlitic structure of a normalized 1080 steel cannot be seen on an as-polished specimen, it suddenly becomes visible after the sample is "etched" with an acid.
  c. Binary Al-Cu alloys with low Cu content are more corrosion-resistant than similar alloys with higher Cu concentrations.

Solution

The explanations are based on microscopic galvanic cells.

  a. Stamping a serial number on metal is a cold-working operation, and the cold-worked region extends well below the bottom of the indentation. If the cold-worked material below the indentation is not removed by grinding, it becomes anodic when etched by an acid and reveals the shape of the prior indentations.
  b. Pearlite is a two-phase microstructure of α-Fe and Fe_3C. The phase boundaries cannot be resolved on a polished surface. Because of the different corrosion potentials of the phases, however, etching results in preferential attack of one phase. (The anodic phase depends on the etchant.) The accelerated material loss in one phase causes surface perturbations (peaks and valleys) that act as light-scattering sites. The two phases can be distinguished easily in the etched condition.
  c. When the Cu content is high, the microstructure consists of two phases. The corrosion resistance of two-phase alloys is generally lower than that of a comparable single-phase alloy because of the potential for microscopic galvanic cell formation.

Concentration Cells

Local anodic regions can also form because of concentration gradients in the electrolyte. For example, if the reduction reaction involves oxygen, regions with higher oxygen concentration become cathodic with respect to the surrounding material. (See reactions 15.2–6 and 15.2–8.) This can be deduced from the Nernst equation, which, for reduction reaction 15.2–8, takes the form:

$$E_{red} = E_{red}^\circ + \frac{0.0592}{n} \log \frac{C_O}{(C_{H^+})^4} \qquad (15.2\text{–}20)$$

Note that a higher oxygen concentration results in a more positive half-cell potential and the formation of a local cathodic region.

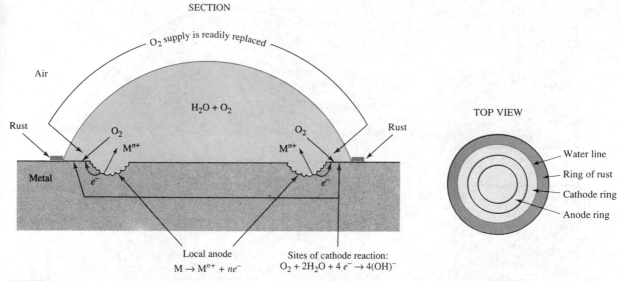

FIGURE 15.2–7   A drop of water on a metal surface results in the formation of an oxygen concentration cell. The regions near the edge of the drop are able to replenish their O_2 supply more easily than the interior regions and become cathodic. The highest rate of corrosion occurs for a ring of metal just inside the cathodic region. In addition, a ring of rust forms just outside the waterline.

Consider the **oxygen concentration cell** shown in Figure 15.2–7, in which a drop of water is partially covering a steel plate. We may assume that the anodic reaction is the oxidation of the metal and the reduction reaction is $0.5O_2 + H_2O + 2e^- \rightarrow 2(OH)^-$. Initially the entire metal surface under the water is anodic, and uniform corrosion will begin. Almost immediately, however, an oxygen concentration gradient develops. Because of their closer proximity to the air, the regions near the edge of the drop are able to replenish their O_2 supply more easily than the interior regions. Thus, the metal just inside the waterline will become cathodic while the remaining covered metal will become anodic. The highest rate of corrosion occurs in a ring of metal just inside the cathodic region. Electrons must flow from the anodic regions through the metal to the cathodic regions, and electron transfer can occur more rapidly for anodic regions located closer to the cathodic regions where there is less resistance.

Careful examination of the oxygen concentration cell shows that a ring of familiar rust forms just outside the waterline. This can be mistakenly interpreted as the location of the anodic reaction. Instead, it is the location of the deposition of the by-product of the reaction between the Fe^{2+} ions, OH^- ions, and additional dissolved oxygen. The form of the by-product depends on the amount of dissolved O_2 and can be $Fe(OH)_3$, $FeO(OH)$, or Fe_3O_4. The removal of Fe^{2+} and OH^- ions decreases the natural concentration polarization of the reaction and increases the corrosion rate.

Pit and Crevice Corrosion

Both **pit** and **crevice corrosion** are highly localized forms of electrochemical attack (usually) involving oxygen concentration gradients. The presence of Cl^- ions increases the severity of the problem. Examples of the geometry of pitting and crevice corrosion are shown in Figure 15.2–8. We focus our attention on crevice corrosion and note that the mechanism for pit corrosion is similar.

The electrolyte within a crevice will be quickly depleted of oxygen and will become anodic with respect to the metal surface just outside the crevice. The metal ions react with

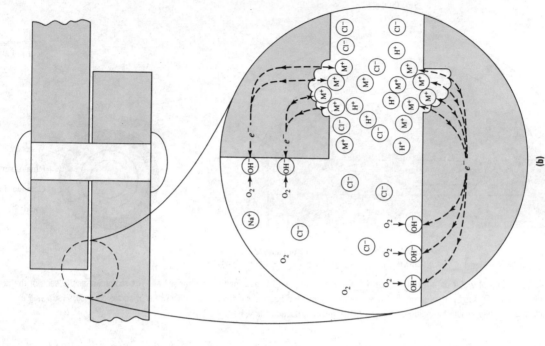

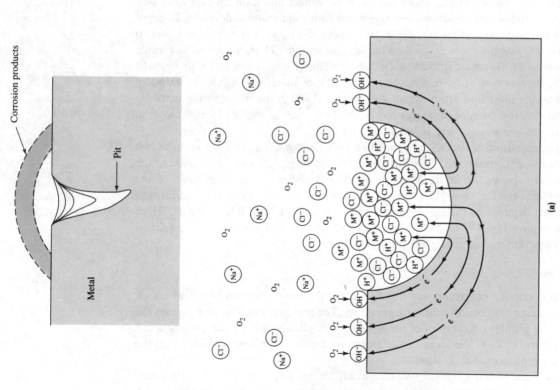

FIGURE 15.2–8 The geometry and mechanisms of pitting and crevice corrosion. **(a)** Pits often form on horizontal surfaces under debris. **(b)** Crevice corrosion can occur in any type of restricted space but is especially common between metal sheets that have been joined using bolts, screws, or rivets. The presence of Cl⁻ ions accelerates both pitting and crevice corrosion. *(Source: Mars G. Fontana and Norbert D. Greene, Corrosion Engineering, 2nd ed. Copyright 1978, McGraw-Hill, Inc. Reprinted with permission of McGraw-Hill, Inc.)*

Cl^- to form a metal chloride, which can then hydrolyze, according to the reactions:

$$M^+ + Cl^- \rightarrow MCl \tag{15.2–21}$$

and

$$MCl + H_2O \rightarrow MOH + H^+Cl^- \tag{15.2–22}$$

These reactions increase the corrosion rate for several reasons. First, the removal of the metal ions by Cl^- decreases the concentration polarization at the anode. Second, the hydrolysis reaction increases the H^+ concentration (acidity), which in turn decreases the cathode concentration polarization. Third, the hydrolysis reaction replenishes the supply of Cl^- ions, so the process is self-sustaining. It is also believed that Cl^- ions disrupt the passive layers formed on certain alloys. In particular, stainless steels are susceptible to pitting and crevice corrosion in Cl-containing environments such as seawater and brine.

Since both pitting and crevice corrosion are highly localized forms of attack, they may be difficult to detect in service. Therefore, they are best controlled by precluding their formation by clever design. For example, joints formed using mechanical connectors such as screws, bolts, or rivets are natural locations for crevice corrosion. Thus, welded joints are often specified for structures designed to be used in aggressive environments.

Empirical observations show that many pits begin under surface deposits of contaminates occurring in stagnant pools of electrolyte located on horizontal sections of metal. Examples of design concepts directed at eliminating conditions favorable to pitting corrosion are are shown in Figure 15.2–9. They include eliminating sharp corners and other geometries that favor the formation of stagnant pools of liquid, and designing to permit free drainage and easy washing of equipment to avoid sediment buildup.

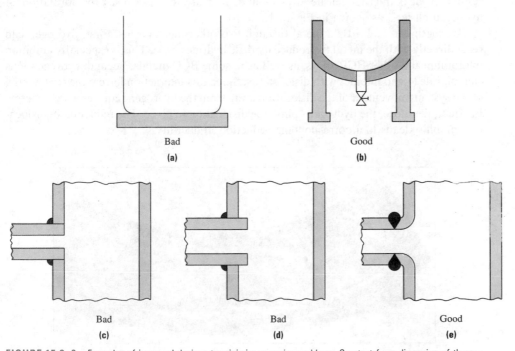

FIGURE 15.2–9 Examples of improved designs to minimize corrosion problems. See text for a discussion of these design features.

EXAMPLE 15.2–9

Consider the pitting corrosion of an iron component in salt water. If you are told that this reaction results in the formation of iron hydroxide, $Fe(OH)_2$, suggest a possible form for the hydrolysis reaction involving Cl.

Solution

In the first part of the reaction, metal ions combine with the chlorine ions to form a metal chloride:

$$Fe^{2+} + 2Cl^- \rightarrow FeCl_2$$

Next, the metal chloride reacts with water to form the corrosion by-product and replenish both the H^+ and Cl^- supply:

$$FeCl_2 + 2H_2O \rightarrow Fe(OH)_2 + 2H^+Cl^-$$

Hydrogen Embrittlement

The presence of hydrogen in the service environment accelerates the reduction reaction and has a deleterious effect on the corrosion resistance of many engineering alloys. In addition, hydrogen can degrade the mechanical properties of some metals in a more direct manner. The family of degradation mechanisms known collectively as **hydrogen embrittlement** is related to the high diffusivity of H^+ in BCC metals. Several related phenomena can occur when H^+ ions enter a metal lattice. First, elemental hydrogen can diffuse to open-volume regions such as voids, where it can recombine to form hydrogen gas. The H_2 molecules have much lower diffusivities than H^+ and are, therefore, trapped in the void. The internal pressure generated by the trapped gas can approach 1 MPa. The pressure results in local internal tensile stresses and, in many cases, causes the formation of hydrogen blisters, as shown in Figure 15.2–10.

Hydrogen can embrittle a metal through two other mechanisms. First, hydrogen can react directly with the metal to produce a brittle hydride phase. This is especially common in titanium and other HCP alloys, as well as in some BCC metals. Second, hydrogen ions can migrate to and interact with dislocations. Since this interaction lowers the free energy of the system, movement of the dislocation away from the hydrogen requires added energy. In effect, therefore, the hydrogen "pins" the dislocation in place. The reduction in dislocation mobility leads to a corresponding reduction in ductility.

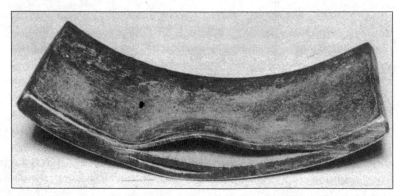

FIGURE 15.2–10   Cross section of a carbon steel plate removed from a petroleum process stream showing a large hydrogen blister. *(Source: Mars G. Fontana and Norbert D. Greene,* Corrosion Engineering, *2nd ed. Copyright 1978, McGraw-Hill, Inc. Reprinted with permission of McGraw-Hill, Inc.)*

Hydrogen embrittlement is unique in that it results in degradation of the cathode. Consider a situation in which a decrease in mechanical strength is the result of hydrogen embrittlement but is mistakenly diagnosed as a result of corrosion. Corrective measures are taken to make the metal more cathodic; however, the problem may be aggravated by this approach.

Good design practice requires avoidance of the unintentional introduction of hydrogen into a metal structure, especially when welding ferrous or titanium alloys. It may be necessary to specify the use of special welding rods that contain low concentrations of hydrogen and to be sure that dry welding conditions prevail. In addition, if hydrogen contamination is suspected, a postfabrication heat treatment, known as bakeout, promotes the diffusion of hydrogen out of the metal component.

....

DESIGN EXAMPLE 15.2–10

The mechanical properties of a stirring blade used in a wood pulp–processing tank are found to degrade rapidly. The blade is fabricated from a plain carbon steel, and the pulp solution has a high concentration of hydrogen sulfide, H_2S. All efforts to minimize the suspected corrosion problem have only increased the rate of degradation of the stirrer blade. Offer a solution for this problem.

Solution

The loss of mechanical strength may be related to hydrogen embrittlement at the cathode rather than corrosion at the anode. This suggestion is supported by the presence of a H-containing solution in contact with a BCC alloy. A common reaction in this situation is $Fe + H_2S \rightarrow FeS + 2H$, resulting in the production of atomic hydrogen, which can then diffuse into the BCC metal. One possible solution is to replace the blade with an alloy that is less sensitive to hydrogen embrittlement.

....

Stress-Assisted Corrosion

An applied stress can interact with an electrochemical corrosion process to accelerate the corrosion rate in many ways. Metal-environment pairs that combine the influence of tensile stress and corrosion can lead to catastrophic failure. Two aspects of this problem, known as **stress corrosion cracking (SCC),** make prevention difficult for design engineers. First, SCC can occur in environments in which the metal is inert in the absence of stress. Second, the stress can result from more subtle sources than a directly applied tensile stress. For example, it may be a residual stress introduced during processing or a thermally induced stress resulting from differential expansion of the constituent phases in a multiphase alloy. The tensile stress levels necessary to cause SCC can be as low as 10% of the yield stress of the metal.

There are two classic examples of SCC. One is "season cracking" of cold-drawn brass cartridge cases in the tropics. The environmental agent is ammonia produced from decaying organic material. The second is caustic (NaOH) embrittlement of the cold-worked regions near rivet holes in boilers of early steam-driven locomotives. The presence of NaOH in boilers is a result of the contamination of the boiler feed water with salt. Photographs of the catastrophic failures resulting from these two examples of SCC are shown in Figure 15.2–11.

Fortunately, only a few environments are capable of leading to SCC for each of the common structural alloys. The specific metal-environment pairs listed in Table 15.2–4 should be avoided if possible. It is important to note that stainless steel is susceptible to SCC in chloride or caustic (NaOH) environments. When a metal must be used in an environment where it is known to be susceptible to SCC, care should be taken to ensure

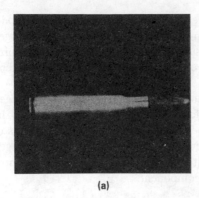
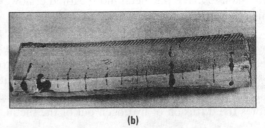

(a)   (b)

FIGURE 15.2–11 Two examples of SCC are **(a)** season cracking of cold-drawn brass cartridge cases in the tropics (ammonia from decaying organic material), and **(b)** caustic (NaOH) embrittlement of steel. *(Source: Mars G. Fontana and Norbert D. Greene, Corrosion Engineering, 2nd ed. Copyright 1978, McGraw-Hill, Inc. Reprinted with permission of McGraw-Hill, Inc.)*

TABLE 15.2–4 Some cases of stress corrosion cracking.

Alloy	Environment	Alloy	Environment
High-strength Al alloys	NaCl solutions Seawater Water vapor	300 series stainless steels (austenitic S.S.)	Metal chloride solutions (NaCl, MgCl$_2$, etc.) Caustic solutions Seawater NaCl + H$_2$O$_2$ solutions
Mg alloys	Solutions of NaCl + K$_2$CrO$_4$ Marine atmosphere Distilled water	Nickel	Strong solutions of NaOH or KOH Fused caustic
Cu alloys (brass)	Ammonia (vapors/solutions) Mercury salt solutions Amines Water	Monel (Ni-Cu alloy)	Solutions of HF or H$_2$SiF$_6$ Mercury salt solutions
Low-carbon steel	NaOH solutions Nitrate solutions [Ca(NO$_3$), NaNO$_3$] Acidic H$_2$S Seawater	Lead Ti alloys	Solutions of lead acetate High-temperature chlorides (fused salt mixtures) Solutions of HCl + methanol Red-fuming HNO$_3$ Chlorinated hydrocarbons
400 series stainless steels	NaCl + H$_2$O$_2$ solutions Seawater NaOH + H$_2$S solutions Caustic solutions Nitric + sulfuric acids		

Source: Z. Jastrzebski, *The Nature and Properties of Engineering Materials*, 3rd ed., p. 592. Copyright © 1959 by John Wiley & Sons, Inc. Reprinted by permission of John Wiley & Sons, Inc.

tensile stresses are not present. This can be achieved by selecting homogeneous single-phase microstructures, fully annealing the structure prior to placing it in the service environment, and eliminating tensile stress during service.

Another form of stress-assisted corrosion is **erosion corrosion.** This type of corrosion is common in elbows of pipes and similar locations where the fluid is impinging upon the metal at high velocities or high angles of attack. The friction and wear caused by the fluid impingement can accelerate the corrosion rate by either decreasing the local polarization or breaking down the protective (passive) oxide coating on the metal surface. Problems related to this phenomenon are common in nuclear reactor piping systems, which carry pressurized water at high temperature.

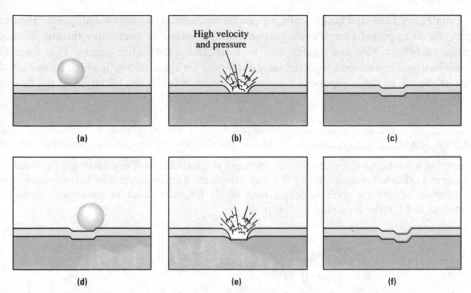

FIGURE 15.2–12   Schematic illustration of the cavitation damage mechanism: **(a)** bubble lands on surface; **(b)** bubble implodes, creating shock wave on surface, which fractures the protective film; **(c)** protective film reforms after some loss of metal; **(d)** repetition of process.   *(Source: Mars G. Fontana and Norbert D. Greene,* Corrosion Engineering, *2nd ed. Copyright 1978, McGraw-Hill, Inc. Reprinted with permission of McGraw-Hill, Inc.)*

A third form of stress-assisted corrosion, known as **cavitation damage,** occurs when vapor bubbles "pop" on a metal surface. (See Figure 15.2–12.) Cavitation is common in many rotating devices immersed in fluid, including ship propellers, pump impellers, and hydraulic turbines. The shock waves associated with the popping of the bubbles can result in pressures exceeding 345 MPa (50 ksi). Such pressure pulses may fracture a brittle oxide layer and expose unprotected metal to the corrosive environment. The nearly constant velocity and pressure gradients associated with rotating components cause the bubbles to continually form and pop at the same locations, leading to a rapid degradation of properties. A photograph of a failed impeller is shown in Figure 15.2–13. Appropriate designs to prevent cavitation damage entail eliminating or minimizing fluid pressure gradients across the surface of the component or coating the metallic parts with a resilient material such as rubber or plastic.

FIGURE 15.2–13   Photograph of an impeller showing severe cavitation damage.   *(Source:* Metals Handbook: Failure Analysis and Prevention, *Vol. 10, 8th ed., 1989, ASM International, Materials Park, OH.)*

Finally, we note that fatigue failures can be initiated or accelerated at anodic regions (pits, for example). Examples of components susceptible to **corrosion fatigue** include gears, propellers, turbines, shafts, and biomedical implants (hip joints). This form of stress-assisted degradation is particularly damaging for applications in which a substantial part of the life of the component is normally associated with the initiation of the fatigue crack.

DESIGN EXAMPLE 15.2–11

A hospital has a high failure rate for stainless steel surgical clamps. The clamps are sterilized by rinsing in a saline solution and drying in a high-temperature environment. The failed clamps show no evidence of uniform corrosion but exhibit highly localized attack in the regions shown in Figure 15.2–14. Offer a solution for this problem.

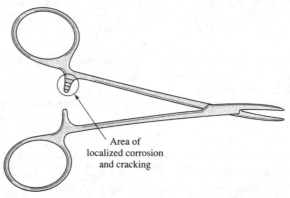

Area of
localized corrosion
and cracking

FIGURE 15.2–14   An illustration of a surgical instrument showing the location of a corrosion-induced failure.

Solution

The combination of stainless steel and a Cl-containing environment suggests the possibility of stress corrosion cracking. SCC, however, requires a tensile stress, which at first glance appears to be absent in this situation. Figure 15.2–14 shows that the failures begin in regions that were heavily cold-worked during the formation of the instrument. Thus, the as-formed part may contain residual tensile stresses. Under the assumption that the instruments are failing because of SCC, there are two ways to minimize this problem: (1) replace the saline rinse with another equivalent process that does not contain Cl^-, or (2) fully anneal the instruments to eliminate the residual tensile stresses due to cold working.

15.2.5 Corrosion Prevention

In this section we discuss strategies for eliminating or minimizing corrosion damage.

Temperature

Since corrosion involves one or more chemical reactions and the rate of most reactions increases with temperature, the rate of corrosion can often be decreased by reducing the operating temperature of the component.

Material Selection

The selection of appropriate materials for a specific environment is one of the principal weapons the design engineer can use to fight corrosion. While some aspects of materials selection are straightforward—avoid alloys known to degrade rapidly or result in SCC in a given environment—other aspects of the selection process are more subtle.

Microscopic galvanic cells can occur as a result of any of the following conditions: alloy segregation, existence of multiple phases, local cold-worked regions, or small grain size (i.e., large number of grain boundaries). Many of these potential problems can be eliminated by selecting appropriate microstructures and heat treatments. The ideal heat treatment yields a homogeneous distribution of the alloying elements, eliminates all residual stresses, and gives a fully recrystallized large grain structure.

The resistance of an alloy to corrosion can also be improved by slight compositional changes, as was demonstrated by the addition of Nb for improving the quality of welds in stainless steel. This method should be used with caution, however, since changes in the alloy composition can significantly impact all other properties of the alloy. Alloy compositions can be modified by incorporating elements that produce protective oxide films (Cr, Ni, Ti) or removing unnecessary grain boundary precipitates. In the extreme, the corrosion resistance of most nonpassivated alloys can be improved by increasing purity. This approach is often unsatisfactory because of cost or because strength decreases as alloy content is reduced. One example of a clever use of the high-purity approach is the laminar composite Alclad, which consists of a surface layer of high-purity Al covering a much higher-strength aluminum alloy sheet.

Design Modifications

As with the design of microstructures, one of the guiding principles in component design is to avoid heterogeneity—that is, to attempt to minimize thermal and velocity gradients so as to eliminate concentration cells. In addition, designers should avoid the use of dissimilar metals in electrical contact. If dissimilar metals must be used, then the anode-to-cathode area ratio should be as large as possible or the materials should be electrically insulated from one another. For example, Teflon washers are often used between steel and Cu pipe, but caution must be used since improper washer design can lead to preferential sites for crevice corrosion. Other ideas include using closed, but not stagnant, fluid systems to minimize the dissolved oxygen content. Use of welded or polymer-filled joints rather than only bolts or rivets minimizes crevice corrosion. Component designs should also minimize the number of sharp corners and recesses where stagnant pools of liquid can develop. Such recesses may be filled with polymer additives.

Cathodic Protection

Cathodic protection is based on reversing the direction of the metal oxidation reaction. That is, corrosion can be prevented if the metal is "forced" to be the cathode. The most straightforward way of forcing the desired cathodic response is to use an external voltage source to supply the metal needing protection with extra electrons. The metal to be protected is simply connected to the negative terminal of a DC power supply, such as a battery. This method is illustrated in Figure 15.2–15, in which a buried steel pipe is protected by an impressed voltage. Note that this method requires that "something else" be connected to the other side of the voltage source. In this application a piece of scrap

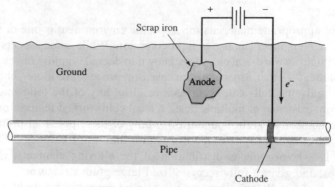

FIGURE 15.2–15 Cathodic protection of a buried steel pipe using the method of impressed voltage. Note that the metal to be protected is supplied with electrons by connecting it to the negative terminal of the voltage supply.

iron is used as the anode in the impressed voltage cell. The design suffers from an unfavorable anode-to-cathode area ratio, and the scrap iron will have to be frequently replaced. This problem can be reduced by wrapping the steel pipe in protective plastic to minimize the amount of exposed metal and thereby minimize cathodic area.

Another version of cathodic protection involves the use of a **sacrificial anode.** In this method a galvanic cell is formed intentionally by placing another, more anodic metal in electrical contact with the metal to be protected. The formation of the galvanic cell results in the preferential corrosion of the sacrificial anode. As with the use of impressed voltages, the protected metal is again supplied with electrons. The common choices for the sacrificial anode material are Zn or more often Mg. This technique is sometimes used on ocean vessels, where large Zn plates are added below the waterline as sacrificial anodes, or in car radiators, where Zn slugs are used.

As with the impressed voltage method, sacrificial anodes corrode rapidly, so they must be replaced periodically. It must also be recognized that the anode only protects the cathodic material that is physically close to the anode. When a large amount of metal is to be protected using this method, a significant number of sacrificial anodes must be used.

Use of Protective Coatings

Another important, simple, and common method for minimizing corrosion damage is to coat the metal. Two general classes of protective coatings are those based on cathodic (galvanic) protection and those that simply prevent contact between metal and fluid. In general, metal coatings can protect based on either or both mechanisms while ceramic and organic coatings are used for their insulating properties.

The use of metal coatings such as Zn or Cr on steel has been discussed previously. While both coatings protect the steel when they offer continuous coverage, only the Zn continues to protect the steel after the coating is scratched. An interesting extension of the idea that scratches will always be present is that if only one of two dissimilar metals is to be painted, it should be the cathodic metal. This guarantees a highly favorable anode-to-cathode area ratio.

There are a variety of types of protective coatings. Some provide only temporary protection while others are meant to protect the material throughout its useful life. Examples of common temporary coatings include grease, oil, and waxes. Many metallic parts are coated with these materials before being placed in long-term storage or as they are prepared for transportation.

Common examples of longer lasting protective coatings include porcelain enamels, tar, Teflon, paints, and lacquers. Many paints permit limited diffusion of oxygen and water, so they must be modified with additives that block diffusion. The major problem with the protective coating is incomplete coverage of the anode, usually resulting from the presence of "pinholes" or other defects in the original coating or damage to the coating during service. This can result in the formation of corrosion cells with unfavorable anode-to-cathode area ratios. To partially combat this problem, many modern paints designed for use on steels contain Zn particles. An added advantage of using paints or adhesives at lap seams is that corrosion can be prevented by physically preventing the fluid from condensing or running into the crevice.

Passive oxide layers, formed in specific metal-environment systems, can be regarded as protective coatings. One common application of this concept is the anodizing of aluminum, which involves applying a high corrosion potential to the aluminum metal to form an oxide layer that is thicker and tougher than the naturally occurring layer. Note, however, that materials that depend on the formation of a protective oxide layer may fail catastrophically in a reducing environment.

Inhibitors

Specific chemicals can be added to an electrolyte that reduce diffusion of ions to the metal-electrolyte interface and enhance the polarization of one or both electrical reactions. Perhaps the most well-known example is the addition of chromate salts to radiator antifreeze (ethylene glycol).

The details of the mechanism by which specific inhibitors limit corrosion vary. Some inhibitors, such as the chromate salts mentioned above, form adsorbed layers on either the anode or cathode. These chemical films act like the coatings described above in that they break the electrical circuit. Other inhibitors are scavengers that decrease the amount of dissolved oxygen in the electrolyte. For example, sodium sulfate removes dissolved oxygen according to the reaction

$$2Na_2SO_3 + O_2 \rightarrow 2Na_2SO_4 \qquad (15.2\text{--}23)$$

Still other additives, such as arsenic or antimony, increase the activation polarization of the system by retarding hydrogen evolution.

Two aspects of the use of inhibitors require caution. First, some of the chemicals (e.g., arsenic) are toxic. Second, if the action of the inhibitor is coating the anode, then not adding enough of the inhibitor to the system may increase the corrosion rate. This results from incomplete coverage of the anode, which yields an unfavorable anode-to-cathode area ratio. As a result of this potential problem, inhibitors that work by coating the anode are sometimes referred to as "dangerous" inhibitors. Since it is difficult if not impossible to maintain the required concentration of the inhibitor in an open-engineering system, the utility of these additives is usually restricted to closed or frequently replenished systems such as in car radiators, as mentioned before.

15.3 DIRECT ATMOSPHERIC ATTACK (GAS-SOLID REACTIONS)

Many engineering materials are used in gaseous rather than liquid environments. In this section we will investigate the characteristics and mechanisms of direct atmospheric attack and explore some design strategies to minimize this form of environmental degradation. We will find a great many similarities between the concepts discussed in this section and those in the previous section on corrosion. For example, while thermodynamic calculations are useful for predicting the relative tendency of a material to interact with the surrounding gases, one must also investigate the kinetics of the resulting reaction to evaluate the potential of various materials for specific environments.

Several commonly encountered gases that can degrade material properties include nitrogen, sulfurous gases, chlorine, and water vapor. The most common reaction, however, involves oxygen and is known as **oxidation.** In contrast to the previous discussion, this section will use the term *oxidation* only to describe reactions in which a material combines with oxygen to produce an oxide. That is, oxidation will refer to a complete chemical reaction of the form:

$$n\,M(\text{solid}) + p\,O_2(\text{gas}) \rightarrow M_nO_{2p}(\text{solid or gas}) \tag{15.3-1}$$

rather than to the general class of electron-producing reactions to which it referred in the previous section.

The distinction between corrosion and direct atmospheric attack is not always clear. In fact, oxidation is often referred to as dry corrosion. An example of the similarities between the two types of degradation mechanisms is provided by the interaction between stainless steels and hydrocarbon gases. Under certain high-temperature conditions, the carbon from the hydrocarbons can diffuse into stainless steel and react with the chromium to form chromium carbides. The result is identical to that associated with the formation of carbides because of improper weld treatment. That is, the regions in the vicinity of the carbides are depleted of Cr and are no longer protected by the passive chromium oxide film. This accelerates degradation either through corrosion or one of the oxidation mechanisms described below.

The remainder of this section is organized as follows. After a brief introduction to the mechanisms by which gases, particularly oxygen, alter the bond structure of materials, we discuss the formation of both gaseous and solid reaction products. Next, we describe the similarities and differences between protective and nonprotective coatings and relate them to the kinetics of the gas-solid reactions. Finally, after describing some of the ways gas-solid reactions can be harnessed to become useful processing operations, we offer some design strategies to prevent or minimize the negative impact of direct atmospheric attack on the properties of engineering materials. Most of the following discussions are couched in terms of oxygen, yet many of the principles can be extended to reactions with other gases.

....

EXAMPLE 15.3–1

Determine the appropriate form of Equation 15.3–1 for the oxidation of Al and the reaction of Al and Cl_2 gas.

Solution

When aluminum reacts with oxygen to form an oxide, the reaction will be

$$4Al(s) + 3O_2(g) \rightarrow 2Al_2O_3(s)$$

When aluminum reacts with chlorine, the reaction will be

$$2Al(s) + 3Cl_2(g) \rightarrow 2AlCl_3(s)$$

....

15.3.1 Alteration of Bond Structures by Atmospheric Gases

As with many phenomena in materials science and engineering, the key to understanding direct atmospheric attack is to examine how the atoms in the gas interact with and alter the bond structure of the solid. For example, if an elastomer contains unsaturated double

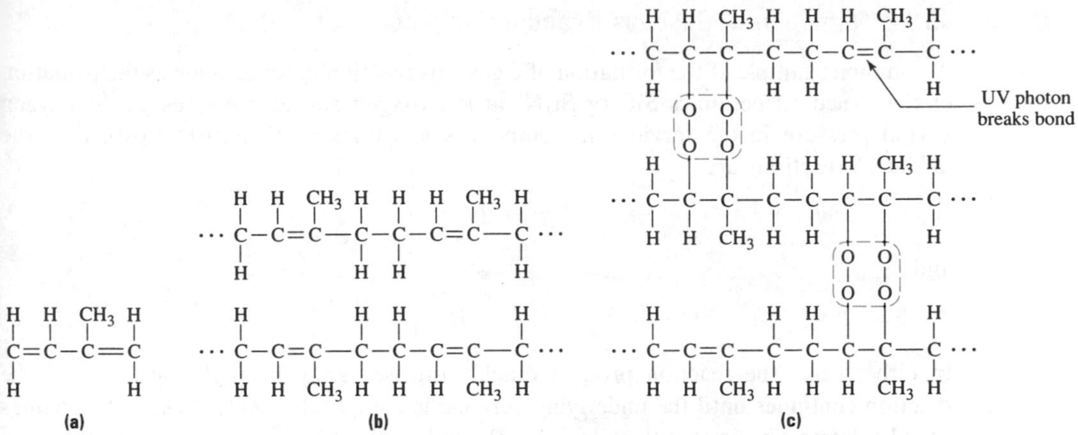

FIGURE 15.3–1   A schematic illustration of the oxygen-induced crosslinking of an elastomer. Exposure to UV light can accelerate the process by supplying the energy to break a carbon-carbon double bond.

bonds in its carbon backbone chain, the oxygen or sulfur in the environment may react with these bonds to produce additional crosslinks. (See Figure 15.3–1.) The increase in crosslink density decreases the ductility of the rubber. A common example is the cracking of automobile tire sidewalls in "normal" environments.

Several factors influence the rate of atmospheric-induced crosslinking. For example, ozone, O_3, is more reactive than O_2. In addition, bonds are broken or altered more readily when an elastomer is under stress, providing the equivalent of SCC in polymers. UV light can accelerate oxidative degradation by providing a portion of the energy to open the carbon-carbon double bonds. The combination of oxidation and radiation damage is so important that it is given a name, photo-oxidation. The carbon black added to stiffen, strengthen, and toughen tires also serves to partially block UV light. While the additive does not prevent degradation, it slows the process so that tires usually wear out before the sidewalls crack excessively. The product Armorall was developed to spray onto tires to slow the process in especially serious environments (such as in Los Angeles).

Virtually all polymers are liable to oxidize at sufficiently high temperature. Oxidation can become a serious problem in melt processing (to be described in Chapter 16) of thermoplastic polymers such as polypropylene, which is often processed at temperatures far above the melting temperature to reduce the melt viscosity. To minimize the problem, antioxidants and stabilizers are commonly added to organic materials. In polymers the antioxidants are added to the melt. Many antioxidants are quinones. An example is butlyated hydroxytoluol (BHT), which you are all familiar with since it is often added to food products to prevent oxidation.

The oxygen crosslink mechanism is characteristic of most atmospheric attack mechanisms in that it involves breaking of primary bonds within the solid and formation of new bonds between the once gaseous species and the atoms of the solid. It differs, however, in that no atoms are removed from the host solid. Returning to the general oxidation equation 15.3–1, we see that atoms are removed from the solid and incorporated into the new oxide phase. The loss of atoms from the host structure is the primary cause of the degradation in physical and mechanical properties.

The resulting oxide phase can be either solid, liquid, or gaseous. Solid oxides can be protective or nonprotective. Materials with the inherent ability to resist high-temperature oxidation from solid protective oxides. In the next subsection we discuss some of the gas-solid reactions that result in the formation of vapor-phase oxides. In Section 15.3.3 we describe the protective and nonprotective solid-phase oxides.

15.3.2 Formation of Gaseous Reaction Products

A common example of the formation of a gaseous reaction product involves the oxidation of the structural ceramics SiC or Si_3N_4 at low oxygen partial pressures. If the oxygen partial pressure in the service environment is less than ~140 Pa (0.02 psi), then the oxidation reactions are:

$$2SiC(s) + 3O_2(g) \rightarrow 2SiO(g) + 2CO_2(g) \qquad (15.3\text{--}2)$$

and

$$2Si_3N_4(s) + 3O_2(g) \rightarrow 6SiO(g) + 4N_2(g) \qquad (15.3\text{--}3)$$

In either case, the reaction product quickly diffuses away from the surface and the reaction continues until the underlying ceramic is completely gone. These two ceramics are also interesting in that they do not melt; rather, they dissociate at elevated temperatures—SiC at 2500°C and Si_3N_4 at 1900°C in air. Dissociation occurs at lower temperatures in vacuum or inert environments. Recognition of this fact is important when sintering these ceramics, since either a vacuum or a high-purity inert gas is usually employed to minimize contamination. In the case of Si_3N_4, sintering is performed under a positive pressure of high-purity nitrogen. This limits contamination and suppresses the dissociation reactions.

Earlier in this chapter we discussed the static fatigue of silicate glasses. The operative chemistry is a hydrolysis reaction,

$$SiO_2 + 2H_2O \rightarrow Si(OH)_4 \qquad (15.3\text{--}4)$$

which leads to degradation of the silica. In a manner analogous to the degradation of SiC and Si_3N_4, the silicates can degrade in vacuum or inert gas according to the reaction

$$2SiO_2(s) \rightarrow 2SiO(g) + O_2(g) \qquad (15.3\text{--}5)$$

In hydrogen gas, SiO_2 can be lost by

$$SiO_2(s) + H_2(g) \rightarrow SiO(g) + H_2O(l) \qquad (15.3\text{--}6)$$

This type of reaction can also degrade the mechanical properties of metals. For example, steel can be decarburized in an oxygen environment according to the reaction

$$2C \text{ (in steel)} + O_2(g) \rightarrow 2CO(g) \qquad (15.3\text{--}7)$$

The resulting carbon-deficient zone near the surface is soft compared with the underlying steel, lowering wear resistance.

In almost all cases the formation of a vapor-phase oxide is undesirable, since it results in the loss of underlying structural material without the possibility of forming a protective coating.

15.3.3 Protective and Nonprotective Solid Oxides

For a solid oxide to protect the underlying material, it must be continuous, adherent, and completely cover the material. For metals, the last requirement can be estimated using the **Pilling-Bedworth (PB) ratio.** This ratio is defined as the volume occupied by one mole of oxide divided by the volume of the metal consumed to create one mole of oxide.

Recalling that density is mass divided by volume, we find that for a reaction of the form given in Equation 15.3–1, the PB ratio is:

$$PB = \frac{M_O \rho_M}{n M_M \rho_O}$$

(15.3–8)

where the subscripts O and M refer to the oxide and metal, M is atomic mass, ρ is density, and n is the number of metal atoms in the oxide formula (e.g., $M_n O_2$).

As shown in Figure 15.3–2, when the PB ratio is less than 1, the volume of the oxide is insufficient to cover the metal, and the oxide will not protect the underlying metal. Table 15.3–1, which lists the PB ratio for many common metals, shows good agreement between the predictions based on PB and experimental observations. Careful examination

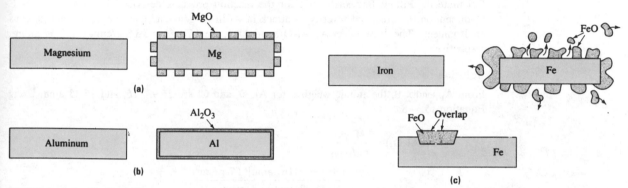

FIGURE 15.3–2   The Pilling-Bedworth (PB) ratio represents the volume occupied by the oxide divided by the volume of the metal consumed to create the oxide. **(a)** If PB < 1, there is insufficient oxide to cover the surface of the metal. **(b)** If 1 < PB < 2.3, the oxide is able to completely cover the metal. **(c)** If PB > 2.3, the relative volume of the oxide is so large that it spalls off the metal due to the development of compressive stresses.

TABLE 15.3–1   Pilling-Bedworth ratios for various metal oxides.

Protective oxides	Nonprotective oxides
Be—1.59	Li—0.57
Cu—1.68	Na—0.57
Al—1.34	K—0.45
Si—2.27	Ag—1.59
Cr—1.99	Cd—1.21
Mn—1.79	Ti—1.95
Fe—1.77	Mo—3.40
Co—1.99	Hf—2.61
Ni—1.52	Sb—2.35
Pd—1.60	W—3.40
Pb—1.40	Ta—2.33
Ce—1.16	U—3.05
Zr—1.50	V—3.18
	Mg—0.80

Source: Adapted from B. Chalmers, *Physical Metallurgy.*
Copyright © 1959 by John Wiley & Sons, Inc. Reprinted
by permission of John Wiley & Sons, Inc.

of the table, however, shows several materials with PB greater than 1 in the nonprotective column. Note that most of these metals have PB > 2.3. This is because if the specific volume of the oxide is much greater than that of the metal, then a substantial compressive stress may develop during oxide formation. The compressive stress may lead to cracking or spalling of the brittle solid oxide layer, as shown in Figure 15.3–2c. Thus, the general guideline is if 1 < PB < 2.3, then the oxide is likely to protect the underlying metal. Caution is warranted, however, since there are several exceptions to this rule, including Ag and Ti.

EXAMPLE 15.3–2

Calculate the Pilling–Bedworth ratio for the reaction products described in Example 15.3–1. Comment on the predicted severity of attack of Al in oxygen- and chlorine-containing gaseous environments. The densities for Al, Al_2O_3, and $AlCl_3$ are 2.70 g/cm^3, 3.8 g/cm^3, and 2.44 g/cm^3, respectively.

Solution

From Appendix B, the atomic weights, for Al, O, and Cl are 26.98, 16, and 35.45 amu. Using Equation 15.3–8:

$$PB(Al_2O_3) = \frac{M_O \rho_M}{n M_M \rho_O}$$

$$= \frac{[2(26.98) + 3(16)\ \text{amu}](2.7\ \text{g/cm}^3)}{(2)(26.98\ \text{amu})(3.8\ \text{g/cm}^3)}$$

$$= 1.34$$

Similarly,

$$PB(AlCl_3) = \frac{M_{Cl} \rho_M}{n M_M \rho_{Cl}}$$

$$= \frac{[26.98 + 3(35.45)\ \text{amu}](2.7\ \text{g/cm}^3)}{(1)(26.98\ \text{amu})(2.44\ \text{g/cm}^3)}$$

$$= 5.47$$

Based on these calculations, we anticipate that the oxide is protective and the chloride is non-protective.

While the PB ratio is reasonably successful in predicting solid oxide properties, it is grossly oversimplified. A protective oxide must possess several other characteristics. Components are often cycled through rather large temperature ranges. Hence, it is important for the oxide to have a coefficient of thermal expansion similar to that of the underlying material. If not, the thermally induced stresses may cause the brittle oxide coating to crack. It is also important in some applications that the oxide have a reasonably high melting temperature. If not, a limitation on the service temperature for the component may be controlled by the T_m of the oxide rather than by that of the metal. This is a common problem with the high-T_m refractory metals, which are known to oxidize much more rapidly than their high bond strength would suggest. For example, MoO_3 is volatile above 550°C, which is considerably below the melting temperature of Mo, 2650°C.

5.3.4 Kinetics of Oxidation

Thermodynamic calculations show that all metals except platinum and gold oxidize in air. Several metals, however, can be used for extended periods in air at high temperatures. This is not a contradiction, since, as was the case with liquid-solid corrosion, the kinetics of gas-solid reactions, not just the thermodynamics, control the suitability of a material for service in aggressive environments. As anticipated, the oxidation rate of most materials increases with temperature.

Returning to the observation that there are two general types of solid oxides— protective and nonprotective—one expects at least two forms of the relationship between the amount of metal lost to oxidation and time of exposure. Oxidation rates can be reported in several different but equivalent ways, including weight gain or loss per unit time and change in the thickness of the oxide per unit time (dy/dt). The metal under a nonprotective oxide is continually exposed directly to the oxygen source. This results in a constant oxidation rate. Mathematically,

$$\frac{dy}{dt} = \text{Oxidation rate} = \text{Constant} \qquad (15.3\text{--}9)$$

or

$$y = C_1 t \qquad (15.3\text{--}10)$$

where C_1 is a material-specific constant. This is referred to as a linear growth rate, since the oxide thickness is proportional to exposure time. The rate-limiting step is chemical oxidation at the surface. Note that in the case of solid nonprotective oxides, which either do not adhere to or spall off of the surface, the thickness calculated from Equation 15.3–10 is the total thickness of oxide produced rather than the thickness of oxide on the surface.

While the initial stage of protective oxide growth is similar to that described above, the situation quickly changes. Once a film of finite thickness has developed, further oxidation requires diffusion of either O^{2-} or M^{n+} ions or both through the oxide. (In all cases electrons must also migrate through the oxide.) The location of the oxidation process depends on the relative diffusivities of the ions, as shown in Figure 15.3–3. When metal

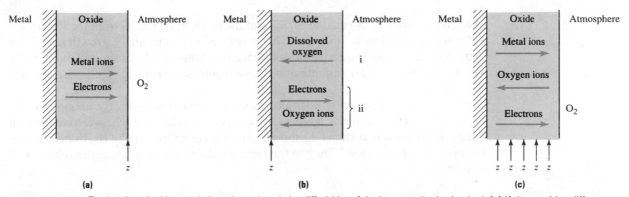

FIGURE 15.3–3 The location of oxide growth depends on the relative diffusivities of the ions or molecules involved. **(a)** If the metal ion diffuses more quickly than the oxygen ions, then growth occurs at the oxide-atmosphere interface. **(b)** Growth occurs at the metal surface either (i) if the oxide is porous (or cracked) such that the O_2 molecule can come in direct contact with the metal surface, or (ii) if the oxide is protective but the oxygen ions diffuse more rapidly than the metal ions. **(c)** If the metal and oxygen ions diffuse at roughly the same rate, then growth can occur anywhere within the oxide layer.

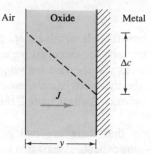

FIGURE 15.3–4   A schematic illustration of the diffusion of oxygen through an oxide of thickness y to a metal surface. The dashed line represents the oxygen concentration profile.

ions diffuse more rapidly than oxygen ions, oxide growth occurs at the oxide-atmosphere interface. However, when the oxygen ions diffuse more rapidly, the oxide growth occurs at the metal-oxide interface. Finally, when both ions diffuse at approximately the same rate, then continued oxidation occurs within the oxide film.

In each case the oxidation rate is limited by the flux of ions to the growth interface. That is,

$$\frac{dy}{dt} \propto J_{ion} \tag{15.3–11}$$

From Fick's first law of diffusion in 1-D (Section 4.4.2):

$$J_{ion} = -D\left(\frac{dc}{dy}\right) \tag{15.3–12}$$

For the situation shown in Figure 15.3–4, we have $dc/dy = C_0/y$. Combining this result with Equations 15.3–11 and 15.3–12 yields:

$$\frac{dy}{dt} \propto -\frac{DC_0}{y} \tag{15.3–13}$$

Separating variables and integrating yields:

$$y^2 = C_2 t + C_3 \tag{15.3–14}$$

where the constant C_2 is equal to $-DC_0$ and the constant C_3 results from the integration and gives the initial oxide thickness. This is referred to as a parabolic growth rate and is characteristic of most protective oxides. Since the diffusivity D increases exponentially with temperature, the growth rate of diffusion-controlled oxides also follows Arrhenius behavior.

In practice, it is generally difficult to measure oxide thickness. By noting that the density of an oxide is uniform, however, we can derive a relationship between weight gain and oxide thickness that allows the former to be used as an indicator of the oxidation rate. The procedure is as follows. The two fundamental definitions concerning the volume V of an oxide are:

$$V = yA \tag{15.3–15}$$

and

$$\rho = \frac{M}{V} \tag{15.3–16}$$

where y is the thickness, A is surface area, M is the mass, and ρ is the density. Combining these two equations by eliminating V and solving for y yields:

$$y = \frac{M}{\rho A} \qquad\qquad (15.3\text{–}17)$$

Since both the oxide density and the surface area are constant, we find $y \propto M$ and, therefore, $\Delta y \propto \Delta M$. That is, weight can be used as a direct measure of change in oxide thickness. Note that if the oxide is not adherent or spalls off, the experiment shows a weight loss rather than a weight gain.

EXAMPLE 15.3–3

A sample of Ni and a sample of Mg are exposed to the same high-temperature oxidation environment. Both metals are initially oxide-free and both have an oxide thickness of 0.17 μm after one minute. Calculate the thickness of the oxide on each material after an exposure of one hour.

Solution

Table 15.3–1 shows that the oxide for Ni is protective while that of Mg is nonprotective. Therefore, the oxide growth rate should be linear for MgO and parabolic for NiO. Mathematically,

$$y_{Mg}(t) = C_1 t \qquad \text{and} \qquad [y_{Ni}(t)]^2 = C_2 t + C_3$$

The value for C_1 can be found by noting that

$$y_{Mg}(60) = 0.17 \ \mu\text{m} = C_1 \times 60 \ \text{s}$$

Solving for C_1 yields:

$$C_1 = 2.833 \times 10^{-3} \ \mu\text{m/s}$$

Therefore, y_{Mg} for a one-hour (3600-s) exposure is

$$y_{Mg}(3600) = (2.833 \times 10^{-3} \ \mu\text{m/s})(3600 \ \text{s}) = 10.2 \ \mu\text{m}$$

For Ni, $y_{Ni}(0) = 0$ implies $C_3 = 0$. Substituting $y_{Ni}(60) = 0.17 \ \mu$m into the expression for $y_{Ni}(t)$ and solving for C_2:

$$C_2 = \frac{y_{Ni}^2(t)}{t} = \frac{(0.17 \ \mu\text{m})^2}{60 \ \text{s}} = 4.82 \times 10^{-4} \ \mu\text{m}^2/\text{s}$$

Therefore, y_{Ni} for a one-hour exposure is

$$[y_{Ni}(3600)]^2 = (4.82 \times 10^{-4} \ \mu\text{m}^2/\text{s})(3600 \ \text{s}) = 1.734 \ \mu\text{m}^2$$
$$y_{Ni}(3600) = \sqrt{1.734 \ \mu\text{m}^2} = 1.32 \ \mu\text{m}$$

EXAMPLE 15.3–4

The experiment in Example 15.3–3 was performed at 300° C, and the growth of NiO is limited by the diffusion of Ni^{2+} ions via a vacancy mechanism ($Q = 210$ kJ/mol). Estimate the oxide film thickness for a sample of Ni subjected to a similar oxidation process for one hour at 500°C.

Solution

From Example 15.3–3, we know $C_2(300°\text{C}) = 4.82 \times 10^{-4} \ \mu\text{m}^2/\text{s}$. Since this constant is proportional to the diffusion coefficient of Ni^{2+} through NiO, the growth rate constant ratio will be the same as the diffusion coefficient ratio:

$$\frac{C_2(500°C)}{C_2(300°C)} = \frac{D(500°C)}{D(300°C)}$$

$$= \frac{D_0 \exp[-Q/R(773\ K)]}{D_0 \exp[-Q/R(573\ K)]}$$

$$= \exp\left[\left(-\frac{210,000\ J/mol}{8.314\ J/mol\text{-}K}\right)\left(\frac{1}{773\ K} - \frac{1}{573\ K}\right)\right]$$

$$= 8.98 \times 10^4$$

Solving for the value of the constant at 500°C:

$$C_2(500°C) = (8.98 \times 10^4)[C_2(300°C)]$$

$$= (8.98 \times 10^4)(4.82 \times 10^{-4}\ \mu m^2/s) = 43.3\ \mu m^2/s$$

This value can be used to calculate the oxide thickness after a one-hour exposure:

$$[y_{Ni}(3600)]^2 = (43.3\ \mu m^2/s)(3600\ s) = 1.56 \times 10^5\ \mu m^2$$

$$y_{Ni}(3600) = \sqrt{1.56 \times 10^5\ \mu m^2} = 395\ \mu m = 0.395\ mm$$

....

While Equation 15.3–14 describes the oxide growth rate for many protective oxides, some of the most resistant oxides, such as those for Al, Zn, Be, and Cr, obey a logarithmic growth law:

$$y = C_4 \log(C_5 t + 1) \tag{15.3–18}$$

The reason for the deviation from parabolic growth for the oxides of Al and Be is related to the extremely low electron mobility in these oxides. As mentioned previously, oxide growth requires not only the diffusion of either the metal or the oxygen ions or both but also the movement of electrons through the oxide. If electron motion is the rate-limiting step, as it is for Al_2O_3 and BeO, then growth rate is logarithmic.

The oxides of Zn and Cr also obey a logarithmic growth law, but for a different reason. In these oxides metal ions build up within the oxide layer, resulting in the formation of an electric field. The field limits further ionic diffusion within the oxide, and subsequent growth is characterized by Equation 15.3–18. Figure 15.3–5 compares the total loss of underlying metal associated with the three oxide growth laws.

An additional characteristic of both Al_2O_3 and Cr_2O_3 gives them their superior protective qualities: their crystal structure is coherent with the underlying metal. This results in strong bonds at the oxide-metal interface and, consequently, excellent oxide adherence.

In a few materials it is possible for the oxygen ions to diffuse more rapidly than both the metal ions and the electrons. In this case the oxygen diffuses through the oxide and continues to diffuse into the metal. The mechanism, known as internal oxidation, leads to the formation of oxide particles within the metal lattice and often results in severe embrittlement of the metal.

15.3.5 Using Atmospheric "Attack" to Advantage

An improved understanding of the mechanisms of atmospheric attack enables engineers to intelligently design materials for use in service environments and to harness these same mechanisms to accomplish processing goals.

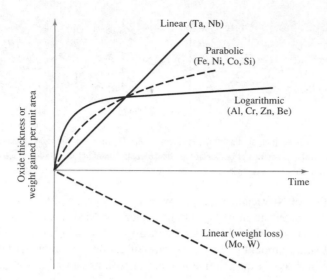

FIGURE 15.3–5

A comparison of linear, parabolic, and logarithmic oxidation rates. The vertical scale represents either oxide thickness or weight gain (loss).

The structural ceramics SiC and Si_3N_4 are susceptible to tensile failure because of strength-limiting surface flaws. A brief exposure of these materials to oxygen at low partial pressures can remove the (damaged) surface layer of the components. (See Equations 15.3–2 and 15.3–3.) This process improves surface finish and, therefore, correspondingly increases the strength of the ceramic component. The component can then be used successfully at higher operating temperatures and stress levels, provided that the oxygen partial pressure in the service environment exceeds 140 Pa (0.02 psi). At the higher oxygen pressure, the favored oxidation reactions are:

$$SiC(s) + O_2(g) \rightarrow SiO_2(s) \tag{15.3–19}$$

and

$$Si_3N_4(s) + O_2(g) \rightarrow SiO_2(s) \tag{15.3–20}$$

The oxide is now solid SiO_2, which is a protective coating.

15.3.6 Methods of Improving Resistance to Atmospheric Attack

Three general methods used to minimize or reduce the rate of atmospheric attack on materials are to select the best material and, if necessary, modify its composition; modify the composition of the service environment; or apply a suitable coating to the material. Examples of each method are described below.

Alloying

In its simplest form, alloying involves adding one or more of the metals (Al, Be, Cr, or Zn) that form tough, adherent, thermally stable, and protective oxides to the material needing protection. The popularity and success of this method can be recognized by noting that most oxidation-resistant alloys contain some Cr.

A slightly more advanced alloying method is to introduce additional cations into the oxide to adjust defect concentrations within the oxide. A change in defect chemistry can substantially alter the ionic diffusion kinetics. Consider the case of NiO (actually $Ni_{1-x}O$) in which the parabolic growth is controlled by the diffusion of Ni ions via a vacancy mechanism. (See Section 4.4.5.) Since the rate of Ni diffusion is directly proportional to

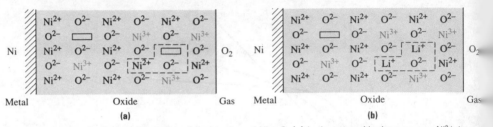

FIGURE 15.3–6   A schematic illustration of the structure of $Ni_{1-x}O$. **(a)** In the pure oxide there are two Ni^{3+} ions for every Ni vacancy. **(b)** The addition of LiO_2 decreases the concentration of vacant Ni sites, since each mole of LiO_2 "replaces" one mole of Ni^{2+}-O^{2-}-V_{Ni} groups.

the concentration of Ni vacancies (V_{Ni}), any compositional modification that results in a decrease in the concentration of V_{Ni} should reduce the Ni diffusion rate.

The structure of $Ni_{1-x}O$ is shown schematically in Figure 15.3–6a. When the NiO is alloyed with a small amount of Li_2O, the ratio of cations to anions within the structure increases. The increase in the relative number of cations can be accommodated in four ways:

1. An increase in the number of cation interstitials
2. A decrease in the number of cation vacancies
3. An increase in the number of anion vacancies
4. A decrease in the number of anion interstitials

The most energetically favorable structural modification is the second option. The decrease in the concentration of V_{Ni} is shown schematically in Figure 15.3–6b. Thus, the result of alloying NiO with Li_2O is a decrease in the oxidation rate of the material.

··

DESIGN EXAMPLE 15.3–5

The growth rate of the oxide Fe_2O_3 is controlled by the vacancy-assisted diffusion of O^{2-} ions through the oxide. Suggest an oxide addition that might reduce the oxidation rate of Fe.

Solution

Since oxide growth rate is controlled by the diffusion of O^{2-} ions through the oxide, the key idea is to try to retard O^{2-} diffusion by reducing the oxygen vacancy concentration in the oxide. To accomplish this objective one might increase the anion: cation ratio by adding an oxide of the form MO_2, where M is a metal ion with valence $+4$. In addition, the new cation should obey the Hume-Rothery rules so that it can substitute for the Fe^{3+} ions. For example, one might try additions of ZrO_2.

··

Modification of the Environment

As with corrosion, the ability to modify the composition of a gaseous environment is limited to reactions occurring in a contained volume. Several environmental modifications have been mentioned previously, including increasing the oxygen partial pressure above 140 Pa if SiC or Si_3N_4 components are in use, sintering Si_3N_4 under a positive partial pressure of nitrogen, and avoiding the presence of hydrocarbons if stainless steels are in use.

Coatings

If a specific material must be used in an aggressive gaseous environment and the material will not form a protective coating naturally, it may be necessary to apply a protective coating to the material before placing it in service. The most common protective coatings include the protective oxides such as Al_2O_3 or Cr_2O_3, the stoichiometric silicates, or a diamond coating. All of these materials share several features. They are adherent, have low ionic diffusivities, are thermally stable, and can withstand thermal shock.

DESIGN EXAMPLE 15.3–6

Suppose you were asked to select a material that will be subjected to both high-temperature oxidation and electrochemical corrosion. Plates of this material will be fastened together with Ni rivets. The design of the structure has narrowed the selection of materials for the plates to Ti, Cu, or plain carbon steel. Which is the best choice?

Solution

We must select a material that will minimize both galvanic corrosion and high-temperature oxidation. From an oxidation point of view, Table 15.3–1 shows that while the oxides of Cu, Ni, and Fe (steel) are all protective, the oxide of Ti is nonprotective. Therefore, Ti plates are not a good choice for this application. Based on the previous discussion of unfavorable area ratios in galvanic couples, one should use a plate material that is at least as anodic as nickel. The galvanic series (Table 15.2–3) shows that of the two remaining candidate materials, Cu and steel, only steel is more anodic than Ni in the active state. Therefore, the best material choice is plain carbon steel plates.

15.4 FRICTION AND WEAR (SOLID-SOLID INTERACTIONS)

Whenever two solid materials in contact experience relative motion, frictional forces can result in a degradation mechanism known as wear. **Wear** can be defined as the removal of surface material as a result of mechanical rather than chemical interaction. The study of the science of friction, wear, and lubrication is called tribology. Lubrication is one of the primary engineering weapons used to control wear.

As with corrosion and oxidation, an understanding of the causes and mechanisms of wear can result in both improved designs to minimize degradation and harnessing of the wear mechanism for use in processing applications. For example, while friction and wear must be minimized in applications like bearings, cams, and gears, they must be maximized in machining operations such as grinding and polishing. Steel wires and nylon fibers in ropes are twisted to maximize interaction among wires through the static frictional transfer of tensile stresses.

Sliding friction generates heat, and the resulting temperature increase can influence wear rates through alteration of the microstructure by annealing, grain coarsening, local melting at the surface of one or both solids, or simply increasing kinetics. Thus, as is frequently the case, temperature plays an important role in determining the wear rate.

15.4.1 Wear Mechanisms

Adhesive Wear

Any surface, no matter how much it has been polished, is in fact rough on a microscopic scale. Consequently, the true area of contact between two solid objects is much less than the apparent contact area. The result is higher local stresses than predicted from macroscopic calculations. Adhesive wear, also known as scoring, galling, or seizing, is a result of the high contact stresses. The sequence of steps occurring during adhesive wear is

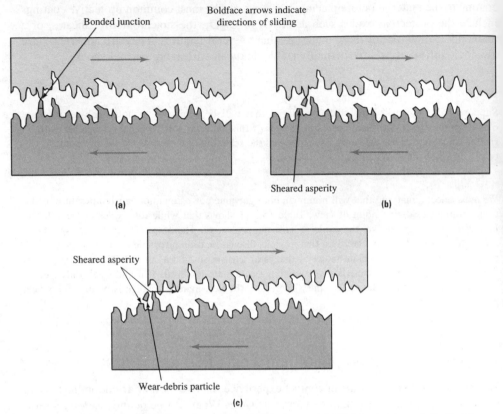

FIGURE 15.4–1 The sequence of steps occurring during adhesive wear. **(a)** High local stresses plastically deform the material in the vicinity of the contact points, resulting in the formation of atomic bonds across the interface. **(b)** As the force causing the relative sliding motion is increased, the shear stress in the joined region increases until it exceeds the shear strength of one of the solids. **(c)** Subsequently, material is lost into the region between the two solids.

shown in Figure 15.4–1. First, the high local stresses plastically deform the material near the contact points, resulting in an increase in contact area and the formation of atomic bonds across the interface (Figure 15.4–1a). As the force causing the sliding motion increases, the shear stress in the bonded region increases until it exceeds the shear strength of one of the solids. This results in transfer of material from one solid to the other (Figure 15.4–1b) and subsequent loss of material into the region between the solids (see Figure 15.4–1c).

The ability of a material to resist adhesive wear has been found to increase as its shear strength decreases and its hardness increases. Unfortunately, the simultaneous requirements of low shear strength and high hardness are difficult to fulfill. It is interesting to note that metals, with their intermediate values of strength and hardness, are generally less resistant to adhesive wear than either the low-strength, high-ductility polymers or the high-hardness, low-ductility ceramics.

The normally incompatible requirements of high hardness and low shear strength make composite materials ideal candidates for this application. For example, a high-hardness substrate able to resist the deformation that increases contact area offers excellent resistance to adhesive wear if coated with a low–shear strength film. Examples include the bearing bronzes (lead particles in a bronze matrix) and polymer-impregnated

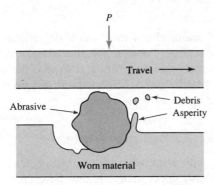

FIGURE 15.4–2 A schematic illustration of the abrasive wear mechanism.

porous bearings such as sintered copper with polytetrafluoroethylene, $(C_2F_4)_n$, infiltrated into the open pore network.

Abrasive Wear

The mechanism of abrasive wear is shown in Figure 15.4–2. The loss of surface material is caused by an interaction with a separate particle "trapped" between the two sliding surfaces. It is not unusual for the abrading particles to be wear debris from an adhesive mechanism. The rate of material loss is related to the relative hardness of the abrading particles and the sliding surfaces. If the surface is harder than the particle, the wear rate is minimal. Materials with high hardness, high toughness, and reasonable temperature stability are good candidates for applications requiring high abrasive wear resistance. Common selections for this application include tempered martensite, surface-hardened steels, cobalt alloys, and many ceramics.

Surface Fatigue

The previous two wear mechanisms are common during sliding motion. If, however, one component is rolling over another, the wear mechanism is often surface fatigue. A quantitative analysis of the state of stress developed during rolling contact shows that a maximum shear stress is developed slightly below the surface. As shown in Figure 15.4–3, the maximum shear stress, known as a Hertz contact stress, may result in the nucleation of subsurface cracks, which can then propagate to the surface and form wear pits. This is a common failure mechanism in railroad wheels.

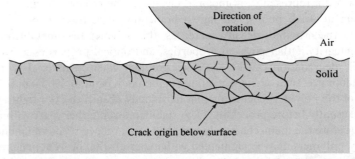

FIGURE 15.4–3 A schematic illustration of the wear fatigue mechanism showing the location of the subsurface crack origin.

Corrosive Wear

The synergistic effects of chemical and mechanical degradation can increase the rate of material loss from a solid surface. For example, mechanical wear associated with sliding or rolling can break down a protective film and expose the underlying material to an aggressive environment. Alternatively, the corrosion reaction products, particularly hard oxide particles, can serve as abrasive particles.

15.4.2 Designing to Minimize Friction and Wear

Common approaches to minimizing wear are the use of lubricants and surface hardening treatments. One can also select materials that have an inherently high wear resistance. In complex designs, however, the last option may not be available. Lubricants provide a protective layer between two moving solid components; thus, direct contact between the surfaces is avoided. They also provide a low–shear resistance layer. Lubricants can be liquids such as oil, or solids such as graphite, boron nitride, or molybdenum disilicide, $MoSi_2$. Surface hardening treatments for reducing wear include case carburizing, commonly used in engine crankshafts; ion implantation, used in surgical instruments; and hard-faced ceramic coatings used in turbine blades and fiber guides in the textile industry.

15.5 RADIATION DAMAGE

Virtually all materials are influenced by high-energy radiation. Polymers are most susceptible to radiation damage, even from the relatively low-energy ultraviolet (UV) radiation. The cracking of vinyl dashboard coverings or seats in automobiles results in large part from UV radiation damage. Without stabilization, most polymers will seriously degrade in a couple of months in the Florida sunshine.

Why are the covalent bonds in polymers so susceptible to radiation damage even though these bonds are about the same strength as those in other material types? Covalent bonds, once broken, generally do not re-form as they were. In addition, impurities often initiate degradative reactions. In polymers, radiation generally leads directly to bond breakage and subsequent crosslinking, so that the molecular structure is dramatically changed.

The degradation begins with the formation of a radical pair, each with a free electron, via the breaking of a covalent bond:

$$R - R' \rightarrow R \cdot + R' \cdot \qquad (15.5\text{--}1)$$

where R and R' represent portions of the polymer chains, the dash represents a covalent bond, and each dot represents an unpaired electron. The radicals $R \cdot$ and $R' \cdot$ attack the polymer chain at weak points, leading to branching, crosslinking, or further radical formation. As the crosslink density increases, the polymer becomes brittle and loses mechanical integrity. Other physical properties, and optical properties as well, are also degraded. UV degradation can be a serious problem when a material is designed for extended outdoor use, such as a tent, caulking compound, paint, or a boat hull. To increase the lifetime of the polymer, chemicals that harmlessly absorb the UV light are added to the polymer, usually before processing. The stabilizers can either react with and quench the radicals (before the main-chain polymer can react), reduce reaction kinetics, or both. Alternatively, polymers may be pigmented with materials that prevent penetration by the radiation. The outer layer of the material then becomes sacrificial. Carbon black is often added to polymers for this purpose. Suntan oils not only reflect a portion of incident

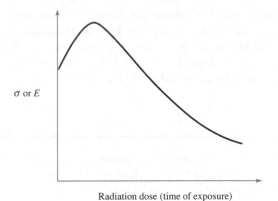

σ or E

Radiation dose (time of exposure)

FIGURE 15.5–1 Variation of the strength and modulus of polyethylene as a function of exposure to UV radiation.

radiation, but also contain UV absorbers to protect the underlying skin (a polymer) from the harmful radiation.

The variation in the strength of polyethylene as a function of time in sunlight is shown in Figure 15.5–1. The strength initially rises as the initial degradation slightly increases crosslink density. Thereafter, the continued free radical attack weakens and embrittles the material.

In metals and ceramics, high-energy radiation, such as neutrons from a nuclear reactor, can displace atoms from their normal crystallographic positions. The result is the formation of vacancy-interstitial pairs. These point defects can diffuse to form defect clusters and eventually result in the generation of dislocation loops. In addition, much of the energy associated with the radiation is dissipated through heating of the crystal along the path of the incoming particle. Since the heating is restricted to a small volume, the high energy density can result in local melting. The large surrounding thermal mass results in a rapid quench rate and a corresponding microstructure.

Metals are better able to resist radiation damage than are ceramics. This is because electrons in a partially filled valence band are able to absorb and dissipate some of the incident energy without significant changes in the properties of the metal. Nevertheless, the radiation-induced increase in the concentration of point defects and dislocations usually increases the metal's yield strength and ultimate tensile strength; however, its ductility, thermal conductivity, and electrical conductivity decrease. In addition, most BCC metals exposed to ionizing radiation experience an unfortunate increase in their ductile-to-brittle transition temperature. The radiation-induced structural and property changes can be at least partially reversed by a thermal annealing treatment.

The extent of radiation damage in ceramics depends in part on the nature of bonding. Covalent ceramics are more readily damaged than their ionic counterparts. The type of damage in the ionic ceramics is similar to that encountered in metals. The increase in the point defect density induces a modest increase in strength, a modest decrease in ductility, a decrease in thermal conductivity, and, in originally transparent compounds, a significant degradation in optical properties. In contrast to metals, however, the electrical conductivity of some ionic ceramics increases with exposure to radiation. Specifically, in compounds that conduct electricity via the motion of ions rather than electrons, the radiation-induced increase in point defect density increases the ionic diffusion rate. As with metals, radiation-induced damage in ionic compounds can be reversed by thermal annealing. Although the damage mechanisms and changes in properties are similar in covalent ceramics, the resulting damage is, for the most part, irreversible.

Semiconductors are affected by radiation in exactly the same way as any other covalent ceramic. However, because of their high purity and low initial defect density, the relative magnitude of the degradation in properties is much larger in these materials. Since radiation exposure seriously degrades both the electrical and magnetic properties of semiconductors, they must be protected from stray ionizing radiation.

SUMMARY

The performance of an engineering material is influenced by interaction with its environment in many ways. The interaction can be beneficial if it is used as part of a materials fabrication or processing operation; it can be detrimental if it results in the degradation of material properties.

The selection of materials for use in aggressive environments requires a knowledge of the thermodynamic driving forces and an understanding of the factors controlling the reaction kinetics.

One of the major classes of environmental interactions involves a solid material in a liquid environment. Most liquid-solid reactions fall into two classes. Those that occur without electron transfer are direct dissolution mechanisms; those that require electron transfer are known as electrochemical corrosion mechanisms. Direct dissolution can affect all classes of materials. Electrochemical corrosion is a phenomenon restricted to metals. As a class, ceramics are generally more resistant to aggressive liquid environments than are either metals or polymers.

All electrochemical reactions can be divided into two or more half-cell reactions. The oxidation, or anodic, reactions produce electrons, while the reduction, or cathodic, reactions consume free electrons. In electrochemical corrosion the oxidation reactions always involve metal going into solution. The reduction reactions can take several forms, depending on the corrosive environment. The relative ability of metals to resist corrosion can be predicted through the use of the standard reduction potential series or the galvanic series. Electrochemical corrosion takes many forms, including uniform corrosion, galvanic corrosion, pitting and crevice corrosion, corrosion due to concentration gradients in the electrolyte, hydrogen embrittlement, and stress-assisted corrosion. General design strategies for corrosion control and prevention include reducing the service temperature, selection of appropriate materials, microstructural modification through alloying or heat treatments, and design modifications. Corrosion problems can be reduced by using sacrificial anodes, protective coatings, impressed voltages, or chemical inhibitors.

The second major class of degradation mechanisms involves gas-solid reactions. Although a variety of gases, including nitrogen, sulfur, and chlorine, can cause degradation, the most common reactions involve oxygen. Oxygen attack can take several forms, ranging from crosslink formation in elastomers to consumption of structural material through the formation of vapor- or solid-phase oxides. Vapor-phase oxide products are almost always undesirable. Solid oxides are protective only if they are continuous, adherent, and of sufficient volume to completely cover the material, but not so much volume that they expand and fracture or spall from the metal. The nature of metal oxides can be predicted using the Pilling-Bedworth (PB) ratio; those oxides with $1 < PB < 2.3$ are generally protective. The strategies used to minimize the rate of atmospheric attack on engineering materials are: (1) modification of alloy composition, (2) modification of service environment, or (3) application of a protective coating.

The third type of material degradation occurs when two solid materials in contact experience relative motion. The degradation mechanisms, known collectively as wear, result from mechanical rather than chemical interaction. Three general approaches for minimizing friction and wear are materials selection and modification, lubrication, and system design. Nonmetals usually offer superior resistance to friction and wear. Since optimum wear resistance requires both high hardness and low shear strength, materials often used in demanding applications contain a high-hardness substrate with a soft surface layer or dispersed second phase.

Virtually all materials are degraded by high-energy radiation, which results in the creation of point defects, the pinning of dislocations, and loss of molecular weight and crystallinity in polymers. These defects degrade the mechanical, optical, electrical, thermal, and magnetic properties of materials. Polymers and semiconductors are especially sensitive to radiation damage.

KEY TERMS

activation polarization	erosion corrosion	oxidation (anodic) reaction	reduction potential
cavitation damage	Faraday's equation	oxygen concentration cell	sacrificial anode
concentration polarization	galvanic corrosion	passivation	static fatigue
corrosion fatigue	hydrogen embrittlement	Pilling-Bedworth (PB) ratio	stress corrosion cracking (SCC)
crevice corrosion	liquid metal embrittlement	pit corrosion	uniform corrosion
direct dissolution mechanisms	Nernst equation	reduction (cathodic) reaction	wear
electrochemical corrosion	oxidation		

HOMEWORK PROBLEMS

1. Ethical citizens consider disposal of products at the time of purchase—Is the product biodegradable? Will it harm the environment? and so on. Many, too many, durable products are currently considered "disposable." Give 10 examples of nondurable disposables and five examples of durable disposables.

SECTION 15.1
Introduction

2. Discuss the similarities and differences between direct dissolution and electrochemical corrosion.

SECTION 15.2
Liquid-Solid Reactions

3. Describe the conditions under which the degradation mechanism known as static fatigue can occur.

4. Explain the driving force for liquid metal embrittlement. That is, why does the liquid penetrate along the metal grain boundaries?

5. Many condoms are composed of elastomers derived from petroleum by-products. Discuss the advisability of using petroleum jelly as a lubricant in this application.

6. This question deals with the dissolution of ceramic materials. Discuss the significance of each of the following variables:
   a. Temperature
   b. Grain size
   c. Use in a stagnant or moving fluid

7. Use your knowledge of "like dissolves like" to describe a good solvent for nylon. Consider ammonia and formic acid in your list of candidate solvents.

8. Will oil mix with water? Why or why not? Can you fool the two into some sort of mixing? (Hint: Consider dishwashing liquid.)

9. Discuss possible solvents for wool, which contains about 10% of the amino acid cysteine. Cysteine contains a disulfide crosslink, which covalently joins neighboring molecules. Also discuss possible solvents for epoxy.

10. A high-quality boat, perhaps 15 to 35 ft long, that someone living near the ocean might own requires various materials. What materials represent good performance and long-lasting quality investments for the following applications:

    a. Metal tie-down points for ropes?
    b. Ropes?
    c. (Sailboat) mast?
    d. Windows?
    e. Hull?

11. You have been hired as a consultant to predict the failure mode for a silicate glass used under a constant stress ($\sigma = 0.3\sigma_f$) in an H_2O environment at a temperature near 25°C. What do you think will be the most likely failure mechanism? Why?

12. Why is it that in the experiment used to determine the standard reduction potential of metals, the Pt/H half cell contains both H^+ and hydrogen gas?

13. In electrochemical corrosion, is the loss of metal associated with the reduction or with the oxidation reaction? Does it occur at the anode or at the cathode?

14. Consider an electrochemical cell composed of a Pb electrode in a 1-molar solution of its ion in electrical contact with a Sn electrode in a solution of its ions. What concentration of Sn ions is required to make the Sn cathodic to the Pb?

15. The current density during zinc electroplating of steel is found to be 1.5×10^3 A/m^2. Estimate the time required to produce a zinc coating of thickness of 50 μm.

16. A steel bridge across a saltwater inlet will contain a critical structural plate that must have minimum thickness of 1.5 cm to withstand the applied loads. If the design life of the bridge is 30 years and a sample of this steel exhibits a corrosion current density of 0.5 A/m^2 in a laboratory test, estimate the required initial thickness of the plate.

17. Consider a zinc sacrificial anode attached to an aluminum plate. If the area of the anode is 10 cm^2, the initial anode thickness is 1 cm, the area of the aluminum plate is 1 m^2, and the cathodic current density is 2.5×10^{-4} A/m^2, estimate the life of the sacrificial anode.

18. Explain why some metals with reduction reaction potentials less than the hydrogen reference value are not always attacked in dilute acids.

19. Two identical metal tanks are used to contain the same acidic solution. One tank is a long-term storage tank and is nearly always filled. The other tank is a short-term storage tank and is constantly filled and drained. Why does the long-term tank suffer less corrosion than the short-term tank?

20. a. Under what conditions will the removal of dissolved oxygen from a corrosive liquid reduce the corrosion rate?
    b. Under what condition might it increase the corrosion rate?

21. Why is it important to know the pH of a solution before selecting a material for use in that solution?

22. If you are designing to minimize erosion corrosion and cavitation damage, should you try to promote laminar or turbulent fluid flow?

23. List three mechanisms by which hydrogen can embrittle a metal.

24. Suggest several methods for protecting cast-iron pipe used to carry water underground.

25. A tank used to contain seawater was fabricated using a plain carbon steel bottom with stainless steel sides. Your boss suggests painting the bottom in an effort to reduce the risk of corrosion. What do you think about this idea?

26. A "tin" can used for food storage is really a laminar composite of Sn and Fe. Do you think the Sn is on the inside or the outside of the can? (Note: Sn is anodic to steel in an oxygen-free environment.)

27. A precipitation-hardened aluminum alloy is found to corrode preferentially along the grain boundaries. Offer an explanation for this observation.

28. Identical sheets of nickel are immersed in four different solutions: one in air-free water, one in aerated water, one in air-free acid, and the final one in aerated acid.
    a. Determine the most likely oxidation and reduction reaction(s) for each electrochemical cell.
    b. Which piece of nickel do you think will corrode most rapidly? Why?

29. Suggest a method to eliminate or reduce the severity of each of the following types of corrosion:
    a. Galvanic corrosion
    b. Pitting
    c. Crevice corrosion
    d. Microscopic galvanic corrosion in the heat-affected zone of stainless steel welds
    e. Hydrogen embrittlement
    f. Uniform corrosion of the external surface of a storage tank
    g. Failure of a brass component in an ammonia environment

30. List three examples of beneficial applications of electrochemical reactions.

31. An Austin-Healey automobile body is constructed of steel, with aluminum hood and trunk shrouds. Discuss the long-term problems associated with using steel bolts to connect the steel and aluminum panels.

32. In some homes built in the 1950s the wiring was Al, since Cu was so expensive. Several homes burnt down. Show why it is a bad idea to use Al rather Cu wiring in a home. (Note: Junction boxes and outlets are steel.)

33. Discuss the problems associated with using Pb-Sn solder to connect Cu water pipes in homes.

34. J. C. Whitney sells Zn sacrificial anodes to mount onto a car radiator (ferrous). Will they work? Is there a better choice?

35. A water heater consists of steel, brass, and copper components. Which element preferentially corrodes? How might the corrosion be prevented or minimized?

36. A ship uses a zinc anode to protect the hull. The initial mass of the anode is 300 lb. The average corrosion current is 12 A. How often should the anode be replaced?

37. Which is more stable in salt water, FeO or Fe_2O_3? In the presence or absence of O_2, will one change to the other?

38. Discuss the electrochemical problems associated with use of carbon fiber-reinforced aluminum.

39. You go to the hardware store to purchase steel fence. There are two options and the price is the same: welded and galvanized fence or galvanized and welded fence. Which do you buy?

40. We have learned how etchants can be used to reveal grain boundaries in polycrystalline samples of metal. Can you propose a similar technique to reveal crystalline and noncrystalline regions in polymers?

41. What is anodizing? Does it offer increased protection?

42. Briefly discuss the characteristics of a protective oxide. What characteristics must an oxide possess if it is to protect the base metal?

SECTION 15.3
Direct Atmospheric Attack (Gas-Solid Reactions)

43. Suppose you are asked to determine if a certain piece of niobium (Nb) is covered with NbO or Nb_2O_5. All you are told is that the oxide exhibits linear growth (thickness increase) with time. Given the information below, determine which form of the oxide is present. Explain your answer.

Material	Molecular weight	Density
Nb	92.9 amu	8.57 g/cm^3
NbO	108.9	7.3
Nb$_2$O$_5$	265.8	4.5

44. List the conditions under which oxide growth occurs at each of the following locations:
    a. The metal-oxide interface
    b. The oxide-atmosphere interface
    c. Within the oxide film

45. Explain the mechanisms that result in each of the following classes of oxidation rate:
    a. Linear
    b. Parabolic
    c. Logarithmic

46. Explain why linear oxide growth can result in either weight gain or weight loss while parabolic oxide growth always results in a weight gain.

47. If the oxidation of Zn occurs by the diffusion of zinc interstitials, suggest the type of oxide addition that might result in a decreased oxidation rate.

48. A turbine blade fabricated from a nickel alloy has an initial oxide thickness of 50 nm. After one hour of operation in a jet engine at 1200°C, the oxide thickness is found to be 500 nm. Estimate the oxide thickness after the blade has been in service for one week at 1200°C.

49. List two ways in which controlled atmospheric attack can be used to improve the properties of engineering materials.

50. You have been asked to select a ceramic component for a structural application in a high-temperature oxygen environment. Recognizing the potential problems associated with an active oxidation reaction, you have performed an experiment in which you measured oxide thickness as a function of time and obtained the data below. Based on these data, do you recommend using this ceramic in this environment?

Oxide thickness (arbitrary units)	5	11	17	32
Time (arbitrary units)	1	3	5	10

51. Explain why metal alloys designed for use in high-temperature environments often contain Cr additions.

52. Although it is possible for a metal undergoing oxidation to experience a linear weight gain, linear weight loss, or parabolic weight gain, it is extremely unlikely for a metal to experience a parabolic weight loss. Explain this observation.

53. The iron ions of FeO can have a valence of either +2 or +3. This leads to a nonstoichiometric oxide described by the formula Fe$_{1-x}$O. Use this information to predict the location of oxide growth in this material.

54. How do you know whether a polymer has oxidized?

55. Suppose you surrender to the fact that the metal you have selected will oxidize in service. To minimize metal loss at short times, what oxidation mechanism do you promote? To minimize oxide thickness at long service times, what mechanism do you want to promote?

56. Why is gold often used as a contact material in electronics?

57. Discuss the pros and cons of the materials used to make bicycle frames. Note that bicyclists sweat profusely during serious training. Bike frames may be steel, aluminum, carbon fiber–epoxy composite, or titanium. Focus on the locations where various materials come in contact, such as the frame and the seat post or the handlebar area.

58. Polymer melts typically have high viscosities. Explain in terms of molecular friction why polymer viscosities are high.

59. Why do sewing machine needles get hot? What material should be used to make them? How can the temperature be minimized?

60. What happens when the wheel bearings in your car run out of grease? Why are the bearings used in the first place?

61. What sports rely on near zero friction underfoot? How is it achieved?

62. Is water a good lubricant? What are the properties of a good lubricant? (Hint: Consider the hydroplaning car.)

63. What solid material indisputably has the lowest coefficient of friction? It is often used as an additive in greases, oils, and the like.

SECTION 15.4
Friction and Wear
(Solid-Solid Interactions)

64. Why do X-ray technicians cover your exposed body parts with lead blankets when a part of your body is being X-rayed?

65. Discuss the possibility of using UV or ionizing radiation to sterilize instruments and garments in operating rooms.

66. Gasoline is stored in large polyethylene (PE) underground tanks at service stations. A homeowner decides to recycle a translucent plastic milk jug (PE) by using it as a gasoline storage tank outdoors. Is this a good idea?

67. A single crystal of quartz is used as a transducer (pressure sensor) in a nuclear power plant. Is it important to shield the crystal? Why or why not?

68. Discuss the problems associated with making lightweight space suits for use outside the spaceship.

69. The glass transition temperature of PVC (polyvinylchloride) is about 40°C. Hence, it is a glass at room temperature. Its use in car seats, etc., is facilitated by plasticizing it with huge amounts of DOP (dioctyl phthalate), which lowers the glass transition temperature of the vinyl to well below room temperature. As the summer sunlight strikes the vinyl, the car interior heats up and diffusion of the DOP out of the vinyl is accelerated. The oily transparent film that condenses on the car windows is DOP. What is the fate of the seats? What might Vinyl-Nu or Armorall do?

70. Your body is composed of a number of polymers, called polypeptides. Simply speaking, they are polyamides, or nylons. What occurs when any one of these polymers, say, your skin, is subjected to prolonged UV radiation?

SECTION 15.5
Radiation Damage

MATERIALS SYNTHESIS AND DESIGN

inished products such as camcorders contain several components. The cost of producing camcorders includes the cost of raw materials, the cost of assembly of components, and also the cost of the several processes involved in manufacturing the components. Often the manufacturing cost is the largest fraction of the total cost. Therefore, finding economical means of producing components often determines whether a product ultimately is commercially successful. Take, for example, Ge and Si, both of which are semiconductors. Ge has been used as a semiconductor for much longer than Si. However, Si is now used much more than Ge and is referred to as the workhorse of the semiconductor industry. Why? One reason is performance. Si has a higher band gap (1.1 eV) than Ge (0.7 eV). Thus, Si devices remain operational to much higher temperatures than Ge devices. However, the overwhelming reason Si devices are favored is economic.

Si is derived from sand, one of the most abundant materials on earth. Therefore, it is inexpensive, even in the single-crystalline form. However, the most important reason Si is successful is not its low raw material cost but its superior processability, which makes it ideal for producing integrated circuits. Modern-day integrated circuits contain millions of devices on a single chip. Device fabrication on chips is facilitated by the excellent characteristics of silicon dioxide (SiO_2). Silicon oxidizes easily, and its oxide, which grows from the substrate, is extremely pure and acts as a high-quality insulator on the chip. It can be etched easily when necessary, and it serves as an excellent mask during implantation of dopants such as B and P. The accompanying figure shows the steps involved in producing an integrated circuit. Several of these steps require expensive equipment and specialized environments such as clean rooms. A high yield (in this case defined as the ratio of acceptable chips to the number of attempts) is desirable. Therefore, processability of the material is extremely important under these circumstances, and this is why Si devices have a big advantage over other semiconductor materials.

The electronic module of the camcorder is only one of its several components. Others include lenses, which are precision ground and require delicate handling during manufacture. The

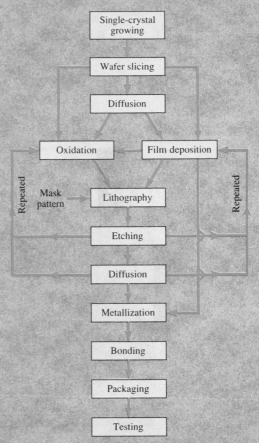

Fabrication sequence for integrated circuits. *(Source: Serope Kalpakjian, Manufacturing Engineering and Technology, 2nd ed., © 1989 by Addison-Wesley. Reprinted by permission of the publisher.)*

housing can be made from a casting of aluminum or steel; however, this choice will add to the weight and cost of the camcorder. Perhaps the least expensive housings are made from polymers by a process called injection molding. The parts are assembled to complete the housing. Thus, producing a camcorder entails a number of manufacturing steps, and the success of these processes determines the cost and quality of the product.

The purpose of Part IV is to describe the different processes available to engineers for making components. These processes are material-dependent. The quality of the product may be related to the processing method(s) used and the control exercised during processing. In the semiconductor industry, yield is dependent on these factors. Therefore, engineers play a critical role in engineering design.

MATERIALS PROCESSING

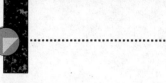

MATERIALS IN ACTION Materials Processing

While a given material may exhibit outstanding properties, unless it can be processed in an economical way into a useful shape, those properties may not be practically attainable. It is, after all, through processing that materials are converted into useful forms. Thus a knowledge of processing approaches is essential to understanding the application of materials.

Many processing techniques have been known for centuries. An example of this is casting of metals in complex shapes via the lost wax process. This technique, which was used by the ancient Egyptians, involved creating a wax model of the part to be cast. The wax model was then embedded in sand and molten metal was poured into a channel connected to the wax part. The wax was burned by the metal and an exact duplicate was created.

It is interesting that a variation of the lost wax technique is the basis of one of the modern triumphs of modern materials science processing: casting of single crystal Ni-base superalloy turbine blades for jet engines. Traditionally, turbine blades were cast in polycrystalline form; however, the grain boundaries served as areas of oxide formation and subsequent cracking. Elimination of the grain boundaries by producing turbine blades in the single crystalline form was very desirable to improve reliability and extend the life. Unfortunately, an economical way to make single crystal blades had not yet been developed. In the mid 1960s a simple process was developed that made use of the lost wax process. In this approach, wax models of turbine blades were produced in "wheels" with each wheel containing as many as 20 turbine blades. The wheels were backed with a ceramic slurry and the "green" ceramic molds were cured in a high temperature furnace to strengthen them and to burn off the wax. Metal was poured into this mold and heat was extracted along the axis of the blades through a " choke" to ensure that a single crystal was formed before reaching the turbine blades. As the heat was extracted, the single crystal grew into the cavities of the turbine blades thereby producing single crystals. This process revolutionized the production of turbine blades. Blades produced in this way last longer, have fewer defects, and can be used at higher temperatures than those produced by conventional casting techniques. Virtually all commercial engines are made today using this technology for the production of turbine blades. In this chapter we will examine many different fabrication processes. However, we should remember that, independent of the process being considered, the goal of all processing is to produce controlled microstructures that lead to the required properties in the manufactured part.

16.1 INTRODUCTION

Successful use of a new or existing material in engineering applications depends on finding reliable, reproducible, and cost-effective fabrication processes to transform materials into the net product shape. Consider, for example, large steam turbines that produce more than 1000 megawatts of energy and run at substantially higher thermal efficiencies than their smaller counterparts. Successfully designing these large turbines requires the ability to produce steel forgings much larger than those used in lower-capacity machines (200 megawatts and lower). Since turbine rotors are highly stressed parts it is important to produce large forgings that are free of strength-limiting defects and have uniform microstructure and properties throughout their volume. Melting practices introduced in the 1960s utilize electrical melting furnaces to achieve better impurity control and are capable of producing ingots from which large forgings can be fabricated. Similarly, microcomputers today have greater capabilities than did large mainframe computers of the 1960s and 1970s. Improvements in speed and size are a direct result of advances in manufacturing processes that allow packaging of a large number of electronic components on small circuit boards.

In this chapter we will discuss the fundamentals of producing, shaping, joining, and surface treating materials to prepare them for service. Although each technique is customized to the specific material application, the underlying principles behind the various processing techniques have much in common. Therefore, we will emphasize the generic aspects of materials processing techniques whenever possible.

16.2 PROCESS SELECTION CRITERIA AND INTERRELATIONSHIP AMONG STRUCTURE, PROCESSING, AND PROPERTIES

Regardless of the material type and application, some common considerations enter into selecting an optimum fabrication process. Primary considerations include meeting performance requirements and the cost of production. Another important consideration is the ability to reproduce quality during mass production so that the inspection procedures result in low rejection rates. In large components it is important that uniform properties be obtained throughout the component.

Factors affecting the cost of production include raw material costs, labor costs, utility costs, and the cost of capital for the required tooling and facilities. For example, if the raw material is expensive, a processing technique that results in near net shape (requiring little or no end machining) is desirable. In a related issue, processing methods that require substantial capital investment are often reserved to produce high-volume parts. Thus, process selection must be an integral part of product design and must be considered in detail at the time of developing design requirements and selecting materials. These aspects will be discussed in more detail in the next chapter.

The relationship between processing and the end properties of the product can hardly be overemphasized. In Chapter 9 we observed that the microstructure and the surface hardness of an alloy steel weldment change as we move across the weldment starting from the base metal region (see Figure 9.2–20). Variation in the welding process and the subsequent postwelding heat treatment substantially affect the microstructure of the weldment and its mechanical strength. Similarly, in ceramic-matrix composites, the fiber-matrix interface strength determines the fracture toughness of the composite. The interfacial strength depends largely on bonding between fiber and matrix, which in turn is related to the processing variables.

Since the properties of the material are linked to its atomic structure, we should not be surprised to learn that the structure of the material can influence the processing methods appropriate for a specific material. For example, the ionic/covalent bond characteristics of ceramics are responsible for both their brittle behavior, which prohibits shaping by plastic deformation and machining, and their refractory nature, which prevents the use of conventional liquid-to-solid casting methods. Therefore, structure, processing, and properties are interdependent.

In the following sections, we describe processes used in producing components of engineering machinery or devices from raw materials. The discussion is limited to some of the most important processes and is primarily concerned with materials issues as opposed to the mechanics of the processes and processing equipment. Shaping processes, including casting, forming, powder processing, and machining, are considered first. These are followed by a discussion of joining processes, surface coatings and treatments, single-crystal and semiconductor processing, fiber manufacturing, and composite manufacturing.

16.3 CASTING

Casting is a manufacturing process dating back to at least 4000 B.C. for inexpensively producing large or complex parts. Components can be directly formed to near final shape with only minor finishing or machining required. The casting process generally involves filling a mold with molten material, which upon solidification takes the shape of the mold. While metals are most frequently used for casting parts, polymers and oxide glasses are often cast into shapes as well. Slip casting is a process by which ceramic materials can be molded into shapes. Injection molding is an example of a casting process that can be applied to polymers, metals, and oxide glasses. Housings of automobile engines, lawn mower engines, and machinery such as turbogenerators are examples of large castings. Piston heads of automobile engines and class rings are examples of intricate castings. Plastic housings, tool handles, bottle caps, and knobs are examples of products made by the injection molding process.

While casting is occasionally used to form large components, it is generally not used for manufacturing large components that bear high stresses during application. Compared with parts made using other shaping processes, such as forging, cast parts tend to contain larger defects that result in lower strength and ductility. In this section, the fundamentals of casting processes are discussed for metals, polymers, ceramics, and oxide glasses. Since the terminology and processes are somewhat different for the different material categories, we will treat them separately.

16.3.1 Metal Casting

The grain structure of a cast metal part is established during solidification. Melting temperatures of metals used for castings range from 650°C to 1550°C. The important parameters are the thermal gradients, which control the heat flow from the interior of the part to the outside surface, and the chemical composition, which affects the temperature range over which solidification occurs. For example, Figure 16.3–1 shows the cast microstructure of a pure metal, a solid solution alloy, and castings produced by eliminating thermal gradients or by using catalysts to induce heterogeneous nucleation of nearly equiaxed grains.

The grain structure of a pure metal casting consists of a mixture of columnar grains in the interior and an exterior shell containing equiaxed grains. The fine equiaxed grains at

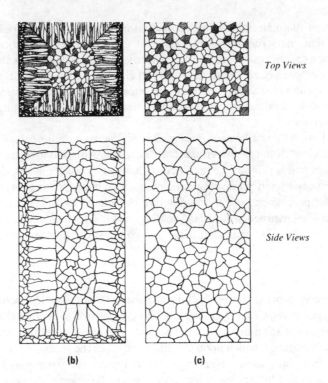

Top Views

Side Views

(b)   **(c)**

FIGURE 16.3–1   Schematic illustration of structures of metals solidified in a square mold: **(a)** a pure metal, **(b)** a solid solution, and **(c)** when no thermal gradients are present during solidification or inoculants (catalysts) are used to induce heterogeneous nucleation of grains.   *(Source: Adapted from G. W. Form, J. F. Wallace, J. L. Walker, and A. Cibula, as found in Serope Kalpakjian,* Manufacturing Engineering and Technology, *2nd ed., © 1989 by Addison-Wesley. Reprinted by permission of the publisher.)*

the surface are a result of the high nucleation rate resulting from the significant undercooling at the relatively cold mold walls. The region of equiaxed grains is also called the chill zone. Pure metals solidify at a fixed temperature while the latent heat of fusion is released. During this process, the liquid-solid interface, or solidification front, moves from the surface near the mold walls to the interior. This leads to the formation of columnar grains with their long axis pointing from the surface toward the interior of the casting.

The solidification of solid-solution alloys also results in a chill zone and a columnar grain structure. In contrast to castings from pure metals, however, the central region of a solid-solution casting is composed of a second region of equiaxed grains. During the solidification process, small crystallites form at the (cold) top of the mold and settle to the bottom, where they act as nucleation sites for further crystal growth. Since such nuclei are present more or less uniformly, the resultant structure consists of equiaxed grains.

Another major difference between the microstructure of the columnar grains in alloys versus pure metals is the presence of the dendritic structure within the columnar grains of alloys. Consider the solidification of an Al-Si alloy. Figure 16.3–2 shows the Al-Si phase diagram, which has a eutectic composition of 12.6 weight percent (wt. %) Si. If an alloy with less than 12.6 wt. % Si is solidifying, primary Al crystals appear near the mold walls. As these crystals grow, they reject Si and enrich the surrounding liquid in Si. Because of the pattern of latent heat rejection and compositional variation, the growth of the Al-rich crystals are in the form of a treelike structure called a **dendrite.** The surround-

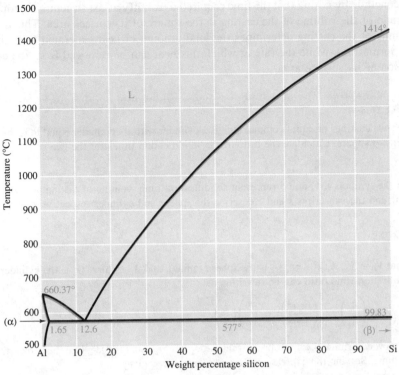

FIGURE 16.3–2   Al-Si phase diagram showing a eutectic point at 12.6 wt. % and limited solubility of Si in Al.

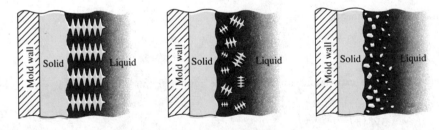

FIGURE 16.3–3   Schematic illustration of three basic types of dendrite structures: **(a)** columnar dendritic,
(b) equiaxed dendritic, and **(c)** equiaxed nondendritic.   *(Source: Serope Kalpakjian, Manufacturing Engineering and Technology, 2nd ed., © 1989 by Addison-Wesley. Reprinted by permission of the publisher.)*

ing liquid has the eutectic composition and freezes at a fixed temperature of 577°C, giving rise to a eutectic microstructure. Figure 16.3–3 shows the structure of three different types of dendrites. In reality, pure cast metals may also have a dendritic structure. Since there is no compositional difference between the dendrites and the interdendritic material, however, the dendritic structure is neither readily apparent nor of practical importance.

With reference to Figure 16.3–1c, a completely equiaxed grain structure can be produced in a casting by eliminating temperature gradients and convection currents within the liquid phase. Convection currents can be reduced considerably in a zero-gravity environment such as in space. A similar result can be obtained by using inoculants (or catalysts) such as sodium, bismuth, or magnesium, which act as heterogeneous nucleation sites throughout the molten metal. In this way grain formation can be initiated and grain growth occurs uniformly throughout the casting, resulting in an equiaxed grain structure.

We conclude by noting that the time required to solidify a cast structure is proportional to the ratio of the volume of the casting to the square of its surface area. This is because the volume of the casting determines the total amount of heat generated (i.e., the latent heat of formation), while the rate at which this heat can be removed from the casting is a function of its surface area.

EXAMPLE 16.3-1

Consider two castings of equal volume. One is a cylinder with its diameter equal to its height, and the other is a sphere. Calculate the ratio of the solidification times for the two castings.

Solution

If we let the symbols t, V, and S represent solidification time, volume of casting, and surface area of casting and the subscripts s and c describe the sphere and cylinder respectively, we find:

$$\frac{t_s}{t_c} = \frac{V_s/A_s^2}{V_c/A_c^2}$$

Given that $V_s = V_c$, $A_s = 4\pi r_s^2$ (r_s is the sphere radius), and $A_c = 6\pi r_c^2$ (r_c is the cylinder radius), the solidification time ratio can be rewritten as:

$$\frac{t_s}{t_c} = \left(\frac{A_c}{A_s}\right)^2 = \left(\frac{6\pi r_c^2}{4\pi r_s^2}\right)^2 = \left(\frac{9}{4}\right)\left(\frac{r_c}{r_s}\right)^4$$

We must now determine the appropriate values for r_s and r_c using the fact that the two castings have equal volume. Solving the expression $V_s = (4/3)\pi r_s^3$ for r_s gives:

$$r_s = \left(\frac{3V_s}{4\pi}\right)^{1/3}$$

A similar calculation for the cylinder yields:

$$r_c = \left(\frac{V_c}{2\pi}\right)^{1/3}$$

Combining these two expressions to determine the ratio r_c/r_s and noting that $V_s = V_c$ gives:

$$\frac{r_c}{r_s} = \left(\frac{V/2\pi}{3V/4\pi}\right)^{1/3} = \left(\frac{2}{3}\right)^{1/3}$$

Substituting this expression into the t_s/t_c expression yields:

$$\frac{t_s}{t_c} = \left(\frac{9}{4}\right)\left(\frac{r_c}{r_s}\right)^4 = \left(\frac{9}{4}\right)\left(\frac{2}{3}\right)^{4/3} = 1.31$$

It takes approximately 31% longer for the spherical casting to cool, since it has a smaller surface-to-volume ratio.

Defects in Metal Castings

Liquid metals generally occupy larger volumes than do their solid counterparts. Thus, shrinkage defects in the form of porosity often occur during solidfication. Shrinkage defects will also occur when the metal in thin sections of the casting freezes and blocks the source of molten metal to other regions of the casting. To avoid such defects, metal heat sinks, known as chills, can be strategically placed in various regions of the mold to control the solidification pattern (i.e., increase the solidification rate in thicker regions of the casting). Several examples of "chill" locations are shown in Figure 16.3–4.

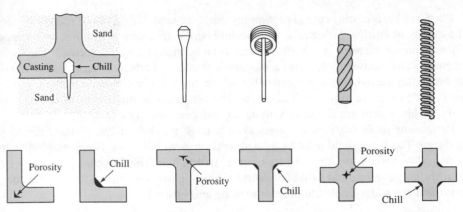

FIGURE 16.3–4 Various types of internal and external chills (dark areas in the corners) placed to avoid shrinkage defects. The figures also show the types of defects that will form if external chills are not used. *(Source: Serope Kalpakjian, Manufacturing Engineering and Technology, 2nd ed., © 1989 by Addison-Wesley. Reprinted by permission of the publisher.)*

Gray cast iron contains graphite, which expands during cooling and compensates for shrinkage of the iron matrix. The result is a slight net expansion of the casting during solidification, which reduces the number and severity of shrinkage defects. Therefore, cast iron is commonly used for large castings such as engine blocks, housings for machine tools, and so on.

Porosity in castings can also be caused by dissolved gases in the molten metal that are expelled in the form of gas bubbles during solidification. The reason for this is that the gases are generally much less soluble in the solid than in the liquid phase. Often the gases are trapped in interdendritic areas, causing microporosity. Dissolved gases can be removed from the molten metal by purging it with inert gases or by pouring the metal in vacuum.

Other defects include hot tearing (or cracking), which occurs if the material is not allowed to contract freely during solidification. Blows, scars, blisters, and scabs are surface defects or cavities that mar the surface finish of the casting product. These defects are often caused by the presence of oxides and contaminants.

Metal Casting Techniques

The primary distinction among the common metal casting methods is the type of mold used. The choice of the correct mold type depends on the properties of the alloy, the shape, size, and contours of the casting, the desired surface finish, and cost.

Sand casting, the oldest of the casting methods, consists of making sand molds using a pattern. The pattern is nearly identical in shape and size to the end piece, with allowances for shrinkage and machining. Patterns are made of wood, plastics, metals, or plaster, depending on the number of times they are to be reused. A mixture of sand and clay for bonding is closely packed around the pattern. The pattern is subsequently removed from the mold, producing the mold cavity into which molten metal is poured. It is common to make the mold in two or more pieces so that patterns can be removed easily without ruining the surface.

In shell molding, a metal pattern is heated to 175–350°C, coated with a parting (lubricating) agent such as silicone, placed in a box, and sprinkled with a mixture of fine sand and a thermosetting resin binder (phenolformaldehyde). When the resin is partially set, the pattern is removed from the shell mold, which is then further baked in an oven. Castings made using this method offer better control of dimensions and improved surface finish compared to those made using conventional sand casting.

Plaster of Paris castings use plaster molds that, like sand molds, can be used only once. The quality of casting, however, is significantly better. In addition, castings with regions of much thinner sections and intricate shapes can be made using this method. The molds are made in two halves by applying a mixture of plaster of Paris and excess water onto the pattern. After partial drying, the two halves of the mold can be separated from the pattern and baked to remove moisture. The baking also leaves many small interconnected pores in the molds, which are useful in venting air and gases during casting.

Permanent mold castings are used when a large number of components are to be produced. The same mold must be used several times to save cost. Compared with sand casting, this method usually provides castings with a fine grain structure, relatively free from shrinkage pores, and with better surface finish. Permanent molds are often made of metal and are water-cooled to prevent melting and erosion.

16.3.2 Casting of Ceramics

Unlike metals and polymers, the high melting temperature of ceramics generally eliminates casting from melts as a viable processing option. Instead, ceramics are "cast," extruded, or pressed from a powder. Therefore, the first step in casting ceramics is crushing to produce powders. The crushing can be either wet or dry. Wet crushing is preferred because it prevents suspension of fine particles in the air, which can be a safety hazard. The ground particles are then mixed with additives that act as binders for the mix, as wetting agents to improve mixing, or to make the mix more plastic and formable.

The most common casting process in ceramics is called **slip casting,** or drain casting. In this process, a suspension of ceramic particles in a liquid is poured into a plaster of Paris mold. The mixture, or slip, must be sufficiently fluid to flow inside the mold. After the moisture from the outer layers of the slip has been absorbed by the porous mold, the remaining slip is emptied out, leaving a ceramic shell behind. The shell is then removed from the mold, dried, and fired (heated to elevated temperatures) to impart the strength and hardness typical of ceramic materials. Solid parts can be produced with this process by omitting the draining step. The slip-casting process is illustrated schematically in Figure 16.3-5a.

As shown in Figure 16.3-5b, thin sheets of ceramics (or polymer films starting from solutions) can be made from a process called the doctor-blade process. In this process the slip is cast over a moving belt that passes under a blade. The distance between the belt and blade determines the thickness of the cast ceramic sheet.

Ceramic parts can also be made by compression molding and injection molding. These techniques are more commonly used for making polymer parts and are described in the next section.

16.3.3 Polymer Molding

Many molding techniques can be used to provide a shaped article from polymers. The simplest and least complex technique is the **compression-molding** process first used in the 1860s. In this process, shown schematically in Figure 16.3-6, the raw material is placed between two heated platens. The platens come together to heat, melt or react, and shape the material. The shaped polymer article is removed hot and then cooled.

Both thermoset and thermoplastic materials can be formed by compression molding. The thermoset materials undergo chemical reaction during the molding process, forming a crosslinked polymer in the press. Thermoplastic polymers form a viscous melt in the press. The formed shape is preserved by rapid cooling.

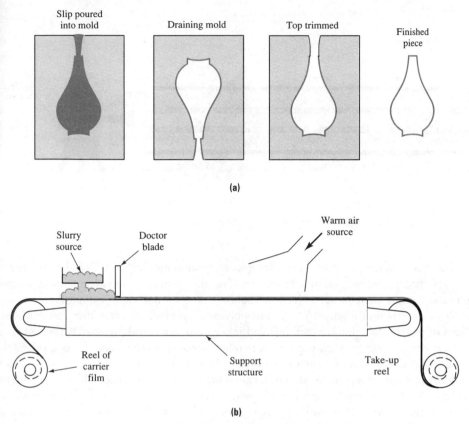

FIGURE 16.3–5   Ceramic processing methods: **(a)** the slip-casting process is used to fabricate ceramic products, and **(b)** thin sheets of ceramics can be made using the doctor-blade process.   *(Source:* **(a)** *F. H. Norton,* Elements of Ceramics, *2nd ed., p. 97, © 1974 Addison-Wesley Publishing Company, Inc. Reprinted by permission of the publisher,* **(b)** *David W. Richerson,* Modern Ceramic Engineering, *1992, Marcel Dekker, Inc. Reprinted by permission of Marcel Dekker, Inc.)*

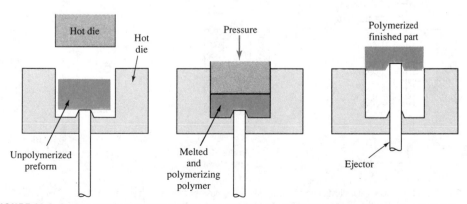

FIGURE 16.3–6   The compression-molding process used to fabricate polymer products.   *(Source: Donald R. Askeland,* The Science and Engineering of Materials, *3rd ed., 1994, p. 517, PWS Publishing Co. Boston, MA. Reprinted by permission of the publisher.)*

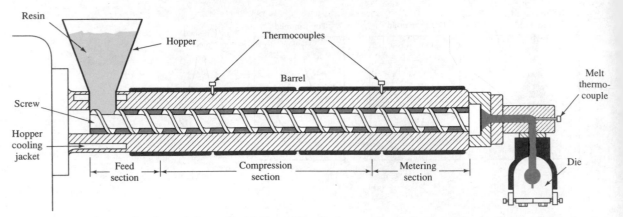

FIGURE 16.3–7   Cross section of an extruder used for injection molding of polymers.   *(Source: Courtesy of Quantum Chemical Company.)*

Transfer molding is a process similar to compression molding in which the polymer is first softened to ensure good flow properties. The molten mass is only then allowed to flow into the compression mold. **Injection molding** is a similar capital-intensive two-step process, as shown in Figure 16.3–7. The polymer is premelted, often using an extruder. Slugs of molten polymer are then injected into a cold mold, and pressure is applied. The formed part is ejected when the two-part mold opens. Injection molding is a relatively rapid process, with cycle times on the order of a few seconds.

Blow molding is a process used to form hollow containers of oxide glass or polymers. Light bulbs, Christmas-tree ornaments, and beverage bottles are examples of products that are amenable to blow molding. A parison, or slug of molten material, is extruded vertically downward. A cold, two-piece mold closes loosely over the parison, and air is blown into the parison from the top, forming a bottle, as shown in Figure 16.3–8. The process is automated and can run at high speeds. All of the incandescent light bulbs in the United States can be produced on just a few high-speed machines.

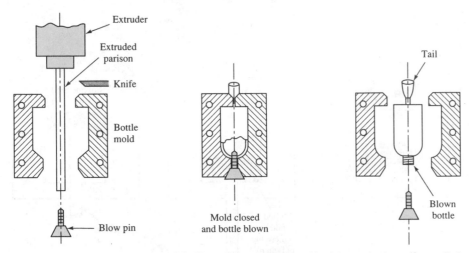

FIGURE 16.3–8   Schematic illustration of the blow-molding process for making beverage bottles.   *(Source: D. C. Miles and J. H. Briston,* Polymer Technology, *as found in Serope Kalpakjian,* Manufacturing Engineering and Technology, *2nd ed.,* © 1989 by Addison-Wesley. Reprinted by permission of the publisher.)

6.4 FORMING

Forming or deformation processes involve changing the shape of a solid by mechanical working and plastic deformation. Therefore, such processes are restricted to materials with sufficient ductility to permit deformation without risk of cracking or fracture. Metals and polymers are particularly suited for these forming processes, while ceramics and composites are typically not shaped in this way. The major exception to this rule is that at temperatures above T_g, oxide glass can be easily formed into shapes. Some important forming processes for each of the three general material categories are described in this section.

16.4.1 Metal Forming

Metals are used extensively as engineering materials in part because of their ability to deform plastically. Several of the forming techniques used to make metal parts are briefly described in this section. These forming operations generally occur after the metal is cast, so it is important to understand how forming operations interact with preexisting casting defects. We will see that most metal-forming operations reduce the severity of casting defects, such as microporosity, and break up coarse particles, such as nonmetallic inclusions, that form during solidification.

Rolling is a common polymer- and metal-forming process. As shown in Figure 16.4–1, rolling is used to reduce the thickness of the workpiece by compressing it between two rolls. When this operation is carried out above the recrystallization temperature of the metal, it is called hot rolling; when the rolling occurs below the recrystallization temperature, it is called cold rolling. Cold rolling increases the dislocation density and alters the shape, but not the average size, of the metal grains. This process generally increases the strength and decreases the ductility of the metal.

Since the metal undergoes recrystallization during hot rolling, this process significantly alters the microstructure. As shown in Figure 16.4–1, the microstructural features affected by the hot-rolling process include the homogeneity of the structure, the size and shape of the grains, and the concentration of point and line defects (dislocations). For example, the hot-rolling process can be used to homogenize the initially nonuniform grain structure in a cast steel resulting from cooling rate gradients during solidification. In addition, the grain size of the hot-rolled product can be controlled accurately by carefully selecting the rolling temperature, the amount of reduction, and the rolling speed.

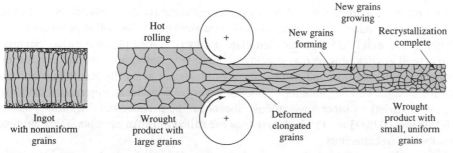

FIGURE 16.4–1 Changes in the grain structure of cast or large-grain wrought metals during hot rolling. *(Source: Serope Kalpakjian, Manufacturing Engineering and Technology, 2nd ed., © 1989 by Addison-Wesley. Reprinted by permission of the publisher.)*

Thus, hot rolling is a much more powerful tool for controlling the microstructure and hence, properties of a metal. The range of microstructures that can be produced by this process extends from very soft (i.e., annealed) structures to high-strength (i.e., fully hardened) microstructures.

Most polymers and some metals such as titanium can develop microstructures in which the grains become rotated during rolling so that certain crystallographic directions are aligned in the rolling direction. This crystallographic alignment, known as texture, can result in anisotropic behavior of the rolled material. That is, properties such as strength and ductility are different depending on whether they are measured parallel or perpendicular to the rolling direction.

CASE STUDY   Process Selection for a Steel Plate

A fabricator of pressure vessels used to store natural gas is shopping for a 10-mm-thick low-carbon-steel plate. The steel plates will be form-rolled into cylindrical shapes and welded along the seam to make pressure vessels. The fabricator is asking for plates that have a relatively low dislocation density to allow successful welding and high ductility and fracture toughness. You, as an engineer with the steel manufacturer, have been asked to recommend a process for making the plate.

Solution

The required microstructure suggests that the end product must be recrystallized and must possess a fine grain size. The recrystallized microstructure has low dislocation density. Additionally, the fine grain size along with low dislocation density will produce higher ductility and fracture toughness. Recall our discussions in Chapter 9 about microstructural variables that influence fracture toughness and also the observation that ductility and fracture toughness increase together. The combination of requirements rules out cold working as a means of producing the plate because it results in high dislocation density and decreased fracture toughness and ductility. On the other hand, hot rolling is a viable processing route because the rolling can be completed above the recrystallization temperature and the plate can then be allowed to cool. The end product will be a recrystallized microstructure with a fine equiaxed grain size of the type shown in Figure 16.4–1. The hot rolling must be conducted at a temperature approximately 15 to 20°C below the eutectoid temperature (723°C; see Figure 7.8–2), to ensure that no solid-state transformations occur during hot rolling. Thus, the rolling temperature should range from 703 to 708°C. By controlling the cooling rates after the hot rolling, a precise control over the resulting microstructure can be maintained.

Flat rolling is a process in which a workpiece of constant thickness enters a set of rolls and exits as a product with a different constant thickness. Flat-rolled products that are thicker than 6 mm (0.25 in) are called plate products. Rolled plates are used extensively in structural applications such as in bridges, pressure vessels, and supertanker hulls, or as armor materials in tanks and planes. Sheet products are generally less than 6 mm thick and are used for making automotive bodies, beverage containers, appliances, desk drawers, and a variety of other applications. Sheets can be produced in thicknesses as small as 0.008 mm (0.0003 in), in which case they are called foils. An example is thin aluminum foils for use in capacitors.

Form rolling or shape rolling is another rolling method for making products such as I-beams, channels, railroad tracks, and large-diameter piping and pressure vessels that are first formed into cylinders from plates and then welded along the longitudinal seams. Even threads can be rolled on cylindrical rods by use of special rolls.

Forging has been used for centuries for making jewelry, horse shoes, hunting weapons, coins, and a variety of other objects. Some modern applications include gas and steam turbine disks and shafts, crankshafts, and valve heads for automobile engines. The basic tools and equipment needed for forging are an anvil, a hammer, and a furnace for heating the workpiece, as in a blacksmith's shop. Modern forging operations, which are extremely sophisticated, make extensive use of computer-controlled dies and mechanical hammers.

In its basic form the forging operation consists of beating the workpiece into the desired shape. When "work" is conducted at room temperature, it is known as cold forging; when done at elevated temperature (above the recrystallization temperature), it is known as warm or hot forging. In an open-die forging operation (Figure 16.4–2) the workpiece is not constrained in the lateral direction and is sandwiched between two flat dies. As deformation proceeds, barreling of the workpiece may occur (see Figure 16.4–2c). In closed-die forging the workpiece is contained in a die, and it assumes the shape of the die at the end of the operation, as shown in Figure 16.4–3. The excess metal, known as flash, is important for proper metal flow during the forging process. The flash is usually machined off after forging is completed.

Forging of complex parts is usually accomplished in several stages, as shown in Figure 16.4–4 for a connecting rod of an internal-combustion engine. The starting material is a blank cylindrical bar forged in a series of dies, which eventually leads to a workpiece with dimensions and shape similar to the final product. In fact, precision forgings are special cases of closed-die forgings that contain fine details. Aluminum and magnesium alloys are better suited for this process because of their relatively low strength

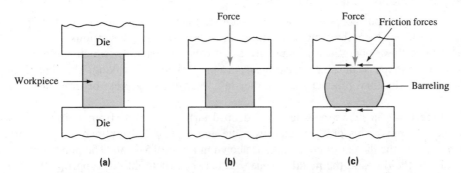

(a)   **(b)**   **(c)**

FIGURE 16.4–2   Schematic of an open-die forging operation: **(a)** solid cylindrical billet being forged between flat dies, **(b)** uniform deformation in the absence of friction, and **(c)** barreling due to friction between the dies and the billet.   *(Source: Serope Kalpakjian,* Manufacturing Engineering and Technology, *2nd ed., © 1989 by Addison-Wesley. Reprinted by permission of the publisher.)*

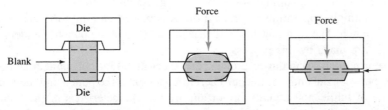

FIGURE 16.4–3   Stages of die forging of a solid round billet. The presence of excess metal known as flash is important for the metal to flow into sharp edges and corners.   *(Source: Serope Kalpakjian,* Manufacturing Engineering and Technology, *2nd ed., © 1989 by Addison-Wesley. Reprinted by permission of the publisher.)*

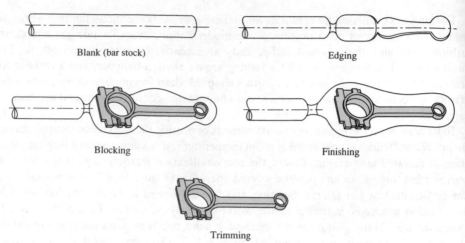

Blank (bar stock)   Edging

Blocking   Finishing

Trimming

FIGURE 16.4–4   Stages in the closed-die forging of an automotive engine connecting rod.   *(Source: Serope Kalpakjian,* Manufacturing Engineering and Technology, *2nd ed., © 1989 by Addison-Wesley. Reprinted by permission of the publisher.)*

and low forging temperatures, which allow them to be forged with less force. Since wear on the expensive dies is related to the magnitude of the applied forces, the use of aluminum and magnesium alloys causes less wear in the dies and makes the process economically feasible.

Deep Drawing and Sheet-Metal Forming

Beverage cans and the exterior bodies of automobiles and appliances are examples of products made from sheet-metal forming and **deep-drawing** operations. The starting material in these operations is typically a flat-rolled sheet from which a blank of an appropriate shape is cut using a shear press, in a process called blanking. The blank is then formed into the final product shape either in a deep-drawing press or in a sheet-metal press.

Figure 16.4–5a shows the stages of a deep-drawn cup starting from a circular blank. Besides the press, this process requires a die and a punch. The punch is a cylindrical mandrel, and the die is a circular ring, as shown in Figure 16.4–5b. The punch draws the metal into the die, and the metal is squeezed between them into the appropriate shape. In this process the metal is subjected to large deformations in certain regions. For example, the metal in the region close to the edge of the circular blank ends up in the rim region of the cup. An estimate of the strain is given by $\ln(D_0/D_f)$, where D_0 is the diameter of the blank and D_f is the diameter of the finished cup. In contrast, the metal in the bottom of the cup is not subjected to any significant plastic deformation.

Sheet-metal forming is similar to deep drawing except the dies are different. It is essentially a stamping operation in which a blank is placed on top of the die cavity and is hit by the punch. The resulting shape is that of the die and the punch. Fenders and oil pans on cars are made by this process.

Large deformations occur during drawing and sheet-metal forming, and large frictional forces exist between die and workpiece. Therefore, design of the punch and die is critical, as is the use of lubricants. Equally important are the characteristics of the metal. The presence of large nonmetallic inclusions can decrease the formability significantly. In a cup-drawing operation, formability can be defined as the ratio of the diameter of the blank to the diameter of the smallest size cup that can be made from it without fracturing.

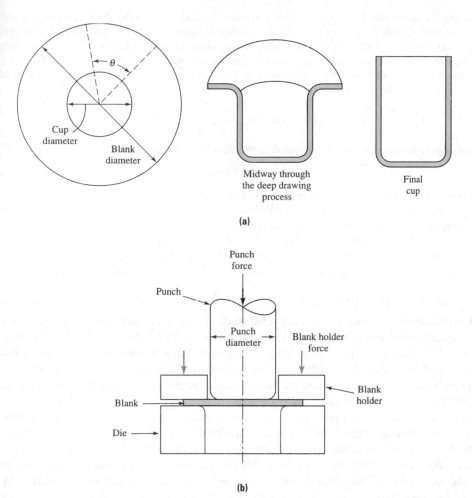

FIGURE 16.4–5 The deep-forming operation: **(a)** stages of forming a cup from a circular blank in a deep-drawing process, and **(b)** pictures of a typical punch and a die used in a deep-drawing operation.

This parameter is called the limiting draw ratio (LDR). LDR is a material property that can be changed by changing the rolling parameters (temperature, speed, etc.) during the production of the sheet. Grain orientation and size can have a profound effect on sheet-metal formability. Metals with high tensile ductilities also have high LDR values.

··

EXAMPLE 16.4–1

A soup can is to be manufactured using a deep-drawing operation. The finished can has a diameter of 50 mm and a height of 150 mm. The sheet thickness is 0.3 mm. Estimate the diameter of the blank needed to make this container and also the minimum number of cup-drawing steps needed to make the container if LDR = 2.0.

Solution

Consider the geometry shown in Figure 16.4–5a. Let r_b represent the blank radius, r_c represent the cup radius, and h equal the cup height. If we consider a segment of the blank having an angle θ, then the arc length at the perimeter of the blank is $r_b\theta$. During drawing this arc length must reduce to $r_c\theta$. Similarly, the distance $r_b - r_c$ must become equal to h. Assuming the sheet thickness does not change significantly, we can write an equation representing volume (or, equivalently, area conservation during plastic deformation. The area of interest is that contained in the blank between concentric circles of radius r_c and r_b and bounded by the angle θ (i.e., $[(r_b\theta - r_c\theta)/2][r_b - r_c]$). During forming this material is transformed into part of the can wall of area $(r_c\theta)h$. Equating the two expressions for area gives:

$$\left(\frac{r_b\theta + r_c\theta}{2}\right)(r_b - r_c) = (r_c\theta)h$$

Solving for r_b and substituting the appropriate values from the problem statement yields:

$$r_b = \sqrt{2r_ch + r_c^2} = \sqrt{[2(25 \text{ mm})(150 \text{ mm})] + (25 \text{ mm})^2} = 90.1 \text{ mm}$$

Therefore, the initial blank diameter must be 180.2 mm.

If the LDR is 2.0, a cup of diameter 90.1 mm can be drawn from a blank of diameter 180.2 mm during the first step. To further reduce the diameter to 50 mm, a second deep-drawing step will be needed. In practice, you may need to begin with a blank larger than 180.2 mm to leave some allowance for trimming at the container rim following the deep-drawing steps.

....

Extrusion and Wire Drawing

Extrusions are produced by pushing a bar stock of material (also called the *billet*) through a die that is shaped to produce the desired cross section (Figure 16.4–6a). The "push" can be accomplished by means of a ram in a direct extrusion process or by means of a fluid (often oil) compressed by a plunger in the hydrostatic extrusion process. In the latter case a small clearance is left between the billet and the chamber containing the billet. The fluid infiltrates this clearance and thereby reduces friction between the billet and the chamber. In the indirect extrusion process, the die moves toward the billet, which remains stationary. The extrusion ratio is defined as the ratio of the area of cross section of the original billet to the area of cross section of the product. Extrusion can be carried out as cold extrusion or as hot extrusion, similar to other metal-working processes. By controlling the extrusion ratio and the temperature, the tensile strength and ductility of the final product can be controlled.

Extrusion can be an inexpensive way of producing several similar parts such as metal gears. These can be produced in long lengths as a single piece of extrusion and then sectioned along the length to produce the individual components, as shown in Figure 16.4–6b. Metal brackets for supporting cabinet shelves are also formed in this way.

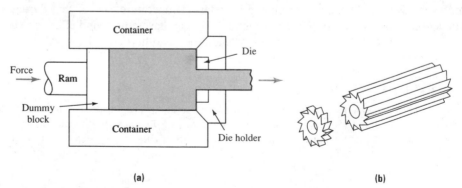

(a)   (b)

FIGURE 16.4–6   The extrusion process: **(a)** a schematic illustration of the process, and **(b)** a gear made by sectioning a long extrusion.   *(Source: Serope Kalpakjian,* Manufacturing Engineering and Technology, *2nd ed., © 1989 by Addison-Wesley. Reprinted by permission of the publisher.)*

Wire drawing is a process used to make wires for several applications including steel ropes, steel-belted tires, or electrical cables. In this process, wires are pulled through a die with typical reductions ranging from 15 to 45%. Several reductions may be required to achieve the final product. Again, by selecting the wire-drawing speed and temperature, the tensile strength of the wire can be controlled.

Kinetics Issues in Metal-Forming Operations

We have mentioned several times that the microstructure and properties of the metal product can be controlled by selecting the processing temperature and speed, whether for a rolling, forging, or drawing process. In this section we investigate the quantitative aspects of this statement. For example, we know from Chapter 8 that recrystallization requires atomic mobility, which in turn depends on temperature. Thus, like several other thermally activated processes, the kinetics of recrystallization can be described by the Arrhenius equation. The time for completing recrystallization, t_r, can be written as:

$$\ln(t_r) = C_0 + \frac{C_1}{T} \tag{16.4-1}$$

where C_0 and C_1 are constants, and T is the absolute temperature. This equation can be used to relate processing times to processing temperatures, as illustrated in the next example.

....

EXAMPLE 16.4-2

Consider a carbon steel wire-drawing operation in which recrystallization must be completed before the wire is wrapped around the take-up spool. If $C_0 = -15.3$ and $C_1 = 1.73 \times 10^4$ K for this steel, calculate the wire-drawing temperature if the distance between the spool and the die is 100 m and the wire-drawing speed is 0.1 m/s. Repeat the problem for a die-to-spool distance of 0.5 m.

Solution

The time available for recrystallization is

$$\text{Time} = \frac{\text{Distance}}{\text{Speed}} = \frac{100 \text{ m}}{0.1 \text{ m/s}} = 1000 \text{ s}$$

Therefore, the minimum temperature for wire drawing is found by solving Equation 16.4–1 for temperature T to yield:

$$T = \frac{C_1}{\ln(t_r) - C_0}$$

$$= \frac{1.73 \times 10^4 \text{ K}}{\ln(1000) + 15.3}$$

$$= 779 \text{ K} = 506°C$$

If the distance is reduced to 0.5 m, the available time = 5 s, with a resulting temperature of:

$$T = 1023 \text{ K} = 750°C$$

....

Note that in the previous example, the required recrystallization temperature of 750°C is greater than the eutectoid temperature of carbon steel (see Figure 7.8–2). Thus, heating to that temperature causes not only recrystallization but also phase transformations (both upon heating and cooling). The resultant microstructure and the strength of the wire are affected significantly by these phase transformations (and by the cooling rate from the austenite phase field). These phase transformations can be viewed either as a problem, i they were not anticipated, or as an advantage, since they provide an engineer with added flexibility. If we take the latter view, we can begin to recognize the importance of materia processing in delivering a product of consistent and known quality. By changing the drawing, speed die-to-spool distances, and drawing temperatures, a variety of microstructures and corresponding property combinations can be produced in steel wire. Similar considerations also apply to extrusion and rolling operations.

16.4.2 Forming of Polymers

In Chapter 6, we described polymers as thermoplastic or thermosetting. This distinction becomes important in selecting processes for forming polymeric materials. Thermoplastics soften upon heating, allowing them to be formed by mechanical working processes such as extrusion, vacuum forming, blow molding, and so on. The thermosetting polymers, on the other hand, must be formed and cured at the same time by compression molding or reaction injection molding.

Polymer Extrusion

Extrusion of polymers is an old technique, dating back to the 1860s, when it was used to process natural rubbers. A variety of polymers can be readily processed into useful shapes by heating polymer granules to above their glass transition temperature and then forcing the viscous liquid through a die into an environment in which it quickly solidifies. Dies of various geometries produce virtually any cross-sectional shapes, just as in the case of metal extrusion. Hollow tubes, rods, beams, films, and fibers are examples of shapes economically and rapidly made using extrusion.

Recall from the previous discussion of polymer molding that an extruder forms the front end of an injection-molding device (see Figure 16.3–7). In injection molding the molten polymer from the extruder is pushed into a mold where it solidifies. In extrusion alone the molten polymer is pushed through a die and hardens in air. In either case, the polymer granules (or pellets) are gravity-fed into the cold end of the extruder. It is important that this end be kept cold so that the granules do not begin to melt, bridge, and cease to flow under gravity. Once inside the barrel, the pellets melt from the transfer of heat from the screw to the polymer and from viscous deformation. A rough estimate of the temperature rise associated with deformation can be made from the following equation:

$$\Delta T = \frac{P\eta}{Q_E C_p} \tag{16.4–2}$$

where ΔT is the temperature rise in °C, η is the motor efficiency, P is power in W, Q_E is the mass of polymer extruded in g/s, and C_p is the polymer heat capacity in J/(g-°C).

....

EXAMPLE 16.4–3

A 10 kW motor operating at 65% efficiency drives an extruder processing polymethylmethacrylate (PMMA). The heat capacity of PMMA can be approximated at the single value of 2.3 J/(g-°C). Calculate the temperature rise of the polymer due to deformation processing when the throughput of polymer is 10 g/s.

Solution

This problem can be solved using Equation 16.4–2:

$$\Delta T = \frac{P\eta}{Q_E C_p}$$

Substituting the appropriate values in the problem statement and noting that 10 kW = 10,000 J/s yield:

$$\Delta T = \frac{(10,000 \text{ J/s})(0.65)}{(10 \text{ g/s})[2.3 \text{ J/(g-}^\circ\text{C})]} = 283^\circ\text{C}$$

In general, most of the temperature rise in an extruder is associated with viscous deformation rather than heat transfer.

The above discussion was concerned with materials that are extruded as viscous liquids and subsequently solidified. It is possible to process some materials in the solid state using hydrostatic extrusion, which is similar to the extrusion technique used for metals. Hydrostatic extrusion of polyethylene at temperatures well below the melting temperature provides transparent, fibrillar, highly oriented, high–elastic modulus samples. Partly because hydrostatic extrusion of polymers is a slow process, it has not been commercialized extensively.

Vacuum and Blow Forming

In vacuum forming, heat-softened sheets of polymers are shaped by creating a vacuum between the mold walls and the polymer sheet (Figure 16.4–7). The polymer deforms and clings to the wall of the molds. Blow forming is a process by which heated polymers are shaped into the form of bottles by means of compressed air. Both processes are inexpensive and, therefore, used extensively.

16.4.3 Forming of Ceramics and Glasses

Plastic forming is not suitable for ceramic materials, which in their final form are hard and brittle. Clay mixtures with about 20–30% water content, however, can be forced through a die by a screw-type machine similar to those used for polymer extrusion. A variety of

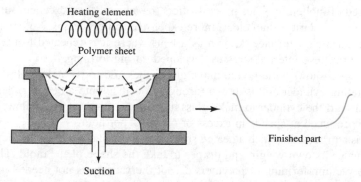

FIGURE 16.4–7 Schematic of the vacuum-forming process used for polymers.   *(Source: Michael F. Ashby and David R. H. Jones, Engineering Materials 2: An Introduction to Microstructures, Processing and Design, 1986, Elsevier Science Ltd., Pergamon Imprint, Oxford, England.)*

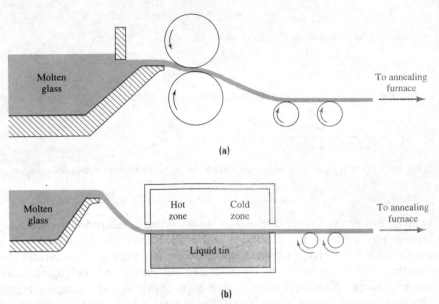

FIGURE 16.4–8   The forming of glass sheets from molten glass using: **(a)** the rolling operation, and **(b)** the Pilkington, or float glass, process.   *(Source: Donald R. Askeland. The Science and Engineering of Materials, 3rd ed., 1994. PWS. Reprinted by permission of the publisher.)*

both solid and hollow shapes can be produced by this method. After extrusion the products are subjected to drying and firing operations to impart strength and hardness. Dimensional changes during drying can cause parts to warp and crack. Therefore, drying is done in a closely controlled environment.

Oxide glass can be formed into sheets (window glass), rods and tubing, bottles, and fibers. The processing of glass fibers will be considered in a later section. The other processes are briefly discussed here.

The glass-forming processes begin with molten glass (i.e., at $T > T_g$). Figure 16.4–8a shows the (now outdated) rolling process for making flat glass sheets. In this process the molten glass passes through a pair of rolls. The viscous liquid is squeezed by the rolls, forming a sheet that is drawn away on a bed of rollers. A pattern can be embossed on the glass surface by the rollers, if desired.

Figure 16.4–8b shows the Pilkington, or float glass, process in which the glass sheet floats on a bed of molten tin. This process offers the advantage of excellent surface finish (and better optical clarity) than older, more expensive roll processes. Note that the glass sheet passes through a furnace, known as a lehr, where it is reheated to the annealing temperature to relieve internal stresses developed during forming.

Figure 16.4–9 shows a method to make glass tubes and rods. Molten glass flows past a hollow rotating cylinder and is drawn through an orifice onto a bed of rollers. Air can be blown through the cylinder to make glass tubes: by cutting off the airflow, solid glass rods can be produced at speeds in excess of 60 miles per hour.

Sagging is a process in which a heated sheet of glass is placed on top of the mold. The sheet sags under its own weight and drapes to take the shape of the mold. This process is similar to vacuum forming of polymers except the vacuum is not needed in this case. The windshields of some automobiles are formed using this process.

Another class of materials known as glass-ceramics can be formed easily in a glassy state. Once the shaping is complete, **devitrification,** which consists of high-temperature nucleation and growth treatments, can transform glass into a crystalline ceramic structure.

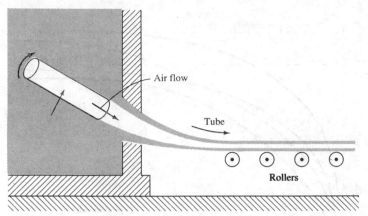

FIGURE 16.4–9 Manufacturing of glass tubing or rods. To make tubes, air is blown through the spinning mandrel. *(Source: Corning Glass Works, as found in Serope Kalpakjian.* Manufacturing Engineering and Technology, *2nd ed., © 1989 by Addison-Wesley. Reprinted by permission of the publisher.)*

Examples of glass ceramics include the lithia-alumino-silicates. Glass ceramics have very low thermal expansion coefficients and are, therefore, resistant to thermal fatigue, making them ideal materials for cookware, heat exchangers, and so on.

16.5 POWDER PROCESSING

Powder processing techniques are used primarily for making parts from metals and ceramics. Powder processing is typically not used with polymers, with the exception of polytetrafluoroethylene, or Teflon®. The underlying principles of powder processing for metals and ceramics are similar, as shown in the following discussion.

16.5.1 Powder Metallurgy

Powder metallurgy (P/M) is a process in which fine metal powders are compacted into intricate shapes. The first known product from this process is the tungsten filament of incandescent light bulbs made early in the 20th century. More recently, the P/M process has been used to produce gears, cams, bushings, cutting tools, piston rings, and valve guides for automotive engines, and aircraft parts such as jet engine disks. The process consists of several steps, including powder production, blending, compaction, sintering, and finishing operations. A brief description of each of these steps is given here.

Metal powders can be produced by a variety of techniques. In the process called atomization, a liquid-metal stream is generated by injecting molten metal through a small orifice. The stream is broken up by a jet of inert gas, air, or even water. Alternatively, fine metal powders can be produced by chemically reducing (removing the oxygen from) fine metal oxide particles using hydrogen or carbon monoxide. It should be pointed out that since metal oxides are brittle, fine powders can be easily produced by grinding. The reduction process is particularly attractive for ductile metals that are difficult to form into small particles by any other means. Powders of brittle metals can be produced either by pulverization in a ball mill or by grinding.

Blending the metal powders is necessary to provide a uniform distribution of powder size, for mixing powders of two metals to make an alloy product, or for mixing lubricants that improve flow of powder metals into dies. Blending is usually done in mechanical mixers under a controlled environment, such as inert gas, or in liquid lubricants.

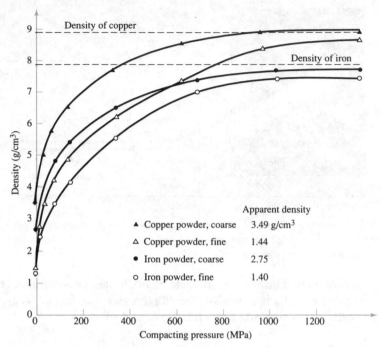

FIGURE 16.5–1 Mass density of copper and iron powder compacts as a function of pressure. The ideal densities are asymptotically approached with increasing pressure. *(Source: F. V. Lenel,* Powder Metallurgy Principles and Applications, *Meta Powder Industries Foundation (Princeton, NJ), 1980, as found in* Manufacturing Engineering and Technology, *2nd ed., by Serope Kalpakjian, © 1989 by Addison-Wesley. Reprinted by permission of the publisher.)*

During the compaction step, powders are pressed into shape using hydraulic or mechanically activated presses. The mass density increases and good particle-to-particle contact is achieved in this step. As the compacting pressure increases, the density approaches the theoretical value, as shown in Figure 16.5–1. After going through the compacting stage, the workpiece is known as a **green compact** and has the strength necessary for the next processing step.

Green compacts may be subjected to additional compaction by processes known as cold isostatic pressing (CIP) or **hot isostatic pressing (HIP).** In CIP, the metal powder or the green compact is placed in a mold made from natural rubber (or other elastomers), and the assembly is pressurized in a water chamber at 400 to 1000 MPa. In the HIP process, the mold is made from a high–melting-point sheet metal, and the pressurizing medium is inert gas. The typical pressures in this process are about 100 MPa, and temperatures up to 1100°C can be used. This process is expensive and is used to make jet engine disks from nickel-based superalloys. Because of the uniformity of pressure, the parts made from isostatic pressing have good uniformity of properties.

Sintering is an important step in powder densification. The driving force for sintering resides in the enormous surface energy stored in powders due to their high surface-area-to-volume ratio. In fact, the stored energy may be a danger during storage because of its potential to explode. During sintering, the green compacts are heated to temperatures between 70 and 90% of their absolute melting temperature to get high diffusion rates along the grain (powder) boundaries. Since sintering is a diffusion-controlled process, the rate of densification, $d\rho/dt$, is given by the following relationship:

$$\frac{d\rho}{dt} = \left(\frac{C}{d^n}\right)\exp\left(-\frac{Q}{RT}\right) \tag{16.5–1}$$

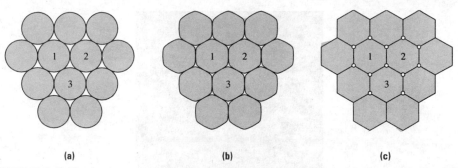

(a)   (b)   (c)

FIGURE 16.5–2 The sintering process: **(a)** powder particles pressed together, **(b)** the first stage of sintering with tetrahedral shape voids, and **(c)** reduction in area by forming spherical pores and hexagonal grains as sintering proceeds. *(Source: Michael F. Ashby and David R. H. Jones. Engineering Materials 2: An Introduction to Microstructures, Processing and Design, 1986, Elsevier Science Ltd., Pergamon Imprint, Oxford, England.)*

where Q is the activation energy for grain boundary diffusion, R is the gas constant, T is absolute temperature, d is the average particle (grain) diameter, and n and C are materials constants.

Theoretical density (i.e., the density associated with the "perfect" packing of individual atoms) may not be achieved during sintering because of residual closed pores. A mechanism that explains the formation of residual porosity is shown in Figure 16.5–2. Hot pressing, such as the HIP process, is required for removing such porosity.

Sintering can be facilitated if there are two adjacent particles of different metals. Alloying can take place at the interface of the two particles, causing local melting. Also, if one metal has a lower melting point than the other, it can melt, and the liquid can surround the other particle because of surface tension. This type of sintering, called liquid-phase sintering, results in reduced porosity.

The P/M process is attractive since it permits the development of intricately shaped metal parts with extremely fine grain sizes (the grain size can approach the original powder diameter). Thus, P/M permits the development of microstructures that cannot be achieved by any other processing method.

CASE STUDY Specification of Powder Size Distribution for Producing Steel Sprockets

A precision steel sprocket is to be made using a powder metallurgy process. In addition to good dimensional tolerances, porosity is an issue because the sprocket will be transmitting cyclic torque during operation. Thus, porosity can cause mechanical failures due to metal fatigue. You are asked to specify the process and the powder size distribution so that low porosity levels can be achieved without resorting to unduly long sintering times.

Solution

If all powder particles are the same size, the maximum achievable density for spherical particles is 74% for either fine or coarse particles (see the discussion of atomic packing factor for either the FCC or HCP structures in Chapter 3). The porosity level can be reduced by compaction, which changes the spherical particles into roughly a hexagonal shape, as shown in Figure 16.5–2c. As shown in Figure 16.5–1, however, theoretical density will never be attained by compaction alone. Therefore, uniformly sized particles will always result in some residual porosity in the sintered part. On the other hand, with the right mix of particle sizes, the small particles can fill the spaces between the close-packed arrangement of large particles. This will considerably increase the density of the green compacts and also significantly reduce the porosity level in the sintered product. As a general rule of thumb the P/M part formed from a bimodal distribution of powder sizes should contain approximately 70% coarse powder and 30% fine powder.

FIGURE 16.5–3   Silicon nitride blade for a gas turbine engine made by the hot isostatic pressing of powder. *(Source: Battelle Columbus Laboratories, as found in Serope Kalpakjian, Manufacturing Engineering and Technology, 2nd ed., © 1989 by Addison-Wesley. Reprinted by permission of the publisher.)*

The finishing operation for P/M parts is intended to give them special surface characteristics or to impart higher strength and higher dimensional accuracy. The latter is accomplished by additional pressing of the sintered part that contains porosity. In other applications, the inherent porosity of P/M parts can be utilized by impregnating them with fluid such as oil. This is accomplished by immersing the parts in heated oil. Internally lubricated bearings and bushings are made by this process.

16.5.2 Powder Processing of Ceramics

Almost all ceramic-processing operations start with powders. Previously we learned that the starting material for molding and extruding ceramic materials was a powder slurry. In this section, we will investigate ceramic powder processing, which involves powder compacting and sintering operations similar to P/M techniques.

Ceramic powders are pressed into shapes in dies similar to those used for the metal powders. Ceramic powders are usually mixed with stearic acids or wax and low–melting temperature fluxes. The stearic acids and wax serve as both lubricants, which aid in the flow of the ceramic powders during the compaction operation, and as temporary binders to hold the green compacts together. After the initial forming operations, the green bodies are placed in a furnace, known as a kiln, and fired. During firing, the water and volatile binders evaporate while the low–melting temperature fluxes bind the powder. Sintering is then conducted at temperatures ranging from 700°C to 2000°C, which facilitates binding among the ceramic powder particles to produce a fine-grained structure. Hot isostatic pressing discussed earlier for P/M parts is also frequently used for making ceramic parts. Figure 16.5–3 shows a silicon nitride gas turbine blade that was processed using this route.

16.6 MACHINING

Machining is the process of removing excess material by plastic shearing. Common machining processes include turning, drilling, milling, shaping, and grinding.

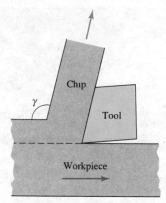

FIGURE 16.6–1   Schematic of a machining process.   *(Source: Michael F. Ashby and David R. H. Jones. Engineering Materials 2: An Introduction to Microstructures, Processing and Design, 1986, Elsevier Science Ltd., Pergamon Imprint, Oxford, England.)*

Figure 16.6–1 shows a schematic of a machining process. Energy is required for several purposes: (1) for creating two new surfaces (one on the workpiece and the other on the chip), (2) for plastic deformation of the chip (quantified by the shear angle γ), and (3) for overcoming the frictional force between the tool and the chip surface. The latter two factors far exceed the energy required for creating new surfaces. Water-soluble cutting fluids are used to reduce the frictional forces and to cool the workpiece and tool.

The ease with which the material can be machined is called machinability. For materials to have high machinability, they must have low shear strength, they must be shock-resistant to prevent fracture from the impact loading between the tool and the workpiece, and the resulting chips must not have a tendency to stick to the tool and thereby increase friction. Ceramics are hard and consequently have a high shear strength; they are also not shock-resistant. Polymers, on the other hand, have low shear strength but may stick to the tool because they tend to melt from the heat generated during the machining process. This requires that the process be slowed down considerably to allow time for heat removal, which makes machining a costly process for these materials. Thus, based on these criteria, it is not surprising that metals possess better machinability than ceramics and polymers.

Machinability varies considerably among metals. For example, precipitation-hardened steels and nickel-based alloys are often difficult to machine because they are relatively hard and have high shear strength. Therefore, components from these alloys are often machined to their near final shape and size before being subjected to precipitation hardening. Only final finishing remains to be performed in the high-strength state, where machinability is poor.

Some alloys that are often used in parts that require heavy machining have built-in lubricants that coat the face of the tool as the chip forms. These are called free-cutting metals. Examples include brass, which contains free lead, and steels, which contain manganese sulfide inclusions to promote chipping. Figure 16.6–2 shows the microstructure of free-cutting brass containing fine lead particles dispersed in the brass matrix.

Machining may be expensive because of labor cost, equipment cost, and the value of the material scrap that is generated. Thus, forming processes may provide a viable alternative.

FIGURE 16.6–2 Microstructure of free-machining brass containing free lead (black dots) dispersed in a two-phase brass microstructure. The dark area is the copper-rich α phase, and the light area is the Zn-rich β phase. The lead is dispersed uniformly in both phases.

16.7 JOINING PROCESSES

Joining materials and parts is essential for building engineering machinery and structures. Joints are made between similar as well as dissimilar materials. The strength of joints is vital in determining the reliability of any structure. Your television set contains numerous joints between copper wires and electronic components. A failure in any one of these joints can degrade your television picture. Thus, these joints must have a lifetime of several years. Similarly, the Alaska pipeline contains many lengths of steel piping that are welded together. In this case, being able to weld with assurance against mechanical failure for several decades was the key factor in selecting the steel for this application.

This section is concerned with four important joining techniques that are based on either diffusive mass transport, pure mechanical bonding, or chemical bonding between the bonding agent and the surfaces being joined.

Irrespective of the bond type in the joint (diffusion, mechanical, or chemical), there will always be an interfacial, or transition, region between the joined components. If the compositions or microstructures of the joined components (or the bonding agent) differ significantly from one another, the interface region will contain sharp chemical and microstructural gradients. When the strength of the interface is less than that of the base material, it becomes the load-limiting feature of the assembly.

16.7.1 Welding, Brazing, and Soldering

A large number of structural parts made from steel contain weldments. Examples include bridges, large pressure vessels, large piping, and ships. The fact that steel is so readily weldable gives it a significant advantage over other alloys such as aluminum for structural applications. Aluminum alloys are not weldable in air because of their tendency to quickly form aluminum oxide, which embrittles the joint. Therefore, specialized techniques are needed to weld aluminum, which make the process expensive.

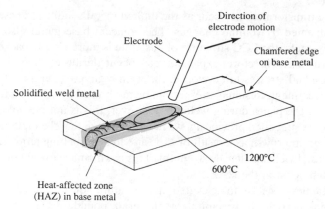

FIGURE 16.7–1   Schematic of a welding operation.   *(Source: Michael F. Ashby and David R. H. Jones,* Engineering Materials 2: An Introduction to Microstructures, Processing and Design, *1986, Elsevier Science Ltd., Pergamon Imprint, Oxford, England.)*

Welding may be carried out by generating an electric arc between the electrode, which contains the weld or filler metal (in this case, the bonding agent), and the workpieces that are to be welded together. The arc generates enough heat to melt both the electrodes and the metal on the edges of the workpiece. As the electrode passes through a region, the molten metal from the electrode mixes with that from the base metal and then solidifies to form the joint. This process is shown in Figure 16.7–1. The purpose of the flux contained within the electrode is to stabilize the arc and to generate gases (CO_2, CO, and H_2O vapor) that act as a shield against the surrounding atmosphere.

Since the region in the base metal adjacent to the weld is heated to temperatures above that at which ferrite (α) transforms to austenite (γ), the microstructure of this region is altered significantly during the welding operation. For example, rapid cooling of austenite can result in the formation of the brittle martensitic phase, or, at the very least, result in a coarsening of the grain structure. Both of these microstructural modifications are undesirable, since they reduce the tensile strength and toughness of the steel. Residual stresses developed during cooling can also weaken welded joints. The region over which the welding operation alters the microstructure of the base metal is called the **heat-affected zone (HAZ)** and is equivalent to the joint interface discussed earlier in this section.

Postwelding heat treatments are often used to develop more uniform microstructures and to relieve residual stresses in the weld region. This added step, which is often necessary, adds to the cost. Desirable weld microstructures may be obtained without postwelding heat treatments by optimizing welding parameters, such as heat input, weld and base metal compositions, and cooling rates.

What factors affect the strength of a welded joint? Suppose you have been asked to select the shielded metal arc welding process for fabricating a pressure vessel used in power plants for containing steam at 550°C. The vessel is made from ferritic steels containing chromium for high-temperature oxidation resistance. What problems might you anticipate?

Since the weldment is subjected to the same conditions of stress and temperature as the base metal, it is important to match the chemical composition of the weld material with that of the base material. If this is done correctly, it is theoretically possible to achieve the same tensile and creep strength at the joint as in the rest of the pressure vessel. However, several other metallurgical concerns need to be addressed.

Several phase transformations occur as the molten metal solidifies. These transformations are accompanied by volume changes. The adjacent base metal also expands and contracts. For example, the FCC structure of austenite is more closely packed than is the BCC ferritic structure. Therefore, expansion can occur during cooling. This expansion causes severe residual stresses to develop in the weld region, which must be relieved by a postweld heat treatment. Residual stresses can be minimized by allowing free expansion and contraction of the plates during welding. However, this is not possible in thick-wall pressure vessels. As suggested previously, we must recognize that the material in the HAZ may experience grain growth as a result of exposure to near melting temperatures, which can result in reduced yield and tensile strengths. This is why mechanical failures in welded components often begin in the HAZ.

Exposing stainless steels to elevated temperature during welding can cause chromium carbide particles to precipitate at the grain boundaries (see Section 15.2.4). When this occurs the strength of the weld may be poor. Additionally, the migration decreases the chromium content of the surrounding region, which in turn decreases the oxidation resistance of the metal in the HAZ. In addition, the corresponding loss of carbon to the precipitates can decrease the creep strength of the weld region.

Because of the above difficulties, several new welding techniques have emerged. They require significantly less heat input per unit weld area. Electron beam welding and laser welding are two such techniques. The heat generated when using these techniques is localized so that the heat-affected zone is extremely small.

Brazing is a joining process in which a filler metal and the flux are sandwiched between the workpieces. The temperature is raised sufficiently to melt the filler material but not the workpieces. The molten metal fills the spaces between closely fitting parts by capillary action. Upon cooling, a strong joint is obtained. Filler metals that melt at temperatures above 450°C are used in brazing applications. The use of flux is essential to prevent oxidation and to remove oxide films from the surfaces of the workpiece so that good metallurgical bonds form.

Brazing is used to join metals to ceramics in the electronics industry. In this process, the mating face of the ceramic part is first coated with a thin film of a refractory metal (molybdenum, tantalum, etc.) by applying it as a powder and then heating it. The metal film is then electroplated with copper. A metal part may then be brazed to the copper plating.

Soldering is a process similar to brazing, except the solder materials melt at much lower temperatures than do brazing filler metals (less than 450°C). Typical soldering materials include tin-lead, tin-zinc, lead-silver, and cadmium-silver alloys. The flux materials, which are used for much the same purpose as fluxes during welding, include inorganic acids or salts, which clean the surfaces rapidly. The residue from the flux materials should be cleaned off after soldering to avoid corrosion during service.

A clever application of the soldering process is the use of gold (Au) to connect silicon chips to chip carriers fabricated from either ceramics or metals. Figure 16.7–2 shows the Au-Si phase diagram, which is a classical eutectic diagram with virtually no solid solubility at low temperatures. The melting point of gold is over 1000°C and that of Si is about 1400°C, but their eutectic composition, consisting of about 20 atomic percent Si, melts at 363°C. The silicon chip is placed in a gold-plated chip cavity in the carrier, which is heated to 425°C. Si diffuses into Au until the eutectic composition is reached, at which time melting occurs. The liquid front advances into the gold and an intimate bond is achieved upon cooling.

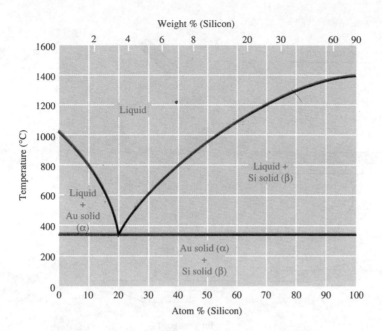

FIGURE 16.7–2

The Au-Si phase diagram.

16.7.2 Adhesive Bonding

Adhesive bonding is used for joining parts made from polymers and polymer-matrix composites. It is also used for joining polymers to metals, metals to metals, and ceramics to metals. Adhesive joints are designed to withstand shear, tensile, and compressive stresses, but are typically weak against peeling forces. An example is adhesive tape, which can be peeled, but not so easily pulled to separation. This joining technique is frequently used in the aerospace, automotive, appliance, and building industries. Because of the weakness against peeling, the joint design for adhesive bonding is critical. Figure 16.7–3 shows examples of poor and good adhesive joint designs.

To achieve high adhesive bond strength, it is desirable to have chemical bonding between the adhesive and the adherent (i.e., the base material). Primary bonds—ionic, covalent, and metallic—require close contact, closer than can be achieved with most adhesives. Therefore, it is necessary that the interfacial gaps between the adhesive and the adherent be reduced to the magnitude of interatomic distances. Even if this can be accomplished, nothing ensures that the significant electronic and molecular rearrangement required for such bonding will occur spontaneously. These demanding conditions can be met only during processes such as welding, brazing, and soldering, as discussed before. On the other hand, secondary bonds have longer range electronic interactions and do not require high-temperature conditions to form. Therefore, this type of chemical bonding may develop in adhesive joints.

The electronic interactions characteristic of secondary bonding, though long range on the atomic scale, still occur only over distances of a few angstroms. Therefore, the adherent and the adhesive must approach each other on an atomic scale. Thus, it is imperative to clean all surfaces thoroughly to allow the adhesive to wet the surface of the adherent. Gases, oils, or fine dust particles at the interface have deleterious effects on the bond strength because they inhibit close contact between bonding surfaces. An atomically rough surface may also promote adhesion via mechanical interlocking. For example, aluminum surfaces can be specially treated by a process called phosphoric acid anodizing,

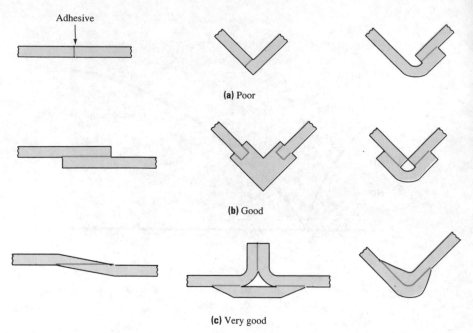

(a) Poor

(b) Good

(c) Very good

FIGURE 16.7–3   Typical designs of adhesively bonded joints rated from poor to very good.   *(Source: Serope Kalpakjian,* Manufacturing Engineering and Technology, *2nd ed., © 1989 by Addison-Wesley. Reprinted by permission of the publisher.)*

which forms strongly bonded columns of aluminum oxide on the surface. A resin adhesive can flow between these columns and provide a strong adhesive bond.

Adhesive can be classified in three categories: natural adhesives, such as starch and its derivatives, and animal products; inorganic adhesives, such as sodium silicate and magnesium oxychloride; and synthetic organic adhesives, which are either thermoplastic or thermosetting polymer resins. Most adhesives are noncrystalline (glasses), so the bond strength is highest when a minimal amount of adhesive is used.

16.7.3 Diffusion Bonding

Diffusion bonding, also known as diffusion welding, is a solid-state bonding process that can be accomplished at temperatures higher than half the absolute melting point of the base material. Since the process involves atomic scale diffusion, the time required to produce a strong bond is a function of temperature. A clamping force is also applied during the process, which helps the mating surfaces to flow plastically and come into intimate contact. This greatly improves the bond quality.

Diffusion bonding works well for joining dissimilar material pairs, for reactive metals such as titanium, beryllium, and the refractory metal alloys, and for composites and ceramic materials. Since ceramics cannot be joined by welding or mechanical fastening, this is an important technique for joining these materials.

Oxide glass is often used as a permanent high-temperature sealant. The quality of sealing depends on the wetting characteristics of the glass on the metal surfaces. The contact angle can be decreased (improved) by saturating the glass with the substrate metal oxide. This can be accomplished by first growing a layer of oxide on the metal surface and then bringing it in contact with the glass. The oxide diffuses into the glass and the metal

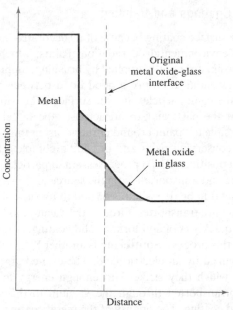

FIGURE 16.7–4   Concentration profile across the glass–metal interface at the beginning and end of the diffusion steps.

oxide–glass interface advances toward the metal, as shown in Figure 16.7–4. This process is continued until all the metal oxide has dissolved in the glass and a direct contact between the metal and the glass is obtained.

16.7.4 Mechanical Joining

Mechanical joining methods often involve threaded holes. These holes act as notches and can cause fatigue fractures, even in the most ductile materials if the joints are subjected to cyclic stresses during service. In brittle materials, fracture can emanate from these holes. Hence, mechanical joints must be designed with consideration of the fatigue and fracture resistance of the material. Therefore, mechanical joining is seldom used with ceramic and composite materials.

Another consideration in mechanical joining is galvanic coupling (see Section 15.2.4). If different materials are being joined or if the rivet or screw material is different from the base materials, galvanic corrosion can occur. For example, if two pieces of copper are joined by a steel screw, steel becomes the anode and all the metal loss is concentrated on the small screw, which will eventually degrade. However, copper screws for joining two pieces of steel are acceptable because the anode has the large area in this case.

16.8 SURFACE COATINGS AND TREATMENTS

Surface coatings and treatments are applied on engineering components for a variety of reasons, including cosmetic, protection from environmental degradation, thermal resistance, wear resistance, fatigue resistance, and electrical insulation. These treatments include: (1) applying thin polymeric, ceramic, or metallic coatings by a variety of processes; (2) changing the chemical composition of the surface through diffusion or by ion implantation; or (3) simple mechanical hardening of the surface by sandblasting or shot peening.

16.8.1 Application of Coatings and Painting

Paint is widely used as a surface coating because of its decorative properties and its ability to protect a part from environmental degradation. Paints, which may be either curing (crosslinking) or noncuring, are usually applied by brushing, dipping, or spraying. Proper surface preparation is essential to ensure a good bond between paint and substrate. In **electrostatic painting** the paint particles and the surface to be painted are given opposite electric charges so that the particles are attracted to the target surface. As shown in Figure 16.8–1, it is possible to "paint around corners" using this technique. The spraying process is effective in conserving paint and providing complete and uniform coverage.

In **physical vapor deposition (PVD)** processes, small particles of the coating material are generated at a source and transported from the source to the workpiece. In the vacuum evaporation process, metal to be deposited is heated to its vaporization point. The metal atoms in the vapor phase are transported through the vacuum to the (room-temperature) workpiece, where they quickly cool and form a solid coating. A uniform and thin coating layer can be applied by this process. **Sputtering** is another PVD process in which an inert gas such as argon is ionized by an electric field. The excited argon ions are accelerated toward a metal target, which they strike with enough energy to displace metal atoms. These atoms are then transported through the vacuum to the workpiece, where they condense to form a solid coating. The quality of the metal coating, as well as the strength of the coating-substrate interface, may be improved by heating the workpiece to promote diffusion of the metal atoms.

In PVD processes the atoms that form the coating are generated by "physical" methods such as evaporation or sputtering. If, however, the coating atoms are generated by a chemical reaction, the process is known as **chemical vapor deposition (CVD).** CVD processes are used to produce a variety of coatings with applications ranging from the microelectronics industry (to be discussed in Section 16.9) to the deposition of wear-resistant surface layers. As an example, consider the CVD process used to deposit a hard titanium nitride coating onto a tool surface to improve its wear resistance. The workpiece is placed in a graphite crucible or tray and heated to 950–1050°C in an inert environment. Titanium tetrachloride vapor, hydrogen, and nitrogen are then introduced in the chamber. The chemical reactions form a titanium nitride coating on the workpiece surface.

Hot dipping is a process used for applying metallic coatings on other higher–melting temperature metals. The workpiece is simply dipped into a molten bath of the coating material. A common example of the hot-dipping process is the coating of steel with zinc.

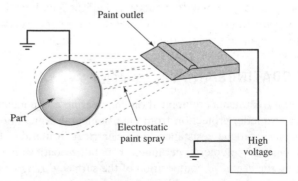

FIGURE 16.8–1   A schematic illustration of the electrostatic painting process.   *(Source: Society of Manufacturing Engineers, as found in Serope Kalpakjian,* Manufacturing Engineering and Technology, *2nd ed., © 1989 by Addison-Wesley. Reprinted by permission of the publisher.)*

The zinc-plated steel, called galvanized steel, is used in pipes, screws and nails, and automotive bodies.

Metal coatings can be applied uniformly by **electrochemical plating,** in which the workpiece is the cathode and the coating material is the anode. Common metals coated by this method include tin, chromium, nickel, copper, cadmium, and zinc. This type of plating can be deposited selectively, which is useful in rebuilding worn or damaged parts.

Anodizing is an oxidation process in which the surface of the workpiece is converted into a hard oxide. The workpiece is the anode in an electrolyte cell containing an acid bath. The workpiece absorbs oxygen from the bath. Since the resulting oxide is porous, this process can be continued until the desired levels of oxide thickness are obtained.

Porcelain enamels and glazes are glassy inorganic coatings that contain various metal oxides. Enamels are used to coat metals and glazes to coat ceramics. Examples of this type of coating include chromium oxide and aluminum oxide for wear resistance, zirconium oxide for thermal resistance, and aluminum oxide for electrical insulation. Such coatings may be applied by spraying, electrodeposition, and dipping.

Hard facing, also known as weld overlay, is a process in which a thick layer of wear-resistant or environment-resistant alloy is deposited on the workpiece by a welding or brazing process. Usually, a number of layers of deposits are made to produce a thick coating. Nuclear pressure vessels, which are made from ferritic steels, have stainless steel cladding on the inside produced by this process.

16.8.2 Surface Treatments

In surface treatment, the surface of the workpiece is hardened by changing its chemical composition, by work hardening, or by heat treatment.

Case hardening involves strengthening the surface layer of the workpiece by either changing its chemical composition or selectively heat-treating the surface differently from the bulk.

The surface hardness of low-carbon steels (less that 0.2 weight percent carbon) can be increased by a process known as carburizing, which raises the surface carbon composition to nearly 1 percent. This is accomplished by heating the workpiece to ~900°C and subjecting it to an atmosphere of carbonaceous gases. By changing the mixture of gases, the carbon content of the outermost surface can be controlled. Since carburization is a diffusion process, the depth of carburizing can be regulated by controlling the time and temperature (see Section 4.4). Carburization is commonly used to harden engine crankshafts, cams, chains, and so on. It is a relatively inexpensive surface-hardening process.

Nitriding is a closely related surface treatment process using nitrogen-containing media such as cyanide baths to produce reactive nitrogen. Since nitrogen is also an interstitial atom in iron, the results of nitriding and carburizing steel are similar. Boron oxide fibers can be nitrided by subjecting them to ammonia at high temperature, thereby producing BN fibers.

Induction hardening is a process in which a workpiece of medium-carbon steel is quickly heated on the surface by induction heating and quenched to produce a martensitic surface layer. (The interior of the workpiece remains as ferrite and pearlite.)

In **ion implantation,** ions of light elements such as boron and nitrogen are accelerated to a high velocity in an electric field and impinge on the workpiece. The ions penetrate to depths of a few microns and substantially increase the surface hardness of the workpiece. This technique is used for surgical tools and implants to develop wear and corrosion resistance. It is also used for doping (introducing controlled amounts of Group III or Group V elements into) the surfaces of semiconductors.

In **shot peening,** the surface of a metal workpiece is barraged with a large number of small glass, ceramic, or steel balls known as shots. These shots plastically deform and work-harden the surface layers, and induce residual compressive stresses on the surface. Both factors increase the resistance to fatigue crack initiation and prolong the fatigue life of the component.

In **explosive hardening,** the workpiece surface is subjected to high pressure pulses by detonating a layer of explosive on the surface. Pressures in the range of 30 GPa can be obtained by this method. The process yields large changes in the surface hardness with little change in component dimension.

CASE STUDY Material and Process Selection for Automobile Engine Crankshafts

Crankshafts in automobile engines are made from medium-carbon steels. In service they are subject to cyclic loading as well as significant friction and wear. Justify the choice of medium-carbon steels and recommend a surface-hardening process for wear and fatigue resistance.

Solution

Since cost is an important consideration in car production, the selected material must be easy to form into the complex shape of the crankshaft. It is relatively inexpensive to forge crankshafts from medium-carbon steels. These steels, however, do not have the necessary hardness for wear resistance. Should the surface be hardened by compositional modification, work hardening, or heat treating (martensite formation)? If the latter option is selected, distortions in shape may occur, since the density of martensite is less than that of austenite. Since crankshafts require strict dimensional control, a heat-treated crankshaft will require large amounts of machining to meet the dimensional tolerance. Also, the shaft will become too brittle for this application. The most inexpensive treatment that does not induce dimensional changes is case carburizing, in which the carbon content of the crankshaft surface is increased to ~1%. For selecting conditions for case carburizing, refer to Section 4.4.7, where this problem was discussed in detail including the kinetic equations that provide the time-temperature-concentration profiles.

16.9 SINGLE-CRYSTAL AND SEMICONDUCTOR PROCESSING

The principal difference between single crystals and polycrystalline materials is the lack of grain boundaries in the former. The absence of grain boundaries is a highly desirable microstructural feature in several materials applications. For example, the use of single-crystal nickel alloys improves the high-temperature resistance of turbine blades by improving their resistance to oxidation and creep deformation, which are both enhanced by the presence of grain boundaries, as explained in Chapters 9 and 15. A preference for single crystals is also apparent in our selection of naturally occurring gemstones such as diamonds, rubies, and sapphires. The vast majority of applications for single crystals, however, occur within the semiconductor industry.

As discussed in Chapter 10, the reason for requiring single crystals in the semiconductor industry is that defects significantly degrade the electrical properties of these materials. Therefore, this industry not only requires materials that are free of grain boundaries, but in addition the single crystals must be free from high concentrations of dislocation, impurity atoms, and other point defects.

In this section, we describe the growth and processing of silicon single crystals and discuss the sequence of processing steps required to fabricate an integrated circuit chip. These operations include oxidation, lithography, etching, diffusion, ion implantation, interconnection, assembly, and packaging.

16.9.1 Growth and Processing of Single Crystals

Most of the silicon crystals used in the semiconductor industry are grown by the **Czochralski (CZ) method.** The CZ process is a simple liquid-solid monocomponent phase transformation system (see Section 7.2). To obtain a single crystal, however, the liquid-to-solid phase transformation must take place at a temperature only slightly below the melting temperature (T_m) in order to ensure a low nucleation rate and a high growth rate. The small undercooling is obtained by holding the melt just above T_m and pulling a seed crystal from the melt. A typical CZ growth system is shown in Figure 16.9–1.

The use of a seed crystal provides two advantages. First, since the seed serves as a heterogeneous nucleation site, it permits the phase transformation to occur at a temperature above that required for the formation of additional homogeneous nucleation sites. Second, the seed crystal can be used to orient the CZ crystal growth axis along a particular crystallographic axis. Many, but not all, Si crystals are grown with their axis in a $\langle 1\ 0\ 0 \rangle$ direction.

After the ingot has been grown, it must be sliced, polished, and etched prior to device fabrication. The slicing operation is usually performed with a diamond-coated stainless steel saw blade. During this process approximately 30% of the ingot is lost as "saw dust."

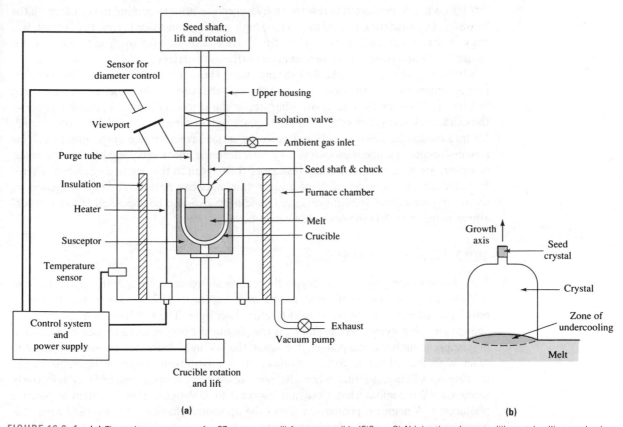

(a)   **(b)**

FIGURE 16.9–1 **(a)** The major components of a CZ system are: (i) furnace: crucible (SiO_2 or Si_3N_4), heating elements; (ii) crystal pulling mechanism: seed shaft and rotation mechanism; (iii) atmosphere control system: pumping system and inert gas source (He or Ar); (iv) control system: sensors to control temperature, pull rate, rotation speed, crystal diameter, and gas pressure. **(b)** A schematic showing the seed crystal and the crystal being grown from the melt. *(Source: VLSI Technology, S. M. Sze, ed., 1988. McGraw-Hill, Inc. Reproduced with permission of McGraw-Hill.)*

The polishing and etching operations ensure that the faces of the wafer are parallel and free from surface contamination.

The most common alternative to CZ crystal growth is **epitaxial growth.** This process involves the formation of a thin crystalline layer on top of a crystalline substrate that serves as the "seed." The phase transformation can be either gas-to-solid (vapor-phase epitaxy), or liquid-to-solid (liquid-phase epitaxy). The low vapor pressure of Si permits condensation of the atoms on a relatively cool substrate (400–800°C). Thus, epilayers can be grown at temperatures well below the melting temperature of Si ($T_m = 1410°C$).

16.9.2 Oxidation

Silicon has emerged as the dominant semiconductor material in part because of the excellent characteristics of silicon dioxide (SiO_2). Some of the desirable characteristics of SiO_2 are (1) it is a good electrical insulator, (2) it provides surface passivation (renders surface defects electrically inactive), and (3) it plays a vital role in the introduction of controlled amounts of dopant atoms in selected regions of the host silicon wafer, as will be discussed later.

The most common method for growing an oxide layer on Si begins with the flow of oxygen over a wafer held at elevated temperatures (900–1200°C). The oxygen can be either "dry" or "wet." In the wet oxidation process the dry oxygen is either bubbled through water or permitted to react with hydrogen gas prior to coming in contact with the Si wafer. Dry oxidation is used to produce high-quality thin oxide layers. In contrast, wet oxidation is faster and is used when thick oxide layers (>0.5 μm) are required, for example, when oxides are needed to act as diffusion barriers.

Oxidation takes place at the SiO_2-Si interface. Thus, at high thicknesses, the oxidation rate is limited by the diffusion of oxygen through the oxide to the interface. As discussed in Chapter 15, since the rate-controlling step in the reaction involves a diffusion process, the oxide thickness (y) increases with the square root of time ($y \propto \sqrt{t}$). On the other hand, for thin oxides the rate-controlling step is the surface reaction rate at the interface. This process results in a linear growth rate ($y \propto t$). Both the linear and parabolic growth rates, however, are proportional to the pressure of the oxygen in the reaction chamber. Therefore, the time to grow an oxide of a given thickness can be substantially decreased by increasing the oxygen partial pressure. Oxide films can also be deposited at low temperatures using the CVD methods discussed above.

16.9.3 Lithography and Etching

The **photolithography** process begins by coating an oxidized wafer with a light-sensitive polymer called a photoresist. Next, as shown in Figure 16.9–2, a mask is used to expose only selected areas of the photoresist to ultraviolet light. The UV light induces chemical changes in the polymer. For example, if the photoresist is composed of a combination of polyisoprene rubber and a photoactive agent, the UV light causes the photoactive agent to react with the rubber to form crosslinks. This crosslinked rubber is insoluble in the developing solution so that when the photoresist is developed, material is selectively removed from regions that were not exposed to UV light. This is called a positive photoresist. A negative photoresist shows the opposite behavior—the exposed areas become more soluble in the developing fluid.

The next step is to etch away the regions of the oxide film that are not protected by the remaining photoresist material. The unprotected regions of the wafer are removed by exposure to hydrofluoric acid, which attacks the SiO_2 without attacking the underlying Si.

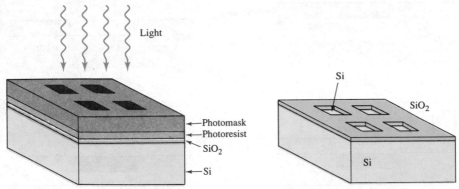

FIGURE 16.9–2   In the photolithography process a mask is used to define selected regions of a light-sensitive photoresist material. HF acid is then used to etch away the regions of the oxide film that were not protected by the UV-hardened photoresist.

Dopant atoms can then be deposited on the portions of the wafer from which the SiO_2 was removed, as discussed in the next subsection.

The present practical limit for the smallest dimension that can be defined by photolithography, also known as the line-width resolution, is ~ 0.3 μm. Electron lithography may eventually permit smaller line widths due to the shorter wavelength of the high-energy electrons. Other possible lithography systems under development include X-ray and ion beam systems.

CASE STUDY   Mask Selection for Doping of Si Wafers

As an engineer in an integrated chip production facility, you are asked to design an appropriate UV mask for doping (with P) a small square region of an oxide-covered circular Si wafer. What should the mask look like?

Solution

The mask must allow the oxide in the target region to be selectively removed by etching. Thus, after applying the photoresist the mask must be positioned so that the target region is protected from exposure to UV light. When the mask is removed and the photoresist is developed, the polymer over the target region will not have hardened and will be easily removed by the developing fluid. The oxide can then be etched away from this selected region in preparation for the doping step. Therefore, the mask should consist of an opaque square piece the size of the region to be doped.

16.9.4 Diffusion and Ion Implantation

In this section we discuss diffusion and ion implantation techniques for incorporating dopant atoms into a Si wafer. The diffusion process involves two stages, as shown in Figure 16.9–3c and d: first the dopant atoms are deposited onto the wafer surface; then they are diffused into the wafer. The dopants can be introduced onto the wafer surface by three methods: solid source, liquid source, or gas source.

The diffusion time and temperature, as well as the dopant concentration on the surface, must be controlled in order to obtain the desired dopant profile. The diffusion mechanism can be either substitutional (Al, Ga, B, P, In, As, Sb) or interstitial (Li, Cu, Fe, Au). Typical diffusion depths are on the order of a few microns.

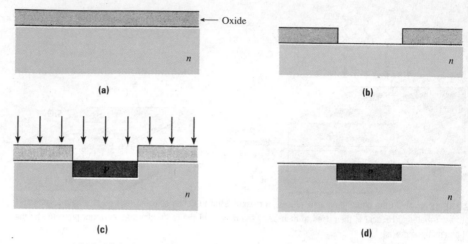

FIGURE 16.9–3   Basic fabrication steps in the silicon doping process: **(a)** oxide formation, **(b)** selective oxide removal, **(c)** deposition of dopant atoms on wafer, and **(d)** diffusion of dopant atoms into exposed regions of silicon.   *(Source: Serope Kalpakjian, Manufacturing Engineering and Technology, 2nd ed., pp. 1017–18, © 1989 by Addison-Wesley. Reprinted by permission of the publisher.)*

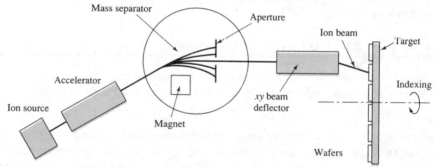

FIGURE 16.9–4   A schematic illustration of a typical ion implantation system.   *(Source: J. A. Schey, as found in Serope Kalpakjian, Manufacturing Engineering and Technology, 2nd ed., © 1989 by Addison-Wesley. Reprinted by permission of the publisher.)*

An alternative doping method is ion implantation (Figure 16.9–4). By manipulating the accelerating voltage, the average implantation depth can be precisely controlled. The dopant concentration can also be tightly controlled by monitoring the current during implantation. Additional advantages of the ion implantation technique include a higher purity of the dopant species and the ability to locate the maximum dopant concentration at a location other than the wafer surface. Furthermore, it is a low-temperature process, and it allows for the incorporation of a variety of impurity atoms (any that can be ionized). The disadvantages of the technique are its high cost and slow speed, and that defects are introduced into the Si crystal during dopant implantation. A postimplantation anneal (typically ~30 minutes at 800–1000°C) is required to "heal" the defects created during ion implantation.

16.9.5 Interconnection, Assembly, and Packaging

The final steps in the fabrication of an integrated circuit chip are:

1. Deposition of metal contacts between doped regions
2. Slicing the wafer into individual "dice"
3. Attachment of dice to a package
4. Attachment of wires from the leads on the package to the metal contacts on the dice
5. Sealing the package
6. Mounting the package, known as an integrated circuit (IC) chip, in electronic systems

Metallic interconnections provide a conductive path between the isolated electrically active regions of the Si wafer. These paths are created by the selective removal of the oxide and subsequent deposit of a metallic overlayer—often aluminum. The metal can be directly deposited by sputtering from a heated source.

After completion of several cycles of oxidation, masking, etching, doping, diffusion, and formation of metallic interconnections, the individual devices, or "dice," are separated and assembled in convenient packages. Since the assembly process is difficult to automate fully, it is often responsible for a substantial fraction of the cost of the finished device.

The dice are separated from one another by fracturing the wafer on lines scribed along specific crystallographic directions. In the next step, known as **die bonding,** either near-eutectic solders or polymers (typically epoxies or polyimides) mixed with silver to enhance conductivity are used to attach the die to the package. In general, eutectic solders (such as Si-Au alloy discussed in Section 16.7.1) are used with ceramic packages, while epoxies are used with plastic packages. Ceramic packages (typically aluminum oxide or occasionally silicon carbide and aluminum nitride) are more reliable and are, therefore, preferred for high-performance devices. The plastic packages are used when cost is paramount. These packages can be produced inexpensively by injection molding, as described earlier in this chapter. The package must perform several functions; it must protect against mechanical and environmental degradation, provide accessible electrical connections, dissipate heat, and lend itself to inspection and repair.

The electrical connections between the die and the package are formed by the wire-bonding technique in which a gold or aluminum wire is attached to the die bond pad and the package leads. The bonding of the Au wire to the pads and leads can be accomplished using a combination of temperature, pressure, and ultrasonic vibration. The resulting wire connections are shown in Figure 16.9–5. After wire bonding, the package is sealed. If it is a ceramic package, the seal can be hermetic.

The final design consideration is the method by which the electronic package will be mounted in an electrical system. As shown in Figure 16.9–6, chips can be attached to printed wire boards by either of two general methods: through hole mounting or surface mounting.

Since the cost of a wafer is determined by the number of processing steps, not by the size of wafer or the number of devices per wafer, there is a strong driving force to place more and more devices on each wafer. This requires the growth of larger-diameter wafers and a reduction in the line-width resolution of the lithography and diffusion techniques. It also requires more rapid heat removal because higher circuit concentration leads to more heat generation. These are some of the issues in the microelectronics industry that represent significant challenges to materials engineers.

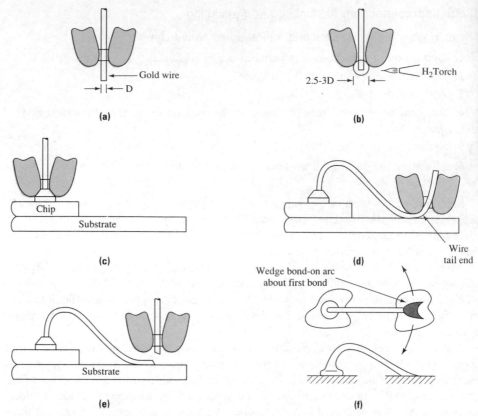

FIGURE 16.9–5   In the wire-bonding process a Au wire is used to make the electrical connections between the die bond pads and the package leads.   *(Source:* VLSI Technology, *S. M. Sze, ed., 1988, McGraw-Hill, Inc. Reproduced with permission of McGraw-Hill.)*

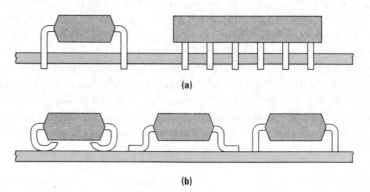

FIGURE 16.9–6   Examples of some of the methods by which the IC chips can be mounted in an electrical system: **(a)** hole mounting, or **(b)** surface mounting.

16.10 FIBER MANUFACTURING

As discussed in detail in Chapter 14, fibers are often used as reinforcement materials in composites because of their high strength and elastic modulus, or for optical transmission in fiber optics; however, most filamentary materials are used in textile applications.

Organic fibers are characterized by low mass density. Fibers are produced in continuous, chopped, or whisker forms with diameters ranging from 5 μm to 1 mm. The small diameters are necessary for minimizing imperfections, thereby allowing a high percentage of the material's theoretical strength to be achieved. Also, small diameters ensure good flexibility for winding on bobbins, and maintaining integrity through tight bends in use.

Fibers can occur naturally, or they can be man-made. Examples of natural fibers of commercial importance include those based on vegetative processes, such as wood fibers, cotton, flax, and hemp, as well as those based on animal or insect processes, such as silk from a silkworm or a spider. History is rich in references to various kinds of textile fibers. In this section we will focus on techniques for formation of man-made fibers such as polymer, oxide glass, crystalline ceramic, and other high-performance fibers for composite applications. Since wire drawing was described in Section 16.4.1, it will not be repeated here. Large-diameter metal reinforcements are used extensively in applications such as steel-belted tires and in reinforced concrete.

16.10.1 Melt Spinning

Melt spinning is an economical process for producing polymer and oxide glass fibers. Typical examples of melt-spun polymer fibers include polyester, nylon, and polyolefins. Since the covalent carbon-carbon bond is very strong, we should expect linear-chain polymers to exhibit high strength provided the molecular chain can be aligned in the direction of the fiber axis. These properties can be achieved using the melt-spinning process.

The back end of a melt-spinning system is usually an extruder. The melt then passes through a spinnerette with several circular holes. The viscous liquid is drawn down from perhaps 400 to 25 μm in diameter and the resulting highly oriented fiber is wound on a bobbin. An example of a melt-spinning apparatus used to form continuous fibers is shown in Figure 16.10–1.

The most common technique for forming oxide glass fibers is also melt spinning. Although the temperatures used are higher, the technique is similar to that described for polymers, with one notable exception—an extruder is not required. The hot viscous melt is simply pressed through the holes in the spinnerette.

16.10.2 Solution Spinning

Some polymers such as cellulose and polyacrylonitrile cannot be heated to the melt because they degrade prior to melting. These polymers can be spun into fibers using solution techniques. In dry spinning, the polymer is dissolved in a suitable volatile solvent, and the solution, known as the dope, is extruded through the spinnerette holes, into air. The evaporation of the solvent facilitates precipitation of polymer in fiber form. Dry spinning contrasts with wet spinning, in which the dope is extruded into a nonsolvent where precipitation is brought about by the inward diffusion of the nonsolvent. After fiber formation, solution-spun fibers, like melt-spun fibers, are drawn to impart molecular orientation. Typical wet- and dry-spinning apparatuses are shown in Figure 16.10–2.

Variations of the melt- and solution-spinning techniques are implemented as required. Liquid crystalline materials, or rodlike polymers that are characterized by molecular orientation in the melt or solution, are processed in a manner that preserves the molecular alignment and directs it along the fiber axis. Fibers produced using liquid crystalline polymer, such as Kevlar or Vectran, have very high strength and modulus. Spectra

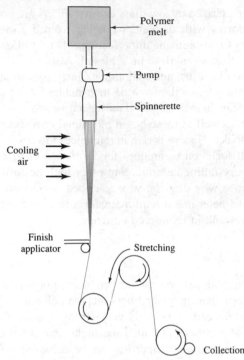

FIGURE 16.10–1 Melt-spinning apparatus: Polymer is melted and forced with pressure through several holes in the spinnerette. Upon extrusion, the filaments are stretched and cooled by air. A finish is applied to the fibers, and they are gathered and wound on a spool or bobbin.

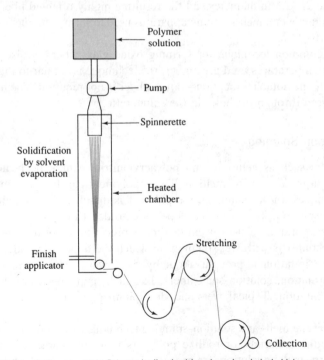

FIGURE 16.10–2 Dry-spinning apparatus. Polymer is dissolved in solvent in relatively high concentrations (greater than 10%) to form a viscous dope. The dope is forced through a spinnerette into either heated air (dry) or a coagulation bath (wet), where the fiber is stretched and the solvent is removed from the fiber, causing precipitation. A finish or lubricant is applied to the gathered bundle or yarn, and the product is wound onto a bobbin or spool.

olyethylene fiber is dry-spun from a gel consisting of only a few percent ultrahigh–
molecular weight polymer (millions of g/mol) in solvent. The extremely high molecular
orientation is produced by collapse of the gel by removal of solvent and by repeated
drawing steps. Like Kevlar and Vectran, Spectra is pound for pound stronger and stiffer
than steel.

16.10.3 Controlled Pyrolysis

Pyrolysis of organic and organometallic precursors (raw materials) has become an important commercial process for producing ceramic and carbon fibers. In this technique a
polymer precursor fiber containing the elements of interest is formed using conventional
techniques. The fiber is then pyrolized (heated to produce a chemical change) without
melting. An example of fibers that can be made using pyrolysis are carbon fibers made
from acrylic precursors. In the pyrolysis steps all the hydrogen, oxygen, and nitrogen
atoms are driven off while the fiber retains its integrity, leaving behind only carbon.
Figure 16.10–3 outlines the various steps required to convert an acrylic fiber to a high-
strength or high-modulus carbon fiber.

Silicon carbide fibers are also made in this fashion. An organosilane polymer fiber is
heated under controlled conditions to remove all atoms except Si and C, as shown in
Figure 16.10–4. Both carbon and SiC fibers are typically used in high-performance
composite materials.

16.10.4 Vapor-Phase Processes

Only a limited number of fibers are made using chemical vapor deposition (CVD) because
of the high cost of the process (upwards of $600/kg). Boron fibers are the most well-
known ones made using CVD. In this process a small resistance-heated tungsten wire is
fed into a chamber containing BCl_3 and H_2 gases. The BCl_3 gas decomposes into elemental boron, which is deposited onto the tungsten substrate, and HCl, which is drained from
the chamber, according to the reaction:

$$2BCl_3 + 3H_2 \rightarrow 2B + 6HCl \qquad (16.10-1)$$

The large fibers that result have high modulus, fair strength, and low density. Silicon
carbide fibers have also been made using CVD.

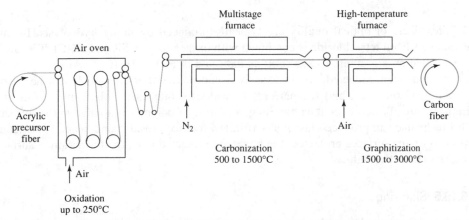

FIGURE 16.10–3 Schematic process for conversion of acrylic to carbon fiber. High-purity, acid-containing acrylic
fiber yarn proceeds under tension into a circulating air oven at about 250°C, where the fiber oxidizes without melting.
The black stabilized material is then carbonized at about 1500°C in an inert atmosphere to remove all atoms except
carbon. Finally, the fiber is heated to temperatures as high as 3000°C to develop high modulus.

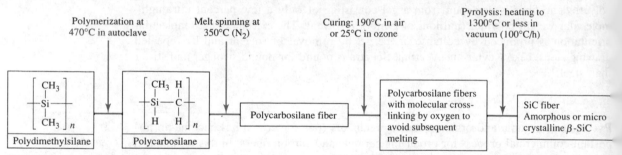

FIGURE 16.10–4 Silicon carbide fiber formation by pyrolysis of organosilane. Polycarbosilane polymer is melt-spun at about 350°C in an inert atmosphere followed by air curing or oxidation in an atmosphere containing a high concentration of reactive oxygen. The crosslinked or stabilized fibers are pyrolyzed at temperatures up to 1300°C in vacuum to yield amorphous or microcrystalline SiC fibers.   *(Source: K. K. Chawla. Composites Material Science and Engineering, 1987, Springer-Verlag New York. Reprinted with permission of Springer-Verlag New York Publishers.)*

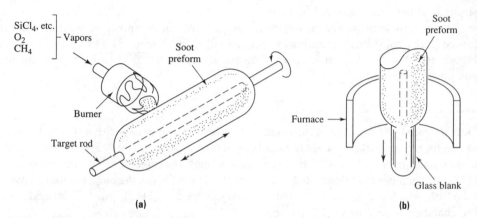

FIGURE 16.10–5 Formation of silica performs for optical waveguides. Silicon tetrachloride is combined with hydrogen in a flame to produce silica, which deposits onto a graphite target rod as a sooty preform. The sooty preform is densified in a second step and eventually drawn into a fiber.   *(Source: Adapted from M. G. Blankenship and C. W. Deneka. "The Outside Vapor Deposition Method of Fabricating Optical Waveguide Fibers," IEEE Journal of Quantum Electronics, 1982, IEEE.)*

Silica fibers of optical quality are typically produced by **flame hydrolysis.** In this technique silicon tetrachloride is combined with oxygen to give SiO_2 and HCl. The silica is deposited on a graphite rod as soot, as indicated in Figure 16.10–5. After a quantity of soot has been collected, the hydrolysis is ended, the soot collapsed with heat, the graphite rod removed, and the perform is melt-spun into single filaments. Before the filaments are allowed to contact anything, a coating of polymer is applied to the surface. This technique can produce glass fibers with low hydroxyl residue and hence low attenuation. With the surfaces protected, the fibers are protected from the formation of surface cracks that scatter light.

16.10.5 Sintering

It is possible to grind virtually any ceramic or metallic material into a fine powder; wet- or dry-spin the powder as a bound slurry into the form of a fiber, which can be dried and subsequently heated to remove the binder; and then sinter the particles. Alumina (Al_2O_3)

bers have been made using this technique; however, such fibers have comparatively low strength because the resulting coarse grain structure contains large defects.

....

EXAMPLE 16.10–1

Use your knowledge of the effect of grain size on strength to show why sintered fibers have poor strength.

Solution

The Griffith equation shows that the strength σ_f of a brittle material decreases with increasing defect size. In this case, the relevant "defect" is the grain boundary itself so the flaw size is in effect the grain size d. Thus, '

$$\sigma_f \propto \frac{1}{\sqrt{d}}$$

The strength of a fiber consisting of grains on the order of a nanometer is nearly two orders of magnitude greater than that of fibers with grain size of 10 μm. Since sintered fibers generally have grain sizes in the latter category, they are comparatively weak.

....

16.10.6 Chemical Reaction

A great deal of creativity has been applied to the development of refractory ceramic fibers. CVD is expensive, pyrolysis seems to be limited to a few compositions, and sintering has problems with regard to good strength. In this section we describe just a few of the numerous techniques developed to produce ceramic fibers.

Borate (B_2O_3) glass fibers can be readily melt-spun. The resulting fibers can be subjected to ammonia to convert them to boron nitride, a refractory material with excellent mechanical and thermal properties:

$$B_2O_3 + 2NH_3 \rightarrow 2BN + 3H_2O \qquad (16.10-2)$$

Alumina fibers (Al_2O_3) can also be nitrided. The product is aluminum nitride (AlN), a material with an excellent combination of thermal and dielectric properties:

$$Al_2O_3 + 2NH_3 \rightarrow 2AlN + 3H_2O \qquad (16.10-3)$$

Fibers can also be made using the **sol-gel technique.** Appropriate metal alkoxides, which are simple liquids and readily available, are mixed with water in an appropriate cosolvent, usually alcohol. Under acid or basic conditions, the solution reacts by condensation polymerization, producing oxides that contain some residual organic groups. The material can be spun into fiber when the reaction mixture reaches an appropriate viscosity. The pure oxide is obtained by heating to drive off the remaining organic molecules. High-purity silica has been made using this technique:

$$Si(OCH_3)_4 + 4H_2O \rightarrow Si(OH)_4 + 4CH_3OH \rightarrow$$
$$SiO_2 + 2H_2O + 4CH_3OH \qquad (16.10-4)$$

In a similar process, a stable sol can be destabilized and allowed to undergo condensation polymerization to form a polymer preceramic containing appropriate metal and nonmetal atoms, such as Al and O. The low–molecular-weight polymer can be dry-spun and thermally treated to form an inorganic fibrous material. This process is a hybrid of pyrolysis and sol-gel techniques. Fibers made in this fashion have superior strength compared with those produced using the sintering technique.

16.11 COMPOSITE-MANUFACTURING PROCESSES

The important composite materials, discussed in Chapter 14, can be classified as polymer matrix, metal-matrix, and ceramic-matrix composites. The manufacturing processes are different for each class of materials and are also quite different from those used for the monolithic materials discussed in previous sections. The cost of these processes is generally high, which drives up the expense of composite components above that of the material costs.

16.11.1 Polymer-Matrix Composites (PMCs)

Hand layup is the simplest technique for making PMC materials. Continuous fibers are laid in a mold, and the resin precursor is either brushed or sprayed on. If chopped fibers are used, a mixture of resin and fibers is sprayed into the mold. The composite is then cured or crosslinked at the appropriate temperature. This technique is most commonly used for composite materials utilizing glass fibers in thermosetting resins, such as shower enclosures.

Hand layup is also used for making laminated composite components. In this technique, prepregs (sheets of partially cured resins with fibers oriented along a single direction) are stacked on top of each other in a mold in any desired fiber orientation and then pressed together in an autoclave to form the composite part. Figure 16.11–1 shows the process of stacking prepregs to make a laminate. Although labor-intensive, this technique is versatile and can be used to make hybrid composites containing more than one type of fiber.

Filament winding is a technique used for making cylindrical objects. Continuous tow or roving (a bundle of fibers) is passed through a resin bath, which impregnates the fiber with the resin. The fibers are then wound on a mandrel with successive layers applied at specific angles until the desired material thickness is attained. The process is schematically shown in Figure 16.11–2. Subsequently, the composite is cured and the mandrel removed.

Pultrusion is an effective way of making long lengths of continuous fiber composites with constant cross section, such as I-beams (see Figure 16.11–3). In this technique, preimpregnated fibers of oxide glass, Kevlar®, or carbon are pulled slowly through a heated die to make the composite. Resins used include polyesters and some epoxies. This technique is similar to the extrusion process described in Section 16.4.1. Pultrusion and filament winding are examples of automated processes that are capable of reliably delivMering high-performance products.

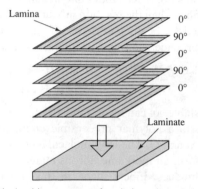

FIGURE 16.11–1 Schematic of the hand layup process of producing polymer matrix laminated composites.

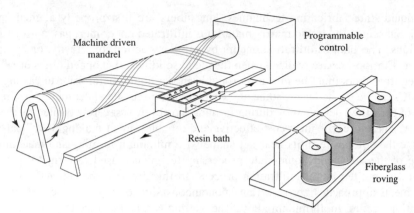

FIGURE 16.11–2   Schematic of the filament-winding method of manufacturing cylindrical parts from polymer-matrix composites.   *(Source: B. D. Agarwal and L. T. Browtman,* Analysis and Performance of Fiber Composites, *2nd ed., Copyright © 1990 by John Wiley & Sons, Inc. Reprinted by permission of John Wiley & Sons, Inc.)*

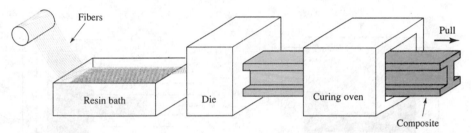

FIGURE 16.11–3   The pultrusion process for fabricating constant–cross section fiber-reinforced composites. *(Source: Donald R. Askeland,* The Science and Engineering of Materials, *3rd ed., 1994, PWS. Reprinted by permission of the publisher.)*

Often fibers are given a surface treatment prior to incorporation into the matrix. For example, silane coupling agents were developed to ensure good wetting between glass and resin and to help form chemical bonds between fiber and matrix.

16.11.2 Metal-Matrix Composites (MMCs)

The most common matrix materials used for making MMCs are aluminum and titanium alloys. Fibers used in MMCs include boron, carbon, and SiC in continuous as well as whisker and particulate forms. Boron and carbon fibers require surface treatments prior to incorporation in MMCs. The surface treatment is necessary to avoid reactions at the fiber-matrix interface at high temperatures during processing and to promote good wetting along interfaces. Boron fibers are coated with a thin layer of silicon carbide or boron carbide by the CVD process. A titanium boride (TiB_2) coating is similarly deployed on carbon fibers.

In the solid-state fabrication technique of MMCs, alternate layers of properly spaced boron fibers and aluminum foil are stacked. Resin-based binders are used to keep the fibers in place. Composites are then made by applying pressure and temperature such that the aluminum flows around the fibers and bonds with the next aluminum layer.

Powder metallurgy techniques described in Section 16.5.1 are used to make silicon carbide whisker–reinforced MMCs. Plasma spray, which is a chemical or physical vapor deposition of matrix material onto properly laid-up fibers, is another solid-state processing technique for making MMCs.

In liquid-state fabrication techniques, long fibers are first properly aligned and arranged. Subsequently, liquid matrix material is infiltrated under inert gas, air, or vacuum conditions. The liquid infiltration occurs by capillary action, by gravity, or by applied pressure. Composites are realized when the metal solidifies. If chopped fibers or whiskers are used, the fibers must be mixed thoroughly in the liquid metal prior to casting.

MMCs can be formed using some of the techniques discussed under metal forming. The extent of permitted deformation during forming, however, is severely limited by their low ductility. Therefore, the most cost-effective and reliable way of making MMC parts is to fabricate the composite in its near net shape with minimum forming and machining.

One of the advanced liquid-state processing techniques used to form MMCs is the pressureless molten metal infiltration process. In this process, the contact angle of the liquid metal approaches zero, so that spontaneous spreading and, hence, spontaneous infiltration occurs. In aluminum alloys, the wetting can be enhanced by adding magnesium or lithium, which can also provide solid solution strengthening in aluminum. Figure 16.11–4a shows a schematic of this process for making a near-net-shape electronic

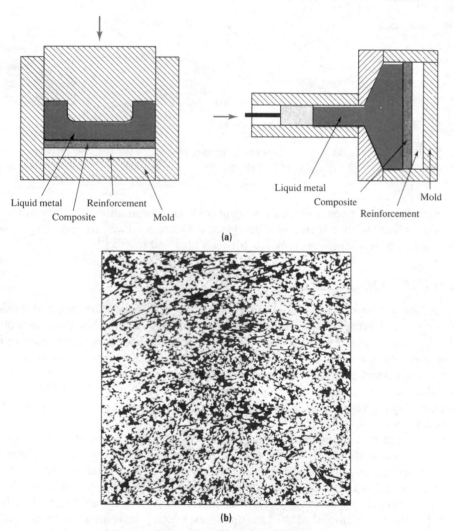

(a)

(b)

FIGURE 16.11–4 **(a)** Schematic of the liquid infiltration process for making metal-matrix composite components (an electronic package casing). **(b)** Microstructure of the resultant composite from the PRIMEX process of the Lanxide Corp. *(Source:* **(a)** *A. Mortensen, V. J. Michaud, and M. C. Flemings,* Journal of Metals, *January 1993, pp. 36–43.)*

package casing composed of SiC particulate in an aluminum matrix. A preform of loosely packed reinforcement material is first made. Then the metal infiltration is carried out under inert environment conditions to complete the process. Figure 16.11–4b shows a microstructure of the resultant composite, which displays excellent metal infiltration.

16.11.3 Ceramic-Matrix Composites (CMCs)

Since ceramic matrices are typically stiff but brittle, the purpose of the fibers is not to increase modulus, but to improve composite toughness. Fiber pullout during fracture requires work and is therefore a desirable feature in these materials. (This mechanism for toughening composites was discussed in Section 14.4.3.) It is crucial to achieve the optimum interface strength between the fiber and matrix in this class of materials.

Another consideration in processing CMCs is that the fabrication techniques require exposure to high temperatures during hot pressing, followed by cooling. Thus, mismatch between the thermal expansion coefficient of the fiber and the matrix must be minimized. Common fibers include silicon carbide in whisker, chopped, or continuous form, and alumina in chopped or continuous forms. Common matrix materials include magnesium oxide, silicon nitride, silicon carbide, alumina, borosilicate glass, and lithia-alumino-silicate glass ceramic.

The most commonly used CMC is concrete. The tensile strength of concrete is orders of magnitude lower than its compressive strength. Reinforced concrete uses steel wires and rods as reinforcements, which add considerably to its tensile strength. Reinforced concrete is made by first laying steel rods or wire meshes in the form of a loosely tied frame. A mixture of concrete consisting of cement, sand, and an aggregate mixed in water is poured on the frame. The cement hardens in a series of complex chemical reactions that involve moisture absorption to form a gel, which crystallizes with time, resulting in a reinforced-concrete product. The composition of the mixture and the water content is extremely important because premature drying can prevent the hydration reaction from forming a gel.

The processing of the more advanced CMC materials is done in two steps. The first involves embedding the fibers in an unconsolidated matrix; the second is the matrix consolidation. The most common process for fiber incorporation is **slurry infiltration,** in which a fiber tow is passed through a slurry of the matrix material. The slurry contains the ceramic powder, a carrier liquid, and often an organic binder. The infiltrated fiber tow is wound on a drum and dried. The coated fibers are then chopped, stacked, and hot-pressed together to form a consolidated matrix. The hot-pressing process is no different from the hot pressing of metal powders discussed as a part of the powder metallurgy technique.

Another technique used for embedding fibers in an unconsolidated ceramic matrix is cold pressing of fiber and ceramic powder followed by sintering. Densification by hot pressing follows.

Carbon fiber CMC materials have also been made by first hot-pressing carbon fibers in a glass matrix followed by a high-temperature treatment to transform the glass matrix into a ceramic. The crystallization of glass is called devitrification, as explained in Section 16.4.3.

SUMMARY

Engineering materials can be processed to produce components of various shapes and sizes with the end product possessing desirable microstructures and physical and mechanical properties. The atomic structure and bonding dictate the processing routes that are available for a given material. For example, ceramic materials, unlike metals and polymers, cannot in practice be cast from the liquid state because of their high melting or decomposition temperatures. Ceramics cannot be shaped plastically because of their intrinsically brittle character. However, they can be easily ground into fine powders enabling the use of powder-processing techniques. In fact, an efficient method of producing fine powder from ductile metals for powder metallurgy processing is to first produce a powder of its oxide and then reduce it to obtain the metal powder.

Polymers can in general be rapidly processed to net or near net shape using low-temperature melt processing, such as injection or compression molding, blow molding, or fiber or film techniques. Sometimes polymers need to be processed from concentrated solutions using techniques similar to those developed for metals.

The basic techniques used for processing materials may be divided into the following categories: (1) casting, (2) forming, (3) powder processing, and (4) machining. Not all processes are equally applicable to all material types. The important results of processing techniques are summarized below:

1. Solidification microstructures are affected significantly by the cooling rates, which depend on the mold geometry and the size of the part. Cooling rates during solidification do not significantly affect the properties of noncrystalline materials.

2. Metals can be shaped by a variety of plastic deformation processes such as rolling, forging, extrusion and wire-drawing, sheet metal forming, deep drawing, and so on. Depending on the temperatures used during the forming operations, large microstructural changes can occur, which can affect the properties of the resulting parts. Plastic deformation during forming can have a significant effect on polymer properties due to the alignment of the molecular chains in the direction of deformation.

3. Metals and ceramics lend themselves to powder processing. Parts with intricate shapes and close dimensional tolerances can be produced via powder-processing routes. The primary concerns about powder-processed parts are the defects that remain and their effect on mechanical strength.

Joining techniques include: (1) welding, brazing, and soldering; (2) adhesive bonding; (3) diffusion bonding; and (4) mechanical joining. Some general observations on joining techniques are:

1. The materials being joined dictate to a large extent what joining approaches may be used.

2. Joints are almost always weaker than the materials being joined.

3. Large residual stresses can develop at weldments, which must be relieved by postweld heat treatment. The grain size of the metal near the weld increases due to exposure to high temperatures, and contributes to joint weakness. This is true for both metals and semicrystalline polymers.

4. Brazing is used for joining metal to metal and metal to ceramic.

5. Intimate contact is necessary to promote good adhesive bonding. This is achieved by starting with clean surfaces and choosing an adhesive that wets the solid surfaces.

6. Diffusion bonding is an effective technique for joining dissimilar materials such as metal to ceramic.

Coatings and surface treatments include old processes such as painting and hot dipping, as well as new methods such as ion implantation. The particular methods used depend on the material, the end application, and the cost.

The steps required to produce integrated circuits include single-crystal growth, deposition of oxide coatings, lithography and etching, diffusion and ion implantation, metallization, and packaging. Each of these steps involves specific technologies.

There are several approaches for producing fibers for various applications, including their use in manufacturing composite materials. The end properties of fibers are dependent on the processing method. The methods include melt-spinning for producing polymer and oxide glass fibers, controlled pyrolysis for carbon fibers, physical and chemical vapor deposition used for producing boron and ceramic fibers, as well as sintering and sol-gel techniques for producing ceramic fibers.

Composite-manufacturing methods are significantly different from those for producing parts from conventional materials. The primary difference is that the composite material and the component are produced simultaneously. Since these materials cannot be machined extensively and the fiber arrangement depends on the geometry of the component, the composite is usually manufactured in near net shape.

KEY TERMS

adhesive bonding	die bonding	heat affected zone (HAZ)	rolling
anodizing	diffusion bonding	hot dipping	shot peening
blow molding	electrochemical plating	hot isostatic pressing	sintering
brazing	electrostatic painting	injection molding	slip casting
case hardening	epitaxial growth	ion implantation	slurry infiltration
casting	explosive hardening	machining	soldering
chemical vapor deposition (CVD)	extrusion	melt spinning	sol-gel technique
compression molding	filament winding	photolithography	sputtering
Czochralski method	flame hydrolysis	physical vapor deposition (PVD)	welding
deep drawing	forging	powder metallurgy	wire drawing
dendrite	green compact	pultrusion	
devitrification	hand layup	pyrolysis	
	hard facing		

HOMEWORK PROBLEMS

1. Explain how the concept of economies of scale influences the selection of a materials-processing method.

2. Explain how the cost of the raw material influences the selection of a materials-processing method.

3. Two of your colleagues are having an argument. One claims that microstructure determines the processing methods that can be used on a material while the other claims that the choice of processing method determines the microstructure of the material. Which colleague would you side with? Why?

4. Processing affects microstructure. Consider the following pairs of products and describe the differences in chemical and physical properties:
   a. Spectra polyethylene fiber and a polyethylene sandwich bag
   b. Ferrous Metglas and ordinary steel of a similar composition
   c. Carbon fiber and pencil graphite

SECTION 16.2
Process Selection Criteria and Interrelationship among Structure, Processing, and Properties

5. A sculptor who wants to make a one-of-a-kind bronze statue is looking for a recommendation for a suitable casting method to get an approximate shape from which she will produce the finished statue. What do you recommend, and what justifies your choice? How would your recommendation change if she wants to mass-produce the statue?

6. An Al-Si alloy containing 11.3 at. % Si (eutectic composition) is used to produce sand castings. Explain the nature of the dendritic structure you would expect in the casting.

7. Two cast rods, one with a circular cross section and the other with an elliptical cross section, have the same length and cross-sectional areas. The major axis of the elliptical rod is twice as long as the minor axis. Derive the ratio for the solidification times needed for the two castings.

8. Consider the casting of a polycarbonate capacitor film: PC is dissolved in solvent, the dope is poured into a flat open mold, a "doctor's knife" is used to ensure a uniform height (or thickness) of dope in the mold, and the solvent is removed by evaporation. The evaporation time is 4 hours. The procedure is repeated, but the doctor's knife is set for half the original thickness so that a film one-half the original thickness will be produced. How long will it take to evaporate the solvent? Note that the volume is one-half that of the original and the surface area is unchanged.

9. What are porosity and shrinkage defects in metal castings? Briefly discuss methods of reducing these defects.

10. Why is wet crushing of powders preferred over dry crushing?

11. Describe the steps in making a flower vase from alumina.

12. Should one be concerned about the solidification structure during compression molding of (a) thermoplastics and (b) thermosetting plastics? Explain your answer.

13. How is a PET beverage bottle made? Compare this with how an oxide glass beverage bottle is made.

14. Briefly describe how the side of a steel can is made. Consider two classes of cans: one has a weld ring at both top and bottom; the other has a weld ring only at the top.

15. The constants in the Arrhenius equation (16.4–1) for an aluminum alloy are $C_0 = -50$ and $C_1 = 30,000$ K. Calculate the maximum speed of wire drawing if complete recrystallization must be achieved before the wire is wrapped on the spool. The distance between the die and the spool is 100 m, and the drawing temperature is 300°C.

16. Discuss the similarities between the rolling and extrusion processes for metals.

17. Explain how ductility is measured in a tensile test and how it may relate to the limiting draw ratio in a deep-drawing operation.

18. Aluminum beverage containers are made by a process that consists of first deep-drawing a cup and then reducing the cup wall thickness by a process known as wall ironing. During this process, the thickness of the cup wall can be reduced considerably and the height of the cup increased correspondingly. Ironing is accomplished by placing the cup on a mandrel and then forcing the mandrel through ring-shaped dies with mandrel-die clearances much less than the wall thickness of the cup. Thus, the height of the cup can be doubled or tripled during this operation. In Example 16.4–1, if the cup wall is ironed to a thickness of 0.1 mm, estimate the blank diameter, the materials savings per container, and the number of deep-drawing steps. The size of the finished container and the LDR is the same as given in Example 16.4–1.

19. What are rolling textures? Describe how textures may affect mechanical properties of flat-rolled sheet.

20. Explain why the blow-molding process is carried out vertically.

21. Describe the process of glass tube and glass rod manufacturing. How does this differ from the glass sheet manufacturing process?

22. Explain how temperature influences the ability of noncrystalline materials to be processed using traditional forming operations.

23. Describe and explain the differences in the microstructures of cold- and hot-rolled metals.

24. Explain why hot rolling offers an engineer more control over the properties of the final metal product than does a similar cold-rolling process.

25. Can cold rolling be done without limit? What about wire drawing?

26. Show with a sketch what happens to two spheres in contact as they sinter.

27. Give reasons why metal powders are easier to produce by reducing powders of their oxide.

28. Why is inert gas used as a medium for pressurizing the green compacts during the hot isostatic pressing (HIP) process?

29. Why is powder metallurgy a good process for fabricating internally lubricated bearings?

30. Why are sintering and hot isostatic pressing necessary for structural parts made from the powder metallurgy process?

31. What is the consequence of igniting a spark in a container filled with fine metallic powder? Explain your answer.

32. What are the roles of stearic acid and wax in powder processing of ceramics?

33. List and briefly describe the five steps involved in the powder metallurgy process.

34. The highest volume fraction of space that can be filled using single-sized spheres is 74%. Calculate the maximum volume fraction of space that can be filled using two different-size spheres (with $r_2 \ll r_1$).

SECTION 16.5
Powder Processing

35. Design a simple test that you could perform to rank the machinability of three steels. Explain your answer.

36. The energy required to perform a machining operation can be grouped into three general categories. List these three categories and describe how the use of a lubricant during the machining process influences the amount of energy required.

37. What are free-cutting brasses, and what is the reason for their good machinability?

38. Explain why metals generally have higher machinability than either ceramics or polymers.

39. Suppose you must design a process for fabricating metal parts that must undergo both a machining operation and an age-hardening treatment. Which of these two operations would you perform first? Why?

40. Consider once again the materials used in bicycle frame construction: high-strength steel, aluminum, titanium alloys, and carbon fiber–reinforced epoxy. What are the various techniques that can be used to join the tubes together to form a bicycle frame? Remember that a modern racing bicycle frame weighs less than 3 pounds!

SECTION 16.6
Machining

41. What are the primary differences between welding, brazing, and soldering?

42. Why are fluxes used during welding, brazing, or soldering?

43. What is a heat-affected zone in connection with welding?

44. Explain what is meant by the term *postweld heat treatment*. What is the purpose of such a treatment?

45. What advantages do electron beam welding and laser welding offer over conventional welding procedures?

46. When designing an adhesive joint, it is important to consider several factors including: (*a*) joint geometry, (*b*) cleanliness of surfaces, and (*c*) magnitude of the gap between the surfaces to be joined together. Briefly explain why each of these factors is important to the success of the adhesive joint.

SECTION 16.7
Joining Processes

47. Suppose you must join two dissimilar ceramic materials and it is not possible to use an adhesive joint. Suggest an appropriate joining method and justify your choice.

48. Are mechanical joints more common in metals, ceramics, polymers, or composites? Why?

SECTION 16.8
Surface Coatings and Treatments

49. What are some of the best ways to keep cars from rusting?

50. What are the differences between the chemical vapor deposition (CVD) and the physical vapor deposition (PVD) processes?

51. What types of materials are used for hard facing, and how are they deposited?

52. What advantages does electrostatic painting offer over conventional painting techniques?

53. The three general classes of surface treatments are compositional modification, cold working, and heat treating. Describe how each method can be used to increase the surface hardness of a steel component.

54. Describe each of the processes listed below:
    a. Ion implantation
    b. Shot peening
    c. Nitriding
    d. Explosive hardening

SECTION 16.9
Single-Crystal and Semiconductor Processing

55. Describe the various steps in producing a Si chip.

56. List three different applications that favor the use of single-crystal materials. For each application explain what advantages the single crystal offers over competing polycrystalline materials.

57. Why are single crystals of silicon generally grown at a temperature just below the melting temperature? What is a seed crystal, and why is it used in the Czochralski crystal growth method?

58. One of the primary reasons that Si is the dominant material in the microelectronics industry is the high quality of its native oxide SiO_2.
    a. Describe the advantageous characteristics of SiO_2.
    b. How would you determine whether to use the "wet" or "dry" procedure for growing SiO_2?

59. Extrinsic semiconductors can be fabricated using either the thermal diffusion process or the ion implantation process. Describe the relative strengths and weaknesses of each doping technique.

60. Explain why decreasing the line-width resolution would lead to substantial reductions in the cost of a semiconductor chip.

61. What limits the line width, now 0.3 μm, on a chip? What techniques might reduce the line width to about 25 angstroms?

62. Explain why the cost of a semiconductor chip would change if the diameter of a single-crystal semiconductor wafer could be increased.

63. What factors would influence your choice of a ceramic or polymer package for a semiconductor chip?

64. Given that one of the common failure mechanisms of an integrated circuit is failure of the die bond resulting from temperature fluctuations, what properties or characteristics are important in the selection of a die-bonding material?

SECTION 16.10
Fiber Manufacturing

65. Calculate the area reduction that occurs when the fiber diameter changes from 400 to 25 μm during extension. What occurs during fiber formation that allows the huge fiber diameter reduction? (Don't forget, plastic deformation cannot go on without limit.)

66. In general, polymers do not have high elastic moduli. Explain how some polymers can be processed into the fiber form so that they have a high elastic modulus.

67. What explains the low strength of sintered ceramic fibers?

68. How would you decide whether to use a melt-spinning or a solution-spinning operation for a particular polymer fiber?

69. Explain why the only two types of fiber that are commonly made by the pyrolysis technique are carbon-based and silicon-based polymer fibers.

70. The Griffith equation [$\sigma = (2E\gamma/\pi c)^{0.5}$, where E is Young's modulus, γ is the surface energy, and c is the crack length] is critically important in materials science. Based on Griffith's equation, how do you anticipate fiber strength will vary with fiber diameter?

71. Suppose you are in a contest to design a light glider plane and you have chosen to make the wings out of a polymer composite. What is the most suitable method for manufacturing the wing?

SECTION 16.11
Composite-Manufacturing
Processes

72. Interface properties between the matrix and the fiber play an important role in ceramic-matrix composites. Is it good to have (*a*) a weak interface, (*b*) a strong interface, or (*c*) one in between? Explain your answer.

73. What is a green body in connection with CMC manufacturing?

74. Describe a method used for manufacturing metal-matrix composites.

75. Describe a composite-manufacturing process that could be used to fabricate a pressure vessel that will contain high-pressure gases.

76. Why is the coefficient of thermal expansion generally a more important consideration in CMCs than in MMCs or PMCs?

C H A P T E R   17

MATERIALS AND ENGINEERING DESIGN

MATERIALS IN ACTION Case Studies

We will see in this chapter that "real-world" problems involve a combination of different materials, phenom-ena, economic considerations, and other factors that must be properly integrated in order to come up with a viable solution.

A good example of the complexity of actual problems and a "systems" approach to solving them is found in a problem that a department store in Cincinnati was having with its parking garage. In order to keep customers, downtown stores must have readily accessible parking, or business will be lost to suburban malls. However, a major department store was forced to close its parking garage because large "chunks" of concrete were coming out of the surface in virtually all areas of the parking deck. Much business was being lost and a fix was needed quickly.

A search of the construction records showed that the floors were poured with reinforced concrete, which was expected for this application. It was also noted that Cincinnati is one of the largest cities in the United State to be located on a major arterial river (the Ohio River) that undergoes frequent freeze/thaw cycles in the winter. As a result of the city's relatively mild climate, use of salts is effective in keeping roads ice-free. Furthermore, salt is economically viable, since it can be brought in easily and inexpensively via the Ohio River. Consequently it is widely used. Thus, during the winter cars would come into the parking deck dripping salt-bearing water. It was hypothesized that the salt-bearing water would seep into the concrete (which is relatively porous) and react with the steel reinforcing bar to form an oxide. As we have seen, iron oxides have a larger volume than the metal from which they form. Consequently, the oxide formation places the concrete under tension. While concrete is superb for compression applications, it is notoriously poor in tension. Thus, the oxides caused cracks. Since the reinforcing bar (known as rebar) was in the form of a crisscrossing network, the cracks intersected, and loose chunks formed. Examination of the rebar showed that it was indeed oxidized as hypothesized and the thickness of the oxide was sufficient to cause stresses that would exceed the tensile fracture stress of the concrete.

Having identified the mechanism leading to the problem, it was relatively easy to implement a solution. The solution involved connecting the embedded reinforcing bar to the negative terminal in a dc circuit. This is an example of cathodic protection and is represented by the equation:

$$M^{n+} + ne^- \rightarrow M$$

This means that the oxidation reaction can't occur in the vicinity of the rebar and there should thus be no stresses generated by the formation of the oxides. Once the cathodic protection scheme was put in place, damage to the parking deck was eliminated.

In this problem, we see that knowledge of economics (salt use), phase transformations (oxidation of rebar), electrochemistry (oxidation reaction, cathodic protection), macrostructure (porosity of the concrete), and mechanical properties (behavior of concrete under tension) was necessary to develop a solution. In this chapter we shall see many such examples where a broadly based approach is needed.

17.1 INTRODUCTION

This chapter is concerned with applying the concepts discussed in the previous chapters to engineering design. We begin by discussing the role of a materials engineer on a design team, which usually consists of members from other engineering disciplines such as mechanical, civil, environmental, chemical, electrical, aerospace, nuclear, textile, or architectural engineering. In several ways, materials engineering acts as a common thread among other engineering disciplines.

If we define engineering design as the process of creating a new part or system, the role of the materials engineer can be easily understood by asking the following question: From what material is the part or the machinery going to be made? To arrive at an answer, a series of related questions must first be answered. Examples include:

1. What are the performance requirements (electrical, mechanical, chemical, thermal) of the part or system?
2. How will it be produced?
3. How long must it last in service?
4. What is the acceptable level of production cost?
5. How frequently will it have to be inspected during service?
6. How frequently will it require repair or maintenance?
7. How will it be disposed of at the end of service?
8. Last, but certainly not least, what are the consequences of an in-service failure?

Thus, materials engineers are directly involved with issues related to product durability, cost, reliability, liability, and environmental impact as a part of material and process selection.

The cost of design and manufacture of transportation equipment such as aircraft increases with increased demand for fuel efficiency while the in-service costs decrease. The cost of manufacture includes the cost of using high-technology materials such as composites, which are needed to boost efficiency by saving weight without compromising payload. The *total* cost is minimized at a certain level of efficiency, which depends on the expected life of the aircraft. If the expected life is longer, a higher front-end cost can be justified. We can also see that if the aircraft or any other equipment can be safely used beyond its projected design life, the economics of continuing to use it are usually very favorable because the initial investment has already been recovered. In finance terms, the costs have been amortized. However, the higher inspection and maintenance costs incurred to offset the added risk of using old equipment must not be excessive. This operating philosophy, called the retirement-for-cause approach, is commonly used for aircraft but is rapidly gaining acceptance in other industries such as in fossil and nuclear power generation and oil refineries. In general this approach is attractive for industries utilizing expensive machinery that cannot be shut down simply because the originally predicted life has been exhausted. Materials engineers play a significant role in plant, or equipment, life extension decisions.

This chapter discusses the analysis methods and the approaches available to systematically arrive at an optimum material choice. The final choice depends on several economic and technical factors. Examples of aids available to engineers for materials and process selection include computerized material databases, computerized material and process specifications, and computer programs that allow quick estimation of material and manufacturing costs. Thus, several materials options can be considered before arriving at the final choice. These tools considerably reduce the turnaround time for introducing new designs. We begin with a discussion of unified life-cycle cost engineering, including manu-facturing and operating cost estimation, followed by materials and process selection, risk assessment and product liability, and, finally, failure analysis and prevention.

17.2 UNIFIED LIFE-CYCLE COST ENGINEERING (ULCE)

Suppose the exhaust pipe in your car has developed a hole that needs repair to avoid incurring a fine of $200 under the state's environmental protection law plus the repair costs. You take your car to the auto repair shop, where the mechanic advises you of the following options. The first is to repair the hole in the pipe by tack-welding a patch around the hole. The labor costs for this service will be $25, and the repair is probably good for three months until another hole appears somewhere else in the exhaust pipe. The second option is to replace the exhaust pipe with a new carbon steel pipe with an anticipated lifetime of five years. The cost of this pipe is $50, and the labor to install it is another $50. The third option is to replace it with a stainless steel pipe that costs $100 in addition to labor cost of $50, but the pipe has a predicted lifetime of 15 years. What is the best decision for you, and how will you arrive at it?

In this problem it is clear that only cost is driving the decision process. Besides the three options given to you by the mechanic, you also have the option of not doing anything. However, when the hole becomes large enough, the excessive noise will attract attention, and the probability of being stopped by an officer and being fined under the state law is high. Therefore, this is not a viable option. The choice among the other options depends on one key question: How long do you intend to keep the car? If the answer is less than three months, the first option is attractive. If the answer is more than three months but less than five years, the second option appears to be the best. However, if your car is relatively new and you are likely to keep it for the next 10 years, the third option may look attractive. This is because if you decide to have the carbon steel pipe installed, you may have to change the exhaust pipe at least one more time and possibly two more times during the lifetime of the car at a total cost of $200 to $300. In this analysis, inflation is not a factor if one assumes that the money saved today will grow to offset additional costs at a future date due to inflation. The third option requires only $150 for an exhaust pipe that almost surely will last through the life of the car.

This was a simple example of unified life-cycle cost engineering (ULCE), which attempts to seek the most economical solution over the projected life of the equipment. It also demonstrates the importance of cost in material choices. Another conclusion from this example is that there is a market for both types of pipes, depending on the circum-stances, finances, and attitude of the car owner.

To perform an accurate ULCE analysis, it is imperative that all cost elements be anticipated as far as possible and a realistic dollar value be associated with each element. The first level of cost breakdown leads to: (1) design and analysis costs, (2) manufacturing costs, (3) operating costs, and (4) cost of disposal. The first two are borne by the man-ufacturer, but the user ultimately pays for the costs of all phases. Therefore, the user benefits the most from a ULCE approach.

17.2.1 Design and Analysis Costs

Design and analysis cost is the front-end engineering cost for making a new product. It includes the cost of (1) conceptual and detailed design and subsequent analysis, (2) pro-totype building and testing, (3) preparing manufacturing specifications, including inspec-tion and quality control criteria, and (4) assessing risk and liability. Evolutionary designs result in considerably less front-end costs because they base material and manufacturing process choices on previous service experience. Besides saving engineering man-hours required for preparing new manufacturing and inspection specifications and for prototype testing, this approach minimizes the risk of in-service failure. Usually, revolutionary designs utilizing new materials and manufacturing processes are first introduced in non-commercial business sectors such as defense or space exploration, where cost is not the most important consideration. Once the technology is proven through service experience, it is introduced in the commercial sector with minimal development costs. For example, the availability of composite materials and novel approaches to the packaging of electronic devices and circuitry are the result of defense and space research. Tennis rackets, golf clubs, skis, and fishing rods made from carbon, boron, and epoxy composites are the outgrowth of developments in the defense and aerospace business leading to new compositive materials.

17.2.2 Manufacturing Costs

Costs of manufacturing include direct production costs, fixed costs, and plant overhead costs. Production costs include costs of raw materials, operating and supervisory labor, utilities, maintenance of equipment, supplies, quality control and inspection, and payments of royalties for any patented process. Fixed costs include depreciation on buildings and equipment, insurance, rents for floor space, investment capital, and taxes. Overhead costs include benefits for employees, advertising and marketing, storage, shipping, inventory, and general plant maintenance.

17.2.3 Operating Costs

Operating costs include labor costs for operating the equipment, fuel or energy costs, buildings, and maintenance and repair costs, including costs due to loss of production during maintenance outages. For example, the ideal aircraft is one that is the most fuel-efficient, never requires maintenance or repair, and lasts forever. Thus, higher availability can reduce operating costs. (Availability is defined as time during which the equipment is operational divided by the elapsed time from the installation of the equipment.) Similarly, longer life also reduces operating costs. The choice of material affects both availability and service life of the equipment and is, therefore, an important consideration in determining operating costs. The example of the exhaust system given earlier illustrated this clearly.

17.2.4 Cost of Disposal

The cost of disposing of scrap can also be an important consideration. For example, in some states a cost is associated with the disposal of car tires. In Germany, manufacturers of certain goods are responsible for their ultimate disposal, and the price of disposal is included in the selling price. Similarly, the cost associated with the disposal of spent nuclear fuel is having a major influence on the survival of the nuclear industry. Therefore, it is desirable to find cost-effective ways to recycle car tires and spent nuclear fuel. As environmental regulations become more stringent, the cost of disposal will play an increasingly important role in determining the overall cost of conducting business. The ability to dispose of a material in a cost-effective and environmentally responsible manner will be a key factor in material selection in the future.

CASE STUDY 🔵 Cost Consideration in Materials Selection

Turbine blades in jet engines are subjected to high stresses along the blade axis and high temperatures (~1000°C) during service at fracture-critical locations. Thus, creep and oxidation effects determine the service life of the blades, which must be replaced several times during the life of the engine, since these blades last an average of two years. Both creep and oxidation occur preferentially along the grain boundaries that are normal to the loading axis. The operating temperature limits are determined by these failure modes. To make the turbine blades more resistant to creep and oxidation, the blades are currently cast from nickel-base superalloys using a directional solidification (DS) casting process. This process results in large and highly elongated grains with their axis aligned along the blade axis and a much smaller grain boundary area normal to the direction of applied stress, thus reducing the potential for creep and oxidation failure.

DS alloy blades have been used in jet engines for quite some time. The cost of one such blade is $300. You are a materials engineer in a company that manufactures DS turbine blades and supplies them to various jet engine manufacturers. One engine manufacturer who is a leader in engine technology has approached your company, asking it to begin manufacturing blades from a silicon carbide (SiC)–reinforced silicon nitride (Si_3N_4) ceramic matrix composite (CMC), which can withstand temperatures up to 1200°C and will considerably boost the fuel efficiency of the engine. Also, the CMC blade will last for four years compared with the two-year life of the DS blade. The customer expects to install the first batch of proven CMC blades in engines in five years and is guaranteeing to purchase 5000 blades per year for five years. The customer has requested that your company submit, within a month, a cost proposal and a firm commitment to supply the blades. You have been asked to study the request and make a suitable recommendation to your management, who will make the final decision whether or not to go ahead with a bid.

Solution

From discussions with the research and development group you know that your company has an ongoing modest effort directed toward developing the CMC blades and a facility to produce prototypes in the laboratory, but no facilities exist for producing them economically in large quantities. Hence, a capital expense of $6 million is needed in the fifth year to develop facilities for large-scale production. The company will need to spend an additional $5 million over the first four years to develop the necessary design data and to conduct the verification testing, the stress analysis, and the heat transfer analysis needed to prove the new CMC blade design. The cost of materials to make the blade and the labor costs plus overhead after production begins are estimated to be $200 per blade, which compares with $150 needed for the DS blade. Therefore, an additional working capital of $1 million (5000 blades × $200) is needed to begin production.

Figure 17.2–1 shows the cash flow as a function of years. If we assume that the depreciated value of the equipment after five years is $3 million, to break even at the end of 10 years (five years after the start of production) your company should have sales worth $9 million. If we divide the $9 million by the number of blades (5000 per year × 5 years = 25,000) we get a cost price of $360. Since the life of CMC blades is double that of DS blades, the engine company should not object to paying twice the price of the DS blade, which is $600. At this price, your company stands to make a profit of $6 million by the end of the first five years of production. At $600 per blade, the engine company still stands to gain. Because of the increased life of the blades, less frequent maintenance will be needed. Also, fuel efficiency will receive a big boost. Both of these factors will result in a higher value for the engine.

If the new blade is successful, your company stands to profit further in future years by continuing to supply blades to this engine company after the capital cost has been recovered as well as by picking up business from other engine companies. Depending on the competition at that time, your company can reduce the price of the blade substantially and still have a highly profitable business. The downside risks include losing the research and development costs if a good blade design cannot be obtaind. If the blades fail during service for some unanticipated reason, the whole idea could be scrapped and the entire $12 million of initial investment would be at risk.

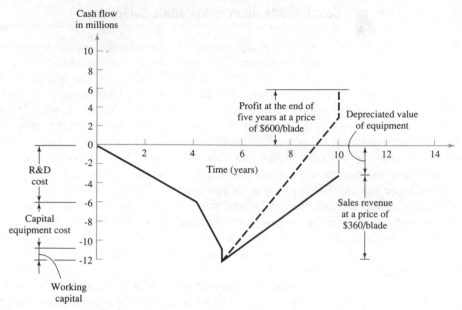

FIGURE 17.2–1   Cash flow (cumulative) for the production of SiC-Si$_3$N$_4$ turbine blades in a 10-year period. Various cost components are considered. Note that sales begin in the fifth year.

Given the above economic data, the decision whether or not your company should commit to manufacturing the CMC blades can be made by the management, depending on the financial circumstances of the company.

The above calculations do not take into account factors such as cost of capital and inflation. These factors can be included in the analysis without much difficulty by making an appropriate mathematical model, usually discussed in texts on engineering economics. It must be added that the cost data of any company are proprietary. Therefore, the above cost figures are not real and should be considered only for illustration purposes.

17.3 MATERIAL AND PROCESS SELECTION

Material and process selection for a part is an important step in the design process. In structural applications, the strength of the material determines the size of the part, as we discussed in Chapter 9. Similarly, the electrical resistivity of the material determines the size of a conducting wire. Both strength and resistivity vary considerably from material to material. Since material selection also in part dictates the fabrication process, an efficient design process must consider material and fabrication process selection simultaneously, taking into account the geometrical and size details of the part. Such a design approach, called concurrent engineering, is rapidly gaining acceptance. The acceptance of this approach has been facilitated by use of computers in the design process. Computers speed up the various design steps and allow the designer to consider several alternatives.

A virtually infinite number of materials can be used in engineering applications if one considers all metal alloys, polymers, ceramics, and composite materials. Therefore, it is important to have the material performance characteristics clearly defined to speed the material selection process. Table 17.3–1 gives a comprehensive list of the relevant physical, mechanical, thermal, chemical, electrical, and fabrication material parameters that determine the engineering usefulness of a material. Because of the large numbers of materials, measuring and compiling even a few relevant performance characteristics of

TABLE 17.3–1 List of material properties commonly used in material selection.

Physical properties	Fatigue crack growth resistance
Crystal structure	Stress corrosion cracking resistance
Melting point	Charpy impact resistance
Density	Creep deformation resistance
Viscosity	Creep rupture
Reflectivity	**Thermal properties**
Transparency	Conductivity
Electrical properties	Coefficient of thermal expansion
Conductivity	Specific heat
Dielectric constant	**Chemical properties**
Mechanical properties	Position in EMF series
Hardness	Corrosion and degradation resistance
Elastic modulus	Oxidation resistance
Poisson's ratio	Stress corrosion resistance
Shear modulus	Ultraviolet radiation resistance
Yield strength	Resistance to various chemicals
Ultimate strength	**Fabrication properties**
Percent elongation	Castability
Strain hardening exponent	Heat-treatability
Strength coefficient	Hardenability
Percent reduction area	Formability
Endurance limit	Machinability
Fracture toughness	Weldability

Source: Table adapted from G. E. Dieter, *Engineering Design—A Materials and Processing Approach.* New York: McGraw-Hill, 1983.

each material is a formidable task. In addition, new materials are being developed at an increasing rate. Hence, material testing and the ability to precisely measure the relevant performance characteristics of the materials are important aspects of the materials engineer's responsibility on a design team. Databases and material and process standards are helpful aids in selecting materials and processes and will be discussed in this section. The impact of the chosen material on the environment is becoming increasingly important. Hence, a separate subsection is devoted to this topic.

17.3.1 Databases for Material Selection

Material properties databases are developed by government organizations such as NASA and the Department of Defense and by major private corporations. Material vendors are an important source of properties data for their own materials. Professional societies such as the ASM International, the Society of Automotive Engineers (SAE), the American Iron and Steel Institute (AISI), and the American Society for Testing and Materials (ASTM) all collect materials data and compile them in computer databases and handbooks. Table 17.3–2 lists some of the sources of material properties.

Materials engineers who regularly participate in new designs also develop files consisting of trade literature, proprietary company reports, and scientific and technical literature containing materials data on their systems of interest. Usually material property measurements obtained from the latest testing techniques and also data for the latest materials are not found in standard databases. Therefore, technical literature in the form of trade articles and scientific journals is a very important source of material data. It must be recognized, however, that some of these data refer to materials that are not exactly like

TABLE 17.3-2 Sources of material properties information.

Metals
1. *Metals Handbook*, Volumes I to III, *Properties and Selection*, ASM International, Metals Park, OH.
2. *SAE Handbook, Materials Parts and Components*, Society of Automotive Engineers, Warrendable, PA.
3. *Metallic Materials and Elements for Aerospace Vehicle Structures*, Military Standardization Handbook, MIL-HDBK-5C, 2 volumes.
4. *Structural Alloys Handbook*, Volumes I and II, Mechanical Properties Data Center, Battelle Memorial Institute, Columbus, OH.
Ceramics and glasses
1. *Engineering Properties of Selected Ceramic Materials*, American Ceramic Society, Columbus, OH.
2. *Engineering Property Data on Selected Ceramics*, Metals and Ceramics Information Center, Battelle Columbus Institute.
3. E. B. Shand, *Glass Engineering Handbook*, McGraw-Hill, New York.
Polymers
1. C. A. Harper, *Handbook of Plastics and Elastomers*, McGraw-Hill, New York.
2. *Encyclopedia of Polymer Science and Technology*, John Wiley & Sons, Inc., New York.
3. *Modern Plastics Encyclopedia*, McGraw-Hill, New York.
4. B. M. Wallace, *Handbook of Thermoplastic Elastomers*, Van Nostrand Reinhold Co., New York.
5. *Engineering Materials Handbook*, Volume 2, *Engineering Plastics*, ASM International, Metals Park, OH.
Electronic materials
1. *Handbook of Electronic Materials*, Volumes 1 to 9, Plenum Publishing Co., New York.
2. *Electronic Materials Handbook*, Volume 1, *Electronic Packaging*, ASM International, Metals Park, OH.
Composites
1. *Engineered Materials Handbook*, Volume 1, *Composites*, ASM International, Metals Park, OH.
2. Y. S. Touloukian, *Thermophysical Properties of High Temperature Solid Materials*, Volumes 1 to 6, Macmillan, New York.

Source: Table adapted from G. E. Dieter, *Engineering Design—A Materials and Processing Approach*. New York: McGraw-Hill, 1983.

the one(s) being considered for implementation. Before a material is introduced into practice, it is important that it be completely evaluated for the intended purpose.

17.3.2 Materials and Process Standards

ASTM, AISI, SAE, the International Standards Organization (ISO), and the American Welding Society (AWS) are among the professional societies promoting material and processing standards. In addition, the military, NASA, and private companies have standards for their own internal use in material selection that also serve as purchasing specifications when the parts must be bought from outside vendors. The purpose of developing standards is to ensure consistency in the quality of products. These standards are usually developed by reaching a consensus within working groups consisting of representatives from the users and producers of materials. Such standards are the result of carefully negotiated balances between the needs of the user and what the material producer can deliver. Therefore, designers prefer to follow consensus standards whenever possible, and manufacturers of equipment like to support standardization efforts. By following industry standards, companies can limit their liability in the event of failure during service.

Standards contain detailed information regarding the composition of the material, relevant physical, mechanical, or other types of characteristics, and the processing methods used for fabricating the material. They are also specific about the acceptable limits of variability on important properties. Standards can be applied to materials, processes, products, and testing. Materials standards typically specify chemical composition, impurity levels, and relevant properties. These standards are followed closely to monitor the quality of the materials at the plant immediately following production. Process standards may include welding, forging, or casting practice, carburizing, application of coatings, or polymer processing. Product standards consist of standardizing items such as nuts, bolts, shafts, gears, and so on. Testing standards address methods for characterizing properties such as tensile strength, fracture strength, fatigue strength, and nondestructive detection of flaws in manufactured components using ultrasonic and X-ray techniques.

17.3.3 Impact of Material Selection on the Environment

For centuries humans have disposed of their waste by burning or casual dumping on available land. In recent years these practices have received considerable negative attention. People are increasingly becoming aware of the effect of materials on the environment: water, land, and air. Available landfills are nearly full, and residents do not want a landfill near their homes. This change in social consciousness will have far-reaching consequences on the nature of consumer and industrial products and what they are made from in upcoming years.

Each person in the United States generates about half a ton of solid waste annually, adding to a total of about 160 million tons produced in 1989. Figure 17.3–1 gives a typical profile of household solid waste. There are a number of ways to deal with solid waste, some more socially responsible than others, including:

1. Dump waste into a landfill, at sea, in space, or elsewhere.
2. Incinerate.
3. Biodegrade (compost).
4. Recycle.

Before we discuss these options, consider the concept of sources reduction. Perhaps one of the best ways to deal with waste is to produce less of it. Industry has already done much to reduce the amount of waste produced in manufacturing. Efforts are under way to reduce the amount of packaging material used, which accounts for more than half the volume of

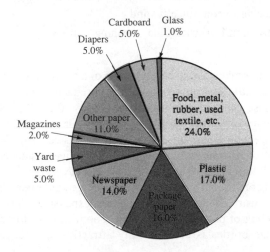

FIGURE 17.3–1

Typical contents of household solid waste.

material in landfills. It seems reasonable to anticipate that there might be a consumer-driven push to replace disposability with durability, so that some items can be used for longer times. Examples of items with needlessly short lifetimes are consumer electronics and automobiles. As discussed in the previous section, very soon, if not already, the cost of a product will reflect disposal costs.

Dumping waste is the traditional disposal route, but it is rapidly becoming obsolete for chemical and other solid waste. Improvements in containment technology have led to the promise of total containment of landfill matter. Although social acceptance of dumping is low, there will probably always be a need to use this method to a limited extent. It is a good technique for disposing of some inorganic wastes, such as concrete, brick, and other ceramics. However, even these materials degrade with time and many contribute undesirable chemicals to the environment.

Incineration is a technique that has its advantages and disadvantages. An advantage is that the material can be used as an energy source to produce heat. A material that has been disposed of in this way for decades is the lignin removed in the pulping process. The lignin is burned to produce heat required for the pulping process. The objective of incineration for disposal of organic materials is to produce relatively innocuous combustion products such as carbon dioxide and water. However, it is often difficult to get the reactions to go to completion and to avoid undesirable by-products. Also, social acceptance of this method is low because no one wants an incinerator in his or her backyard. This technique will continue to have limited use, especially in the chemical industry.

Composting is the best technique for disposal of innocuous organic material, such as garbage, leaves, and grass clippings, which may account for up to 25 vol.% of the material that currently goes to landfills. Composting literally turns trash into valuable organic matter used to build soil. Composting reactions occur in a relatively short period of time with little odor. Many people in rural settings already compost. The difficulty of composting in urban areas is the establishment of a centralized composting facility and minimization of transportation costs.

Recycling is another satisfying, ethical, and socially acceptable alternative for waste disposal. Nearly 30% of all paper and 30% of all plastic soft-drink containers are currently recycled. The easiest products to recycle are those that:

1. Are used in large quantity,
2. Can be remelted and reformed with little or no loss in properties,
3. Are available without elaborate separation from other components,
4. Can be transported economically, and
5. Have relatively high intrinsic value, meaning that the cost of the material is high.

Aluminum in the form of cans is the best example of a recyclable consumer product. The amount of energy that can be saved by not having to reduce aluminum oxide to aluminum is large. Also, because crushed cans have low density, their transportation cost is not prohibitive. The ability to recycle Al is the primary reason for its being the preferred material for beverage containers in comparison with steel. Figure 17.3–2 shows the increase in the usage of Al cans and the decline in the usage of steel containers with time.

Polymers are well known for their low density. However, polymers are more resilient than aluminum, and consequently, used plastic beverage containers cannot be compacted to near theoretical density. Hence, transportation is expensive. Until the balance in economics shifts, recycling of plastics will not be economically viable. Currently, recycling of polyester beverage bottles is economically attractive, but recycling of lower-value

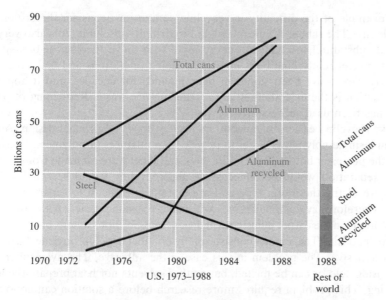

FIGURE 17.3–2 Increase in the usage of recyclable Al beverage containers and the decline in the use of steel containers. *(Source: JOM (formerly Journal of Metals) 42, no. 4 (April 1990), p. 38, a publication of the Minerals, Metals & Materials Society, Warrendale, PA. Reprinted with permission.)*

TABLE 17.3–3 Economics of select polymers.

Polymer	Volume in U.S. (billion lb)	Cost ($/lb)
Polyester	3.43	0.65
Polyethylene	18.75	0.43
Polypropylene	7.25	0.43
Polystyrene	5.19	0.52
Polyvinylchloride	8.31	0.40

Source: *Modern Plastics* (January 1990), p. 168; data for 1989.

polymers such as high-density polyethylene (milk jugs) cannot be justified solely on an economic basis. Table 17.3–3 shows the approximate cost of various polymer resins and the volume sold in the United States per year. Unfortunately for recycling, the large-volume polymers are typically those that use low-cost resins.

Recycling newspaper saves trees, and recycling glass saves considerable energy. However, these recycling efforts are also more socially than financially satisfying. The economics may change substantially in the future when the purchase price of these goods reflects disposal costs.

The most difficult products to recycle or otherwise dispose of are some of the composite structures. Composites often contain crosslinked polymers that are not easily recycled. In fact, most polymers are not compatible, and remelting does not produce a useful material. Hence, the polymers must be separated, and to date no simple method for separation has been discovered. Our inability to separate paper from the polymeric gloss that covers the paper in high-quality catalogs is the reason that these catalogs cannot be recycled. Newspapers that do not have this gloss can be readily recycled.

As an example of the difficulty in recycling composites, consider the disposal of an automobile tire. The rubber is vulcanized and is virtually insoluble; it is also very tough, requiring a substantial mechanical energy input to achieve rheological breakdown. In addition, the steel reinforcing belts hinder shredding operations and make incineration unattractive. Tires have been stored in huge mounds for decades, and it appears that source reduction is the best alternative for reducing tire waste. Retreading of truck-tire carcasses has been used successfully for years.

To make recycling easier, manufacturers need to focus on product design. A product made from a single polymer is easier to recycle than mixtures of polymers. Consider, for example, the polyester bottle. If the bottle, cap, and label were all made from compatible polymers, separation would be unnecessary and recycling would be simplified. Many adhesives are of the thermoset type and are extremely difficult to separate from other materials. Therefore, avoiding adhesives will make recycling easier. An example of this problem is found in the carpet industry, which generates millions of tons of waste that goes directly to landfills. The carpet consists of nylon fibers, polypropylene backing, and thermoset adhesive. The problem in this case is the adhesive. The nylon fibers can be dissolved using acid or can be melted, but neither solvents nor heat prepare the adhesive for recycling. This problem requires more research before a solution can be found.

CASE STUDY   Material Selection for Electronic Package Casing

This problem is concerned with the design of an optimum material for electronic package casing. In the 1970s and 1980s there was a revolution in the electronics industry. This resulted in improved performance of electronic devices, such as highspeed computers, while simultaneously reducing their size and weight. These advances were made possible by the availability of small electronic packages containing multiple integrated circuit chips, each containing thousands of circuits. Examples of such packages are shown in Figure 17.3–3. As the circuit density (number of active devices per unit package area) increases, the power dissipated also increases. This causes the semiconductor junction temperatures to rise and may lead to failure. Therefore, circuit density may be limited by the ability to remove heat. Besides junction failures, thermal-fatigue problems

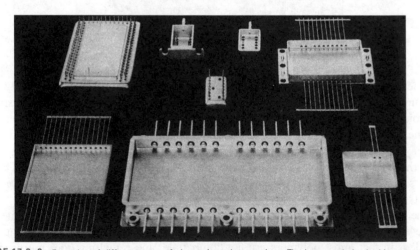

FIGURE 17.3–3 Examples of different types of electronic package casings. The integrated circuit chips are mounted on substrates and attached to the bottom of the casings. The connections with other circuits are made through the pins. The package casing is closed by placing a lid and hermetically sealing the gaps between the lid and the casing. *(Source: Electronic Materials Handbook, Vol. 1, 1989, Packaging, ASM International, Materials Park, OH. Reprinted by permission of the publisher.)*

can also result from temperature increases. The problems include increased failure probability at the thousands of interconnect joints and at the attachments between the integrated circuit chips and the package housing. As a materials engineer working in the electronics assembly industry, you are asked to choose a package casing material for future designs so that: (1) there is at least a factor of 5 improvement in thermal conductivity (for fast heat removal); (2) the coefficient of thermal expansion of the casing material matches as closely as possible that of silicon to reduce thermal-fatigue problems; and (3) the new casing is considerably lighter than the currently used casing, which is made from an iron-cobalt-nickel alloy known by the trade name Kovar.

Solution

The relevant material properties in this example are: (1) thermal conductivity, (2) density, and (3) coefficient of thermal expansion. The package housing should be made from a material that does not oxidize with time in a humid environment. Table 17.3–4 lists some candidate materials that can be used either in monolithic form or in combinations (i.e., as composites). The elastic modulus E, density ρ, coefficient of thermal expansion α, and thermal conductivity K are listed for the various materials, including silicon and Kovar.

If the package has to be significantly lighter than one made from Kovar, none of the metals except aluminum can meet this requirement in the monolithic form. Aluminum has an α value 5 times that of Si. Hence, even though its thermal conductivity is more than 5 times that of Kovar, it is not suitable for this application in the monolithic form. Epoxy by itself has a high α value and low thermal conductivity and, therefore, cannot be considered for this application. All monolithic ceramic materials listed in Table 17.3–4 (Al_2O_3, AlN, BeO) have α values that are in the range of those of Si and Kovar, and low densities. However, the thermal conductivity of Al_2O_3 is comparable to that of Kovar, and it does not meet the requirement for improved heat transfer. Hence, alumina is also dropped from consideration. Be and BeO are toxic and will be ruled out for that reason. Therefore, among the monolithic materials considered (metals, ceramics, and polymers), AlN is the only candidate that meets all the design requirements.

Let us now consider composite systems. Among the reinforcement materials—C fibers, E-glass fibers, and SiC particles—the C-fiber properties immediately stand out for several reasons. The first is the negative coefficient of thermal expansion, which enables us to use it for reinforcing virtually any of the ceramic, metal, or polymer materials listed in Table 17.3–4 and tailoring the expansion to match that of Si. Thus, we can eliminate the risk of thermal fatigue. The second reason is its high thermal conductivity, which, when coupled with its low density, makes it a very attractive reinforcement material. Glass fibers can be combined with either a ceramic or an epoxy matrix. An epoxy-glass composite is not likely to have the needed thermal conductivity,

TABLE 17.3–4 Key properties of materials which may be used in the electronic package housing application. Properties of the fibers are in the axial direction only.

Material	Density ρ (g/cm³)	CTE α (°C⁻¹)	Thermal conductivity K (W/m·°C)	Elastic modulus E (GPa)
Aluminum	2.7	2.3	180	69
Copper	8.9	17.0	400	117
Stainless steel	8.0	17.0	16	193
Alumina (Al_2O_3)	3.9	6.7	20	380
Aluminum nitride (AlN)	3.2	4.5	250	330
Beryllia (BeO)	2.9	6.7	250	330
Epoxy	1.2	54	1.7	3
Pitch-base C fiber	2.2	−1.6	1100	895
E-glass fiber	2.6	5.0	2.1	72
SiC particles [$(SiC)_p$]	2.9	4.9	81	520
Kovar	8.3	5.9	17	131
Silicon	2.3	4.1	150	—

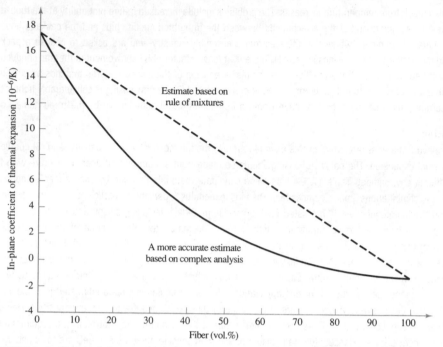

FIGURE 17.3–4 Comparison between the predicted values of the coefficient of thermal expansion of C-Cu composite using the simple rule of mixtures and a considerably more complex equation. *(Source:* Electronic Materials Handbook, *Vol. 1, 1989,* Packaging, *ASM International, Materials Park, OH. Reprinted by permission of the publisher.)*

because thermal conductivities for both materials are low. The properties of glass are comparable to those of AlN, and in fact, its thermal conductivity is lower. Hence, there is no benefit in making an AlN-glass composite for this application. We thus focus on composites with C fibers.

C fibers may be used with an aluminum or copper matrix (because of their high thermal conductivities), with epoxy, or with AlN. To get a first-order estimate of the volume fractions of C fibers required, we use the simple rule of mixtures for estimating composite properties. The rule of mixtures states that:

$$p_c = p_m(1 - V_f) + p_f V_f \tag{17.3–1}$$

where p_c, p_m, and p_f are respectively the composite, matrix, and fiber properties (ρ, K, or α) and V_f is the fiber volume fraction. More accurate expressions for estimating properties deviate somewhat from this linear relationship. Figure 17.3–4 shows α for a C-Cu composite as a function of V_f estimated from a more accurate but considerably more complex relationship. The predicted expansion coefficients from such relationships are somewhat lower than those predicted by the rule of mixtures. For the sake of simplicity, we will limit our analysis to the rule of mixtures.

The approach used in this example is to solve for the volume fraction that will yield an expansion coefficient identical to that of silicon using Equation 17.3–1. Subsequently, this volume fraction is used to predict values of thermal conductivity and density. Table 17.3–5 presents these values for various composites and for monolithic AlN and Kovar. All composite materials appear to meet the design requirements. Among metal-matrix composites, the Al-C system is preferred. The epoxy-C system is marginal because theoretically the maximum V_f that can be attained by stacking cylindrical fibers of equal diameter is 91%. Since the required volume fraction is close to the theoretical limit, it may not be possible to achieve this level. The critical V_f for an AlN matrix is very small, and the additional cost of manufacturing the composite may easily offset the small advantage in properties compared with monolithic AlN.

TABLE 17.3–5   Estimated properties of the various composite systems considered. The properties of Kovar and AIN in monolithic forms are listed for comparison.

Composite type	V_f (%)*	ρ (g/cm³)	Thermal conductivity K (W/m-°C)	Elastic modulus E (GPa)
Al-C	76.8	2.3	886	703
Cu-C	69.3	4.3	885	656
AIN-C	6.5	3.1	305	366
Epoxy-C	89.7	2.1	986	803
Kovar	0	8.3	17	131
AIN	0	3.2	250	330

*V_f = volume fraction of fibers.

In summary, the two optimum materials are AIN in monolithic form and the Al-C composite. The final choice may be based on cost factors, manufacturability, and the extent of hermeticity desired in the application. The AIN package is likely to be more hermetic than the Al-C composite package. Other secondary factors such as impact resistance may also tip the decision in favor of one material over the other. It should be noted that, in comparison with the Kovar casing, the weight savings with AIN and the Al-C composite are 61 and 72%, respectively, if we assume that the same volume of material is used. Both these savings are significant and appear to meet the design objectives.

CASE STUDY   ▶   Material Selection for a Nuclear Waste Container

By the year 2000 it is expected that the world's 500-plus nuclear power plants will be producing 15,300 tons of high-level nuclear waste annually. The problem of what to do with this waste is a matter of considerable debate. Without an acceptable solution, the future of nuclear power is uncertain. An approach that is gaining some consensus is for the waste to be contained in lead "coffins" and buried underground, because it has been demonstrated that lead is a superior radiation shield. However, the containment must function effectively for 10,000 years. Two potential problems are failures caused by creep and corrosion. We will discuss approaches to minimize the risks of such failures.

Solution

The melting point of lead is 327°C, or 600 K. The ambient temperature of 298 K is approximately 0.5 T_m, where T_m is the melting temperature. Therefore, creep under its own weight is a significant design consideration even at room temperature. The radioactive nuclear waste can increase the temperature in the container and accelerate the creep process. Since the containers will be buried underground and may come in contact with groundwater, the possibility of corrosion under aqueous conditions must be minimized.

The corrosion behavior of lead has been studied by conducting corrosion tests and examining archeological lead samples exposed to water or seawater for up to 2000 years. The conclusion is that lead has excellent corrosion resistance. The minimal corrosion that does occur is isolated at grain boundaries. In large-grain materials, the depth of corrosion penetration along a few grain boundaries was substantial, but in fine-grain materials, the progress of corrosion is lost in the maze of grains and damage is limited to the surface. Therefore, fine-grain structure is more suitable for corrosion resistance than large-grain structure.

One approach for design against creep is to place the spent nuclear fuel rods in a thick-walled copper container with lead filling the space between the fuel rods and the container. Since the corrosion resistance of lead is superior to that of copper, the outer copper walls can be lined with sheets of lead to protect the copper from the environment.

In conclusion, despite the potential problems of creep and corrosion, lead still appears to be the most attractive containment material for burying high-level nuclear waste. The problems can be overcome through grain size control and through the provision of support structures for the lead containers made from stronger corrosion-resistant metals. However, considerable corrosion and creep testing of the designs is needed to ensure that these approaches are safe.

CASE STUDY   Development of Lead-Free, Free-Cutting Copper Alloy

As discussed in Chapter 16, the machinability of copper alloys can be enhanced significantly by adding 0.5 to 3 wt. % lead. Lead has virtually no solid solubility in copper but has extensive liquid solubility. During solidification, lead precipitates and forms a dispersion of second-phase particles that are uniformly distributed within the grains and along grain boundaries. During machining, these particles function as chip breakers and prevent the tool from cold-welding onto the workpiece. Many plumbing fixtures are made from these materials. The annual usage of lead-containing copper alloys is approximately 500,000 tons in the United States alone. Medical studies have shown that lead is toxic, and its quantities in drinking water are strictly limited by legislation in both the United States and Europe. Therefore, a substitute for lead in free-cutting copper alloys is needed.

Solution

The binary phase diagrams of Cu-X alloys (where X is Te, Se, Tl, Bi, S) reveal that these elements also have limited solid solubility in Cu and can perform the role of Pb. In the case of Te, Se, and S, intermetallics form that can be uniformly dispersed in the same way as Pb and have the same result on machinability. Unfortunately, these elements have severe limitations. Te, Se, and Tl are toxic and expensive. S has a deleterious reaction with other elements used in copper and its alloys. Bi is nontoxic, but severely embrittles copper even when present in small amounts. Figure 17.3–5 shows the effect of Bi and Pb additions on the ductility of copper. A 0.01 wt. % addition of Bi reduces the ductility by a factor of 6 to 7. If the ductility problem of Bi addition can be overcome, it would make a successful substitute for Pb in free-cutting copper alloys.

Although the Cu-Pb and Cu-Bi phase diagrams are similar, the second-phase dispersions in these binary systems are quite different. In Cu-Pb alloys, the Pb is harmless to the mechanical properties because the globules are distributed uniformly between grain boundaries and the grains. Lead's FCC structure makes it inherently ductile. Hence, even up to 30 wt. %, Pb-Cu alloys have significant ductilities. Bi globules are also found in the gain interiors of Bi-Cu alloys, which is desirable. However, Bi in the grain boundaries is present

FIGURE 17.3–5

The influence of Pb and Bi additions on the ductility of Cu. *(Source: Advanced Materials and Processes, Vol. 14D, no. 4 (October 1991), ASM International, Materials Park, OH. Reprinted by permission of the publisher.)*

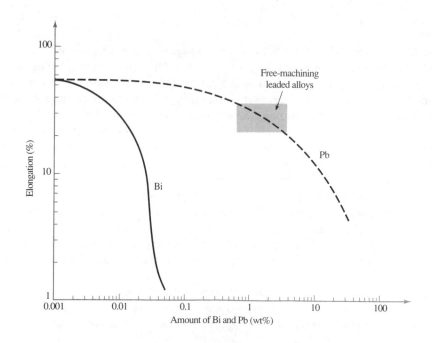

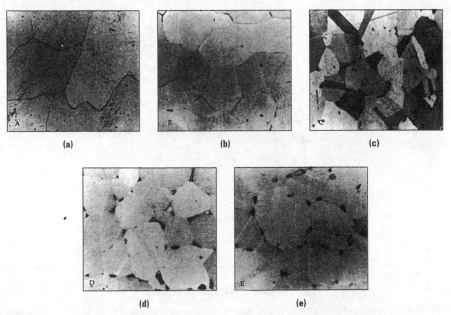

(a)   **(b)**   **(c)**

(d)   **(e)**

FIGURE 17.3–6   Metallography (500 ×) reveals the positive effects on ductility of adding third elements to binary Cu-Bi alloy. **(a)** Extensive grain-boundary wetting by bismuth in Cu–1% Bi embrittles the alloy (tensile elongation is 0.8%). **(b)** Some dewetting has occurred in this Cu–0.54 Bi–0.55 Sn alloy (elongation is 11.6%). **(c)** Grain-boundary dewetting in Cu–0.54 Bi–0.18 P (45% elongation). **(d)** Cu–10 Zn–3 Bi–3 Sn (22% elongation). **(e)** Cu–10 Zn–1 Bi–0.15 P (37% elongation). The Cu-Zn-Bi-Sn alloy (part d) is the bismuth-containing counterpart of leaded semired brass (6 to 8% Pb). Ferric chloride.   *(Source: Advanced Materials and Processes, Vol. 14D, no. 4 (October 1991), ASM International, Materials Park, OH. Reprinted by permission of the publisher.)*

in the form of film instead of globules. The presence of the film embrittles the alloy. The microstructure of the Bi-Cu alloy is shown in Figure 17.3–6. In addition to the film formation, Bi has a rhombohedral microstructure, which is inherently brittle. Thus, a successful modification to the Bi-Cu alloy will be one that changes the grain boundary film into the shape of a globule.

The Bi film formation at the grain boundary is directly related to the differences in surface tensions of liquid Bi and Cu. The surface tensions of various liquid metals are shown in Table 17.3–6. We assume that the

TABLE 17.3–6   Surface tensions of various liquid metals.

Material	Surface tension (dyn/cm^2)
Phosphorus	70
Bismuth	350
Antimony	400
Lead	450
Indium	550
Tin	570
Gemanium	620
Gallium	700
Zinc	760
Aluminum	860
Copper	1300
Nickel	1725

Source: Adapted from J. T. Plewes and D. N. Loiacono, *Advanced Materials and Processes* 14D, no. 4 (October 1991), p. 23.

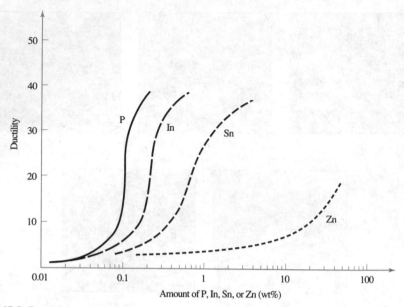

FIGURE 17.3–7 Ductility of a nominal Cu–1% Bi alloy as a function of additions of P, In, Sn, or Zn. These additions reduce the surface tension of Cu, making it closer to that of Bi. This in turn means that the Bi will not as readily form a film along the Cu-Cu grain boundaries but will instead take on a globular shape. *(Source:* Advanced Materials and Processes, *Vol. 14D, no. 4 (October 1991), ASM International, Materials Park, OH. Reprinted by permission of the publisher.)*

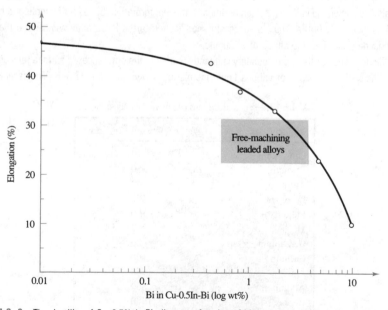

FIGURE 17.3–8 The ductility of Cu–0.5% In-Bi alloy as a function of bismuth content. There is a range of Bi content that will provide similar ductilities to those of free-machining leaded alloys. *(Source:* Advanced Materials and Processes, *Vol. 14D, no. 4 (October 1991), ASM International, Materials Park, OH. Reprinted by permission of the publisher.)*

surface tensions of the solid metals are proportional to those of the liquids. Thus, the problem can be solved either by reducing the surface tension of Cu by alloy addition or by increasing the surface tension of Bi by alloy addition. Pb and Tl are the only elements that are not soluble in Cu but are soluble in Bi and can increase its surface tension. However, both are toxic and must be ruled out. On the other hand, P, In, Sn, Ge, Ga, Zn, and Al are all soluble in Cu and are not soluble in Bi and can reduce the surface tension of Cu considerably. Hence, the solution lies in reducing the surface tension of the Cu. The effects of each of these alloys on the ductility of 1 wt. % Bi-Cu alloy are shown in Figure 17.3–7. Figure 17.3–8 shows the ductility of Cu–0.5 In-Bi alloys as a function of Bi content. The microstructures of the various modified Bi-Cu alloys are shown in Figure 17.3–6b to e. The change in morphology of the second phase at the grain boundaries is apparent.

In conclusion, small quantities of In, Sn, and Bi can serve as an effective substitute to Pb for improving the machinability of copper alloys.

17.4 RISK ASSESSMENT AND PRODUCT LIABILITY

All engineers have an ethical obligation to fully understand, and to alert all potential users about, the negative consequences of any technology they introduce to society. In fact, it is more than an obligation because engineers legally assume, through the companies that they work for, the financial responsibility for negligence during design. This is known as product liability. Therefore, quantifying the risk of a new design or a design modification is a necessary part of the design process. Since failures can result from materials not meeting their expected levels of performance, materials engineers have a major role in risk assessment. In this section we illustrate some methods of quantitative risk assessment. Risk is defined as:

$$
\begin{aligned}
\text{Risk (in dollars)} = \ & \text{(Failure probability)} \\
& \times \text{(Number of components)} \\
& \times \text{(Dollars/failure)}
\end{aligned}
\tag{17.4–1}
$$

The cost of failure must include all direct, consequential, and punitive damages. The risk, in dollars, must then be compared with the potential for profit in making a decision about implementing the new design or design modification. Such analysis is subjective because it involves estimating probability of failure of components that have not been in service as well as predicting consequential and punitive damages.

If we define reliability as the probability that a component will perform its expected function during its design life, then reliability $R = (1 - f)$, where f = failure probability. A high reliability is always desirable, but it can drive up the cost of the product considerably. Figure 17.4–1 shows the cost of design and manufacture, the operating

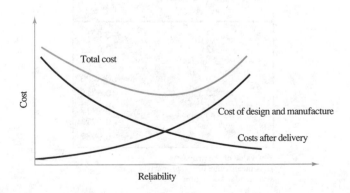

FIGURE 17.4–1

Cost of design and manufacture, cost after delivery, and total cost as a function of reliability.

costs, and the total cost as a function of reliability. The minimum cost is achieved at some value of reliability less than 1. Hence, in some applications an acceptable failure rate may be more cost-effective than a failure-proof design. However, such an analysis does not take into account psychological factors such as customer dissatisfaction, which may have an extreme impact on the business.

17.4.1 Failure Probability Estimation

Equipment failure can occur for a variety of reasons. We will be concerned with material failure only. To estimate risk, an estimate of failure probability is required. Failure probability can be obtained empirically if prior service data are available. If such data are not available, statistical principles can be used to estimate its value.

   When we determine a material property in the laboratory (for example, tensile strength, fracture toughness, etc.), we always conduct multiple tests and report an average value. This is necessary because of the inherent variability in the material; therefore, the property value is not a constant. For example, the strength of ceramic materials can vary considerably because of flaws that vary in size from sample to sample. Tensile strength

FIGURE 17.4–2

The probability distribution as a function of standard normal deviate. The shaded area represents the cumulative probability.

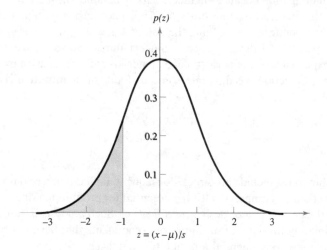

$$z = (x - \mu)/s$$

TABLE 17.4–1   Area under standardized normal frequency curves.

Standard normal deviate $(x - \mu)/s$	Area
−3	0.0013
−2	0.0228
−1	0.1587
0	0.5000
0.5	0.6915
1	0.8413
2	0.9772
3	0.9987

x = value of the property.
μ = mean value of the property.
s = standard deviation.

often follows a normal distribution given by the equation:

$$p\left(\frac{x - \mu}{s}\right) = \left(\frac{1}{s\sqrt{2\pi}}\right) \exp\left[-\frac{1}{2}\left(\frac{x - \mu}{s}\right)^2\right] \qquad (17.4\text{--}2)$$

where μ is the mean value of the variable x, s is the standard deviation, and p is the probability that the property has value x.

A plot of Equation 17.4–2 is shown in Figure 17.4–2. The probability that the value of x will equal the mean is approximately 0.4. The total area under the curve is unity. The cumulative probability that the value of the standard normal deviate, $(x - \mu)/s$ is less than -1 is shown by the shaded area. Table 17.4–1 gives the areas under the standardized normal frequency curve for various values of $(x - \mu)/s$.

....

EXAMPLE 17.4–1

This problem illustrates how failure probabilities can be estimated. Let us assume that a structural ceramic material has a tensile strength σ_u, with a mean value, $\overline{\sigma}_u$ of 100 MPa and a standard deviation σ_{us} of 20 MPa. Let us further assume that in an application, the applied stress σ has a mean value $\overline{\sigma}$ of 55.5 MPa with a standard deviation σ_s of 10 MPa. We are to estimate the probability of failure.

Solution

Failure will occur in the structural ceramic component when the applied stress exceeds the tensile strength of the material. The distributions of the applied stress and the material strength are shown in Figure 17.4–3. Also shown in this figure is the distribution of Q, which is the difference between the ultimate strength and the applied stress (i.e., $Q = \sigma_u - \sigma$). The mean value of Q is $\overline{Q} = (\overline{\sigma}_u - \overline{\sigma}) = 44.5$ MPa. The standard deviation of the distribution of Q is given by $Q_s = \sqrt{\sigma_{us}^2 + \sigma_s^2} \approx 22.5$ MPa. The probability of failure is given by the area under the curve to the left of $Q = 0$. This corresponds to the standard normal deviate of $(0 - 44.5)/22.5 = -2.0$. From Table 17.4–1, the probability of failure is then given by 0.0228, or slightly higher than 2%.

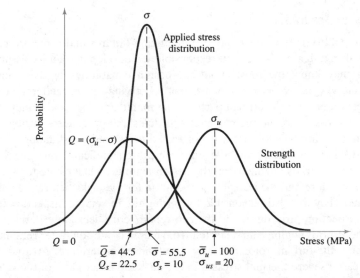

FIGURE 17.4–3 Probability curves for strength (rightmost curve), applied stress (middle curve), and difference between strength and applied stress (leftmost curve). The latter is obtained by subtracting the probability of applied stress from the probability of strength. The probability of failure is the area to the left of $Q = 0$.

The example shows that as we increase the mean applied stress, the probability of failure will increase. To achieve a probability of failure that is approximately 1 in 1000, the standard normal deviate has to be -3.0. This implies that Q must be 67.5 or $\bar{\sigma}$ must be 32.5 MPa. Hence, the mean applied service stress has to decrease from 55.5 MPa to 32.5 MPa to decrease failure probability from 2% to 0.1%. This would imply that the cross-sectional area must increase by $[(55.5 - 32.5)/55.5] \times 100 = 42\%$. This translates into a similar increase in the weight of the part. In an aerospace application for a noncritical part, this penalty would be unacceptable. Thus, this example illustrates the trade-offs between achieving reliability and cost and how this trade-off is affected by the end use.

17.4.2 Liability Assessment

Assessment of liability is necessary to determine the cost of failure as part of risk assessment. The total cost is the sum of direct cost plus consequential damage cost plus punitive damage cost. Direct cost of failure in a machinery part is the cost associated with replacing that part. This is also the total cost when an impending failure is detected during scheduled maintenance, because consequential and punitive damages have been averted. For example, if your car's brake shoes are worn and this problem is detected during scheduled maintenance, the total cost equals the direct cost of replacing the shoes. Consequential damage cost is the cost of the failure of a part during service. For example, if the brake shoes fail during the operation of the car and cause an accident, the costs of the accident are considered consequential damage. This may include damage to other cars or to property, injury, and the repair costs for your own car. Punitive damages are often awarded by courts if the accident is caused by gross negligence on the part of the manufacturer or the operator of the equipment. For example, punitive damages on the order of several hundred million dollars were assessed against the company responsible for a major oil spill off the coast of Alaska in 1989. Since the spill caused extensive environmental damage, the company also had to pay significant cleanup costs, which would be categorized as the consequential damage cost.

17.4.3 Quality Assurance Criteria

Maintaining quality is the surest way of minimizing failures and therefore reducing the dollars at risk. Design engineers are responsible for developing quality assurance criteria because they know the most about the design capabilities of the equipment and factors of safety. Quality assurance consists of specifying the inspection method and the accept/reject criteria for the critical parts during manufacturing. It also includes specifying operating procedures and maintenance schedules and procedures during service. For example, aircraft turbine disks are highly stressed parts, and a disk failure during operation can result in high-speed projectiles that can cause significant consequential damage. Therefore, the materials engineers on the design team are responsible for specifying the inspection procedure and the accept/reject criteria for the disks before they are installed in the engine. They are also responsible for setting the in-service inspection interval, the in-service inspection procedure, and the run/repair/retire decision criteria.

When the aircraft engine is first started from a cold condition (after remaining idle for several hours), the turbine rotor parts are at ambient temperature. A rapid start to full power can cause severe thermal stresses in the disk and other parts of the rotor. It is the design engineer's responsibility to specify the start-up procedure for a cold start. This has to be done with complete knowledge of the material's capability to resist thermal-fatigue failure, which is an input the design engineer will receive from the materials engineer.

While ensuring quality by imposing strict component qualification requirements is highly desirable, the engineering world cannot ignore the economic considerations of maintaining high quality. Imposing materials quality standards that are unnecessarily restrictive can make one's product noncompetitive in the market or cut profit margins. This is illustrated in the following case study.

CASE STUDY   Inspection Criterion for Large Industrial Fans

The rotor shafts of large industrial fans are several meters long and up to 0.75 m in diameter. These shafts are typically made from medium carbon steel heat-treated to a yield strength of about 300 MPa and an ultimate strength of 500 MPa. The shafts are first forged into shape and then machined to final dimensions. Your company, which manufactures fans, buys the shaft forgings from a supplier at a cost of $100,000. These forgings must be inspected by a magnetic particle method (see Section 17.5.3 for a description of inspection methods) prior to shipment. Your company requires that forgings containing any detectable defects be rejected. The procedure employed is capable of detecting defects that are 1 mm deep on the surface. Limiting the defect size to 1 mm can ensure that all defects will be removed during the final machining of the shaft.

The forging supplier tells your company that it has been rejecting 1 out of 3 forgings and has been passing on the cost of the rejected forgings by raising the price of the forgings that pass inspection. Also, rejections have been responsible for several delays in supplying these custom forgings. As a materials engineer you have been asked to assess the suitability of your company's inspection criterion, which has been practiced for two decades.

Solution

The first step in the investigation is to determine if there have been any service failures in these shafts during the past 20 years. Your company produces about 50 fans every year. Therefore, about 1000 fans are in service, and your service department has not received a single complaint about failures or cracks in the fan shafts. These shafts are supported at both ends, and rotate at 1800 rpm. Most of the stresses are due to gravity bending resulting from the shaft weight. The design department does not allow the fatigue stresses to exceed ± 25 MPa due to rotating bending. The endurance limit for the shaft material is approximately 150 MPa. Hence, the probability of fatigue failure initiating from regions with no preexisting defects is negligible. To proceed with this analysis you must use fracture mechanics to establish safe allowable flaw sizes, then base a new inspection criterion on these calculations.

If the shaft is designed for a 30-year life, it experiences a huge number of fatigue cycles during its operation (estimated at 3×10^{10} cycles). Hence, preexisting defects must have a near zero crack growth rate. This implies that the ΔK level must be less than the threshold value of ΔK. Figure 17.4–4 shows the fatigue crack growth behavior of the shaft steel. Since the stresses are completely reversed, we select the fatigue crack growth behavior corresponding to $R = -1$ to estimate the threshold ΔK for this material. That value is 6 MPa-$\sqrt{m}$. The ΔK for surface flaws is given by the equation:

$$\Delta K = 1.12 \, \Delta\sigma \, \sqrt{\pi a}$$

where $\Delta\sigma$ is the fatigue stress range ($= 50$ MPa) and a is crack depth. If the crack depth for which the applied ΔK equals the threshold ΔK is a_0, we can write

$$1.12(50 \text{ MPa})\sqrt{\pi a} = 6 \text{ MPa-}\sqrt{m}$$

Solving for a_0 gives:

$$a_0 = \left(\frac{1}{\pi}\right)\left[\frac{6 \text{ MPa-}\sqrt{m}}{(1.12)(50 \text{ MPa})}\right]^2 = 3.65 \times 10^{-3} \text{ m}$$

$$= 3.65 \text{ mm}$$

Hence, flaws that are more than 3½ times deeper than used in the current inspection criterion will not lead to failures of the shaft. The forging supplier has kept accurate records of sizes of defects that it has found

FIGURE 17.4–4

Fatigue crack growth be-
havior of a medium carbon
steel used in fan shafts.
*(Source: A. Saxena and J.
Opoku, "Evaluating Flaws in
Metal Parts," Machine De-
sign (March 1980), pp. 2–8,
Penton/IPC, Inc., Cleveland,
OH. Reprinted by permis-
sion of the publisher.)*

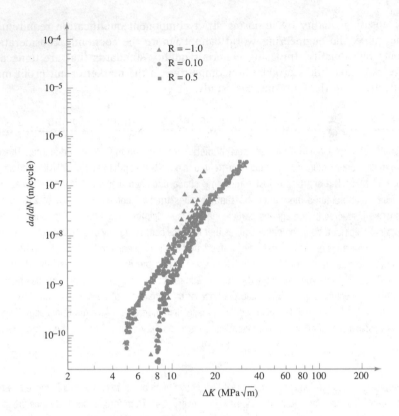

in the shafts over the years. The defects have never exceeded 2 mm in depth. Hence, this calculation shows not only that all rejected shafts would be acceptable but also that there is no need to inspect all the shafts. Hence, the recommendation should be to inspect only one shaft out of several that are supplied to ensure that some checks are maintained on the process and to accept all shafts that have cracks up to 3 mm deep. The machining allowance on forgings is 1 mm. Hence, a defect that is initially 3 mm deep will never be larger than 2 mm deep after machining. This provides additional safety against service failures.

17.5 FAILURE ANALYSIS AND PREVENTION

In Chapters 9 and 15 we discussed a variety of potential failure mechanisms for structural components. By studying the resistance of materials to such failures, we can design to minimize the failure risk. Despite all the testing and analysis, failures do occur—because of unanticipated failure mechanisms, faulty design calculations, faulty material selection, abuse of the machinery during service (for example, not following recommended operating procedure or not following the proper maintenance schedule and procedure), or a critical manufacturing defect escaping detection during the quality assurance procedures. Failure analysis is performed to understand the cause of a failure so that measures can be taken to prevent it in the future and to assign legal responsibility for the damage caused by it. Thus, failure analysis is sometimes called forensic engineering.

    Failure mechanisms can be indentified by conducting a series of examinations that reveal the quality of the material, the quality of the fabrication process, and possible abuse during service. The tools used by failure analysts include optical and scanning electron microscopes, and test machines such as those used for tensile, hardness, and Charpy mpact testing. Failed parts are also subjected to nondestructive testing, chemical analysis, and advanced mechanical tests to verify fracture toughness, fatigue strength, or creep strength. We first discuss the general methods used in failure analysis and then examine some specific failures using these methods.

17.5.1 General Practice in Failure Analysis

The first step in failure analysis is a complete visual examination of the failed parts, even before any cleaning is undertaken. Often the debris found on the part can provide useful evidence of the cause of the failure. For example, the oxide thickness on a crack surface can provide an estimate of how long the crack was present prior to failure. All fragments of the failure are examined and photographed. They must be handled carefully to prevent damage to the fracture surfaces.

The second step is to obtain the service history of the part, including:

1. How long was the part in service?
2. What typical stress levels did it experience during service?
3. What was the service environment?
4. Were there any unusual excursions in the service conditions?

Excursions can take the form of high stress, high temperature, or exposure to an unusually corrosive environment. In the case of such excursions, their intensities, durations, and frequencies should be noted.

Destructive testing is often conducted to determine the mechanical strength, microstructure, and chemical composition of the failed part. The mechanical tests performed may include tensile, hardness, Charpy impact, and, in some cases, creep, fracture toughness, and fatigue tests. Metallography is conducted to examine the microstructure. For destructive testing it is important to use only portions of the failed part that are adjacent to but do not include the failure site. The portions from the failure site are used for fractography.

Fractography consists of examining the fracture surface at magnifications ranging from 10 to 100,000 times using optical and electron microscopes. It is an indispensable tool for identifying fracture mechanisms. The first step in conducting fractography may be cleaning and cutting the fracture surface if fragments of failed parts are larger than the size that can be conveniently examined in the microscope. Cutting must be done at a sufficient distance from the fracture site to avoid altering the microstructure of the material underlying the fracture. Furthermore, before reexamining failures involved in legal cases, permission must be sought before cutting, or penalties could be assessed against the failure analyst. Cleaning is required to remove dirt from the fracture surface and can be performed by a dry air blast or by treating the surface with solvents. The solvent must be able to remove the deposits without attacking the base material. Another effective cleaning method is to strip the deposits using a cellulose acetate tape that has been softened by acetone. The soft tape is pressed on the fracture surface and, after drying, can be peeled off with the deposits adhering to it. The tape samples with the deposits may be saved in case an analysis of the deposits is required.

Examination of fracture surfaces at magnifications from 1 to 100 times may be done with the unaided eye, a handheld magnifying glass, or a low-power optical microscope. This step is known as macroscopic examination. Figure 17.5–1 shows a fractured polyethylene pipe clearly indicating the point of origin of the fracture, which is a manufacturing defect. In almost all cases, the fracture origin has a different texture from the rest of the fracture surface and can be readily identified during macroscopic examination.

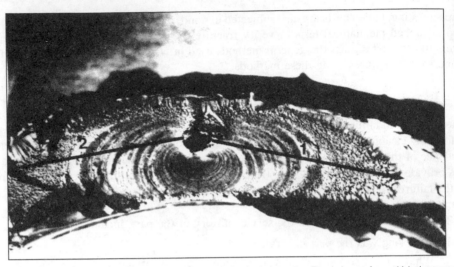

FIGURE 17.5-1 Fracture surface appearance of a cracked polyethylene pipe. The dark spot from which the arrows are emanating is the origin of the fracture and could be a manufacturing defect. The arrows point in the direction of crack propagation. The ends of the arrows indicate the crack profile at the time of the burst. *(Source: Metal Handbook, 9th ed., Vol. 11, 1986, Failure Analysis and Prevention, ASM International, Materials Park. OH. Reprinted by permission of the publisher.)*

Microscopic fracture surface examination can be performed with an optical microscope, a transmission electron microscope, or a scanning electron microscope. Figure 17.5–2 shows a crack nucleating in a weldment of a high-temperature, high-pressure steam pipe. This photograph was taken directly by a camera after the specimen was polished and etched using a mild solution of nitric acid. Optical microscopes may be used to obtain photographs that are magnified 50 to 500 times. Scanning electron microscopes (SEMs) are most commonly used for studying fracture surfaces at magnifications ranging

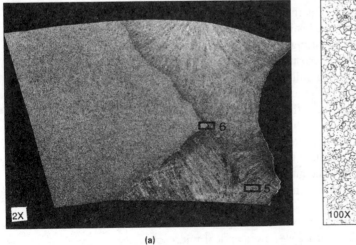

(a)

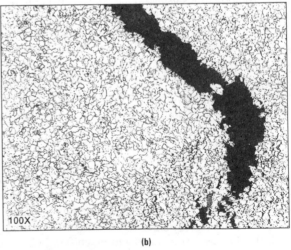

(b)

FIGURE 17.5-2 Failure of a welded steam pipe: **(a)** A creep crack (region marked 6) that has developed in a steam pipe operated at high temperatures. The crack is located in the weldment at the interface between the base metal and the weld metal in the creep region of the weldment. **(b)** The appearance of the cracked region at a higher magnification showing extensive creep damage in the form of cavities. *(Source: R. H. Norris and A. Saxena.)*

Chapter 17   Materials and Engineering Design   **751**

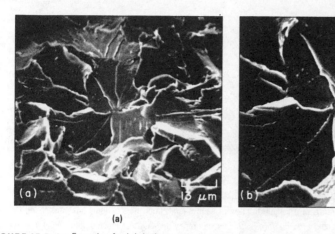

(a)   (b)

FIGURE 17.5–3   Example of a brittle fracture in a steel sample: **(a)** microscopic view showing fracture origin in the middle of the picture, and **(b)** a higher-magnification picture of the region near the particle (marked A). Typically, brittle fracture propagates along crystallographic planes and is, therefore, very flat in appearance.   *(Source:* Metals Handbook, *8th ed., Vol. 9, 1974,* Fractography and Atlas of Fractographs, *ASM International, Materials Park, OH. Reprinted by permission of the publisher.)*

from 100 to 100,000 times. Plastic replicas of the fracture surfaces can be made and examined at high magnifications in a transmission electron microscope (TEM). However, modern SEMs can provide resolutions comparable to the TEM/plastic replica techniques. Use of the SEM has become dominant because sample preparation is much simpler.

Figure 17.5–3a shows a brittle cleavage fracture originating from a carbide particle at the intersection of four grains in a steel specimen. The particle can be seen more clearly in Figure 17.5–3b at a slightly higher magnification. Figure 17.5–4 shows an example of ductile fracture in which microscopic voids grow by plastic deformation and link together to form the fracture surface. Figure 17.5–5 shows a fractograph containing fatigue

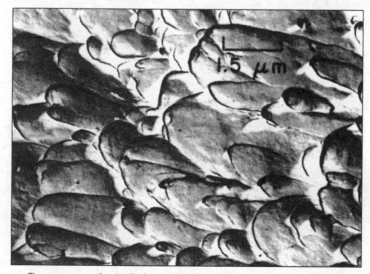

FIGURE 17.5–4   The appearance of a ductile fracture in which voids nucleate at foreign particles but then grow considerably to form the elongated dimples on the fracture surface.   *(Source:* Metals Handbook, *8th ed., Vol. 9, 1974,* Fractography and Atlas of Fractographs, *ASM International, Materials Park, OH. Reprinted by permission of the publisher.)*

FIGURE 17.5–5 Typical appearance of a fatigue fracture at high magnification. The periodic lines are called fatigue striations. *(Source: Metals Handbook, 8th ed., Vol. 9, 1974, Fractography and Atlas of Fractographs, ASM International, Materials Park, OH. Reprinted by permission of the publisher.)*

striations, which are related to the amount of crack extension during each fatigue cycle. These striations are a clear indication of fatigue loading. These are a few examples illustrating the use of fractography in identifying the mechanisms of failure. Understanding the failure mechanism provides an important clue for the cause of failure.

Most failure analysts have access to an atlas of typical fractographs prepared from laboratory-generated fractures under precisely controlled loading and environmental conditions. The fractographs obtained from the failed parts can then be compared with the ones from laboratory specimens. Thus, service-loading conditions, which are sometimes difficult to establish, can be determined.

C A S E   S T U D Y   Failure Analysis of Seam-Welded Steam Pipes

Large steel pipes are used for carrying supersaturated steam to the turbines in power plants. These pipes have diameters of about 0.75 m and are made from steel plates that are approximately 37.5 mm thick. They carry steam at a pressure of 5 MPa and a temperature of 538°C (811 K), which is approximately 44% of the melting point of steel on the absolute temperature scale. At this temperature, creep deformation and rupture is of concern in Cr-Mo steels used for this application.

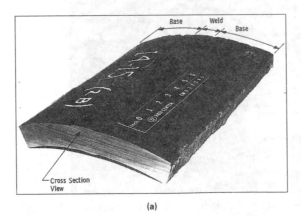

(a)

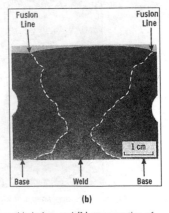

(b)

FIGURE 17.5–6 (a) Picture of the weld in a piece taken from a long seam-welded pipe, and (b) cross section of the weldment showing the various regions. *(Source: P. K. Liaw, A. Saxena, and J. Schaffer, "Estimating Remaining Life of Elevated-Temperature Steam Pipes—Part I," Material Properties, Engineering Fracture Mechanics 32, no. 5 (1989), pp. 675–708, Pergamon Press PLC, Great Britain. Reprinted by permission.)*

Solution

These pipes are fabricated from flat plates that are form-rolled to a cylindrical shape and then welded along the pipe length using a submerged arc-welding process and making multiple passes. This process leaves a welded seam along the entire length of the pipe. Figure 17.5–6 shows a picture of a typical weld. The fusion lines between the different weld passes and also the ones between the weld metal and the base metal can be seen in the picture. In the mid-1980s ruptures in such steam pipes in two separate power plants caused considerable damage in both plants and multiple fatalities and injuries in one of them. Figure 17.5–7 shows a picture of a failed pipe. As a result of the incidents, inspections in similar steam pipes were carried out in other power plants, and a number of cracking problems were discovered in pipes that had been in service for approximately 15 years. These failures needed to be understood in order to ensure the safety of workers in the plants and to prevent costly damage to equipment.

Figure 17.5–8 shows a higher-magnification photomicrograph of the cracked region in a pipe that had been in service for several years. Voids known as creep cavities in front of the crack tip are a clear indication of the creep process causing the failure. Hence, it is reasonable to attempt a creep rupture analysis to determine if the design was faulty.

The internal pressure in a steam pipe applies a tangential stress σ_θ on the pipe wall, given by the following approximate equation (for $r/t > 10$):

$$\sigma_\theta = \frac{pr}{t} \qquad\qquad (17.5\text{–}1)$$

where p is the internal pressure, r is the pipe radius, and t is the pipe wall thickness. Thus, $\sigma_\theta = 50$ MPa in this example. The relationship between the creep rupture time t_R and the applied stress for 2.25 Cr–1 Mo steel used in this piping is given by:

$$\sigma = f[T(20 + \log t_R)] \qquad\qquad (17.5\text{–}2)$$

FIGURE 17.5–7

A picture of a burst in a large high-temperature welded steam pipe. The opening at the center of the burst is approximately 2 meters. *(Source: Courtesy of Detroit Edison Co. Reprinted with permission.)*

FIGURE 17.5–8

An optical micrograph of the end of a crack in a steam pipe clearly showing the damage in the form of creep cavities that first form at the grain boundaries and then subsequently grow in size and ultimately join together to form larger cracks.

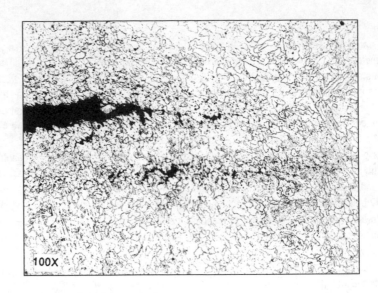

FIGURE 17.5–9

A plot of the Larsen-Miller parameter for creep rupture as a function of the applied stress for a Cr-Mo steel used in making steam pipes.   (Source: P. K. Liaw, A. Saxena, and J. Schaffer, "Estimating Remaining Life of Elevated-Temperature Steam Pipes—Part I," Material Properties, Engineering Fracture Mechanics 32, no. 5 (1989), pp. 675–708, Pergamon Press PLC, Great Britain. Reprinted by permission of the author.)

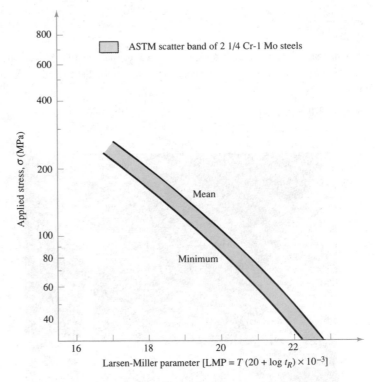

where T is absolute temperature and σ is the applied stress. The term in the brackets, $T(20 + \log t_R)$, called the Larsen-Miller parameter (LMP), was described in Section 9.6.3. Plotting the applied stress against the LMP yields a temperature-independent relationship that can be described by the function f in Equation 17.5–2. From Figure 17.5–9, the minimum value of LMP for a stress of 50 MPa is 21,100. Hence, at 811 K, this corresponds to a t_R that can be calculated as follows:

$$811(20 + \log t_R) = 21{,}100$$

or

$$\log t_R = 6.0172$$

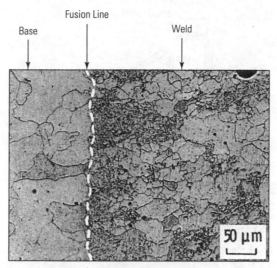

FIGURE 17.5–10   The microstructure of the weldment in a Cr-Mo steel pipe. The large change in the microstructure in going from the base metal to weld metal is apparent.   *(Source: P. K. Liaw, A. Saxena, and J. Schaffer, "Estimating Remaining Life of Elevated-Temperature Steam Pipes—Part I," Material Properties, Engineering Fracture Mechanics 32, no. 5 (1989), pp. 675–708, Pergamon Press PLC, Great Britain. Reprinted by permission of the author.)*

or

$$t_R = 1.04 \times 10^6 \text{ h} = 118.7 \text{ years}$$

Therefore, based on this calculation, the minimum expected life of the pipe should be over 118 years. However, the cracking has been found at times almost an order of magnitude smaller. The creep rupture data used in the above calculation pertained to the base metal, but the problems occur in the weldments. Therefore, a reasonable conclusion would be to assume that the weldments have a lower creep strength than the base metal. Figure 17.5–10 shows the microstructure of the base metal and the weld metal in the region of the fusion line. The density of inclusions appears to be larger along the fusion line as compared with the base metal. This most likely reduces the creep rupture strength of the weldment, which was not taken into account in the above estimate of rupture life.

Since several pipes of this type are operating in service, it is important to inspect these pipes for cracks on a regular basis. The pipes in which cracks are found must be replaced. This additional maintenance requirement increases the operating costs of the plant. Another remedy would be to reduce the temperature or pressure of the steam so that the weldments are subjected to less severe conditions. However, this requires accepting a penalty in the amount of power generated and the efficiency of energy conversion. Since safety cannot be compromised, one of the two actions is required, both of which result in economic penalties. However, these economic penalties are preferred over risking a rupture. In future power plant designs, the welding process must be changed so that the inclusion density is reduced, or the use of welded steam pipes should be discontinued. Seamless pipes can be used in their place. Another approach would be to reduce the stress by choosing a thicker pipe.

CASE STUDY   Failure in Wire Bonds in Electronic Circuits

Modern electronic circuits contain hundreds of bonds between gold wire and aluminum pads that form the electrical connection between the silicon chip and the package housing. These bonds are made by a thermo-compression or thermosonic process. Their geometry is illustrated schematically in Figure 17.5–11. The process consists of an Au ball pressed onto the Al pad on one side and a wedge-shaped bond on the other side. Failure have been found in several of these wires during service. We will examine the reasons for the failure.

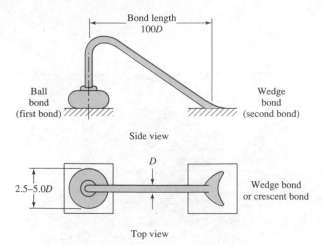

Bond length
100D

Ball
bond
(first bond)

Wedge
bond
(second bond)

Side view

D

2.5–5.0D

Wedge bond
or crescent bond

Top view

FIGURE 17.5–11   Schematic illustration of the wire bond in modern high-performance electronic packages. A ball-shaped joint results on one side and a wedge-shaped joint on the other.   *(Source:* Electronic Materials Handbook, *Vol. I, 1989,* Packaging, *ASM International, Materials Park, OH. Reprinted by permission of the publisher.)*

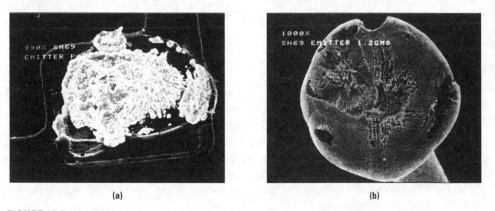

(a)   (b)

FIGURE 17.5–12   (a) Failures in wire bonds from thermal fatigue of the brittle intermetallic compound, and **(b)** formation of voids at the interface due to increase in the amount of intermetallic with time.   *(Source:* Electronic Materials Handbook, *Vol. 1, 1989,* Packaging, *ASM International, Materials Park, OH. Reprinted by permission of the publisher.)*

Solution

Al and Au can interdiffuse and form two types of brittle intermetallic compounds. These intermetallic compounds, Au_2Al and Au_5Al_2, can gradually form in the temperature range of 23 to 50°C, which is the service temperature range for these circuits. The brittle intermetallic layer formed on the bonding surface causes fatigue failures due to thermal cycling, as shown in Figure 17.5–12a. The loss of the gold to intermetallic compounds can also cause void formation in the bonds, which progresses with time and eventually causes fracture (see Figure 17.5–12b).

C A S E   S T U D Y   Failure in a Polyethylene Pipe

A fracture surface of a polyethylene pipe was shown in Figure 17.5–1. During service, this pipe was subjected to an estimated axial stress of 6.9 MPa and a hoop stress of 6.2 MPa from water pressure. In addition, the pipe experienced severe cyclic bending strains in the failed region on the order of 6%. Bending strains in long piping can occur because of failures in hangers that mechanically support the pipe. Six percent strain is far more severe than normal for such pipes.

Solution

The pipe contained a subsurface imperfection (the dark, diamond-shaped spot in Figure 17.5–1) that acted as a crack starter. Concentric circular striations seemed to originate from the crack starter and grew simultaneously in the radial and circumferential directions. These striations represent crack arrest lines, where the distance between two striations represents the increase in crack length during a single excursion (or burst). The spacing between the striations (or the band width) increased with crack length in the circumferential direction, clearly showing that the amount of crack extension increased with each additional burst. This indicates the increase in the stress intensity parameter K with increase in crack size (see Chapter 9). Crack propagation continued until the defect grew sufficiently in the radial and circumferential directions to reach critical size and cause failure. In summary, the cause of the failure was a combination of the severe bending strains in service and the manufacturing defect in the pipe wall.

17.5.2 Failure Analysis in Composite Materials

The principles of failure analysis described in the preceding sections for monolithic materials apply equally to composite materials. The fracture modes, however, are quite different for composites. In this section, we will briefly examine some prominent composite failure modes.

Figure 17.5–13 shows a cross-sectional optical micrograph of laminated composite with fibers oriented at different fixed angles with respect to the major loading direction. The fixed angles commonly used in making laminated composites are $0°$, $45°$, $-45°$, and $90°$ to the major load axis. The ply angles are chosen on the basis of the desired mechanical properties in the different directions. Because the stresses can vary significantly within adjacent plies, high shear stresses are generated between the piles that can cause interlaminar shear failure. The resulting fracture surface is parallel to the plane of fibers, as shown schematically in Figure 17.5–14a. Figure 17.5–14b shows an SEM fractograph of an interlaminar shear failure at a high magnification. The large deformation in the matrix

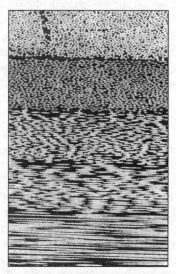

FIGURE 17.5–13 The microstructure of a laminated composite with fibers oriented at various angles from $0°$ in the top layer to $90°$ in the bottom layer. The picture also shows the defects in stacking of fibers and the interlaminar regions with low fiber concentration making them weak. *(Source:* Metals Handbook, *9th ed., Vol. 11, 1986,* Failure Analysis and Prevention, *ASM International, Materials Park, OH. Reprinted by permission of the publisher.)*

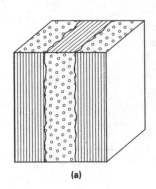

(a)

(b)

FIGURE 17.5–14

(a) Schematic of an interlaminar fracture in a laminate composite, and **(b)** a high-magnification photograph of an interlaminar fracture surface.
(Source: **(b)** Metals Handbook, *9th ed., Vol. 11, 1986,* Failure Analysis and Prevention, *ASM International, Materials Park, OH. Reprinted by permission of the publisher.)*

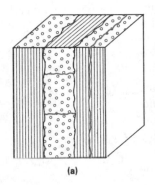

(a)

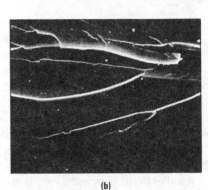

(b)

FIGURE 17.5–15

(a) Schematic of intralaminar fractures in composites, and **(b)** an SEM high-magnification picture of an intralaminar fracture. *(Source:* **(b)** Metals Handbook, *9th ed., Vol. 11, 1986,* Failure Analysis and Prevention, *ASM International, Materials Park, OH. Reprinted by permission of the publisher.)*

indicates a fracture accompanied by considerable energy absorption, making the material behave like a ductile material.

Intralaminar fractures occur within the plies in the weak matrix, as shown in Figure 17.5–15a. As in the case of interlaminar shear fracture, the fracture surface in intralaminar fracture is parallel to the plane of the fibers. However, intralaminar fracture is a result of normal stresses acting in a direction perpendicular to the fiber planes. Little energy is absorbed during such fracture. At the microscopic level, these fractures are brittle in appearance as shown by the flat features in the SEM fractograph shown in Figure 17.5–15b.

The third fracture mode is translaminar fracture, which occurs because fibers break, as shown schematically in Figure 17.5–16. This occurs when enough fibers break on a given cross section of the specimen that the load carried by the broken fibers cannot be transferred to the unfractured fibers. In these fractures, the amount of energy absorbed depends on the amount of fiber pullout. For example, Figure 17.5–17a shows a fracture in a polymer matrix composite characterized by a small amount of energy absorbed. On the other hand, Figure 17.5–17b shows a fracture with a considerable amount of fiber pullout in a B-Al system. The energy required for fiber pullout must be supplied by the external load. Hence, fiber pullout can add considerably to the fracture toughness of the composites and makes them behave like ductile materials. This is a common mechanism of toughening in structural ceramics. To achieve more fiber pullout in ceramic-matrix composite materials, short, discontinuous fibers are often used. By using short fibers, less tensile force is required on the fibers to pull them out, thus reducing the chances of breaking the fibers.

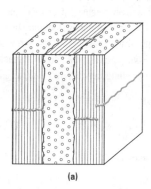

(a)   (b)

FIGURE 17.5–16   (a) Schematic of a translaminar fracture in a laminate composite, and **(b)** high-magnification SEM picture of a translaminar fracture.   *(Source:* Metals Handbook, *9th ed., Vol. 11, 1986,* Failure Analysis and Prevention, *ASM International Materials Park, OH. Reprented by permission of the publisher.)*

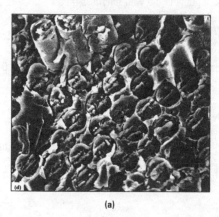

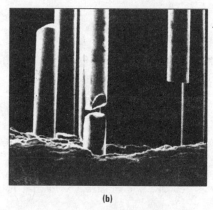

(a)   (b)

FIGURE 17.5–17   (a) Low-energy fiber fracture in a polymer-matrix composite showing fibers breaking almost in one plane, and **(b)** picture of fiber pullout in an Al-B composite resulting in considerable energy absorption, making the behavior of the composite appear ductile.   *(Source:* Metals Handbook, *9th ed, Vol. 11, 1986,* Failure Analysis and Prevention, *ASM International, Materials Park, OH. Reprinted by permission of the publisher.)*

17.5.3 Failure Prevention

Responsible operating practice, good maintenance, and thorough periodic inspection of the fracture-critical components are the best ways to avoid failures. Materials engineers are frequently involved in choosing inspection intervals and techniques.

Earlier in this chapter and in Chapter 9, we discussed the fracture mechanics approach for determining inspection criteria during manufacturing. The same principles can be used for determining safe inspection intervals and the criteria for run/retire decisions during service. A closely related consideration is the selection of the inspection technique. The technique must be able to detect flaws, or cracks, that are larger than the size that will cause failure. In-service inspection intervals must then be determined on the basis of the largest flaw that can be missed by this technique, as illustrated in the following case study.

An aerospace pressure vessel is made from Al. The outside diameter of the vessel is 20 cm, and it is 1 cm thick. This vessel is subjected to 10 MPa hydrogen pressure, which is gradually released over four hours to 2 MPa. The vessel is then repressurized. The design life of the vessel is five years. The manufacturing process leaves the possibility of longitudinal flaws running along the entire length of the pressure vessel. The vessel is inspected using an ultrasonic method (described later in this section). You are to recommend an inspection criterion during manufacturing and an inspection interval to be following during service. The fatigue crack growth behavior of this Al alloy is given by the equation:

$$\frac{da}{dN} = (8.5 \times 10^{-12})(\Delta K)^4 \tag{17.5-3}$$

where ΔK is in MPa-$\sqrt{m}$ and da/dN is in m/cycle. The fracture toughness K_{Ic} of this alloy is 30 MPa-$\sqrt{m}$.

Solution

The ΔK for this flaw is given approximately by the following simplified equation:

$$\Delta K = 1.12 \, \Delta\sigma \, \sqrt{\pi a} \, f\left(\frac{a}{t}\right) \tag{17.5-4}$$

where $\Delta\sigma$ is the cyclic hoop stress range, a is the crack depth, t is the wall thickness, and

$$f\left(\frac{a}{t}\right) = 1 + 0.361\left(\frac{a}{t}\right) + 2.28\left(\frac{a}{t}\right)^2 + 2.9\left(\frac{a}{t}\right)^3 \tag{17.5-5}$$

The function $f(a/t)$ is shown in Figure 17.5–18a. The hoop stress can be estimated using Equation 17.5–1 (i.e., $\sigma = pr/t$). Hence, $\Delta\sigma$ is estimated as:

$$\Delta\sigma = \frac{(10 \text{ MPa} - 2 \text{ MPa})(0.1 \text{ m})}{0.01 \text{ m}} = 80 \text{ MPa}$$

The expected design life is (6 cycles/day)(365 day/yr)(5 yr) = 10,950 cycles. The critical crack size at fracture is given by a_f, which can be obtained as follows:

$$30 \text{ MPa-}\sqrt{m} = 1.12\sigma_{max}\sqrt{\pi a_f} \, f\left(\frac{a_f}{t}\right)$$

where σ_{max} is the maximum stress (= 100 MPa) that occurs at maximum pressure (= 10 MPa). Solving for $\sqrt{a_f} \, f(a_f/t)$ gives:

$$\sqrt{a_f} \, f\left(\frac{a_f}{t}\right) = 0.151 \, \sqrt{m}$$

Figure 17.5–18a shows a plot of $\sqrt{a} f(a/t)$ versus crack size a. From this figure, a value of $a_f = 0.005$ m is obtained.

Next, we estimate an initial crack size a that will permit a design life of 10,950 cycles. Integrating Equation 17.5–3 after substituting for ΔK from Equation 17.5–4 gives:

$$\int_0^{10,950} dN = \int_{a_0}^{0.005} \frac{da}{8.5 \times 10^{-12}(1.12 \, \Delta\sigma \, \sqrt{\pi a} \, f(a/t))^4}$$

or

$$59.2 = \int_{a_0}^{0.005} \frac{da}{a^2 f^4}$$

From Figure 17.5–18d, $a_0 = 2.2 \times 10^{-3}$ m = 2.2 mm. The ultrasonic system selected to detect flaws that are 2.2 mm deep can miss flaws that are 3.5 mm deep with a finite probability. Therefore, in-service inspection

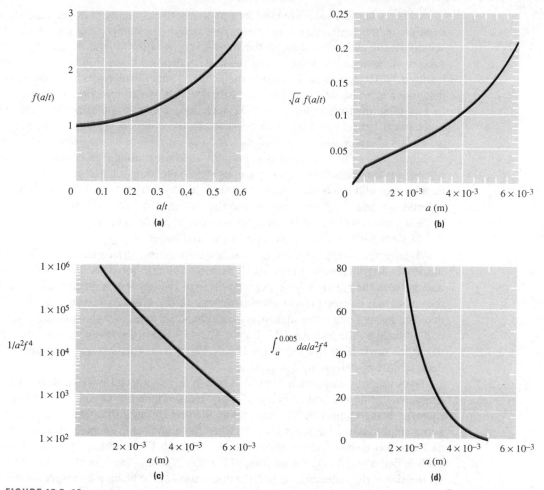

FIGURE 17.5–18 (a) The K-calibration factor, f, as a function of normalized crack size; (b) the function $\sqrt{a}\,f(a/t)$ versus a; (c) the function $1/\{a^2[f(a/t)]^4\}$ versus a; and (d) a versus the function $\int_a^{0.005}[da/(a^2 f^4)]$.

should be scheduled at intervals such that a flaw 3.5 mm deep does not become critical between successive inspections. To estimate the number of cycles between inspections, N_i, the following equation should be solved:

$$\int_0^{N_i} dN = \int_{3.5\times10^{-3}}^{5\times10^{-3}} \frac{da}{8.5\times10^{-12}(1.12\,\Delta\sigma\,\sqrt{\pi a}\,f)^4}$$

or

$$5.406\times10^{-3}\,N_i = \int_{3.5\times10^{-3}}^{5\times10^{-3}} \frac{da}{a^2 f^4}$$

From Figure 17.5–18d, the value of the integral on the right-hand side is 10.0. Solving for N_i gives $N_i = 1849$ cycles. This corresponds to every 10 months. In summary, the recommendation is to specify an ultrasonic inspection process capable of detecting flaws deeper than approximately 2.0 mm and not missing flaws any deeper than 3.5 mm. Also, the pressure vessel must be inspected using the same procedure every 10 months.

A variety of techniques are used for nondestructive component inspection. The choice depends on the material, the geometry and size of the component, the desired level of sensitivity for detecting defects, the most likely location of the defects (for example, surface versus internal defects), and the cost. A detailed discussion of all nondestructive testing (NDT) techniques is outside the scope of this introductory text. We restrict our discussion to some simple techniques; others will be mentioned only in passing.

Hardness measurement is an NDT technique for ensuring that the right thermal-mechanical treatments have been imparted to the component. This technique was described in Chapter 9.

Liquid penetrant inspection is commonly used for finding surface flaws in essentially nonporous materials. This technique consists of spreading a liquid that wets the component surface and soaks into cracks or defects by capillary action. If the liquid contains a colored dye or a flourescent compound that is easily visible under fluorescent light, the surface outlines of the cracks or defects become visible. This technique reportedly can detect flaws with surface openings of 4 μm and larger.

Magnetic particle inspection is a method for locating surface and subsurface discontinuities in ferromagnetic materials. It is most suitable for detecting discontinuities or defects with their planes lying perpendicular to the applied magnetic field because the defect surface causes magnetic fields to leak to the surface. The leakage field is then detected by applying finely divided ferromagnetic particles over the surface, which gather and are held by the leakage field. Figure 17.5–19 shows the magnetic particles gathered to outline two defects in a forged steel hook. Paints and other coatings adversely affect the detectability of defects by this technique.

The ultrasonic inspection method is widely used to detect internal flaws and surface flaws. With this method, a beam of high-frequency sound waves is introduced in the material being inspected. The sound waves travel through the good material, suffering some amplitude attenuation. However, when the sound beam encounters an interface such as a crack surface, a portion of the beam is reflected. The reflected beam can be analyzed both to detect and locate the position of the flaw. The amplitude of the reflection depends on the size of the reflecting surface, the orientation of the beam with respect to the surface (the strongest reflection is realized when the beam is normal to the reflecting surface), and the physical state of matter on the opposite side of the interface. For instance, sound waves are almost completely reflected at metal-gas interfaces and are only partially reflected at

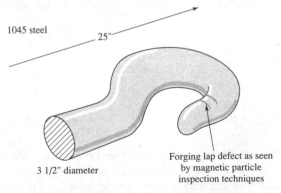

1045 steel

25"

3 1/2" diameter

Forging lap defect as seen by magnetic particle inspection techniques

FIGURE 17.5–19 Magnetic particle inspection technique used to outline flaws in a forged steel hook. *(Source: Metals Handbook, 8th ed., Vol. 11, 1976,* Nondestructive Inspection, *ASM International, Materials Park, OH. Reprinted by permission of the publisher.)*

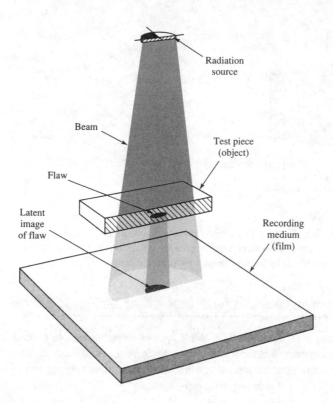

Radiation
source

Beam

Test piece
(object)

Flaw

Latent
image
of flaw

Recording
medium
(film)

FIGURE 17.5–20

A schematic of the X-ray radiography method for detecting internal flaws in components. *(Source: Metals Handbook, 8th ed., Vol. 11, 1976,* Nondestructive Inspection, *ASM International, Materials Park, OH. Reprinted by permission of the publisher.)*

metal-liquid interfaces. Cracks, laminations, shrinkage cavities, pores, welding defects, and so on act as metal-gas interfaces. Hence, this technique is particularly suited for detecting such defects.

Radiography, used for nondestructive detection of internal defects in components or assemblies, is based on differential absorption of penetrating electromagnetic radiation in the form of X rays or gamma rays. The component is bombarded with a beam of X rays, and the radiation transmitted through the component falls upon an X-ray film, exposing the film emulsion similar to the way light exposes a photographic film. The developed film provides a two-dimensional image of the defect. The region around the defect has different absorption characteristics compared with the rest of the material, which give rise to the image. Figure 17.5–20 shows the essential principle behind this technique. Cracks lying normal to an X-ray beam may go undetected by this technique. This is because the strength of the beam exiting the material depends only on the thickness of material traversed. The presence of a crack does not generally alter the thickness normal to the crack plane. For this reason, several angles should be used when making radiographs (see Figure 17.5–21).

Other nondestructive methods include eddy current techniques, optical holography, and neutron radiography. The acoustic emission technique is based on detecting stress waves emitted from growing cracks due to release of strain energy and can be used for monitoring crack growth during service. This technique is quite suitable for composite and other nonmetallic materials that are otherwise difficult to inspect. Optical fibers may be embedded in composite structures, and their characteristics are monitored during service. The signals can be used to detect damage and impending failure. This a new technology with strong potential for composite materials. These components/materials are often referred to as smart structures or smart materials.

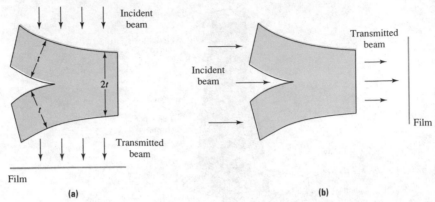

FIGURE 17.5–21 The effect of angle on the detectability of a crack. **(a)** The crack is undetected because the amount of material traversed by the beam is the same for the cracked and uncracked portions. **(b)** The crack is visible, since less material is traversed by the X-ray beam in the cracked region.

CASE STUDY Choosing Optimum Locations for Probes during Ultrasonic Testing

In a previous example we discussed creep failure in seam-welded pipes. One effective method of failure prevention in these systems is to periodically inspect for flaws or cracks. If defects are oriented in the axial and circumferential directions, the transmitter and receiver probes have to be carefully located and arranged to realize the best chance of detecting these defects.

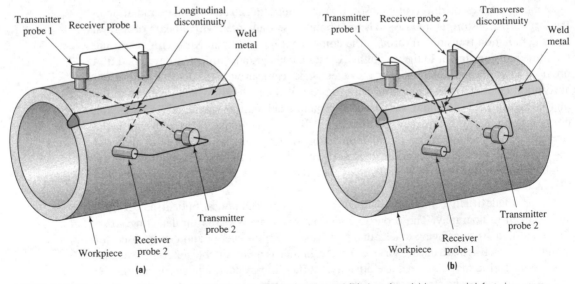

FIGURE 17.5–22 Location of ultrasonic probes for detecting **(a)** longitudinal and **(b)** circumferential (transverse) defects in a seam-welded steel pipe. *(Source: Metals Handbook, 8th ed., Vol. 11, 1976,* Nondestructive Inspection, *ASM International, Materials Park, OH. Reprinted by permission of the publisher.)*

Solution

The ultrasonic inspection system depends on detecting the sound wave reflected from the defect surface. Since the angle between the reflected wave and the surface normal to the defect is the same as the angle between the incident wave and the normal, the most suitable arrangement is as shown in Figure 17.5–22. This figure shows an arrangement consisting of two transmitter probes and two receiver probes, which maximizes the chances of detecting longitudinally and circumferentially oriented defects.

SUMMARY

We have seen that materials are an important aspect of engineering design because the choice of material affects the cost, the manufacturing process, the component design life, the inspection method and frequency, and the risk of failure. Unified life-cycle cost engineering is a useful aid in the materials selection process because it includes the total cost of an engineering component throughout its design life.

Material and process selection methods must be chosen in view of the specific property and cost requirements for each application. The requirements for a "package" for an advanced electronic device and a "package" to contain nuclear waste are different, and, therefore, the appropriate materials and processing methods also differ. The selection of appropriate materials and processes can be guided by the use of industry standards developed jointly by the manufacturers and users of engineering components. It is important to recognize that material disposal costs will continue to increase. Thus, the design process should include an analysis of waste minimization and address the issue of recycled, or recyclable, materials.

Failure risks, including product liability, can be quantified, and the results show that there are always trade-offs between minimizing risk of failure and the cost of the product. The optimum conditions lie in acceptance of a finite failure probability. However, the level of acceptable risk (or failure probability) varies with the application.

Good operating practices, including periodic nondestructive inspection, can help to prevent failures. Fracture mechanics can be used to determine inspection criteria and inspection intervals. If a component does eventually fail, failure analysis techniques can and should be used to determine the cause of the failure so that the component can be redesigned to minimize the future risks of using the component.

HOMEWORK PROBLEMS

1. Explain what is meant by unified life-cycle cost engineering (ULCE). Find from your own experience an example similar to the exhaust pipe discussed in Section 17.2, and go through the ULCE steps.

2. What are the major cost elements during the life cycle of an engineering component?

3. Construct a cash flow diagram to compare the economics of operating car A, which costs $30,000 and consumes three gallons of gas every 100 kilometers, with that of car B, which costs $15,000 and consumes three gallons of gas every 150 kilometers. Assume that the lifetime of car A is 10 years, that of car B is 5 years, and the owner drives 25,000 kilometers/year. At the end of their lifetime, the residual worth of the cars is 20% of their original value. Neglect inflation and the cost of capital, and assume that maintenance costs for the two cars are similar.

SECTION 17.2
Unified Life-Cycle Cost
Engineering (ULCE)

4. Explain the benefits of developing materials and processing standards.

5. You are asked to select a material for a highly stressed part in aircraft landing gear. Make a list of all mechanical properties you will check to determine the suitability of various materials.

6. What are the advantages of concurrent engineering? Use the example of the design of a ceramic-matrix composite gas turbine blade to illustrate the approach of concurrent engineering.

7. List the reasons why aluminum beverage containers are easier to recycle than steel or plastic containers.

8. Discuss how the retreading of car tires will contribute to protecting the environment.

9. Suppose you are asked to select a material for carrying liquid oxygen in a spacecraft. Thoroughly discuss the considerations involved in choosing the right material.

10. Corrosion resistance and creep deformation of lead containers for storing spent nuclear fuel rods were discussed in this chapter. If the sites for disposing of nuclear waste are selected in remote areas, what additional material problems do you foresee during transportation of these containers from nuclear plants to the disposal site?

11. Discuss why lead in free-cutting copper alloys does not embrittle the alloy and why even small amounts of bismuth additions to copper significantly reduce its ductility.

12. What is the role of In and Sn in developing lead-free, free-cutting copper alloys?

13. The fatigue endurance limit of a motor shaft steel is 200 MPa with a standard deviation of 20 MPa. Calculate the failure probability of shafts if the average fatigue stress the shafts encounter during service is 120 MPa and the standard deviation is 50 MPa.

14. You are asked to recommend an inspection criterion for a structural component and a realistic inspection interval during service. How will you go about developing a sound recommendation?

15. Why is it necessary to accept a calculated risk of failure in components in situations where human lives are not involved?

16. What are the steps involved in analyzing failures in field components?

17. What are fatigue striations, and what quantitative information can you obtain by photographing them?

18. A seam-welded pipe is carrying supersaturated steam at 550°C in a power plant. The pipeline goes through a high-traffic region. What precautions will you recommend to minimize the risk of failure?

19. A 30-cm-thick pressure vessel wall made from steel has to be inspected for surface and internal defects. Discuss the suitability of the various NDT techniques for finding flaws in this vessel.

20. Explain why flaws can sometimes escape detection during ultrasonic and radiographic inspection.

APPENDICES

APPENDIX A **Periodic Table**

GROUP →

I A	II A	III A	IV A	V A	VI A	VII A	← VIII A →		I B	II B	III B	IV B	V B	VI B	VII B	VIII B

Metals ▌Nonmetals

1 **H** 1.0079																2 **He** 4.0026
3 **Li** 6.941	4 **Be** 9.012										5 **B** 10.811	6 **C** 12.011	7 **N** 14.007	8 **O** 15.999	9 **F** 18.998	10 **Ne** 20.180
11 **Na** 22.99	12 **Mg** 24.305				*d* Transition Elements						13 **Al** 26.982	14 **Si** 28.086	15 **P** 30.974	16 **S** 32.066	17 **Cl** 35.453	18 **Ar** 39.948
19 **K** 39.098	20 **Ca** 40.078	21 **Sc** 44.956	22 **Ti** 47.88	23 **V** 50.942	24 **Cr** 51.996	25 **Mn** 54.938	26 **Fe** 55.847	27 **Co** 58.933	28 **Ni** 58.69	29 **Cu** 63.546	30 **Zn** 65.39	31 **Ga** 69.723	32 **Ge** 72.610	33 **As** 74.921	34 **Se** 78.960	35 **Br** 79.904	36 **Kr** 83.80
37 **Rb** 85.468	38 **Sr** 87.620	39 **Y** 88.906	40 **Zr** 91.224	41 **Nb** 92.906	42 **Mo** 95.940	43 **Tc** (97.907)	44 **Ru** 101.07	45 **Rh** 102.906	46 **Pd** 106.42	47 **Ag** 107.87	48 **Cd** 112.41	49 **In** 114.82	50 **Sn** 118.71	51 **Sb** 121.75	52 **Te** 127.60	53 **I** 126.90	54 **Xe** 131.29
55 **Cs** 132.91	56 **Ba** 137.33	57 **La*** 138.91	72 **Hf** 178.49	73 **Ta** 180.95	74 **W** 183.85	75 **Re** 186.21	76 **Os** 190.20	77 **Ir** 192.22	78 **Pt** 195.08	79 **Au** 196.97	80 **Hg** 200.59	81 **Tl** 204.38	82 **Pb** 207.20	83 **Bi** 208.98	84 **Po** (208.99)	85 **At** (209.99)	86 **Rn** (222.02)
87 **Fr** (223.02)	88 **Ra** (226.03)	89 **Ac*** (227.03)	104 **Unq** (261.11)	105 **Unp** (262.11)	106 **Unh** (262.12)												

Gas ─┐ ┌─── Atomic number
      | 34
      | **Se**
Liquid ─┘ └─── Atomic mass (g mol⁻¹)
         78.96

f Transition Elements

*Lanthanides (Rare Earths)	58 **Ce** 140.12	59 **Pr** 140.91	60 **Nd** 144.24	61 **Pm** (144.91)	62 **Sm** 150.36	63 **Eu** 151.97	64 **Gd** 157.25	65 **Tb** 158.93	66 **Dy** 162.50	67 **Ho** 164.94	68 **Er** 167.26	69 **Tm** 168.93	70 **Yb** 173.04	71 **Lu** 174.97
**Actinides	90 **Th** 232.04	91 **Pa** (231.04)	92 **U** (238.05)	93 **Np** (237.05)	94 **Pu** (244.06)	95 **Am** (243.06)	96 **Cm** (247.07)	97 **Bk** (247.07)	98 **Cf** (242.06)	99 **Es** (252.08)	100 **Fm** (257.10)	101 **Md** (258.10)	102 **No** (259.10)	103 **Lr** (260.11)

APPENDIX B Physical and Chemical Data for the Elements

Atomic number	Element	Symbol	1s	2s	2p	3s	3p	3d	4s	4p	4d	4f	5s	5p	5d	5f	6s	6p	6d	7s	Atomic mass (a) (amu)	Density of solid (b) (at 20°C) (Mg/m^3 = g/cm^3)	Crystal structure (c) (at 20°C)	Melting point (d) (°C)	Atomic number	Electronegativity*
1	Hydrogen	H	1																		1.008	—	—	−259.34 (T.P.)	1	2.10
2	Helium	He	2																		4.003	—	—	−271.69	2	—
3	Lithium	Li		1																	6.941	0.533	BCC	180.6	3	0.98
4	Beryllium	Be		2																	9.012	1.85	HCP	1289	4	1.57
5	Boron	B		2	1																10.81	2.47		2092	5	2.04
6	Carbon	C		2	2																12.01	2.27	hex.	3826 (S.P.)	6	2.55
7	Nitrogen	N		2	3																14.01	—		−210.0042 (T.P.)	7	3.04
8	Oxygen	O		2	4																16.00	—		−218.789 (T.P.)	8	3.44
9	Fluorine	F		2	5																19.00	—		−219.67 (T.P.)	9	3.98
10	Neon	Ne		2	6																20.18	—		−248.587	10	
11	Sodium	Na				1															22.99	0.966	BCC	97.8	11	0.93
12	Magnesium	Mg				2															24.31	1.74	HCP	650	12	1.31
13	Aluminum	Al				2	1														26.98	2.70	FCC	660.452	13	1.61
14	Silicon	Si				2	2														28.09	2.33	dia. cub.	1414	14	1.90
15	Phosphorus	P				2	3														30.97	1.82 (white)	ortho.	44.14 (white)	15	2.19
16	Sulfur	S				2	4														32.06	2.09 (white)	ortho.	115.22	16	2.58
17	Chlorine	Cl				2	5														35.45	—		−100.97 (T.P.)	17	3.16
18	Argon	Ar				2	6														39.95	—		−189.352 (T.P.)	18	
19	Potassium	K							1												39.10	0.862	BCC	63.71	19	0.82
20	Calcium	Ca							2												40.08	1.53	FCC	842	20	1.00
21	Scandium	Sc						1	2												44.96	2.99	FCC	1541	21	1.36
22	Titanium	Ti						2	2												47.90	4.51	HCP	1670	22	1.54
23	Vanadium	V						3	2												50.94	6.09	BCC	1910	23	1.63
24	Chromium	Cr						5	1												52.00	7.19	BCC	1863	24	1.66
25	Manganese	Mn						5	2												54.94	7.47	cubic	1246	25	1.55
26	Iron	Fe						6	2												55.85	7.87	BCC	1538	26	1.83
27	Cobalt	Co						7	2												58.93	8.8	HCP	1495	27	1.88
28	Nickel	Ni						8	2												58.71	8.91	FCC	1455	28	1.91
29	Copper	Cu						10	1												63.55	8.93	FCC	1084.87	29	1.90
30	Zinc	Zn						10	2												65.38	7.13	HCP	419.58	30	1.65
31	Gallium	Ga						10	2	1											69.72	5.91	ortho.	29.7741 (T.P.)	31	1.81
32	Germanium	Ge						10	2	2											72.59	5.32	dia.cub.	938.3	32	2.01
33	Arsenic	As						10	2	3											74.92	5.78	rhomb.	603 (S.P.)	33	2.18
34	Selenium	Se						10	2	4											78.96	4.81	hex.	221	34	2.55
35	Bromine	Br						10	2	5											79.90	—		−7.25 (T.P.)	35	2.96
36	Krypton	Kr						10	2	6											83.80	—		−157.385	36	
37	Rubidium	Rb											1								85.47	1.53	BCC	39.48	37	0.82
38	Strontium	Sr											2								87.62	2.58	FCC	769	38	0.95
39	Yttrium	Y									1		2								88.91	4.48	HCP	1522	39	1.22
40	Zirconium	Zr									2		2								91.22	6.51	HCP	1855	40	1.33
41	Niobium	Nb									4		1								92.91	8.58	BCC	2469	41	1.6

Electronic configuration[a] (number of electrons in each group)

Z	Element	Symbol	Atomic weight	Crystal structure	Melting point	Density	Electronegativity
42	Molybdenum	Mo	95.94	BCC	2623	10.22	2.16
43	Technetium	Tc	98.91	HCP	2204	11.50	1.9
44	Ruthenium	Ru	101.07	HCP	2334	12.36	2.2
45	Rhodium	Rh	102.91	FCC	1963	12.42	2.28
46	Palladium	Pd	106.4	FCC	1555	12.00	2.20
47	Silver	Ag	107.87	FCC	961.93	10.50	1.93
48	Cadmium	Cd	112.4	HCP	321.108	8.65	1.69
49	Indium	In	114.82	FCT	156.634	7.29	1.78
50	Tin	Sn	118.69	BCT	231.9681	7.29	1.96
51	Antimony	Sb	121.75	rhomb.	630.755	6.69	2.05
52	Tellurium	Te	127.60	hex.	449.57	6.25	2.1
53	Iodine	I	126.90	hex.	113.6 (T.P.)	4.95	2.66
54	Xenon	Xe	131.30	ortho	−111.7582 (T.P.)		2.6
55	Cesium	Cs	132.91	BCC	28.39	1.91 (−10°)	0.79
56	Barium	Ba	137.33	BCC	729	3.59	0.89
57	Lanthanum	La	138.91	hex.	918	6.17	1.10
58	Cerium	Ce	140.12	FCC	798	6.77	1.12
59	Praseodymium	Pr	140.91	hex.	931	6.78	1.13
50	Neodymium	Nd	144.24	hex.	1021	7.00	1.14
51	Promethium	Pm	(145)	hex.	1042		1.13
62	Samarium	Sm	150.4	rhomb.	1074	7.54	1.17
63	Europium	Eu	151.96	BCC	822	5.25	1.2
64	Gadolinium	Gd	157.25	HCP	1313	7.87	1.20
65	Terbium	Tb	158.93	HCP	1356	8.27	1.1
66	Dysprosium	Dy	162.50	HCP	1412	8.53	1.22
67	Holmium	Ho	164.93	HCP	1474	8.80	1.23
68	Erbium	Er	167.26	HCP	1529	9.04	1.24
69	Thulium	Tm	168.93	HCP	1545	9.33	1.25
70	Ytterbium	Yb	173.04	FCC	819	6.97	1.1
71	Lutetium	Lu	174.97	HCP	1663	9.84	1.27
72	Hafnium	Hf	178.49	HCP	2231	13.28	1.3
73	Tantalum	Ta	180.95	BCC	3020	16.67	1.5
74	Tungsten	W	183.85	BCC	3422	19.25	2.36
75	Rhenium	Re	186.2	HCP	3186	21.02	1.9
76	Osmium	Os	190.2	HCP	3033	22.58	2.2
77	Iridium	Ir	192.22	FCC	2447	22.55	2.20
78	Platinum	Pt	195.09	FCC	1769.0	21.44	2.28
79	Gold	Au	196.97	FCC	1064.43	19.28	2.54
80	Mercury	Hg	200.59		−38.836	13.6	2.00
81	Thallium	Tl	204.37	HCP	304	11.87	2.04
82	Lead	Pb	207.2	FCC	327.502	11.34	2.33
83	Bismuth	Bi	208.98	rhomb.	271.442	9.80	2.02

The central portion of the chart gives the electron configuration (number of electrons in each subshell; the "Xenon core" is indicated) for each element.

*Percent ionic character of a single chemical bond

Difference in electronegativity	0.1	0.2	0.3	0.4	0.5	0.6	0.7	0.8	0.9	1.0	1.1	1.2	1.3	1.4	1.5	1.6	1.7	1.8	1.9	2.0	2.1	2.2	2.3	2.4	2.5	2.6	2.7	2.8	2.9	3.0	3.1	3.2
Percent ionic character %	0.5	1	2	4	6	9	12	15	19	22	26	30	34	39	43	47	51	55	59	63	67	70	74	76	79	82	84	86	88	89	91	92

Sources: Data from

(a) *Handbook of Chemistry and Physics*, 58th ed., R. C. Weast, ed. (Boca Raton, Fla.: CRC Press, 1977).

(b) X-ray diffraction measurements are tabulated in B. D. Cullity, *Elements of X-ray Diffraction*, 2nd ed. (Reading, Mass.: Addison-Wesley Publishing Co., 1978).

(c) R. W. G. Wyckoff, *Crystal Structure*, 2nd ed., Vol. 1, Interscience Publishers, New York, 1963; and *Metals Handbook*, 9th ed., Vol. 2 (Metals Park, Ohio: American Society for Metals, 1979).

(d) *Binary Alloy Phase Diagrams*, Vols. 1 and 2, T. B. Massalski, ed. (Metals Park, Ohio: American Society for Metals, 1986)

T.P. = triple point. S.P. = sublimation point at atmospheric pressure.

APPENDIX B   Physical and Chemical Data for the Elements (concluded)

Electronic configuration[a]
(number of electrons in each group)

Note: Columns 1s, 2s, 2p, 3s, 3p, 3d, 4s, 4p, 4d, 5s, 5p comprise the **Radon core**.

Atomic number	Element	Symbol	4f	5d	5f	6s	6p	6d	7s	Atomic mass (a) (amu)	Density of solid (b) (at 20°C) (Mg/m³ = g/cm³)	Crystal structure (c) (at 20°C)	Melting point (d) (°C)	Atomic number	Electronegativity*
84	Polonium	Po	14	10		2	4			(~210)	9.2	monoclinic	254	84	2.0
85	Astatine	At	14	10		2	5			(210)			≈302	85	2.2
86	Radon	Rn	14	10		2	6			(222)			−71	86	—
87	Francium	Fr							1	(223)			≈27	87	0.7
88	Radium	Ra							2	226.03		BCC	700	88	0.89
89	Actinium	Ac						1	2	(227)		FCC	1051	89	1.1
90	Thorium	Th						2	2	232.04	11.72	FCC	1755	90	1.3
91	Protoactinium	Pa			2			1	2	231.04		BCT	1572	91	1.5
92	Uranium	U			3			1	2	238.03	19.05	ortho.	1135	92	1.38
93	Neptunium	Np			4			1	2	237.05		ortho.	639	93	1.36
94	Plutonium	Pu			6				2	(244)	19.81	monoclinic	640	94	1.28
95	Americium	Am			7				2	(243)		hex.	1176	95	1.3
96	Curium	Cm			7			1	2	(247)		hex.	1345	96	1.3
97	Berkelium	Bk			9				2	(247)		hex.	1050	97	1.3
98	Californium	Cf			10				2	(251)			900	98	1.3
99	Einsteinium	Es			11				2	(254)			860	99	1.3
100	Fermium	Fm			12				2	(257)			≈1527	100	1.3
101	Mendelevium	Md			13				2	(258)			≈827	101	1.3
102	Nobelium	No			14				2	(259)			≈827	102	1.3
103	Lawrencium	Lr			14			1	2	(260)			≈1627	103	1.3

APPENDIX C   Atomic and Ionic Radii of the Elements

Atomic number	Symbol	Atomic radius (nm)	Ion	Ionic radius (nm)
1	H	0.046	H^-	0.154
2	He	—	—	—
3	Li	0.152	Li^+	0.078
4	Be	0.114	Be^{2+}	0.054
5	B	0.097	B^{3+}	0.02
6	C	0.077	C^{4+}	< 0.02
7	N	0.071	N^{5+}	0.01–0.02
8	O	0.060	O^{2-}	0.132
9	F	—	F^-	0.133
10	Ne	0.160	—	—
11	Na	0.186	Na^+	0.098
12	Mg	0.160	Mg^{2+}	0.078
13	Al	0.143	Al^{3+}	0.057
14	Si	0.117	Si^{4-}	0.198
			Si^{4+}	0.039
15	P	0.109	P^{5+}	0.03–0.04
16	S	0.106	S^{2-}	0.174
			S^{6+}	0.034
17	Cl	0.107	Cl^-	0.181
18	Ar	0.192	—	—
19	K	0.231	K^+	0.133
20	Ca	0.197	Ca^{2+}	0.106
21	Sc	0.160	Sc^{2+}	0.083
22	Ti	0.147	Ti^{2+}	0.076
			Ti^{3+}	0.069
			Ti^{4+}	0.064
23	V	0.132	V^{3+}	0.065
			V^{4+}	0.061
			V^{5+}	~0.04
24	Cr	0.125	Cr^{3+}	0.064
			Cr^{6+}	0.03–0.04
25	Mn	0.112	Mn^{2+}	0.091
			Mn^{3+}	0.070
			Mn^{4+}	0.052
26	Fe	0.124	Fe^{2+}	0.087
			Fe^{3+}	0.067
27	Co	0.125	Co^{2+}	0.082
			Co^{3+}	0.065
28	Ni	0.125	Ni^{2+}	0.078
29	Cu	0.128	Cu^+	0.096
			Cu^{2+}	0.072
30	Zn	0.133	Zn^{2+}	0.083
31	Ga	0.135	Ga^{3+}	0.062
32	Ge	0.122	Ge^{4+}	0.044
33	As	0.125	As^{3+}	0.069
			As^{5+}	~0.04
34	Se	0.116	Se^{2-}	0.191
			Se^{6+}	0.03–0.04
35	Br	0.119	Br^-	0.196
36	Kr	0.197	—	—
37	Rb	0.251	Rb^+	0.149
38	Sr	0.215	Sr^{2+}	0.127
39	Y	0.181	Y^{3+}	0.106
40	Zr	0.158	Zr^{4+}	0.087
41	Nb	0.143	Nb^{4+}	0.074
			Nb^{5+}	0.069
42	Mo	0.136	Mo^{4+}	0.068
			Mo^{6+}	0.065

Source: After a tabulation by R. A. Flinn and P. K. Trojan, *Engineering Materials and Their Applications* (Boston: Houghton Mifflin, 1986). The ionic radii are based on the calculations of V. M. Goldschmidt, who assigned radii based on known interatomic distances in various ionic crystals. Reprinted with permission.

Atomic number	Symbol	Atomic radius (nm)	Ion	Ionic radius (nm)
43	Tc	—	—	—
44	Ru	0.134	Ru^{4+}	0.065
45	Rh	0.134	Rh^{3+}	0.068
			Rh^{4+}	0.065
46	Pd	0.137	Pd^{2+}	0.050
47	Ag	0.144	Ag^{+}	0.113
48	Cd	0.150	Cd^{2+}	0.103
49	In	0.157	In^{3+}	0.092
50	Sn	0.158	Sn^{4-}	0.215
			Sn^{4+}	0.074
51	Sb	0.161	Sb^{3+}	0.090
52	Te	0.143	Te^{2-}	0.211
			Te^{4+}	0.089
53	I	0.136	I^{-}	0.220
			I^{5+}	0.094
54	Xe	0.218	—	—
55	Cs	0.265	Cs^{+}	0.165
56	Ba	0.217	Ba^{2+}	0.143
57	La	0.187	La^{3+}	0.122
58	Ce	0.182	Ce^{3+}	0.118
			Ce^{4+}	0.102
59	Pr	0.183	Pr^{3+}	0.116
			Pr^{4+}	0.100
60	Nd	0.182	Nd^{3+}	0.115
61	Pm	—	Pm^{3+}	0.106
62	Sm	0.181	Sm^{3+}	0.113
63	Eu	0.204	Eu^{3+}	0.113
64	Gd	0.180	Gd^{3+}	0.111
65	Tb	0.177	Tb^{3+}	0.109
			Tb^{4+}	0.089
66	Dy	0.177	Dy^{3+}	0.107
67	Ho	0.176	Ho^{3+}	0.105
68	Er	0.175	Er^{3+}	0.104
69	Tm	0.174	Tm^{3+}	0.104
70	Yb	0.193	Yb^{3+}	0.100
71	Lu	0.173	Lu^{3+}	0.099
72	Hf	0.159	Hf^{4+}	0.084
73	Ta	0.147	Ta^{5+}	0.068
74	W	0.137	W^{4+}	0.068
			W^{6+}	0.065
75	Re	0.138	Re^{4+}	0.072
76	Os	0.135	Os^{4+}	0.067
77	Ir	0.135	Ir^{4+}	0.066
78	Pt	0.138	Pt^{2+}	0.052
			Pt^{4+}	0.055
79	Au	0.144	Au^{+}	0.137
80	Hg	0.150	Hg^{2+}	0.112
81	Tl	0.171	Tl^{+}	0.149
			Tl^{3+}	0.106
82	Pb	0.175	Pb^{4-}	0.215
			Pb^{2+}	0.132
			Pb^{4+}	0.084
83	Bi	0.182	Bi^{3+}	0.120
84	Po	0.140	Po^{6+}	0.067
85	At	—	At^{7+}	0.062
86	Rn	—	—	—
87	Fr	—	Fr^{+}	0.180
88	Ra	—	Ra^{+}	0.152
89	Ac	—	Ac^{3+}	0.118
90	Th	0.180	Th^{4+}	0.110
91	Pa	—	—	—
92	U	0.138	U^{4+}	0.105

APPENDIX D Mechanical Properties

TABLE D–1 Tensile properties of selected carbon and alloy steels in hot-rolled, normalized, and annealed conditions.

AISI No.*	Treatment	Austenitizing temperature (°C)	Tensile strength (MPa)	Yield strength (MPa)	Elongation (%)†	Reduction in area (%)	Hardness, BHN
1015	As-rolled	—	420.6	313.7	39.0	61.0	126
	Normalized	925	424.0	324.1	37.0	69.6	121
	Annealed	870	386.1	284.4	37.0	69.7	111
1020	As-rolled	—	448.2	330.9	36.0	59.0	143
	Normalized	870	441.3	346.5	35.8	67.9	131
	Annealed	870	394.7	294.8	36.5	66.0	111
1030	As-rolled	—	551.6	344.7	32.0	57.0	179
	Normalized	925	520.6	344.7	32.0	60.8	149
	Annealed	845	463.7	341.3	31.2	57.9	126
1040	As-rolled	—	620.5	413.7	25.0	50.0	201
	Normalized	900	589.5	374.0	28.0	54.9	170
	Annealed	790	518.8	353.4	30.2	57.2	149
1050	As-rolled	—	723.9	413.7	20.0	40.0	229
	Normalized	900	748.1	427.5	20.0	39.4	217
	Annealed	790	636.0	365.4	23.7	39.9	187
1060	As-rolled	—	813.6	482.6	17.0	34.0	241
	Normalized	900	775.7	420.6	18.0	37.2	229
	Annealed	790	625.7	372.3	22.5	38.2	179
1080	As-rolled	—	965.3	586.1	12.0	17.0	293
	Normalized	900	1010.1	524.0	11.0	20.6	293
	Annealed	790	615.4	375.8	24.7	45.0	174
1095	As-rolled	—	965.3	572.3	9.0	18.0	293
	Normalized	900	1013.5	499.9	9.5	13.5	293
	Annealed	790	656.7	379.2	13.0	20.6	192
1117	As-rolled	—	486.8	305.4	33.0	63.0	143
	Normalized	900	467.1	303.4	33.5	63.8	137
	Annealed	855	429.5	279.2	32.8	58.0	121
1118	As-rolled	—	521.2	316.5	32.0	70.0	149
	Normalized	925	477.8	319.2	33.5	65.9	143
	Annealed	790	450.2	284.8	34.5	66.8	131
1137	As-rolled	—	627.4	379.2	28.0	61.0	192
	Normalized	900	668.8	396.4	22.5	48.5	197
	Annealed	790	584.7	344.7	26.8	53.9	174
1141	As-rolled	—	675.7	358.5	22.0	38.0	192
	Normalized	900	706.7	405.4	22.7	55.5	201
	Annealed	815	598.5	353.0	25.5	49.3	163
1144	As-rolled	—	703.3	420.6	21.0	41.0	212
	Normalized	900	667.4	399.9	21.0	40.4	197
	Annealed	790	584.7	346.8	24.8	41.3	167
1340	Normalized	870	836.3	558.5	22.0	62.9	248
	Annealed	800	703.3	436.4	25.5	57.3	207
3140	Normalized	870	891.5	599.8	19.7	57.3	262
	Annealed	815	689.5	422.6	24.5	50.8	197
4130	Normalized	870	668.8	436.4	25.5	59.5	197
	Annealed	865	560.5	360.6	28.2	55.6	156
4140	Normalized	870	1020.4	655.0	17.7	46.8	302
	Annealed	815	655.0	417.1	25.7	56.9	197
4150	Normalized	870	1154.9	734.3	11.7	30.8	321
	Annealed	815	729.5	379.2	20.2	40.2	197
4320	Normalized	895	792.9	464.0	20.8	50.7	235
	Annealed	850	579.2	609.5	29.0	58.4	163

TABLE D–1   (*concluded*)

AISI No.*	Treatment	Austenitizing temperature (°C)	Tensile strength (MPa)	Yield strength (MPa)	Elongation (%)[†]	Reduction in area (%)	Hardness, BHN
4340	Normalized	870	1279.0	861.8	12.2	36.3	363
	Annealed	810	744.6	472.3	22.0	49.9	217
4620	Normalized	900	574.3	366.1	29.0	66.7	174
	Annealed	855	512.3	372.3	31.3	60.3	149
4820	Normalized	860	75.0	484.7	24.0	59.2	229
	Annealed	815	681.2	464.0	22.3	58.8	197
5140	Normalized	870	792.9	472.3	22.7	59.2	229
	Annealed	830	572.3	293.0	28.6	57.3	167
5150	Normalized	870	870.8	529.5	20.7	58.7	255
	Annealed	825	675.7	357.1	22.0	43.7	197
5160	Normalized	855	957.0	530.9	17.5	44.8	269
	Annealed	815	722.6	275.8	17.2	30.6	197
6150	Normalized	870	939.8	615.7	21.8	61.0	269
	Annealed	815	667.4	412.3	23.0	48.4	197
8620	Normalized	915	632.9	357.1	26.3	59.7	183
	Annealed	870	536.4	385.4	31.3	62.1	149
8630	Normalized	870	650.2	429.5	23.5	53.5	187
	Annealed	845	564.0	372.3	29.0	58.9	156
8650	Normalized	870	1023.9	688.1	14.0	40.4	302
	Annealed	795	715.7	386.1	22.5	46.4	212
8740	Normalized	870	929.4	606.7	16.0	47.9	269
	Annealed	815	695.0	415.8	22.2	46.4	201
9255	Normalized	900	932.9	579.2	19.7	43.4	269
	Annealed	845	774.3	486.1	21.7	41.1	229
9310	Normalized	890	906.7	570.9	18.8	58.1	269
	Annealed	845	820.5	439.9	17.3	42.1	241

*All grades are fine-grained except for those in the 1100 series, which are coarse-grained. Heat-treated specimens were oil-quenched unless otherwise indicated.
[†]Elongation in 50 mm.
Source: Adapted from *Metals Handbook, Desk Edition,* p. 4.20, ASM International, Materials Park, Ohio, 1984. Reprinted by permission of the publisher.

TABLE D–2   Tensile properties of selected carbon and alloy steels in the quenched-and-tempered condition.

AISI No.*	Tempering temperature (°C)	Tensile strength (MPa)	Yield strength (MPa)	Elongation (%)[†]	Reduction in area (%)	Hardness, HB
1040	205	779	593	19	48	262
	315	779	593	20	53	255
	425	758	552	21	54	241
	540	717	490	26	57	212
	650	634	434	29	65	192
1050	205	—	—	—	—	—
	315	979	724	14	47	321
	425	938	655	20	50	277
	540	876	579	23	53	262
	650	738	469	29	60	223
1060	205	1103	779	13	40	321
	315	1103	779	13	40	321
	425	1076	765	14	41	311
	540	965	669	17	45	277
	650	800	524	23	54	229

ABLE D–2 *(continued)*

AISI No.*	Tempering temperature (°C)	Tensile strength (MPa)	Yield strength (MPa)	Elongation (%)[†]	Reduction in area (%)	Hardness, HB
1080	205	1310	979	12	35	388
	315	1303	979	12	35	388
	425	1289	951	13	36	375
	540	1131	807	16	40	321
	650	889	600	21	50	255
1095	205	1289	827	10	30	401
	315	1262	813	10	30	375
	425	1213	772	12	32	363
	540	1089	676	15	37	321
	650	896	552	21	47	269
1137	205	1082	938	5	22	352
	315	986	841	10	33	285
	425	876	731	15	48	262
	540	758	607	24	62	229
	650	655	483	28	69	197
1141	205	1634	1213	6	17	461
	315	1462	1282	9	32	415
	425	1165	1034	12	47	331
	540	896	765	18	57	262
	650	710	593	23	62	217
1144	205	876	627	17	36	277
	315	869	621	17	40	262
	425	848	607	18	42	248
	540	807	572	20	46	235
	650	724	503	23	55	217
1340	205	1806	1593	11	35	505
	315	1586	1420	12	43	453
	425	1262	1151	14	51	375
	540	965	827	17	58	295
	650	800	621	22	66	252
4037	205	1027	758	6	38	310
	315	951	765	14	53	295
	425	876	731	20	60	270
	540	793	655	23	63	247
	650	696	421	29	60	220
4042	205	1800	1662	12	37	516
	315	1613	1455	13	42	455
	425	1289	1172	15	51	380
	540	986	883	20	59	300
	650	793	689	28	66	238
4140	205	1772	1641	8	38	510
	315	1551	1434	9	43	445
	425	1248	1138	13	49	370
	540	951	834	18	58	285
	650	758	655	22	63	230
4150	205	1931	1724	10	39	530
	315	1765	1593	10	40	495
	425	1517	1379	12	45	440
	540	1207	1103	15	52	370
	650	958	841	19	60	290
4340	205	1875	1675	10	38	520
	315	1724	1586	10	40	486
	425	1469	1365	10	44	430
	540	1172	1076	13	51	360
	650	965	855	19	60	280

TABLE D-2 *(continued)*

AISI No.*	Tempering temperature (°C)	Tensile strength (MPa)	Yield strength (MPa)	Elongation (%)[†]	Reduction in area (%)	Hardness, HB
5046	205	1744	1407	9	25	482
	315	1413	1158	10	37	401
	425	1138	931	13	50	336
	540	938	765	18	61	282
	650	786	655	24	66	235
50B46	205	—	—	—	—	560
	315	1779	1620	10	37	505
	425	1393	1248	13	47	405
	540	1082	979	17	51	322
	650	883	793	22	60	273
50B60	205	—	—	—	—	600
	315	1882	1772	8	32	525
	425	1510	1386	11	34	435
	540	1124	1000	15	38	350
	650	896	779	19	50	290
5130	205	1613	1517	10	40	475
	315	1496	1407	10	46	440
	425	1275	1207	12	51	379
	540	1034	938	15	56	305
	650	793	689	20	63	245
5140	205	1793	1641	9	38	490
	315	1579	1448	10	43	450
	425	1310	1172	13	50	365
	540	1000	862	17	58	280
	650	758	662	25	66	235
5150	205	1944	1731	5	37	525
	315	1737	1586	6	40	475
	425	1448	1310	9	47	410
	540	1124	1034	15	54	340
	650	807	814	20	60	270
5160	205	2220	1793	4	10	627
	315	1999	1772	9	30	555
	425	1606	1462	10	37	461
	540	1165	1041	12	47	341
	650	896	800	20	56	269
51B60	205	—	—	—	—	600
	315	—	—	—	—	540
	425	1634	1489	11	36	460
	540	1207	1103	15	44	355
	650	965	869	20	47	290
6150	205	1931	1689	8	38	538
	315	1724	1572	8	39	483
	425	1434	1331	10	43	420
	540	1158	1069	13	50	345
	650	945	841	17	58	282
81B45	205	2034	1724	10	33	550
	315	1765	1572	8	42	475
	425	1407	1310	11	48	405
	540	1103	1027	16	53	338
	650	896	793	20	55	280
8630	205	1641	1503	9	38	465
	315	1482	1392	10	42	430
	425	1276	1172	13	47	375
	540	1034	896	17	54	310
	650	772	689	23	63	240

TABLE D–2 (concluded)

AISI No.*	Tempering temperature (°C)	Tensile strength (MPa)	Yield strength (MPa)	Elongation (%)[†]	Reduction in area (%)	Hardness, HB
8640	205	1862	1669	10	40	505
	315	1655	1517	10	41	460
	425	1379	1296	12	45	400
	540	1103	1034	16	54	340
	650	896	800	20	62	280
86B45	205	1979	1641	9	31	525
	315	1696	1551	9	40	475
	425	1379	1317	11	41	395
	540	1103	1034	15	49	335
	650	903	876	19	58	280
8650	205	1937	1675	10	38	525
	315	1724	1551	10	40	490
	425	1448	1324	12	45	420
	540	1172	1055	15	51	340
	650	965	827	20	58	280
8660	205	—	—	—	—	580
	315	—	—	—	—	535
	425	1634	1551	13	37	460
	540	1310	1213	17	46	370
	650	1068	951	20	53	315
8740	205	1999	1655	10	41	578
	315	1717	1551	11	46	495
	425	1434	1358	13	50	415
	540	1207	1138	15	55	363
	650	986	903	20	60	302
9255	205	2103	2048	1	3	601
	315	1937	1793	4	10	578
	425	1606	1489	8	22	477
	540	1255	1103	15	32	352
	650	993	814	20	42	285
9260	205	—	—	—	—	600
	315	—	—	—	—	540
	425	1758	1503	8	24	470
	540	1324	1131	12	30	390
	650	979	814	20	43	295
94B30	205	1724	1551	12	46	475
	315	1600	1420	12	49	445
	425	1344	1207	13	57	382
	540	1000	931	16	65	307
	650	827	724	21	69	250

*All grades are fine-grained except for those in the 1100 series, which are coarse-grained. Heat-treated specimens were oil-quenched.
[†]Elongation in 50 mm.

Source: Adapted from *Metals Handbook, Desk Edition*, p. 4.21, ASM International, Materials Park, Ohio, 1984. Reprinted by permission of the publisher.

TABLE D–3 Tensile properties of non-heat-treatable wrought aluminium alloys.

Alloy	Temper	Tensile strength (MPa)	Yield strength (MPa)*	Elongation (%)[†]	Hardness, HB[‡]
1100	O	90	35	35	23
	H14	125	115	9	32
	H18	165	150	5	44

TABLE D-3 (*continued*)

Alloy	Temper	Tensile strength (MPa)	Yield strength (MPa)*	Elongation (%)[†]	Hardness, HB[‡]
3003	O	110	40	30	28
	H14	150	145	8	40
	H18	200	185	4	55
Alclad 3003	O	110	40	30	—
	H14	150	145	8	—
	H18	200	185	4	—
3004	O	180	70	20	45
	H34	240	200	9	63
	H38	285	250	5	77
	H19	295	285	2	
Alclad 3004	O	180	70	20	—
	H34	240	200	9	—
	H38	285	250	5	—
	H19	295	285	2	—
3104	H19	290	260	4	—
3005	O	130	55	25	—
	H14	180	165	7	—
	H18	240	225	4	—
3105	O	115	55	24	—
	H25	180	160	8	—
	H18	215	195	3	—
5005	O	125	40	25	28
	H34	160	140	8	41
	H38	200	185	5	55
5042	H19	360	345	4.5	—
5050	O	145	55	24	36
	H34	190	165	8	53
	H38	200	200	6	63
5052	O	195	90	25	47
	H34	260	215	10	68
	H38	290	255	7	77
5252	O	180	85	23	46
	H25	235	170	11	68
	H28	285	240	5	75
5154	O	240	115	27	58
	H34	290	230	13	73
	H38	330	270	10	80
	H112	240	115	25	63
5454	O	250	115	22	62
	H34	305	240	10	81
	H111	260	180	14	70
	H112	250	125	18	62
5056	O	290	150	35	65
	H18	435	405	10	105
	H38	310	345	15	100
5456	O	310	160	24	—
	H112	310	165	22	—
	H116	350	255	16	90
5457	O	130	50	22	32
	H25	180	160	12	48
	H28	205	185	6	55
5657	O	110	40	25	28
	H25	160	140	12	40
	H28	195	165	7	50

TABLE D-3   (concluded)

Alloy	Temper	Tensile strength (MPa)	Yield strength (MPa)*	Elongation (%)†	Hardness, HB‡
5082	H19	395	370	4	—
5182	O	275	130	21	—
	H19	420	395	4	—
5083	O	290	145	22	—
	H116	315	230	16	—
5086	O	260	115	22	—
	H34	325	255	10	—
	H112	270	130	14	—
	H116	290	205	12	—
7072	O	70	—	15	—
	H113	75	—	15	—
8001	O	110	40	30	—
	H18	200	185	4	—
8081	H25	165	145	13	—
	H112	195	170	10	—
8280	O	115	50	28	—
	H18	220	205	4	—

*At 0.2% offset.
†In 50 mm or 2 in.
‡500-kg load, 10-mm ball, 30 s.
Source: Adapted from *Metals Handbook, Desk Edition*, p. 6.33, ASM International, Materials Park, Ohio, 1984. Reprinted by permission of the publisher.

TABLE D-4   Tensile properties of heat-treatable wrought aluminium alloys.

Alloy	Temper	Tensile strength (MPa)	Yield strength (MPa)*	Elongation (%)†	Hardness, HB‡
2011	T3	380	295	15	95
	T6	395	270	17	97
	T8	405	310	12	100
2014	O	185	95	18	45
	T4, T451	425	290	20	105
	T6, T651	485	415	13	135
Alclad 2014	O	170	70	21	—
	T3	435	275	20	—
	T4, T451	420	255	22	—
	T6, T651	470	415	10	—
2017	O	180	70	22	45
	T4, T451	425	275	22	105
2117	T4	300	165	27	70
2218	T72	330	255	11	95
2618	T61	435	370	10	—
2219	O	170	70	18	—
	T42	360	185	20	—
	T31, T351	360	250	17	100
	T37	395	315	11	117
	T62	415	290	10	115
	T81, T85x	455	350	10	130
	T87	475	395	10	130
2024	O	185	75	20	47
	T3	485	345	18	120

TABLE D-4   (*continued*)

Alloy	Temper	Tensile strength (MPa)	Yield strength (MPa)*	Elongation (%)[†]	Hardness, HB[‡]
2024	T361	495	395	13	130
	T4, T351	470	325	20	120
	T81, T851	485	450	6	128
	T86	515	490	6	135
Alclad 2024	O	180	76	20	—
	T3	450	310	18	—
	T36	460	365	11	—
	T4, T351	440	290	19	—
	T8, T851	450	415	6	—
	T86	485	455	6	—
2124	T351	470	325	20	120
	T851	485	450	6	128
2025	T6	400	255	19	110
2036	T4	340	195	24	—
4032	T6	380	315	9	120
6101	T6	220	195	15	71
6009	T4	230	125	25	62
6010	T4	290	170	24	78
6151	T6	330	295	17	100
6351	T4, T451	290	185	20	60
	T6, T651	340	295	13	95
6951	O	110	40	30	28
	T6	270	230	13	82
6053	O	110	55	35	26
	T6	255	220	13	80
6063	O	90	50	—	25
	T1	150	90	20	42
	T4	170	90	22	—
	T5	185	145	12	60
	T6	240	215	12	73
	T83	255	240	9	82
	T831	205	185	10	70
	T832	290	270	12	95
	T835	330	295	8	105
6463	O	90	50	—	25
	T1	150	90	20	42
	T4	170	90	22	—
	T5	185	145	12	60
	T6	240	215	12	74
6061	O	125	55	25	30
	T4, T451	240	145	22	65
	T6, T651	310	275	12	95
	T91	405	395	12	—
	T913	460	455	10	—
Alclad 6061	O	115	50	25	—
	T4, T451	230	130	22	—
	T6, T651	290	255	12	—
6262	T9	400	380	10	120
6066	O	150	85	18	43
	T4, T451	360	205	18	90
	T6, T651	395	360	12	120
6070	O	145	70	20	35
	T6	380	350	10	120

TABLE D-4   (concluded)

Alloy	Temper	Tensile strength (MPa)	Yield strength (MPa)*	Elongation (%)†	Hardness, HB‡
7001	O	255	150	14	60
	T6, T651	675	625	9	160
	T75	580	495	12	—
7005	O	195	80	20	—
	W	345	205	20	—
	T6	350	290	13	—
7016	T5	360	315	15	96
7021	T62	420	380	13	—
7029	T5	430	380	15	—
7049	T73	540	475	10	146
7050	T74, T7451, T7452	510	450	13	142
7075	O	230	105	17	60
	T6, T651	570	505	11	150
	T73, T735x	505	435	13	—
Alclad 7075	O	220	95	17	—
	T6	525	460	11	—
	T651	525	460	11	—
7175	T736, T7365x	550	485	12	145
	T7351	505	435	13	—
7475	T7351	505	435	14	—
7076	T61	510	470	14	150
7178	O	230	105	15	60
	T6, T651	605	540	10	160
	T76, T7651	570	505	9	—
Alclad 7178	O	220	95	16	—
	T6	560	490	10	—
	T651	560	490	10	—

*At 0.2% offset.
†In 50 mm or 2 in.
‡500-kg load, 10-mm ball.
Source: Adapted from *Metals Handbook, Desk Edition*, p. 6.34–6.35, ASM International, Materials Park, Ohio, 1984.
Reprinted by permission of the publisher.

TABLE D-5   Tensile properties of die-cast aluminum alloys.*

AA No.	Temper	Tensile strength (MPa)	Yield strength (MPa)†	Elongation in 50 mm (%)	Hardness, HB‡	Endurance limit (MPa)§	Modulus of elasticity (kPa × 10⁶)§
360.0	F	324	172	3.0	75	131	71
A360.0	F	317	165	5.0	75	124	—
364.0	F	296	159	7.5	—	124	—
380.0	F	331	165	3.0	80	145	71
A380.0	F	324	159	4.0	80	138	—
384.0	F	324	172	1.0	—	145	71
390.0	F	279	241	1.0	120	138	82
	T5	296	265	1.0	—	—	—
392.0	F	290	262	< 0.5	—	103(f)	—
413.0	F	296	145	2.5	80	131	71

TABLE D–5 (*concluded*)

AA No.	Temper	Tensile strength (MPa)	Yield strength (MPa)[†]	Elongation in 50 mm (%)	Hardness, HB[‡]	Endurance limit (MPa)[§]	Modulus of elasticity (kPa × 10⁶)[§]
A413.0	F	241	110	3.5	80	131	—
443.0	F	228	110	9.0	50	117	71
513.0	F	276	152	10.0	—	124	—
515.0	F	283	—	10.0	—	—	—
518.0	F	310	186	8.0	80	138	—

* Tension properties are average values determined from ASTM standard 6-mm (1/4-in) diam test specimens cast on a cold-chamber (high-pressure) die casting machine.
† At 0.2% offset.
‡ 500-kg load on 10-mm ball.
§ Average of tension and compression moduli; compression modulus is about 2% greater than tension modulus.
Source: Adapted from *Metals Handbook, Desk Edition*, p. 6.60, ASM International, Materials Park, Ohio, 1984. Reprinted by permission of the publisher.

TABLE D–6   Tensile properties of wrought, cast, and powder metallurgy titanium alloys.

Product and condition	Tensile strength (MPa)	Yield strength (MPa)	Elongation (%)	Reduction in area (%)
Unalloyed Ti				
Wrought bar, annealed	550	480	18	33
Cast bar, as cast	635	510	20	31
P/M compact, annealed*	480	370	18	22
Ti–5Al–2.5Sn–ELI				
Wrought bar, annealed	815	710	19	34
Cast bar, as cast	795	725	10	17
P/M compact, annealed and forged†	795	715	16	27
Ti–6Al–4V				
Wrought bar, annealed	1000	925	16	34
Cast bar:				
As cast	1025	880	12	19
Annealed	1015	890	10	16
Solution-treated and aged‡	1180	1085	6	11
P/M compact:				
Annealed*	825–855	740–785	5–8	8–14
Annealed and forged†	925	840	12	27
Solution-treated and aged‡	965	895	4	6

* About 94% dense.
† Almost 100% dense.
‡ Aging treatment not specified.
Source: Adapted from *Metals Handbook, Desk Edition*, p. 9.8, ASM International, Materials Park, Ohio, 1984. Reprinted by permission of the publisher.

TABLE D–7 Tensile properties of wrought nickel-base superalloys:

Astroloy, bar

Temperature °C	°F	Tensile strength (MPa)	Yield strength (MPa)	Elongation (%)
21	70	1410	1050	16
540	1000	1240	965	16
650	1200	1310	965	18
760	1400	1160	910	21
870	1600	770	690	25

Hastelloy X, sheet

Temperature °C	°F	Tensile strength (MPa)	Yield strength (MPa)	Elongation (%)
21	70	785	360	43
540	1000	650	290	45
650	1200	570	275	37
760	1400	435	260	37
870	1600	255	180	50

Inconel X 600, bar

Temperature °C	°F	Tensile strength (MPa)	Yield strength (MPa)	Elongation (%)
21	70	620	250	47
540	1000	580	195	47
650	1200	450	180	39
760	1400	185	115	46
870	1600	105	62	80

Inconel 601, sheet

Temperature °C	°F	Tensile strength (MPa)	Yield strength (MPa)	Elongation (%)
21	70	740	340	45
540	1000	725	150	38
650	1200	525	180	45
760	1400	290	200	73
870	1600	160	140	92

Inconel 625, bar

Temperature °C	°F	Tensile strength (MPa)	Yield strength (MPa)	Elongation (%)
21	70	855	490	50
540	1000	745	405	50
650	1200	710	420	35
760	1400	505	420	42
870	1600	285	475	125

Inconel 718, bar

Temperature °C	°F	Tensile strength (MPa)	Yield strength (MPa)	Elongation (%)
21	70	1430	1190	21
540	1000	1280	1060	18
650	1200	1230	1020	19
760	1400	950	740	25
870	1600	340	330	88

Inconel 718, sheet

Temperature °C	°F	Tensile strength (MPa)	Yield strength (MPa)	Elongation (%)
21	70	1280	1050	22
540	1000	1140	945	26
650	1200	1030	870	15
760	1400	675	625	8

Inconel X 750, bar

Temperature °C	°F	Tensile strength (MPa)	Yield strength (MPa)	Elongation (%)
21	70	1120	635	24
540	1000	965	580	22
650	1200	825	565	9
760	1400	485	455	9
870	1600	235	165	47

René 95, bar

Temperature °C	°F	Tensile strength (MPa)	Yield strength (MPa)	Elongation (%)
21	70	1620	1310	15
540	1000	1540	1250	12
650	1200	1460	1220	14
760	1400	1170	1100	15

Udimet 500, bar

Temperature °C	°F	Tensile strength (MPa)	Yield strength (MPa)	Elongation (%)
21	70	1310	840	32
540	1000	1240	795	28
650	1200	1210	760	28
760	1400	1040	730	39
870	1600	640	495	20

Udimet 520, bar

Temerature °C	°F	Tensile strength (MPa)	Yield strength (MPa)	Elongation (%)
21	70	1310	860	21
540	1000	1240	825	20
650	1200	1170	795	17
760	1400	725	725	15
870	1600	515	515	20

Udimet 700, bar

Temerature °C	°F	Tensile strength (MPa)	Yield strength (MPa)	Elongation (%)
21	70	1410	965	17
540	1000	1280	895	16
650	1200	1240	855	16
760	1400	1030	825	20
870	1600	690	635	27

Waspaloy, bar

Temerature °C	°F	Tensile strength (MPa)	Yield strength (MPa)	Elongation (%)
21	70	1280	795	25
540	1000	1170	725	23
650	1200	1120	690	34
760	1400	795	675	28
870	1600	525	515	35

Source: Adapted from *Metals Handbook, Desk Edition*, p. I6.15, ASM International, Materials Park, Ohio, 1984. Reprinted by permission of the publisher.

TABLE D–8   Rupture strengths of selected wrought nickel-base superalloys.

Temperature		For stress rupture at:	
°C	°F	100 h (MPa)	1000 h (MPa)
Inconel 600			
815	1500	 55	39
870	1600	 37	24
Inconel 601*			
540	1000	.... —	400
870	1600	 48	30
980	1800	 23	14
Inconel 625*			
650	1200	440	370
815	1500	130	93
870	1600	 72	48
Inconel 718[†]			
540	1000	.... —	951
595	1100	860	760
650	1200	690	585

* Solution-treated at 1150°C (2100°F).
[†] Heat-treated to 980°C (1800°F) plus 720°C (1325°F) hold for 8 h, F.C. to 620°C (1150°F), hold for 8 h.

Source: Adapted from *Metals Handbook, Desk Edition*, p. 16.16, ASM International, Materials Park, Ohio, 1984. Reprinted by permission of the publisher.

TABLE D–9   Average strength of selected engineering ceramics.*

	Compressive strength, MPa	Tensile strength, MPa	Flexural strength, MPa
Alumina 85%	1620	124	293
90%	2413	138	317
95%	2413	193	338
99%	2586	207	344
Alumina silicate	275	172	62
ZrO_2–Al_2O_3	2413	—	—
3% Y_2O_3–PSZ[†]	2965	—	1172
TTZ[‡]	1758	352	634
9% MgO–PSZ[†]	1861	—	690
Slip-cast Si_3N_4	138	24	69
Reaction-bonded SiC	690	138	255
Pressureless-sintered SiC	3861	172	551
Sintered SiC + free Si	1034	165	324
Sintered SiC + graphite	413	34	55
Reaction-bonded Si_3N_4	772	—	207
Hot-pressed Si_3N_4	3447	—	861

* Strength is dependent on test method, sample preparation, and sample size.
[†] Partially stabilized zirconia.
[‡] Transformation-toughened zirconia.

Source: "Advanced Materials and Processes," *ASM Journal*, June 1990.

TABLE D–10 Mechanical properties of selected polymers.

Material	Tensile strength (MPa)	Tensile modulus (GPa)	Flexural strength (MPa)	Impact strength (J/m)	Hardness, Rockwell
ABS	41	2.3	72.4	347	R103
ABS-PC	59	2.6	89.6	560	R117
DAP	48	10.3	117	37	E80
POM	69.0	3.2	98.6	133	R120
PMMA	72.4	3.0	110	21	M68
PAR	68	2.1	82.7	288	R122
LCP	120	11.0	124	101	R80
MF	52	9.65	93.1	16	M120
Nylon 6	81.4	2.76	113	59	R119
Nylon 6/6	82.7	2.83	110	53	R121
Nylon 12	81.4	2.3	113	64	R122
PAE	121	8.96	138	64	M85
PBT	52	2.3	82.7	53	R117
PC	69.0	2.3	96.5	694	R118
PBT-PC	55	2.2	86.2	800	R115
PEEK	93.8	3.5	110	59	R120
PEI	105	3.0	152	53	M109
PESV	84.1	2.6	129	75	M88
PET	159	8.96	245	101	R120
PF	41	5.9	62	21	M105
PPO	54	2.5	88.3	267	R115
PPS	138	11.7	179	69	R123
PSU	73.8	2.5	106	64	M69
SMA	31	1.9	55	133	R95
UP	41	5.5	82.7	32	M88

Source: Adapted with permission from Donald V. Rosato, D. P. DiMattia, and Dominick V. Rosato, *Designing with Plastics and Composites: A Handbook* (New York: Chapman & Hall, 1991), p. 533.

TABLE D–11 Fracture toughness of selected metals at room temperature.

Alloy	Form	Specimen orientation	K_{1c} (MPa-$\sqrt{m}$)
Aluminum alloys			
2024-T351	Plate	L-T	31–44
2024-T351	Plate	T-L	30–37
2024-T851	Plate	L-T	23–28
2024-T851	Plate	T-L	21–24
7050-T73651	Plate	L-T	33–41
7050-T73651	Plate	T-L	29–38
7050-T73651	Plate	S-L	25–28
7075-T651	Plate	L-T	27–31
7075-T651	Plate	T-L	25–28
7075-T651	Plate	S-L	16–21
7075-T7351	Plate	L-T	31–35
7075-T7351	Plate	T-L	26–41
7475-T651	Plate	T-L	33–37
7475-T7351	Plate	T-L	39–44

TABLE D–11 (*concluded*)

Alloy	Form	Specimen orientation	K_{1c} (MPa-$\sqrt{m}$)
Ferrous alloys			
4330V (275°C temper)	Forging	L-T	86–94
4330V (425°C temper)	Forging	L-T	103–110
4340 (205°C temper)	Forging	L-T	44–66
4340 (260°C temper)	Plate	L-T	50–63
4340 (425°C temper)	Forging	L-T	79–91
D6AC (540°C temper)	Plate	L-T	102
D6AC (540°C temper)	Plate	L-T	62
9-4-20 (550°C temper)	Plate	L-T	132–154
18 Ni(200) (480°C 6 hr)	Plate	L-T	110
18 Ni(250) (480°C 6 hr)	Plate	L-T	88–97
18 Ni(300) (480°C)	Plate	L-T	50–64
18 Ni(300) (480°C 6 hr)	Forging	L-T	83–105
Titanium alloys			
Ti–6 Al–4 V	(Mill anneal plate)	L-T	123
Ti–6 Al–4 V	(Mill anneal plate)	T-L	106
Ti–6 Al–4 V	(Recryst. anneal plate)	L-T	85–107
Ti–6 Al–4 V	(Recryst. anneal plate)	T-L	77–116

Source: Adapted from MCIC-HB-01, *Damage Tolerance Design Handbook*, MCIC, Battelle Columbus Labs, 1975, p. 369.

TABLE D–12 Fracture toughness of selected ceramics at room temperature.*

Material	Comments	Fracture toughness K_{1c} (MPa-$\sqrt{m}$)
Medium-strength steel		50.0
NaCl	Monocrystal	0.4
Soda-lime glass[†]	Amorphous	0.74 DCB
Aluminosilicate glass	Amorphous	0.91 DCB
ZnSe	Vapor-deposited	0.9
WC	Cobonded	13.0
ZnS	Vapor-deposited	1.0
Si_3N_4	Hot-pressed	5.0
Al_2O_3	MgO-doped	4.0
Al_2O_3 (sapphire)	Monocrystal	2.1
SiC	Hot-pressed	4.0
$SiC-ZrO_2$	Hot-pressed[‡]	5.0
MgF_2	Hot-pressed	0.9
MgO	Hot-pressed	1.2
B_4C	Hot-pressed	6.0
Si	Monocrystal	0.6
ZrO_2	Ca-stabilized	7.6 DCB

*Double torsion measurement technique, except where double cantilever beam test (DCB) indicated.
[†]Commercial sheet glass.
[‡]20% ZrO_2, 14 wt. % mullite. ZrO_2 present in monoclinic form; no transformation toughening.
Source: "Advanced Materials Processes," *ASM Journal*, June 1990.

TABLE D–13   Fracture toughness of selected polymers and composites at room temperature.

Material	Fracture toughness $(\text{MPa-}\sqrt{\text{m}})$
PMMA*	0.8–1.75
PS*	0.8–1.1
PC*	2.75–3.3
E-glass cloth fabric, 50 vol. % epoxy[†]	4.6–5.4
T300 carbon fiber, 934 epoxy[†]	5.5–6.8
Kevlar 49,181 fabric, 50 vol. % [±45]$_2$, Hexcel F-155 epoxy[†]	17.4–27.2

Sources: *Adapted from R. W. Hertzberg, *Defraction and Fracture Mechanics of Engineering Materials* (New York: John Wiley & Sons, 1976), pp. 369–70. Copyright © 1976 by John Wiley & Sons. Reprinted by permission of John Wiley & Sons, Inc.
[†]Adapted from N. L. Harcox and R. M. Mayer, *Design Data for Re-inforced Plastics* (New York: Chapman & Hall, 1994), p. 148.

APPENDIX E: ANSWERS TO SELECTED PROBLEMS

CHAPTER 2

1. 3, 5
5. e = 29, p = 29
9. 2.91
23. A
27. 10 times
29. Ti = 2.9454 Å, Cu = 2.57 Å
31. $CN(Ca^{2+})$ = 8
37. $0.225 \leq (r/R) < 0.414$
39. Same size

CHAPTER 3

9. Hexagonal lattice
11. 8.94 g/cm^3
15. 14.45%
17. $(2\,\bar{2}\,\bar{1})$, $[2\,\bar{2}\,\bar{1}]$
19. 109.5 degrees
23. $(4\,\bar{2}\,1)$, $[4\,\bar{2}\,1]$
25. $0.25/r^2$, $0.18/r^2$, $0.29/r^2$
29. 1.63 r
31. 91 vol %
35. APF = 0.67
37. 0.74
39. 2 in BCC, 6 in FCC
43. 1, 2
45. 5.43 Å
51. NaCl structure
53. CsCl structure
57. 3.37 g/cm^3
79. $d_{(130)}$ $10^{0.5}$
83. 1.15 Å, does not occur
85. Copper

CHAPTER 4

3. 140°C
5. No
7. Cu is solvent, substitutional S. S., Yes
9. Interstitial S. S.
13. 2 moles of oxygen ion vacancies
15. C_{CV} = 3.108 × 10^{-4}, C_{AV} = 9.6 × 10^{-8}, C_{CI} = 5.48 × 10^{-6}
19. N, Ag
21. Interstitial, 5%, slower
23. J = 8.5 × 10^9 atoms/(cm^2-s)
27. 1531 K
33. 25.3 μm
35. 1 hr
41. increase, increase, increase
43. 1.36 hr
45. 1298 K

CHAPTER 5

1. 0.125 MPa
3. 0.407 nm
5. BCC
7. 0.289 nm
9. −4572 psi (compression)
17. 0.296 nm
19. 202.8 MPa, does not change
23. −69.4 MPa, BCC
29. ASTM#10.3, ASTM#6.3, 7620 in^2/in^3, 1905 in^2/in^3
31. 1.76 × 10^6 dislocations/cm
41. 4.76 × 10^{-2} at%

CHAPTER 6

5. 10 × 10^{-6} °C^{-1}
9. range of temperatures

13. unusual
19. 79 °C
21. 312.2 g/mole, approx. 29 mers joined
29. Glassy
31. Yes
35. Silicone rubbers
39. 218 Å

CHAPTER 7

3. FCC
5. 1110, 1070 degrees Celsius
13. Critical temperature
19. 0.15 B, 0.1 A
23. 0.70
27. 10.5 wt% Ag
29. 10.5 to 66.3 wt% Ag
37. monotectic at 955°C, eutectic at 326°C

CHAPTER 8

1. nucleation and growth
15. Yes
17. 75,600 J/mole
19. 38,400 J/mole*K
33. Alloy III-hypo, Alloy I-eutectoid, Alloy II-hyper
35. hypo = 0.7% C, hyper = 1.1% C
39. 0.072
49. F, F, F, F
51. 16°C/s
61. 1.50×10^5 J/mole
65. 119 kJ/mole
67. 1 hour
71. 5.7%
77. n = 3.1960, 1.8145, 2.3807 k = 0.1299, 0.0462, 0.0264

CHAPTER 9

3. 0.05%, 0.5%, 5%
5. Al = 7.09×10^{-4} m, Cu = 4.08×10^{-4} m, W = 1.29×10^{-4} m
7. 3.1 GPa
9. 172 MPa, 0.00448
15. 8.70×10^5 J/m^3
19. $E \propto T_g$
23. 2945 N

37. 7075-T6 Al = 4.07 mm; 300 Maraging = 2.26 cm; $Al_2O_3 = 3.25 \times 10^{-5}$ m
39. Kelvar/Epoxy = 0.102m; E-glass/Epoxy = 0.006 m
41. 192 MPa
43. 125.9 MPa
45. 0.67 MPa-m$^{1/2}$
55. 1.6×10^6 cycles
59. 500 cycles
63. 3150 cycles
67. 1.39×10^{-8} hr^{-1}
69. 1509 hours
75. 658, −62, 100, 417°C

CHAPTER 10

7. 1.49 m^2/(V-s)
9. 0.20 ohms
15. 4.25
29. Electron conductors: 0, 0, <0, <0
    Ionic conductors: >0, >0, >0, >0
    Semiconductors: >0, >0, <0, <0
33. 32.4%
35. −13.4°C, +1.3°C, semiconductor
39. 0.288 eV, 12.5 ohms
45. 0.1 eV
47. 1.58×10^{-4} (ohm-cm)$^{-1}$
49. III-V semiconductor, zinc-blend
57. 3
63. 553°C
65. Dopped with P
67. −163.6 K
69. pn diode

CHAPTER 11

1. No band gap
3. BN, epoxy, diamond
5. C-F
7. 2.04×10^{-9} Å
9. 2.25×10^{-6} %
13. 0.22 d$_{PE}$
19. 8.7×10^3 V/m
21. 0.71 V, 0.36 V
25. 0
27. Bakelite
31. 1.470
33. 4%

35. 1.570

37. n = 1.33

49. 3.5 km

51. 15.5 μm

61. Intensity decreases 19%

67. 14.4s

69. LED

CHAPTER 12

11. 1.0000131, 1.257 $\times$ 10^{-6} (T-m/A)

29. 2.5 $\times$ 10^6 A/m, 3.14 T

33. 2.17 atom^{-1}

35. 0 T, 1.5 $\times$ 10^{-4}(T-m/A); 1,1 T, 6 $\times$ 10^{-6} (T-m/A)

37. 7.54 A/m, -4.15×10^{-5} A/m, 9.47 $\times$ 10^{-6} T

39. 1.7 $\times$ 10^6 A/m; 0 A/m; 2.56 $\times$ 10^5 A/m

45. Cubic ferrites, rare earth ferrites, hexagonal ferrites

51. Co

57. 1.2 $\times$ 10^{-4} watts

63. 300 T, 202 T

CHAPTER 13

7. 572°C

11. 40, 15.2, 208°C

13. W

17. 1307°C, No

23. 121.4 J; 5250 J; 7.14 $\times$ 10^6 J

25. PE

27. MgO

29. 850 W

31. Metals

41. 5.88 $\times$ 10^7 psi

49. 0.737 mm

CHAPTER 14

11. Spectra

15. glass fiber, glass fiber

23. PMCs

27. 14.3%

35. 0.048

39. No, Yes

45. 29.36 GPa, 35 MPa, fiber

49. 31%, 19.6%

51. 145.5 W/m-K

CHAPTER 15

11. static fatigue

15. 11.69 minutes

17. 26.7 years

37. Fe_2O_3

39. welded and galvanized

43. Nb_2O_5

47. Li_2O

53. oxide/atmosphere interface

55. linear, logarithmic

63. PTFE (Teflon)

CHAPTER 16

7. t_c/t_e = 1.06

15. 9.48 m/s

25. No, No

39. machine first

47. diffusion bonding

51. metals

61. etching process, use laser vaporization of etching

63. cost, protection, amount of heated needed to be dissipated

65. 256

67. defects

71. hand lay-up technique

CHAPTER 17

13. 9.3% chance of failure

15. economically favorable

17. obtain rate at which crack grew

G L O S S A R Y

absorption coefficient   A measure of the amount of light absorbed by a transparent material, expressed in units of reciprocal length.

acceptor level   The impurity level within the band gap of a semiconductor associated with the electron holes introduced by the incorporation of a Group II or III element into a Group IV semiconductor.

activation polarization   A mechanism by which the corrosion rate is reduced in some systems. It occurs when one of the steps in the corrosion process has a high activation barrier so that the whole process is slowed considerably.

adhesive bonding   A process by which two parts are joined by an adhesive. This process is used to join polymers to other polymers, metals, and ceramics and also for bonding composites.

aggregate composites   Composites containing reinforcements in the aggregate form such as sand and gravel.

aging   A heat treatment that precipitates a second phase from a supersaturated solid solution.

amorphous material   A material that lacks the long-range order that is characteristic of a crystalline solid. Amorphous materials are either rubbers or glasses.

anisotropic   Having properties that vary with direction.

anodizing   A process by which the surface of a component is converted into a hard oxide by an electrochemical action.

antiferromagnetic   A class of materials in which adjacent magnetic dipoles of equal strength point in opposite directions. These materials show weak response to external magnetic fields.

Arrhenius equation   Any equation of the form $C = C_0 \times \exp(-Q/RT)$. This equation states that the variable C increases exponentially as the temperature increases.

artificial aging   The precipitation of a second phase from a supersaturated solid solution at a temperature above room temperature.

atactic   Configuration of a vinyl polymer whose side groups are randomly positioned during polymerization.

athermal   Describing a phase transformation that is not thermally activated. That is, the extent of the reaction is not a function of time, but depends on temperature. The martensite transformations in steels are athermal.

atomic packing factor (APF)   The fraction of space filled by spherical atoms.

atomic scale structure   The structure of a material on the size scale of an atom (i.e., on the scale of about 10^{-8} to 10^{-10} m).

austenite   The face-centered cubic (FCC) form of iron. The FCC phase in iron alloys is also referred to as austenite.

bainite   A two-phase microstructure consisting of ferrite and cementite. Bainite forms when an austenitic steel is quenched to a temperature below the pearlite region but above the martensite start (M_s) temperature.

band gap   The forbidden range of electron energy levels located between the top of the valence band and the bottom of the conduction band in a semiconductor or insulator.

basal plane   The top or bottom plane in a crystal. In hexagonal crystals the basal plane is defined by the atoms placed on the six corners of hexagons. It is normal to the c axis.

basal slip   Deformation that occurs on the basal plane of HCP crystals as a result of dislocation motion.

basis   The number of atoms or ions on each lattice position.

bipolar junction transistor   A solid-state semiconductor device composed of three layers of doped material—either npn or pnp—often used as an amplifier.

block copolymer   A polymer consisting of two or more mers in which each mer occurs in long sequences in the molecule.

blow molding   A process by which a slug of molten oxide glass or polymer is extruded vertically downward by blowing air from the top.

body-centered cubic (BCC)   A cubic structure in which atoms are located in the corners of the unit cell plus one in the center of the cubic unit cell.

bond angles   The characteristic angle between an atom or ion and two of its nearest neighbors. The term is most often used to describe specific bond angles in covalent structures.

bond energy   The amount of energy required to move two atoms or ions to an infinite separation distance. It is equivalent to the depth of the bond-energy well.

bond-energy curve (bond-energy well)   The curve that describes the energy associated with a pair of atoms or ions as a function of the distance between the two atoms or ions.

bond-force curve   The curve that describes the relationship between the total force between two atoms or ions as a function of the distance between the two atoms or ions. This curve is equivalent to the derivative of the bond-energy curve.

bond length   The equilibrium separation distance between two atoms or ions. It can be estimated from either the bond-energy curve (the point where the energy is a minimum) or the bond-force curve (the point where the total force is zero).

Bragg's law   The defining equation of X-ray diffraction, $n\lambda = 2d \sin \theta$.

brazing   A process of joining two materials in which a low-temperature filler metal is sandwiched between the two pieces being joined and the temperature is raised sufficiently to melt the filler metal but not high enough to melt the materials being joined.

brittle   Term used to describe materials that are unable to absorb energy by "bending" but instead fracture (break into pieces) when subjected to external loads. For example, the glass in a car window is brittle, while the metal or plastic bumper is not. More precisely, brittle materials exhibit a low failure strain.

Burgers circuit   A circuit drawn around a dislocation through defect-free material that would normally close if there were no dislocation inside of the circuit.

Burgers vector b   The vector that joins the end to the start of the Burgers circuit. The Burgers vector is an invariant property of a dislocation.

carbon-carbon composites   Composites in which carbon is used both in the fibers as well as the matrix.

case hardening   A process by which the surface of a component is hardened by changing the chemical composition of the surface or by subjecting the surface to a hardening heat treatment.

casting   A process in which a mold is filled with molten material that upon solidification takes the shape of the mold.

cavitation   A form of stress-assisted corrosion in which vapor bubbles form and then burst on a surface, causing damage to the underlying material.

cementite   An iron carbide phase, Fe_3C.

ceramic-matrix composites   Composites in which ceramics are used as the matrix material.

chemical vapor deposition (CVD)   A process by which two gases are chemically reacted to produce a thin coating of the reaction product, a solid, on the surface of the workpiece.

close-packed direction   A crystallographic direction in a crystal along which atoms touch.

close-packed plane   A crystallographic plane in a crystal on which each atom contacts six neighbors.

close-packed structures   A crystal of a single atom type in which the spherical atoms are packed as tightly as possible, giving a coordination number of 12 and an APF of 0.74. HCP and FCC are close-packed structures.

coefficient of thermal expansion α_{th} Term defined through the equation $\epsilon_{th} = \alpha_{th} \Delta T$, which states that the coefficient of thermal expansion is the constant that relates thermal strain to changes in temperature.

coherent interface   An interface separating two phases where there is a one-to-one match of lattice planes across the boundary.

cold working   Plastic deformation of a metal or alloy at a temperature where dislocations are created faster than they are annihilated.

component   A chemical substance (element or compound) used to specify the composition of an alloy.

composites   Materials formed by combining two or more basic materials in which one material, called the reinforcing phase, is in the form of fibers, sheets, or particles embedded in another material, called the matrix.

compression molding   A process by which a raw material is shaped by heating and squeezing into the desired shape between two platens that act as molds.

concentration polarization   A mechanism by which the corrosion rate is reduced due to low ion diffusion rates within the electrolyte.

concrete   A mixture (or aggregate composite) containing small sand particles and gravel stones in a matrix of cement.

conduction band   The band of permissible electron energy levels located above, but separated by the band gap from, the valence band.

conductivity σ An intrinsic materials property that describes the ease with which electric charge is transported through a material in response to an external electric field.

conductor   A material, often a metal, with a high conductivity. These materials are able to transport electrical charge efficiently.

conformation   The arrangement of the rotatable bonds within a molecule.

congruent melting   The equilibrium melting reaction that occurs when a solid phase transforms to a liquid phase having identical composition.

coordination number   The number of nearest neighbors that surround an atom or ion.

core band   A band of electron energy levels that is (usually) completely filled and is located closer to the nucleus than either the valence or conduction bands.

core electrons   Those electrons that are contained within the filled inner shells of an atom.

coring   Variation in the solute distribution of a single phase due to nonequilibrium solidification. Coring occurs because the composition of the solid and the liquid are not the same at a specific temperature.

corrosion fatigue   A degradation mechanism in which corrosion and mechanical fatigue occur simultaneously and their synergistic effects are much more serious than would be predicted from their separate effects.

coulombic force   Name given to the electrostatic force that develops between charged species. The magnitude of the force is proportional to the charges of the two species and inversely related to the square of their separation distance. This force pulls the species closer together if they have opposite charge and pushes them apart if they have charges of the same sign.

covalent bond   A type of primary bond formed between two electronegative elements when their average number of valence electrons is greater than or equal to 4.

creep   Time-dependent deformation that occurs at relatively high temperatures and low stresses.

crevice corrosion   A highly localized form of corrosion (usually) involving oxygen concentration gradients in regions where the flow of the electrolyte is limited.

critical fiber length   The minimum fiber length-to-diameter ratio for load to be effectively transferred to the strong fibers in a discontinuous fiber composite.

critical nucleus   Nucleus of the size such that either growth or dissolution will decrease its free energy.

critical resolved shear stress (τ_{CRSS})   The shear stress resolved in the slip plane in the slip direction at which plastic deformation occurs.

crosslinks   Primary bonds formed between adjacent polymer chains. These atomic "bridges" are often composed of small chains of either oxygen or sulfur but can be composed of many other small groups of atoms.

crystal lattice   The framework on which atoms are placed in periodic three-dimensional structures.

crystalline   Having atoms or ions arranged on a three-dimensional lattice having long-range order.

crystalline material   A material that contains a regular and repeating atomic or molecular arrangement such that long-range order is established within the structure.

Curie temperature   The temperature above which a ferromagnetic material exhibits paramagnetic characteristics.

Czochralski method   A process by which a single crystal of Si of a predetermined orientation is fabricated from a melt of Si.

deep drawing   A process by which a flat circular piece of metal is shaped into a cup by mechanical deformation.

degree of polymerization   The number of mers linked to form a polymer; the molecular weight of the polymer divided by the molecular weight of the mer.

degrees of freedom   The number of independent variables necessary to specify equilibrium.

dendrite   A treelike structure in a solidified metal formed due to the pattern of latent heat rejection during solidification.

devitrification   A high-temperature treatment given to certain oxide glasses to convert the structure into crystalline ceramics.

diamagnetic   A class of magnetic materials in which the internal field is antiparallel to the external field. These materials show a weak response to external magnetic fields.

die bonding   A process by which an integrated circuit chip is bonded to the housing of an electronic package.

dielectric constant   Ability of a material to store charge when subject to an electric field. The dielectric constant normalized to that of vacuum is the relative dielectric constant. Also called *permittivity*.

dielectric strength   The maximum electric field strength that a material can withstand without failure.

diffraction   A specific interaction of radiation with materials. Diffraction appears as reflections from parallel planes and is governed by Bragg's law.

diffusion bonding   A process by which two pieces are bonded or joined together by atomic mixing at the interface accomplished by solid-state diffusion under conditions of high temperature and pressure.

diffusion coefficient   The constant in the equation relating mass flux per unit area to the concentration gradient. It obeys an Arrhenius-type equation and as such is strongly temperature-dependent. The diffusion coefficient describes the ease of atomic or ionic motion in the solid state.

direct dissolution mechanism   A degradation mechanism in which the solid dissolves directly into the surrounding liquid without substantial charge transfer.

dislocation   A linear defect in a crystalline solid that is responsible for plastic deformation.

dislocation climb   The motion of an edge dislocation that occurs when vacancies migrate to the edge of the extra half plane. Climb results in the dislocation being displaced to a parallel slip plane. Climb requires mass transport.

dislocation glide   Motion of a dislocation on its slip plane (defined by the Burgers vector and the unit tangent vector). This kind of dislocation motion does not involve mass transport.

dispersion   The variation in refractive index with wavelength.

displacive transformation   Term used by ceramists and geologists to identify martensitic, or athermal, transformation in nonmetallic materials.

donor level   Impurity level within the band gap of a semiconductor associated with the extra electrons introduced by the incorporation of a Group V or VI element into a Group IV semiconductor.

drift velocity   The average velocity with which an electrical charge carrier moves through a solid in response to an external electric field.

ductile   Term used to describe materials that are able to absorb energy by "bending" rather than by fracture (breaking into pieces) when subjected to external loads. For example, the metal or plastic bumper on a car is ductile, while the glass in a car window is not. More precisely, ductile materials exhibit a high failure strain.

ductile-to-brittle transition temperature (DBTT)   The temperature at which 50% of the fracture surface in a Charpy specimen shows crystallographic facets. Alternatively, it is the temperature corresponding to the midpoint between the lower and upper shelf energies in a Charpy test.

edge dislocation   A dislocation that is formed (figuratively) by the insertion of an extra half plane into an otherwise perfect region of a crystal.

effective penetration distance   The distance at which the concentration of a solute in a solid solution reaches a value of approximately the average of the surface concentration and the initial bulk concentration.

elastic deformation   Deformation that is recoverable when a load is removed. This means that the part will return to its original size when the load is removed. Deformation of a rubber band is typically elastic.

elastic limit   The stress beyond which there is permanent deformation. Below the elastic limit all the deformation is recovered when the load is removed.

elastomer   A rubber that has been imparted a memory, usually by crosslinking or incorporation of a hard segment, so that up to about 600% extension the polymer recovers substantially completely. Only polymers can be elastomers.

electrochemical corrosion   A liquid-solid degradation mechanism in which the transfer of electrical charge plays an important role.

electrochemical plating   An electrochemical process of depositing a thin metal coating. The workpiece is the cathode and the coating material is the anode.

electron affinity   The energy released when an isolated neutral electronegative atom gains an electron.

electron configuration   The distribution of electrons within the permissible energy levels in an atom.

electron transfer   A process that occurs in ionic bonds in which an electropositive element gives up an electron to become a cation and this electron is then added to an electronegative element that becomes an anion.

electronegativity   The relative tendency of an element to gain, or attract, an electron.

electronic polarization   The temporary displacement of the center of positive and negative charge in atoms in the presence of an electric field.

electrostatic painting   A spray-painting process in which the paint particles and the surface being painted are given opposite electrical charges so that the particles accelerate toward the target surface.

electrostriction   The change in dimensions of a material brought on by an electric field.

endurance limit S_e The value of the stress amplitude in fatigue below which failure will not occur regardless of the number of repeated load applications.

energy gap   See *band gap*.

engineering strain ϵ The change in length of a specimen or component divided by the original length.

engineering strain at fracture ϵ_f The value of the strain when failure occurs.

engineering stress σ The load divided by the original cross-sectional area normal to the applied load.

epitaxial growth   A process in which a thin layer of crystalline material is grown on a substrate that acts as a seed crystal.

equilibrium state   The characteristics of the system remain constant indefinitely. Equilibrium occurs when the free energy of the system is at its minimum value.

erosion corrosion   A form of degradation common in elbows of pipes and similar locations where the fluid impinges upon the material at high velocities or high angles of attack.

eutectic reaction   The transformation, during cooling, of a liquid phase isothermally and reversibly into two solid phases.

eutectoid reaction   The transformation, during cooling, of a solid phase isothermally and reversibly into two new solid phases.

extrinsic conduction   Electrical conduction resulting from the incorporation of dopant atoms into a semiconductor such as pure Si, Ge, or GaAs.

extrusion   A process in which a billet is mechanically pushed through a die that is shaped to produce the desired cross section.

face-centered cubic (FCC)   A cubic crystal in which the atoms or ions are positioned at each corner and in the center of each of the six faces.

failure   Usually refers to separation into two (or more) pieces of a specimen or component that is loaded beyond its capacity. It may also refer to changes in dimension that are large enough to render the component unsuitable to fulfill its intended purpose.

families of directions   All the equivalent directions in a crystal, which are all the permutations of the Miller indices of any direction in the family.

families of planes   All the equivalent planes in a crystal, which are all the permutations of the Miller indices of any plane in the family.

Faraday's equation   The equation that relates the weight loss due to corrosion to the corrosion current.

fatigue   The process of repeated load or strain application to a specimen or component. A part subjected to repeated loading is said to be fatigued.

Fermi-Dirac distribution function   A mathematical function that describes the probability that a specific electron energy level is occupied at a given temperature.

Fermi energy E_f The electron energy level that has a 50% chance of containing an electron at any temperature.

ferrimagnetic   A class of magnetic materials in which adjacent magnetic dipoles of unequal strength point in opposite directions. The result is a material that shows a significant response to external magnetic fields.

ferrite   The body-centered cubic (BCC) form of iron.

ferromagnetic   A class of magnetic materials in which adjacent magnetic dipoles spontaneously align with one another to create a strong internal magnetic field. These materials show a significant response to external magnetic fields.

fiber optics   Use of transparent fibers to carry light or transport information via light signals.

Fick's first law   An equation which states that the mass flux per unit area of a diffusing species under steady-state conditions is proportional to the concentration gradient.

Fick's second law   An equation which states that for non-steady-state conditions the rate of change of concentration with respect to time at a given location is proportional to the second derivative of concentration with respect to position.

filament winding   A process used to make hollow cylindrical parts from composites in which a resin impregnated fiber is wound around a mandrel in a definite pattern and then subsequently cured to make the composite part.

flame hydrolysis   A process by which silica fibers are produced by oxidizing silicon tetrachloride, resulting in the fiber and HCl.

flow stress   Stress required to continue plastic deformation. The stress at which plastic deformation first occurs is a specific value of the flow stress and is referred to as the yield stress.

fluorescence   Luminescence that ceases when the stimulus is removed.

forging   A process in which a metal is first heated and subsequently hammered into the desired shape.

fracture   The physical separation of a part or component. In many instances it is synonymous with failure.

fracture toughness   The property of a material that has to do with its ability to absorb energy before fracturing. If a material can absorb much energy it is said to have a high fracture toughness. It is also common to refer to the value of the stress intensity parameter at which fracture occurs as the fracture toughness.

Frenkel defect   A defect in an ionic solid consisting of a vacancy/interstitial pair of the same ionic species. Due to size considerations, most Frenkel defects involve cations.

galvanic corrosion   A form of corrosion in which the local anodic and cathodic regions are a result of compositional variations.

Gibbs free energy   A thermodynamic variable that is a function of the enthalpy and entropy (randomness) of the system.

Gibbs phase rule   For a system at equilibrium, an equation that relates the number of phases present to the number of externally controllable variables.

glass   A brittle noncrystalline solid below its glass transition temperature at the temperature of interest, usually room temperature.

glass-ceramic   A fine-grained crystalline ceramic that results from the devitrification of an oxide glass.

glass transition temperature   Temperature at which motion on the time scale of the experiment and on the size scale of the repeat unit becomes frozen upon cooling.

grain boundary   Boundary that occurs when two crystals of arbitrary orientation to each other are joined along an arbitrary surface. A grain boundary is a planar defect.

green compact   A powder compaction product in which some densification has been achieved so that the piece has sufficient strength for the next stage of processing such as isostatic pressing.

ground state　Electron configuration in which an atom's electrons occupy the lowest-energy subshells consistent with the Pauli exclusion principle.

Guinier-Preston (GP) zones　Solute-rich coherent precipitates. The terminology is commemorative and recognizes the efforts of A. Guinier and G. D. Preston, who independently demonstrated the occurrence of these precipitates during the initial stages of decomposition in an aluminum-rich Al-Cu alloy. The general use of the term GP zones has widened to include solute-rich coherent precipitates in other aluminum alloys as well as other alloy systems.

hand layup　A simple process of manufacturing laminate composites in which sheets of resin-impregnated fibers are stacked on top of each other by hand in a predetermined orientation and cured to form the composite.

hard facing　A process in which a thick layer of hard material is deposited on the surface of the workpiece by either a welding or a brazing process.

hard magnets　Ferro- or ferrimagnetic materials with large hysteresis loops.

hardness　Ability of a material to resist penetration. A material is said to be hard if large forces are required to cause a permanent indentation mark.

heat-affected zone (HAZ)　A zone of material on the base metal side that is adjacent to the weld metal–base metal interface in which microstructural changes occur due to the intense heat of the welding process.

heat capacity　A materials property that represents the amount of heat energy required to raise the temperature of one mole of the material by one degree.

heterogeneous nucleation　Specific site-induced nucleation.

hexagonal close-packed (HCP)　A crystal that has a regular hexagonal basal plane and rectangular side planes normal to the basal plane. Individual atoms or ions are located at each corner of the basal plane and in the center of the basal plane, and three more atoms are on the midplane, located over the interstices of the atoms in the basal plane. Each atom has 12 neighbors that touch, giving an APF of 0.74.

highest-density plane　A crystallographic plane with the largest number of atoms or ions centered on the plane per unit area.

hole　An empty electron level within a nearly filled band of electron levels.

homogeneous nucleation　Random nucleation process.

hot dipping　A process used for applying coatings of low–melting-point metals to higher–melting-point metals by dipping the latter in a molten bath of the former.

hot isostatic pressing　A process carried out at a high temperature and pressure to increase the density of green compacts during powder processing.

Hume-Rothery rules　A set of rules for the formation of substitutional solid solutions. They incorporate relative sizes, bond character, and electronegativity of the atomic species being considered.

hydrogen bond　A type of secondary bond in which a hydrogen atom is shared between two strongly electronegative atoms such as N, O, F, or Cl.

hydrogen embrittlement　A family of degradation mechanisms related to the high diffusivity of H^+ in BCC metals that results in a decrease in ductility.

hypereutectic　An alloy whose composition is greater than the eutectic composition.

hypereutectoid　An alloy whose composition is greater than the eutectoid composition.

hypoeutectic　An alloy whose composition is less than the eutectic composition.

hypoeutectoid alloy　An alloy whose composition is less than the eutectoid composition.

impurity　Any atom or ion species that differs from the host or solvent species. While considered to be defects, impurities may confer desirable properties on the solid solution.

impurity diffusion　The diffusion of an impurity species at very low concentrations within the solvent or host material.

incoherent interface　An interface separating two phases that do not have similar atomic arrangements across the interface.

incongruent melting　The equilibrium melting reaction that occurs when a solid phase transforms into a solid and a liquid phase both having different compositions than the original solid.

index of absorption　The imaginary part of the complex index of refraction.

injection molding　A process in which slugs of molten polymer are injected into a cold mold and pressure is applied to make the polymer conform to the shape of the mold.

insulator　A material that is a poor conductor of electricity—that is, one with a low value of conductivity.

integrated circuits　Collections of pn junction diodes, bipolar junction transistors, and other devices linked together to form microelectronic circuits ranging in complexity from a simple amplifier to a complete miniature digital computer known as a microprocessor.

interfacial polarization The accumulation of mobile charge in the regions near grain boundaries or surfaces.

intermetallics Stoichiometric compounds with characteristic metal atom ratios such as AlLi, Ni_3Al, Al_3V, AlSb, CuZn, Ti_3Al, and Mg_2S.

interstices The regions between atoms that are defined by at least four atoms or ions. Interstices in crystals are usually tetrahedral or octahedral.

interstitial A point defect consisting of an atom or ion that is located in the interstices of the structure under consideration. Interstitials may be either the same or a different species than the solvent.

interstitial solid solution A solid solution in which atoms or ions of a foreign species are located in the interstitial positions.

intrinsic conduction Electrical conduction in pure semiconductors such as Si, Ge, and GaAs.

invariant reaction A reaction is which there are zero degrees of freedom.

ion implantation A process in which ions of light elements such as boron and nitrogen are made to impinge on the surface of the workpiece at a high velocity. The surface hardens due to penetration of the ions.

ionic bond A type of primary bond involving electronegative and electropositive atoms that display a significant difference in their electronegativity values (usually $\Delta EN \geq 1.7$).

ionic (atomic) polarization The temporary displacement of positive and negative ions in the presence of an electric field.

ionization potential The energy required to remove an electron from an isolated neutral atom.

isomorphous Having the same structure. When applied to a phase diagram, indicating that the solid phase has the same structure and hence complete solubility at every composition.

isostrain An assumption in the analysis of composites in which the fiber and the matrix are considered to be subjected to identical strain levels. It is the result of a load being applied parallel to the fibers.

isostress An assumption in the analysis of composites in which the fiber and the matrix are considered to be subjected to identical stress. It is the result of a load being applied perpendicular to the fibers.

isotactic Configuration of a vinyl polymer established during polymerization corresponding to all side groups positioned on the same side of the backbone.

isotropic Characterized by properties that are independent of the direction in which they are measured.

kinetics The study of the rate of reactions and the factors that affect them, including the influence of time on phase transformations.

laminate composites Composites made from stacking several laminae in a predetermined sequence of fiber orientation.

Larsen-Miller parameter A parameter employed in the analysis of creep. It may be used at a given stress to determine how long the part will last at a given temperature or to determine the maximum allowable temperature for a fixed time duration.

lattice Three-dimensional space-filling repeating pattern on which atoms or ions are placed to form a crystal.

lattice point One position in the three-dimensional space-filling repeating pattern that forms a crystal. One or more atoms or ions are placed at each lattice point.

lever rule An equation that enables one to calculate the relative amounts of two phases present in a two-phase mixture in terms of the compositions of the alloy and the phases present.

line compound A compound whose composition is independent of temperature.

linear density The number of atoms centered on a crystallographic direction per unit length.

liquid crystal An anisotropic fluid, usually a melt or solution characterized by molecular alignment.

liquid metal embrittlement A degradation mechanism in which a liquid metal preferentially attacks the grain boundaries in a solid.

liquidus The temperature at which a liquid begins to freeze during equilibrium cooling conditions.

long-range order Existence of a regular repeating arrangement of atoms, ions, or molecules within a crystalline region of a material.

luminescence Absorption of light or electromagnetic radiation at high frequencies with subsequent reradiation at lower frequencies.

machining A process by which excess material is removed by plastic shearing.

magnetic domains Microscopic regions of a ferri- or ferromagnetic crystal in which all of the internal magnetic dipoles are aligned with one another.

magnetic field strength H A measure of the intensity of the external magnetic field.

magnetic induction B A measure of the degree of magnetism resulting from a given external magnetic field strength.

magnetic susceptibility χ The constant of proportionality that relates the magnetization **M** to the magnetic field strength **H**.

magnetic vector M   A measure of the intensity of the internal magnetic field within a solid.

martensite   Originally used to describe the metastable body-centered tetragonal (BCT) iron phase that is supersaturated in carbon. Iron-carbon martensite forms when austenite is rapidly quenched to low temperatures. A wide variety of metallic and nonmetallic materials· have also been shown to exhibit phases having characteristics similar to iron-carbon martensite; hence, the more recent generic use of the term martensite to describe these phases as well.

martensite finish temperature M_f In iron-carbon martensites, the temperature at which the transformation of austenite to martensite is complete. For other systems, M_f is the temperature at which all the parent phase has transformed to the product phase.

martensite start temperature M_s In iron-carbon martensites, the temperature at which the transformation of austenite to martensite is just beginning. For other systems, M_s is the temperature at which the parent phase is just beginning to transform to the product phase.

martensitic transformation   A shearlike transformation with no change in composition when the parent phase transforms into the product phase.

matrix phase   The material used in a composite to fill the space between fibers or other reinforcement with the primary purpose of bonding them together and to provide them with lateral strength.

mean time between collisions   The average time between scattering events during the motion of charge carriers in response to an external electric field.

melt spinning   A process of making fibers in which a viscous melt of the material passes through a spinnerette with hundreds of holes and turns into a fiber as it cools.

metal-matrix composites   Composites utilizing metals as the matrix material.

metallic bonds   A type of primary bond formed between two electropositive elements when their average number of valence electrons is less than or equal to 3.

metastable state   A nonequilibrium state, but one that can persist for a very long time. The free energy has a local minimum, but is not the lowest free energy.

Miller indices   A shorthand for expressing directions and planes in crystals.

mobility μ A materials property that describes the ease of charge carrier motion in response to an external electric field. It is defined as the average charge carrier velocity divided by the electric field strength.

molecular (dipole) polarization   The temporary alignment of permanent dipoles in molecules in the presence of an electric field.

monomers   Relatively small molecules that combine with other similar molecules to form much larger polymer molecules.

monotectic reaction   The transformation, during cooling, of a liquid phase isothermally and reversibly into another liquid and a solid phase having different compositions from the initial liquid.

monotectoid reaction   The transformation, during cooling, of a solid phase isothermally and reversibly into two solids having different compositions from the initial solid.

n-type   A class of extrinsic semiconductors containing dopant atoms with more than four valence electrons.

natural aging   The precipitation of a second phase from a supersaturated solution at room temperature.

Nernst equation   An equation that relates the electrochemical cell potential to the activities of the reactants and products.

network former   An element in a chain capable of forming at least three bonds with its neighbors (trifunctionality) or an ion that satisfies Zachariasen's rules. They are good glass formers.

network modifier   An ion that is not capable of forming three or more bonds with its neighbors, usually in the context of an oxide glass. Network modifiers disrupt three-dimensional networks. They are not good glass formers.

noncrystalline   Amorphous, lacking three-dimensional order. Amorphous materials are either rubbers or glasses.

normal stress   The force divided by the area normal to the force.

nucleation   The first step in a phase transformation.

octahedral site   A position in a crystal defined by six neighbors, usually nearest neighbors.

0.2% offset yield strength   The stress at which a line starting at 0.2% on the strain axis and drawn parallel to the initial elastic portion of a stress-strain curve intersects the stress-strain curve. This is the conventional yield strength for materials that do not exhibit a "sharp" transition from elastic to plastic behavior.

Ohm's law   The relationship between voltage, current, and resistance which has the general form $V = IR$.

oxidation   A gas-phase degradation mechanism in which a material combines with oxygen to produce an oxide.

oxidation (anodic) reaction   An electrochemical reaction that results in the production of electrons.

oxygen concentration cell   An electrochemical cell resulting from oxygen concentration gradients in the electrolyte.

p-type   A class of extrinsic semiconductors containing dopant atoms with fewer than four valence electrons.

paramagnetic   A class of magnetic materials in which the internal field is parallel to the external field. These materials show a weak response to external magnetic fields.

particulate composites   Composite systems in which the reinforcement phase consists of small particles.

passivation   A corrosion surface condition under which a material displays a significant positive shift in its reduction potential.

Pauli exclusion principle   A quantum mechanics concept stating that no two interacting electrons can have the same four quantum numbers.

pearlite   ·A two-phase microstructure of alternate ferrite and cementite lamellae occurring in some steels. Pearlite forms by the decomposition of austenite.

percent reduction in area (%RA)   The change in area divided by the original area (expressed in percent) of a tensile specimen that has been fractured.

peritectic reaction   The transformation, during cooling, of a solid and a liquid phase isothermally and reversibly into a solid having a different composition from either the initial liquid or solid phase.

peritectoid reaction   The transformation, during cooling, of two solid phases isothermally and reversibly into a solid having a different composition from either of the initial solids.

permanent dipole   A molecule in which the spatial center of positive charge is always different from that of the negative charge.

permeability μ The materials property that relates the magnetic induction $\mathbf{B}$, to the magnetic field strength $\mathbf{H}$.

permittivity   See *dielectric constant*.

Petch-Hall equation   Equation that provides the relationship between the grain size and the yield stress. The yield stress increases with decreasing grain size according to the Petch-Hall equation.

phase   A homogeneous portion of matter bounded by a surface so that it is mechanically separate from any other portion.

phase diagram   A graphical representation showing the phase or phases present for a given composition as a function of temperature.

phosphorescence   Luminescence that persists more than 10^{-8} s after the stimulus is removed.

photoconductivity   An increase in electrical conductivity brought about by photostimulation.

photolithography   A process that utilizes ultraviolet-light–sensitive polymers to etch small devices on an integrated chip.

photovoltaic   Describing production of a voltage by stimulation with light or electromagnetic radiation, usually in solid-state devices.

physical vapor deposition (PVD)   A process by which vapor-phase particles of the coating material are first created and then deposited on the workpiece to produce a thin, uniform coating.

piezoelectric effect   The production of electrical current brought about by changing a material's sample dimensions, usually with stress.

Pilling-Bedworth (PB) ratio   The volume occupied by one-mole of oxide divided by the volume of the metal consumed to create one mole of oxide.

pitting corrosion   A highly localized form of corrosion usually involving oxygen concentration gradients that result from surface deposits of contaminants occurring in stagnant pools of electrolyte located on horizontal sections of metal.

planar density   The number of atoms or ions centered on a crystallographic plane per unit area of the plane.

plastic deformation   Deformation that is permanent; for example, a metal part that has been permanently bent is said to have undergone plastic deformation.

pn junction diode   A microelectronics device composed of adjacent layers of p-type and n-type semiconductors that serves to pass current in one direction while (essentially) blocking current in the opposite direction.

point defect   A "mistake" in the material that involves a single atom, ion, lattice site, or interstitial position. For example, if an atom is missing at a regular lattice site, a point defect is said to have been formed.

Poisson's ratio ν The negative ratio of the transverse strain divided by the longitudinal strain in a tensile test.

polarization   Separation of positive and negative charge in a material, usually by an electric field. The vector sum of all the dipoles in a material per unit volume is the polarization.

polycrystalline   Describing a solid, usually isotropic, of joined crystals or grains.

polymer   Engineering material composed of high–molecular-weight molecules. These molecules usually have either a linear (chain) structure or a three-dimensional network structure.

polymer-matrix composites   Composite systems in which the matrix material is a polymer.

polymorphic   Allotropic; capable of more than one crystal structure.

powder metallurgy   A process consisting of producing very small particles of metallic powders and then consolidating them in the form of near-net-shape components.

precipitate   A small region of a crystal that generally has a different chemical composition from the matrix and may also have a different crystal structure.

precipitation hardening   Hardening or strengthening of an alloy by the presence of a fine, uniformly distributed precipitate.

primary solid phase   The first solid that appears upon cooling an alloy from the liquid phase.

proeutectoid cementite   The first cementite that forms when a hypereutectoid steel is cooled through the austenite and cementite two-phase field and held at a temperature below the eutectoid transformation temperature.

proeutectoid ferrite   The first ferrite that forms when a hypoeutectoid steel is quenched through the austenite and ferrite two-phase field and held at a temperature below the eutectoid transformation temperature.

pultrusion   A process in which resin-impregnated fibers are pulled through heated dies to make long lengths of polymer composite products such as I-beams.

pyrolysis   A process for making carbon and ceramic fibers in which a polymer precursor fiber containing the elements of interest is heated to temperatures high enough to cause chemical changes resulting in the desired fiber composition.

quantum number   One of four values, three of which are integers and the other is $\pm\frac{1}{2}$, that together determine the energy and many other important characteristics of an electron within an atom.

recombination   An interaction between an electron and a hole that results in the loss of both charge carriers.

recovery   The reduction of some of the stored energy introduced by cold-working a material.

recrystallization   Process involving the nucleation and growth of new, strain-free grains in a deformed material.

recrystallization temperature   The temperature at which recrystallization occurs within some specified time, frequently one hour.

reduction potential   A measure of the reactivity of a metal that is useful for predicting the severity of some types of corrosion.

reduction (cathodic) reaction   An electrochemical reaction that results in a decrease in the concentration of electrons.

refraction   The bending of a light ray as it passes from one medium into another.

refractive index   Speed of light in vacuum divided by the speed of light in a medium; sine of the angle of incidence divided by the sine of the angle of refraction; a measure of the amount of light reflected from a dielectric surface.

reinforcing phase   The component in a composite that is usually strong and can be present in the form of particles, fibers, sheets, or whiskers.

relative permeability μ_r The ratio of the permeability of a material to the permeability of a vacuum.

resistance R The extrinsic materials property that describes the ability of a material to resist, or oppose, the transport of electrical charge in response to an external electric field. It is defined as $R = \rho(L/A)$.

resistivity ρ The intrinsic materials property that describes the ability of a material to resist, or oppose, the transport of electrical charge in response to an external electric field.

rolling   A process in which the thickness of a flat plate is reduced by compressing it between two cylindrical rolls.

rubber   A noncrystalline or amorphous fluidlike material that is above its glass transition temperature at the temperature of interest, usually room temperature. Only polymers can be rubbers.

rule of mixtures   A formula that expresses the composite properties as a linear sum of the constituent properties weighted in proportion to their respective volume fractions.

sacrificial anode   A version of cathodic protection involving the intentional formation of a galvanic cell by placing another, more anodic metal in electrical contact with the metal to be protected.

Schmid's law   The equation that relates the orientation of a crystal and the normal stress to the shear stress at which plastic deformation occurs.

Schottky defect   Small cation and anion vacancy clusters that are formed in ionic solids. The cation : anion ratio in these clusters is adjusted to maintain electroneutrality.

screw dislocation   A dislocation that may be formed by making a cut in a crystal and displacing the top half of the cut region relative to the bottom half in a direction parallel to the cut. The Burgers vector of a screw dislocation is parallel to the dislocation line.

self-diffusion   The mechanism by which a species diffuses in itself.

semicoherent interface   An interface separating two phases where the differences in the atomic arrangements of the atoms forming the interface can be accommodated by dislocations.

semiconductor   A class of materials that show an increase in electrical conductivity as the temperature increases.

shear modulus G The slope of the shear stress versus shear strain curve in the elastic region.

shear stress   Force divided by the area of the plane in which the force acts. Shear stresses are responsible for the motion of dislocations.

short-range order   The local arrangement of nearest-neighbor atoms or ions around a centrally located atom or ion.

simple cubic (SC)   A crystal with cubic symmetry in which the atoms or ions are located at all eight corner positions.

sintering   The coalescence of powders at an elevated temperature via extensive solid-state diffusion.

slip casting   A process for making hollow ceramic parts in which a suspension of ceramic particles called slip is poured into a plaster of Paris mold and the contents are emptied out after the moisture from the outer layers has been absorbed by the porous mold.

slip direction   The direction in which atoms are displaced across the slip plane when a dislocation passes over the slip plane. The slip direction is parallel to the Burgers vector of the dislocation responsible for the slip being considered.

slip plane   The plane on which a dislocation moves by glide. It is defined by the Burgers vector and the unit tangent vector.

slip system   Any combination of slip direction and slip plane on which deformation can occur. For example, in FCC metals $[1\ 1\ 0]\ (\bar{1}\ 1\ 1)$ is a slip system and the family of all slip systems is represented by $\langle 1\ 1\ 0 \rangle \{1\ 1\ 1\}$.

slurry infiltration   A process by which fibers are coated with ceramic matrix material by passing the fiber tow through a ceramic slurry containing the ceramic powder, a carrier liquid, and an organic binder.

soft magnets   Ferro- or ferrimagnetic materials with small hysteresis loops.

sol-gel technique   A technique for making ceramics in which metal alkoxides are reacted with water to make a gel that can be spun into a fiber when the reaction mixture achieves a certain viscosity. The pure oxide is achieved by subsequently driving off the remaining organic molecules by heating.

soldering   A process used to join metals by sandwiching a very low–melting-point alloy (less than 450°C) and melting it by heating to form a good bond.

solidus boundary   The temperature at which a liquid phase disappears during equilibrium cooling conditions.

solute   An atom or ion species that is dissolved in a host or solvent.

solution heat treatment   A high-temperature heat treatment designed to take into solution the soluble elements in an alloy.

solvent   The host species for a solute.

solvus boundary   The phase boundary that separates single-phase solid regions from two-phase solid regions.

specific heat   Heat capacity per unit mass.

specific modulus   Elastic modulus divided by the density of the material.

specific strength   Ultimate tensile strength divided by the density of the material.

spheroidite   A two-phase microstructure, consisting of spheroidized carbide in ferrite, which is formed by heat-treating pearlite, bainite, or martensite at a temperature below the eutectoid temperature.

spheroidizing   For a steel, heat-treating below the eutectoid temperature to produce spheroidite.

spherulite   A spherical aggregate of crystalline and noncrystalline materials that grows radially outward when polymers are crystallized from melts.

sputtering   A process in which inert gas is ionized and then accelerated toward a target from which metal atoms are released and transported to a workpiece where they are deposited.

stacking fault   A variation in the normal stacking sequences of close-packed planes. For example, in FCC crystals the normal stacking sequence is ABCABC . . . , and if in some region of the crystal the stacking is ABCBCABC . . . , that region is said to contain a stacking fault.

state point   Temperature, pressure, and composition of a phase.

static fatigue   A chemical degradation mechanism involving the rupture of Si-O bonds due to an interaction with water at room temperature.

strain hardening   Hardening that occurs as a result of deforming a metal. During strain hardening, dislocations are generated and the high dislocation density makes it difficult for other dislocations to move.

striations   Parallel ridges that are seen on the surfaces of many fatigue failures at high magnifications. In some cases one striation forms for each fatigue cycle.

stress corrosion cracking (SCC)   A type of corrosion that results from the application of a tensile stress in conjunction with a specific material-electrolyte pair.

stress intensity parameter K The parameter that characterizes the stresses and strains at a given location ahead of a crack. It is a valid parameter for situations in which plastic deformation ahead of the crack tip is nonexistent or very limited.

stress relaxation   Reduction in stress that occurs when a component is subjected to a constant value of strain. Typically polymers and metals at relatively high temperatures exhibit stress relaxation.

substitutional solid solution   A solution formed by substituting species B directly for species A on the lattice sites normally occupied by A. The solution so formed is said to be a substitutional solid solution of B in A.

superalloys   FCC alloys containing nickel, cobalt, iron, and chromium strengthened with a fine dispersion of $Ni_3(Al, Ti)$ precipitates.

superconductivity   A state of matter in which there is no resistance to the motion of electrical charge carriers.

supercooled liquid   A liquid that is below its thermo-dynamic melting temperature. A supercooled liquid is metastable at best and may crystallize at any time.

supercooling   Cooling below a transformation temperature without the transformation occurring.

surface tension   Energy per unit area associated with internal or external surfaces. It has units of energy/area, which is equivalent to force/length.

syndiotactic   Configuration of a vinyl polymer established during polymerization corresponding to side groups positioned on alternating sides of the backbone.

temperature coefficient of resistivity   The materials property that describes the magnitude of the change in the resistivity of a conductor resulting from a change in temperature.

tempered martensite   The result of heat-treating a microstructure containing martensite. Tempered martensite consists of a fine dispersion of Fe_3C particles in an α-ferrite matrix. The ductility and toughness of a martensitic microstructure are enhanced by tempering.

temporary dipole   An atom or molecule in which the spatial center of positive charge is momentarily different from the center of negative charge.

tetrahedral site   A position in a crystal defined by four nearest neighbors.

thermal capacitance   The extrinsic counterpart of specific heat obtained by multiplying specific heat by the mass of the object under investigation.

thermal conductivity   The constant of proportionality that relates the magnitude of a thermal gradient to the resulting heat flow rate.

thermal diffusivity   The constant of proportionality in the non-steady-state heat flow equation. It is defined as the thermal conductivity divided by the product of the heat capacity at constant pressure multiplied by density.

thermal fatigue   A mechanism of cracking caused by fatigue stresses that are induced by thermal fluctuations.

thermal shock   A failure mechanism that can result when a brittle material is subjected to a sudden change (usually a decrease) in temperature.

thermodynamics   The science dealing with the relationships between the thermal properties of matter and the external system variables such as pressure, temperature, and composition.

thermoplastic polymer   A polymer composed of long-chain molecules often composed of a covalently bonded backbone and various side groups. These macromolecules can be heated to form a melt repeatedly.

thermoset polymer   A polymer (usually composed of a three-dimensional network of covalently bonded atoms) that does not melt when reheated.

tie line   Horizontal line drawn on a phase diagram in a two-phase field connecting the composition of the two phases in equilibrium.

tilt boundary   A boundary composed of edge dislocations of the same sign stacked vertically above one another. As a result of all of the extra half planes being in one portion of the crystal, a tilt is created.

time-dependent deformation   Deformation that occurs as a function of time, independent of whether such deformation is elastic or plastic.

transference numbers   A set of values ranging from 0 to 1 that describe the relative contributions of anions, cations, electrons, and holes in the transport of electrical charge through a solid.

true strain ϵ_t The natural logarithm of the instantaneous length divided by the original length. Numerically, it is essentially equivalent to the engineering strain for strains up to about 0.1.

true stress σ_t The force divided by the instantaneous area normal to the applied force.

twin   A portion of a crystal in which all of the atoms are in mirror-image positions compared with atoms in the bulk of the crystal.

twist boundary   A boundary that is composed of a grid of screw dislocations of like sign on the same plane. As a result of the displacements around the grid, a twist is created.

ultimate tensile strength σ_{uts} The maximum value of the engineering stress in a tensile test.

unidirectional fiber–reinforced composite   A composite material in which all fibers are aligned parallel to each other.

unified life-cycle cost engineering   A cost analysis procedure that considers the overall cost of a piece of equipment including its manufacturing cost and also the operating cost during service.

uniform corrosion   A common type of electrochemical corrosion resulting from a time-varying change in the location of the local anodic regions.

uniform strain ϵ_u The strain in a specimen that occurs before reaching the ultimate tensile strength.

Deformation up to this point is uniformly distributed throughout the gage section so the strain is also uniform.

unit cell   The smallest representation of a material. In crystals the unit cell is the smallest patterned collection of atoms or ions that repeats in space.

vacancy   A missing atom or ion at a lattice site that would normally be occupied.

valence band   The highest-energy electron band that is at least partially filled at 0 K.

valence electrons   Electrons located in the shell that is furthest from the atomic nucleus.

van der Waals bond   A type of secondary bond in which a temporary dipole induces another dipole in an adjacent atom. The two dipoles then experience a coulombic force of attraction.

viscoelastic deformation   A combination of time-dependent and time-independent deformation. Such behavior is generally seen most prominently in polymers. Mechanically it is equivalent to a combination of spring and dashpot.

viscosity   The proportionality constant between shear stress and shear strain rate, which is a measure of the energy dissipated in a flowing material.

void   A volumetric defect in which there is no matter.

volumetric density   The number of atoms centered in a unit volume of crystal. Often the volume considered is the unit cell.

volumetric thermal expansion coefficient   The fractional change in volume of a material per unit change in temperature.

wear   The removal of surface material as a result of mechanical rather than chemical interaction.

welding   A process used to join two pieces of metal in which the parent metal near the joint and a filler metal are heated until they melt and mix together to form a strong joint upon freezing.

wire drawing   A process in which a metal wire is pulled through a die to reduce its diameter.

yield point strain ϵ_{yp}   The strain at which deformation becomes permanent. The yield point strain is well defined in metals such as steel, which exhibit a "sharp" transition from elastic to plastic deformation.

yield stress σ_{ys}   The stress at which deformation becomes permanent in a tensile test.

Young's modulus   The elastic property of solids that describes the inherent stiffness of the material. It is also the slope of the stress-strain curve in the elastic region.

REFERENCES

Ashby, Michael F., and David R. H. Jones. *Engineering Materials 1: An Introduction to Their Properties and Applications.* Oxford: Pergamon Press, 1980.

Ashby, Michael F., and David R. H. Jones. *Engineering Materials 2: An Introduction to Microstructure, Processing and Design.* Oxford: Pergamon Press, 1986.

Askeland, Donald R. *The Science and Engineering of Materials,* 3rd ed. Boston: PWS Publishing Company, 1994.

ASM Handbook, Vol. 8, *Mechanical Testing.* Materials Park, OH: ASM International, 1985.

ASM Handbook, Vol. 4, *Heat Treating.* Materials Park, OH: ASM International, 1991.

ASM Handbook, Vol. 33, *Alloy Phase Diagrams.* Materials Park, OH: ASM International, 1992.

Baer, E. "Advanced Polymers." *Scientific American* 255, no. 4 (October 1986), pp. 178–90.

Barrett, C. S., and T. B. Massalski. *Structure of Metals,* 3rd ed. Oxford: Pergamon Press, 1980.

Borg, R. J., and G. J. Dienes, eds. *An Introduction to Solid State Diffusion.* San Diego: Academic Press, 1988.

Bowen, H. K. "Advanced Ceramics." *Scientific American* 225, no. 4 (October 1986), pp. 168–76.

Callister, William D. *Materials Science and Engineering: An Introduction,* 3rd ed. New York: John Wiley, 1994.

Chaudhari, P. "Electronic and Magnetic Materials." *Scientific American* 255, no. 4 (October 1986), pp. 136–44.

Collins, J. A. *Failure of Materials in Mechanical Design.* New York: John Wiley, 1981.

Cullity, B. D. *Introduction to Magnetic Materials.* Reading, MA: Addison-Wesley, 1972.

Du Pont Design Handbooks. Du Pont de Nemours and Co., Polymer Products Department, Wilmington, DE, 1981.

Edwards, K. S., Jr., and R. B. McKee. *Fundamentals of Mechanical Component Design.* New York: McGraw-Hill, 1991, Chap. 18.

Engineered Materials Handbook, Vol. 4, *Ceramics and Glasses.* Materials Park, OH: ASM International, 1991.

Flinn, Richard A., and Paul K. Trojan. *Engineering Materials and Their Applications,* 4th ed. Boston: Houghton Mifflin, 1990.

Grayson, M., ed. *Encyclopedia of Composite Materials and Components.* New York: John Wiley, 1983.

Hench, L. L., and J. K. West. *Principles of Electronic Ceramics.* New York: John Wiley, 1990.

Hertzberg, R. *Deformation and Fracture Mechanics of Engineering Materials.* New York: John Wiley, 1976.

Hull, D. *An Introduction to Composite Materials.* Cambridge: Cambridge University Press, 1981.

Hull, D. *Introduction to Dislocations,* 3rd ed. Elmsford, NY: Pergamon Press, 1984.

Jastrzebski, Zbigniew D. *The Nature and Properties of Engineering Materials,* 3rd ed. New York: John Wiley, 1987.

Keffer, F. "The Magnetic Properties of Materials." *Scientific American* 217, no. 3 (September 1967), pp. 222–34.

Kingery, W. D. *Property Measurements at High Temperatures.* New York: John Wiley, 1959.

Kingery, W. D., H. K. Bowen, and D. R. Uhlmann. *Introduction to Ceramics.* New York: John Wiley, 1976.

Kotz, J. C., and K. F. Purcell. *Chemistry and Chemical Reactivity,* 2nd ed. Philadelphia: Saunders College Publishing, 1991.

Lewis, G. *Selection of Engineering Materials.* Englewood Cliffs, NJ: Prentice Hall, 1990.

Marin, J. *Mechanical Behavior of Engineering Materials.* Englewood Cliffs, NJ: Prentice Hall, 1962.

McClintock, F. A., and A. S. Argon. *Mechanical Behavior of Materials.* Reading, MA: Addison-Wesley, 1966.

Metals Handbook, 9th ed., Vol. 6, *Welding, Brazing, and Soldering.* Metals Park, OH: American Society for Metals, 1983.

Metals Handbook, 9th ed., Vol. 1, *Properties and Selection: Irons, Steels, and High-Performance Alloys.* Materials Park, OH: ASM International, 1990.

Mysen, B. O., ed. *Phase Diagrams for Ceramists,* Vol. VIII. Columbus, OH: American Ceramic Society, 1990.

Orowan. *Fatigue and Fracture of Metals.* Cambridge, MA: MIT Press, 1950.

Papanek. *Design for the Real World.* New York: Random House, 1971.

Richerson, D. W. *Modern Ceramic Engineering,* 2nd ed. New York: Marcel Dekker, 1992.

Rosen, S. L. *Fundamental Principles of Polymeric Materials.* New York: John Wiley, 1976.

Schweitzer, P. A. *Corrosion Resistance Tables,* 3rd ed. New York: Marcel Dekker, 1991.

Shackelford, James F. *Introduction to Materials Science for Engineers,* 3rd ed. New York: Macmillan, 1992.

Tadokoro, H. *Structure of Crystalline Polymers,* New York: John Wiley, 1979.

Van Vlack, Lawrence. *Elements of Materials Science and Engineering,* 6th ed. Reading, MA: Addison-Wesley, 1989.

Vander Voort, G. F. *Metallography: Principles and Practice.* New York: McGraw-Hill 1984.

Vander Voort, G., ed. *Atlas of Time-Temperature Diagrams for Nonferrous Alloys.* Materials Park, OH: ASM International, 1991.

Welding Handbook, 7th ed. Miami, FL: American Welding Society, 1976. In five volumes.

Young, R. J., and P. Lovell. *Introduction to Polymers,* 2nd ed. London: Chapman & Hall, 1991.

INDEX

UNIT CONVERSION FACTORS

Temperature	$K = °C + 273$
	$°C = 1.8(°F - 32)$
	$°R = °F + 460$
Length	$1 \text{ m} = 10^{10} \text{ Å} = 3.28 \text{ ft} = 39.4 \text{ in}$
Mass	$1 \text{ kg} = 2.2 \text{ lbm}$
Force	$1 \text{ N} = 1 \text{ kg-m/s}^2 = 0.225 \text{ lbf}$
Pressure (stress)	$1 \text{ Pa} = 1 \text{ N/m}^2 = 1.45 \times 10^{-4} \text{ psi}$
Energy	$1 \text{ J} = 1 \text{ W-s} = 1 \text{ N-m} = 1 \text{ V-C}$
	$1 \text{ J} = 0.239 \text{ cal} = 6.24 \times 10^{18} \text{ eV}$
Current	$1 \text{ A} = 1 \text{ C/s} = 1 \text{ V}/\Omega$

CONSTANTS

Avogadro's number	$6.02 \times 10^{23} \text{ mol}^{-1}$
Gas constant, R	8.314 J/(mol-K)
Boltzmann's constant, k	$8.62 \times 10^{-5} \text{ eV/K}$
Planck's constant, h	$6.63 \times 10^{34} \text{ J-s}$
Speed of light in a vacuum, c	$3 \times 10^8 \text{ m/s}$
Electron charge, q	$1.6 \times 10^{-19} \text{C}$

SI PREFIXES

giga, G	10^9
mega, M	10^6
kilo, k	10^3
centi, c	10^{-2}
milli, m	10^{-3}
micro, μ	10^{-6}
nano, n	10^{-9}